Handbook of
Hazardous Chemical Properties

Nicholas P. Cheremisinoff, Ph.D.

N&P Limited
United States – Ukraine – Russia – Central Eastern Europe

BUTTERWORTH
HEINEMANN

Boston Oxford Auckland Johannesburg Melbourne New Delhi

Library of Congress Cataloging-in-Publication Data

ISBN 0-7506-7209-9

British Library Cataloguing-in-Publication Data
A catalogue record for this book is available from the British Library.

The publisher offers special discounts on bulk orders of this book.
For information, please contact:
Manager of Special Sales
Butterworth-Heinemann
225 Wildwood Avenue
Woburn, MA 01801-2041
Tel: 781-904-2500
Fax: 781-904-2620

For information on all Butterworth–Heinemann publications available, contact our World Wide Web home page at: http://www.bh.com

10 9 8 7 6 5 4 3 2 1

Printed in the United States of America

CONTENTS

iii

PREFACE

This volume has been prepared as a reference source on the hazardous properties of industrial and consumer chemicals. It is designed to assist chemical handling specialists, emergency responders, and health and safety engineers and technicians in the safe handling and shipping practices of chemicals.

To use the volume effectively, the reader should first review the Glossary of Terms section immediately preceding the first chemical entry. This section contains precise definitions used for certain parameters where data have been obtained for each chemical. A review of these terms will help the reader interpret certain information. In addition, a list of abbreviations used throughout the volume is also provided in the front section of the handbook.

Chemical information is compiled in this volume in accordance with an alphabetical listing based on the most commonly used chemical name. The most common chemical name designation is based either on (1) that designation specified in the Code of Federal Regulations (CFR), Titles 46 and 49, or (2) a common name for those chemicals known to be hazardous during shipment. As such, for most common names, the shipping name recommended by the U.S. Department of Transportation (DOT) is used as it appears in Title 49 of the CFRs. For each chemical entry, there are five data or information fields that are provided. These information fields are as follows:

- Chemical Designation - A list of common synonyms is given. Synonym names are alternative systematic chemical names and commonly used trivial names for chemicals. An index of synonyms is provided at the end of the handbook to assist the reader in identifying a particular chemical and researching chemical hazards information in the event that the common name of the chemical is not known. The data field also contains the chemical formula. The chemical formula is limited to a commonly used one-line formula. In the case of some organic chemical compounds it has not been possible to represent the chemical structure within such limitation.

- Observable Characteristics - This includes the physical state of the chemical under normal conditions of handling and shipping, its characteristic color and odor. Typical designations for the physical state of a chemical include liquefied gas, liquefied compressed gas, liquid, and solid. Where a compound may be shipped or handled as either a liquid or solid, both designations are given. The color description provided is that for pure liquid. The reader should recognize that occasionally the color of a chemical changes when it dissolves in water or becomes a gas. Similarly, the odor description is that for pure material. The term "characteristic" is used in those cases when no other reasonable description of the chemical's odor could be found.

- Physical and Chemical Properties - Information provided for each chemical include the material's physical state, its molecular weight, boiling point, freezing point, critical properties (temperature and pressure), specific gravity, vapor (gas) density, the ratio of specific heats of vapor, and various thermodynamic properties. The following are more detailed explanations of the information field entries. The *Physical State at 15 °C and 1 atm* is provided, which indicates whether the chemical is a solid, liquid, or gas after it has reached equilibrium with its surroundings at "ordinary" conditions of temperature and pressure. The *Molecular Weight* is the weight of a molecule of the chemical relative to a value of 12 for one atom of carbon. The molecular weight is useful in converting from molecular units to weight units, and in calculating the pressure, volume and temperature relationships of gaseous substances. The *Boiling Point at 1 atm*, the *Freezing Point*, and the *Critical Temperature* data are each given in three sets of units as follows: °F, °C, °K. As an example - for the chemical ACETALDEHYDE, the boiling point at 1 atm is 68.7 °F, 20.4 °C, and 293.6 °K. Entries for *Critical Pressure* are given in three sets of units: psia, atm, MN/m². As an example - for acetaldehyde, the critical pressure data in three units are 820 psia, 56 atm, and 5.7 MN/m² The entries for *Specific Gravity* are typically based on 4 °C unless otherwise specified, and the entry for *Vapor (Gas) Density* is described in the Glossary of Terms section. Thermodynamic properties include the *Ratio of Specific Heats of Vapor (Gas)*, The *Latent Heat of Vaporization*, *Heat of Combustion*, and *Heat of Decomposition*. These data are given in the following three sets of units: Btu/lb, cal/g, J/kg. As an example - for acetaldehyde, the latent heat of vaporization is 245 Btu/lb, 136 cal/g, and 5.69 x 10⁵ J/kg.

- Health Hazards Information - Information included are recommended personal protective equipment for hazard materials handling specialist, typical symptoms following exposure to the chemical, general first aid treatment procedures, and various toxicological information including toxicity by ingestion, inhalation and short term exposures. Additional information included are the liquid or solid irritant characteristics and odor threshold data.

- Fire Hazards - Information compiled includes flash point temperature, flammable limits (explosivity range) in air, a list of fire extinguishing agents to be used, along with a list of fire extinguishing units not to be used, special by-products or hazards of combustion, a description of the chemical's behavior under a fire situation, the ignition temperature, its electrical hazard rating, and its burning rate (if applicable). The burning rate is based on experimentally reported literature data for a standing pool of liquid chemical. If a data field has the entry - "no data", it means that none could be found. If the entry "not pertinent" is given , it means that the property or characteristic does not apply. As an example, for a non-volatile chemical, the flash point temperature has no significance.

- Chemical Reactivity - Information provided includes the chemical's reactivity when in contact with water, as well as its chemical reactivity with common materials such as metals, plastics and organic matter. Information of the chemical's stability during transport is also given. Where appropriate, information on recommended neutralizing agents for acids and caustics are provided. Finally, information on whether the material polymerizes, along with a recommended inhibitor of polymerization are given where appropriate.

The reader should note that although the author has made every reasonable attempt to verify the accuracy of the information compiled in this volume by a review of multiple open literature sources, there are no guarantees as to the accuracy of information, and we do not recommend or endorse the application of this information for design purposes or emergency response procedures. This handbook provides guidance only, and much of the information and data will require interpretation and prudent judgement on the part of a knowledgeable reader with training in chemistry, engineering, and safe handling procedures for hazardous chemicals.

Nicholas P. Cheremisinoff, Ph.D.

ABOUT THE AUTHOR

Nicholas P. Cheremisinoff is President of N&P Limited, a consulting firm specializing in environmental management and privatization issues in Central & Eastern Europe and the Newly Independent States of the former Soviet Union. He has nearly twenty years of industry, applied research, and business development experience, and he has been on multi-year assignments in Russia and Ukraine addressing environmental and industrial health & safety problems as they relate to privatization and industry sustainability in economies in transition. Among his clients are the World Bank Organization, the United States Agency for International Development, Chemonics International, Booz-Allen & Hamilton Corporation, K&M Engineering and Consulting Company, the United States Department of Energy, and others. Dr. Cheremisinoff has contributed extensively to the industrial press with hundreds of articles and as the author, co-author or editor of more than 150 technical books, including Butterworth-Heinemann's *Liquid Filtration - 2nd edition*. He received his B.S., M.S. and Ph.D. degrees in chemical engineering from Clarkson College of Technology.

ABBREVIATIONS

ACGIH	American Conference of Governmental Hygienists
ANSI	American National Standards Institute
atm	atmospheres
Btu/lb	British thermal units per pound
CC	closed cup method
CFR	Code of Federal Regulations
CPC	chemical protective clothing
deg. C	degrees Celsius
deg. F	degrees Fahrenheit
DOT	Department of Transportation
est.	estimated value
g/kg	grams per kilogram
IDLH	immediately dangerous to life and health
ISO	International Standards Organization
LC_{50}	lethal concentration at 50th percentile
LD_{50}	lethal dose at 50th percentile
LEL	lower explosion limit
LFL	lower flammabiliy limit
mg/m^3	milligrams per cubic meter
Min.	minimum value
mm/min	millimeters per minute
mol. wt.	molecular weight
MSA	Mine Safety Administration
n -	normal
NFPA	National Fire Protection Association
NIOSH	National Institute of Occupational Safety and Health
OC	open cup method
OSHA	Occupational Safety and Health Administration
ppm	parts per million
psia	pounds per square inch - absolute
SCBA	self-contained breathing apparatus
STEL	short term exposure limit
tech. grades	technical grades
TLV	threshold limit value
UEL	upper explosion limit
UFL	upper flammability limit

GLOSSARY OF TERMS

Boiling Point at 1 atm - Defined as the characteristic temperature of a liquid when its vapor pressure is 1 atm. As an example, when water is heated to 100°C (212°F), its vapor pressure rises to 1 atm and the liquid boils. The boiling point at 1 atm indicates whether the liquid will boil and become a gas at any particular temperature and at sea-level atmospheric pressure.

Burning Rate - Defined as the rate (in millimeters per minute) at which a pool of liquid decreases as the liquid burns. Details of measurements are provided by D. S. Burgess, A. Strasser, and J. Grumer, "Diffusive Burning of Liquid Fuels in Open Trays," Fire Research Abstracts and Reviews, 3, 177 (1961).

Flammable Limits in Air - This is a concentration expressed as percent by volume of the chemical in air, whereby spontaneous combustion will be supported. The lowest concentration where combustion will be supported is known as the lower flammability limit (LFL) or lower explosion limit (LEL). LEL and LFL are considered interchangeable. The upper concentration limit is the UFL (Upper Flammability Limit) or UEL (Upper Explosion Limit).

Flammability Range - Defined as the difference between the UEL and LEL. This difference is an indication of how wide the flammability limits of a chemical are. The wider this range, the more hazardous the chemical may be considered from a fire standpoint.

Flash Point - The flash point of a material is the lowest temperature at which vapors above a volatile substance will ignite in air when exposed to a flame. Depending on the test method used, the value of flash point temperature is either Tag Closed Cup (CC) based on ASTM D56 test method, or Cleveland Open Cup (CC) based on ASTM 093. The value provides a relative indication of the flammability of the chemical.

Freezing Point - Defined as the temperature at which a liquid changes from liquid to solid state. For example, liquid water changes to solid ice at 0 °C (32 °F). Some liquids solidify very slowly even when cooled below their freezing point. When liquids are not pure, their freezing points are lowered slightly.

Heat of Combustion - Defined as the amount of heat liberated when the specific weight is burned in oxygen at 25 °C. The products of combustion are assumed to remain as gases, and the value given is referred to as the "lower heat value". A negative sign before the value indicates that heat is given off when the chemical burns. Three sets of units are given: Btu per pound, calories per gram, and joules per kilogram.

Heat of Decomposition - Defined as the amount of heat liberated when the specified weight decomposes to more stable substances. This value is given for very few chemicals , because most are stable and do not decompose under the conditions of temperature and pressure they are normally handled under. A negative sign before a value indicates that heat is given off during the decomposition. The value does not include the heat that is given off when the chemical burns. Three sets of units are given: Btu per pound, calories per gram, and joules per kilogram.

Ignition Temperature - This is defined as the minimum temperature at which a chemical substance will ignite without a spark or flame being present. Along with the values of flash point and flammability range, it provides and indication of the relative fire potential for the chemical.

Late Toxicity - Where there is evidence that a chemical can cause cancer, mutagenic effects, teratogenic effects, or delayed injury to vital organs such as the liver or kidney, a qualitative description of the chemical is given. The term implies long-term or chronic effects due to exposure to the chemical.

Latent Heat of Vaporization - Defined as the heat that must be added to the specified weight of a liquid before it can change to vapor (gas). The value varies with temperature. The value given in the handbook is that derived at the chemical's boiling point at 1 atm.. Three sets of units are given: Btu per pound, calories per gram, and joules per kilogram. No value is given for chemicals with very high boiling points at 1 atm, because such chemical substances are considered essentially nonvolatile.

Molecular Weight - Defined as the weight of a molecule of the chemical relative to a value of 12 for one atom of carbon. The molecular weight is useful in converting from molecular units to weight units, and in calculating the pressure, volume and temperature relationships of gaseous substances. The ratio of the densities of any two gases is approximately equal to the ratio of their molecular weights. The molecular weights of mixtures can be calculated if both the identity and quantity of each component of the mixture are known. Because the composition of mixtures described in this handbook are not known, or because they vary from chemical suppliers, no molecular weights are given for mixtures.

Short Term Exposure Limits - Defined as the parts of vapor (gas) per million parts of contaminated air by volume at 25 °C (77 °F) and atmospheric pressure. The limits are given in milligrams per cubic meter for chemicals that can form a fine mist or dust. The values are the maximum permissible average exposures for the time periods specified. The term Short Term Exposure Limit, or STEL, is also used and is considered interchangeable with Short Term Inhalation Limit. The STEL designation is derived from the OSHA standards.

Synonyms - These are alternative systematic chemical names and commonly used trivial names for chemicals. An index of synonyms is provided at the end of the handbook to assist the reader in researching chemical hazards information.

Toxicity by Ingestion - The designation LD_{50} is commonly used in the handbook. The LD_{50} values are those defined in most cases by the national Academy of Sciences, but actual data were collected from various sources such as company specific material safety data sheets. The term LD_{50} (meaning lethal dose at the 50th percentile population) indicates that about 50 percent of the test animals given a specified dose by mouth will die. Thus, for a chemical whose LD_{50} is below 50 mg/kg, the toxic dose for 50 % of animals weighing 70 lb (150 kg) is 70 x 50 = 3500 mg = 3.5 g, which is less than one teaspoon. For a chemical with an LD_{50} value of between 5 to 15 g/kg, the amount would be between a pint and a quart for a 150 lb man.

Threshold Limit Value - The term refers to toxicity by inhalation. The abbreviation used is TLV. The TLV is usually expressed in units of parts per million (ppm) - i.e., the parts of vapor (gas) per million parts of contaminated air by volume at 25 °C (77°F) and atmospheric pressure. For chemicals that form a fine mist or dust, the concentration is given in milligrams per cubic meter (mg/m^3). The TLV is defined as the concentration of the chemical in air that can be breathed for five consecutive eight-hour workdays (i.e., 40 hours per week) by most people without suffering adverse health effects. This is the definition given by the American Conference of Governmental Industrial Hygienists.

Vapor (Gas) Specific Gravity - Defined as the ration of the weight of the vapor to an equal volume of dry air at the same conditions of temperature and pressure. Buoyant vapors have a vapor specific gravity less than one. The value may be approximated by the ratio M/29, where M is the molecular weight of the chemical and 29 is the molecular weight of air. In some instances, the vapor may be at a temperature that is different from that of the surrounding air. For example, the vapor form a container of boiling methane at -172 °F sinks in warm air, enough though the vapor specific gravity of methane at 60 °F is about 0.6.

A

ACETALDEHYDE

Chemical Designations - *Synonyms*: Acetic Aldehyde; Ethanal, Ethyl Aldehyde; *Chemical Formula*: CH_3CHO.

Observable Characteristics - *Physical State (as normally shipped)*: Liquid; *Color*: Colorless; *Odor*: Penetrating, fruity; sharp pungent.

Physical and Chemical Properties - *Physical State at 15 °C and 1 atm.*: Liquid; *Molecular Weight*: 44.05; *Boiling Point at 1 atm.*: 68.7, 20.4, 293.6; *Freezing Point*: -189, -123, 150; *Critical Temperature*: 370, 188, 461; *Critical Pressure*: 820, 56, 5.7; *Specific Gravity*: 0.780 at 20 °C (liquid); *Vapor (Gas) Density*: 1.5; *Ratio of Specific Heats of Vapor (Gas)*: 1.182; *Latent Heat of Vaporization*: 245, 136, 5.69; *Heat of Combustion*: -10600, -5890, -246.4; *Heat of Decomposition*: Not pertinent.

Health Hazards Information - *Recommended Personal Protective Equipment*: Rubber gloves, eye goggles, and other equipment to prevent contact with the body. Organic canister or air pack as required; *Symptoms Following Exposure*: Breathing vapors will be irritating and may cause nausea, vomiting, headache, and unconsciousness. Contact with eyes may cause burns. Skin contact from clothing wet with the chemical causes burns or severe irritation; *General Treatment for Exposure*: INHALATION: remove victim to fresh air; if breathing has stopped, give artificial respiration; if breathing is difficult, give oxygen; call a physician at once. SKIN: wash with soap and water. EYES: flush with water; *Toxicity by Inhalation (Threshold Limit Value)*: 100 ppm; *Short-Term Exposure Limits*: 50 ppm for 60 min.; *Toxicity by Ingestion*: LD_{50} 0.5 to 5 g/kg (cat); *Late Toxicity*: No data found; *Vapor (Gas) Irritant Characteristics*: Vapor is moderately irritating such that workers will not usually tolerate moderate to high concentrations; *Liquid or Solid Irritant Characteristics*: Minimum hazard. If spilled on clothing and allowed to remain, may cause smarting and reddening of the skin; *Odor Threshold*: 0.21 ppm.

Fire Hazards - *Flash Point (deg. F)*: -36 CC; -59 OC; *Flammable Limits in Air (%)*: 4 - 60; *Fire Extinguishing Agents*: Dry chemical, alcohol foam, carbon dioxide; *Fire Extinguishing Agents Not To Be Used*: Water may be ineffective; *Special Hazards of Combustion Products*: Produces irritating vapors when heated; *Behavior in Fire*: Vapors are heavier than air and may travel to a considerable distance for a source of ignition and flash back; *Ignition Temperature (deg. F)*: 365; *Electrical Hazard*: Class 1, Group C; *Burning Rate*: 3.3 mm/min.

Chemical Reactivity - *Reactivity with Water*: No reaction; *Reactivity with Common Materials*: No reaction; *Stability During Transport*: Stable; *Neutralizing Agents for Acids and Caustics*: Not Pertinent; *Polymerization*: May occur. Avoid contact with heat, dust, strong oxidizing and reducing agents, strong acids and bases; *Inhibitor of Polymerization*: None.

ACETIC ACID

Chemical Designations - *Synonyms*: Ethanoic Acid, Glacial Acetic Acid, Vinegar acid; *Chemical Formula*: CH_3COOH.

Observable Characteristics - *Physical State (as normally shipped)*: Liquid; *Color*: Colorless; *Odor*: Characteristic vinegar, pungent; vinegar-like; sharp.

Physical and Chemical Properties - *Physical State at 15 °C and 1 atm.*: Liquid; *Molecular Weight*: 60.05; *Boiling Point at 1 atm.*: 244, 117.9, 391.1; *Freezing Point*: 62.1, 16.7, 290; *Critical Temperature*: 611, 321.6, 594.8; *Critical Pressure*: 839, 57.1, 5.78; *Specific Gravity*: 1.051 at 20 °C(liquid); *Vapor (Gas) Density*: Not pertinent; *Ratio of Specific Heats of Vapor (Gas)*: 1.145; *Latent Heat of Vaporization*: 17.1, 96.7, 4.05; *Heat of Combustion*: -5645, -3136, -131.3; *Heat of Decomposition*: Not pertinent.

Health Hazards Information - *Recommended Personal Protective Equipment*: Protective clothing should be worn when skin contact can occur. Respiratory protection is necessary when exposed to vapor. Complete eye protection is recommended; *Symptoms Following Exposure*: Breathing of vapors

causes coughing, chest pains, and irritation of the nose and throat; may cause nausea and vomiting. Contact with skin and eyes causes burns; *General Treatment for Exposure*: INHALATION: Move the victim immediately to fresh air. If breathing becomes difficult, give oxygen and get medical attention immediately. INGESTION: If the victim is conscious, have him drink water or milk. Do not induce vomiting. SKIN OR EYE CONTACT: Flush immediately with lots of clean running water; wash eyes for at least 15 min. and get medical attention as quickly as possible; remove contaminated clothing; *Toxicity by Inhalation (Threshold Limit Value)*: 10 ppm; *Short-Term Exposure Limits*: 40 ppm for 5 min.; *Toxicity by Ingestion*: LD_{50} 0.5 to 5.0 g/kg (rat); *Late Toxicity*: No data; *Vapor (Gas) Irritant Characteristics*: Vapors cause moderate irritation such that workers will find high concentrations very unpleasant. Effects are temporary; *Liquid or Solid Irritant Characteristics*: This is a fairly severe skin irritant; may cause pain and secondary burns after a few minutes of contact; *Odor Threshold*: 1.0 ppm.
Fire Hazards - *Flash Point (deg. F)*: 112 OC; 104 CC; *Flammable Limits in Air (%)*: 5.4 - 16.0; *Fire Extinguishing* Agents: Water, alcohol foam, dry chemical or carbon dioxide; *Fire Extinguishing Agents Not To Be Used*: None; *Special Hazards of Combustion Products*: Irritating vapors produced when heated; *Behavior in Fire*: Not Pertinent; *Ignition Temperature (deg. F)*: 800; *Electrical Hazard*: Not Pertinent; *Burning Rate*: 1.6 mm/min.
Chemical Reactivity - *Reactivity with Water*: No reaction; *Reactivity with Common Materials*: Corrosive, particularly when diluted. Attacks most common metals, including most stainless steels. Excellent solvent for many synthetic resins or rubber; *Stability During Transport*: Stable; *Neutralizing Agents for Acids and Caustics*: Dilute with water, rinse with sodium bicarbonate solution; *Polymerization*: Not pertinent; *Inhibitor of Polymerization*: Not pertinent.

ACETIC ANHYDRIDE
Chemical Designations - *Synonyms*: Ethanoic Anhydride; *Chemical Formula*: $CH_3CO-O-COCH_3$.
Observable Characteristics - *Physical State (as normally shipped)*: Liquid; *Color*: Colorless; *Odor*: Very strong; pungent; vinegar-like characteristic odor.
Physical and Chemical Properties - *Physical State at 15 °C and 1 atm.*: Liquid; *Molecular Weight*: 102.09; *Boiling Point at 1 atm.*: 282, 139, 412; *Freezing Point*: -101, -74.1, 199.1; *Critical Temperature*: 565, 296, 569; *Critical Pressure*: 679, 46.2, 4.68; *Specific Gravity*: 1.08 at 20 °C (liquid); *Vapor (Gas) Density*: Not pertinent; *Ratio of Specific Heats of Vapor (Gas)*: 1.093; *Latent Heat of Vaporization*: 119, 66.2, 2.77; *Heat of Combustion*: -7058, -3921, -164.2; *Heat of Decomposition*: Not pertinent.
Health Hazards Information - *Recommended Personal Protective Equipment*: Protective clothing when skin contact might occur; respiratory protection is necessary for all exposures; complete eye protection is recommended; *Symptoms Following Exposure*: Liquid is volatile and causes little irritation on unprotected skin. However, causes severe burns when cloning is wet with the chemical, or if it enters gloves or shoes; causes skin and eye burns and irritation of the respiratory tract. Nausea and vomiting may develop after exposure; *General Treatment for Exposure*: INHALATION: Move the victim immediately to fresh air; if breathing becomes difficult, give oxygen, and seek medical attention immediately. INGESTION: Do not induce vomiting. SKIN OR EYE CONTACT WITH LIQUID OR VAPOR: Flush immediately with clean, running water; wash eyes for at least 15 minutes; seek medical attention immediately; *Toxicity by Inhalation (Threshold Limit Value)*: 5 ppm; *Short-Term Exposure Limits*: No data found; *Toxicity by Ingestion*: 0.5 to 5.0 g/kg (rat); *Late Toxicity*: Not pertinent; *Vapor (Gas) Irritant Characteristics*: Vapor is moderately irritating such that personnel will not usually tolerate moderate or high concentrations; *Liquid or Solid Irritant Characteristics*: Fairly severe skin irritant; may cause pain and second degree burns; burns skin after a few minutes of contact; *Odor Threshold*: 0.14 ppm.
Fire Hazards - *Flash Point (deg. F)*: 136 OC; 120 CC; *Flammable Limits in Air (%)*: 2.7 - 10.0; *Fire Extinguishing Agents*: Water spray, dry chemical, alcohol foam, or carbon dioxide; *Fire Extinguishing Agents Not To Be Used*: Water and foam react, but heat liberated is not enough to create a hazard. Dry chemical forced below the surface can cause foaming and boiling; *Special Hazards of Combustion Products*: Irritating vapors generated upon heating; *Behavior in Fire*: Not pertinent; *Ignition

Temperature (deg. F): 600; *Electrical Hazard*: Not pertinent; *Burning Rate*: 3.3 mm/min.

Chemical Reactivity - *Reactivity with Water*: Reacts slowly with water, but considerable heat is liberated when contacted with spray water; *Reactivity with Common Materials*: Corrodes iron, steel and other metals; *Stability During Transport*: Stable; *Neutralizing Agents for Acids and Caustics*: Dilute with water and use sodium bicarbonate solution to rinse; *Polymerization*: Not pertinent; *Inhibitor of Polymerization*: Not pertinent.

ACETONE

Chemical Designations - *Synonyms*: Dimethyl Ketone, 2-Propanone; *Chemical Formula*: CH_3COCH_3.

Observable Characteristics - *Physical State (as normally shipped)*: Liquid; *Color*: Colorless; *Odor*: Sweetish; pleasant, resembling that of mint or fruit; pungent, sharp, penetrating, ketonic pleasant, non-residual.

Physical and Chemical Properties - *Physical State at 15 °C and 1 atm.*: Liquid; *Molecular Weight*: 58.08; *Boiling Point at 1 atm.*: 133, 56.1, 329.3; *Freezing Point*: -138, -94.7, 178.5; *Critical Temperature*: 455, 235, 508; *Critical Pressure*: 682, 46.4, 4.70; *Specific Gravity*: 0.971 at 20 °C (liquid); *Vapor (Gas) Density*: 2.0; *Ratio of Specific Heats of Vapor (Gas)*: 1.127; *Latent Heat of Vaporization*: 220, 122, 5.11; *Heat of Combustion*: -12,250, -6,808, -285.0; *Heat of Decomposition*: Not pertinent.

Health Hazards Information - *Recommended Personal Protective Equipment*: Organic vapor canister or air-supplied respirator; synthetic rubber gloves; chemical safety goggles or face splash shield; *Symptoms Following Exposure*: INHALATION: vapor irritating to eyes and mucous membranes; acts as an anesthetic in very high concentrations. INGESTION: low order of toxicity but very irritating to the mucous membranes. SKIN: prolonged excessive contact causes defatting of the skin, possibly leading to dermatitis; *General Treatment for Exposure*: INHALATION: if victim is overcome, remove to fresh air and call a physician; administer artificial respiration if breathing is irregular or stopped. INGESTION: if victim has swallowed large amounts and is conscious and not having convulsions, induce vomiting and seek medical help immediately. SKIN: wash with clean running water. EYES: flush with water immediately for at least 15 min. and consult a physician; *Toxicity by Inhalation (Threshold Limit Value)*: 1000 ppm; *Short-Term Exposure Limits*: 1000 ppm for 30 minutes; *Toxicity by Ingestion*: LD_{50} 5 to 15 g/kg (dog); *Late Toxicity*: Not pertinent; *Vapor (Gas) Irritant Characteristics*: If present in high concentrations, vapors cause moderate irritation of the eyes or respiratory system. Effects are temporary; *Liquid or Solid Irritant Characteristics*: No appreciable hazard. Practically harmless to the skin because it is very volatile and evaporates quickly from the skin; *Odor Threshold*: 100 ppm.

Fire Hazards - *Flash Point (deg. F)*: 4 OC, 0 CC; *Flammable Limits in Air (%)*: 2.6 - 12.8; *Fire Extinguishing Agents*: Alcohol foam, dry chemical, carbon dioxide; *Fire Extinguishing Agents Not To Be Used*: Water in straight hose streams will scatter fire and is not recommended; *Special Hazards of Combustion Products*: Not pertinent; *Behavior in Fire*: Not pertinent; *Ignition Temperature (deg. F)*: 869; *Electrical Hazard*: Class I, Group D; *Burning Rate*: 3.9 mm/min.

Chemical Reactivity - *Reactivity with Water*: No reaction; *Reactivity with Common Materials*: No reaction; *Stability During Transport*: Stable; *Neutralizing Agents for Acids and Caustics*: Not pertinent; *Polymerization*: Not pertinent; *Inhibitor of Polymerization*: Not pertinent.

ACETONE CYANOHYDRIN

Chemical Designations - *Synonyms*: alpha-Hydroxyisobutyronitrile, 2-Methyllactonitrile; *Chemical Formula*: $(CH_3)_2C(OH)CN$.

Observable Characteristics - *Physical State (as normally shipped)*: Liquid; *Color*: Colorless; *Odor*: Characteristic; distinct, strong cyanide.

Physical and Chemical Properties - *Physical State at 15 °C and 1 atm.*: Liquid; *Molecular Weight*: 85.11; *Boiling Point at 1 atm.*: Decomposes; *Freezing Point*: -5.8, -21, 252; *Critical Temperature*: Not pertinent; *Critical Pressure*: Not pertinent; *Specific Gravity*: 0.925 at 25 °C (liquid); *Vapor (Gas) Density*: Not pertinent; *Ratio of Specific Heats of Vapor (Gas)*: 1.074; *Latent Heat of Vaporization*: No data; *Heat of Combustion*: Not pertinent; *Heat of Decomposition*: Not pertinent.

Health Hazards Information - *Recommended Personal Protective Equipment*: Air-supplied respirator or chemical cartridge respirator approved for use with acrylonitrile in less than 2% concentrations; rubber or plastic gloves; cover goggles or face mask; rubber boots; chemical protective suit; safety helmet; *Symptoms Following Exposure*: At low dosages the earliest symptoms may be weakness, headaches, confusion, sporadic nausea and vomiting. Respiratory rate and depth will usually be increased at the beginning and at later stages become slow and gasping; *General Treatment for Exposure*: Call a physician for all cases of over-exposure. INHALATION: Remove victim to fresh air. First responders/rescuers should wear suitable respiratory protection. If breathing has stopped, give artificial respiration until physician arrives. CHRIS advises that if victim is unconscious, administer amyl nitrate by crushing an ampule in a cloth and holding it under the nose for 15 seconds in every minute. Do not interrupt artificial respiration during the process. Replace the ampule when its strength is spent; continue treatment until victim's condition improves or physician arrives. INGESTION: If the victim is conscious, induce vomiting by having him drink strong salt water. SKIN: Remove contaminated clothing and wash affected skin thoroughly with soap and water. Use copious amount of water. EYES: Hold eyelids apart and wash with continuous, gentle stream of water for at least 15 min.; *Toxicity by Inhalation (Threshold Limit Value)*: No data; *Short-Term Exposure Limits*: No data; *Toxicity by Ingestion*: < 50 mg/kg (rats); *Late Toxicity*: Causes liver damage in rats; *Vapor (Gas) Irritant Characteristics*: Vapors irritate eyes and respiratory tract when present in high concentrations; however effects are temporary; *Liquid or Solid Irritant Characteristics*: Causes smarting of the skin and first-degree chemical burns on short exposure, and second-degree chemical burns on long exposure; *Odor Threshold*: No data found.

Fire Hazards - *Flash Point (deg. F)*: 165 CC; *Flammable Limits in Air (%)*: 2.2 - 12; *Fire Extinguishing Agents*: Water spray, dry chemical, alcohol foam, carbon dioxide; *Fire Extinguishing Agents Not To Be Used*: Not pertinent; *Special Hazards of Combustion Products*: Toxic hydrogen cyanide is generated upon heating; *Behavior in Fire*: Not pertinent; *Ignition Temperature (deg. F)*: 1270; *Electrical Hazard*: Not pertinent; *Burning Rate*: No data.

Chemical Reactivity - *Reactivity with Water*: No reaction; *Reactivity with Common Materials*: No reactions; *Stability During Transport*: Stable; *Neutralizing Agents for Acids and Caustics*: Not pertinent; *Polymerization*: Not pertinent; *Inhibitor of Polymerization*: Not pertinent.

ACETONITRILE

Chemical Designations - *Synonyms*: Ethanenitrile, Ethyl Nitrate, Cyanomethane, Methyl cyanide; *Chemical Formula*: CH_3CN.

Observable Characteristics - *Physical State (as normally shipped)*: Liquid; *Color*: Colorless; *Odor*: Sweet; ethereal.

Physical and Chemical Properties - *Physical State at 15 °C and 1 atm.*: Liquid; *Molecular Weight*: 41.05; *Boiling Point at 1 atm.*: 179, 81.6, 354.8; *Freezing Point*: -50.3, -45.7, 227.5; *Critical Temperature*: 526.5, 274.7, 547.9; *Critical Pressure*: 701, 47.7, 4.83; *Specific Gravity*: 0.787 at 20 °C (liquid); *Vapor (Gas) Density*: 1.4; *Ratio of Specific Heats of Vapor (Gas)*: 1.192; *Latent Heat of Vaporization*: 313, 174, 7.29; *Heat of Combustion*: -13360, -7420, -310.7; *Heat of Decomposition*: Not pertinent.

Health Hazards Information - *Recommended Personal Protective Equipment*: Must wear self contained breathing apparatus (SCBA); *Symptoms Following Exposure*: Exposure to 160 ppm for 4 hours causes flushing of the face and feeling of constriction in the chest. Exposure to 500 ppm for brief periods is irritating to the nose and throat. Severe exposure results in irritability, skin eruptions, confusion, delirium, convulsions, paralysis, and death due to central nervous system depression; *General Treatment for Exposure*: Remove victim from contaminated atmosphere. Apply artificial respiration and oxygen if respiration is impaired; *Toxicity by Inhalation (Threshold Limit Value)*: 40 ppm; *Short-Term Exposure Limits*: 40 ppm for 60 minutes; *Toxicity by Ingestion*: 500 mg/kg (guinea pig); *Late Toxicity*: Not pertinent; *Vapor (Gas) Irritant Characteristics*: Vapors cause slight smarting of the eyes or respiratory system if present in high concentrations. Effects are temporary; *Liquid or*

Solid Irritant Characteristics: Represents a minimum hazard. If spilled on clothing and allowed to remain, may cause smarting and reddening of the skin; *Odor Threshold*: 40 ppm.

Fire Hazards - *Flash Point (deg. F)*: 42 OC; *Flammable Limits in Air (%)*: 4.4 - 16; *Fire Extinguishing Agents*: Alcohol foam, dry chemical, carbon dioxide; *Fire Extinguishing Agents Not To Be Used*: Water may be ineffective; *Special Hazards of Combustion Products*: Toxic vapors generated during heating; *Behavior in Fire*: Vapor is heavier than air and may travel considerable distance to ignition source and flash back; *Ignition Temperature (deg. F)*: 975; *Electrical Hazard*: Not pertinent; *Burning Rate*: 2.7 mm/min.

Chemical Reactivity - *Reactivity with Water*: No reaction; *Reactivity with Common Materials*: No reactions; *Stability During Transport*: Stable; *Neutralizing Agents for Acids and Caustics*: Not pertinent; *Polymerization*: Not pertinent; *Inhibitor of Polymerization*: Not pertinent.

ACETYL BROMIDE

Chemical Designations - *Synonyms*: No common synonyms; *Chemical Formula*: CH_3COBr.

Observable Characteristics - *Physical State (as normally shipped)*: Liquid; *Color*: Colorless; *Odor*: Acrid and sharp.

Physical and Chemical Properties - *Physical State at 15 °C and 1 atm.*: Liquid; *Molecular Weight*: 122.95; *Boiling Point at 1 atm*: 169, 76, 349; *Freezing Point*: -141.7, -96.5, 176.7; *Critical Temperature*: Not pertinent; *Critical Pressure*: Not pertinent; *Specific Gravity*: 1.66 at 16 °C (liquid); *Vapor (Gas) Density*: 4.24; *Ratio of Specific Heats of Vapor (Gas)*: 1.44; *Latent Heat of Vaporization*: 106, 59, 2.5; *Heat of Combustion*: No data; *Heat of Decomposition*: Not pertinent.

Health Hazards Information - *Recommended Personal Protective Equipment*: Safety goggles; gloves; adequate ventilation; provisions for flushing eyes or skin with water; *Symptoms Following Exposure*: Inhalation results in primary irritation of the respiratory tract; symptoms of lung damage may be delayed. Contact with liquid produces primary irritation of eyes and severe skin damage; delayed blistering of the skin often occurs; *General Treatment for Exposure*: INHALATION: Remove the victim from the contaminated area. If breathing has stopped, give artificial respiration. If breathing is difficult, give oxygen. Watch victim carefully for any signs of delayed lung damage. EYES: Flush with water for at least 15 min. and seek medical attention. SKIN: Flush with water and treat chemical burns as needed; *Toxicity by Inhalation (Threshold Limit Value)*: No data; *Short-Term Exposure Limits*: No data; *Toxicity by Ingestion*: Oral rat LD50 3,310 mg/kg (acetic acid). Note that this chemical decomposes violently in water, forming bromic acid and acetic acid; *Late Toxicity*: No data; *Vapor (Gas) Irritant Characteristics*: No data; *Liquid or Solid Irritant Characteristics*: No data; *Odor Threshold*: 5.0×10^{-4} ppm.

Fire Hazards - *Flash Point*: Data not available; *Flammable Limits in Air (%)*: Data not available; *Fire Extinguishing Agents*: Carbon dioxide; *Fire Extinguishing Agents Not To Be Used*: Water; *Special Hazards of Combustion Products*: Toxic and irritating hydrogen bromide fumes may form in fires; *Behavior in Fire*: Do not apply water to adjacent fires. Reacts with water to produce toxic and irritating gases; *Ignition Temperature*: Data not available; *Electrical Hazard*: Data not available; *Burning Rate*: Data not available.

Chemical Reactivity - *Reactivity with Water*: Reacts violently, forming corrosive and toxic fumes of hydrogen bromide; *Reactivity with Common Materials*: Attacks and corrodes wood and most metals in the presence of moisture. Flammable hydrogen gas may collect in enclosed spaces; *Stability During Transport*: Stable if protected from moisture; *Neutralizing Agents for Acids and Caustics*: Flood with water, rinse with dilute sodium bicarbonate or soda ash solution; *Polymerization*: Not pertinent; *Inhibitor of Polymerization*: Not pertinent.

ACETYL CHLORIDE

Chemical Designations - *Synonyms*: No common synonyms; *Chemical Formula*: CH_3COCl.

Observable Characteristics - *Physical State (as normally shipped)*: Liquid; *Color*: Colorless; *Odor*: Pungent.

Physical and Chemical Properties - *Physical State at 15 °C and 1 atm.*: Liquid; *Molecular Weight*:

78.5; *Boiling Point at 1 atm.*: 124, 51, 324; *Freezing Point*: -170, -112, 161; *Critical Temperature*: 475, 246, 519; *Critical Pressure*: 845, 57.5, 5.83; *Specific Gravity*: 1.1039 at 21 °C (liquid); *Vapor (Gas) Density*: 3; *Ratio of Specific Heats of Vapor (Gas)*: 1.1467; *Latent Heat of Vaporization*: 160, 88, 3.7; *Heat of Combustion*: -6000, -3300, -140; *Heat of Decomposition*: Not pertinent.

Health Hazards Information - *Recommended Personal Protective Equipment*: Safety goggles, rubber or plastic gloves, self contained breathing apparatus (SCBA); *Symptoms Following Exposure*: Vapor irritates mucous membranes. Ingestion of liquid or contact with eyes or skin causes severe irritation; *General Treatment for Exposure*: INHALATION; Remove victim from exposure and seek immediate medical attention. EYES: Flush with copious amounts of fresh running water. INGESTION: Administer plenty of water; do not induce vomiting; *Toxicity by Inhalation (Threshold Limit Value)*: No data; *Short-Term Exposure Limits*: No data; *Toxicity by Ingestion*: Readily hydrolyzes to form hydrochloric and acetic acids. Oral human LD_{LO} 1470 mg/kg (acetic acid). oral rat LD_{50} 3310 mg/kg (acetic acid); *Late Toxicity*: None; *Vapor (Gas) Irritant Characteristics*: Vapors cause severe irritation of eyes and throat and can cause eye and lung injury. Cannot be tolerated even at low concentrations; *Liquid or Solid Irritant Characteristics*: Severe skin irritant. Causes second- and third-degree chemical burns on short contact and is very damaging to eyes; *Odor Threshold*: Acetic acid - 1 ppm; hydrochloric acid - 1 ppm.

Fire Hazards - *Flash Point (deg. F)*: 40 CC; *Flammable Limits in Air (%)*: Data not available; *Fire Extinguishing Agents*: Carbon dioxide, dry chemical; *Fire Extinguishing Agents Not To Be Used*: Water, foam; *Special Hazards of Combustion* Products: When heated to decomposition, hydrogen chloride and phosgene, extremely poisonous gases, are involved; *Behavior in Fire*: Vapor is heavier than air and may travel a considerable distance to a source of ignition and flash back; *Ignition Temperature (deg. F)*:734; *Electrical Hazard*: Data not available; *Burning Rate: 2.6 mm/min.*

Chemical Reactivity - *Reactivity with* Water: Reacts vigorously with water, involving hydrogen chloride fumes (hydrochloric acid); *Reactivity with Common Materials*: Is highly corrosive to most metals in the presence of moisture; *Stability During Transport*: Stable; *Neutralizing Agents for Acids and Caustics*: Following dilution with water, limestone or sodium bicarbonate can be used; *Polymerization*: Not pertinent; *Inhibitor of Polymerization*: Not pertinent.

ACETYL PEROXIDE

Chemical Designations — *Synonyms*: Diacetyl Peroxide Solution; *Chemical Formula*: $CH_3CO(O_2)OCCH_3$ in dimethyl phthalate.

Observable Characteristics — *Physical State (as normally shipped)*: Liquid; *Color*: Colorless; *Odor*: Pungent.

Physical and Chemical Properties — *Physical State at 15 °C and 1 atm.*: Liquid; *Molecular Weight*: Mixture; *Boiling Point at 1 atm.*: Decomposes; *Freezing Point*: 17, -8, 265; *Critical Temperature*: Not pertinent; *Critical Pressure*: Not pertinent; *Specific Gravity*: 1.2 at 20 °C (liquid); *Vapor (Gas) Density*: Not pertinent; *Ratio of Specific Heats of Vapor (Gas)*: Not pertinent; *Latent Heat of Vaporization*: Not pertinent; *Heat of Combustion*: -15,700, -8750, -366; *Heat of Decomposition*: -50, -28, -1.2.

Health Hazards Information - *Recommended Personal Protective Equipment*: Protective goggles, rubber apron, and gloves; *Symptoms Following Exposure*: Contact with liquid causes irritation of eyes and skin. If ingested, irritates mouth and stomach; *General Treatment for Exposure*: EYES: Wash with plenty of water and seek medical attention. SKIN: Flush with water and wash thoroughly with soap and water; seek medical attention. INGESTION: Induce vomiting and seek immediate medical attention; *Toxicity by Inhalation (Threshold Limit Value)*: No data; *Short-Term Exposure Limits*: No data; *Toxicity by Ingestion*: No data; *Late Toxicity*: No data; *Vapor (Gas) Irritant Characteristics*: No data; *Liquid or Solid Irritant Characteristics*: No data; *Odor Threshold*: No data.

Fire Hazards - *Flash Point (deg. F)*: 113 OC; *Flammable Limits in Air (%)*: Not pertinent; *Fire Extinguishing Agents*: Water, dry chemical, carbon dioxide; *Fire Extinguishing Agents Not To Be Used*: Not pertinent; *Special Hazards of Combustion Products*: Not pertinent; *Behavior in Fire*: May explode. Burns with accelerating intensity; *Ignition Temperature (deg. F)*: Explodes; *Electrical Hazard*: Data not available; *Burning Rate*: Data not available.

Chemical Reactivity - *Reactivity with Water*: No reaction; *Reactivity with Common Materials*: May ignite combustible materials such as wood; *Stability During Transport*: Heat-and-shock-sensitive crystals may separate at very low temperature during transport; *Neutralizing Agents for Acids and Caustics*: Not pertinent; *Polymerization*: Not pertinent; *Inhibitor of Polymerization*: Not pertinent.

ACRIDINE

Chemical Designations - *Synonyms*: 10-Azaanthracene, Benzo (b) Quinoline, Dibenzo [b,e] Pyridine; *Chemical Formula*: $C_{13}H_9N$.

Observable Characteristics — *Physical State (as normally shipped)*: Solid; *Color*: Yellow; *Odor*: Weak, somewhat irritating.

Physical and Chemical Properties - *Physical State at 15 °C and 1 atm.*: Solid; *Molecular Weight*: 179.08; *Boiling Point at 1 atm.*: 655, 346, 619; *Freezing Point*: 230, 110, 383; *Critical Temperature*: Not pertinent; *Critical Pressure*: Not pertinent; *Specific Gravity*: 1.2 at 20 °C (solid); *Vapor (Gas) Density*: Not pertinent; *Ratio of Specific Heats of Vapor (Gas)*: Not pertinent; *Latent Heat of Vaporization*: Not pertinent; *Heat of Combustion*: -15800, -8790, -368; *Heat of Decomposition*: Not pertinent.

Health Hazards Information - *Recommended Personal Protective Equipment*: Dust respirator, chemical goggles, rubber gloves; *Symptoms Following Exposure*: Inhalation irritates respiratory system and causes sneezing. Contact with liquid causes eye irritation, irritation of skin, and mucous membranes. At high temperatures and during sun exposure, damage to the cornea, skin and mucous membranes may occur following the liberation of acridine vapor; *General Treatment for Exposure*: INHALATION: Remove victim to fresh air; if breathing has stopped, give artificial respiration; if breathing is difficult, give oxygen. EYES: wash with copious amounts of water for 20 minutes and seek immediate medical attention. SKIN: wash with large amounts of water for 20 min.; *Toxicity by Inhalation (Threshold Limit Value)*: No data; *Short-Term Exposure Limits*: No data; *Toxicity by Ingestion*: oral rat LD_{50} 2,000 mg/kg; *Late Toxicity*: No data; *Vapor (Gas) Irritant Characteristics*: No data; *Liquid or Solid Irritant Characteristics*: No data; *Odor Threshold*: No data.

Fire Hazards - *Flash Point*: Not pertinent (combustible solid); *Flammable Limits in Air (%)*: Not pertinent; *Fire Extinguishing Agents*: Water, foam, monoammonium phosphate, dry chemical; *Fire Extinguishing Agents Not To Be Used*: Carbon dioxide and other dry chemicals may not be effective; *Special Hazards of Combustion Products*: Toxic oxides of nitrogen may form in fire; *Behavior in Fire*: Sublimes before melting; *Ignition Temperature*: Data not available; *Electrical Hazard*: Not pertinent; *Burning Rate*: Not pertinent.

Chemical Reactivity - *Reactivity with Water*: No reaction; *Reactivity with Common Materials*: Data not available; *Stability During Transport*: Stable; *Neutralizing Agents for Acids and Caustics*: Not pertinent; *Polymerization*: Not pertinent; *Inhibitor of Polymerization*: Not pertinent.

ACROLEIN

Chemical Designations - *Synonyms*: Acraldehyde, Acrylic Aldehyde, 2-Propenal, Acrylaldehyde; *Chemical Formula*: $CH_2=CH•CHO$.

Observable Characteristics - *Physical State (as normally shipped)*: Liquid; *Color*: Colorless to slight yellow; *Odor*: Extremely sharp - lachrymator; piercing and disagreeable; extremely **Physical and Chemical Properties** - *Physical State at 15 °C and 1 atm.*: Liquid; *Molecular Weight*: 56.1; *Boiling Point at 1 atm.*: 127, 53, 326; *Freezing Point*: -125, -87, 186; *Critical Temperature*: 489, 254, 527; *Critical Pressure*: 737, 50.0, 5.08; *Specific Gravity*: 0.843 at 20 °C (liquid); *Vapor (Gas) Density*: 1.94; *Ratio of Specific Heats of Vapor (Gas)*: 1.1487; *Latent Heat of Vaporization*: 216, 120, 5.02; *Heat of Combustion*: -12500, -6950, -290; *Heat of Decomposition*: Not pertinent.

Health Hazards Information - *Recommended Personal Protective Equipment*: Chemical safety goggles and full face shield, self-contained breathing apparatus (SCBA), positive pressure hose mask, airline mask; rubber safety shoes, chemical protective clothing; *Symptoms Following Exposure*: Inhalation causes irritation of nose and throat, a feeling of pressure in the chest, and shortness of breath. Nausea and vomiting occur. Loss of consciousness can occur if exposure has been sufficiently great. Congestion in the chest may be present in varying amounts, and fluid may collect in the lungs (pulmonary edema) of severely exposed victims. Vapor also causes severe eye irritation (redness, weeping, and swelling of lids; liquid burns eyes, contact with skin causes reddening or blistering. Ingestion causes severe irritation of mouth and stomach; *General Treatment for Exposure*: Keep patient warm and quiet; if conscious, give coffee and call a physician immediately after all types of exposures to this chemical. INHALATION: Remove patient to fresh air; if breathing becomes difficult, give oxygen. If breathing has stopped, give artificial respiration. EYES: Immediately flush with plenty of water for at least 15 min. If medical attention is not immediately available, continue eye irrigation for another 15 min. period. Upon completion of first 15 min. eye irrigation period, it is permissible to instill 2 or 3 drops of an effective aqueous local eye anesthetic for relief of pain. No oils or ointments should be used unless so instructed by a physician. SKIN: Flush at once with large amounts of water. Wash thoroughly with soap and large amounts of water. INGESTION: Have victim drink large amounts of water. Induce vomiting and keep patient warm and quiet until physician arrives; *Toxicity by Inhalation (Threshold Limit Value)*: 0.1 ppm; *Short-Term Exposure Limits*: 0.5 ppm - 5 min.; 0.2 ppm - 60 min.; *Toxicity by Ingestion*: LD_{50} < 50 mg/kg; *Late Toxicity*: oral rat LD_{50} 46 mg/kg; oral rabbit LD_{50} 7 mg/kg; *Vapor (Gas) Irritant Characteristics*: Vapors cause severe irritation of eyes and throat, and can cause eye and lung injury. There is no tolerance even at very low concentrations; *Liquid or Solid Irritant Characteristics*: Causes smarting of the skin and first-degree chemical burns on short exposure; may cause second degree chemical burns on short exposure; *Odor Threshold*: 0.21 ppm.

Fire Hazards - *Flash Point (deg. F)*: <0 OC; -13 CC; *Flammable Limits in Air (%)*: 2.8 - 31; *Fire Extinguishing Agents*: Foam, dry chemical, carbon dioxide; *Fire Extinguishing Agents Not To Be Used*: Water may be ineffective; *Special Hazards of Combustion Products*: Poisonous vapor of acrolein is formed from hot liquid; *Behavior in Fire*: Vapor is heavier than air and may travel a considerable distance to a source of ignition and flash back. Polymerization may take place, and containers may explode in fire; *Ignition Temperature (deg. F)*: 453; *Electrical Hazard*: Data not available; *Burning Rate*: 3.8 mm/min.

Chemical Reactivity - *Reactivity with Water*: No reaction; *Reactivity with Common Materials*: No reaction; *Stability During Transport*: Stable when inhibited; *Neutralizing Agents for Acids and Caustics*: Not pertinent; *Polymerization*: Undergoes uncatalyzed polymerization reaction around 200°C. Light promotes polymerization; *Inhibitor of Polymerization*: Hydroquinone: 0.10 to 0.25 %.

ACRYLAMIDE

Chemical Designations - *Synonyms*: Acrylic Amide 50%, Propenamide 50%; *Chemical Formula*: $CH_2 = CHCONH_2 - H_2O$.

Observable Characteristics - *Physical State (as normally shipped)*: Liquid; *Color*: Clear; *Odor*: None.

Physical and Chemical Properties - *Physical State at 15 °C and 1 atm.*: Liquid; *Molecular Weight*: 71 (solute only); *Boiling Point at 1 atm.*: Data not available (Vapor Pressure 0.033 atm at 125 °C; *Freezing Point*: 183, 84, 357; *Critical Temperature*: Not pertinent; *Critical Pressure*: Not pertinent; *Specific Gravity*: 1.05 at 25 °C; *Vapor (Gas) Density*: Not pertinent; *Ratio of Specific Heats of Vapor (Gas)*: Not pertinent; *Latent Heat of Vaporization*: Not pertinent; *Heat of Combustion*: Not pertinent; *Heat of Decomposition*: Not pertinent.

Health Hazards Information - *Recommended Personal Protective Equipment*: Safety glasses with side shields; clean body-covering clothing; rubber gloves, boots, apron as dictated by circumstances; in absence of proper environmental control, use approved dust respirator; *Symptoms Following Exposure*: Has produced central nervous system damage, which is partly reversible. Effects can be produced by oral or skin contact as well as by injection. Chronic acrylamide poisoning can cause midbrain disturbance and peripheral neuropathy. Contact with liquid can cause moderate irritation of eyes and skin and may cause moderate transient corneal injury; *General Treatment for Exposure*: INHALATION: if ill effects occur, immediately get patient to fresh air, keep him quiet and warm, and get medical help. INGESTION: if ingested, immediately give large amounts of water (or milk if immediately available), then induce vomiting and get medical help. EYES: immediately flush with plenty of water for at least 15 min. and get medical promptly. SKIN: immediate, continuous, and thorough washing in flowing water is imperative, preferably deluge shower with abundant soap; if burns are present, get medical help; discard all contaminated clothing and wearing accessories; *Toxicity by Inhalation (Threshold Limit Value)*: 0.3 mg/m^3; *Short-Term Exposure Limits*: Data not available; *Toxicity by Ingestion*: Grade 3; oral rat LD_{50} 170 mg/kg; *Late Toxicity*: Repeated exposure to small amounts may cause essentially reversible neurological effects; *Vapor (Gas) Irritant Characteristics*: Data not available; *Liquid or Solid Irritant Characteristics*: Data not available; *Odor Threshold*: Not pertinent.

Fire Hazards - *Flash Point*: Not flammable; *Flammable Limits in Air (%)*: Not flammable; *Fire Extinguishing Agents*: Not pertinent; *Fire Extinguishing Agents Not To Be Used*: Not pertinent; *Special Hazards of Combustion Products*: Toxic oxides of nitrogen may form in fire; *Behavior in Fire*: Sealed containers may burst as a result of polymerization; *Ignition Temperature*: Not pertinent; *Electrical Hazard*: Not pertinent; *Burning Rate*: Not pertinent.

Chemical Reactivity - *Reactivity with Water*: No reaction; *Reactivity with Common Materials*: Data not available; *Stability During Transport*: Stable; *Neutralizing Agents for Acids and Caustics*: Not pertinent; *Polymerization*: May occur at temperature above 50°C (120°F); *Inhibitor of Polymerization*: Oxygen (air) plus 50 ppm of copper as copper sulfate.

ACRYLIC ACID

Chemical Designations - *Synonyms*: Propenoic Acid; *Chemical Formula*: $CH_2=CHCOOH$.

Observable Characteristics - *Physical State (as normally shipped)*: Liquid; *Color*: Colorless; *Odor*: Acrid.

Physical and Chemical Properties - *Physical State at 15 °C and 1 atm.*: Liquid; *Molecular Weight*: 72.06; *Boiling Point at 1 atm.*: 286.3, 141.3, 414.5; *Freezing Point*: 54.1, 12.3, 285.5; *Critical Temperature*: 648, 342, 615; *Critical Pressure*: 840, 57, 5.8; *Specific Density*: 1.0497 at 20 °C (liquid); *Vapor (Gas) Gravity*: Not pertinent; *Ratio of Specific Heats of Vapor (Gas)*: 1.121; *Latent Heat of Vaporization*: 272.7, 151.5, 6.343; *Heat of Combustion*: -8100, -4500, -188.4; *Heat of Decomposition*: Not pertinent.

Health Hazards Information - *Recommended Personal Protective Equipment*: Chemical respirator at ambient temperatures to avoid inhalation of noxious fumes; rubber gloves if exposed to wet material; acid goggles or face shield for splash exposure; safety shower and/or eye fountain may be required; *Symptoms Following Exposure*: May burn skin or eyes upon short contact. INHALATION: eye and nasal irritation and lacrimation. INGESTION: may cause severe damage to the gastrointestinal tract; *General Treatment for Exposure*: Get medical attention promptly for all exposures. INHALATION: remove victim to fresh air. INGESTION: do NOT induce vomiting. SKIN OR EYES: flush with water for at least 15 min.; *Toxicity by Inhalation (Threshold Limit Value)*: Data not available; *Short-Term Exposure Limits*: Data not available; *Toxicity by Ingestion*: Grade 2; LD_{50} 0.5 to 5 g/kg (rat); *Late Toxicity*: Not pertinent; *Vapor (Gas) Irritant Characteristics*: Vapor is moderately irritating such that personnel will not usually tolerate moderate or high vapor concentrations; *Liquid or Solid Irritant Characteristics*: Fairly severe skin irritant; may cause pain second-degree burns after a few minutes of contact; *Odor Threshold*: Data not available.

Fire Hazards - *Flash Point (deg. F)*: (Glacial) 118 OC; *Flammable Limits in Air (%)*: (Tech.) 2.4 LEL; *Fire Extinguishing Agents*: Water spray, alcohol foam, dry chemical, or carbon dioxide; *Fire Extinguishing Agents Not To Be Used*: Not pertinent; *Special Hazards of Combustion Products*: Toxic vapor are generated when heated; *Behavior in Fire*: May polymerize and explode; *Ignition Temperature (deg. F)*: 374; *Electrical Hazard*: Not pertinent; *Burning Rate*: 1.6 mm/min.

Chemical Reactivity - *Reactivity with Water*: No reaction; *Reactivity with Common Materials*: No reaction; *Stability During Transport*: Normally unstable but will be detonate; *Neutralizing Agents for Acids and Caustics*: Wash with water, rinse with sodium bicarbonate solution; *Polymerization*: May occur in contact with acids, iron salts, or at elevated temperatures and release high energy rapidly; may cause explosion under confinement; *Inhibitor of Polymerization*: Monomethyl ether of hydroquinone 180-200 ppm; phenothiazine (for tech. grades) 1000 ppm; hydroquinone (0.1%); methylene blue (0.5

%); N, N'-diphenyl-p-phenylenediamine (0.05%).

ACRYLONITRILE

Chemical Designations - *Synonyms*: Cyanoethylene, Fumigrain, Ventox, Vinyl Cyanide; *Chemical Formula*: $CH_2=CHCN$.

Observable Characteristics - *Physical State (as normally shipped)*: Liquid; *Color*: Colorless; *Odor*: Mild; pungent, resembling that of peach seed kernels.

Physical and Chemical Properties - *Physical State at 15 °C and 1 atm.*: Liquid; *Molecular Weight*: 53.06; *Boiling Point at 1 atm.*: 171, 77.4, 350.6; *Freezing Point*: -118, -83.6, 189.6; *Critical Temperature (°F,, °K)*: 505, 263, 536; *Critical Pressure*: 660, 45, 4.6; *Specific Gravity*: 0.8075 at 20°C (liquid); *Vapor (Gas) Density*: 1.8; *Ratio of Specific Heats of Vapor (Gas)*: 1.151; *Latent Heat of Vaporization*: 265, 147, 6.16; *Heat of Combustion*: -14,300, -7930, -332; *Heat of Decomposition*: Not pertinent.

Health Hazards Information - *Recommended Personal Protective Equipment*: Air-supplied mask, industrial chemical type, with approved canister for acrilonitrile in low (less than 2%) concentrations; rubber or plastic gloves; cover goggles or face mask; rubber boots; slicker suit; safety helmet; *Symptoms Following Exposure*: Similar to those of hydrogen cyanide. Vapor inhalation may cause weakness, headache, sneezing, abdominal pain, and vomiting. Similar symptoms shown if large amounts of liquid are absorbed through the skin; lesser amounts cause stinging and sometime blisters; contact with eyes causes severe irritation. Ingestion produced nausea, vomiting and abdominal pain; *General Treatment for Exposure*: Skilled medical treatment is necessary; call physician for all cases of exposure. INHALATION: remove victim to fresh air. (Wear an oxygen or fresh-air-supplied mask when entering contaminated area). INGESTION: induce vomiting by administering strong solution of salt water, but only if victim is conscious. SKIN: remove contaminated clothing and wash affected area thoroughly with soap and water. EYES: hold eyelids apart and wash with continuous gentle stream of water for at least 15 min.. If victim is not breathing, give artificial respiration until physician arrives.. If he is unconscious, crush an amyl nitrite ampule in a cloth and hold it under his nose for 15 seconds in every minute. Do not interrupt artificial respiration while doing this. Replace ampule when its strength is spent and continue treatment until condition improves or physician arrives; *Toxicity by Inhalation (Threshold Limit Value)*: 20 ppm; *Short-Term Exposure Limits*: 40 ppm for 30 min.; *Toxicity by Ingestion*: Grade 3; LD_{50} 50 to 500 mg/kg (rat, guinea pig); *Late Toxicity*: Data not available; *Vapor (Gas) Irritant Characteristics*: Vapor is moderately irritating such that personnel will not usually tolerate moderate or high vapor concentrations; *Liquid or Solid Irritant Characteristics*: If spilled on clothing and allowed to remain, may cause smarting and reddening of the skin. Large amounts may be absorbed through the skin and cause poisoning; *Odor Threshold*: 21.4 ppm (Sense of smell fatigues rapidly).

Fire Hazards - *Flash Point (deg. F)*: 30 CC; 31 OC; *Flammable Limits in Air (%)*: 3.05 - 17.0; *Fire Extinguishing Agents*: Dry chemical, alcohol foam, carbon dioxide; *Fire Extinguishing Agents Not To Be Used*: Water or foam may cause frothing; *Special Hazards of Combustion Products*: When heated or burned, ACN may evolve toxic hydrogen cyanide gas and oxides of nitrogen; *Behavior in Fire*: Vapor is heavier than air and may travel a considerable distance to a source of ignition and flash back. May polymerize and explode; *Ignition Temperature (deg. F)*: 898; *Electrical Hazard*: Class I, Group D; *Burning Rate*: Data not available.

Chemical Reactivity - *Reactivity with Water*: No reaction; *Reactivity with Common Materials*: Attacks copper and copper alloys; these metals should not be used. Penetrates leather, so contaminated leather shoes and gloves should be destroyed. Attacks aluminum in high concentrations; *Stability During Transport*: Stable; *Neutralizing Agents for Acids and Caustics*: Not pertinent; *Polymerization*: May occur spontaneously in absence of oxygen or on exposure to visible light or excessive heat, violently in the presence of alkali. Pure ACN is subject to polymerization with rapid pressure development. The commercial product is inhibited and not subject to this reaction; *Inhibitor of Polymerization*: Methylhydroquinone (35 - 45 ppm).

ALDRIN

Chemical Designations - *Synonyms*: endo-, exo-,1,2,3,4,10,10-Hexachloro-1,4,4a,5,8,8a-Hexahydro-1,4:5,8-Dimethanonaphtalene, HHDN; *Chemical Formula*: $C_{12}H_8Cl_6$.

Observable Characteristics — *Physical State (as normally shipped)*: Solid; *Color*: Tan to dark brown; *Odor*: Mild chemical.

Physical and Chemical Properties - *Physical State at 15 °C and 1 atm.*: Solid; *Molecular Weight*: 364.93; *Boiling Point at 1 atm.*: Not pertinent; *Freezing Point*: 219, 104, 377; *Critical Temperature*: Not pertinent; *Critical Pressure*: Not pertinent; *Specific Gravity*: 1.6 at 20°C (solid); *Vapor (Gas) Density*: Not pertinent; *Ratio of Specific Heats of Vapor (Gas)*: Not pertinent; *Latent Heat of Vaporization*: Not pertinent; *Heat of Combustion*: Not pertinent; *Heat of Decomposition*: Not pertinent.

Health Hazards Information - *Recommended Personal Protective Equipment*: During prolonged exposure to mixing and loading operations, wear clean synthetic rubber gloves and mask or respirator of the type passed by the U.S. Bureau of Mines for aldrin protection; *Symptoms Following Exposure*: Ingestion, inhalation, or skin absorption of a toxic dose will induce nausea, vomiting, hyperexcitability, tremors, epileptiform convulsions, and ventricular fibrillation. Aldrin may cause temporary reversible kidney and liver injury. Symptoms may be seen after ingestion of less than 1 gram in an adult; ingestion of 25 mg has caused death in children; *General Treatment for Exposure*: SKIN CONTACT: wash with soap and running water. If material gets into eyes, wash immediately with running water for at least 15 min.; get medical attention. INGESTION: call physician immediately; induce vomiting. Repeat until vomit fluid is clear. Never give anything by mouth to an unconscious person. Keep patient prone and quiet. PHYSICIAN: administer barbiturates as anti-convulsant therapy. Observe patient carefully because repeated treatment may be necessary; *Toxicity by Inhalation (Threshold Limit Value)*: 0.25 mg/m³; *Short-Term Exposure Limits*: 1 mg/m³ for 30 min.; *Toxicity by Ingestion*: Grade 3; LD 50 to 500 mg/kg (rat); *Late Toxicity*: Chronic exposure produces benign tumors in mice; *Vapor (Gas) Irritant Characteristics*: Vapors cause slight smarting of the eyes or respiratory system if present in high concentration. Effects is temporary; *Liquid or Solid Irritant Characteristics*: Minimum hazard. If spilled on clothing and allowed to remain, may cause smarting and reddening if the skin; *Odor Threshold*: Data not available.

Fire Hazards - *Flash Point*: Not flammable; *Flammable Limits in Air (%)*: Not pertinent; *Fire Extinguishing Agents*: Water spray, dry chemical, foam or carbon dioxide for fire involving solutions of aldrin in hydrocarbon solvents; *Fire Extinguishing Agents Not To Be Used*: Not pertinent; *Special Hazards of Combustion Products*: Irritating fumes of hydrochloric acid and chlorinated decomposition products are given off; *Behavior in Fire*: Not pertinent; *Ignition Temperature*: Not pertinent; *Electrical Hazard*: Not pertinent; *Burning Rate*: Not pertinent.

Chemical Reactivity - *Reactivity with Water*: No reaction; *Reactivity with Common Materials*: No reaction; *Stability During Transport*: Stable; *Neutralizing Agents for Acids and Caustics*: Not pertinent; *Polymerization*: Not pertinent; *Inhibitor of Polymerization*: Not pertinent.

ALLYL ALCOHOL

Chemical Designations - *Synonyms*: 2-Propen-1-ol-Vinylcarbinol; *Chemical Formula*: $CH_2=CHCH_2OH$.

Observable Characteristics - *Physical State (as normally shipped)*: Liquid; *Color*: Colorless; *Odor*: Characteristic, pungent; sharp; causes tears.

Physical and Chemical Properties - *Physical State at 15 and 1 atm.*: Liquid; *Molecular Weight*: 58.08; *Boiling Point at 1 atm.*: 206, 96.9, 370.1; *Freezing Point*: -200, -129, 144; *Critical Temperature*: 521.4, 271.9, 545.1; *Critical Pressure*: 840, 57, 5.8; *Specific Gravity*: 0.852 at 20°C (liquid); *Vapor (Gas) Density*: 2.0; *Ratio of Specific Heats of Vapor (Gas)*: 1.12; *Latent Heat of Vaporization*: 295, 164, 6.87; *Heat of Combustion*: -13,720, -7620, -319.0; *Heat of Decomposition*: Not pertinent.

Health Hazards Information - *Recommended Personal Protective Equipment*: Organic canister or air pack; rubber gloves, goggles; other protective equipment as required to prevent all body contact; *Symptoms Following Exposure*: Vapors are quite irritating to eyes, nose and throat. Eye irritation may be accompanied by complaints of photophobia and pain in the eyeball; pain may not begin until 6 hours after exposure. Liquid may cause first- and second-degree burns of the skin, with blister formation; underlying part will become swollen and painful, and local muscle spasms may occur; *General Treatment for Exposure*: INHALATION: remove victim from contaminated area and administer oxygen; get medical attention immediately. SKIN: remove liquid with soap and water. EYES: flush with continuous stream of water for 15 min.; *Toxicity by Inhalation (Threshold Limit Value)*: 2 ppm; *Short-Term Exposure Limits*: 5 ppm for 30 min; *Toxicity by Ingestion*: Grade 3; LD_{50} 50 to 500 mg/kg (mouse, rat); *Late Toxicity*: Data not available; *Vapor (Gas) Irritant Characteristics*: Vapor is moderately irritating such that personnel will not usually tolerate moderate or high vapor concentration; *Liquid or Solid Irritant Characteristics*: Causes smarting of the skin and first-degree burns on short exposure; may cause secondary burns on a long exposure; *Odor Threshold*: 0.78 ppm.

Fire Hazards - *Flash Point (deg. F)*: 72 CC; 90 OC; *Flammable Limits in Air (%)*: 2.5 - 18; *Fire Extinguishing Agents*: Dry chemical, alcohol foam, carbon dioxide; *Fire Extinguishing Agents Not To Be Used*: Water may be ineffective; *Special Hazards of Combustion Products*: Toxic vapor is generated when heated; *Behavior in Fire*: Vapor heavier than air and may travel a considerable distance to a source of ignition and flash back; *Ignition Temperature (deg. F)*:829; *Electrical Hazard*: Not pertinent; *Burning Rate*: 2.7 mm/min.

Chemical Reactivity - *Reactivity with Water*: No reaction; *Reactivity with Common Materials*: No reaction; *Stability During Transport*: Stable at ordinary temperatures and pressures; *Neutralizing Agents for Acids and Caustics*: Not pertinent; *Polymerization*: Not pertinent; *Inhibitor of* Polymerization: Not pertinent.

ALLYL CHLOROFORMATE

Chemical Designations - *Synonyms*: Allyl Chlorocarbonate; *Chemical Formula*: $CH_2=CH\bullet CH_2\bullet O\bullet COCl$.

Observable Characteristics - *Physical State (as normally shipped)*: Liquid; *Color*: Colorless; *Odor*: Extremely irritating, causes tears; pungent.

Physical and Chemical Properties - *Physical State at 15 °C and 1 atm.*: Liquid; *Molecular Weight*: 120.5; *Boiling Point at 1 atm.*: 113, 45, 318; *Freezing Point*: -112, -80, 193; *Critical Temperature*: Not pertinent; *Critical Pressure*: Not pertinent; *Specific Gravity:* 1.139 at 20°C (liquid); *Vapor (Gas) Density*: 4.15; *Ratio of Specific Heats of Vapor (Gas)*: 1.0804; *Latent Heat of Vaporization*: 100, 56, 2.3; *Heat of Combustion*: -7,800, -4,300, -180; *Heat of Decomposition*: Not pertinent.

Health Hazards Information - *Recommended Personal Protective Equipment*: Vapor-proof protective goggles and face shield; plastic or rubber gloves, shoes and clothing; gas mask or self-contained breathing apparatus; *Symptoms Following Exposure*: Vapor irritates eyes and respiratory tract. Contact with liquid causes eye and skin irritation, and ingestion irritates mouth and stomach; *General Treatment for Exposure*: INHALATION: remove from exposure; support respiration if necessary; call physician. EYES: if irritated by either vapor or liquid, flush with water for at least 15 min. SKIN: wash with large amounts of water for at least 15 min. INGESTION: do NOT induce vomiting; give water; call physician; *Toxicity by Inhalation (Threshold Limit Value)*: Data not available; *Short-Term Exposure Limits*: Data not available; *Toxicity by Ingestion*: Grade 3; LD_{50} 50 to 500 mg/kg; *Late Toxicity*: Data not available; *Vapor (Gas) Irritant Characteristics*: Vapors are moderately irritating such that personnel will not usually tolerate moderate or high vapor concentrations; *Liquid or Solid Irritant Characteristics*: Fairly severe skin irritant. May cause pain and second-degree burns after a few minutes' contact; *Odor Threshold*: 1.4 ppm.

Fire Hazards - *Flash Point (deg. F)*: 92 OC; 88 CC; *Flammable Limits in Air (%)*: Data not available; *Fire Extinguishing Agents*: Dry chemical, foam, carbon dioxide; *Fire Extinguishing Agents Not To Be Used*: Water may be ineffective; *Special Hazards of Combustion Products*: When heated to decomposition, emits highly toxic phosgene gas; *Behavior in Fire*: Vapor heavier than air and may travel a considerable distance to a source of ignition and flash back; *Ignition Temperature*: Data not available; *Electrical Hazard*: Data not available; *Burning Rate*: 4.9 mm/min.

Chemical Reactivity - *Reactivity with Water*: Reacts slowly generating hydrogen chloride; *Reactivity* *with Common Materials*: Corrosive metals; *Stability During Transport*: Stable; *Neutralizing Agents for Acids and Caustics*: Flush with water, rinse with sodium bicarbonate solution; *Polymerization*: Not pertinent; *Inhibitor of Polymerization*: Not pertinent.

ALLYLTRICHLOROSILANE

Chemical Designations - *Synonyms*: Allylsilicone Trichloride; *Chemical Formula*: $CH_2=CH \cdot CH_2 \cdot SiCl$.

Observable Characteristics - *Physical State (as normally shipped)*: Liquid; *Color*: Colorless; *Odor*: Sharp; pungent, irritating, like hydrochloric acid.

Physical and Chemical Properties - *Physical State at 15 °C and 1 atm.*: Liquid; *Molecular Weight*: 175.5; *Boiling Point at 1 atm.*: 241, 116, 389; *Freezing Point*: Not pertinent; *Critical Temperature*: Not pertinent; *Critical Pressure*: Not pertinent; *Specific Gravity*: 1.215 at 20°C (liquid); *Vapor (Gas) Density*: 6; *Ratio of Specific Heats of Vapor (Gas)*: 1.0863; *Latent Heat of Vaporization*: 97, 54, 2.3; *Heat of Combustion*: -5,200, -2,900, -120; *Heat of Decomposition*: Not pertinent.

Health Hazards Information - *Recommended Personal Protective Equipment*: Acid-vapor-type respiratory protection; rubber gloves; chemical goggles; other equipment necessary to protect skin and eyes; *Symptoms Following Exposure*: Inhalation of vapor irritates mucous membranes. Liquid causes severe burns of eyes and skin and severe internal burns if ingested; *General Treatment for Exposure*: Get medical attention after all exposures of this compound. INHALATION: remove from exposure; support respiration. EYES: flush with water 15 min. SKIN: flush with water. INGESTION: do NOT induce vomiting; give water; *Toxicity by Inhalation (Threshold Limit Value)*: Data not available; *Short-Term Exposure Limits*: Data not available; *Toxicity by Ingestion*: Grade 3; LD_{50} 50 to 500 mg/kg; *Late Toxicity*: Data not available; *Vapor (Gas) Irritant Characteristics*: Vapors cause severe irritation of eyes and throat and can cause eye or lung injury. They can not be tolerated even at low concentrations; *Liquid or Solid Irritant Characteristics*: Severe skin irritant. Causes second- and third-degree burns on short contact and is very injurious to the eyes; *Odor Threshold*: Data not available.

Fire Hazards - *Flash Point (deg. F)*: 100 OC; 95 CC; *Flammable Limits in Air (%)*: Data not available; *Fire Extinguishing Agents*: Dry chemical, carbon dioxide; *Fire Extinguishing Agents Not To Be Used*: Water; *Special Hazards of Combustion Products*: Irritating vapor of hydrogen chloride and phosgene may form; *Behavior in Fire*: Difficult to extinguish. Re-ignition may occur; *Ignition Temperature*: Data not available; *Electrical Hazard*: Data not available; *Burning Rate*: 2.2 mm/min.

Chemical Reactivity - *Reactivity with Water*: Reacts vigorously, generating hydrogen chloride (hydrochloric acid); *Reactivity with Common Materials*: Corrodes metals because of hydrochloric acid formed; *Stability During Transport*: Stable; *Neutralizing Agents for Acids and Caustics*: Flush with water, rinse with sodium bicarbonate; *Polymerization*: Not pertinent; *Inhibitor of Polymerization*: Not pertinent.

ALUMINUM CHLORIDE

Chemical Designations - *Synonyms*: Anhydrous Aluminum Chloride; *Chemical Formula*: $AlCl_3$.

Observable Characteristics - *Physical State (as normally shipped)*: Solid; *Color*: Orange to yellow through gray to white; *Odor*: Like hydrogen chloride; like hydrochloric acid.

Physical and Chemical Properties - *Physical State at 15 °C and 1 atm.*: Solid; *Molecular Weight*: 133.34; *Boiling Point at 1 atm.*: Not pertinent; *Freezing Point*: Not pertinent; *Critical Temperature*: Not pertinent; *Critical Pressure*: Not pertinent; *Specific Gravity*: 2.44 to 25°C (solid); *Vapor (Gas) Density*: Not pertinent; *Ratio of Specific Heats of Vapor (Gas)*: Not pertinent; *Latent Heat of Vaporization*: Not pertinent; *Heat of Combustion*: Not pertinent; *Heat of Decomposition*: Not pertinent.

Health Hazards Information - *Recommended Personal Protective Equipment*: All personnel in the area should wear safety clothing, including fully closed goggles, rubber or plastic-coated gloves, rubber shoes and coverall of acid-resistant material. An acid-vapor canister mask should be carried in case of emergency. In certain applications, it may be advisable to wear this equipment on a routine basis; *Symptoms Following Exposure*: Contact with the skin or eyes in the presence of moisture causes thermal and acid burns; *General Treatment for Exposure*: INGESTION: if victim is conscious have him drink water or milk. Do NOT induce vomiting. SKIN: flush immediately with plenty of water. For eye contact, flush with water for at least 15 min. and get medical attention immediately; *Toxicity by Inhalation (Threshold Limit Value)*: 5 ppm (hydrogen chloride); *Short-Term Exposure Limits*: 5 ppm for 5 min.; 30 ppm for 10 min.; 20 ppm for 20 min.; 10 ppm for 60 min.; *Toxicity by Ingestion*: No systemic effects, but severe burns of mouth; *Late Toxicity*: None recognized; *Vapor (Gas) Irritant Characteristics*: Vapor (or hydrogen chloride) is moderately irritating such that personnel will not usually tolerate moderate or high vapor concentrations; *Liquid or Solid Irritant Characteristics*: Fairly severe skin irritant; may cause pain and second-degree burns after a few minutes' contact; *Odor Threshold*: 1-5 ppm (hydrogen chloride).

Fire Hazards - *Flash Point*: Not flammable; *Flammable Limits in Air* (%): Not flammable; *Fire Extinguishing* Agents: Not pertinent; *Fire Extinguishing Agents Not To Be Used*: Do not use water on adjacent fires; *Special Hazards of Combustion Products*: Not pertinent; *Behavior in Fire*: Reacts violently with water used in extinguishing adjacent fires; *Ignition Temperature*: Not flammable; *Electrical Hazard*: Not pertinent; *Burning Rate*: Not flammable.

Chemical Reactivity - *Reactivity with Water*: Reacts violently with water, liberating hydrogen chloride gas and heat; *Reactivity with Common Materials*: None if dry. If wet it attacks metals because of hydrochloric acid formed; flammable hydrogen is formed; *Stability During Transport*: Stable if kept dry and protected from atmospheric moisture; *Neutralizing Agents for Acids and Caustics*: Hydrochloric acid formed by reaction with water can be flushed away with water. Rinse with sodium bicarbonate or lime solution; *Polymerization*: Not pertinent; *Inhibitor of Polymerization*: Not pertinent.

ALUMINUM NITRATE

Chemical Designations - *Synonyms*: Aluminum Nitrate Nonahydrate; Nitric Acid, Aluminum Salt; *Chemical Formula*: $Al(NO_3)_3 \cdot 9H_2O$.

Observable Characteristics - *Physical State (as normally shipped)*: Solid; *Color*: White; *Odor*: None.

Physical and Chemical Properties - *Physical State at 15 °C and 1 atm.*: Solid; *Molecular Weight*: 375.13; *Boiling Point at 1 atm.*: Not pertinent (decomposes); *Freezing Point*: 163, 73, 346; *Critical Temperature*: Not pertinent; *Critical Pressure*: Not pertinent; *Specific Gravity:* >1 at 20°C (solid); *Vapor (Gas) Density*: Not pertinent; *Ratio of Specific Heats of Vapor (Gas)*: Not pertinent; *Latent Heat of Vaporization*: Not pertinent; *Heat of Combustion*: Not pertinent; *Heat of Decomposition*: Not pertinent.

Health Hazards Information - *Recommended Personal Protective Equipment*: Goggles or face shield; dust respirator; rubber gloves; *Symptoms Following Exposure*: Ingestion of large doses causes gastric irritation, nausea, vomiting, and purging. Contact with dust irritates eyes and skin; *General Treatment for Exposure*: EYES: flush with water for at least 15 min. SKIN: flush with water; wash with soap and water; *Toxicity by Inhalation (Threshold Limit Value)*: Data not available; *Short-Term Exposure Limits*: Data not available; *Toxicity by Ingestion*: Grade 3; oral rat LD50 264 mg/kg (nonahydrate); *Late Toxicity*: Data not available; *Vapor (Gas) Irritant Characteristics*: Data not available; *Liquid or Solid Irritant Characteristics*: Data not available; *Odor Threshold*: Odorless.

Fire Hazards - *Flash Point*: Not flammable; *Flammable Limits in Air* (%): Not flammable; *Fire Extinguishing Agents*: Not pertinent; *Fire Extinguishing Agents Not To Be Used*: Not pertinent; *Special Hazards of Combustion Products*: Toxic oxides of nitrogen may form in fire; *Behavior in Fire*: May increase the intensity if fire when used with combustible material; *Ignition Temperature*: Not pertinent; *Electrical Hazard*: Not pertinent; *Burning Rate*: Not pertinent.

Chemical Reactivity - *Reactivity with Water*: Dissolves and forms a weak solution if nitric acid. The reaction is not hazardous; *Reactivity with Common Materials*: May corrode metals in presence of moisture; *Stability During Transport*: Stable; *Neutralizing Agents for Acids and Caustics*: Flush with water; *Polymerization*: Not pertinent; *Inhibitor of Polymerization*: Not pertinent.

AMMONIA, ANHYDROUS

Chemical Designations - *Synonyms*: Liquid Ammonia; *Chemical Formula*: NH_3.

Observable Characteristics - *Physical State (as normally shipped)*: Compressed liquified gas; *Color*: Colorless; *Odor*: Pungent; extremely pungent.

Physical and Chemical Properties - *Physical State at 15 °C and 1 atm.*: Gas; *Molecular Weight*: 17.03; *Boiling Point at 1 atm.*: -28.1, -33.4, 239.8; *Freezing Point*: -108, -77.7, 265.5; *Critical Temperature*: 271, 133, 406; *Critical Pressure*: 1,636, 11.3, 11.27; *Specific Gravity:* 0.682 at -33.4°C (liquid); *Vapor (Gas) Density*: 0.6; *Ratio of Specific Heats of Vapor (Gas)*: 1.3 at 20°C; *Latent Heat of Vaporization*: 589, 327, 13.7; *Heat of Combustion*: -7992, -4440, -185.9; *Heat of Decomposition*: Not pertinent.

Health Hazards Information - *Recommended Personal Protective Equipment*: Gas-tight chemical goggles, self-contained breathing apparatus, rubber boots, rubber gloves, emergency shower and eye bath; *Symptoms Following Exposure*: 700 ppm causes eye irritation, and permanent injury may result if prompt remedial measures are not taken; 5000 ppm can cause immediate death from spasm, inflammation, or edema of the larynx. Contact of the liquid with skin freezes the tissue; causes a caustic burn; *General Treatment for Exposure*: INHALATION: move victim to fresh air, give artificial respiration if necessary. Oxygen may be useful. Observe for laryngeal spasm and perform tracheostomy if indicated. SKIN OR EYES: flood immediately with running water for 15 min. Treat as thermal burn; *Toxicity by Inhalation (Threshold Limit Value)*: 25 ppm; *Short-Term Exposure Limits*: 50 ppm for 5 min.; *Toxicity by Ingestion*: Not pertinent; *Late Toxicity*: Not pertinent; *Vapor (Gas) Irritant Characteristics*: Vapors cause severe eye or throat irritation and may cause eye or lung injury; vapors cannot be tolerated even at low concentrations; *Liquid or Solid Irritant Characteristics*: Causes smarting of the skin and first-degree burns on short exposure; may cause secondary burns on long exposure; *Odor Threshold*: 46.8 ppm.

Fire Hazards - *Flash Point:* Generally not flammable; *Flammable Limits in Air (%)*: 15.50 - 27.00; *Fire Extinguishing Agents*: Stop flow of gas or liquid. Let fire burn; *Fire Extinguishing Agents Not To Be Used*: None; *Special Hazards of Combustion Products*: Not pertinent; *Behavior in Fire*: Not pertinent; *Ignition Temperature (deg. F)*: 1204; *Electrical Hazard*: Class I, Group D; *Burning Rate*: 1 mm/min.

Chemical Reactivity - *Reactivity with Water*: Dissolves with mild heat effect; *Reactivity with Common Materials*: Corrosive to copper and galvanized surfaces; *Stability During Transport*: Stable; *Neutralizing Agents for Acids and Caustics*: Dilute with water; *Polymerization*: Not pertinent; *Inhibitor of Polymerization*: Not pertinent.

AMMONIUM BIFLUORIDE

Chemical Designations - *Synonyms*: Acid Ammonium Fluoride, Ammonium Acid Fluoride, Ammonium Hydrogen Fluoride; *Chemical Formula*: NH_4NF_2.

Observable Characteristics - *Physical State (as normally shipped)*: Solid; *Color*: White; *Odor*: None.

Physical and Chemical Properties - *Physical State at 15 °C and 1 atm.*: Solid; *Molecular Weight*: 57.04; *Boiling Point at 1 atm.*: 463.1, 239.5, 512.7; *Freezing Point*: 258, 125.6, 398.8; *Critical Temperature*: Not pertinent; *Critical Pressure*: Not pertinent; *Specific Gravity:* 1.5 at 20°C (solid); *Vapor (Gas) Density*: Not pertinent; *Ratio of Specific Heats of Vapor (Gas)*: Not pertinent; *Latent Heat of Vaporization*: Not pertinent; *Heat of Combustion*: Not pertinent; *Heat of Decomposition*: Not pertinent.

Health Hazards Information - *Recommended Personal Protective Equipment*: Bu. Mines approved respirator; rubber gloves; safety goggles; *Symptoms Following Exposure*: Inhalation of dust may cause irritation of respiratory system. Ingestion causes irritation of mouth and stomach, vomiting, abdominal pain, convulsions, collapse, acute toxic nephritis. Contact with dust irritates eyes and may cause burns or rash on skin. High concentrations of fluorine in the urine have been reported following skin contact; *General Treatment for Exposure*: Begin first aid as quickly as possible. INHALATION: remove victim to fresh air. INGESTION: perform gastric lavage with lime water or 1% calcium chloride solution; support respiration; call a physician. EYES: flush with water for at least 15 min.; consult physician. SKIN: flush with water; treat burns. OTHER: remove all contaminated clothing in the shower at once; *Toxicity by Inhalation (Threshold Limit Value)*: 2.5 mg/m^3 (as fluorine); *Short-Term Exposure Limits*: Data not available; *Toxicity by Ingestion*: Grade 3; LD_{50} 50 mg/kg (guinea pig), 60 mg/kg (rat); *Late Toxicity*: Data not available; *Vapor (Gas) Irritant Characteristics*: Data not available; *Liquid or Solid*

Irritant Characteristics: Data not available; *Odor Threshold*: Data not available.

Fire Hazards: *Flash Point:* Not flammable; *Flammable Limits in Air* (%): Not flammable; *Fire Extinguishing Agents*: Not pertinent; *Fire Extinguishing Agents Not To Be Used*: Do not apply water to adjacent fires; *Special Hazards of Combustion Products*: Toxic ammonia and hydrogen fluoride gases may form in fire; *Behavior in Fire*: Not pertinent; *Ignition Temperature*: Not pertinent; *Electrical Hazard*: Not pertinent; *Burning Rate*: Not pertinent.

Chemical Reactivity - *Reactivity with Water*: Dissolves and forms a weak solution of hydrofluoric acid; *Reactivity with Common Materials*: In presence of moisture will corrode glass, cement and most metals. Flammable hydrogen gas may collect in enclosed spaces; *Stability During Transport*: Stable; *Neutralizing Agents for Acids and Caustics*: Flush with water, rinse with dilute solution of sodium of sodium bicarbonate or soda ash; *Polymerization*: Not pertinent; *Inhibitor of Polymerization*: Not pertinent.

AMMONIUM CARBONATE

Chemical Designations - *Synonyms*: Hartshorn; Salt Volatile; *Chemical Formula*: $(NH_4)_2CO_3$.

Observable Characteristics - *Physical State (as normally shipped)*: Solid; *Color*: White; *Odor*: Strong ammonia.

Physical and Chemical Properties - *Physical State at 15 °C and 1 atm.*: Solid; *Molecular Weight*: 157.1; *Boiling Point at 1 atm.*: Not pertinent; *Freezing Point*: Not pertinent; *Critical Temperature*: Not pertinent; *Critical Pressure*: Not pertinent; *Specific Gravity:* 1.5 at 20°C (solid); *Vapor (Gas) Density*: Not pertinent; *Ratio of Specific Heats of Vapor (Gas)*: Not pertinent; *Latent Heat of Vaporization*: Not pertinent; *Heat of Combustion*: Not pertinent; *Heat of Decomposition*: Not pertinent.

Health Hazards Information - *Recommended Personal Protective Equipment*: Dust respirator; protection against ammonia vapors; *Symptoms Following Exposure*: Inhalation causes irritation of nose and throat. Ingestion may cause gastric irritation. Contact with eyes or skin causes irritation; *General Treatment for Exposure*: INHALATION: leave contaminated area. INGESTION: give large amount of water. EYES: flush with copious amounts of water. SKIN: flush with water; *Toxicity by Inhalation (Threshold Limit Value)*: Data not available; *Short-Term Exposure Limits*: Data not available; *Toxicity by Ingestion*: Data not available; *Late Toxicity*: Data not available; *Vapor (Gas) Irritant Characteristics*: Data not available; *Liquid or Solid Irritant Characteristics*: Data not available; *Odor Threshold*: < 1.5 ppm (as ammonia gas).

Fire Hazards - *Flash Point:* Not pertinent; *Flammable Limits in Air* (%): Not pertinent; *Fire Extinguishing Agents*: Water; *Fire Extinguishing Agents Not To Be Used*: Not pertinent; *Special Hazards of Combustion Products*: Toxic ammonia gas will form in fires; *Behavior in Fire*: Decomposes, but reaction is not explosive. Ammonia gas is formed; *Ignition Temperature*: Not pertinent; *Electrical Hazard*: Not pertinent; *Burning Rate*: Not pertinent.

Chemical Reactivity - *Reactivity with Water*: No reaction; *Reactivity with Common Materials*: No reaction; *Stability During Transport*: Stable; *Neutralizing Agents for Acids and Caustics*: Not pertinent; *Polymerization*: Not pertinent; *Inhibitor of Polymerization*: Not pertinent.

AMMONIUM DICHROMATE

Chemical Designations - *Synonyms*: Ammonium Bichromate; *Chemical Formula*: $(NH_4)_2Cr_2O_7$.

Observable Characteristics - *Physical State (as normally shipped)*: Solid; *Color*: Orange-yellow; bright red-orange; orange to red; *Odor*: None.

Physical and Chemical Properties - *Physical State at 15 °C and 1 atm.*: Solid; *Molecular Weight*: 252.06; *Boiling Point at 1 atm.*: Not pertinent; *Freezing Point*: Not pertinent; *Critical Temperature*: Not pertinent; *Critical Pressure*: Not pertinent; *Specific Gravity:* 2.15 at 25°C (solid); *Vapor (Gas) Density*: Not pertinent; *Ratio of Specific Heats of Vapor (Gas)*: Not pertinent; *Latent Heat of Vaporization*: Not pertinent; *Heat of Combustion*: Not pertinent; *Heat of Decomposition*: Not pertinent.

Health Hazards Information - *Recommended Personal Protective Equipment*: Dust respirator; protective goggles, gloves. clothing; *Symptoms Following Exposure*: Inhalation causes irritation or ulceration of the mucous membranes of the nose, throat or respiratory tract. Respiratory irritation can produce symptoms resembling those of asthma. Continuing irritation of the nose may lead to perforation of the nasal septum. External contact can cause eye irritation and conjunctivitis, irritation and ulceration of skin wounds, and rash or external ulcers. If ingested, irritates mucous membrane and causes vomiting; *General Treatment for Exposure*: INHALATION: remove to clean air and summon medical attention. EYES: immediately flush with water for at least 15 min. and consult a physician. SKIN: flush with water; if skin irritation develops, get medical attention. INGESTION: vomiting should occur; follow with an emetic of soapy water; give large amounts of water; *Toxicity by Inhalation (Threshold Limit Value)*: 0.1 mg/m^3; *Short-Term Exposure Limits*: Not pertinent; *Toxicity by Ingestion*: Data not available; *Late Toxicity*: Data not available; *Vapor (Gas) Irritant Characteristics*: Not pertinent; *Liquid or Solid Irritant Characteristics*: Data not available; *Odor Threshold*: Not pertinent.

Fire Hazards - *Flash Point:* Flammable solid; *Flammable Limits in Air (%)*: Not pertinent; *Fire Extinguishing Agents*: Water; *Fire Extinguishing Agents Not To Be Used*: Not pertinent; *Special Hazards of Combustion Products*: Greenish chromic oxide smoke may cause irritation of lungs and mucous membranes; *Behavior in Fire*: Decomposes at about 180° C with spectacular swelling and evolution of heat and nitrogen, leaving chromic oxide residue. Pressure of confined gases can burst closed containers explosively; *Ignition Temperature (deg. F)*: 437; *Electrical Hazard*: Data not available; *Burning Rate*: Not pertinent.

Chemical Reactivity - *Reactivity with Water*: No reaction; *Reactivity with Common Materials*: Can ignite combustible material such as wood shavings; *Stability During Transport*: Stable; *Neutralizing Agents for Acids and Caustics*: Not pertinent; *Polymerization*: Not pertinent; *Inhibitor of Polymerization*: Not pertinent.

AMMONIUM FLUORIDE

Chemical Designations - *Synonyms*: Neutral Ammonium Fluoride; *Chemical Formula*: NH$_4$F.

Observable Characteristics - *Physical State (as normally shipped)*: Solid; *Color*: White; *Odor*: None.

Physical and Chemical Properties - *Physical State at 15 °C and 1 atm.*: Solid; *Molecular Weight*: 37.04; *Boiling Point at 1 atm.*: Not pertinent (decomposes); *Freezing Point*: Not pertinent; *Critical Temperature*: Not pertinent; *Critical Pressure*: Not pertinent; *Specific Gravity:* 1.32 at 25°C (solid); *Vapor (Gas) Density*: Not pertinent; *Ratio of Specific Heats of Vapor (Gas)*: Not pertinent; *Latent Heat of Vaporization*: Not pertinent; *Heat of Combustion*: Not pertinent; *Heat of Decomposition*: Not pertinent.

Health Hazards Information - *Recommended Personal Protective Equipment*: Dust mask; goggles or face shield; rubber gloves; *Symptoms Following Exposure*: Inhalation of dust may cause irritation of respiratory system. Ingestion is harmful; readily soluble fluorides may be fatal if relatively small quantities are swallowed. Contact with eyes may causes local irritation of the mucous membrane. Contact with skin may cause burns. High concentrations. of fluorine in the urine have been reported following skin contact; *General Treatment for Exposure*: Begin first aid as quickly as possible. INHALATION: remove to fresh air. INGESTION: perform gastric lavage with limewater or 1% calcium chloride solution; support respiration if necessary; call a physician. EYES: flush with water for at least 15 min.; consult physician. SKIN: shower immediately with large quantities of water; remove all contaminated clothing in the shower at once; consult physician; *Toxicity by Inhalation (Threshold Limit Value)*: 2.5 mg/m^3 (as fluorine); *Short-Term Exposure Limits*: Data not available; *Toxicity by Ingestion*: Data not available; *Late Toxicity*: Data not available; *Vapor (Gas) Irritant Characteristics*: Data not available; *Liquid or Solid Irritant Characteristics*: Data not available; *Odor Threshold*: Data not available.

Fire Hazards - *Flash Point:* Not flammable; *Flammable Limits in Air (%)*: Not flammable; *Fire Extinguishing Agents*: Not pertinent; *Fire Extinguishing Agents Not To Be Used*: Not pertinent; *Special Hazards of Combustion Products*: Toxic ammonia and hydrogen fluoride gases are formed in fires; *Behavior in Fire*: May sublime when hot and condense on cool surfaces; *Ignition Temperature*: Not pertinent; *Electrical Hazard*: Not pertinent; *Burning Rate*: Not pertinent.

Chemical Reactivity - *Reactivity with Water*: Dissolves and forms dilute solution of hydrofluoric acid; *Reactivity with Common Materials*: May corrode glass, cement and most metals; *Stability During*

Transport: Stable; *Neutralizing Agents for Acids and Caustics*: Not pertinent; *Polymerization*: Not pertinent; *Inhibitor of Polymerization*: Not pertinent.

AMMONIUM HYDROXIDE (< 20 % AQUEOUS AMMONIA)

Chemical Designations - *Synonyms*: Ammonia Water, Aqueous Ammonia, Household Ammonia; *Chemical Formula*: NH_4OH-H_2O.

Observable Characteristics — *Physical State (as normally shipped)*: Liquid; *Color*: Colorless; *Odor*: Pungent.

Physical and Chemical Properties - *Physical State at 15 °C and 1 atm.*: Liquid; *Molecular Weight*: Not pertinent; *Boiling Point at 1 atm.*: Not pertinent; *Freezing Point*: Not pertinent; *Critical Temperature*: Not pertinent; *Critical Pressure*: Not pertinent; *Specific Gravity*: 0.89 at 20°C (liquid); *Vapor (Gas) Density*: Not pertinent; *Ratio of Specific Heats of Vapor (Gas)*: Not pertinent; *Latent Heat of Vaporization*: Not pertinent; *Heat of Combustion*: Not pertinent; *Heat of Decomposition*: Not pertinent.

Health Hazards Information - *Recommended Personal Protective Equipment*: Rubber boots, gloves. apron, and coat; broad-brimmed rubber or felt hat; safety goggles. Use of protective oil will reduce skin irritation from ammonia; *Symptoms Following Exposure*: Contact of liquid or vapor with skin, mucous membranes, lungs, or gastroenteric tract causes marked local irritation. Ingestion causes burning pain in mouth, throat, stomach, and thorax, constriction of throat, and coughing. This is soon followed by vomiting of blood or by passage of loose stools containing blood. Breathing difficulty, convulsions, and shock may result. Brief exposure to 5000 ppm of ingestion of 3-4 ml may be fatal; *General Treatment for Exposure*: INHALATION: give artificial respiration and oxygen if needed; enforce rest. INGESTION: do NOT induce vomiting; lavage stomach with water or lemon juice, milk, or demulcents; delay may cause perforation of esophagus or stomach; swelling of glottis may necessitate tracheostomy. EYES OR SKIN: wash with plenty of water; *Toxicity by Inhalation (Threshold Limit Value)*: 1 ppm; *Short-Term Exposure Limits*: (ammonia gas) 100 ppm for 30 min.; 500 ppm for 10 min.; *Toxicity by Ingestion*: Grade 3; oral rat, LD_{50} 350 mg/kg; *Late Toxicity*: Data nat available; *Vapor (Gas) Irritant Characteristics*: Vapors cause moderate irritation such that personnel will find high concentrations intolerable. The effect is temporary; *Liquid or Solid Irritant Characteristics*: Causes smarting of the skin and first-degree burns on short exposure; may cause second-degree burns on long exposure; *Odor Threshold*: 50 ppm.

Fire Hazards - *Flash Point:* Not flammable; *Flammable Limits in Air (%)*: Not flammable; *Fire Extinguishing Agents*; Not pertinent; *Fire Extinguishing Agents Not To Be Used*: Not pertinent; *Special Hazards of Combustion Products*: Not pertinent; *Behavior in Fire*: Not pertinent; *Ignition Temperature*: Not flammable; *Electrical Hazard*: Data not available; *Burning Rate*: Not flammable.

Chemical Reactivity - *Reactivity with Water*: Mild liberation of heat; *Reactivity with Common Materials*: Corrosive to copper, copper alloys, aluminum alloys, galvanized surfaces; *Stability During Transport*: Stable; *Neutralizing Agents for Acids and Caustics*: Dilute with water; *Polymerization*: Not pertinent; *Inhibitor of Polymerization*: Not pertinent.

AMMONIUM LACTATE

Chemical Designations - *Synonyms*: Ammonium Lactate Syrup; *dl*-Lactic Acid, Ammonium Salt; *Chemical Formula*: $CH_3CH(OH)COONH_4$.

Observable Characteristics - *Physical State (as normally shipped)*: Solid or liquid; *Color*: White; *Odor*: None.

Physical and Chemical Properties - *Physical State at 15 °C and 1 atm.*: Solid; *Molecular Weight*: 107.11; *Boiling Point at 1 atm.*: Not pertinent; *Freezing Point*: Not pertinent; *Critical Temperature*: Not pertinent; *Critical Pressure*: Not pertinent; *Specific Gravity:* 1.2 at 15°C (solid); *Vapor (Gas) Density*: Not pertinent; *Ratio of Specific Heats of Vapor (Gas)*: Not pertinent; *Latent Heat of Vaporization*: Not pertinent; *Heat of Combustion*: Not pertinent; *Heat of Decomposition*: Not pertinent.

Health Hazards Information - *Recommended Personal Protective Equipment*: Dust mask; goggles of face shield; rubber gloves; *Symptoms Following Exposure*: Inhalation causes irritation of nose and throat. Contact with eyes causes irritation; *General Treatment for Exposure*: INHALATION: remove to fresh air. EYES: flush with water; *Toxicity by Inhalation (Threshold Limit Value)*: Data not available; *Short-Term Exposure Limits*: Data not available; *Toxicity by Ingestion*: Data not available; *Late Toxicity*: Data not available; *Vapor (Gas) Irritant Characteristics*: Data not available; *Liquid or Solid Irritant Characteristics*: Data not available; *Odor Threshold*: Data not available.

Fire Hazards - *Flash Point*: Not pertinent (combustible solid); *Flammable Limits in Air* (%): Not pertinent; *Fire Extinguishing Agents*: Water, foam; *Fire Extinguishing Agents Not To Be Used*: Not pertinent; *Special Hazards of Combustion Products*: Toxic oxides of nitrogen may be formed in a fire; *Behavior in Fire*: Not pertinent; *Ignition Temperature*: Not pertinent; *Electrical Hazard*: Not pertinent; *Burning Rate*: Not pertinent.

Chemical Reactivity - *Reactivity with Water*: No reaction; *Reactivity with Common Materials*: No reaction; *Stability During Transport*: Stable; *Neutralizing Agents for Acids and Caustics*: Not pertinent; *Polymerization*: Not pertinent; *Inhibitor of Polymerization*: Not pertinent.

AMMONIUM NITRATE

Chemical Designations - *Synonyms*: Nitram; *Chemical Formula*: NH_4NO_3.

Observable Characteristics - *Physical State (as normally shipped)*: Solid; *Color*: Colorless (pure) to gray or brown (fertilizer grades); *Odor*: None.

Physical and Chemical Properties - *Physical State at 15 °C and 1 atm.*: Solid; *Molecular Weight*: 80.05; *Boiling Point at 1 atm.*: Not pertinent; *Freezing Point*: 337.8, 169.9, 443.1; *Critical Temperature*: Not pertinent; *Critical Pressure*: Not pertinent; *Specific Gravity*: 1.72 at 20°C (solid); *Vapor (Gas) Density*: Not pertinent; *Ratio of Specific Heats of Vapor (Gas)*: Not pertinent; *Latent Heat of Vaporization*: Not pertinent; *Heat of Combustion*: Not pertinent; *Heat of Decomposition*: Not pertinent.

Health Hazards Information - *Recommended Personal Protective Equipment*: Wear self-contained breathing apparatus; *Symptoms Following Exposure*: Irritation of eyes and mucous membranes. Absorption via ingestion or inhalation causes urination and acid urine. Large amount causes systemic acidosis and metheglobinemia (abnormal hemoglobin); *General Treatment for Exposure*: Remove from exposure - symptoms reversible; *Toxicity by Inhalation (Threshold Limit Value)*: Not pertinent; *Short-Term Exposure Limits*: Not pertinent; *Toxicity by Ingestion*: Data not available; *Late Toxicity*: Data not available; *Vapor (Gas) Irritant Characteristics*: Not pertinent; *Liquid or Solid Irritant Characteristics*: None; *Odor Threshold*: Not pertinent.

Fire Hazards - *Flash Point:* Not flammable; *Flammable Limits in Air* (%): Not flammable; *Fire Extinguishing Agents*: Use flooding amount of water in early stages of fire. When large quantities are involved in massive fires, control efforts should be confined to protecting from explosion; *Fire Extinguishing Agents Not To Be Used*: Not pertinent; *Special Hazards of Combustion Products*: Decomposes, giving off extremely toxic oxides of nitrogen; *Behavior in Fire*: May explode in fires. Supports combustion of common organic fuel; *Ignition Temperature*: Not flammable; *Electrical Hazard*: Not pertinent; *Burning Rate*: Not flammable.

Chemical Reactivity - *Reactivity with Water*: No reaction; *Reactivity with Common Materials*: No reaction; *Stability During Transport*: If heated strongly, decomposes, giving off toxic gases which support combustion. Undergoes detonation if heated under confinement; *Neutralizing Agents for Acids and Caustics*: Not pertinent; *Polymerization*: Not pertinent; *Inhibitor of Polymerization*: Not pertinent.

AMMONIUM NITRATE-SULFATE MIXTURE

Chemical Designations - *Synonyms*: No common synonyms; *Chemical Formula*: $NH_4NO_3—(NH_4)_2SO_4$.

Observable Characteristics - *Physical State (as normally shipped)*: Solid; *Color*: Grayish-white; *Odor*: Characteristic.

Physical and Chemical Properties - *Physical State at 15 °C and 1 atm.*: Solid; *Molecular Weight*: Not pertinent; *Boiling Point at 1 atm.*: Not pertinent; *Freezing Point*: Not pertinent; *Critical Temperature*: Not pertinent; *Critical Pressure*: Not pertinent; *Specific Gravity*: 1.8 at 20°C (solid); *Vapor (Gas) Density*: Not pertinent; *Ratio of Specific Heats of Vapor (Gas)*: Not pertinent; *Latent Heat of*

Vaporization: Not pertinent; *Heat of Combustion*: Not pertinent; *Heat of Decomposition*: No data.

Health Hazards Information - *Recommended Personal Protective Equipment*: Self-contained breathing apparatus must be used when fighting fires. At other times a dust mask is adequate; *Symptoms Following Exposure*: Inhalation causes irritation of nose and throat. Contact with eyes causes irritation; *General Treatment for Exposure*: INHALATION: move to fresh air. EYES: flush with water for 15 min.; *Toxicity by Inhalation (Threshold Limit Value)*: Data not available; *Short-Term Exposure Limits*: Data not available; *Toxicity by Ingestion*: Grade 3; oral rat LD_{50} 58 mg/kg (ammonium sulfate); *Late Toxicity*: Data not available; *Vapor (Gas) Irritant Characteristics*: Not pertinent; *Liquid or Solid Irritant Characteristics*: Data not available; *Odor Threshold*: Data not available.

Fire Hazards - *Flash Point:* Not pertinent; *Flammable Limits in Air (%)*: Not pertinent; *Fire Extinguishing Agents*: Water; *Fire Extinguishing Agents Not To Be Used*: Steam, inert gases, foam, dry chemical; *Special Hazards of Combustion Products*: Toxic and irritating oxides of nitrogen may form in fires; *Behavior in Fire*: Will increase intensity of fire when in contact with combustible material. Containers may explode; *Ignition Temperature*: Not pertinent; *Electrical Hazard*: Not pertinent; *Burning Rate*: Not pertinent.

Chemical Reactivity - *Reactivity with Water*: No reaction; *Reactivity with Common Materials*: Corrodes metals to same degree as ordinary fertilizer; the reaction is not hazardous; *Stability During Transport*: Stable; *Neutralizing Agents for Acids and Caustics*: Not pertinent; *Polymerization*: Not pertinent; *Inhibitor of Polymerization*: Not pertinent.

AMMONIUM OXALATE

Chemical Designations - *Synonyms*: Ammonium Oxalate Hydrate; Diammonium Oxalate; Oxalic Acid, Diammonium Salt; *Chemical Formula*: $(NH_4)_2C_2O_4 \cdot H_2O$.

Observable Characteristics - *Physical State (as normally shipped)*: Solid; *Color*: White; *Odor*: None.

Physical and Chemical Properties - *Physical State at 15 °C and 1 atm.*: Solid; *Molecular Weight*: 142.11; *Boiling Point at 1 atm.*: Not pertinent (decomposes at 70°C); *Freezing Point*: Not pertinent (decomposes at 70°C); *Critical Temperature*: Not pertinent; *Critical Pressure*: Not pertinent; *Specific Gravity:* 1.50 at 18.5°C (solid); *Vapor (Gas) Density*: Not pertinent; *Ratio of Specific Heats of Vapor (Gas)*: Not pertinent; *Latent Heat of Vaporization*: Not pertinent; *Heat of Combustion*: Data not available; *Heat of Decomposition*: Not pertinent.

Health Hazards Information - *Recommended Personal Protective Equipment*: Approved dust respirator; rubber or plastic-coated gloves; chemical goggles; *Symptoms Following Exposure*: Ingestion or excessive inhalation of dust causes systemic poisoning; possible symptoms include pain in throat, esophagus, and stomach; mucous membranes turn white; vomiting, severe purging, weak pulse, cardiovascular collapse, neuromuscular symptoms. Contact with eyes causes irritation. Contact with skin causes irritation of severe burns; *General Treatment for Exposure*: Speed is essential. INHALATION: remove to fresh air. INGESTION: call physician immediately; induce vomiting. EYES: flush with water and seek medical attention. SKIN: flush with water. OTHER: watch for swelling of the glottis and delayed constriction of the esophagus; *Toxicity by Inhalation (Threshold Limit Value)*: Data not available; *Short-Term Exposure Limits*: Data not available; *Toxicity by Ingestion*: Data not available; *Late Toxicity*: Kidney damage; *Vapor (Gas) Irritant Characteristics*: Data not available; *Liquid or Solid Irritant Characteristics*: Data not available; *Odor Threshold*: Odorless.

Fire Hazards - *Flash Point:* Not pertinent (combustible solid); *Flammable Limits in Air (%)*: Not pertinent; *Fire Extinguishing Agents*: Water, foam; *Fire Extinguishing Agents Not To Be Used*: Not pertinent; *Special Hazards of Combustion Products*: Toxic oxides of nitrogen may form in fire; *Behavior in Fire*: Not pertinent; *Ignition Temperature*: Not pertinent; *Electrical Hazard*: Not pertinent; *Burning Rate*: Not pertinent.

Chemical Reactivity - *Reactivity with Water*: No reaction; *Reactivity with Common Materials*: No reaction; *Stability During Transport*: Stable; *Neutralizing Agents for Acids and Caustics*: Not pertinent; *Polymerization*: Not pertinent; *Inhibitor of Polymerization*: Not pertinent.

AMMONIUM PERCHLORATE

Chemical Designations - *Synonyms*: No common synonyms; *Chemical Formula*: NH_4ClO_4; **Observable Characteristics** - *Physical State (as normally shipped)*: Solid; *Color*: White; *Odor*: None.
Physical and Chemical Properties - *Physical State at 15 °C and 1 atm.*: Solid; *Molecular Weight*: 117.49; *Boiling Point at 1 atm.*: Not pertinent; *Freezing Point*: Not pertinent; *Critical Temperature*: Not pertinent; *Critical Pressure*: Not pertinent; *Specific Gravity:* 1.95 at 15°C (solid); *Vapor (Gas) Density*: Not pertinent; *Ratio of Specific Heats of Vapor (Gas)*: Not pertinent; *Latent Heat of Vaporization*: Not pertinent; *Heat of Combustion*: Not pertinent; *Heat of Decomposition*: Not pertinent.
Health Hazards Information - *Recommended Personal Protective Equipment*: Data not available; *Symptoms Following Exposure*: Irritating to skin and mucous membranes; *General Treatment for Exposure*: Data not available; *Toxicity by Inhalation (Threshold Limit Value)*: Not pertinent; *Short-Term Exposure Limits*: Not pertinent; *Toxicity by Ingestion*: Grade 2; oral rat LD_{50} 3500 mg/kg; *Late Toxicity*: Not pertinent; *Vapor (Gas) Irritant Characteristics*: Not pertinent; *Liquid or Solid Irritant Characteristics*: Data not available; *Odor Threshold*: Not pertinent.
Fire Hazards - *Flash Point:* Not flammable; *Flammable Limits in Air (%)*: Not flammable; *Fire Extinguishing Agents*: Water (from protected location); *Fire Extinguishing Agents Not To Be Used*: Not pertinent; *Special Hazards of Combustion Products*: Toxic gases are produced in a fire; *Behavior in Fire*: May explode when involved in a fire or exposed to shock or friction; *Ignition Temperature*: Not flammable; *Electrical Hazard*: Not pertinent; *Burning Rate*: Not flammable.
Chemical Reactivity - *Reactivity with Water*: No reaction; *Reactivity with Common Materials*: No reaction; *Stability During Transport*: If contaminated with carbonaceous materials, can become an explosive which is sensitive to shock and friction. Ready detonates or explodes; *Neutralizing Agents for Acids and Caustics*: Not pertinent; *Polymerization*: Not pertinent; *Inhibitor of Polymerization*: Not pertinent.

AMMONIUM PERSULFATE

Chemical Designations - *Synonyms*: Ammonium Peroxydisulfate; Peroxydisulfuric Acid, Diammonium Salt; *Chemical Formula*: $(NH_4)_2S_2O_8$.
Observable Characteristics - *Physical State (as normally shipped)*: Solid; *Color*: Light straw to colorless; *Odor*: Slight acid.
Physical and Chemical Properties - *Physical State at 15 °C and 1 atm.*: Solid; *Molecular Weight*: 228.20; *Boiling Point at 1 atm.*: Not pertinent (decomposes at 120°C); *Freezing Point*: Not pertinent(decomposes at 120°C); *Critical Temperature*: Not pertinent; *Critical Pressure*: Not pertinent; *Specific Gravity:* 1.98 at 20°C (solid); *Vapor (Gas) Density*: Not pertinent; *Ratio of Specific Heats of Vapor (Gas)*: Not pertinent; *Latent Heat of Vaporization*: Not pertinent; *Heat of Combustion*: Not pertinent; *Heat of Decomposition*: Data not available.
Health Hazards Information - *Recommended Personal Protective Equipment*: U.S. Bu. Mines approved toxic dust mask; chemical goggles; rubber gloves; neoprene-coated shoes; *Symptoms Following Exposure*: Inhalation produces slight toxic effects. Contact with eyes irritates eyes and causes skin rash; *General Treatment for Exposure*: INHALATION: remove to fresh air. EYES: wash with water for 20 min.; call a physician. SKIN: wash with water; *Toxicity by Inhalation (Threshold Limit Value)*: Data not available; *Short-Term Exposure Limits*: Data not available; *Toxicity by Ingestion*: Grade 2; oral rat LD_{50} 820 mg/kg; *Late Toxicity*: Data not available; *Vapor (Gas) Irritant Characteristics*: Data not available; *Liquid or Solid Irritant Characteristics*: Data not available; *Odor Threshold*: Data not available.
Fire Hazards - *Flash Point:* Not pertinent; *Flammable Limits in Air (%)*: Not pertinent; *Fire Extinguishing Agents*: Water; *Fire Extinguishing Agents Not To Be Used*: Data not available; *Special Hazards of Combustion Products*: Toxic oxides of nitrogen and sulfuric acid fumes are formed in fire; *Behavior in Fire*: Decomposes with loss of oxygen that increases intensity of fire; *Ignition Temperature*: Not pertinent; *Electrical Hazard*: Not pertinent; *Burning Rate*: Not pertinent.
Chemical Reactivity - *Reactivity with Water*: No reaction; *Reactivity with Common Materials*: Contact with grease, wood and other combustibles may result in fire; *Stability During Transport*: Stable;

Neutralizing Agents for Acids and Caustics: Not pertinent; *Polymerization*: Not pertinent; *Inhibitor of Polymerization*: Not pertinent.

AMMONIUM SILICOFLUORIDE
Chemical Designations - *Synonyms*: Ammonium Fluosilicate; *Chemical Formula*: $(NH_4)_2SiF_6$.
Observable Characteristics - *Physical State (as normally shipped)*: Solid; *Color*: White; *Odor*: None.
Physical and Chemical Properties - *Physical State at 15 °C and 1 atm.*: Solid; *Molecular Weight*: 178.14; *Boiling Point at 1 atm.*: Not pertinent (decomposes); *Freezing Point*: Not pertinent; *Critical Temperature*: Not pertinent; *Critical Pressure*: Not pertinent; *Specific Gravity:* 2.0 at 20°C (solid); *Vapor (Gas) Density*: Not pertinent; *Ratio of Specific Heats of Vapor (Gas)*: Not pertinent; *Latent Heat of Vaporization*: Not pertinent; *Heat of Combustion*: Not pertinent; *Heat of Decomposition*: Not pertinent.
Health Hazards Information — *Recommended Personal Protective Equipment*: Dust respirator; acid resistant clothing and hat; rubber gloves; goggles and safety shoes; *Symptoms Following Exposure*: Inhalation of dust can cause pulmonary irritation and can be fatal in some cases. Ingestion may be fatal. Contact with dust causes irritation of eyes and irritation or ulceration of skin; *General Treatment for Exposure*: INHALATION: remove patient to fresh air. INGESTION: cause vomiting by giving soapy water or mustard water; have patient drink large quantities of lime water; if necessary, give stimulant such as strong coffee; keep patient warm. EYES: flush with water for 20 min., holding eyelids open. SKIN: wash with soap and water; *Toxicity by Inhalation (Threshold Limit Value)*: 2.5 mg/m³; *Short-Term Exposure Limits*: Data not available; *Toxicity by Ingestion*: Grade 3; LD_{50} 100 mg/kg (rat); *Late Toxicity*: Data not available; *Vapor (Gas) Irritant Characteristics*: Data not available; *Liquid or Solid Irritant Characteristics*: Data not available; *Odor Threshold*: Data not available.
Fire Hazards - *Flash Point:* Not flammable; *Flammable Limits in Air (%)*: Not flammable; *Fire Extinguishing Agents*: Not pertinent; *Fire Extinguishing Agents Not To Be Used*: Not pertinent; *Special Hazards of Combustion Products*: Toxic and irritating hydrogen fluoride, silicon tetrafluoride, and oxides of nitrogen may form in fires; *Behavior in Fire*: Not pertinent; *Ignition Temperature (deg. F)*: Not pertinent; *Electrical Hazard*: Not pertinent; *Burning Rate*: Not pertinent.
Chemical Reactivity - *Reactivity with Water*: No reaction; *Reactivity with Common Materials*: No reaction; *Stability During Transport*: Stable; *Neutralizing Agents for Acids and Caustics*: Not pertinent; *Polymerization*: Not pertinent; *Inhibitor of Polymerization*: Not pertinent.

AMYL ACETATE
Chemical Designations - *Synonyms*: Amyl Acetate, Mixed Isomers; Pentyl Acetates; *Chemical Formula*: $CH_3COOC_5H_{11}$.
Observable Characteristics - *Physical State (as normally shipped)*: Liquid; *Color*: Colorless to yellow; *Odor*: Pleasant banana-like; mild; characteristic banana- or pear- like odor;
Physical and Chemical Properties — *Physical State at 15 °C and 1 atm.*: Liquid; *Molecular Weight*: 130.19; *Boiling Point at 1 atm.*: 295, 146, 419; *Freezing Point*: <-148, <-100, <173; *Critical Temperature*: Not pertinent; *Critical Pressure*: Not pertinent; Specific Gravity: 0.876 at 20 °C (liquid); *Vapor (Gas) Density*: Not pertinent; *Ratio of Specific Heats of Vapor (Gas)*: Not pertinent; *Latent Heat of Vaporization*: 140, 75, 3.1; *Heat of Combustion*: -13,360, -7423, -310.8; *Heat of Decomposition*: Not pertinent.
Health Hazards Information - *Recommended Personal Protective Equipment*: Air-supplied mask or chemical cartridge respirator, protective gloves, goggles, safety shower, and eye bath; *Symptoms Following Exposure*: Irritation of eyes, nose and throat. Dizziness, nausea, headache; *General Treatment for Exposure*: INHALATION: move victim to fresh air; call physician; administer oxygen. SKIN OR EYES: flush with water; *Toxicity by Inhalation (Threshold Limit Value)*: 100 ppm; *Short-Term Exposure Limits*: 200 ppm for 30 min.; *Toxicity by Ingestion*: Grade 1; LD_{50} 6.5 g/kg (rat); *Late Toxicity*: None; *Vapor (Gas) Irritant Characteristics*: Vapors cause a slight smarting of the eyes or respiratory system if present in high concentration. The effect is temporary; *Liquid or Solid Irritant Characteristics*: o appreciable hazard. Practically harmless to the skin. If spilled on clothing and

allowed to remain, may cause smarting and reddening of the skin; *Odor Threshold*: 0.067 ppm.
Fire Hazards - *Flash Point (deg. F):* (iso-): 69 CC (n-); 91 CC; *Flammable Limits in Air* (%): 1.1 -
7.5; *Fire Extinguishing Agents*: alcohol foam, dry chemical, carbon dioxide; *Fire Extinguishing Agents
Not To Be Used*: Water in straight hose stream will scatter and spread fire and should not be used;
Special Hazards of Combustion Products: Not pertinent; *Behavior in Fire*: Not pertinent; *Ignition
Temperature (deg. F)*: 572 (n); *Electrical Hazard*: Not pertinent; *Burning Rate*: 4.1 mm/min.
Chemical Reactivity - *Reactivity with Water*: No reaction; *Reactivity with Common Materials*: No
reaction; *Stability During Transport*: Stable; *Neutralizing Agents for Acids and Caustics*: Not pertinent;
Polymerization: Not pertinent; *Inhibitor of Polymerization*: Not pertinent.

ANILINE
Chemical Designations - *Synonyms*: Aminobenzene; Aniline Oil; Blue Oil; Phenylamine; *Chemical
Formula*: $C_6H_5NH_2$.
Observable Characteristics - *Physical State (as normally shipped)*: Liquid; *Color*: Colorless to pale
brown; *Odor*: Aromatic amine like; characteristic, peculiar; strongly amine-like.
Physical and Chemical Properties - *Physical State at 15 °C and 1 atm.*: Liquid; *Molecular Weight*:
93.13; *Boiling Point at 1 atm.*: 363.6, 184.2, 457.4; *Freezing Point*: 21, -6.1, 267.1; *Critical
Temperature*: 798.1, 425.6, 698.8; *Critical Pressure*: 770, 52.4, 5.31; *Specific Gravity:* 1.022 at 20°C
(liquid); *Vapor (Gas) Density*: Not pertinent; *Ratio of Specific Heats of Vapor (Gas)*: 1.1; *Latent Heat
of Vaporization*: 198, 110, 4.61; *Heat of Combustion*: -14,980, -8320, -348,3; *Heat of Decomposition*:
Not pertinent.
Health Hazards Information - *Recommended Personal Protective Equipment*: Respirator for organic
vapors, splashproof goggles, rubber gloves, boots; *Symptoms Following Exposure*: ACUTE
EXPOSURE: blue discoloration of finger-tips, cheeks, lips and nose; nausea, vomiting, headache and
drowsiness followed by delirium, coma and shock. CHRONIC EXPOSURE: loss of appetite, loss of
weight, headaches,, visual disturbances; skin lesions; *General Treatment for Exposure*: Remove victim
to fresh air and call a physician at once. SKIN OR EYES: immediately flush skin or eyes with plenty
of water for at least 15 min. If cyanosis is present, shower with soap and warm water, with special
attention to scalp and fingernails; *Toxicity by Inhalation (Threshold Limit Value)*: 5 ppm; *Short-Term
Exposure Limits*: 50 ppm for 30 min.; 5 ppm for 8 hr.; *Toxicity by Ingestion*: Grade 3; LD_{50} 5 to 500
mg/kg; *Late Toxicity*: None recognized; *Vapor (Gas) Irritant Characteristics*: Vapors cause a slight
smarting of the eyes or respiratory system if present in high concentrations. The effect is temporary;
Liquid or Solid Irritant Characteristics: If spilled on clothing and allowed to remain, may cause
smarting and reddening of the skin.; *Odor Threshold*: 0.5 ppm.
Fire Hazards - *Flash Point (deg. F):* 168 OC; 158 CC; *Flammable Limits in Air* (%): 1.3 (LEL); *Fire
Extinguishing Agents*: Water, foam, dry chemical or carbon dioxide; *Fire Extinguishing Agents Not To
Be Used*: Not pertinent; *Special Hazards of Combustion Products*: Toxic vapors are generated when
heated; *Behavior in Fire*: Not pertinent; *Ignition Temperature (deg. F)*: 1418; *Electrical Hazard*: Not
pertinent; *Burning Rate*: 3.0 mm/min.
Chemical Reactivity - *Reactivity with Water*: No reaction; *Reactivity with Common Materials*: No
reaction; *Stability During Transport*: Stable; *Neutralizing Agents for Acids and Caustics*: Flush with
water and rinse with dilute acetic acid; *Polymerization*: Not pertinent; *Inhibitor of Polymerization*: Not
pertinent.

ANISOYL CHLORIDE
Chemical Designations - *Synonyms*: p-Anisoyl Chloride; *Chemical Formula*: $p-CH_3OC_6H_5COCl$.
Observable Characteristics - *Physical State (as normally shipped)*: Liquid; *Color*: Yellow; brown;
Odor: Sharp, penetrating.
Physical and Chemical Properties - *Physical State at 15 °C and 1 atm.*: Liquid; *Molecular Weight*:
171.6; *Boiling Point at 1 atm.*: 504, 262, 535; *Freezing Point*: 72, 22, 295; *Critical Temperature*: Not
pertinent; *Critical Pressure*: Not pertinent; *Specific Gravity:* 1.26 at 20°C (liquid); *Vapor (Gas)
Density*: Not pertinent; *Ratio of Specific Heats of Vapor (Gas)*: Not pertinent; *Latent Heat of*

Vaporization: Not pertinent; *Heat of Combustion*: -10,500, -5,830, -244; *Heat of Decomposition*: Not pertinent.

Health Hazards Information - *Recommended Personal Protective Equipment*: Goggles or face shield; plastic gloves; protective clothing; *Symptoms Following Exposure*: Vapor irritates mucous membranes. Contact of liquid with eyes or skin causes severe irritation. Ingestion causes severe irritation of mouth and stomach; *General Treatment for Exposure*: INHALATION: remove to fresh air. EYES: flush with water for at least 15 min.; get medical attention. SKIN: flush with water; wash well with soap and water. INGESTION: do NOT induce vomiting; give large amounts of water; *Toxicity by Inhalation (Threshold Limit Value)*: Data not available; *Short-Term Exposure Limits*: Data not available; *Toxicity by Ingestion*: Data not available; *Late Toxicity*: Data not available; *Vapor (Gas) Irritant Characteristics*: Data not available; *Liquid or Solid Irritant Characteristics*: Data not available; *Odor Threshold*: Data not available.

Fire Hazards - *Flash Point:* Data not available; *Flammable Limits in Air (%)*: Not pertinent; *Fire Extinguishing Agents*: Carbon dioxide, dry chemical; *Fire Extinguishing Agents Not To Be Used*: Water, foam; *Special Hazards of Combustion Products*: Irritating hydrogen chloride fumes may be formed; *Behavior in Fire*: Not pertinent; *Ignition Temperature*: Data not available; *Electrical* Hazard: Data not available; *Burning Rate*: Data not available.

Chemical Reactivity - *Reactivity with Water*: Reacts slowly to generate hydrogen chloride (hydrochloric acid). The reaction is not hazardous; *Reactivity with Common Materials*: Corrodes metal slowly; *Stability During Transport*: Stable; *Neutralizing Agents for Acids and Caustics*: Flush with water, rinse with sodium bicarbonate or lime solution; *Polymerization*: Not pertinent; *Inhibitor of Polymerization*: Not pertinent.

ANTHRACENE

Chemical Designations - *Synonyms*: Anthracin; Green Oil; Paranaphtalene; *Chemical Formula*: $C_{14}H_{10}$.

Observable Characteristics - *Physical State (as normally shipped)*: Solid; *Color*: White to yellow; *Odor*: Weak aromatic.

Physical and Chemical Properties - *Physical State at 15 °C and 1 atm.*: Solid; *Molecular Weight*: 178.23; *Boiling Point at 1 atm.*: 646.2, 341.2, 614.4; *Freezing Point*: 421.7, 216.5, 489.7; *Critical Temperature*: Not pertinent; *Critical Pressure*: Not pertinent; *Specific Gravity:* 1.24 at 20°C (solid); *Vapor (Gas) Density*: Not pertinent; *Ratio of Specific Heats of Vapor (Gas)*: Not pertinent; *Latent Heat of Vaporization*: Not pertinent; *Heat of Combustion*: -17,100, -9,510, -398; *Heat of Decomposition*: Not pertinent.

Health Hazards Information — *Recommended Personal Protective Equipment*: Dust mask; goggles or face shield, rubber gloves; *Symptoms Following Exposure*: Inhalation of dust irritates nose and throat. Contact with eyes causes irritation; *General Treatment for Exposure*: INHALATION: move to fresh air. EYES: flush with water for 15 min; *Toxicity by Inhalation*: Data not available; *Short-Term Exposure Limits*: Data not available; *Toxicity by Ingestion*: Data not available; *Late Toxicity*: Data not available; *Vapor (Gas) Irritant Characteristics*: Data not available; *Liquid or Solid Irritant Characteristics*: Data not available; *Odor Threshold*: Data not available.

Fire Hazards - *Flash Point:* Combustible Solid; *Flammable Limits in Air (%)*: Not pertinent; *Fire Extinguishing Agents*: Water, foam, dry chemical, carbon dioxide; *Fire Extinguishing Agents Not To Be Used*: None; *Special Hazards of Combustion Products*: Data not available; *Behavior in Fire*: Data not available; *Ignition Temperature (deg. F)*: 1004; *Electrical Hazard*: Not pertinent; *Burning Rate*: Not pertinent.

Chemical Reactivity - *Reactivity with Water*: No reaction; *Reactivity with Common Materials*: No reaction; *Stability During Transport*: Stable; *Neutralizing Agents for Acids and Caustics*: Not pertinent; *Polymerization*: Not pertinent; *Inhibitor of Polymerization*: Not pertinent.

ANTIMONY PENTACHLORIDE

Chemical Designations - *Synonyms*: Antimony (V) Chloride; Antimony Perchloride; *Chemical Formula*: $SbCl_5$.

Observable Characteristics - *Physical State (as normally shipped)*: Liquid; *Color*: Colorless to medium brown; yellow; red-brown; *Odor*: Pungent; offensive.

Physical and Chemical Properties - *Physical State at 15 °C and 1 atm.*: Liquid; *Molecular Weight*: 299.05; *Boiling Point at 1 atm.*: 347, 175, 448; *Freezing Point*: 37, 3, 276; *Critical Temperature*: Not pertinent; *Critical Pressure*: Not pertinent; *Specific Gravity:* 2.354 at 20°C (liquid); *Vapor (Gas) Density*: Not pertinent; *Ratio of Specific Heats of Vapor (Gas)*: Not pertinent; *Latent Heat of Vaporization*: 68.9, 38.3, 1.60; *Heat of Combustion*: Not pertinent; *Heat of Decomposition*: Not pertinent.

Health Hazards Information - *Recommended Personal Protective Equipment*: Organic vapor-acid gas type canister mask; rubber, neoprene, vinyl, etc. gloves; chemical safety goggles, plus face shield where appropriate; acid resistant clothing, plus apron for splash protection; rubber safety shoes or boots; hard hat; *Symptoms Following Exposure*: Inhalation causes irritation of nose and throat. Contact of liquid with eyes or skin causes severe burns. Ingestion causes vomiting and severe burns of mouth and stomach. Overexposure by any route can cause bloody stools, slow pulse, low blood pressure, coma, convulsions, cardiac arrest; *General Treatment for Exposure*: INHALATION: remove to clean air; rinse mouth and gargle with water; if overexposure is serious, get prompt medical attention. EYES: flush eyes and eye-lids thoroughly with large amounts of water; get prompt medical attention. SKIN: flush thoroughly with water; remove contaminated clothing; wash affected area with soap and water; if overexposure is serious, get prompt medical attention. INGESTION: dilute by drinking water; if vomiting occurs, administer more water. If overexposure is serious, get prompt medical attention; *Toxicity by Inhalation (Threshold Limit Value)*: 0.5 mg/m^3 as antimony; *Short-Term Exposure Limits*: Data not available; *Toxicity by Ingestion*: Grade 2; oral LD50 1,115 mg/kg (rat), 900 mg/kg (guinea pig); *Late Toxicity*: Antimony poisoning may result; *Vapor (Gas) Irritant Characteristics*: Vapors are moderately irritating such that personnel will not usually tolerate moderate or high vapor concentrations; *Liquid or Solid Irritant Characteristics*: Severe skin irritant; causes second- and third-degree burns on short contact and is very injurious to the eyes; *Odor Threshold*: Data not available.

Fire Hazards - *Flash Point:* Not flammable; *Flammable Limits in Air (%)*: Not flammable; *Fire Extinguishing Agents*: Not pertinent; *Fire Extinguishing Agents Not To Be Used*: Do not use water or foam on adjacent fires; *Special Hazards of Combustion Products*: Not pertinent; *Behavior in Fire*: Irritating fumes of hydrogen chloride given off when water or foam is used to extinguish adjacent fire; *Ignition Temperature*: Not pertinent; *Electrical Hazard*: Not pertinent; *Burning Rate*: Not pertinent.

Chemical Reactivity - *Reactivity with Water*: Reacts to form hydrogen chloride gas (hydrochloric acid); *Reactivity with Common Materials*: Causes corrosion on metal; *Stability During Transport*: Stable; *Neutralizing Agents for Acids and Caustics*: Soda ash or soda ash-lime mixture; *Polymerization*: Not pertinent; *Inhibitor of Polymerization*: Not pertinent.

ANTIMONY PENTAFLUORIDE

Chemical Designations - *Synonyms*: No common synonyms; *Chemical Formula*: SbF_5.

Observable Characteristics - *Physical State (as normally shipped)*: Liquid; *Color*: Colorless; *Odor*: Pungent.

Physical and Chemical Properties - *Physical State at 15 °C and 1 atm.*: Liquid; *Molecular Weight*: 216.7; *Boiling Point at 1 atm.*: 289, 143, 416; *Freezing Point*: 45, 7, 280; *Critical Temperature*: Not pertinent; *Critical Pressure*: Not pertinent; *Specific Gravity:* 2.340 at 30°C (liquid); *Vapor (Gas) Density*: Not pertinent; *Ratio of Specific Heats of Vapor (Gas)*: Not pertinent; *Latent Heat of Vaporization*: 79, 44, 1.8; *Heat of Combustion*: Not pertinent; *Heat of Decomposition*: Not pertinent.

Health Hazards Information - *Recommended Personal Protective Equipment*: Acid-gas-type canister mask; rubber gloves, protective clothing; safety goggles and face shield; *Symptoms Following Exposure*: Inhalation causes irritation of nose and throat. Contact of liquid with eyes or skin causes severe burns. Ingestion causes vomiting and severe burns of mouth and throat. Overexposure by any route can cause bloody stools, slow pulse, low blood pressure, coma, convulsions, cardiac arrest; *General Treatment for Exposure*: INHALATION: remove to fresh air; rinse mouth with water; give oxygen if necessary to assist breathing; get medical attention. EYES: irrigate with copious amounts of water for at least 15 min.; get medical attention. SKIN: flush with copious amounts of water; wash well with soap and water. INGESTION: dilute by drinking water; if vomiting occurs, drink more water; get medical attention promptly; *Toxicity by Inhalation (Threshold Limit Value)*: 0.5 mg/m^3 as antimony; *Short-Term Exposure Limits*: Data not available; *Toxicity by Ingestion*: Data not available; *Late Toxicity*: Antimony poisoning may result; *Vapor (Gas) Irritant Characteristics*: Data not available; *Liquid or Solid Irritant Characteristics*: Data not available; *Odor Threshold*: Data not available.

Fire Hazards - *Flash Point:* Not flammable; *Flammable Limits in Air (%)*: Not flammable; *Fire Extinguishing Agents*: Not pertinent; *Fire Extinguishing Agents Not To Be Used*: Do not use water or foam on adjacent fire; *Special Hazards of Combustion Products*: Not pertinent; *Behavior in Fire*: Gives off toxic hydrogen fluoride fumes when water is used to extinguish adjacent fire; *Ignition Temperature*: Not pertinent; *Electrical Hazard*: Not pertinent; *Burning Rate*: Not pertinent.

Chemical Reactivity - *Reactivity with Water*: Reacts vigorously to form toxic hydrogen fluoride (hydrofluoric acid); *Reactivity with Common Materials*: When moisture is present, causes severe corrosion of metals (except steel) and glass. If confined and wet can cause explosion. May cause fire in contact with combustible material; *Stability During Transport*: Stable; *Neutralizing Agents for Acids and Caustics*: Flush with water, rinse with sodium bicarbonate or lime solution; *Polymerization*: Not pertinent; *Inhibitor of Polymerization*: Not pertinent.

ANTIMONY POTASSIUM TARTRATE

Chemical Designations - *Synonyms*: Potassium Antimonyl Tartrate; Tartar Emetic; Tartarized Antimony; Tartrated Antimony; *Chemical Formula*: KOOC•CHOH•CHOH•COO(SbO)•½H$_2$O.

Observable Characteristics - *Physical State (as normally shipped)*: Solid; *Color*: White; *Odor*: None.

Physical and Chemical Properties - *Physical State at 15 °C and 1 atm.*: Solid; *Molecular Weight*: 334; *Boiling Point at 1 atm.*: Not pertinent; *Freezing Point*: Not pertinent; *Critical Temperature*: Not pertinent; *Critical Pressure*: Not pertinent; *Specific Gravity:* 2.60 at 20°C (solid); *Vapor (Gas) Density*: Not pertinent; *Ratio of Specific Heats of Vapor (Gas)*: Not pertinent; *Latent Heat of Vaporization*: Not pertinent; *Heat of Combustion*: Not pertinent; *Heat of Decomposition*: Not pertinent.

Health Hazards Information - *Recommended Personal Protective Equipment*: Dust respirator; rubber or plastic-coated gloves; chemical goggles; tightly woven, close fitting clothes; Bu. Mines approved respirator; *Symptoms Following Exposure*: Inhalation causes inflammation of membranes of nose and throat, upper respiratory irritation, headache, dizziness. Ingestion causes gastrointestinal upset, strong irritation, vomiting. Contact with eyes or skin causes irritation. Further symptoms of exposure include nervous complaints (i.e., irritability, dizziness, muscular and neurological pain); *General Treatment for Exposure*: INHALATION: move to fresh air. INGESTION: call physician immediately; use water (plain, soapy, or salty) or milk (3-4 glasses) to provoke vomiting. EYES: flush with water for 15 min.; consult a physician. SKIN: flush with water; wash well with soap and water; *Toxicity by Inhalation (Threshold Limit Value)*: 0.5 mg/m^3 (as antimony); *Short-Term Exposure Limits*: Data not available; *Toxicity by Ingestion*: Grade 3; oral rat LD$_{50}$ 115 mg/kg; *Late Toxicity*: Data not available; *Vapor (Gas) Irritant Characteristics*: Data not available; *Liquid or Solid Irritant Characteristics*: Data not available; *Odor Threshold*: Odorless.

Fire Hazards - *Flash Point:* Not flammable; *Flammable Limits in Air (%)*: Not flammable; *Fire Extinguishing Agents*: Not pertinent; *Fire Extinguishing Agents Not To Be Used*: Not pertinent; *Special Hazards of Combustion Products*: Not pertinent; *Behavior in Fire*: Not pertinent; *Ignition Temperature*: Not pertinent; *Electrical Hazard*: Not pertinent; *Burning Rate*: Not pertinent.

Chemical Reactivity - *Reactivity with Water*: No reaction; *Reactivity with Common Materials*: No reactions; *Stability During Transport*: Stable; *Neutralizing Agents for Acids and Caustics*: Not pertinent; *Polymerization*: Not pertinent; *Inhibitor of Polymerization*: Not pertinent.

ANTIMONY TRICHLORIDE

Chemical Designations - *Synonyms*: Antimony Butter; Antimony (iii) Chloride; Butter of Antimony; *Chemical Formula*: SbCl$_3$.

Observable Characteristics - *Physical State (as normally shipped)*: Solid; *Color*: White to pale yellow;

Odor: Sharp, acrid.

Physical and Chemical Properties - *Physical State at 15 °C and 1 atm.*: Solid; *Molecular Weight*: 228; *Boiling Point at 1 atm.*: 433, 223, 496; *Freezing Point*: 163, 73, 346; *Critical Temperature*: Not pertinent; *Critical Pressure*: Not pertinent; *Specific Gravity:* 3.14 at 20°C (solid); *Vapor (Gas) Density*: Not pertinent; *Ratio of Specific Heats of Vapor (Gas)*: Not pertinent; *Latent Heat of Vaporization*: Not pertinent; *Heat of Combustion*: Not pertinent; *Heat of Decomposition*: Not pertinent.

Health Hazards Information - *Recommended Personal Protective Equipment*: Bu. Mines approved respirator; chemical safety goggles; face shield; leather or rubber safety shoes; rubber apron; rubber gloves; *Symptoms Following Exposure*: Inhalation of small amounts may cause only irritation of the nose, throat, and air passages; large exposures result in severe air-passage irritation. Ingestion causes vomiting, purging with bloody stools, slow pulse and low blood pressure; slow, shallow breathing; coma and convulsions sometimes followed by death. Contact with eyes causes severe eye burns or at least severe eye irritation. Contact of dry chemical with skin may result in deep chemical burns; *General Treatment for Exposure*: INHALATION: move victim at once to fresh air and keep him warm, but not hot; call a physician immediately; nasal passages may be irrigated from a gently flowing hose. INGESTION: induce vomiting by giving large quantities of warm salt water; have a physician see the patient at once. SKIN: flush with large quantities of flowing water following by washing of skin surfaces with soap and water; remove all contaminated clothing promptly; *Toxicity by Inhalation (Threshold Limit Value)*: 0.5 mg/m^3 (as antimony); *Short-Term Exposure Limits*: Data nat available; *Toxicity by Ingestion*: Grade 2; oral rat LD$_{50}$ 675 mg/kg; *Late Toxicity*:; *Vapor (Gas) Irritant Characteristics*: Data nat available; *Liquid or Solid Irritant Characteristics*: Data nat available; *Odor Threshold*: Data nat available.

Fire Hazards - *Flash Point:* Not flammable; *Flammable Limits in Air* (%): Not flammable; *Fire Extinguishing Agents*: Not pertinent; *Fire Extinguishing Agents Not To Be Used*: Do not apply water on adjacent fires; *Special Hazards of Combustion Products*: Toxic and irritating antimony oxide and hydrogen chloride may form in fires; *Behavior in Fire*: No data; *Ignition Temperature*: Not pertinent; *Electrical Hazard*: Not pertinent; *Burning Rate*: Not pertinent.

Chemical Reactivity - *Reactivity with Water*: Reacts vigorously to form a strong solution of hydrochloric acid; *Reactivity with Common Materials*: Corrodes may metals in the presence of moisture and flammable hydrogen gas may collect in confined spaces; *Stability During Transport*: Stable; *Neutralizing Agents for Acids and Caustics*: Large amounts of water followed by sodium bicarbonate or soda ash solution; *Polymerization*: Not pertinent; *Inhibitor of Polymerization*: Not pertinent.

ANTIMONY TRIFLUORIDE

Chemical Designations - *Synonyms*: No common synonyms; *Chemical Formula*: SbF_3.

Observable Characteristics - *Physical State (as normally shipped)*: Solid; *Color*: White; *Odor*: None.

Physical and Chemical Properties - *Physical State at 15 °C and 1 atm.*: Solid; *Molecular Weight*: 178.75; *Boiling Point at 1 atm.*: Not pertinent; *Freezing Point*: 558, 292, 565; *Critical Temperature*: Not pertinent; *Critical Pressure*: Not pertinent; *Specific Gravity:* 4.38 at 21°C (solid); *Vapor (Gas) Density*: Not pertinent; *Ratio of Specific Heats of Vapor (Gas)*: Not pertinent; *Latent Heat of Vaporization*: Not pertinent; *Heat of Combustion*: Not pertinent; *Heat of Decomposition*: Not pertinent.

Health Hazards Information - *Recommended Personal Protective Equipment*: Approved respirator; rubber gloves; *Symptoms Following Exposure*: Resemble those of lead and arsenic poisoning. ACUTE POISONING: irritation of the mouth, nose, stomach and intestines; vomiting, purging with bloody stools; slow pulse and low blood pressure; slow, shallow breathing; coma and convulsions sometimes followed by death from cardiac and respiratory exhaustion. CHRONIC POISONING: dryness of throat; pain on swallowing; occasional vomiting and persistent nausea; susceptibility to fainting; diarrhea, loss of appetite and weight; giddiness; dermatitis, either pustulating or ulcerative; anemia; *General Treatment for Exposure*: If any symptoms, however slight, are noticed, the affected individual should be removed from contact with chemical and placed under the care of the physician who is versed in the treatment necessary; *Toxicity by Inhalation (Threshold Limit Value)*: 0.5 mg/m^3; *Short-Term Exposure Limits*: Not pertinent; *Toxicity by Ingestion*: Grade 3; LD$_{50}$ 50 to 500 mg/kg (guinea pig); *Late Toxicity*:

Data not available; *Vapor (Gas) Irritant Characteristics*: Not pertinent; *Liquid or Solid Irritant Characteristics*: Fairly severe skin irritant. May cause pain and second-degree burns after a few minutes contact; *Odor Threshold*: Not pertinent.

Fire Hazards - *Flash Point:* Not flammable; *Flammable Limits in Air (%)*: Not flammable; *Fire Extinguishing Agents*: Not pertinent; *Fire Extinguishing Agents Not To Be Used*: Not pertinent; *Special Hazards of Combustion Products*: Not flammable; *Behavior in Fire*: Not flammable; *Ignition Temperature*: Not flammable; *Electrical Hazard*: Not pertinent; *Burning Rate*: Not pertinent.

Chemical Reactivity - *Reactivity with Water*: No reaction; *Reactivity with Common Materials*: No reaction; *Stability During Transport*: Stable; *Neutralizing Agents for Acids and Caustics*: Not pertinent; *Polymerization*: Not pertinent; *Inhibitor of Polymerization*: Not pertinent.

ANTIMONY TRIOXIDE

Chemical Designations - *Synonyms*: Diantimony Trioxide; Exitelite; Flowers of antimony; Senarmontite; Valentinite; Weisspiessglanz; *Chemical Formula*: Sb_2O_3.

Observable Characteristics - *Physical State (as normally shipped)*: Solid; *Color*: White; *Odor*: None.

Physical and Chemical Properties - *Physical State at 15 °C and 1 atm.*: Solid; *Molecular Weight*: 291.50; *Boiling Point at 1 atm.*: Not pertinent; *Freezing Point*: Not pertinent; *Critical Temperature*: Not pertinent; *Critical Pressure*: Not pertinent; *Specific Gravity*: 5.2 at 25°C (solid); *Vapor (Gas) Density*: Not pertinent; *Ratio of Specific Heats of Vapor (Gas)*: Not pertinent; *Latent Heat of Vaporization*: Not pertinent; *Heat of Combustion*: Not pertinent; *Heat of Decomposition*: Not pertinent.

Health Hazards Information - *Recommended Personal Protective Equipment*: Rubber gloves; safety goggles; dust mask; *Symptoms Following Exposure*: Inhalation causes inflammation of upper and lower respiratory tract, including pneumonitis. Ingestion causes irritation of the mouth, nose, stomach and intestines; vomiting, purging with bloody stools; slow pulse and low pressure; slow, shallow breathing; coma and convulsions sometimes followed by death. Contact with eyes causes conjunctivitis. Contact with skin causes dermatitis and rhinitis; *General Treatment for Exposure*: If any of the symptoms of poisoning, even slight, are noticed, the affected individual should be removed from contact with the chemical and placed under care of a physician. INGESTION: induce vomiting. EYES: flush with water for at least 15 min. SKIN: wash well with soap and water; *Toxicity by Inhalation (Threshold Limit Value)*: 0.5 mg/m^3 (as antimony); *Short-Term Exposure Limits*: Data not available; *Toxicity by Ingestion*: Grade 0; oral rat LD$_{50}$ 20,000 mg/kg; *Late Toxicity*: Data not available; *Vapor (Gas) Irritant Characteristics*: Data not available; *Liquid or Solid Irritant Characteristics*: Data not available; *Odor Threshold*: Data not available.

Fire Hazards - *Flash Point:* Not flammable; *Flammable Limits in Air (%)*: Not flammable; *Fire Extinguishing Agents*: Not pertinent; *Fire Extinguishing Agents Not To Be Used*: Not pertinent; *Special Hazards of Combustion Products*: Not pertinent; *Behavior in Fire*: Not pertinent; *Ignition Temperature*: Not flammable; *Electrical Hazard*: Not pertinent; *Burning Rate*: Not pertinent.

Chemical Reactivity - *Reactivity with Water*: No reaction; *Reactivity with Common Materials*: No reactions; *Stability During Transport*: Stable; *Neutralizing Agents for Acids and Caustics*: Not pertinent; *Polymerization*: Not pertinent; *Inhibitor of Polymerization*: Not pertinent.

ARSENIC ACID

Chemical Designations - *Synonyms*: Arsenic Pentoxide; Orthoarsenic Acid; *Chemical Formula*: As_2O_5 or $H_3AsO_4 \cdot \frac{1}{2}H_2O$.

Observable Characteristics - *Physical State (as normally shipped)*: Solid; a concentrated water solution is sometimes shipped; *Color*: White; *Odor*: None.

Physical and Chemical Properties - *Physical State at 15 °C and 1 atm.*: Solid; *Molecular Weight*: 229.8; *Boiling Point at 1 atm.*: Not pertinent; *Freezing Point*: Not pertinent; *Critical Temperature*: Not pertinent; *Critical Pressure*: Not pertinent; *Specific Gravity*: 2.2 at 20°C (solid); *Vapor (Gas) Density*: Not pertinent; *Ratio of Specific Heats of Vapor (Gas)*: Not pertinent; *Latent Heat of Vaporization*: Not pertinent; *Heat of Combustion*: Not pertinent; *Heat of Decomposition*: Not pertinent.

Health Hazards Information - *Recommended Personal Protective Equipment*: Calamine lotion and zinc oxide powder on hands and other skin areas; rubber gloves; Bu. Mines approved dust respirator; *Symptoms Following Exposure*: Ingestion causes irritations of stomach, weakness, other gastrointestinal symptoms. Overdose can cause arsenic poisoning, but symptoms are delayed; *General Treatment for Exposure*: Get medical attention after all exposures to this compound. Be alert for arsenic poisoning symptoms. SKIN: wash well with soap and water. INGESTION: induce vomiting; drink freely lime water, milk, or raw egg; give a cathartic; *Toxicity by Inhalation (Threshold Limit Value)*: 0.5 mg/m^3 as arsenic; *Short-Term Exposure Limits*: Data not available; *Toxicity by Ingestion*: Grade 4; oral LD$_{50}$ 48 mg/kg (young rats); *Late Toxicity*: Arsenic compounds may be carcinogenic; *Vapor (Gas) Irritant Characteristics*: Vapors are nonirritating to eyes and throat; *Liquid or Solid Irritant Characteristics*: Minimum hazard. If spilled on clothing and allowed to remain, may cause smarting and reddening of the skin; *Odor Threshold*: Odorless.

Fire Hazards - *Flash Point:* Not flammable; *Flammable Limits in Air (%)*: Not flammable; *Fire Extinguishing Agents*: Not pertinent; *Fire Extinguishing Agents Not To Be Used*: Not pertinent; *Special Hazards of Combustion Products*: Not pertinent; *Behavior in Fire*: Not pertinent; *Ignition Temperature*: Not pertinent; *Electrical Hazard*: Not pertinent; *Burning Rate*: Not pertinent.

Chemical Reactivity - *Reactivity with Water*: No reaction; *Reactivity with Common Materials*: Will corrode metal and give off toxic arsine gas; *Stability During Transport*: Stable; *Neutralizing Agents for Acids and Caustics*: Flush with water and rinse with sodium bicarbonate or lime solution; *Polymerization*: Not pertinent; *Inhibitor of Polymerization*: Not pertinent.

ARSENIC DISULFIDE

Chemical Designations - *Synonyms*: Realgar; Red Arsenic Glass; Red Arsenic Sulfide; Red Ointment; Ruby Arsenic; *Chemical Formula*: As$_2$S$_2$.

Observable Characteristics - *Physical State (as normally shipped)*: Solid; *Color*: Red-brown; *Odor*: None.

Physical and Chemical Properties - *Physical State at 15 °C and 1 atm.*: Solid; *Molecular Weight*: 214; *Boiling Point at 1 atm.*: 1.049, 565, 838; *Freezing Point*: 585, 307, 580; *Critical Temperature*: Not pertinent; *Critical Pressure*: Not pertinent; *Specific Gravity:* 3.5 at 20°C (solid); *Vapor (Gas) Density*: Not pertinent; *Ratio of Specific Heats of Vapor (Gas)*: Not pertinent; *Latent Heat of Vaporization*: Not pertinent; *Heat of Combustion*: Not pertinent; *Heat of Decomposition*: Not pertinent.

Health Hazards Information - *Recommended Personal Protective Equipment*: Approved respirator; goggles; rubber gloves; clean protective clothing; *Symptoms Following Exposure*: Repeated inhalation causes irritation of nose, laryngitis, mild bronchitis. Ingestion causes weakness, loss of appetite, gastrointestinal disturbances, peripheral neuritis, occasional hepatitis. Contact with eyes causes irritation. Irritates skin, especially where moist; if not treated, may cause ulceration; *General Treatment for Exposure*: Consult physician after all overexposure to this compound. INHALATION: move to fresh air. INGESTION: induce vomiting by giving warm salt water; repeat until vomit is clear. EYES: flush with water for at least 15 min. SKIN: wash well with water; *Toxicity by Inhalation (Threshold Limit Value)*: 0.5 mg/m^3 (as arsenic); *Short-Term Exposure Limits*: Data not available; *Toxicity by Ingestion*: Grade 4; LD$_{50}$ <50 mg/kg; *Late Toxicity*: Possible skin and lung cancer; *Vapor (Gas) Irritant Characteristics*: Data not available; *Liquid or Solid Irritant Characteristics*: Data not available; *Odor Threshold*: Odorless.

Fire Hazards - *Flash Point:* Not pertinent; *Flammable Limits in Air (%)*: Not pertinent; *Fire Extinguishing Agents*: Water; *Fire Extinguishing Agents Not To Be Used*: Not pertinent; *Special Hazards of Combustion Products*: Poisonous fumes of the compound may be formed during fires. If ignited, will form sulfur dioxide gas; *Behavior in Fire*: May ignite at very high temperatures; *Ignition Temperature*: Not pertinent; *Electrical Hazard*: Not pertinent; *Burning Rate*: Not pertinent.

Chemical Reactivity - *Reactivity with Water*: No reaction; *Reactivity with Common Materials*: No data; *Stability During Transport*: Stable; *Neutralizing Agents for Acids and Caustics*: Not pertinent; *Polymerization*: Not pertinent; *Inhibitor of Polymerization*: Not pertinent.

ARSENIC TRICHLORIDE

Chemical Designations - *Synonyms*: Arsenic (iii) Trichloride; Arsenic Chloride; Arsenous Chloride; Butter of Arsenic; Caustic Arsenic Chloride; Caustic Oil of Arsenic; Fuming Liquid Arsenic; *Chemical Formula*: $AsCl_3$.

Observable Characteristics - *Physical State (as normally shipped)*: Liquid; *Color*: Colorless; *Odor*: Acrid.

Physical and Chemical Properties - *Physical State at 15 °C and 1 atm.*: Liquid; *Molecular Weight*: 181.3; *Boiling Point at 1 atm.*: 266.4, 130.2, 403.4; *Freezing Point*: 9, -13, 260; *Critical Temperature*: Not pertinent; *Critical Pressure*: Not pertinent; *Specific Gravity:* 2.156 at 25°C (liquid); *Vapor (Gas) Density*: Not pertinent; *Ratio of Specific Heats of Vapor (Gas)*: Not pertinent; *Latent Heat of Vaporization*: 88.31, 49.06, 2.054; *Heat of Combustion*: Not pertinent; *Heat of Decomposition*: Not pertinent.

Health Hazards Information - *Recommended Personal Protective Equipment*: Safety goggles and face shield; acid-type canister gas mask; rubber gloves; protective clothing; *Symptoms Following Exposure*: Inhalation causes irritation of nose and throat. Contact of liquid with eyes or skin causes severe irritation. Ingestion causes weakness and severe irritation of mouth and stomach. Overdose can cause arsenic poisoning, but symptoms are delayed; *General Treatment for Exposure*: Get medical attention after all exposures to the compound. Be alert fro arsenic poisoning symptoms. INHALATION: remove to fresh air; give artificial respiration if needed. EYES: flush with water for at least 15 min. SKIN: flush with water. INGESTION: give large amounts of water, then induce vomiting; give lime water, milk, or raw egg; give a cathartic; *Toxicity by Inhalation (Threshold Limit Value)*: 0.5 mg/m^3 (as arsenic); *Short-Term Exposure Limits*: Data not available; *Toxicity by Ingestion*: Grade 3; oral rat LD50 138 mg/kg; fatal human dose 70-180 mg, depending on weight; *Late Toxicity*: Arsenic compounds may be carcinogenic; *Vapor (Gas) Irritant Characteristics*: Data not available; *Liquid or Solid Irritant Characteristics*: Data not available; *Odor Threshold*: Data not available.

Fire Hazards - *Flash Point:* Not flammable; *Flammable Limits in Air (%)*: Not flammable; *Fire Extinguishing Agents*: Not pertinent; *Fire Extinguishing Agents Not To Be Used*: Avoid water on adjacent fires; *Special Hazards of Combustion Products*: Irritating and toxic hydrogen chloride formed when involved in fires; *Behavior in Fire*: Becomes gaseous and causes irritation. Forms hydrogen chloride (hydrochloric acid) by reaction with water used to fight adjacent fires; *Ignition Temperature*: Not pertinent; *Electrical Hazard*: Not pertinent; *Burning Rate*: Not pertinent.

Chemical Reactivity - *Reactivity with Water*: Reacts with water to form hydrogen chloride (hydrochloric acid); *Reactivity with Common Materials*: Corrodes metal; *Stability During Transport*: Stable; *Neutralizing Agents for Acids and Caustics*: Flush with water, rinse with sodium bicarbonate or lime solution; *Polymerization*: Not pertinent; *Inhibitor of Polymerization*: Not pertinent.

ARSENIC TRIOXIDE

Chemical Designations - *Synonyms*: Arsenous Acid; Arsenous Acid Anhydride; Arsenous Oxide; Arsenic Sesquioxide; White Arsenic; *Chemical Formula*: As_2O_3.

Observable Characteristics - *Physical State (as normally shipped)*: Solid; *Color*: White; *Odor*: Like garlic; none.

Physical and Chemical Properties - *Physical State at 15 °C and 1 atm.*: Solid; *Molecular Weight*: 197.8; *Boiling Point at 1 atm.*: 855, 457, 730; *Freezing Point*: 599, 315, 588; *Critical Temperature*: Not pertinent; *Critical Pressure*: Not pertinent; *Specific Gravity:* 3.7 at 20°C (solid); *Vapor (Gas) Density*: Not pertinent; *Ratio of Specific Heats of Vapor (Gas)*: Not pertinent; *Latent Heat of Vaporization*: Not pertinent; *Heat of Combustion*: Not pertinent; *Heat of Decomposition*: Not pertinent.

Health Hazards Information - *Recommended Personal Protective Equipment*: Chemical cartridge approved respirator; protective gloves, eye protection; full protective coveralls; *Symptoms Following Exposure*: Ingestion causes irritation of mucous membrane, weakness, loss of appetite, gastrointestinal disturbances. Overdose can cause arsenic poisoning, but symptoms are delayed; *General Treatment for Exposure*: Get medical attention after all exposures to this compound. Be alert for arsenic poisoning symptoms. SKIN: wash thoroughly with soap and water; remove contaminated clothing and shower with soap and water; irritations, except for milder cases which disappear in a day or two, should have medical attention. INGESTION: vomiting should be induced and a physician should be called at once; drink freely of lime water, sweet milk, or raw eggs, followed by castor oil or any brisk cathartic; *Toxicity by Inhalation (Threshold Limit Value)*: 0.5 mg/m^3 (as arsenic); *Short-Term Exposure Limits*: Data not available; *Toxicity by Ingestion*: Grade 4; oral mouse LD$_{50}$ 45 mg/kg; *Late Toxicity*: Arsenic compounds may be carcinogenic; *Vapor (Gas) Irritant Characteristics*: Data not available; *Liquid or Solid Irritant Characteristics*: Data not available; *Odor Threshold*: Odorless.

Fire Hazards - *Flash Point:* Not flammable; *Flammable Limits in Air* (%): Not flammable; *Fire Extinguishing Agents*: Not pertinent; *Fire Extinguishing Agents Not To Be Used*: Not pertinent; *Special Hazards of Combustion Products*: Toxic fumes of arsenic trioxide and arsine may form in fire situations; *Behavior in Fire*: Can volatilize forming toxic fumes of arsenic trioxide; *Ignition Temperature*: Not pertinent; *Electrical Hazard*: Not pertinent; *Burning Rate*: Not pertinent.

Chemical Reactivity - *Reactivity with Water*: No reaction; *Reactivity with Common Materials*: No reaction; *Stability During Transport*: Stable; *Neutralizing Agents for Acids and Caustics*: Flush with water; *Polymerization*: Not pertinent; *Inhibitor of Polymerization*: Not pertinent.

ARSENIC TRISULFIDE

Chemical Designations - *Synonyms*: Arsenic Yellow; King's Gold; King's Yellow; Ointment; Yellow Arsenic Sulfide; *Chemical Formula*: As$_2$S$_3$.

Observable Characteristics - *Physical State (as normally shipped)*: Solid; *Color*: Yellow - orange; *Odor*: None.

Physical and Chemical Properties - *Physical State at 15 °C and 1 atm.*: Solid; *Molecular Weight*: 246; *Boiling Point at 1 atm.*: Not pertinent; *Freezing Point*: 572, 300, 573; *Critical Temperature*: Not pertinent; *Critical Pressure*: Not pertinent; *Specific Gravity:* 3.43 at 20°C (solid); *Vapor (Gas) Density*: Not pertinent; *Ratio of Specific Heats of Vapor (Gas)*: Not pertinent; *Latent Heat of Vaporization*: Not pertinent; *Heat of Combustion*: Not pertinent; *Heat of Decomposition*: Not pertinent.

Health Hazards Information - *Recommended Personal Protective Equipment*: Self-contained breathing apparatus; goggles; rubber gloves; clean protective clothing; *Symptoms Following Exposure*: (Acute and sub-acute poisoning are not common.) Repeated inhalation causes irritation of nose, laryngitis, mild bronchitis. Ingestion causes weakness, loss of appetite, gastrointestinal disturbances, peripheral neuritis, occasional hepatitis. Contact with eyes causes irritation. Irritates skin, especially where moist; if not treated, may cause ulceration; *General Treatment for Exposure*: Consult physician after all over exposures to this compound. INHALATION: move to fresh air. INGESTION: induce vomiting by giving warm salt water; repeat until vomit is clear. EYES: flush with water for at least 15 min. SKIN: wash well with water; *Toxicity by Inhalation (Threshold Limit Value)*: 0.5 mg/m^3 (as arsenic); *Short-Term Exposure Limits*: Data not available; *Toxicity by Ingestion*: Grade 4; LD$_{50}$ < 50 mg/kg; *Late Toxicity*: Possible skin and lung cancer; *Vapor (Gas) Irritant Characteristics*: Data not available; *Liquid or Solid Irritant Characteristics*: Data not available; *Odor Threshold*: Odorless.

Fire Hazards - *Flash Point:* Not pertinent; *Flammable Limits in Air* (%): Not pertinent; *Fire Extinguishing Agents*: Water; *Fire Extinguishing Agents Not To Be Used*: Not pertinent; *Special Hazards of Combustion Products*: Poisonous fumes of compound may be formed in fire situations; *Behavior in Fire*: May ignite at very high temperatures; *Ignition Temperature*: Not pertinent; *Electrical Hazard*: Not pertinent; *Burning Rate*: Not pertinent.

Chemical Reactivity - *Reactivity with Water*: No reaction; *Reactivity with Common Materials*: No reactions; *Stability During Transport*: Stable; *Neutralizing Agents for Acids and Caustics*: Not pertinent; *Polymerization*: Not pertinent; *Inhibitor of Polymerization*: Not pertinent.

ASPHALT

Chemical Designations - *Synonyms*: Asphalt Cements; Asphaltic Bitumen; Bitumen; Petroleum Asphalt; *Chemical Formula*: Not pertinent.

Observable Characteristics - *Physical State (as normally shipped)*: Liquid; *Color*: Dark brown to black; *Odor*: Tarry.

Physical and Chemical Properties - *Physical State at 15 °C and 1 atm.*: Liquid; *Molecular Weight*: Not pertinent; *Boiling Point at 1 atm.*: Not pertinent; *Freezing Point*: Not pertinent; *Critical Temperature*: Not pertinent; *Critical Pressure*: Not pertinent; *Specific Gravity*: 1.00 at 20°C (liquid); *Vapor (Gas) Density*: Not pertinent; *Ratio of Specific Heats of Vapor (Gas)*: Not pertinent; *Latent Heat of Vaporization*: Not pertinent; *Heat of Combustion*: Data not available; *Heat of Decomposition*: Not pertinent.

Health Hazards Information - *Recommended Personal Protective Equipment*: Protective clothing; face and eye protection when handling hot material; *Symptoms Following Exposure*: Contact with skin may cause dermatitis. Inhalation of vapors may cause moderate irritation of nose and throat. Hot liquid burns skin; *General Treatment for Exposure*: Severe burns may result from contact with hot asphalt. If molten asphalt strikes the exposed skin, coll the skin immediately by quenching with cold water. A burn should be covered with a sterile dressing, and the patient should be taken immediately to a hospital; *Toxicity by Inhalation (Threshold Limit Value)*: 5 mg/m^3; *Short-Term Exposure Limits*: Data not available; *Toxicity by Ingestion*: Grade 1; LD$_{50}$ 5 to 15 g/kg; *Late Toxicity*: None observed; *Vapor (Gas) Irritant Characteristics*: Vapors cause a slight smarting of the eyes or respiratory system if present in high concentrations. The effect is temporary; *Liquid or Solid Irritant Characteristics*: Causes smarting of the skin and first-degree burns on short exposure; may cause secondary burns on long exposure; *Odor Threshold*: Data not available.

Fire Hazards - *Flash Point (deg. F)*: 300 - 350 OC; *Flammable Limits in Air (%)*: Not pertinent; *Fire Extinguishing Agents*: Water spray, dry chemical, foam or carbon dioxide; *Fire Extinguishing Agents Not To Be Used*: Water or foam may cause foaming; *Special Hazards of Combustion Products*: Not pertinent; *Behavior in Fire*: Not pertinent; *Ignition Temperature (deg. F)*: 400 - 700; *Electrical Hazard*: Not pertinent; *Burning Rate*: No data.

Chemical Reactivity - *Reactivity with Water*: No reaction; *Reactivity with Common Materials*: No reactions; *Stability During Transport*: Stable; *Neutralizing Agents for Acids and Caustics*: Not pertinent; *Polymerization*: Not pertinent; *Inhibitor of Polymerization*: Not pertinent.

ATRAZINE

Chemical Designations - *Synonyms*: 2-Chloro-4-Ethylamino-6-Isopropylamino-S-Triazine, Aatrex Herbicide; *Chemical Formula*: $C_8H_{14}N_5Cl$.

Observable Characteristics - *Physical State (as normally shipped)*: Solid; *Color*: White; *Odor*: No data.

Physical and Chemical Properties - *Physical State at 15 °C and 1 atm.*: Solid; *Molecular Weight*: 215.7; *Boiling Point at 1 atm.*: Decomposes; *Freezing Point*: 347, 175, 348; *Critical Temperature*: Not Pertinent; *Critical Pressure*: Not Pertinent; Specific Gravity: 20°C: 1.2 at 20 °C (solid); *Vapor (Gas) Density*: Not Pertinent; *Ratio of Specific Heats of Vapor (Gas)*: Not Pertinent; *Latent Heat of Vaporization*: Not Pertinent; *Heat of Combustion*: -9500, -5300, -220; *Heat of Decomposition*: Not Pertinent.

Health Hazards Information - *Recommended Personal Protective Equipment*: Dust mask, goggles and rubber gloves; *Symptoms Following Exposure*: Irritation of eyes and skin. If ingested, irritates mouth and stomach.; *General Treatment for Exposure*: EYES - Flush with copious amounts of water for at least 15 to 20 minutes. SKIN - Wash with large amounts of water. INGESTION - Induce vomiting and give saline laxative and supportive therapy; *Toxicity by Inhalation (Threshold Limit Value)*: No data; *Short-Term Exposure Limits*: No data; *Toxicity by Ingestion*: Oral rat LD$_{50}$ = 3080 mg/kg; *Late Toxicity*: No data; *Vapor (Gas) Irritant Characteristics*: Vapors are nonirritating to eyes and throat.; *Liquid or Solid Irritant Characteristics*: Causes smarting of the skin and first-degree chemical burns on short exposure and may cause second-degree burns on long exposure; *Odor Threshold*: No data.

Fire Hazards - *Flash Point*: Not flammable; *Flammable Limits in Air (%)*: Not flammable; *Fire Extinguishing Agents*: Not pertinent; *Fire Extinguishing Agents Not To Be Used*: Not pertinent; *Special Hazards of Combustion Products*: Irritating hydrogen chloride and toxic oxides of nitrogen may be formed; *Behavior in Fire*: Not pertinent; *Ignition Temperature*: Not pertinent; *Electrical Hazard*: Not pertinent; *Burning Rate*: Not pertinent.

Chemical Reactivity - *Reactivity with Water*: No reaction; *Reactivity with Common Materials*: No

reactions; *Stability During Transport*: Stable; *Neutralizing Agents for Acids and Caustics*: Not pertinent; *Polymerization*: Not pertinent; *Inhibitor of Polymerization*: Not pertinent.

AZINPHOSMETHYL

Chemical Designations - *Synonyms*: O-O-Dimethyl S-[(4-Oxo-1,2,3-Benzotriazine-3(4H)-yl)Methyl] Phosphorodithioate, Gurthion Insecticide, Gusathion Insecticide; *Chemical Formula*: $C_{10}H_{12}N_3O_3PS_2$.
Observable Characteristics — *Physical State (as normally shipped)*: Solid; *Color*: Brown; *Odor*: No data.
Physical and Chemical Properties - *Physical State at 15 °C and 1 atm.*: Solid; *Molecular Weight*: 317; *Boiling Point at 1 atm.*: Decomposes; *Freezing Point*: 163, 73, 346; *Critical Temperature*: Not Pertinent; *Critical Pressure*: Not Pertinent; Specific Gravity: 1.4 at 20°C (solid); *Vapor (Gas) Density*: Not Pertinent; *Ratio of Specific Heats of Vapor (Gas)*: Not Pertinent; *Latent Heat of Vaporization*: Not Pertinent; *Heat of Combustion*: -8600, -4800, -200; *Heat of Decomposition*: Not Pertinent.
Health Hazards Information - *Recommended Personal Protective Equipment*: Dust mask, protective goggles, and rubber gloves; *Symptoms Following Exposure*: Dust irritates eyes. Inhalation or ingestion causes sweating, constriction of pupils of the eyes, asthmatic conditions, cramps, weakness, convulsions, collapse; *General Treatment for Exposure*: INHALATION - Remove victim to fresh air; keep warm, and seek physician. EYES - Flush with fresh running water for at least 15 min. SKIN - Flush with water and wash with soap and water; *Toxicity by Inhalation (Threshold Limit Value)*: 0.2 mg/m^3; *Short-Term Exposure Limits*: No data; *Toxicity by Ingestion*: Oral rat LD_{50} = 11 ~ 18.5 mg/kg; *Late Toxicity*: No data; *Vapor (Gas) Irritant Characteristics*: No data; *Liquid or Solid Irritant Characteristics*: No data; *Odor Threshold*: No data.
Fire Hazards - *Flash Point:* Not flammable; *Flammable Limits in Air (%)*: Not flammable; *Fire Extinguishing Agents*: Not pertinent; *Fire Extinguishing Agents Not To Be Used*: Not pertinent; *Special Hazards of Combustion Products*: Oxides of sulfur and phosphorous may be formed when exposed to a fire situation; *Behavior in Fire*: Data not available; *Ignition Temperature*: Not pertinent; *Electrical Hazard*: Not pertinent; *Burning Rate*: Not pertinent.
Chemical Reactivity - *Reactivity with Water*: No reaction; *Reactivity with Common Materials*: No reactions; *Stability During Transport*: Stable; *Neutralizing Agents for Acids and Caustics*: Not pertinent; *Polymerization*: Not pertinent; *Inhibitor of Polymerization*: Not pertinent.

B

BARIUM CARBONATE

Chemical Designations - *Synonyms*: No common synonyms; *Chemical Formula*: $BaCo_3$.
Observable Characteristics - *Physical State*: Solid; *Color*: White; *Odor*: None.
Physical and Chemical Properties - *Physical State at 15 °C and 1 atm.*: Solid; *Molecular Weight*: 197.35; *Boiling Point at 1 atm.*: Not pertinent; *Freezing Point*: Not pertinent; *Critical Temperature*: Not pertinent; *Critical Pressure*: Not pertinent; *Specific Gravity*: 4.3 at 20°C(solid); *Vapor (Gas) Density*: Not pertinent; *Ratio of Specific Heats of Vapor (Gas)*: Not pertinent; *Latent Heat of Vaporization*: Not pertinent; *Heat of Combustion*: Not pertinent; *Heat of Decomposition*: Not pertinent.
Health Hazards Information - *Recommended Personal Protective Equipment*: Dust respirator; *Symptoms Following Exposure*: (Ingestion only): excessive salvation, vomiting, severe abdominal pain, and violent purging with watering and bloody stools; a slow and often irregular pulse and a transient elevation in arterial blood pressure; tinnitus, giddiness and vertigo; muscle twitching, progressing to convulsions and/or paralysis; dilated pupils with impaired accommodation; confusion and increasing somnolence, without coma; collapse and death from respiratory failure and cardiac arrest; *General Treatment for Exposure*: Rapid oral administration of a soluble sulfate in water, such as magnesium or sodium sulfate(2 oz),alum (4 mg), or very dilute sulfuric acid (30 ml of a 10 % solution diluted to 1

qt). These agents precipitate barium as the insoluble sulfate. Seek medical attention; *Toxicity by Inhalation (Threshold Limit Value)*: Not pertinent; *Short-Term Exposure Limits*: Not pertinent; *Toxicity by Ingestion*:LD$_{50}$ 0.5 to 5 g/kg (rabbit, rat, guinea pig); *Late Toxicity*: None observed; *Vapor (Gas) Irritant Characteristics*: Not pertinent; *Liquid or Solid Irritant Characteristics*: None.

Fire Hazards - *Flash Point:* Not flammable; *Flammable Limits in Air (%)*: Not flammable; *Fire Extinguishing Agents*: Not pertinent; *Fire Extinguishing Agents Not To Be Used*: Not pertinent; *Special Hazards of Combustion Products*: Not pertinent; *Behavior in Fire*: Not pertinent; *Ignition Temperature*: Not flammable; *Electrical Hazard*: Not pertinent; *Burning Rate*: Not flammable.

Chemical Reactivity - *Reactivity with Water*: No reaction; *Reactivity with Common Materials*: No reactions; *Stability During Transport*: Stable; *Neutralizing Agents for Acids and Caustics*: Not pertinent; *Polymerization*: Not pertinent; *Inhibitor of Polymerization*: Not pertinent.

BARIUM CHLORATE

Chemical Designations - *Synonyms*: Barium Chlorate Monohydrate; *Chemical Formula*: Ba(ClO$_3$)$_2$ H$_2$O.

Observable Characteristics - *Physical State (as normally shipped)*: Solid; *Color*: White; *Odor*: None.

Physical and Chemical Properties - *Physical State at 15 °C and 1 atm.*: Solid; *Molecular Weight*: 332 (monohydrate); *Boiling Point at 1 atm.*: Not pertinent; *Freezing Point*: 777, 414,687; *Critical Temperature*: Not pertinent; *Critical Pressure*: Not pertinent; *Specific Gravity*: 3,18 at 20°C (solid); *Vapor (Gas) Density*: Not pertinent; *Ratio of Specific Heats of Vapor (Gas)*: Not pertinent; *Latent Heat of Vaporization*: Not pertinent; *Heat of Combustion*: Not pertinent; *Heat of Decomposition*: Not pertinent.

Health Hazards Information - *Recommended Personal Protective Equipment*: Goggles or face shield; dust respirator (U.S. Bureau of Mines approved); rubberized shoes and gloves; coveralls or other suitable outer clothing; *Symptoms Following Exposure*: Inhalation causes irritation of upper respiratory system. Contact with eyes or skin causes irritation. Ingestion causes abdominal pain, nausea and vomiting diarrhea, pallor, blueness, shortness of breath, excessive salvation, convulsive tremors, slow, hard pulse, elevated blood pressure, unconsciousness. Hemorrhages may occur in the stomach, intestines, and kidneys. Muscular paralysis may follow; *General Treatment for Exposure*: Get medical attention. Alert doctor to possibility of barium poisoning, particularly if compound was swallowed. INHALATION: remove to fresh air. EYES: flush with copious quantities of water for at least 15 min; get medical attention. SKIN: flush with water. INGESTION: induce vomiting and call a physician; have victim drink aqueous 10% solution of magnesium or sodium sulfate; for severe intoxication, calcium or a magnesium salt may have to be given I.V. with caution; treatment otherwise is supportive and symptomatic; *Toxicity by Inhalation (Threshold Limit Value)*: 0,5 mg/m^3; *Short-Term Exposure Limits*: Data not available; *Toxicity by Ingestion*: Data not available; *Late Toxicity*: Barium poisoning; *Vapor (Gas) Irritant Characteristics*: Not pertinent; *Liquid or Solid Irritant Characteristics*: Data not available; *Odor Threshold*: Not pertinent.

Fire Hazards - *Flash Point:* Not flammable but may cause explosions when involved in fires; *Flammable Limits in Air (%)*: Not pertinent; *Fire Extinguishing Agents*: Not pertinent; *Fire Extinguishing Agents Not To Be Used*: No data; *Special Hazards of Combustion Products*: Produces toxic fumes when involved in a fire; *Behavior in Fire*: May cause an explosion when involved in a fire; *Ignition Temperature*: Not pertinent; *Electrical Hazard*: Not pertinent; *Burning Rate*: Not pertinent.

Chemical Reactivity - *Reactivity with Water*: No reaction; *Reactivity with Common Materials*: Can form explosive mixtures with combustible materials such as wood, oil - these mixtures can be ignited readily by friction; *Stability During Transport*: Stable; *Neutralizing Agents for Acids and Caustics*: Not pertinent; *Polymerization*: Not pertinent; *Inhibitor of Polymerization*: Not pertinent.

BARIUM NITRATE

Chemical Designations - *Synonyms*: No common synonyms; *Chemical Formula*: Ba(NO$_3$)$_2$.

Observable Characteristics - *Physical State (as normally shipped)*: Solid; *Color*: White; *Odor*: None.

Physical and Chemical Properties - *Physical State at 15 °C and 1 atm.*: Solid; *Molecular Weight*:

261.35; *Boiling Point at 1 atm.*: Decomposes; *Freezing Point*: 1,098, 592, 865; *Critical Temperature*: Not pertinent; *Critical Pressure*: Not pertinent; *Specific Gravity*: 3.24 at 23°C; *Vapor (Gas) Density*: Not pertinent; *Ratio of Specific Heats of Vapor (Gas)*: Not pertinent; *Latent Heat of Vaporization*: Not pertinent; *Heat of Combustion*: Not pertinent; *Heat of Decomposition*: Not pertinent.

Health Hazards Information — *Recommended Personal Protective Equipment*: Goggles or face shield; dust respirator; rubber gloves and shoes; suitable coveralls; *Symptoms Following Exposure*: Inhalation or contact with eyes or skin causes irritation. Ingestion causes excessive salivation, vomiting, colic, diarrhea, convulsive tremors, slow, hard pulse, elevated blood pressure. Hemorrhages may occur in the stomach, intestines, and kidneys, Muscular paralysis may follow; *General Treatment for Exposure*: Get medical attention. Alert doctor to possibility of barium poisoning, particularly if compound was swallowed. Inhalation: remove to fresh air. Eyes: flush with water for at least 15 min. Skin: flush with water. Ingestion: oral administration of a aqueous 10% solution of magnesium or sodium sulfate; in severe intoxication, calcium or magnesium salt may have to be given iv with caution; treatment otherwise is supportive and symptomatic; *Toxicity by Inhalation (Threshold Limit Value)*: 0.5 mg/m^3; *Short-Term Exposure Limits*: Data not available; *Toxicity by Ingestion*: Grade 3; oral rat LD_{50}=355 mg/kg; *Late Toxicity*: Barium poisoning; *Vapor (Gas) Irritant Characteristics*: Data not available; *Liquid or Solid Irritant Characteristics*: Data not available; *Odor Threshold*: Not pertinent.

Fire Hazards - *Flash Point:* Not flammable but can aggravate fires; *Flammable Limits in Air (%)*: Not flammable; *Fire Extinguishing Agents*: Not pertinent; *Fire Extinguishing Agents Not To Be Used*: Not pertinent; *Special Hazards of Combustion Products*: Produces toxic gaseous oxides of nitrogen when involved in fires; *Behavior in Fire*: Mixtures with combustible materials are readily ignited and burn fiercely. Containers may explode; *Ignition Temperature*: Not pertinent; *Electrical Hazard*: Not pertinent; *Burning Rate*: Not pertinent.

Chemical Reactivity - *Reactivity with Water*: No reaction; *Reactivity with Common Materials*: Fire can result by contact of this material with combustibles; *Stability During Transport*: Stable; *Neutralizing Agents for Acids and Caustics*: Not pertinent; *Polymerization*: Not pertinent; *Inhibitor of Polymerization*: Not pertinent.

BARIUM PERCHLORATE

Chemical Designations - *Synonyms*: Barium Perchlorate Trihydrate; *Chemical Formula*: Ba(ClO$_4$)-3H$_2$O.

Observable Characteristics - *Physical State (as normally shipped)*: Solid; *Color*: White; *Odor*: None.

Physical and Chemical Properties - *Physical State at 15 °C and 1 atm.*: Solid; *Molecular Weight*: 390.35; *Boiling Point at 1 atm.*: Decomposes; *Freezing Point*: 941, 505, 778; *Critical Temperature*: Not pertinent; *Critical Pressure*: Not pertinent; *Specific Gravity*: Not pertinent; *Vapor (Gas) Density*: Not pertinent; *Ratio of Specific Heats of Vapor (Gas)*: Not pertinent; *Latent Heat of Vaporization*: Not pertinent; *Heat of Combustion*: Not pertinent; *Heat of Decomposition*: Not pertinent.

Health Hazards Information - *Recommended Personal Protective Equipment*: Goggles or face shield; dust respirator; rubber gloves and shoes; suitable coveralls; *Symptoms Following Exposure*: Inhalation or contact with eyes or skin causes irritation. Ingestion causes excessive salivation, vomiting, colic, diarrhea, convulsive tremors, slow, hard pulse, and elevated blood pressure; hemorrhages may occur in the stomach, intestines and kidneys; muscular paralysis may follow; *General Treatment for Exposure*: Get medical attention. Alert doctor to possibility of barium poisoning, particularly if compound was swallowed. INHALATION: remove to fresh air. EYES: flush with water for at least 15 min. Skin: flush with water. Ingestion: oral administration of aqueous 10% solution of magnesium or sodium sulfate; for severe intoxication, calcium or magnesium salt may have to be given and i.v. with caution; *Toxicity by Inhalation (Threshold Limit Value)*: 0.5 mg/m^3; *Short-Term Exposure Limits*: Data not available; *Toxicity by Ingestion*: Data not available; *Late Toxicity*: Barium poisoning; *Vapor (Gas) Irritant Characteristics*: Data not available;

Fire Hazards - *Flash Point:* Not flammable but can aggravate fire intensity; *Flammable Limits in Air (%)*: Not flammable; *Fire Extinguishing Agents*: Not pertinent; *Fire Extinguishing Agents Not To Be Used*: Not pertinent; *Special Hazards of Combustion Products*: Not pertinent; *Behavior in Fire*:

Increases the intensity of fires. Containers may burst or explode; *Ignition Temperature*: Not pertinent; *Electrical Hazard*: Not pertinent; *Burning Rate*: Not pertinent.

Chemical Reactivity - *Reactivity with Water*: No reaction; *Reactivity with Common Materials*: When mixed with combustible materials or finely divided metals, can become explosive; *Stability During Transport*: Stable; *Neutralizing Agents for Acids and Caustics*: Not pertinent; *Polymerization*: Not pertinent; *Inhibitor of Polymerization*: Not pertinent.

BARIUM PERMANGANATE

Chemical Designations - *Synonyms*: No common synonyms; *Chemical Formula*: $Ba(MnO_4)_2$.

Observable Characteristics - *Physical State (as normally shipped)*: Solid; *Color*: Dark purple to red; *Odor*: None.

Physical and Chemical Properties - *Physical State at 15 °C and 1 atm.*: Solid; *Molecular Weight*: 375; *Boiling Point at 1 atm.*: Decomposes; *Freezing Point*: Not pertinent; *Critical Temperature*: Not pertinent; *Critical Pressure*: Not pertinent; *Specific Gravity*: 3.77 at 20°C; *Vapor (Gas) Density*: Not pertinent; *Ratio of Specific Heats of Vapor (Gas)*: Not pertinent; *Latent Heat of Vaporization*: Not pertinent; *Heat of Combustion*: Not pertinent; *Heat of Decomposition*: Not pertinent.

Health Hazards Information - *Recommended Personal Protective Equipment*: Goggles or face shield; dust respirator; rubber gloves and shoes; *Symptoms Following Exposure*: Inhalation or contact with eyes or skin causes irritation. Ingestion causes abdominal pain, nausea, vomiting, pallor, shortness of breath; *General Treatment for Exposure*: Get medical attention. Alert doctor to possibility of barium poisoning, particularly if compound was swallowed. INHALATION: remove to fresh air. EYES: flush with copious amount of water for 15 min. SKIN: wash with copious amount of water. INGESTION: induce vomiting, give a 10% water solution of Epsom salt; *Toxicity by Inhalation (Threshold Limit Value)*: 0.5 mg/m^3; *Short-Term Exposure Limits*: Data not available; *Toxicity by Ingestion*: Data not available; *Late Toxicity*: Barium poisoning; *Vapor (Gas) Irritant Characteristics*: Data not available; *Liquid or Solid Irritant Characteristics*: Data not available; *Odor Threshold*: Not pertinent.

Fire Hazards - *Flash Point:* Not flammable; *Flammable Limits in Air (%)*: Not flammable; *Fire Extinguishing Agents*: Not pertinent; *Fire Extinguishing Agents Not To Be Used*: Not pertinent; *Special Hazards of Combustion Products*: Not Pertinent; *Behavior in Fire*: Can increase the intensity of fires; *Ignition Temperature*: Not pertinent; *Electrical Hazard*: Not pertinent; *Burning Rate*: Not pertinent.

Chemical Reactivity - *Reactivity with Water*: No reaction; *Reactivity with Common Materials*: When mixed with combustible materials, can ignite by friction or in an acidic state and may become spontaneously combustible; *Stability During Transport*: Stable; *Neutralizing Agents for Acids and Caustics*: Not pertinent; *Polymerization*: Not pertinent; *Inhibitor of Polymerization*: Not pertinent.

BARIUM PEROXIDE

Chemical Designations - *Synonyms*: Barium Dioxide; Barium Superoxide; Barium Binoxide; *Chemical Formula*: BaO_2.

Observable Characteristics - *Physical State (as normally shipped)*: Solid; *Color*: Light grayish-tan; grayish-white; *Odor*: None.

Physical and Chemical Properties - *Physical State at 15 °C and 1 atm.*: Solid; *Molecular Weight*: 169.4; *Boiling Point at 1 atm.*: Decomposes; *Freezing Point*: 842, 450, 723; *Critical Temperature*: Not pertinent; *Critical Pressure*: Not pertinent; *Specific Gravity*: 4,96 at 20°C; *Vapor (Gas) Density*: Not pertinent; *Ratio of Specific Heats of Vapor (Gas)*: Not pertinent; *Latent Heat of Vaporization*: Not pertinent; *Heat of Combustion*: Not pertinent; *Heat of Decomposition*: -194, -108, -4.52.

Health Hazards Information - *Recommended Personal Protective Equipment*: Toxic gas respirator; liquid-proof PVC gloves; chemical safety goggles, full cover clothing; *Symptoms Following Exposure*: Inhalation causes irritation of mucous membranes, throat and nose. Contact with eyes or skin causes severe burns. Ingestion causes severe salivation, vomiting, colic, diarrheas, convulsive tremors, slow, hard pulse, and elevated blood pressure; hemorrhages may occur in the stomach, intestines, and kidneys; muscular paralysis may follow; *General Treatment for Exposure*: Get medical attention. Alert doctor to possibility of barium poisoning, particularly if compound was swallowed. INHALATION:

remove to fresh air. EYES: flush with copious amount of water for 15 min. SKIN: flush with water. INGESTION: oral administration of aqueous 10% solution of magnesium or sodium sulfate; for severe intoxication, calcium or magnesium salt may have to be given i.v. with caution; treatment otherwise is supportive and symptomatic; *Toxicity by Inhalation (Threshold Limit Value)*: 0.5 mg/m^3; *Short-Term Exposure Limits*: Data not available; *Toxicity by Ingestion*: Data not available; *Late Toxicity*: Barium poisoning; *Vapor (Gas) Irritant Characteristics*: Data not available; *Liquid or Solid Irritant Characteristics*: Data not available; *Odor Threshold*: Not pertinent.

Fire Hazards - *Flash Point:* Not flammable but may cause fires upon contact with combustible materials; *Flammable Limits in Air (%)*: Not pertinent; *Fire Extinguishing Agents*: Flood with water, dry powder (e.g., graphite or powdered limestone); *Fire Extinguishing Agents Not To Be Used*: Not pertinent; *Special Hazards of Combustion Products*: Not pertinent; *Behavior in Fire*: Can increase the intensity of fires; *Ignition Temperature*: Not pertinent; *Electrical Hazard*: Not pertinent; *Burning Rate*: Not pertinent.

Chemical Reactivity - *Reactivity with Water*: Decomposes slowly but the reaction is not hazardous; *Reactivity with Common Materials*: Corrodes metals slowly. If mixed with combustible materials or finely divided metals, mixture can spontaneously ignite or become unstable by friction; *Stability During Transport*: Sable; *Neutralizing Agents for Acids and Caustics*: Not pertinent; *Polymerization*: Not pertinent; *Inhibitor of Polymerization*: Not pertinent.

BENZALDEHYDE

Chemical Designations - *Synonyms*: Benzoic Aldehyde, Oil of Bitter Almond; *Chemical Formula*: C_6H_5CHO.

Observable Characteristics - *Physical State (as normally shipped)*: Liquid; *Color*: Colorless to pure yellow; *Odor*: Like almonds.

Physical and Chemical Properties - *Physical State at 15 °C and 1 atm.*: Liquid; *Molecular Weight*: 106.12; *Boiling Point at 1 atm.*: 354, 179, 452; *Freezing Point*: Not pertinent; *Critical Temperature*: 666, 352, 625; *Critical Pressure*: 316, 21.5, 2.18; *Specific Gravity*: 1.046 at 20°C (liquid); *Vapor (Gas) Density*: Not pertinent; *Ratio of Specific Heats of Vapor (Gas)*: 1.1; *Latent Heat of Vaporization*: 156, 86.5, 3.62; *Heat of Combustion*: -13,730, -7,630, -319.5; *Heat of Decomposition*: Not pertinent.

Health Hazards Information - *Recommended Personal Protective Equipment*: Chemical goggles and chemical protective clothing.; *Symptoms Following Exposure*: Inhalation of concentrated vapor can irritate eyes, nose and throat. Liquid is irritating to the eyes. Prolonged contact with skin causes irritation.; *General Treatment for Exposure*: SKIN, EYE CONTACT: Move the victim to fresh air and contact doctor immediately. Wash contaminated skin area with water. Flush eyes with fresh running water for at least 15 minutes. INGESTION: Induce vomiting and call a doctor.; *Toxicity by Inhalation (Threshold Limit Value)*: No data; *Short-Term Exposure Limits*: No data; *Toxicity by Ingestion*: LD_{50} 0.5 to 5 g/kg; *Late Toxicity*:; *Vapor (Gas) Irritant Characteristics*: Vapors can cause sever irritation of eyes and throat and can cause eye and lung injury. They cannot be tolerated even at very low concentrations.; *Liquid or Solid Irritant Characteristics*: Represents minimum hazard. If spilled on clothing and allowed to remain, may cause smarting and reddening of the skin.; *Odor Threshold*: 0.042 ppm.

Fire Hazards - *Flash Point (deg. F):* 148 CC, 163 OC; *Flammable Limits in Air (%)*: No data; *Fire Extinguishing Agents*: Water spray, foam, carbon dioxide, or dry chemical; *Fire Extinguishing Agents Not To Be Used*: Not pertinent; *Special Hazards of Combustion Products*: Not pertinent; *Behavior in Fire*: Not pertinent; *Ignition Temperature (deg. F)*: 378; *Electrical Hazard*: Not pertinent; *Burning Rate*: 3.8 mm/min.

Chemical Reactivity - *Reactivity with Water*: No reaction; *Reactivity with Common Materials*: No reactions; *Stability During Transport*: Stable; *Neutralizing Agents for Acids and Caustics*: Not pertinent; *Polymerization*: Not pertinent; *Inhibitor of Polymerization*: Not pertinent.

BENZENE

Chemical Designations - *Synonyms*: Benzol, Benzole; *Chemical Formula*: C_6H_6.

Observable Characteristics - *Physical State (as normally shipped)*: Liquid; *Color*: Colorless; *Odor*: Aromatic; Pleasant aromatic odor, characteristic odor.

Physical and Chemical Properties - *Physical State at 15 °C and 1 atm.*: Liquid; *Molecular Weight*: 78.11; *Boiling Point at 1 atm.*: 176, 80.1, 353.3; *Freezing Point*: 42.0, 5.5, 278.7; *Critical Temperature*: 552.0, 288.9, 562.1; *Critical Pressure*: 710, 48.3, 4.89; *Specific Gravity*: 0.879 at 20 °C (liquid); *Vapor (Gas) Density*: 2.7; *Ratio of Specific Heats of Vapor (Gas)*: 1.061; *Latent Heat of Vaporization*: 169, 94.1, 3.94; *Heat of Combustion*: -17460, -9698, -406.0; *Heat of Decomposition*: Not pertinent.

Health Hazards Information - *Recommended Personal Protective Equipment*: Hydrocarbon vapor canister, supplied air respirator or a hose mask; hydrocarbon insoluble rubber or plastic gloves; chemical goggles or face splash shield; hydrocarbon-insoluble apron such as neoprene.; *Symptoms Following Exposure*: Dizziness, excitation, pallor, followed by flushing, weakness, headache, breathlessness, chest constriction. Coma and possible death.; *General Treatment for Exposure*: SKIN: Flush with water followed by soap and water; remove contaminated clothing and wash skin. EYES: Flush with plenty of water until irritation subsides. INHALATION: Remove from exposed environment immediately. Call a physician. If breathing is irregular or stopped, start resuscitation, administer oxygen.; *Toxicity by Inhalation (Threshold Limit Value)*: 25 ppm; *Short-Term Exposure Limits*: 75 ppm for 30 minutes; *Toxicity by Ingestion*: LD_{50} 50 ~ 500 mg/kg; *Late Toxicity*: Leukemia; *Vapor (Gas) Irritant Characteristics*: If present in high concentrations, vapors may cause irritation of eyes or respiratory system. Effect is usually temporary.; *Liquid or Solid Irritant Characteristics*: Represents minimum hazard. If spilled on clothing and allowed to remain, may cause smarting and reddening of skin.; *Odor Threshold*: 4~7 ppm.

Fire Hazards - *Flash Point (deg. F)*: 12 CC; *Flammable Limits in Air (%)*: 1.3 - 7.9; *Fire Extinguishing Agents*: Dry chemical, foam and carbon dioxide; *Fire Extinguishing Agents Not To Be Used*: Water may be ineffective; *Special Hazards of Combustion Products*: Not pertinent; *Behavior in Fire*: Vapor is heavier than air and can travel considerable distance to source of ignition and flash back; *Ignition Temperature (deg. F)*: 1,097; *Electrical Hazard*: Class I, Group D; *Burning Rate*: 6.0 mm/min.

Chemical Reactivity - *Reactivity with Water*: No reaction; *Reactivity with Common Materials*: No reactions; *Stability During Transport*: Stable; *Neutralizing Agents for Acids and Caustics*: Not pertinent; *Polymerization*: Not pertinent; *Inhibitor of Polymerization*: Not pertinent.

BENZENE HEXACHLORIDE

Chemical Designations - *Synonyms*: BHC, 1,2,3,4,5,6-Hexachlorocyclohexane Lindane; *Chemical Formula*: $C_6H_6Cl_6$.

Observable Characteristics - *Physical State (as normally shipped)*: Solid; *Color*: Light tan to dark brown; *Odor*: Characteristic.

Physical and Chemical Properties - *Physical State at 15 °C and 1 atm.*: Solid; *Molecular Weight*: 290.83; *Boiling Point at 1 atm.*: Not pertinent; *Freezing Point*: Not pertinent; *Critical Temperature*: Not pertinent; *Critical Pressure*: Not pertinent; *Specific Gravity*: 1.891 at 19 °C (solid); *Vapor (Gas) Density*: Not pertinent; *Ratio of Specific Heats of Vapor (Gas)*: Not pertinent; *Latent Heat of Vaporization*: Not pertinent; *Heat of Combustion*: Not pertinent; *Heat of Decomposition*: Not pertinent.

Health Hazards Information - *Recommended Personal Protective Equipment*: Respiratory protection; ensure handling in a well ventilated area.; *Symptoms Following Exposure*: Hyperirritability and central nervous system excitation; notably vomiting, restlessness, muscle spasms, ataxia, clonic and tonic convulsions. Occasional dermatitis and urticaria.; *General Treatment for Exposure*: Gastric lavage and saline cathartics (not oil laxatives because they promote abortion). Sedatives: pentobarbital or phenobarbitol in amounts adequate to control convulsions. Calcium gluconate intravenously may be used in conjunction with sedatives to control convulsions. Keep patient quiet. Do not use epinephrine because ventricular fibrillation may result; *Toxicity by Inhalation (Threshold Limit Value)*: 0.5 mg/m^3;

Short-Term Exposure Limits: 1 mg/m^3 for 30 minutes; *Toxicity by Ingestion*: LD$_{50}$ 0.5 ~ 5 g/kg (Technical Mixture); LD$_{50}$ 50 ~ 500 mg/kg (rat) (Gamma Isomer - Lindane); *Late Toxicity*: Mutagen to human lymphocytes; *Vapor (Gas) Irritant Characteristics*: Moderately irritating. Workers will not usually tolerate moderate to high concentrations.; *Liquid or Solid Irritant Characteristics*: Minimum hazard. If spilled on clothing and allowed to remain, the chemical may cause smarting or reddening of skin.; *Odor Threshold*: No data.

Fire Hazards - *Flash Point:* Not flammable; *Flammable Limits in Air (%)*: Not flammable; *Fire Extinguishing Agents*: Not pertinent; *Fire Extinguishing Agents Not To Be Used*: Not pertinent; *Special Hazards of Combustion Products*: Toxic gases are generated when solid is heated or when solution exposed to intense heat; *Behavior in Fire*: Not pertinent; *Ignition Temperature*: Not flammable; *Electrical Hazard*: Not pertinent; *Burning Rate*: Not flammable.

Chemical Reactivity - *Reactivity with Water*: No reaction; *Reactivity with Common Materials*: No reactions; *Stability During Transport*: Stable; *Neutralizing Agents for Acids and Caustics*: Not pertinent; *Polymerization*: Not pertinent; *Inhibitor of Polymerization*: Not pertinent.

BENZENE PHOSPHOROUS DICHLORIDE

Chemical Designations - *Synonyms*: Phenyl Phosphonous Dichloride, Phenylphosphine Dichloride, Dichlorophenylphosphine; *Chemical Formula*: $C_6H_5PCl_2$.

Observable Characteristics - *Physical State (as normally shipped)*: Liquid; *Color*: Colorless; *Odor*: Acrid and pungent.

Physical and Chemical Properties - *Physical State at 15 °C and 1 atm.*: Liquid; *Molecular Weight*: 179.0; *Boiling Point at 1 atm.*: 430, 221, 494; *Freezing Point*: -60, -51, 222; *Critical Temperature*: Not pertinent; *Critical Pressure*: Not pertinent; *Specific Gravity*: 1.140 at 25 °C (liquid); *Vapor (Gas) Density*: Not pertinent; *Ratio of Specific Heats of Vapor (Gas)*: Not pertinent; *Latent Heat of Vaporization*: Not pertinent; *Heat of Combustion*: -8,200, -4,500, -190; *Heat of Decomposition*: Not pertinent.

Health Hazards Information - *Recommended Personal Protective Equipment*: Self-contained breathing apparatus (SCBA); acid-type canister mask; goggles and face shield, rubber gloves and chemical protective clothing.; *Symptoms Following Exposure*: Inhalation causes irritation of the nose and throat; pulmonary edema may develop following severe exposure. Contact with skin or eyes causes sever burns. Ingestion causes sever burns of mouth and stomach.; *General Treatment for Exposure*: Seek immediate medical attention following all exposures to this chemical. INHALATION: Remove victim to fresh air; if breathing has stopped, start mouth to mouth resuscitation. EYES: Flush with clean running water for at least 15 minutes. Do not use any oils or ointments. SHIN: Flush with water; wash with soap and water. INGESTION: Give victim large amounts of milk or water. Do not induce vomiting. If victim begins to vomit, give milk or beaten eggs at one-hour intervals.; *Toxicity by Inhalation (Threshold Limit Value)*: No data; *Short-Term Exposure Limits*: No data; *Toxicity by Ingestion*: No data; *Late Toxicity*: No data; *Vapor (Gas) Irritant Characteristics*: No data; *Liquid or Solid Irritant Characteristics*: No data; *Odor Threshold*: No data.

Fire Hazards - *Flash Point (deg. F):* 215 OC; This value may be lower because of the presence of phosphorus; *Flammable Limits in Air (%)*: Not pertinent; *Fire Extinguishing Agents*: Large amounts of water; *Fire Extinguishing Agents Not To Be Used*: Not pertinent; *Special Hazards of Combustion Products*: Toxic fumes include oxides of phosphorous and hydrogen chloride; *Behavior in Fire*: Containers may rupture. The hot liquid is spontaneously flammable because of the presence of dissolved phosphorus; *Ignition Temperature (deg. F)*: 319; *Electrical Hazard*: No data; *Burning Rate*: No data.

Chemical Reactivity - *Reactivity with Water*: Reacts vigorously to form hydrogen chloride (hydrochloric acid); *Reactivity with Common Materials*: Corrodes metal except 316 stainless steel, nickel, and Hastelloy; *Stability During Transport*: Stable; *Neutralizing Agents for Acids and Caustics*: Flush with water and rinse with sodium bicarbonate or lime solution; *Polymerization*: Not pertinent; *Inhibitor of Polymerization*: Not pertinent.

BENZENE PHOSPHOROUS THIODICHLORIDE

Chemical Designations - *Synonyms*: Benzenethiophosphonyl Chloride, Phenylphosphonothioic Dichloride, Phenylphosphine Thiodichloride; *Chemical Formula*: $C_6H_5PSCl_2$.

Observable Characteristics - *Physical State (as normally shipped)*: Liquid; *Color*: Colorless to light yellow; *Odor*: Acrid and pungent.

Physical and Chemical Properties - *Physical State at 15 °C and 1 atm.*: Liquid; *Molecular Weight*: 211; *Boiling Point at 1 atm*: 518, 270, 543; *Freezing Point*: -11.2, -24.0, 249.2; *Critical Temperature*: Not pertinent; *Critical Pressure*: Not pertinent; *Specific Gravity*: 1.378 at 20 °C (liquid); *Vapor (Gas) Density*: Not pertinent; *Ratio of Specific Heats of Vapor (Gas)*: Not pertinent; *Latent Heat of Vaporization*: Not pertinent; *Heat of Combustion*: -7700, -4300, -180; *Heat of Decomposition*: Not pertinent.

Health Hazards Information - *Recommended Personal Protective Equipment*: SCBA or acid-type canister mask; *Symptoms Following Exposure*: Inhalation of vapor irritates nose and throat; pulmonary edema may result. Contact with eyes or skin causes severe irritation. Ingestion causes severe irritation of mouth and stomach.; *General Treatment for Exposure*: Get medical attention following all exposures. INHALATION: Remove victim to fresh air. EYES: Flush with fresh running water for at least 15 minutes.; do not apply oils. SKIN: Flush with water, followed by washing with soap and water. INGESTION: Give large amounts of water or milk, eggs or olive oil.; *Toxicity by Inhalation*: No data; *Short-Term Exposure Limits*: No data; *Toxicity by Ingestion*: No data; *Late Toxicity*: No data; *Vapor (Gas) Irritant Characteristics*: No data; *Liquid or Solid Irritant Characteristics*: No data; *Odor Threshold*: No data.

Fire Hazards - *Flash Point (deg. F)*: 252 OC; *Flammable Limits in Air (%)*: Not pertinent; *Fire Extinguishing Agents*: Water; *Fire Extinguishing Agents Not To Be Used*: Not pertinent; *Special Hazards of Combustion Products*: Toxic fumes include oxides of phosphorous, sulfur and hydrogen chloride; *Behavior in Fire*: Containers may rupture; *Ignition Temperature (deg. F)*: 338; *Electrical Hazard*: No data; *Burning Rate*: No data.

Chemical Reactivity - *Reactivity with Water*: Forms hydrogen chloride fumes. Reaction is slow unless the water is hot; *Reactivity with Common Materials*: Corrodes metals slowly; *Stability During Transport*: Stable; *Neutralizing Agents for Acids and Caustics*: Flush with water and rinse with sodium bicarbonate solution; *Polymerization*: Not pertinent; *Inhibitor of Polymerization*: Not pertinent.

BENZOIC ACID

Chemical Designations - *Synonyms*: Benzenecarboxylic Acid, Carboxybenzene, Dracyclic Acid; *Chemical Formula*: C_6H_5COOH.

Observable Characteristics - *Physical State (as normally shipped)*: Solid; *Color*: White; *Odor*: Faint, pleasant; slightly aromatic.

Physical and Chemical Properties - *Physical State at 15 °C and 1 atm.*: Solid; *Molecular Weight*: 122.12; *Boiling Point at 1 atm.*: 480.6, 249.2, 522.4; *Freezing Point*: 252.1, 122.3, 395.5; *Critical Temperature*: 894, 479, 752; *Critical Pressure*: 660, 45, 4.6; *Specific Gravity*: 1.316 at 28 °C (solid); *Vapor (Gas) Density*: Not pertinent; *Ratio of Specific Heats of Vapor (Gas)*: Not pertinent; *Latent Heat of Vaporization*: Not pertinent; *Heat of Combustion*: Not pertinent; *Heat of Decomposition*: Not pertinent.

Health Hazards Information - *Recommended Personal Protective Equipment*: Dust respirator; when melted material is present, use eye protection and organic respirator for fumes; *Symptoms Following Exposure*: Dust is irritating to eyes and nose. At elevated temperatures, fumes may cause irritation of eyes, respiratory system and skin.; *General Treatment for Exposure*: Remove victim to fresh air. EYE CONTACT: Flush eyes with water.; *Toxicity by Inhalation (Threshold Limit Value)*: Not pertinent; *Short-Term Exposure Limits*: Not pertinent; *Toxicity by Ingestion*: LD_{50} 0.5 ~ 5 g/kg; *Late Toxicity*: None; *Vapor (Gas) Irritant Characteristics*: Not pertinent; *Liquid or Solid Irritant Characteristics*: Minimum hazard. If spilled on clothing and allowed to remain, the chemical may cause smarting or reddening of skin. Dust may irritate nose and eyes; *Odor Threshold*: Not pertinent.

Fire Hazards - *Flash Point (deg. F)*: 250 CC; *Flammable Limits in Air (%)*: Not pertinent; *Fire

Extinguishing Agents: Dry chemical, carbon dioxide, water fog, chemical foam; *Fire Extinguishing Agents Not To Be Used*: None; *Special Hazards of Combustion Products*: Not pertinent; *Behavior in Fire*: Vapor from molten benzoic acid may form explosive mixture with air. Concentrated dust may form explosive mixture in air; *Ignition Temperature (deg. F)*: 1,063; *Electrical Hazard*: Not pertinent; *Burning Rate*: Not pertinent.

Chemical Reactivity - *Reactivity with Water*: No reaction; *Reactivity with Common Materials*: No reactions; *Stability During Transport*: Stable; *Neutralizing Agents for Acids and Caustics*: Not pertinent; *Polymerization*: Not pertinent; *Inhibitor of Polymerization*: Not pertinent.

BENZONITRILE

Chemical Designations - *Synonyms*: Benzoic Acid Nitrile, Cyanobenzene, Phenylcyanide; *Chemical Formula*: C_6H_5CN.

Observable Characteristics - *Physical State (as normally shipped)*: Liquid; *Color*: Colorless; *Odor*: Almond-like.

Physical and Chemical Properties - *Physical State at 15 °C and 1 atm.*: Liquid; *Molecular Weight*: 103.12; *Boiling Point at 1 atm.*: 376, 191, 464; *Freezing Point*: 9.0, -12.8, 260.4; *Critical Temperature*: 799.2, 426.2, 699.4; *Critical Pressure*: 611, 41.6, 4.22; *Specific Gravity*: 1.01 at 25 °C (liquid); *Vapor (Gas) Density*: 3.6; *Ratio of Specific Heats of Vapor (Gas)*: 1.091; *Latent Heat of Vaporization*: 157.7, 87.6, 3.67; *Heat of Combustion*: -15,100, -8,400, -351; *Heat of Decomposition*: Not pertinent.

Health Hazards Information - *Recommended Personal Protective Equipment*: Rubber gloves; chemical resistant splash-proof goggles; rubber boots; chemical protective clothing for splash protection, chemical cartridge type respirator or other suitable protection against vapor must be worn when working in poorly ventilated areas or where overexposure by inhalation could occur.; *Symptoms Following Exposure*: Personnel can be overexposed to this chemical by ingestion, absorption through the skin, or inhalation. The earliest symptoms of cyano-compound intoxication may be weakness, headaches, confusion, and occasionally nausea and vomiting. The respiratory rate and depth will usually be increased at the beginning and at later stages become slow and gasping. Blood pressure is usually normal, especially in mild or moderately severe cases, although the pulse rate is usually more rapid than normal.; *General Treatment for Exposure*: INHALATION: Remove patient to fresh air; seek immediate medical attention. INGESTION: Call physician immediately. Until doctor arrives, take the following steps: a) Provide for inhalation by amyl nitrate vapor from ampules crushed in a handkerchief and held to nose of victim. b) Induce vomiting unless patient is unconscious. (Gastric lavage should be employed by, or under the supervision of a physician). c) Keep patient warm and quiet until medical attention arrives. EYES: Immediately flush with large volumes of fresh water for ate least 15 minutes. SKIN: Wash thoroughly at once, without scrubbing, with large amounts of soap and water. OTHER: Exposed personnel should be checked periodically for chronic toxic effects.; *Toxicity by Inhalation (Threshold Limit Value)*: No data; *Short-Term Exposure Limits*: No data; *Toxicity by Ingestion*: LD_{50} = 800 mg/kg (rat); *Late Toxicity*: No data; *Vapor (Gas) Irritant Characteristics*: No data; *Liquid or Solid Irritant Characteristics*: No data; *Odor Threshold*: No data.

Fire Hazards - *Flash Point (deg. F)*: 167 CC, This material is combustible but burns with difficulty; *Flammable Limits in Air (%)*: No data; *Fire Extinguishing Agents*: Foam, dry chemical, carbon dioxide; *Fire Extinguishing Agents Not To Be Used*: Water may be ineffective; *Special Hazards of Combustion Products*: Toxic hydrogen cyanide and oxides of nitrogen form; *Behavior in Fire*: No data; *Ignition Temperature*: No data; *Electrical Hazard*: No data; *Burning Rate*: Difficult to burn.

Chemical Reactivity - *Reactivity with Water*: No reaction; *Reactivity with Common Materials*: Will attack some plastics; *Stability During Transport*: Stable; *Neutralizing Agents for Acids and Caustics*: Not pertinent; *Polymerization*: Not pertinent; *Inhibitor of Polymerization*: Not pertinent.

BENZOPHENONE

Chemical Designations — *Synonyms*: Benzoylbenzene, Diphenyl Ketone, Diphenyl methanone, alpha-Oxodiphenylmethane, alpha-Oxoditane; *Chemical Formula*: $C_6H_5COC_6H_5$.

Observable Characteristics - *Physical State (as normally shipped)*: Liquid or solid; *Color*: White; *Odor*: Characteristic.

Physical and Chemical Properties - *Physical State at 15 °C and 1 atm.*: Solid; *Molecular Weight*: 182; *Boiling Point at 1 atm.*: 581.9, 305.5, 578.7; *Freezing Point*: 118.2, 47.9, 321.1; *Critical Temperature*: Not pertinent; *Critical Pressure*: Not pertinent; *Specific Gravity*: 1.085 at 50°C (liquid); *Vapor (Gas) Density*: Not pertinent; *Ratio of Specific Heats of Vapor (Gas)*: Not pertinent; *Latent Heat of Vaporization*: 126.0, 70.0, 2.93; *Heat of Combustion*: -15,400, -8,550, -358; *Heat of Decomposition*: Not pertinent.

Health Hazards Information - *Recommended Personal Protective Equipment*: Goggles or face shield, rubber gloves; *Symptoms Following Exposure*: Ingestion causes gastrointestinal disturbances. Contact causes eye irritation and, if prolonged, irritation of skin; *General Treatment for Exposure*: INHALATION: remove to fresh air. Ingestion: get medical attention. EYES: flush with water for at least 15 min; get medical attention if irritation persists. SKIN: flush with water, wash with soap and water; *Toxicity by Inhalation (Threshold Limit Value)*: Data not available; *Short-Term Exposure Limits*: Data not available; *Toxicity by Ingestion*: Grade 1; acute oral rat $LD_{50} \geq 10,000$ mg/kg; *Late Toxicity*: Data not available; *Vapor (Gas) Irritant Characteristics*: Data not available; *Liquid or Solid Irritant Characteristics*: Data not available; *Odor Threshold*: Data not available.

Fire Hazards - *Flash Point:* This is a combustible product; *Flammable Limits in Air (%)*: No data; *Fire Extinguishing Agents*: Foam, dry chemical, carbon dioxide; *Fire Extinguishing Agents Not To Be Used*: Water may be ineffective; *Special Hazards of Combustion Products*: No data; *Behavior in Fire*: No data; *Ignition Temperature*: No data; *Electrical Hazard*: No data; *Burning Rate*: Not pertinent.

Chemical Reactivity - *Reactivity with Water*: No reaction; *Reactivity with Common Materials*: Will attack some plastics; *Stability During Transport*: Stable; *Neutralizing Agents for Acids and Caustics*: Not pertinent; *Polymerization*: Not pertinent; *Inhibitor of Polymerization*: Not pertinent.

BENZOYL CHLORIDE

Chemical Designations - *Synonyms*: Benzenecarbonyl Chloride; *Chemical Formula*: C_6H_5COCL.

Observable Characteristics - *Physical State (as normally shipped)*: Liquid; *Color*: Colorless; may become slightly brownish on standing; *Odor*: Pungent characteristic.

Physical and Chemical Properties - *Physical State at 15 °C and 1 atm.*: Liquid; *Molecular Weight*: 140.57; *Boiling Point at 1 atm.*: 387, 197.3, 470.5; *Freezing Point*: 30.9, -0.6, 272.6; *Critical Temperature*: Not pertinent; *Critical Pressure*: Not pertinent; *Specific Gravity*: 1.211 at 25° C (liquid); *Vapor (Gas) Density*: Not pertinent; *Ratio of Specific Heats of Vapor (Gas)*: Not pertinent; *Latent Heat of Vaporization*: Data not available; *Heat of Combustion*: -10,030, -5570, -233.2; *Heat of Decomposition*: Not pertinent.

Health Hazards Information - *Recommended Personal Protective Equipment*: Full protecting clothing, including full-face respirator for acid gases and organic vapors (yellow GMC canister), close fitting goggles, nonslip rubber gloves, plastic apron, face shield; *Symptoms Following Exposure*: INHALATION: may irritate eyes, nose and throat. INGESTION: causes acute discomfort. SKIN: causes irritation and burning; *General Treatment for Exposure*: INHALATION: remove to fresh air; administer oxygen with patient in sitting position. INGESTION: give water; call physician at once; give milk. EYES: flush with water for 15 min, get medical attention; SKIN: wash with plenty of soap and water; *Toxicity by Inhalation (Threshold Limit Value)*: Data not available; *Short-Term Exposure Limits*: Data not available; *Toxicity by Ingestion*: Data not available; *Late Toxicity*: Data not available; *Vapor (Gas) Irritant Characteristics*: Vapors cause severe irritation of eyes and throat and can cause eye and lung injury. They cannot be tolerated even at low concentrations; *Liquid or Solid Irritant Characteristics*: Severe skin irritant. Cause second- and third-degree burns on short contact and is very injurious to the eyes; *Odor Threshold*: Data not available.

Fire Hazards - *Flash Point (deg. F):* 162 OC; *Flammable Limits in Air (%)*: 1.2 - 4.9; *Fire Extinguishing Agents*: Foam, carbon dioxide, dry chemical, water fog; *Fire Extinguishing Agents Not To Be Used*: Water spray. Do not allow water to enter containers; *Special Hazards of Combustion Products*: Highly poisonous phosgene gas forms during fires; *Behavior in Fire*: At fire temperature the compound may react violently with water or steam; *Ignition Temperature (deg. F)*: 185; *Electrical Hazard*: Not pertinent; *Burning Rate*: No data.

**Chemical Reactivity - *Reactivity with Water*: Slow reaction with water to produce hydrochloric acid fumes. The reaction is more rapid with steam; *Reactivity with Common Materials*: Slow corrosion of metals but no immediate danger; *Stability During Transport*: Not pertinent; *Neutralizing Agents for Acids and Caustics*: Soda ash and water, lime; *Polymerization*: Does not occur; *Inhibitor of Polymerization*: Not pertinent.

BENZYL ALCOHOL

**Chemical Designations - *Synonyms*: Benzenecarbinol,; alpha-Hydroxytoluene; Phenylcarbinol; Phenylmethanol; Phenylmethyl Alcohol; *Chemical Formula*: $C_6H_5CH_2OH$.

**Observable Characteristics - *Physical State (as normally shipped)*: Liquid; *Color*: Colorless; *Odor*: Mild, pleasant.

**Physical and Chemical Properties - *Physical State at 15 °C and 1 atm.*: Liquid; *Molecular Weight*: 108.13; *Boiling Point at 1 atm.*: 401, 205, 478; *Freezing Point*: 4.5, -15.3, 257.9; *Critical Temperature*: 757. 403, 676; *Critical Pressure*: 663, 45.0, 4.57; *Specific Gravity*: 1.050 at 15/15°C(liquid); *Vapor (Gas) Density*: 3.73; *Ratio of Specific Heats of Vapor (Gas)*: 1.070; *Latent Heat of Vaporization*: 193, 107, 4.48; *Heat of Combustion*: -14,850, -8,260, -345; *Heat of Decomposition*: Not pertinent.

**Health Hazards Information - *Recommended Personal Protective Equipment*: Rubber gloves, chemical safety goggles; *Symptoms Following Exposure*: Inhalation of vapor may cause irritation of upper respiratory tract. Prolonged or excessive inhalation may result in headache, nausea, vomiting and diarrhea. In severe cases, respiratory stimulation followed by respiratory and muscular paralysis, convulsions, narcosis and death may result. Ingestion may produce severe irritation of the gastrointestinal tract, followed by nausea, vomiting, cramps and diarrhea; tissues ulceration may result. Contact with eyes causes local irritation, Material can be absorbed through skin with anesthetic or irritant effect; *General Treatment for Exposure*: INHALATION: remove victim from contaminated atmosphere; call physician immediately. INGESTION: induce vomiting and contact a physician. EYES: flush with plenty of water for 15 min, and contact a physician. SKIN: flush with water, wash with soap and water, obtain medical attention in case of irritation or central nervous system depression; *Toxicity by Inhalation (Threshold Limit Value)*: Data not available; *Short-Term Exposure Limits*: Data not available; *Toxicity by Ingestion*: Grade 2; oral rat LD_{50} = 1,230 mg/kg; *Late Toxicity*: Data not available; *Vapor (Gas) Irritant Characteristics*: Data not available; *Liquid or Solid Irritant Characteristics*: Data not available; *Odor Threshold*: 5.5 ppm.

**Fire Hazards - *Flash Point (deg. F)*: 220 OC, 213 CC; *Flammable Limits in Air (%)*: No data; *Fire Extinguishing Agents*: Alcohol foam. dry chemical, carbon dioxide; *Fire Extinguishing Agents Not To Be Used*: Water or foam may cause foaming; *Special Hazards of Combustion Products*: No data; *Behavior in Fire*: No data; *Ignition Temperature (deg. F)*: 817; *Electrical Hazard*: No data; *Burning Rate*: 3.74 mm/min.

**Chemical Reactivity - *Reactivity with Water*: No reaction; *Reactivity with Common Materials*: Will attack some plastics; *Stability During Transport*: Stable; *Neutralizing Agents for Acids and Caustics*: Not pertinent; *Polymerization*: Not pertinent; *Inhibitor of Polymerization*: Not pertinent.

BENZYLAMINE

**Chemical Designations - *Synonyms*: Alpha-Aminotoluene, Phenylmethyl Amine; *Chemical Formula*: $C_6H_5CH_2NH_2$

**Observable Characteristics - *Physical State (as normally shipped)*: Liquid; *Color*: Colorless to light yellow; *Odor*: Strong ammonia.

**Physical and Chemical Properties - *Physical State at 15 °C and 1 atm.*: Liquid; *Molecular Weight*: 107.16; *Boiling Point at 1 atm.*: 364.1, 184.5, 457.7; *Freezing Point*: -51, -46, 227; *Critical Temperature*: Not pertinent; *Critical Pressure*: Not pertinent; *Specific Gravity*: 0.98 at 20°C (liquid); *Vapor (Gas) Density*: 3.70; *Ratio of Specific Heats of Vapor (Gas)*: 1.070; *Latent Heat of Vaporization*:

164, 91, 3.8; *Heat of Combustion*: -16,260; -9,040; -378; *Heat of Decomposition*: Not pertinent.

Health Hazards Information - *Recommended Personal Protective Equipment*: Self-contained breathing apparatus; goggles of face shield; rubber gloves; *Symptoms Following Exposure*: Inhalation of vapor causes irritation of mucous membranes of the nose and throat, and lung irritation with respiratory distress and cough. Headache, nausea, faintness and anxiety can occur. Exposure to vapor produces eye irritation with lachrymation, conjunctivitis, and corneal edema resulting in halos around lights. Direct local contact with liquid is known to produce severe and sometimes permanent eye damage and skin burns. Vapors may also produce primary skin irritation and dermatitis; *General Treatment for Exposure*: INHALATION: remove victim from exposure, if breathing is difficult, administer oxygen, if breathing has stopped, begin artificial respiration. EYES or SKIN: wash with copious amounts of water for 15 min; *Toxicity by Inhalation (Threshold Limit Value)*: Data not available; *Short-Term Exposure Limits*: Data not available; *Toxicity by Ingestion*: Data not available; *Late Toxicity*: Data not available; *Vapor (Gas) Irritant Characteristics*: Data not available; *Liquid or Solid Irritant Characteristics*: Data not available; *Odor Threshold*: Data not available.

Fire Hazards - *Flash Point (deg. F)*: 168 OC; *Flammable Limits in Air* (%): No data; *Fire Extinguishing Agents*: Alcohol foam, dry chemical, carbon dioxide; *Fire Extinguishing Agents Not To Be Used*: Water may be ineffective; *Special Hazards of Combustion Products*: Toxic nitrogen oxides form in fire situations; *Behavior in Fire*: No data; *Ignition Temperature*: No data; *Electrical Hazard*: No data; *Burning Rate*: 4.13 mm/min.

Chemical Reactivity - *Reactivity with Water*: No reaction; *Reactivity with Common Materials*: In presence of moisture may severely corrode some metals. In liquid state this chemical will attack some plastics; *Stability During Transport*: Stable; *Neutralizing Agents for Acids and Caustics*: Flush with water; *Polymerization*: Not pertinent; *Inhibitor of Polymerization*: Not pertinent.

BENZYL BROMIDE

Chemical Designations - *Synonyms*: Alpha-Bromotoluene; Omega-Bromotoluene; Bromotoluene, Alpha; *Chemical Formula*: $C_6H_5CH_2Br$.

Observable Characteristics - *Physical State (as normally shipped)*: Liquid; *Color*: Colorless to yellow; *Odor*: Very sharp, pungent, like tear gas.

Physical and Chemical Properties - *Physical State at 15 °C and 1 atm.*: Liquid; *Molecular Weight*: 171.0; *Boiling Point at 1 atm.*: 388, 198, 471; *Freezing Point*: 25.0, -3.9, 269.3; *Critical Temperature*: Not pertinent; *Critical Pressure*: Not pertinent; *Specific Gravity*: 1.441 at 22°C (liquid); *Vapor (Gas) Density*: 5.9; *Ratio of Specific Heats of Vapor (Gas)*: Data not available; *Latent Heat of Vaporization*: 120, 66.4, 2.78; *Heat of Combustion*: -9,000; -5,000; -210; *Heat of Decomposition*: Not pertinent.

Health Hazards Information - *Recommended Personal Protective Equipment*: Self-contained breathing apparatus; goggles; rubber gloves; protective clothing; *Symptoms Following Exposure*: Inhalation causes irritation of nose and throat; severe exposure may cause pulmonary edema. Vapors cause severe eyes irritation; liquid can burn eyes. Skin contact cause irritation of mouth and stomach; *General Treatment for Exposure*: INHALATION: remove to fresh air. EYES: irrigate with copious amount of water for 15 min. SKIN: flush with water. INGESTION: do NOT induce vomiting; give large amounts of water; *Toxicity by Inhalation (Threshold Limit Value)*: Data not available; *Short-Term Exposure Limits*: Data not available; *Toxicity by Ingestion*: Data not available; *Late Toxicity*: Data not available; *Vapor (Gas) Irritant Characteristics*: Data not available; *Liquid or Solid Irritant Characteristics*: Data not available; *Odor Threshold*: Data not available.

Fire Hazards - *Flash Point (deg. F)*: 174 CC; *Flammable Limits in Air* (%): Not pertinent; *Fire Extinguishing Agents*: Water, dry chemical, foam, carbon dioxide; *Fire Extinguishing Agents Not To Be Used*: Not pertinent; *Special Hazards of Combustion Products*: Irritating and toxic hydrogen bromide gas is formed; *Behavior in Fire*: Forms vapor that is powerful tear gas; *Ignition Temperature*: No data; *Electrical Hazard*: No data; *Burning Rate*: 2.6 mm/min.

Chemical Reactivity - *Reactivity with Water*: Reacts slowly generating hydrogen bromide (hydrobromic acid); *Reactivity with Common Materials*: Decomposes rapidly in the presence of all common metals except nickel and lead, liberating heat and hydrogen bromide; *Stability During Transport*: Stable; *Neutralizing Agents for Acids and Caustics*: Rinse with sodium bicarbonate or lime solution; *Polymerization*: Polymerizes with evolution of heat and hydrogen bromide when in presence with all common metals except nickel and lead; *Inhibitor of Polymerization*: None.

BENZYL CHLORIDE

Chemical Designations - *Synonyms*: alpha-Chlorotoluene; omega-Chlorotoluene; Chlorotoluene, alpha; *Chemical Formula*: $C_6H_5CH_2Cl$.

Observable Characteristics - *Physical State (as normally shipped)*: Liquid; *Color*: Colorless to pale yellow; *Odor*: Pungent, irritating.

Physical and Chemical Properties - *Physical State at 15 °C and 1 atm.*: Liquid; *Molecular Weight*: 126.6; *Boiling Point at 1 atm.*: 354.9, 179.4, 452.6; *Freezing Point*: -38.6, -39.2, 234.0; *Critical Temperature*: 772, 411, 684; *Critical Pressure*: 567, 38.5, 3.91; *Specific Gravity*: 1.10 at 25°C (liquid); *Vapor (Gas) Density*: 4.36; *Ratio of Specific Heats of Vapor (Gas)*: 1.0689; *Latent Heat of Vaporization*: 130, 70, 2.9; *Heat of Combustion*: -12,000, -6,700, -280; *Heat of Decomposition*: Not pertinent.

Health Hazards Information - *Recommended Personal Protective Equipment*: Chemical safety goggles or face shield, self-contained breathing apparatus, positive-pressure hose mask, industrial canister-type gas mask, or chemical cartridge respirator; rubber gloves, protective clothing; *Symptoms Following Exposure*: Inhalation causes severe irritation of upper respiratory tract with coughing, burning of the throat, headache, dizziness, and weakness; lung damage and pulmonary edema may occur after severe exposure; chronic irritation of the upper respiratory tract may occur after prolonged and repeated exposure to vapors. Immediate and severe eye irritation may result from contact with the liquid or vapor; prolonged or permanent eye damage may result. Vapors irritate skin and liquid may cause severe burns. Ingestion may cause immediate and severe burns of the mouth and throat, and gastrointestinal tract; nausea, vomiting, cramps, and diarrhea may follow; gastrointestinal damage and systemic effects may result; *General Treatment for Exposure*: INHALATION: remove victim from contaminated atmosphere; if breathing has ceased, start mouth-to-mouth resuscitation; oxygen, if available, should be administered only by an experienced person when authorized by a physician; keep patient warm and comfortable; call a physician immediately. EYES: immediately flush with large quantities of running water for a minimum of 15 min; hold eyelids apart during irrigation to ensure flushing of the entire surface of the eye and lids with water; do not attempt to neutralize with chemical agents; obtain medical attention as soon as possible; oils or ointments should not be used unless directed by a physician; continue irrigation for an additional 15 min, if physician is not available. SKIN: immediately flush affected area with water; remove contaminated clothing under shower, continue washing with water, do not attempt to neutralize with chemical agents; obtain medical attention if irritation persists. INGESTION: give large amounts of water; do not induce vomiting; *Toxicity by Inhalation (Threshold Limit Value)*: 1 ppm; *Short-Term Exposure Limits*: Data not available; *Toxicity by Ingestion*: Grade 2; oral rat LD_{50} = 1,231 mg/kg; *Late Toxicity*: Data not available; *Vapor (Gas) Irritant Characteristics*: Vapors cause severe irritation of eyes and throat and cause eye and lung injury. They cannot be tolerated even at low concentrations; *Liquid or Solid Irritant Characteristics*: Severe skin irritant. Causes second- and third-degree burns on short contact and is very injurious to the eyes; *Odor Threshold*: 0.047 ppm.

Fire Hazards - *Flash Point (deg. F)*: 165 OC, 140 CC; *Flammable Limits in Air (%)*: 1.1 (LEL); *Fire Extinguishing Agents*: Water, dry chemical, foam, and carbon dioxide; *Fire Extinguishing Agents Not To Be Used*: Not pertinent; *Special Hazards of Combustion Products*: Irritating hydrogen chloride gas forms; *Behavior in Fire*: Forms vapor that is a powerful tear gas; *Ignition Temperature (deg. F)*: 1,161; *Electrical Hazard*: No data; *Burning Rate*: 4.2 mm/min.

Chemical Reactivity - *Reactivity with Water*: Undergoes slow hydrolysis, liberating hydrogen chloride (hydrochloric acid); *Reactivity with Common Materials*: Decomposes rapidly in the presence of all common metals (with the exception of nickel and lead), liberating heat and hydrogen chloride; *Stability During Transport*: Stable; *Neutralizing Agents for Acids and Caustics*: Rinse with sodium bicarbonate or lime solution; *Polymerization*: Polymerizes with evolution of heat and hydrogen chloride when in contact with all common metals except nickel and lead; *Inhibitor of Polymerization*: Triethylamine, propylene oxide or sodium carbonate.

BENZYL CHLOROFORMATE

Chemical Designations - *Synonyms*: Carbobenzoxy Chloride; Chloroformic Acid, Benzyl Ester; Benzylcarbonyl Chloride; Benzyl Chlorocarbonate; *Chemical Formula*: $C_6H_5CH_2OCOCl$.

Observable Characteristics - *Physical State (as normally shipped)*: Liquid; *Color*: Colorless; *Odor*: Irritating; sharp, penetrating.

Physical and Chemical Properties - *Physical State at 15 °C and 1 atm.*: Liquid; *Molecular Weight*: 170.6; *Boiling Point at 1 atm.*: 306, 152, 425; *Freezing Point*: Not pertinent; *Critical Temperature*: Not pertinent; *Critical Pressure*: Not pertinent; *Specific Gravity*: 1.22 at 20°C; *Vapor (Gas) Density*: Not pertinent; *Ratio of Specific Heats of Vapor (Gas)*: Not pertinent; *Latent Heat of Vaporization*: 90, 50, 2.1; *Heat of Combustion*: -10,000, -5,700, -240; *Heat of Decomposition*: Not pertinent.

Health Hazards Information - *Recommended Personal Protective Equipment*: Self-contained breathing apparatus or acid-type canister mask; goggles or face shield; rubber gloves; protective clothing; *Symptoms Following Exposure*: Inhalation causes mucous membrane irritation. Eyes are irritated by excessive exposure to vapor. Liquid causes severe irritation of eyes and irritates skin. Ingestion causes irritation of mouth and stomach; *General Treatment for Exposure*: INHALATION: remove from exposure, support respiration, call physician. EYES: irrigate with copious amounts of water for 15 min. SKIN: flush with large quantities of water; wash with soap and water. INGESTION: give large amounts of water, do NOT induce vomiting; *Toxicity by Inhalation (Threshold Limit Value)*: Data not available; *Short-Term Exposure Limits*: Data not available; *Toxicity by Ingestion*: Grade 3; LD_{50} to 500 mg/kg; *Late Toxicity*: Data not available; *Vapor (Gas) Irritant Characteristics*: Vapors cause moderate irritation such that personnel will find high concentrations unpleasant. The effect is temporary; *Liquid or Solid Irritant Characteristics*: Causes smarting of the skin and first-degree burns on short exposure and may cause second-degree on long exposure; *Odor Threshold*: Data not available.

Fire Hazards - *Flash Point (deg. F)*: 176 OC, 227 CC; Vigorous decomposition occurs at these temperatures. These values are anomalous due to the effect of the decomposition products of benzyl chloride and CO_2; *Flammable Limits in Air (%)*: Not pertinent; *Fire Extinguishing Agents*: Dry chemical, foam, carbon dioxide; *Fire Extinguishing Agents Not To Be Used*: No data; *Special Hazards of Combustion Products*: Toxic phosgene, hydrogen chloride, and benzyl chloride vapors may form; *Behavior in Fire*: Containers may explode; *Ignition Temperature*: No data; *Electrical Hazard*: No data; *Burning Rate*: 4.0 mm/min.

Chemical Reactivity - *Reactivity with Water*: Forms hydrogen chloride (hydrochloric acid). Reaction not very vigorous in cold water; *Reactivity with Common Materials*: Slow corrosion of metals; *Stability During Transport*: Stable; *Neutralizing Agents for Acids and Caustics*: Flush with and rinse with sodium bicarbonate or lime solution; *Polymerization*: Not pertinent; *Inhibitor of Polymerization*: Not pertinent.

BERYLLIUM CHLORIDE

Chemical Designations - *Synonyms:* No common synonyms; *Chemical Formula*: $BeCl_2$; **(ii)**

Observable Characteristics - *Physical State (as normally shipped)*: Solid; *Color*: White to green; *Odor*: Sharp, acrid.

Physical and Chemical Properties - *Physical State at 15 °C and 1 atm.*: Solid; *Molecular Weight*: 79.9; *Boiling Point at 1 atm.*: 968, 520, 793; *Freezing Point*: 824, 440, 713; *Critical Temperature*: Not pertinent; *Critical Pressure*: Not pertinent; *Specific Gravity*: 1.90 at 25°C; *Vapor (Gas) Density*: Not pertinent; *Ratio of Specific Heats of Vapor (Gas)*: Not pertinent; *Latent Heat of Vaporization*: Not pertinent; *Heat of Combustion*: Not pertinent; *Heat of Decomposition*: Not pertinent.

Health Hazards Information - *Recommended Personal Protective Equipment*: Respiratory protection; gloves; freshly laundered clothing; chemical safety goggles; *Symptoms Following Exposure*: Inhalation causes pneumonitis, nasopharyngitis, tracheobronchitis, dyspnea, chromic cough. Ingestion causes irritation of mouth and stomach. Contact with dust causes conjunctival inflamation of eyes and irritation of skin. Any dramatic, unexplained weigh loss should be considered as a possible first indication of beryllium disease; *General Treatment for Exposure*: INHALATION: chest x-ray should be taken immediately for evidence of pneumonitis. EYES: flush with water for at least 15 min; if irritation persists, get medical attention. SKIN: cuts or puncture wounds in which beryllium may be embedded under the skin should be thoroughly cleansed immediately by a physician; *Toxicity by Inhalation (Threshold Limit Value)*: 0.002 mg/m^3 (as beryllium); *Short-Term Exposure Limits*: 0.025 mg/m^3 less than 30 min; *Toxicity by Ingestion*: Grade 3; oral rat LD = 86 mg/kg; *Late Toxicity*: Be produces a chromic systematic disease that primarily affects the lung but also can involve other organs such as lymph nodes, liver, bones, and kidney; *Vapor (Gas) Irritant Characteristics*: Data not available; *Liquid or Solid Irritant Characteristics*: Data not available; *Odor Threshold*: Data not available.

Fire Hazards - *Flash Point:* Not flammable; *Flammable Limits in Air (%)*: Not flammable; *Fire Extinguishing Agents*: Not pertinent; *Fire Extinguishing Agents Not To Be Used*: Do not use water on adjacent fires; *Special Hazards of Combustion Products*: Toxic and irritating beryllium oxide fumes and hydrogen chloride may form in fires; *Behavior in Fire*: No data; *Ignition Temperature*: Not pertinent; *Electrical Hazard*: Not pertinent; *Burning Rate*: Not pertinent.

Chemical Reactivity - *Reactivity with Water*: Reacts vigorously as an exothermic reaction. Forms beryllium oxide and hydrochloric acid solution; *Reactivity with Common Materials*: Corrodes most metals in the presence of moisture. Flammable and explosive hydrogen gas may collect in confined spaces; *Stability During Transport*: Stable; *Neutralizing Agents for Acids and Caustics*: Flush with water and rinse with dilute solution of sodium bicarbonate or soda ash; *Polymerization*: Not pertinent; *Inhibitor of Polymerization*: Not pertinent.

BERYLLIUM FLUORIDE

Chemical Designations - *Synonyms*: No common synonyms; *Chemical Formula*: BeF_2.

Observable Characteristics - *Physical State (as normally shipped)*: Solid; *Color*: White; *Odor*: None.

Physical and Chemical Properties - *Physical State at 15 °C and 1 atm.*: Solid; *Molecular Weight*: 47; *Boiling Point at 1 atm.*: Not pertinent; *Freezing Point*: Not pertinent; *Critical Temperature*: Not pertinent; *Critical Pressure., MN/m^2)*: Not pertinent; *Specific Gravity*: 1.99 at 20°C (solid); *Vapor (Gas) Density*: Not pertinent; *Ratio of Specific Heats of Vapor (Gas)*: Not pertinent; *Latent Heat of Vaporization*: Not pertinent; *Heat of Combustion*: Not pertinent; *Heat of Decomposition*: Not pertinent.

Health Hazards Information - *Recommended Personal Protective Equipment*: Respiratory protection; gloves; goggles; *Symptoms Following Exposure*: Any dramatic weight loss should be considered as possible first indication of beryllium disease. Inhalation causes irritation of nose, throat and lungs, severe pneumonitis, and/or pulmonary edema. Ingestion causes fatigue, weakness, loss of appetite. Contact with eyes causes severe irritation and burns. Contact with skin causes dermatitis and non-healing ulcers; *General Treatment for Exposure*: INHALATION: move to fresh air, chest x-ray should be taken immediately to detect pneumonitis, if exposure has been severe. INGESTION: induce vomiting, get medical attention. EYES: flush with water for at least 15 min; get medical attention. SKIN: flush with water; get medical attention if skin has been broken; *Toxicity by Inhalation (Threshold Limit Value)*: 0.002 mg/m^3 (as beryllium); *Short-Term Exposure Limits*: 0.025 mg/m^3, less than 30 min.; *Toxicity by Ingestion*: Grade 3; oral LD = 100 mg/kg (mouse); *Late Toxicity*: Berylliosis of lungs may occur from 3 months to 15 years after exposure. Chromic systemic diseases of the liver, spleen, kidney, and other organs may also occur; *Vapor (Gas) Irritant Characteristics*: Data not available; *Liquid or Solid Irritant Characteristics*: Data not available; *Odor Threshold*: Data not available.

Fire Hazards - *Flash Point:* Not flammable; *Flammable Limits in Air (%)*: Not flammable; *Fire Extinguishing Agents*: Not pertinent; *Fire Extinguishing Agents Not To Be Used*: Not pertinent; *Special Hazards of Combustion Products*: Toxic and irritating vapors may form from unburned material in a fire situation; *Behavior in Fire*: Not pertinent; *Ignition Temperature*: Not flammable; *Electrical Hazard*: Not pertinent; *Burning Rate*: Not flammable.

Chemical Reactivity - *Reactivity with Water*: No reaction; *Reactivity with Common Materials*: No reactions; *Stability During Transport*: Stable; *Neutralizing Agents for Acids and Caustics*: Not pertinent; *Polymerization*: Not pertinent; *Inhibitor of Polymerization*: Not pertinent.

BERYLLIUM METALLIC

Chemical Designations — *Synonyms*: No common synonyms; *Chemical Formula*: Be.

Observable Characteristics — *Physical State (as normally shipped)*: Solid; *Color*: White; *Odor*: None.

Physical and Chemical Properties - *Physical State at 15 °C and 1 atm.*: Solid; *Molecular Weight*: 9.01; *Boiling Point at 1 atm.*: Not pertinent; *Freezing Point*: Not pertinent; *Critical Temperature*: Not pertinent; *Critical Pressure*: Not pertinent; *Specific Gravity*: 1.85 at 20 °C (solid); *Vapor (Gas) Density*: Not pertinent; *Ratio of Specific Heats of Vapor (Gas)*: Not pertinent; *Latent Heat of Vaporization*: Not pertinent; *Heat of Combustion*: -28000, -15560, -652; *Heat of Decomposition*: Not pertinent.

Health Hazards Information - *Recommended Personal Protective Equipment*: Chemical cartridge respirator; clean work clothes daily; gloves and eye protection.; *Symptoms Following Exposure*: Any dramatic, unexplained weight loss should be considered as possible first indication of beryllium disease. The dust is extremely toxic when inhaled; symptoms include coughing, shortness of breath, and acute or chronic lung disease. There is no record of illness from ingestion of beryllium. Contact with dust causes conjuctival inflammation of the eyes and dermatitis; *General Treatment for Exposure*: INHALATION: acute disease may require hospitalization with administration of oxygen; chest x-ray should be taken immediately. EYES: Flush with water followed by washing with soap and water; all cuts, scratches or other injuries should receive prompt medical attention.; *Toxicity by Inhalation (Threshold Limit Value)*: 0.002 mg/m^3; *Short-Term Exposure Limits*: 0.025 mg/m^3 for 5 minutes; *Toxicity by Ingestion*: LD$_{50}$ 50 ~ 500 mg/kg; *Late Toxicity*: Beryllium disease may occur in the lungs, lymph nodes, liver, spleen, kidney, and other organs.; *Vapor (Gas) Irritant Characteristics*: No data; *Liquid or Solid Irritant Characteristics*: No data; *Odor Threshold*: No data.

Fire Hazards - *Flash Point:* Not pertinent. This is a combustible solid; *Flammable Limits in Air (%)*: Not pertinent; *Fire Extinguishing Agents*: Graphite, sand, or any other inert dry powder; *Fire Extinguishing Agents Not To Be Used*: Water; *Special Hazards of Combustion Products*: Combustion results in beryllium oxide fumes which are toxic to inhalation; *Behavior in Fire*: Powder may form explosive mixture in air; *Ignition Temperature (deg. F)*: Not pertinent; *Electrical Hazard*: Not pertinent; *Burning Rate*: Not pertinent.

Chemical Reactivity - *Reactivity with Water*: No reaction; *Reactivity with Common Materials*: No reactions; *Stability During Transport*: Stable; *Neutralizing Agents for Acids and Caustics*: Not pertinent; *Polymerization*: Not pertinent; *Inhibitor of Polymerization*: Not pertinent.

BERYLLIUM NITRATE

Chemical Designations - *Synonyms*: Beryllium Nitrate Trihydrate; *Chemical Formula*: Be(NO$_3$)$_2$•3H$_2$O.

Observable Characteristics - *Physical State (as normally shipped)*: Solid; *Color*: White; *Odor*: None.

Physical and Chemical Properties - *Physical State at 15 °C and 1 atm.*: Solid; *Molecular Weight*: 205.1; *Boiling Point at 1 atm.*: Not pertinent; *Freezing Point*: Not pertinent; *Critical Temperature*: Not pertinent; *Critical Pressure*: Not pertinent; *Specific Gravity*: 1.56 at 20°C (solid); *Vapor (Gas) Density*: Not pertinent; *Ratio of Specific Heats of Vapor (Gas)*: Not pertinent; *Latent Heat of Vaporization*: Not pertinent; *Heat of Combustion*: Not pertinent; *Heat of Decomposition*: Not pertinent.

Health Hazards Information - *Recommended Personal Protective* Equipment: Respiratory protection; gloves; freshly laundered clothing; chemical safety goggles; *Symptoms Following Exposure*: Any dramatic, unexplained weight loss should be considered as possible first indication of beryllium disease. Inhalation causes pneumonitis, nasopharyngitis, tracheobronchitis, dyspnea, chronic cough. Ingestion causes anorexia, fatigue, weakness, malaise. Contact with eyes causes dermatitis and non-healing ulcers; *General Treatment for Exposure*: INHALATION: remove to fresh air; take chest x-ray immediately to check the pneumonitis. INGESTION: induce vomiting; get medical attention. EYES: flush with water for at least 15 min.; get medical attention. SKIN: cuts or puncture wounds in which beryllium may be embedded under the skin should be thoroughly cleansed immediately by a physician; *Toxicity by Inhalation (Threshold Limit Value)*: 0.02 mg/m^3 (as beryllium); *Short-Term Exposure Limits*: 0.025 mg/m^3 less then 30 min.; *Toxicity by Ingestion*: Data not available; *Late Toxicity*: May cause chronic systemic disease of the lung as well as other organs such as liver, spleen, lymph nodes, bone and kidney; *Vapor (Gas) Irritant Characteristics*: Data not available; *Liquid or Solid Irritant Characteristics*: Data not available; *Odor Threshold*: Data not available.

Fire Hazards - *Flash Point:* Not combustible; *Flammable Limits in Air* (%): Not combustible; *Fire Extinguishing Agents*: Water; *Fire Extinguishing Agents Not To Be Used*: Not pertinent; *Special Hazards of Combustion Products*: Toxic and irritating beryllium oxide and oxides of nitrogen may form in fires; *Behavior in Fire*: May increase the intensity of fire when in contact with combustible materials; *Ignition Temperature*: Not pertinent; *Electrical Hazard*: Not pertinent; *Burning Rate*: Not pertinent.

Chemical Reactivity - *Reactivity with Water*: Reacts to form weak solution of nitric acid, however the reaction is usually not considered hazardous; *Reactivity with Common Materials*: In presence of moisture will attack and damage wood and corrode most metals; *Stability During Transport*: Stable; *Neutralizing Agents for Acids and Caustics*: Not pertinent; *Polymerization*: Not pertinent; *Inhibitor of Polymerization*: Not pertinent.

BERYLLIUM OXIDE

Chemical Designations - *Synonyms*: Beryllia; Bromellite; *Chemical Formula*: BeO.

Observable Characteristics - *Physical State (as normally shipped)*: Solid; *Color*: White; *Odor*: None.

Physical and Chemical Properties - *Physical State at 15 °C and 1 atm.*: Solid; *Molecular Weight*: 25; *Boiling Point at 1 atm.*: Not pertinent; *Freezing Point*: Not pertinent; *Critical Temperature*: Not pertinent; *Critical Pressure*: Not pertinent; *Specific Gravity*: 3.0 at 20°C (solid); *Vapor (Gas) Density*: Not pertinent; *Ratio of Specific Heats of Vapor (Gas)*: Not pertinent; *Latent Heat of Vaporization*: Not pertinent; *Heat of Combustion*: Not pertinent; *Heat of Decomposition*: Not pertinent.

Health Hazards Information - *Recommended Personal Protective Equipment*: Respiratory protection; gloves; freshly laundered clothing; chemical safety goggles; *Symptoms Following Exposure*: Dramatic, unexplained weight loss may be considered as first indication of beryllium disease. Symptoms include anorexia, fatigue, weakness, malaise. Inhalation causes pneumonitis, nasopharyngitis, tracheobronchitis, dyspnea, chronic cough. Dust contact causes conjunctival inflammation of eyes and skin irritation; *General Treatment for Exposure*: INHALATION: take chest x-ray immediately to check the pneumonitis. INGESTION: induce vomiting; get medical attention. EYES: flush with water for at least 15 min.; get medical attention. SKIN: cuts or puncture wounds in which beryllium may be embedded under the skin should be thoroughly cleansed immediately by a physician; *Toxicity by Inhalation (Threshold Limit Value)*: 0.002 mg/m^3 (as beryllium); *Short-Term Exposure Limits*: 0.025 mg/m^3 less than 30 min.; *Toxicity by Ingestion*: No data; *Late Toxicity*: Beryllium disease may occur in lymph nodes, liver, spleen, kidney, and lungs; *Vapor (Gas) Irritant Characteristics*: No data; *Liquid or Solid Irritant Characteristics*: No data; *Odor Threshold*: No data.

Fire Hazards - *Flash Point:* Not flammable; *Flammable Limits in Air* (%): Not flammable; *Fire Extinguishing Agents*: Not pertinent; *Fire Extinguishing Agents Not To Be Used*: Not pertinent; *Special Hazards of Combustion Products*: Toxic beryllium oxide fume may form in fires; *Behavior in Fire*: Not pertinent; *Ignition Temperature*: Not flammable; *Electrical Hazard*: Not pertinent; *Burning Rate*: Not flammable.

Chemical Reactivity - *Reactivity with Water*: No reaction; *Reactivity with Common Materials*: No reactions; *Stability During Transport*: Stable; *Neutralizing Agents for Acids and Caustics*: Not pertinent; *Polymerization*: Not pertinent; *Inhibitor of Polymerization*: Not pertinent.

BERYLLIUM SULFATE

Chemical Designations - *Synonyms*: Beryllium Sulfate Tetrahydrate; *Chemical Formula*: BeSO$_4$•4H$_2$O.

Observable Characteristics - *Physical State (as normally shipped)*: Solid; *Color*: White; *Odor*: None.

Physical and Chemical Properties - *Physical State at 15 °C and 1 atm.*: Solid; *Molecular Weight*: 177.14; *Boiling Point at 1 atm.*: Not pertinent (decomposes); *Freezing Point*: Not pertinent; *Critical Temperature*: Not pertinent; *Critical Pressure*: Not pertinent; *Specific Gravity*: 1.71 at 11°C (solid); *Vapor (Gas) Density*: Not pertinent; *Ratio of Specific Heats of Vapor (Gas)*: Not pertinent; *Latent Heat of Vaporization*: Not pertinent; *Heat of Combustion*: Not pertinent; *Heat of Decomposition*: Not pertinent.

Health Hazards Information - *Recommended Personal Protective Equipment*: Respiratory protection;

gloves; freshly laundered clothing; chemical safety goggles; *Symptoms Following Exposure*: Any dramatic, unexplained weight loss should be considered as possible first indication of beryllium disease. Other symptoms include anorexia, fatigue, weakness, malaise. Inhalation causes pneumonitis, nasopharyngitis, tracheobronchitis, dyspnea, chronic cough. Contact with eyes causes conjunctival inflammation. Contact with skin causes dermatitis of primary irritant or sensitization type; causes ulcer formation when in contact wit cuts; *General Treatment for Exposure*: INHALATION: take chest x-ray immediately to check for evidence of pneumonitis. INGESTION: induce vomiting; get medical attention. EYES: flush with water for at least 15 min.; get medical attention. SKIN: cuts or puncture wounds in which beryllium may be embedded under the skin should be thoroughly cleansed immediately by a physician; *Toxicity by Inhalation (Threshold Limit Value)*: 0.002 mg/m^3 (as beryllium); *Short-Term Exposure Limits*: 0.025 mg/m^3 less than 30 min.; *Toxicity by Ingestion*: Grade 3; oral rat LD$_{50}$ 82 mg/kg; *Late Toxicity*: Beryllium disease may occur in the lymph nodes, liver, spleen, kidney, etc., as well as lung; *Vapor (Gas) Irritant Characteristics*: Data not available; *Liquid or Solid Irritant Characteristics*: Data not available; *Odor Threshold*: Data not available.

Fire Hazards - *Flash Point:* Not flammable; *Flammable Limits in Air (%)*: Not flammable; *Fire Extinguishing Agents*: Not pertinent; *Fire Extinguishing Agents Not To Be Used*: Not pertinent; *Special Hazards of Combustion Products*: Toxic beryllium oxide and sulfuric acid fumes may form in fire situations; *Behavior in Fire*: Not pertinent; *Ignition Temperature*: Not flammable; *Electrical Hazard*: Not pertinent; *Burning Rate*: Not flammable.

Chemical Reactivity - *Reactivity with Water*: No reaction; *Reactivity with Common Materials*: No reactions; *Stability During Transport*: Stable; *Neutralizing Agents for Acids and Caustics*: Not pertinent; *Polymerization*: Not pertinent; *Inhibitor of Polymerization*: Not pertinent.

BISMUTH OXYCHLORIDE

Chemical Designations - *Synonyms*: Basic Bismuth Chloride; Bismuth Chloride Oxide; Bismuth Subchloride; Bismuthyl Chloride; Pearl White; *Chemical Formula*: BiOCl.

Observable Characteristics - *Physical State (as normally shipped)*: Solid; *Color*: White; *Odor*: None.

Physical and Chemical Properties - *Physical State at 15 °C and 1 atm.*: Solid; *Molecular Weight*: 260.4; *Boiling Point at 1 atm.*: Not pertinent (decomposes); *Freezing Point*: Not pertinent; *Critical Temperature*: Not pertinent; *Critical Pressure*: Not pertinent; *Specific Gravity*: 7.7 at 20°C (solid); *Vapor (Gas) Density*: Not pertinent; *Ratio of Specific Heats of Vapor (Gas)*: Not pertinent; *Latent Heat of Vaporization*: Not pertinent; *Heat of Combustion*: Not pertinent; *Heat of Decomposition*: Not pertinent.

Health Hazards Information - *Recommended Personal Protective Equipment*: Goggles or face shield; protective gloves; dust mask; *Symptoms Following Exposure*: Contact with yes causes mild eye irritation and can cause skin rashes; *General Treatment for Exposure*: EYES: flush with water; *Toxicity by Inhalation (Threshold Limit Value)*: Data not available; *Short-Term Exposure Limits*: Data not available; *Toxicity by Ingestion*: Grade 0; LD$_{50}$ > 21.5 g/kg (rat); *Late Toxicity*:; *Vapor (Gas) Irritant Characteristics*: Data not available; *Liquid or Solid Irritant Characteristics*: Data not available; *Odor Threshold*: Data not available.

Fire Hazards - *Flash Point:* Not flammable; *Flammable Limits in Air (%)*: Not flammable; *Fire Extinguishing Agents*: Not pertinent; *Fire Extinguishing Agents Not To Be Used*: Not pertinent; *Special Hazards of Combustion Products*: Irritating hydrogen chloride gas may form in fires; *Behavior in Fire*: Not pertinent; *Ignition Temperature*: Not flammable; *Burning Rate*: Not flammable.

Chemical Reactivity - *Reactivity with Water*: No reaction; *Reactivity with Common Materials*: No reactions; *Stability During Transport*: Stable; *Neutralizing Agents for Acids and Caustics*: Not pertinent; *Polymerization*: Not pertinent; *Inhibitor of Polymerization*: Not pertinent.

BISPHENOL A

Chemical Designations - *Synonyms*: 2,2-Bis(4-Hydroxyphenyl)Propane; p,p'-Dihydroxy-diphenyldimethylmethane; 4,4'-Isopropylidenediphenol; Ucar Bisphenol HP; *Chemical Formula*: p-HOC$_6$H$_4$C(CH$_3$)$_2$C$_6$H$_4$OH-p.

Observable Characteristics - *Physical State (as normally shipped)*: Solid; *Color*: White to cream; *Odor*: Very weak phenolic.

Physical and Chemical Properties - *Physical State at 15 °C and 1 atm.*: Solid; *Molecular Weight*: 228.28; *Boiling Point at 1 atm*: Not pertinent; *Freezing Point*: 315, 157, 430; *Critical Temperature*: Not pertinent; *Critical Pressure*: Not pertinent; *Specific Gravity*: 1.195 at 25°C (solid); *Vapor (Gas) Density*: Not pertinent; *Ratio of Specific Heats of Vapor (Gas)*: Not pertinent; *Latent Heat of Vaporization*: Not pertinent; *Heat of Combustion*: Not pertinent; *Heat of Decomposition*: Not pertinent.

Health Hazards Information - *Recommended Personal Protective Equipment*: Approved dust mask and clean, body-covering clothing sufficient to prevent excessive or repeated exposure to dust, fumes, or solutions. Safety glasses with side shields; *Symptoms Following Exposure*: Dust irritating to upper respiratory passages; may cause sneezing; *General Treatment for Exposure*: SKIN: wash with soap and plenty of water. Avoid wearing contaminated clothing. EYES: promptly flush with plenty of water for at least 15 min. and get medical help. INGESTION: if large amounts are swallowed, induce vomiting promptly and get medical help promptly. No known antidote; *Toxicity by Inhalation (Threshold Limit Value)*: Not pertinent; *Short-Term Exposure Limits*: Not pertinent; *Toxicity by Ingestion*: Grade 2; LD_{50} 0.5 to 5 g/kg (rat); *Late Toxicity*: Lowered hemoglobin and erythrocyte (red blood cell) counts below normal in rats; *Vapor (Gas) Irritant Characteristics*: Not pertinent; *Liquid or Solid Irritant Characteristics*: Minimum hazard. If spilled on clothing and allowed to remain, may cause smarting and reddening of the skin; *Odor Threshold*: Not pertinent.

Fire Hazards - *Flash Point (deg. F)*: 415 OC; *Flammable Limits in Air (%)*: Not pertinent; *Fire Extinguishing Agents*: Foam, dry chemical, carbon dioxide; *Fire Extinguishing Agents Not To Be Used*: No data; *Special Hazards of Combustion Products*: Not pertinent; *Behavior in Fire*: Nor pertinent; *Ignition Temperature*: No data; *Electrical Hazard*: Not pertinent; *Burning Rate*: Not pertinent.

Chemical Reactivity - *Reactivity with Water*: No reaction; *Reactivity with Common Materials*: No reactions; *Stability During Transport*: Stable; *Neutralizing Agents for Acids and Caustics*: Not pertinent; *Polymerization*: Not pertinent; *Inhibitor of Polymerization*: Not pertinent.

BORIC ACID

Chemical Designations - *Synonyms*: Boracic Acid; Orthoboric Acid; *Chemical Formula*: H_3BO_3.

Observable Characteristics - *Physical State (as normally shipped)*: Solid; *Color*: White; *Odor*: None.

Physical and Chemical Properties - *Physical State at 15 and 1 atm.*: Solid; *Molecular Weight*: 61.83; *Boiling Point at 1 atm.*: Not pertinent (decomposes); *Freezing Point*: Not pertinent; *Critical Temperature*: Not pertinent; *Critical Pressure*: Not pertinent; *Specific Gravity*: 1.51 at 14°C (solid); *Vapor (Gas) Density*: Not pertinent; *Ratio of Specific Heats of Vapor (Gas)*: Not pertinent; *Latent Heat of Vaporization*: Not pertinent; *Heat of Combustion*: Not pertinent; *Heat of Decomposition*: Not pertinent.

Health Hazards Information - *Recommended Personal Protective Equipment*: Chemical goggles; *Symptoms Following Exposure*: Although no adverse effects have been reported from inhaling boric acid dust, it is absorbed through mucous membranes. Ingestion of 5 grams or more may irritate gastrointestinal tract and affect central nervous system. Contact with dust or aqueous solutions may irritate eyes; no chronic effects have been recognized, but continued contact should be avoided. Dust and solutions are absorbed through burns and open wounds but not through unbroken skin; *General Treatment for Exposure*: INHALATION: remove from contaminated atmosphere. INGESTION: obtain medical attention as soon as possible; if the patient is conscious, induce vomiting by giving warm salty water (2 tablespoons of table salt to a pint of water) or warm soapy water; if this measure is unsuccessful, vomiting may be induced by ticking the back of the patient's throat with a finger; vomiting should be encouraged about three times or until the vomitus is clear; additional water may be given to wash out the stomach. EYES: immediately flush the eyes with large quantities of running water for a minimum of 15 min.; hold the eyelids apart during the irrigation to ensure flushing of the entire surface of the eye and lids with water; obtain medical attention as soon as possible; continue the irrigation for an additional 15 min. if the physician is not available. SKIN: immediately flush affected area with water; remove contaminated clothing under the shower; continue washing with water - do not attempt to neutralize with chemical agents; obtain medical attention unless burn is minor; *Toxicity by Inhalation (Threshold Limit Value)*: 10 mg/m^3 (as boric oxide); *Short-Term Exposure Limits*: Data not available; *Toxicity by Ingestion*: Grade 2; oral rat LD$_{50}$ 2.660 mg/kg; *Late Toxicity*: Data not available; *Vapor (Gas) Irritant Characteristics*: Data not available; *Liquid or Solid Irritant Characteristics*: Data not available; *Odor Threshold*: Odorless.

Fire Hazards - *Flash Point:* Not flammable; *Flammable Limits in Air (%)*: Not flammable; *Fire Extinguishing Agents*: Not pertinent; *Fire Extinguishing Agents Not To Be Used*: Not pertinent; *Special Hazards of Combustion Products*: Not pertinent; *Behavior in Fire*: Not pertinent; *Ignition Temperature*: Not flammable; *Electrical Hazard*: Not pertinent; *Burning Rate*: Not flammable.

Chemical Reactivity - *Reactivity with Water*: No reaction; *Reactivity with Common Materials*: No reactions; *Stability During Transport*: Stable; *Neutralizing Agents for Acids and Caustics*: Not pertinent; *Polymerization*: Not pertinent; *Inhibitor of Polymerization*: Not pertinent.

BORON TRICHLORIDE

Chemical Designations - *Synonyms*: Boron Chloride; *Chemical Formula*: Bcl$_3$.

Observable Characteristics - *Physical State (as normally shipped)*: Liquid; *Color*: Colorless; *Odor*: Acrid and irritating.

Physical and Chemical Properties - *Physical State at 15 °C and 1 atm.*: Gas; *Molecular Weight*: 117.2; *Boiling Point at 1 atm.*: 54.3, 12.4, 285.6; *Freezing Point*: -161, -107, 166; *Critical Temperature*: 352, 178, 451; *Critical Pressure*: 566, 38.5, 3.90; *Specific Gravity*: 1.35 at 11°C (liquid); *Vapor (Gas) Density*: 4; *Ratio of Specific Heats of Vapor (Gas)*: 1.1470; *Latent Heat of Vaporization*: 68.8, 38.2, 1.60; *Heat of Combustion*: Not pertinent; *Heat of Decomposition*: Not pertinent.

Health Hazards Information - *Recommended Personal Protective Equipment*: Chemical goggles; rubber protective clothing and gloves; self-contained breathing apparatus; *Symptoms Following Exposure*: Inhalation causes edema and severe irritation of the upper respiratory system. Contact with liquid causes acid burns of eyes and severe burns of skin. Ingestion causes severe burns of mouth and stomach; *General Treatment for Exposure*: INHALATION: remove to fresh air; give oxygen or apply artificial respiration; keep warm; call a doctor at once; observe for pulmonary edema. EYES: wash with plenty of water for 15 min.; consult an eye specialist. SKIN: wash off with plenty of water. INGESTION: do not induce vomiting; give large amount of water; *Toxicity by Inhalation (Threshold Limit Value)*: Data not available; *Short-Term Exposure Limits*: Data not available; *Toxicity by Ingestion*: Grade 2; LD$_{50}$ 0.5 to 5 g/kg; *Late Toxicity*: Data not available; *Vapor (Gas) Irritant Characteristics*: Vapors cause severe irritation of eyes and throat and cause eye or lung injury. They cannot be tolerated even at low concentrations; *Liquid or Solid Irritant Characteristics*: Severe skin irritant. Causes second- and third-degree burns on short contact and is very injurious to the eyes; *Odor Threshold*: Decomposes in moist air, releasing hydrochloric acid and decomposition products. Hydrochloric acid - 1 ppm.

Fire Hazards - *Flash Point:* Not flammable; *Flammable Limits in Air (%)*: Not flammable; *Fire Extinguishing Agents*: Not pertinent; *Fire Extinguishing Agents Not To Be Used*: Not pertinent; *Special Hazards of Combustion Products*: Not pertinent; *Behavior in Fire*: Toxic fumes of hydrogen chloride are generated upon contact with water used to fight adjacent fires; *Ignition Temperature*: Not pertinent; *Electrical Hazard*: Not pertinent; *Burning Rate*: Not pertinent.

Chemical Reactivity - *Reactivity with Water*: Reacts vigorously, liberating heat and forming hydrogen chloride fumes (hydrochloric acid) and boric acid; *Reactivity with Common Materials*: Vigorously attacks elastomers and various packaging materials. Viton, Tygon, silastic elastomers, natural rubber, some synthetic rubbers are not recommended for service. Avoid lead and graphite impregnated asbestos. In the presence of moisture this chemical will aggressively attack most metals; *Stability During Transport*: Stable; *Neutralizing Agents for Acids and Caustics*: Flush with water and rinse with sodium bicarbonate and lime solution; *Polymerization*: Not pertinent; *Inhibitor of Polymerization*: Not pertinent.

BROMINE

Chemical Designations - *Synonyms*: No common synonyms; *Chemical Formula*: Br_2

Observable Characteristics - *Physical State (as normally shipped)*: Liquid; *Color*: Dark red; red-brown; *Odor*: Sharp, harsh, penetrating.

Physical and Chemical Properties - *Physical State at 15 °C and 1 atm.*: Liquid; *Molecular Weight*: 159.81; *Boiling Point at 1 atm.*: 138, 58.8, 332; *Freezing Point*: 19, -7.2, 266; *Critical Temperature*: Not pertinent; *Critical Pressure*: Not pertinent; *Specific Gravity*: 3.12 at 20°C (liquid); *Vapor (Gas) Density*: 5.5 at 20°C; *Ratio of Specific Heats of Vapor (Gas)*: 1.3; *Latent Heat of Vaporization*: 80.6, 44.8, 1.88; *Heat of Combustion*: Not pertinent; *Heat of Decomposition*: Not pertinent.

Health Hazards Information - *Recommended Personal Protective Equipment*: Chemical safety goggles, face shield; self-contained air-line canister mask; rubber suit; *Symptoms Following Exposure*: SKIN: contact with liquid or vapor may cause acne and slow-healing ulcers. INHALATION: induces severe irritation of the respiratory passages and pulmonary edema. Probable lethal oral dose fro an adult is 1 ml. A brief exposure to 1000 ppm may be fatal; *General Treatment for Exposure*: SKIN AND EYES: wash well with water and sodium bicarbonate solution. RESPIRATORY SYSTEM: if there is obstruction to breathing establish airway by pulling tongue forward, inserting an airway tube, or doing a tracheotomy; begin artificial respiration; if difficulty in breathing is a result of pulmonary edema, treatment should be carried out with the patient in the sitting position. Administration of oxygen is most important; INGESTION: do not induce vomiting. Have victim drink water and milk; *Toxicity by Inhalation (Threshold Limit Value)*: 0.1 ppm; *Short-Term Exposure Limits*: 0.4 ppm for 30 min.; *Toxicity by Ingestion*: Not pertinent; *Late Toxicity*: None; *Vapor (Gas) Irritant Characteristics*: Causes severe eye or throat irritations which cause eye or lung injury; cannot be tolerated even at low concentrations; *Liquid or Solid Irritant Characteristics*: Severe skin irritant. Causes second- and third-degree burns on short contact; very injurious to the eyes; *Odor Threshold*: 3.5 ppm.

Fire Hazards - *Flash Point:* Not flammable; *Flammable Limits in Air (%)*: Not flammable; *Fire Extinguishing Agents*: Use water spray to cool exposed containers and to wash spill away from a safe distance; *Fire Extinguishing Agents Not To Be Used*: Not pertinent; *Special Hazards of Combustion Products*: Toxic and irritating gases are formed when heated or in fires; *Behavior in Fire*: Not pertinent; *Ignition Temperature*: Not flammable; *Electrical Hazard*: Not pertinent; *Burning Rate*: Not flammable.

Chemical Reactivity - *Reactivity with Water*: No reaction; *Reactivity with Common Materials*: Reacts violently with aluminum. May cause fire on contact with common materials such as wood, cotton, straw. Iron, steel, stainless steel, and copper are corroded by bromine and will undergo severe corrosion when in contact with wet bromine. Plastics are also degraded/ attacked by bromine except for highly fluorinated plastics which resist attack; *Stability During Transport*: Stable; *Neutralizing Agents for Acids and Caustics*: Not pertinent; *Polymerization*: Not pertinent; *Inhibitor of Polymerization*: Not pertinent.

BROMINE TRIFLUORIDE

Chemical Designations - *Synonyms*: No common synonyms; *Chemical Formula*: BrF_3.

Observable Characteristics - *Physical State (as normally shipped)*: Liquid; *Color*: Colorless to gray-yellow; *Odor*: Extremely irritating.

Physical and Chemical Properties - *Physical State at 15 °C and 1 atm.*: Liquid; *Molecular Weight*: 136.9; *Boiling Point at 1 atm.*: 258.4, 125.8, 399; *Freezing Point*: 47.8, 8.8, 282; *Critical Temperature*: 621, 327, 600; *Critical Pressure*: Not pertinent; *Specific Gravity*: 2.81 at 20°C (liquid); *Vapor (Gas) Density*: 4.7; *Ratio of Specific Heats of Vapor (Gas)*: 1.1428; *Latent Heat of Vaporization*: 130, 74, 3.1; *Heat of Combustion*: Not pertinent; *Heat of Decomposition*: Not pertinent.

Health Hazards Information - *Recommended Personal Protective Equipment*: Self-contained breathing apparatus; complete protective clothing; safety glasses; face shield; *Symptoms Following Exposure*: Inhalation causes severe irritation of upper respiratory system. Contact with liquid or vapor causes severe burns of eyes and can cause ulcers and blindness. Contact with skin causes severe burns. Ingestion causes severe burns of mucous membranes; *General Treatment for Exposure*: Get immediate medical attention for all exposures. INHALATION: remove from exposure; support respiration. EYES: irrigate with copious amounts of water for at least 15 min. SKIN: wash with large amounts of water for at least 15 min., then rinse with sodium bicarbonate or lime solution; *Toxicity by Inhalation (Threshold Limit Value)*: 0.1 ppm (suggested); *Short-Term Exposure Limits*: 50 ppm/30 min.; 100 ppm/3 min.; *Toxicity by Ingestion*: Data not available; *Late Toxicity*: Data not available; *Vapor (Gas) Irritant Characteristics*: Data not available; *Liquid or Solid Irritant Characteristics*: Data not available; *Odor Threshold*: Data not available.

Fire Hazards - *Flash Point:* Not flammable but can cause fire on contact with combustibles; *Flammable Limits in Air* (%): Not flammable; *Fire Extinguishing Agents*: Dry chemical, carbon dioxide; *Fire Extinguishing Agents Not To Be Used*: Water, foam; *Special Hazards of Combustion Products*: No data; *Behavior in Fire*: Forms highly toxic and irritating fumes; *Ignition Temperature*: Not pertinent; *Electrical Hazard*: Not pertinent; *Burning Rate*: Not pertinent.

Chemical Reactivity - *Reactivity with Water*: Reacts vigorously generating toxic hydrogen fluoride gas (hydrofluoric acid); *Reactivity with Common Materials*: Causes severe corrosion of common metals and glass. May cause fire when in contact with organic materials such as wood, cotton or straw; *Stability During Transport*: Stable; *Neutralizing Agents for Acids and Caustics*: Flush with water and rinse with sodium bicarbonate or lime solution; *Polymerization*: Not pertinent; *Inhibitor of Polymerization*: Not pertinent.

BROMOBENZENE

Chemical Designations - *Synonyms*: Monobromobenzene; Phenyl Bromide; Bromobenzol; *Chemical Formula*: C_6H_5Br.

Observable Characteristics - *Physical State (as normally shipped)*: Liquid; *Color*: Colorless; *Odor*: Pleasant.

Physical and Chemical Properties - *Physical State at 15 °C and 1 atm.*: Liquid; *Molecular Weight*: 157; *Boiling Point at 1 atm.*: 313, 156, 429; *Freezing Point*: -23.1, -30.6, 242.6; *Critical Temperature*: 747, 397, 670; *Critical Pressure*: 655, 44.6, 4.52; *Specific Gravity*: 1.49 at 25°C (liquid); *Vapor (Gas) Density*: 5.4; *Ratio of Specific Heats of Vapor (Gas)*: 1.0931; *Latent Heat of Vaporization*: 104, 58, 2.4; *Heat of Combustion*: -8,510, -4,730, -198; *Heat of Decomposition*: Not pertinent.

Health Hazards Information - *Recommended Personal Protective Equipment*: goggles or face shield; rubber gloves and apron; *Symptoms Following Exposure*: Contact with liquid causes irritation of eyes and mild irritation of skin. Ingestion causes mild irritation of mouth and stomach; *General Treatment for Exposure*: EYES: flush with water for at least 15 min. SKIN: wipe off, wash with soap and water. INGESTION: induce vomiting; consult a doctor; *Toxicity by Inhalation (Threshold Limit Value)*: Data not available; *Short-Term Exposure Limits*: Data not available; *Toxicity by Ingestion*: Data not available; *Late Toxicity*: Data not available; *Vapor (Gas) Irritant Characteristics*: Data not available; *Liquid or Solid Irritant Characteristics*: Data not available; *Odor Threshold*: Data not available.

Fire Hazards - *Flash Point (deg. F):* 124 CC; *Flammable Limits in Air* (%): Not pertinent; *Fire Extinguishing Agents*: Water, dry chemical, foam, carbon dioxide; *Fire Extinguishing Agents Not To Be Used*: Not pertinent; *Special Hazards of Combustion Products*: Irritating hydrogen bromide and other gases form in a fire situation; *Behavior in Fire*: Not pertinent; *Ignition Temperature (deg. F)*: 1,049; *Electrical Hazard*: No data; *Burning Rate*: 3.8 mm/min.

Chemical Reactivity - *Reactivity with Water*: No reaction; *Reactivity with Common Materials*: No reactions; *Stability During Transport*: Stable; *Neutralizing Agents for Acids and Caustics*: Not pertinent; *Polymerization*: Not pertinent; *Inhibitor of Polymerization*: Not pertinent.

BUTADIANE, INHIBITED

Chemical Designations - *Synonyms*: Biethylene; 1,3-Butadiene; Bivinyl; Divinyl; Vinylethylene; *Chemical Formula*: $CH_2=CHCH=CH_2$.

Observable Characteristics - *Physical State (as normally shipped)*: Gas; *Color*: Colorless; *Odor*: Mildly aromatic.

Physical and Chemical Properties - *Physical State at 15 °C and 1 atm.*: Gas; *Molecular Weight*:

54.09; *Boiling Point at 1 atm.*: 24.1, -4.4, 268.8; *Freezing Point*: -164, -108.9, 164.3; *Critical Temperature*: 306, 152, 425; *Critical Pressure*: 628, 42.7, 4.32; *Specific Gravity*: 0.621 at 20°C (liquid); *Vapor (Gas) Density*: 1.9 at 20°C; *Ratio of Specific Heats of Vapor (Gas)*: 1.1; *Latent Heat of Vaporization*: 180, 100, 4.19; *Heat of Combustion*: -19,008, -10,560, -442.13; *Heat of Decomposition*: Not pertinent.

Health Hazards Information - *Recommended Personal Protective Equipment*: Chemical-type safety goggles; rescue harness and life line for those entering a tank or enclosed storage space; hose mask with hose inlet in a vapor-free atmosphere; self-contained breathing apparatus; rubber suit; *Symptoms Following Exposure*: Slight anesthetic effect at high concentrations; causes "frostbite" from skin contact; slight irritation to eyes and nose in high concentrations; *General Treatment for Exposure*: remove from exposure immediately. Call a physician. INHALATION: if breathing is irregular or stopped, start resuscitation, administer oxygen. SKIN: remove contaminated clothing and wash affected skin area. EYES: irrigate with water for at 15 min.; *Toxicity by Inhalation (Threshold Limit Value)*: 1,000 ppm; *Short-Term Exposure Limits*: Data not available; *Toxicity by Ingestion*: Data not available; *Late Toxicity*: None; *Vapor (Gas) Irritant Characteristics*: Vapors cause a slight smarting of the eyes or respiratory system if present in high concentrations. The effect is temporary; *Liquid or Solid Irritant Characteristics*: Minimum hazard. If spilled on clothing and allowed to remain, may cause smarting and reddening of the skin because of frostbite; *Odor Threshold*: 4 mg/m^3.

Fire Hazards - *Flash Point (deg. F):* -105 (est.); *Flammable Limits in Air (%)*: 2.0 - 11.5; *Fire Extinguishing Agents*: Stop flow of gas; *Fire Extinguishing Agents Not To Be Used*: Not pertinent; *Special Hazards of Combustion Products*: Not pertinent; *Behavior in Fire*: Vapor is heavier than air and can travel distances to ignition source and flash back. Containers may explode in a fire due to polymerization; *Ignition Temperature (deg. F)*: 788; *Electrical Hazard*: Class I, Group B; *Burning Rate*: 8.0 mm/min.

Chemical Reactivity - *Reactivity with Water*: No reaction; *Reactivity with Common Materials*: No reactions; *Stability During Transport*: Explosive decomposition when contaminated with peroxides formed by reaction with air; *Neutralizing Agents for Acids and Caustics*: Not pertinent; *Polymerization*: Polymerization inhibited when stabilizer is used; *Inhibitor of Polymerization*: tert-Butylcatehol (0.01 - 0.02%).

BUTANE

Chemical Designations - *Synonyms*: n-Butane; *Chemical Formula*: n-C_4H_{10}.

Observable Characteristics - *Physical State (as normally shipped)*: Compressed gas; *Color*: Colorless; *Odor*: Like gasoline.

Physical and Chemical Properties - *Physical State at 15 °C and 1 atm.*: Gas; *Molecular Weight*: 58.12; *Boiling Point at 1 atm.*: 31.1, -0.48, 272.72; *Freezing Point*: -216, -138, 135; *Critical Temperature*: 306, 152, 425; *Critical Pressure*: 550.8, 37.47, 3.796; *Specific Gravity*: 0.60 at 0°C (liquid); *Vapor (Gas) Density*: 20 at 20°C; *Ratio of Specific Heats of Vapor (Gas)*: 1.092; *Latent Heat of Vaporization*: 170, 92, 3.9; *Heat of Combustion*: -19,512, -10,840, -453.85; *Heat of Decomposition*: Not pertinent.

Health Hazards Information - *Recommended Personal Protective Equipment*: Self-contained breathing apparatus and safety goggles; *Symptoms Following Exposure*: high exposure produces drowsiness but no other evidence of systemic effect; *General Treatment for Exposure*: ORAL AND ASPIRATION: no treatment required. INHALATION: guard against self-injury if stuporous, confused, or anesthetized. Apply artificial respiration if not breathing. Avoid administration of epinephrine or other sympathomimetic amines. Prevent aspirations of vomitus by proper positioning of the head. Give symptomatic and supportive treatment; *Toxicity by Inhalation (Threshold Limit Value)*: 500 ppm; *Short-Term Exposure Limits*: Data not available; *Toxicity by Ingestion*: Not pertinent; *Late Toxicity*: None; *Vapor (Gas) Irritant Characteristics*: None; *Liquid or Solid Irritant Characteristics*: No appreciable hazard. Practically harmless to the skin because it is very volatile and evaporates quickly from the skin. Some frostbite possible; *Odor Threshold*: 6.16 ppm.

Fire Hazards - *Flash Point (deg. F):* -100 (est.); *Flammable Limits in Air (%)*: 1.8 - 8.4; *Fire

Extinguishing Agents: Stop flow of gas; *Fire Extinguishing Agents Not To Be Used*: Not pertinent; *Special Hazards of Combustion Products*: Not pertinent; *Behavior in Fire*: Not pertinent; *Ignition Temperature (deg. F)*: 807; *Electrical Hazard*: Class I, Group D; *Burning Rate*: 7.9 mm/min.
Chemical Reactivity - *Reactivity with Water*: No reaction; *Reactivity with Common Materials*: No reactions; *Stability During Transport*: Stable; *Neutralizing Agents for Acids and Caustics*: Not pertinent; *Polymerization*: Not pertinent; *Inhibitor of Polymerization*: Not pertinent.

N-BUTYL ACETATE
Chemical Designations - *Synonyms*: Acetic Acid, Butyl Ester; Butyl Acetate; Butyl Ethanoate; *Chemical Formula*: $CH_3COO(CH_2)_3CH_3$.
Observable Characteristics - *Physical State (as normally shipped)*: Liquid; *Color*: Colorless; *Odor*: Characteristic; agreeable fruity (in low concentrations); non residual.
Physical and Chemical Properties - *Physical State at 15 °C and 1 atm.*: Liquid; *Molecular Weight*: 116.16; *Boiling Point at 1 atm.*: 259, 126, 399; *Freezing Point*: -100, -73.5, 199.7; *Critical Temperature*: 582.6, 305.9, 579.1; *Critical Pressure*: 455, 31, 3.1; *Specific Gravity*: 0.875 at 20°C (liquid); *Vapor (Gas) Density*: Not pertinent; *Ratio of Specific Heats of Vapor (Gas)*: 1.058; *Latent Heat of Vaporization*: 133, 73.9, 3.09; *Heat of Combustion*: -13,130, -7294, -305.4; *Heat of Decomposition*: Not pertinent.
Health Hazards Information - *Recommended Personal Protective Equipment*: All-purpose canister mask, chemical safety goggles, rubber gloves; *Symptoms Following Exposure*: SKIN: prolonged or frequently repeated exposures may lead to drying. INHALATION: headaches, dizziness, nausea, irritation of respiratory passages and eyes; *General Treatment for Exposure*: EYES: in cause of contact, flush with water for at least 15 min. INHALATION: remove from exposure immediately. Call a physician. If breathing is irregular or stopped, start resuscitation, administer oxygen. INGESTION: induce vomiting and call a physician; *Toxicity by Inhalation (Threshold Limit Value)*: 150-200 ppm; *Short-Term Exposure Limits*: 300 ppm for 30 min.; *Toxicity by Ingestion*: Grade 2; LD_{50} 0.5 to 5 g/kg; *Late Toxicity*: None; *Vapor (Gas) Irritant Characteristics*: Vapors cause a slight smarting of the eyes or respiratory system if present in high concentrations. The effect is temporary; *Liquid or Solid Irritant Characteristics*: Minimum hazard. If spilled on clothing and allowed to remain, may cause smarting and reddening of the skin; *Odor Threshold*: 10 ppm.
Fire Hazards - *Flash Point (deg. F)*: 99 OC, 75 CC; *Flammable Limits in Air (%)*: 1.7 - 7.6; *Fire Extinguishing Agents*: Foam, dry chemical, carbon dioxide; *Fire Extinguishing Agents Not To Be Used*: Water in straight hose stream will scatter and spread fire and should be avoided; *Special Hazards of Combustion Products*: Not pertinent; *Behavior in Fire*: Not pertinent; *Ignition Temperature (deg. F)*: 760; *Electrical Hazard*: Class I, Group D; *Burning Rate*: 4.4 mm/min.
Chemical Reactivity - *Reactivity with Water*: No reaction; *Reactivity with Common Materials*: No reactions; *Stability During Transport*: Stable; *Neutralizing Agents for Acids and Caustics*: Not pertinent; *Polymerization*: Not pertinent; *Inhibitor of Polymerization*: Not pertinent.

SEC-BUTYL ACETATE
Chemical Designations - *Synonyms*: Acetic Acid, sec-Butyl Ester; *Chemical Formula*: $CH_3COOCH(CH_3)CH_2CH_3$.
Observable Characteristics - *Physical State (as normally shipped)*: Liquid; *Color*: Colorless; *Odor*: Pleasant.
Physical and Chemical Properties - *Physical State at 15 °C and 1 atm.*: Liquid; *Molecular Weight*: 116.16; *Boiling Point at 1 atm.*: 234, 112, 385; *Freezing Point*: -100, -73.5, 199.7; *Critical Temperature*: 550, 288, 561; *Critical Pressure*: 469, 32, 3.2; *Specific Gravity*: 0.872 at 20°C (liquid); *Vapor (Gas) Density*: Not pertinent; *Ratio of Specific Heats of Vapor (Gas)*: 1.061; *Latent Heat of Vaporization*: 130, 74, 3.1; *Heat of Combustion*: -13,100, -7300, -305; *Heat of Decomposition*: Not pertinent.
Health Hazards Information - *Recommended Personal Protective Equipment*: Organic vapor canister or air-supplied mask; chemical goggles or face splash shield; *Symptoms Following Exposure*:

Headaches, dizziness, nausea, irritation of respiratory passage and eyes; *General Treatment for Exposure*: INHALATION: if victim is overcome by vapors, remove from exposure immediately; call a physician; if breathing is irregular or stopped, start resuscitation and administer oxygen. EYES: flush with water for at least 15 min.; *Toxicity by Inhalation (Threshold Limit Value)*: 200 ppm; *Short-Term Exposure Limits*: Data not available; *Toxicity by Ingestion*: Data not available; *Late Toxicity*: None; *Vapor (Gas) Irritant Characteristics*: Vapors cause a slight smarting of the eyes or respiratory system if present in high concentrations. The effect is temporary; *Liquid or Solid Irritant Characteristics*: Minimum hazard. If spilled on clothing and allowed to remain, may cause smarting and reddening of the skin; *Odor Threshold*: Data not available.

Fire Hazards - *Flash Point (deg. F):* 62 CC, 88 OC; *Flammable Limits in Air (%)*: 1.7 - 9.8; *Fire Extinguishing Agents*: Foam, carbon dioxide or dry chemical; *Fire Extinguishing Agents Not To Be Used*: Water may be ineffective; *Special Hazards of Combustion Products*: Not pertinent; *Behavior in Fire*: Not pertinent; *Ignition Temperature (deg. F)*: No data; *Electrical Hazard*: Not pertinent; *Burning Rate*: 4.4 mm/min.

Chemical Reactivity - *Reactivity with Water*: No reaction; *Reactivity with Common Materials*: No reactions; *Stability During Transport*: Stable; *Neutralizing Agents for Acids and Caustics*: Not pertinent; *Polymerization*: Not pertinent; *Inhibitor of Polymerization*: Not pertinent.

ISO-BUTYL ACRYLATE

Chemical Designations - *Synonyms*: Acrylic Acid, Isobutyl Ester; *Chemical Formula*: $CH_2=CHCOOCH_2CH(CH_3)_2$.

Observable Characteristics - *Physical State (as normally shipped)*: Liquid; *Color*: Colorless; *Odor*: Sharp, fragrant.

Physical and Chemical Properties - *Physical State at 15 °C and 1 atm.*: Liquid; *Molecular Weight*: 128.17; *Boiling Point at 1 atm.*: 280.2, 137.9, 411.1; *Freezing Point*: -78.0, -61.1, 212.1; *Critical Temperature*: 599, 315, 588; *Critical Pressure*: 440, 30, 3.0; *Specific Gravity*: 0.889 at 20°C (liquid); *Vapor (Gas) Density*: Not pertinent; *Ratio of Specific Heats of Vapor (Gas)*: 1.044; *Latent Heat of Vaporization*: 130, 71, 3.0; *Heat of Combustion*: -13,500, -7500, -314; *Heat of Decomposition*: Not pertinent.

Health Hazards Information - *Recommended Personal Protective Equipment*: Self-contained breathing apparatus; rubber gloves; chemical goggles; *Symptoms Following Exposure*: Moderate toxicity when swallowed. Contact with eyes causes minor irritation no worse than that caused by hand soap; *General Treatment for Exposure*: INHALATION: move victim to fresh air at once; give oxygen if breathing is difficult or artificial respiration if breathing has stopped; call a doctor. INGESTION: make victim vomit by sticking a finger down the throat or by giving strong, warm water to drink; get medical attention. SKIN AND EYES: remove chemical by flushing with plenty of clean, running water; remove contaminated clothing and wash exposed skin with soap and water; *Toxicity by Inhalation (Threshold Limit Value)*: Data not available; *Short-Term Exposure Limits*: Data not available; *Toxicity by Ingestion*: Data not available; *Late Toxicity*: Not pertinent; *Vapor (Gas) Irritant Characteristics*: Vapors cause a slight smarting of the eyes or respiratory system if present in high concentrations. The effect is temporary; *Liquid or Solid Irritant Characteristics*: Minimum hazard. If spilled on clothing and allowed to remain, may cause smarting and reddening of the skin; *Odor Threshold*: Data not available.

Fire Hazards - *Flash Point (deg. F):* 94 OC; *Flammable Limits in Air (%)*: 1.9 - 8.0; *Fire Extinguishing Agents*: Dry chemical, foam or carbon dioxide; *Fire Extinguishing Agents Not To Be Used*: Not pertinent; *Special Hazards of Combustion Products*: Not pertinent; *Behavior in Fire*: Not pertinent; *Ignition Temperature (deg. F)*: 644; *Electrical Hazard*: Not pertinent; *Burning Rate*: 4.8 mm/min.

Chemical Reactivity - *Reactivity with Water*: No reaction; *Reactivity with Common Materials*: No reactions; *Stability During Transport*: Stable; *Neutralizing Agents for Acids and Caustics*: Not pertinent; *Polymerization*: Polymerizes upon exposure to heat; uncontrolled bulk polymerization can be explosive; *Inhibitor of Polymerization*: Methyl ether of hydroquinone: 10 - 100 ppm; Hydroquinone: 5 ppm.

N-BUTYL ACRYLATE
Chemical Designations - *Synonyms*: Acrylic Acid, Butyl Ester, Butyl Acrylate, Butyl 2-Propenoate; *Chemical Formula*: $CH_2=CHCOO(CH_2)_3CH_3$.
Observable Characteristics - *Physical State (as normally shipped)*: Liquid; *Color*: Colorless; *Odor*: Characteristic acrylic.
Physical and Chemical Properties - *Physical State at 15 °C and 1 atm.*: Liquid; *Molecular Weight*: 128.17; *Boiling Point at 1 atm.*: 299.8, 148.8, 422; *Freezing Point*: -83, -64, 209; *Critical Temperature*: 621, 327, 600; *Critical Pressure*: 426, 29, 2.9; *Specific Gravity*: 0.899 at 20°C (liquid); *Vapor (Gas) Density*: Not pertinent; *Ratio of Specific Heats of Vapor (Gas)*: 1.08; *Latent Heat of Vaporization*: 120, 66.4, 2.78; *Heat of Combustion*: -13,860, -7,700, -322.4; *Heat of Decomposition*: Not pertinent.
Health Hazards Information - *Recommended Personal Protective Equipment*: Self-contained breathing apparatus, rubber gloves, acid goggles; *Symptoms Following Exposure*: Vapor is irritating when breathed at high concentrations. Contact with liquid causes irritation of skin and burning of eyes; *General Treatment for Exposure*: INHALATION: remove to fresh air; administer artificial respiration or oxygen if indicated; call a physician. SKIN AND EYES: wash with plenty of water; *Toxicity by Inhalation (Threshold Limit Value)*: Data not available; *Short-Term Exposure Limits*: LD_{50} 100 ppm, 4 hr; *Toxicity by Ingestion*: Grade 2; 0.5 to 5 g/kg (rat); *Late Toxicity*: Data not available; *Vapor (Gas) Irritant Characteristics*: Vapors cause a slight smarting of the eyes or respiratory system if present in high concentrations. The effect is temporary; *Liquid or Solid Irritant Characteristics*: Minimum hazard. If spilled on clothing and allowed to remain, may cause smarting and reddening of the skin; *Odor Threshold*: Data not available.
Fire Hazards - *Flash Point (deg. F)*: 118 OC; *Flammable Limits in Air (%)*: 1.4 - 9.4; *Fire Extinguishing Agents*: Dry chemical, foam or carbon dioxide; *Fire Extinguishing Agents Not To Be Used*: Not pertinent; *Special Hazards of Combustion Products*: Not pertinent; *Behavior in Fire*: Not pertinent; *Ignition Temperature (deg. F)*: 534; *Electrical Hazard*: Not pertinent; *Burning Rate*: 4.7 mm/min.
Chemical Reactivity - *Reactivity with Water*: No reaction; *Reactivity with Common Materials*: No reactions; *Stability During Transport*: Stable; *Neutralizing Agents for Acids and Caustics*: Not pertinent; *Polymerization*: Polymerizes upon exposure to heat; uncontrolled bulk polymerization can be explosive; *Inhibitor of Polymerization*: Methyl ether of hydroquinone: 15 - 100 ppm. Store in contact with air.

N-BUTYL ALCOHOL
Chemical Designations - *Synonyms*: Butanol; Butyl Alcohol; 1-Butanol; 1-Hydroxybutane; -Propylcarbinol; *Chemical Formula*: $CH_3(CH_2)_2CH_2OH$.
Observable Characteristics - *Physical State (as normally shipped)*: Liquid; *Color*: Colorless; *Odor*: Alcohol-like; pungent; strong; characteristic; mildly alcoholic, non residual.
Physical and Chemical Properties - *Physical State at 15 °C and 1 atm.*: Liquid; *Molecular Weight*: 74.12; *Boiling Point at 1 atm.*: 243.9, 117.7, 390.9; *Freezing Point*: -129, -89.3, 183.9; *Critical Temperature*: 553.6, 289.8, 563; *Critical Pressure*: 640.2, 43.55, 4.412; *Specific Gravity*: 0.810 at 20°C (liquid); *Vapor (Gas) Density*: Not pertinent; *Ratio of Specific Heats of Vapor (Gas)*: 1.083; *Latent Heat of Vaporization*: 256, 142, 5.95; *Heat of Combustion*: -14,230, -7,906, -331.0; *Heat of Decomposition*: Not pertinent.
Health Hazards Information - *Recommended Personal Protective Equipment*: Organic vapor canister or air-supplied mask; chemical goggles or face splash shield; *Symptoms Following Exposure*: Anesthesia, nausea, headache, dizziness, irritation of respiratory passages. Mildly irritating to the skin and eyes; *General Treatment for Exposure*: INHALATION: remove from exposure immediately; call a physician; if breathing is irregular or has stopped, start resuscitation and administer oxygen. INGESTION: induce vomiting and call a physician. EYES: flush with water for at least 15 min.; *Toxicity by Inhalation (Threshold Limit Value)*: 100 ppm; *Short-Term Exposure Limits*: 150 ppm for 30 min.; *Toxicity by Ingestion*: Grade 2; LD_{50} 0.5 to 5 g/kg (rat); *Late Toxicity*: None; *Vapor (Gas) Irritant Characteristics*: Vapors cause a slight smarting of the eyes or respiratory system if present in

high concentrations. The effect is temporary; *Liquid or Solid Irritant Characteristics*: Minimum hazard. If spilled on clothing and allowed to remain, may cause smarting and reddening of the skin; *Odor Threshold*: 2.5 ppm.

Fire Hazards - *Flash Point (deg. F)*: 84 CC, 97 OC; *Flammable Limits in Air (%)*: 1.4 - 11.2; *Fire Extinguishing Agents*: Carbon dioxide, dry chemicals; *Fire Extinguishing Agents Not To Be Used*: Not pertinent; *Special Hazards of Combustion Products*: Not pertinent; *Behavior in Fire*: Not pertinent; *Ignition Temperature (deg. F)*: 650; *Electrical Hazard*: Class I, Group D; *Burning Rate*: 3.2 mm/min.

Chemical Reactivity - *Reactivity with Water*: No reaction; *Reactivity with Common Materials*: No reactions; *Stability During Transport*: Stable; *Neutralizing Agents for Acids and Caustics*: Not pertinent; *Polymerization*: Not pertinent; *Inhibitor of Polymerization*: Not pertinent.

SEC-BUTYL ALCOHOL

Chemical Designations - *Synonyms*: 2-Butanol; Butylene Hydrate; 2-Hydroxybutane; Methylethylcarbinol; *Chemical Formula*: $CH_3CH_2CH(OH)CH_3$.

Observable Characteristics - *Physical State (as normally shipped)*: Liquid; *Color*: Colorless; *Odor*: Strong, pleasant.

Physical and Chemical Properties - *Physical State at 15 °C and 1 atm.*: Liquid; *Molecular Weight*: 74.12; *Boiling Point at 1 atm.*: 211, 99.5, 372.7; *Freezing Point*: -174.5, -114.7, 158.5; *Critical Temperature*: 505.0, 262.8, 536; *Critical Pressure*: 608.4, 41.39, 4.193; *Specific Gravity*: 0.807 at 20°C (liquid); *Vapor (Gas) Density*: Not pertinent; *Ratio of Specific Heats of Vapor (Gas)*: 1.080; *Latent Heat of Vaporization*: 243, 135, 5.65; *Heat of Combustion*: -15,500, -8,600, -360; *Heat of Decomposition*: Not pertinent.

Health Hazards Information - *Recommended Personal Protective Equipment*: Organic vapor canister or air-supplied mask; chemical goggles or face splash shield; *Symptoms Following Exposure*: Headache, dizziness, and respiratory irritation. Liquid is severely irritating to the eyes and may cause eyeburn; *General Treatment for Exposure*: INHALATION: remove from exposure immediately; calla physician; if breathing is irregular or has stopped, start resuscitation and administer oxygen. INGESTION: induce vomiting and call a physician. EYES: flush with water for at least 15 min.; *Toxicity by Inhalation (Threshold Limit Value)*: 150 ppm; *Short-Term Exposure Limits*: 200 ppm for 60 min.; *Toxicity by Ingestion*: Grade 1; 5 -15 g/kg (rat-single oral dose); *Late Toxicity*: None; *Vapor (Gas) Irritant Characteristics*: Vapors cause a slight smarting of the eyes or respiratory system if present in high concentrations. The effect is temporary; *Liquid or Solid Irritant Characteristics*: No appreciable hazard. Practically harmless to the skin; *Odor Threshold*: Data not available.

Fire Hazards - *Flash Point (deg. F)*: 75 CC; *Flammable Limits in Air (%)*: 1.7 - 9.0; *Fire Extinguishing Agents*: Carbon dioxide, dry chemical; *Fire Extinguishing Agents Not To Be Used*: Not pertinent; *Special Hazards of Combustion Products*: Not pertinent; *Behavior in Fire*: Not pertinent; *Ignition Temperature (deg. F)*: 763; *Electrical Hazard*: Class I, Group D; *Burning Rate*: 3.1 mm/min.

Chemical Reactivity - *Reactivity with Water*: No reaction; *Reactivity with Common Materials*: No reactions; *Stability During Transport*: Stable; *Neutralizing Agents for Acids and Caustics*: Not pertinent; *Polymerization*: Not pertinent; *Inhibitor of Polymerization*: Not pertinent.

TERT-BUTYL ALCOHOL

Chemical Designations - *Synonyms*: 2-Methyl-2-Propanol; Trimethylcarbinol; *Chemical Formula*: $(CH_3)_3COH$.

Observable Characteristics - *Physical State (as normally shipped)*: Liquid. Sometimes freezes below 75°F; *Color*: Colorless; *Odor*: Characteristic; camphor-like; pungent.

Physical and Chemical Properties - *Physical State at 15 °C and 1 atm.*: Liquid; *Molecular Weight*: 74.12; *Boiling Point at 1 atm.*: 181, 82.6, 355.8; *Freezing Point*: 78.3, 25.7, 298.9; *Critical Temperature*: 451, 233, 506; *Critical Pressure*: 576, 39.2, 3.97; *Specific Gravity*: 0.78 at 26°C (liquid); *Vapor (Gas) Density*: 2.6; *Ratio of Specific Heats of Vapor (Gas)*: 1.080; *Latent Heat of Vaporization*: 234, 130, 5.44; *Heat of Combustion*: -14,000, -7,780, -325.7; *Heat of Decomposition*: Not pertinent.

Health Hazards Information - *Recommended Personal Protective Equipment*: Air pack or organic canister mask, rubber gloves, and goggles; *Symptoms Following Exposure*: Vapor is narcotic in action and irritating to respiratory passages. Liquid is irritating to skin and eyes; *General Treatment for Exposure*: INHALATION: remove victim from exposure and restore breathing. SKIN AND EYES: remove liquid from skin with water. Flush eyes with water; *Toxicity by Inhalation (Threshold Limit Value)*: 100 ppm; *Short-Term Exposure Limits*: 150 ppm for 30 min.; *Toxicity by Ingestion*: Grade 2; 0.5 to 5 g/kg (rat); *Late Toxicity*: Data not available; *Vapor (Gas) Irritant Characteristics*: Vapors cause a slight smarting of the eyes or respiratory system if present in high concentrations. The effect is temporary; *Liquid or Solid Irritant Characteristics*: No appreciable hazard. Practically harmless to the skin; *Odor Threshold*: Data not available.

Fire Hazards - *Flash Point (deg. F)*: 52 CC, 61 OC; *Flammable Limits in Air (%)*: 2.35 - 8.00; *Fire Extinguishing Agents*: Dry chemical, carbon dioxide; *Fire Extinguishing Agents Not To Be Used*: Not pertinent; *Special Hazards of Combustion Products*: Not pertinent; *Behavior in Fire*: Not pertinent; *Ignition Temperature (deg. F)*: 896; *Electrical Hazard*: Class I, Group D; *Burning Rate*: 3.4 mm/min.

Chemical Reactivity - *Reactivity with Water*: No reaction; *Reactivity with Common Materials*: No reactions; *Stability During Transport*: Stable; *Neutralizing Agents for Acids and Caustics*: Not pertinent; *Polymerization*: Not pertinent; *Inhibitor of Polymerization*: Not pertinent.

N-BUTYLAMINE

Chemical Designations - *Synonyms*: 1-Aminobutane; Butylamine; Mono-n-butylamine; Norvalamine; *Chemical Formula*: $CH_3(CH_2)_3NH_4$.

Observable Characteristics - *Physical State (as normally shipped)*: Liquid; *Color*: Colorless; *Odor*: Fish-like; ammonia-like.

Physical and Chemical Properties - *Physical State at 15 °C and 1 atm.*: Liquid; *Molecular Weight*: 73.14; *Boiling Point at 1 atm.*: 171.3, 77.4, 350.6; *Freezing Point*: -56, -49, 224; *Critical Temperature*: 484, 251, 524; *Critical Pressure*: 603, 41, 4.16; *Specific Gravity*: 0.741 at 20°C (liquid); *Vapor (Gas) Density*: 2.5; *Ratio of Specific Heats of Vapor (Gas)*: 1.071; *Latent Heat of Vaporization*: 180, 100, 4.2; *Heat of Combustion*: -17,595, -9,775, -409; *Heat of Decomposition*: Not pertinent.

Health Hazards Information - *Recommended Personal Protective Equipment*: Air-supplied mask; rubber gloves; coverall goggles; face shield; butyl rubber apron; *Symptoms Following Exposure*: Inhalation causes irritation, nausea, vomiting, headache, faintness, severe coughing and chest pains; can cause lung edema. Ingestion causes severe irritation of mouth and stomach. Contact with eyes causes severe irritation and edema of the cornea. Contact with skin causes burns; absorption through skin may cause nausea, vomiting and shock; *General Treatment for Exposure*: INHALATION: remove victim to fresh air; call a physician; give oxygen if breathing is difficult; if not breathing, give artificial respiration. INGESTION: give large amounts of water; get medical attention. EYES: flush with water for at least 15 min.; get medical care. SKIN: remove contaminated clothing; flush skin with plenty of water at least 15 min.; *Toxicity by Inhalation (Threshold Limit Value)*: 5 ppm; *Short-Term Exposure Limits*: 5 ppm, 5 min.; *Toxicity by Ingestion*: Grade 3; oral LD_{50} 500 mg/kg (rat); *Late Toxicity*: Data not available; *Vapor (Gas) Irritant Characteristics*: Data not available; *Liquid or Solid Irritant Characteristics*: Data not available; *Odor Threshold*: Data not available.

Fire Hazards - *Flash Point (deg. F)*: 30 OC, 10 CC; *Flammable Limits in Air (%)*: 1.7 - 9.8; *Fire Extinguishing Agents*: Alcohol foam, dry chemical, carbon dioxide; *Fire Extinguishing Agents Not To Be Used*: Water may be ineffective; *Special Hazards of Combustion Products*: Toxic oxides of nitrogen may form during fires; *Behavior in Fire*: Vapor is heavier than air and can travel distances to ignition source and flash back. Containers may explode; *Ignition Temperature (deg. F)*: 594; *Electrical Hazard*: No data; *Burning Rate*: 5.8 mm/min.

Chemical Reactivity - *Reactivity with Water*: No reaction; *Reactivity with Common Materials*: May corrode some metals in presence of water; *Stability During Transport*: Stable; *Neutralizing Agents for Acids and Caustics*: Flush with water; *Polymerization*: Not pertinent; *Inhibitor of Polymerization*: Not pertinent.

SEC-BUTYLAMINE

Chemical Designations - *Synonyms*: No common synonyms; *Chemical Formula*: $CH_3CH_2CH(CH_3)NH_2$

Observable Characteristics - *Physical State (as normally shipped)*: Liquid; *Color*: Colorless; *Odor*: Ammonia-like.

Physical and Chemical Properties - *Physical State at 15 °C and 1 atm.*: Liquid; *Molecular Weight*: 73.1; *Boiling Point at 1 atm.*: 145, 63, 336; *Freezing Point*: -155, -104, 169; *Critical Temperature*: Data not available; *Critical Pressure*: Data not available; *Specific Gravity*: 0.721 at 20°C (liquid); *Vapor (Gas) Density*: 2.52; *Ratio of Specific Heats of Vapor (Gas)*: 1.073 at 20°C; *Latent Heat of Vaporization*: 178.09, 98.94, 4.160; *Heat of Combustion*: -17,600, -9,780, -409; *Heat of Decomposition*: Not pertinent.

Health Hazards Information - *Recommended Personal Protective Equipment*: Chemical safety goggles; rubber gloves and apron; respiratory protective equipment; non-sparking shoes; *Symptoms Following Exposure*: Inhalation causes irritation of burns of the respiratory system; exposure to concentrated vapors and cause asphyxiation. Ingestion causes burns of mouth and stomach. Contact with eyes causes lachrymation, conjunctivitis, burns, corneal edema. Contact with skin causes irritation or burns, dermatitis; *General Treatment for Exposure*: INHALATION: remove patient from exposure; keep him quiet; contact physician. INGESTION: give large amount of water; induce vomiting; consult a physician. EYES: flush thoroughly with water for 15 min.; call physician immediately. SKIN: remove all contaminated clothing; flood affected area with large quantities of water; consult a physician; *Toxicity by Inhalation (Threshold Limit Value)*: Data not available; *Short-Term Exposure Limits*: Data not available; *Toxicity by Ingestion*: Grade 3; oral LD_{50} 380 mg/kg (rat); *Late Toxicity*: Data not available; *Vapor (Gas) Irritant Characteristics*: Vapors are moderately irritating such that personnel will not usually tolerate moderate or high concentrations; *Liquid or Solid Irritant Characteristics*: Severe skin irritant. Causes second- and third-degree burns on short contact and is very injurious to the eyes; *Odor Threshold*: Data not available.

Fire Hazards - *Flash Point (deg. F)*: 16 CC; *Flammable Limits in Air (%)*: No data; *Fire Extinguishing Agents*: Alcohol foam, dry chemical, carbon dioxide; *Fire Extinguishing Agents Not To Be Used*: Water may be ineffective; *Special Hazards of Combustion Products*: Toxic oxides of nitrogen may be formed; *Behavior in Fire*: Vapor is heavier than air and can travel distances to ignition source and flash back. Containers may explode; *Ignition Temperature (deg. F)*: 712; *Electrical Hazard*: No data; *Burning Rate*: 6.2 mm/min.

Chemical Reactivity - *Reactivity with Water*: No reaction; *Reactivity with Common Materials*: May corrode some metals in presence of water; *Stability During Transport*: Stable; *Neutralizing Agents for Acids and Caustics*: Flush with water; *Polymerization*: Not pertinent; *Inhibitor of Polymerization*: Not pertinent.

TERT-BUTYLAMINE

Chemical Designations - *Synonyms*: 2-Aminoisobutane; 2-Amino-2-Methylpropane; 1,1-Dimethylethylamine; TBA; Trimethylaminomethane; *Chemical Formula*: $(CH_3)_3CNH_2$

Observable Characteristics - *Physical State (as normally shipped)*: Liquid; *Color*: Colorless; *Odor*: Like ammonia.

Physical and Chemical Properties - *Physical State at 15 °C and 1 atm.*: Liquid; *Molecular Weight*: 73.14; *Boiling Point at 1 atm.*: 113, 45, 318; *Freezing Point*: Not pertinent; *Critical Temperature*: Data not available; *Critical Pressure*: Data not available; *Specific Gravity*: 0.696 at 20°C (liquid); *Vapor (Gas) Density*: 8.13; *Ratio of Specific Heats of Vapor (Gas)*: Data not available; *Latent Heat of Vaporization*: 167, 92.8, 3.88; *Heat of Combustion*: -17,600, -9,790, -410; *Heat of Decomposition*: Data not available.

Health Hazards Information - *Recommended Personal Protective Equipment*: Self-contained breathing apparatus; goggles or face shield; rubber gloves; *Symptoms Following Exposure*: Inhalation causes irritation of nose, mouth, and lung. Ingestion causes irritation of mouth and stomach. Contact with liquid causes severe irritation of eyes and moderate irritation of skin; *General Treatment for Exposure*:

INHALATION: move to fresh air; give artificial respiration if breathing has stopped. INGESTION: give large amounts of water and induce vomiting. EYES: immediately flush with water for at least 15 min.; get medical attention. SKIN: flush with water; wash with soap and water; *Toxicity by Inhalation (Threshold Limit Value)*: Data not available; *Short-Term Exposure Limits*: Data not available; *Toxicity by Ingestion*: Grade 3; oral LD_{50} 180 mg/kg (rat); *Late Toxicity*: Data not available; *Vapor (Gas) Irritant Characteristics*: Data not available; *Liquid or Solid Irritant Characteristics*: Data not available; *Odor Threshold*: Data not available.

Fire Hazards - *Flash Point (deg. F):* 16 CC; *Flammable Limits in Air (%)*: 1.7 - 8.9 (at 212 of); *Fire Extinguishing Agents*: Dry chemical, alcohol foam, carbon dioxide; *Fire Extinguishing Agents Not To Be Used*: Water may be ineffective; *Special Hazards of Combustion Products*: Toxic oxides of nitrogen may form; *Behavior in Fire*: No data; *Ignition Temperature (deg. F)*: 716; *Electrical Hazard*: No data; *Burning Rate*: 7 mm/min.

Chemical Reactivity - *Reactivity with Water*: No reaction; *Reactivity with Common Materials*: Liquid will attack some plastics; *Stability During Transport*: Stable; *Neutralizing Agents for Acids and Caustics*: Flush with water; *Polymerization*: Not pertinent; *Inhibitor of Polymerization*: Not pertinent.

BUTYLENE

Chemical Designations - *Synonyms*: 1-Butene; *Chemical Formula*: $CH_3CH_2CH{=}CH_2$.

Observable Characteristics - *Physical State (as normally shipped)*: Compressed gas; *Color*: Colorless; *Odor*: Sweetish.

Physical and Chemical Properties - *Physical State at 15 °C and 1 atm.*: Gas; *Molecular Weight*: 56.10; *Boiling Point at 1 atm.*: 20.7, -6.3, 266.9; *Freezing Point*: -297, -183, 90; *Critical Temperature*: 295.5, 146.4, 419.6; *Critical Pressure*: 584, 39.7, 4.02; *Specific Gravity*: 0.595 at 20°C (liquid); *Vapor (Gas) Density*: 1.9; *Ratio of Specific Heats of Vapor (Gas)*: 1.104; *Latent Heat of Vaporization*: 168, 93.4, 3.91; *Heat of Combustion*: -19,487, -10,826, -453.26; *Heat of Decomposition*: Not pertinent.

Health Hazards Information - *Recommended Personal Protective Equipment*: Chemical goggles, gloves, self-contained breathing apparatus or organic canister; *Symptoms Following Exposure*: May act as an asphyxiate or slight anesthetic at high vapor concentrations. Vapor concentrations are not usually a hazard at room temperature except in enclosed spaces; *General Treatment for Exposure*: INHALATION: remove victim to fresh air and supply resuscitation. Call a doctor. EYES AND SKIN: flush with water for at least 15 min.; *Toxicity by Inhalation (Threshold Limit Value)*: Data not available; *Short-Term Exposure Limits*: Data not available; *Toxicity by Ingestion*: Not pertinent; *Late Toxicity*: None; *Vapor (Gas) Irritant Characteristics*: Vapors are non-irritating to the eyes and throat; *Liquid or Solid Irritant Characteristics*: No appreciable hazard. Practically harmless to the skin because it is very volatile and evaporates quickly. May cause frostbite; *Odor Threshold*: Data not available.

Fire Hazards - *Flash Point:* Not pertinent; *Flammable Limits in Air (%)*: 1.6 - 10; *Fire Extinguishing Agents*: Stop flow of gas; *Fire Extinguishing Agents Not To Be Used*: Not pertinent; *Special Hazards of Combustion Products*: Not pertinent; *Behavior in Fire*: Containers may explode in fires. Vapor is heavier than air and may travel considerable distance to ignition source and flash back; *Ignition Temperature (deg. F)*: 725; *Electrical Hazard*: Not pertinent; *Burning Rate*: No data.

Chemical Reactivity - *Reactivity with Water*: No reaction; *Reactivity with Common Materials*: No reactions; *Stability During Transport*: Stable; *Neutralizing Agents for Acids and Caustics*: Not pertinent; *Polymerization*: Not pertinent; *Inhibitor of Polymerization*: Not pertinent.

BUTYLENE OXIDE

Chemical Designations - *Synonyms*: 1-Butene Oxide; 1,2-Butylene Oxide; *alpha*-Butylene Oxide; 1,2-Epoxybutane; *Chemical Formula*: $C_2H_5CHCH_2O$.

Observable Characteristics - *Physical State (as normally shipped)*: Liquid; *Color*: Colorless; *Odor*: Pungent.

Physical and Chemical Properties - *Physical State at 15 °C and 1 atm.*: Liquid; *Molecular Weight*: 72; *Boiling Point at 1 atm.*: 145 63, 336; *Freezing Point*: < -58, < -50, < 223; *Critical Temperature*: Data not available; *Critical Pressure*: Data not available; *Specific Gravity*: 0.826 at 25°C (liquid);

Vapor (Gas) Density: 2.49; *Ratio of Specific Heats of Vapor (Gas)*: Data not available; *Latent Heat of Vaporization*: 180, 100, 4.2; *Heat of Combustion*: -15,200, -8,470, -354; *Heat of Decomposition*: Not pertinent.

Health Hazards Information - *Recommended Personal Protective Equipment*: Clean protective clothing; rubber gloves; chemical worker's goggles; self-contained breathing apparatus; *Symptoms Following Exposure*: Inhalation: intolerable odor and irritation; respiratory injury may occur at higher levels. Ingestion causes irritation of mouth and stomach. Contact with either liquid or vapor may cause burns of eyes. Liquid produces frostbite-type of skin burn if free to evaporate; if confined to skin, burn may cause skin sensitization; not readily absorbed in toxic amounts; *General Treatment for Exposure*: INHALATION: if any ill effects occur, immediately remove person to fresh air and get medical help; if breathing stops, start artificial respiration. INGESTION: induce vomiting promptly and get medical help. EYES: promptly flush with plenty of water for at least 15 min. and get medical help. SKIN: promptly flush with plenty of water; remove all contaminated clothing and wash before reuse; *Toxicity by Inhalation (Threshold Limit Value)*: Data not available; *Short-Term Exposure Limits*: Data not available; *Toxicity by Ingestion*: Grade 2; oral LD_{50} 1,410 mg/kg (rat); *Late Toxicity*: Data not available; *Vapor (Gas) Irritant Characteristics*: Data not available; *Liquid or Solid Irritant Characteristics*: Data not available; *Odor Threshold*: Data not available.

Fire Hazards - *Flash Point (deg. F)*:20 OC; *Flammable Limits in Air (%)*: 1.5 - 18.3; *Fire Extinguishing Agents*: Dry chemical, alcohol foam, carbon dioxide; *Fire Extinguishing Agents Not To Be Used*: Water may be ineffective; *Special Hazards of Combustion Products*: No data; *Behavior in Fire*: Containers may explode in fires. Apply water to cool containers from a safe distance; *Ignition Temperature (deg. F)*: 959; *Electrical Hazard*: No data; *Burning Rate*: No data.

Chemical Reactivity - *Reactivity with Water*: No reaction; *Reactivity with Common Materials*: No data; *Stability During Transport*: Stable; *Neutralizing Agents for Acids and Caustics*: Not pertinent; *Polymerization*: May occur when the product is in contact with strong acids and bases; *Inhibitor of Polymerization*: No data.

N-BUTYL MERCAPTAN

Chemical Designations - *Synonyms*: 1-Butanethiol; Thiobutyl Alcohol; *Chemical Formula*: $CH_3CH_2CH_2CH_2SH$.

Observable Characteristics - *Physical State (as normally shipped)*: Liquid; *Color*: Colorless to yellow; *Odor*: Strong skunk-like.

Physical and Chemical Properties - *Physical State at 15 °C and 1 atm.*: Liquid; *Molecular Weight*: 90.2; *Boiling Point at 1 atm.*: 209.3, 98.5, 371.7; *Freezing Point*: -176.2, -115.7, 157.5; *Critical Temperature*: 554, 290, 563; *Critical Pressure*: 572, 38.9, 3.94; *Specific Gravity*: 0.841 at 20°C (liquid); *Vapor (Gas) Density*: 3.1; *Ratio of Specific Heats of Vapor (Gas)*: 1.0770 at 16°C; *Latent Heat of Vaporization*: 154.0, 85.58, 3.583; *Heat of Combustion*: -16,601, -9,223, -386; *Heat of Decomposition*: Not pertinent.

Health Hazards Information - *Recommended Personal Protective Equipment*: Plastic gloves, goggles; self-contained breathing apparatus; *Symptoms Following Exposure*: Inhalation causes loss of sense of smell; muscular weakness, convulsions, and respiratory paralysis may follow prolonged exposure. Contact of liquid with eyes or ski causes slight irritation. Ingestion causes nausea; *General Treatment for Exposure*: INHALATION: remove victim from contaminated atmosphere; give artificial respiration and oxygen if needed; observe for signs of pulmonary edema. EYES: wash with plenty of water; see a physician. SKIN: wash with soap and water. INGESTION: induce vomiting and follow with gastric lavage; *Toxicity by Inhalation (Threshold Limit Value)*: 0.5 ppm; *Short-Term Exposure Limits*: Data not available; *Toxicity by Ingestion*: Grade 2; oral LD_{50} 1,500 mg/kg (rat); *Late Toxicity*: Data not available; *Vapor (Gas) Irritant Characteristics*: Data not available; *Liquid or Solid Irritant Characteristics*: Data not available; *Odor Threshold*: 0.001 ppm.

Fire Hazards - *Flash Point (deg. F)*: 53 OC; *Flammable Limits in Air (%)*: No data; *Fire Extinguishing Agents*: Dry chemical, alcohol foam, carbon dioxide; *Fire Extinguishing Agents Not To Be Used*: Water; *Special Hazards of Combustion Products*: Irritating sulfur dioxide; *Behavior in Fire*: Vapors

are heavier than air and may travel considerable distance to ignition source and flash back; *Ignition Temperature*: No data; *Electrical Hazard*: No data; *Burning Rate*: 7.4 mm/min.
Chemical Reactivity - *Reactivity with Water*: No reaction; *Reactivity with Common Materials*: No reactions; *Stability During Transport*: Stable; *Neutralizing Agents for Acids and Caustics*: Not pertinent; *Polymerization*: Not pertinent; *Inhibitor of Polymerization*: Not pertinent.

N - BUTYL METHACRYLATE

Chemical Designations - *Synonyms*: Methacrylic Acid, Butyl Ester; Butyl Methacrylate; Butyl 2-Methacrylate; n-Butyl alpha-Methyl Acrylate; Butyl 2-Methyl-2-Propenoate; *Chemical Formula*: $CH_2 = C(CH_3)COOCH_2CH_2CH_2CH_3$.

Observable Characteristics - *Physical State (as normally shipped)*: Liquid; *Color*: Colorless; *Odor*: Moderate acrylate.

Physical and Chemical Properties - *Physical State at 15 °C and 1 atm.*: Liquid; *Molecular Weight*: 142.2; *Boiling Point at 1 atm.*: 325, 163, 436; *Freezing Point*: < 32, < 0, < 273; *Critical Temperature*: Not pertinent; *Critical Pressure*: Not pertinent; *Specific Gravity*: 0.8975 at 20°C; *Vapor (Gas) Density*: Not pertinent; *Ratio of Specific Heats of Vapor (Gas)*: Not pertinent; *Latent Heat of Vaporization*: Not pertinent; *Heat of Combustion*: -14,800, -8,230, -344; *Heat of Decomposition*: Not pertinent.

Health Hazards Information - *Recommended Personal Protective Equipment*: Self-contained breathing apparatus; impervious gloves; chemical splash goggles; *Symptoms Following Exposure*: Inhalation may cause nausea because of offensive odor. Contact with liquid causes irritation of eyes and mild irritation of skin. Ingestion causes irritation of mouth and stomach; *General Treatment for Exposure*: INHALATION: remove to fresh air; give oxygen or artificial respiration as required. EYES: flush with copious amounts of water for 15 min. and consult physician. SKIN: wash with soap and water. INGESTION: induce vomiting; call a physician; *Toxicity by Inhalation (Threshold Limit Value)*: Data not available; *Short-Term Exposure Limits*: Data not available; *Toxicity by Ingestion*: Grade 0; LD_{50} > 15 g/kg; *Late Toxicity*: Birth defects in rats (gross and skeletal abnormalities); *Vapor (Gas) Irritant Characteristics*: Data not available; *Liquid or Solid Irritant Characteristics*: Data not available; *Odor Threshold*: Data not available.

Fire Hazards - *Flash Point (deg. F)*: 150 OC; *Flammable Limits in Air* (%): 2 - 8; *Fire Extinguishing Agents*: Dry chemical, foam, carbon dioxide; *Fire Extinguishing Agents Not To Be Used*: Water may be ineffective; *Special Hazards of Combustion Products*: Not pertinent; *Behavior in Fire*: Containers may explode; *Ignition Temperature (deg. F)*: 562; *Electrical Hazard*: No data; *Burning Rate*: 4.8 mm/min.

Chemical Reactivity - *Reactivity with Water*: No reaction; *Reactivity with Common Materials*: No reactions; *Stability During Transport*: Stable; *Neutralizing Agents for Acids and Caustics*: Not pertinent; *Polymerization*: May occur upon exposure to heat; *Inhibitor of Polymerization*: 9 - 15 ppm monomethyl ether of hydroquinone; 90 - 120 ppm hydroquinone.

1,4-BUTYNEDIOL

Chemical Designations - *Synonyms*: 2-Butyne-1,4-Diol; 1,4-Dihydroxy-2-Butyne; *Chemical Formula*: $HOCH_2C \equiv CCH_2OH$.

Observable Characteristics - *Physical State (as normally shipped)*: Solid or 35% aqueous solution; *Color*: Solid: colorless to pale yellow; Solution: straw to amber color; *Odor*: Data not available.

Physical and Chemical Properties - *Physical State at 15 °C and 1 atm.*: Solid; *Molecular Weight*: 36.09; *Boiling Point at 1 atm.*: 460, 238, 511; *Freezing Point*: 140, 58, 331; *Critical Temperature*: Not pertinent; *Critical Pressure*: Not pertinent; *Specific Gravity*: 1.07 at 20°C (solid); *Vapor (Gas) Density*: Not pertinent; *Ratio of Specific Heats of Vapor (Gas)*: Not pertinent; *Latent Heat of Vaporization*: Not pertinent; *Heat of Combustion*: -11,020, -6,120, -256.2; *Heat of Decomposition*: Not pertinent.

Health Hazards Information - *Recommended Personal Protective Equipment*: Neoprene rubber gloves and safety goggles or face shield; *Symptoms Following Exposure*: May cause dermatitis; *General Treatment for Exposure*: SKIN: wash affected skin area thoroughly with water. EYES: immediately wash with water for at least 15 min. and get medical attention; *Toxicity by Inhalation (Threshold Limit*

Value): Not pertinent; *Short-Term Exposure Limits*: Not pertinent; *Toxicity by Ingestion*: Grade 3; LD_{50} 50 to 500 mg/kg; *Late Toxicity*: Data not available; *Vapor (Gas) Irritant Characteristics*: Not pertinent; *Liquid or Solid Irritant Characteristics*: Minimum hazard. If spilled on clothing and allowed to remain, may cause smarting and reddening of the skin; *Odor Threshold*: Not pertinent.

Fire Hazards - *Flash Point (deg. F):* 263 OC (pure butynediol); *Flammable Limits in Air (%)*: Not pertinent; *Fire Extinguishing Agents*: Water, alcohol foam, dry chemical or carbon dioxide; *Fire Extinguishing Agents Not To Be Used*: Not pertinent; *Special Hazards of Combustion Products*: Not pertinent; *Behavior in Fire*: Not pertinent; *Ignition Temperature (deg. F)*: No data; *Electrical Hazard*: Not pertinent; *Burning Rate*: No data.

Chemical Reactivity - *Reactivity with Water*: No reaction; *Reactivity with Common Materials*: No reactions; *Stability During Transport*: Stable; *Neutralizing Agents for Acids and Caustics*: Not pertinent; *Polymerization*: Not pertinent; *Inhibitor of Polymerization*: Not pertinent.

ISO-BUTYRALDEHYDE

Chemical Designations - *Synonyms*: Isobutyric Aldehyde; Isobutyraldehyde; Isobutylaldehyde; 2-Methylpropanal; *Chemical Formula*: $(CH_3)_2CHCHO$.

Observable Characteristics - *Physical State (as normally shipped)*: Liquid; *Color*: Colorless; *Odor*: Pungent.

Physical and Chemical Properties - *Physical State at 15 °C and 1 atm.*: Liquid; *Molecular Weight*: 72.11; *Boiling Point at 1 atm.*: 147, 64.1, 337.3; *Freezing Point*: -112, -80, 193; *Critical Temperature*: 464, 240, 513; *Critical Pressure*: 600, 41, 4.2; *Specific Gravity*: 0.791 at 20°C (liquid); *Vapor (Gas) Density*: 2.5; *Ratio of Specific Heats of Vapor (Gas)*: 1.093; *Latent Heat of Vaporization*: 180, 98, 4.1; *Heat of Combustion*: -13,850, -7,693, -322.1; *Heat of Decomposition*: Not pertinent.

Health Hazards Information - *Recommended Personal Protective Equipment*: Appropriate protective clothing, including rubber gloves, rubber shoes and protective eyewear; *Symptoms Following Exposure*: Vapor is irritating to the eyes and mucous membranes; *General Treatment for Exposure*: EYES: immediately flush with plenty of water for at least 15 min.; *Toxicity by Inhalation (Threshold Limit Value)*: Data not available; *Short-Term Exposure Limits*: Data not available; *Toxicity by Ingestion*: Grade 2; 0.5 to 5 g/kg; *Late Toxicity*: Data not available; *Vapor (Gas) Irritant Characteristics*: Vapors cause moderate irritation such that personnel will find high concentrations unpleasant. The effect is temporary; *Liquid or Solid Irritant Characteristics*: Minimum hazard. If spilled on clothing and allowed to remain, may cause smarting and reddening of the skin; *Odor Threshold*: 0.047.

Fire Hazards - *Flash Point (deg. F):* 13 OC, -40 CC; *Flammable Limits in Air (%)*: 2.0 -10.0; *Fire Extinguishing Agents*: Foam, dry chemical or carbon dioxide; *Fire Extinguishing Agents Not To Be Used*: Data not available; *Special Hazards of Combustion Products*: Not pertinent; *Behavior in Fire*: Vapors are heavier than air and may travel considerable distances to a source of ignition and flash back. Fires are difficult to control because of reignition; *Ignition Temperature (deg. F)*: 385; *Electrical Hazard*: Not pertinent; *Burning Rate*: 4.8 mm/min.

Chemical Reactivity - *Reactivity with Water*: No reaction; *Reactivity with Common Materials*: No reactions; *Stability During Transport*: Stable; *Neutralizing Agents for Acids and Caustics*: Not pertinent; *Polymerization*: Not pertinent; *Inhibitor of Polymerization*: Not pertinent.

N-BUTYRALDEHYDE

Chemical Designations - *Synonyms*: Butanal; Butyraldehyde; Butyric Aldehyde; Butyl Aldehyde; *Chemical Formula*: $CH_3CH_2CH_2CHO$.

Observable Characteristics - *Physical State (as normally shipped)*: Liquid; *Color*: Colorless; *Odor*: Pungent aldehyde; pungent and intense.

Physical and Chemical Properties - *Physical State at 15 °C and 1 atm.*: Liquid; *Molecular Weight*: 72.11; *Boiling Point at 1 atm.*: 167, 74.8, 348; *Freezing Point*: -142, -96.4, 176.8; *Critical Temperature*: 484, 251, 524; *Critical Pressure*: 590, 40, 4.1; *Specific Gravity*: 0.803 at 20°C; *Vapor (Gas) Density*: 2.5; *Ratio of Specific Heats of Vapor (Gas)*: 1.089; *Latent Heat of Vaporization*: 184, 102, 4.27; *Heat of Combustion*: -15,210, -8,450, -353.8; *Heat of Decomposition*: Not pertinent.

Health Hazards Information - *Recommended Personal Protective Equipment*: Protective goggles, gloves, and organic canister gas mask; *Symptoms Following Exposure*: Inhalation will cause irritation and possibly nausea, vomiting, headache, and loss of consciousness. Contact with eyes causes burns. Skin contact may be irritating; *General Treatment for Exposure*: INHALATION: remove victim to fresh air; if breathing has stopped, give artificial respiration; if breathing is difficult, give oxygen; call a doctor at once. SKIN AND EYES: immediately flush with water for at least 15 min.; get medical attention for eyes; remove contaminated clothing and wash underlying skin; *Toxicity by Inhalation (Threshold Limit Value)*: Data not available; *Short-Term Exposure Limits*: Data not available; *Toxicity by Ingestion*: Grade 1; 5-15 g/kg (rat); *Late Toxicity*: Data not available; *Vapor (Gas) Irritant Characteristics*: Vapors cause moderate irritation such that personnel will find high concentrations unpleasant. The effect is temporary; *Liquid or Solid Irritant Characteristics*: Minimum hazard. If spilled on clothing and allowed to remain, may cause smarting and reddening of the skin; *Odor Threshold*: 0.0046 ppm.

Fire Hazards - *Flash Point (deg. F):* 15 OC, 20 CC; *Flammable Limits in Air (%)*: 2.5 - 10.6; *Fire Extinguishing Agents*: Dry chemical, carbon dioxide, foam; *Fire Extinguishing Agents Not To Be Used*: Not pertinent; *Special Hazards of Combustion Products*: Not pertinent; *Behavior in Fire*: Vapors are heavier than air and may travel considerable distances to a source of ignition and flash back. Fires are difficult to control because of recognition; *Ignition Temperature (deg. F)*: 446; *Electrical Hazard*: Not pertinent; *Burning Rate*: 4.4 mm/min.

Chemical Reactivity - *Reactivity with Water*: No reaction; *Reactivity with Common Materials*: No reactions; *Stability During Transport*: Stable; *Neutralizing Agents for Acids and Caustics*: Not pertinent; *Polymerization*: May occur in the presence of heat, acids or alkalis; *Inhibitor of Polymerization*: None.

N-BUTYRIC ACID

Chemical Designations - *Synonyms*: Butanic Acid; Butanoic Acid; Butyric Acid; Ethylacetic Acid; Propanecarboxylic Acid; *Chemical Formula*: $CH_3CH_2CH_2COOH$.

Observable Characteristics - *Physical State (as normally shipped)*: Liquid; *Color*: Clear; *Odor*: Rancid, disagreeable; strong, penetrating, like rancid butter.

Physical and Chemical Properties - *Physical State at 15 and 1 atm.*: Liquid; *Molecular Weight*: 88.1; *Boiling Point at 1 atm.*: 327, 164, 437; *Freezing Point*: 23, -5, 268; *Critical Temperature*: 671, 355, 628; *Critical Pressure*: 764, 52, 5.3; *Specific Gravity*: 0.958 at 20°C; *Vapor (Gas) Density*: 3.0; *Ratio of Specific Heats of Vapor (Gas)*: 1.079 at 20°C; *Latent Heat of Vaporization*: 167, 92.7, 3.88; *Heat of Combustion*: -10,620, -5,900, -247; *Heat of Decomposition*: Not pertinent.

Health Hazards Information - *Recommended Personal Protective Equipment*: Self-contained breathing apparatus; rubber gloves; vapor-proof plastic goggles; impervious apron and boots; *Symptoms Following Exposure*: Inhalation causes irritation of mucous membrane and respiratory tract; may cause nausea and vomiting. Ingestion causes irritation of mouth and stomach. Contact with eyes may cause serious injury. Contact with skin may cause burns; chemical is readily absorbed through the skin and may cause damage by this route; *General Treatment for Exposure*: INHALATION: remove victim to fresh air; give oxygen if breathing is difficult; call a physician. INGESTION: give large amount of water and induce vomiting. EYES: irrigate with water for 15 min. and get medical attention. SKIN: flush affected areas immediately and thoroughly with water; *Toxicity by Inhalation (Threshold Limit Value)*: Data not available; *Short-Term Exposure Limits*: Data not available; *Toxicity by Ingestion*: Grade 2; oral LD_{50} 2,940 mg/kg (rat); *Late Toxicity*: Data not available; *Vapor (Gas) Irritant Characteristics*: Vapors cause moderate irritation. The effect is temporary; *Liquid or Solid Irritant Characteristics*: Fairly severe skin irritant. May cause pain and second-degree burns after a few minutes of contact; *Odor Threshold*: 0.001 ppm.

Fire Hazards - *Flash Point (deg. F):* 166 OC, 160 CC; *Flammable Limits in Air (%)*: 2.19 - 13.4; *Fire Extinguishing Agents*: Dry chemical, alcohol foam, carbon dioxide; *Fire Extinguishing Agents Not To Be Used*: Water may be ineffective; *Special Hazards of Combustion Products*: No data; *Behavior in Fire*: No data; *Ignition Temperature (deg. F)*: 842; *Electrical Hazard*: No data; *Burning Rate*: 2.7 mm/min.

Chemical Reactivity - *Reactivity with Water*: No reaction; *Reactivity with Common Materials*: May attack aluminum or other light metals with the formation of flammable hydrogen gas; *Stability During Transport*: Stable; *Neutralizing Agents for Acids and Caustics*: Flush with water; *Polymerization*: Not pertinent; *Inhibitor of Polymerization*: Not pertinent.

C

CACODYLIC ACID

Chemical Designations - *Synonyms*: Hydroxydimethylarsine Oxide; Dimethylarsenic Acid; Ansar; Silvisar 510; *Chemical Formula*: $(CH_3)_2AsOOH$.

Observable Characteristics - *Physical State (as normally shipped)*: Solid; *Color*: White; water solutions may be dyed blue; *Odor*: None.

Physical and Chemical Properties - *Physical State at 15 °C and 1 atm.*: Solid; *Molecular Weight*: 138; *Boiling Point at 1 atm.*: > 392, >200, >473; *Freezing Point*: Not pertinent; *Critical Temperature*: Not pertinent; *Critical Pressure*: Not pertinent; *Specific Gravity*: >1.1 (est.) At 20°C (solid); *Vapor (Gas) Density*: Not pertinent; *Ratio of Specific Heats of Vapor (Gas)*: Not pertinent; *Latent Heat of Vaporization*: Not pertinent; *Heat of Combustion*: -6,000, -3,300, -140; *Heat of Decomposition*: Not pertinent.

Health Hazards Information - *Recommended Personal Protective Equipment*: Dust respirator; goggles, protective clothing; *Symptoms Following Exposure*: Chemical is essentially non-irritating in contact with skin or eyes. Ingestion causes arsenic poisoning, but symptoms are delayed; *General Treatment for Exposure*: Be alert for delayed arsenic poisoning symptoms. EYES or SKIN: flush with water. INGESTION: induce vomiting and call a physician at once; *Toxicity by Inhalation (Threshold Limit Value)*: Data not available; *Short-Term Exposure Limits*: Data not available; *Toxicity by Ingestion*: Grade 2; oral rat $LD_{50}=700$ mg/kg; *Late Toxicity*: Arsenic poisoning; *Vapor (Gas) Irritant Characteristics*: Data not available; *Liquid or Solid Irritant Characteristics*: Data not available; *Odor Threshold*: Not pertinent.

Fire Hazards - *Flash Point:* Not flammable; *Flammable Limits in Air (%)*: Not flammable; *Fire Extinguishing Agents*: Not pertinent; *Fire Extinguishing Agents Not To Be Used*: Not pertinent; *Special Hazards of Combustion Products*: Not pertinent; *Behavior in Fire*: May form toxic oxides of arsenic when heated; *Ignition Temperature*: Not pertinent; *Electrical Hazard*: Not pertinent; *Burning Rate*: Not pertinent.

Chemical Reactivity - *Reactivity with Water*: No reaction; *Reactivity with Common Materials*: No reactions; *Stability During Transport*: Stable; *Neutralizing Agents for Acids and Caustics*: Not pertinent; *Polymerization*: Not pertinent; *Inhibitor of Polymerization*: Not pertinent.

CADMIUM ACETATE

Chemical Designations - *Synonyms*: Cadmium Acetate Dihydrate; *Chemical Formula*: $Cd(C_2H_3O_2)_2 2H_2O$.

Observable Characteristics - *Physical State (as normally shipped)*: Solid; *Color*: White; *Odor*: None.

Physical and Chemical Properties - *Physical State at 15 °C and 1 atm.*: Solid; *Molecular Weight*: 266.52; *Boiling Point at 1 atm.*: Not pertinent; *Freezing Point*: Not pertinent; *Critical Temperature*: Not pertinent; *Critical Pressure*: Not pertinent; *Specific Gravity*: 2.34 at 20° (solid); *Vapor (Gas) Density*: Not pertinent; *Ratio of Specific Heats of Vapor (Gas)*: Not pertinent; *Latent Heat of Vaporization*: Not pertinent; *Heat of Combustion*: Not pertinent; *Heat of Decomposition*: Not pertinent.

Health Hazards Information - *Recommended Personal Protective Equipment*: Dust mask; goggles or face shield; rubber gloves; *Symptoms Following Exposure*:; *General Treatment for Exposure*: INHALATION: remove victim to fresh air, seek medical attention. INGESTION: induce vomiting; allay gastrointestinal irritation by swallowing milk or egg whites at frequent intervals; perform gastric lavage; seek medical attention. EYES: flush with water for at least 15 min; *Toxicity by Inhalation (Threshold Limit Value)*: 0.2 mg/m (as cadmium); *Short-Term Exposure Limits*: Data not available; *Toxicity by Ingestion*: Grade 4; LD_{50} < 50 mg/kg; *Late Toxicity*:; *Vapor (Gas) Irritant Characteristics*: Data not available; *Liquid or Solid Irritant Characteristics*: Data not available; *Odor Threshold*: Data not available.

Fire Hazards - *Flash Point:* Not flammable; *Flammable Limits in Air (%)*: Not flammable; *Fire Extinguishing Agents*: Not pertinent; *Fire Extinguishing Agents Not To Be Used*: Not pertinent; *Special Hazards of Combustion Products*: Toxic cadmium oxide fumes may form; *Behavior in Fire*: No data; *Ignition Temperature*: Not pertinent; *Electrical Hazard*: Not pertinent; *Burning Rate*: Not pertinent.

Chemical Reactivity - *Reactivity with Water*: No reaction; *Reactivity with Common Materials*: No reactions; *Stability During Transport*: Stable; *Neutralizing Agents for Acids and Caustics*: Not pertinent; *Polymerization*: Not pertinent; *Inhibitor of Polymerization*: Not pertinent.

CADMIUM BROMIDE

Chemical Designations - *Synonyms*: Cadmium Bromide Tetrahydrate; *Chemical Formula*: CdBr4H$_2$O.

Observable Characteristics - *Physical State (as normally shipped)*: Solid; *Color*: White; *Odor*: None.

Physical and Chemical Properties - *Physical State at 15 °C and 1 atm.*: Solid; *Molecular Weight*: 344.27; *Boiling Point at 1 atm.*: Not pertinent; *Freezing Point*: Not pertinent; *Critical Temperature*: Not pertinent; *Critical Pressure*: Not pertinent; *Specific Gravity*: > 1.1 at 20°C (solid); *Vapor (Gas) Density*: Not pertinent; *Ratio of Specific Heats of Vapor (Gas)*: Not pertinent; *Latent Heat of Vaporization*: Not pertinent; *Heat of Combustion*: Not pertinent; *Heat of Decomposition*: Not pertinent.

Health Hazards Information - *Recommended Personal Protective Equipment*: Dust mask; goggles, or face shield; rubber gloves; *Symptoms Following Exposure*: Inhalation causes coughing, sneezing symptoms of lung damage. Ingestion produces severe toxic symptoms, both kidney and liver injures may occur. Contact with dust causes eye irritation; *General Treatment for Exposure*: INHALATION: remove victim to fresh air, seek medical attention. INGESTION: induce vomiting; allay gastrointestinal irritation by swallowing milk or egg whites at frequent intervals; perform gastric lavage; seek medical attention. EYES: flush with water for at least 15 min; *Toxicity by Inhalation (Threshold Limit Value)*: 0.2 mg/m^3; (as cadmium) *Short-Term Exposure Limits*:; *Toxicity by Ingestion*: Grade 4; LD_{50}; > 50 mg/kg; *Late Toxicity*: Delayed liver, lung and kidney damage has followed respiratory exposures to cadmium salts in industry; *Vapor (Gas) Irritant Characteristics*: Data not available; *Liquid or Solid Irritant Characteristics*: Data not available; *Odor Threshold*: Data not available.

Fire Hazards - *Flash Point:* Not flammable; *Flammable Limits in Air (%)*: Not flammable; *Fire Extinguishing Agents*: Not pertinent; *Fire Extinguishing Agents Not To Be Used*: Not pertinent; *Special Hazards of Combustion Products*: Toxic cadmium oxide fumes can form; *Behavior in Fire*: No data; *Ignition Temperature*: Not pertinent; *Electrical Hazard*: Not pertinent; *Burning Rate*: Not pertinent.

Chemical Reactivity - *Reactivity with Water*: No reaction; *Reactivity with Common Materials*: No reactions; *Stability During Transport*: Stable; *Neutralizing Agents for Acids and Caustics*: Not pertinent; *Polymerization*: Not pertinent; *Inhibitor of Polymerization*: Not pertinent.

CADMIUM CHLORIDE

Chemical Designations - *Synonyms*: No common synonyms; *Chemical Formula*: CdCl$_2$

Observable Characteristics - *Physical State (as normally shipped)*: Crystalline solid; *Color*: White; *Odor*: Odorless.

Physical and Chemical Properties - *Physical State at 15 °C and 1 atm.*: Solid; *Molecular Weight*: 228.35; *Boiling Point at 1 atm.*: Not pertinent; *Freezing Point*: Not pertinent; *Critical Temperature*: Not pertinent; *Critical Pressure*: Not pertinent; *Specific Gravity*: 4.05 at 25°C (solid); *Vapor (Gas) Density*: Not pertinent; *Ratio of Specific Heats of Vapor (Gas)*: Not pertinent; *Latent Heat of Vaporization*: Not pertinent; *Heat of Combustion*: Not pertinent; *Heat of Decomposition*: Not pertinent.

Health Hazards Information - *Recommended Personal Protective Equipment*: Safety glasses, rubber gloves, and respirator with proper filter; *Symptoms Following Exposure*: Ingestion causes gastroenteric distress, pain and prostration. Sensory disturbances, liver injury, and convulsions have been observed in severe intoxications; *General Treatment for Exposure*: INGESTION: induce vomiting and follow with gastric lavage, a saline catharic, and demulcents. Consider using atropine, opiates, and fluid therapy. CaNa$_2$ EDTA has been effective in acutely poisoned animals and in a few humans. BAL has been found sufficiently effective in animal experiments, to justify its use in human intoxication. Since the BAL-cadmium complex has a nephrotoxic action, the physician will have to decide whether or not to use this drug.; *Toxicity by Inhalation (Threshold Limit Value)*: 2 ppm; *Short-Term Exposure Limits*: Not pertinent; *Toxicity by Ingestion*: Grade 4; LD$_{50}$ below 50 mg/kg; *Late Toxicity*: Data not available; *Vapor (Gas) Irritant Characteristics*: Not pertinent; *Liquid or Solid Irritant Characteristics*: Causes smarting of the skin and first degree burns on short exposure; may cause second degree burns on long exposure; *Odor Threshold*: Not pertinent.

Fire Hazards - *Flash Point:* Not flammable; *Flammable Limits in Air (%)*: Not flammable; *Fire Extinguishing Agents*: Not pertinent; *Fire Extinguishing Agents Not To Be Used*: Not pertinent; *Special Hazards of Combustion Products*: Not pertinent; *Behavior in Fire*: Not pertinent; *Ignition Temperature*: Not pertinent; *Electrical Hazard*: Not pertinent; *Burning Rate*: Not pertinent.

Chemical Reactivity - *Reactivity with Water*: No reaction; *Reactivity with Common Materials*: No reactions; *Stability During Transport*: Stable; *Neutralizing Agents for Acids and Caustics*: Not pertinent; *Polymerization*: Not pertinent; *Inhibitor of Polymerization*: Not pertinent.

CADMIUM FLUOROBORATE

Chemical Designations - *Synonyms*: Cadmium Fluoborate; *Chemical Formula*: Cd(BF$_4$)$_2$-H$_2$O.

Observable Characteristics - *Physical State (as normally shipped)*: Liquid; *Color*: Colorless; *Odor*: None.

Physical and Chemical Properties - *Physical State at 15 °C and 1 atm.*: Liquid; *Molecular Weight*: 286; *Boiling Point at 1 atm.*: Not pertinent; *Freezing Point*: Not pertinent; *Critical Temperature*: Not pertinent; *Critical Pressure*: Not pertinent; *Specific Gravity*: 1.60 at 20°C (liquid); *Vapor (Gas) Density*: Not pertinent; *Ratio of Specific Heats of Vapor (Gas)*: Not pertinent; *Latent Heat of Vaporization*: Not pertinent; *Heat of Combustion*: Not pertinent; *Heat of Decomposition*: Not pertinent.

Health Hazards Information - *Recommended Personal Protective Equipment*: Rubber gloves and apron; safety glasses and face shield; *Symptoms Following Exposure*: Ingestion produces severe toxic symptoms; both kidney and liver injures may occur; may be fatal. Contact with eyes or skin causes irritation; *General Treatment for Exposure*: INHALATION: remove patient to fresh air; seek medical attention. INGESTION: call a physician at once; if victim is conscious, induce vomiting by giving a tablespoon of salt in a glass of warm water and repeat until vomit is clear; give milk or whites of eggs beaten with water; keep patient warm and quiet. EYES: flush with plenty of water and get medical attention. SKIN: flush with plenty of water; *Toxicity by Inhalation (Threshold Limit Value)*: 0.2 mg/m^3 (as cadmium); *Short-Term Exposure Limits*: Data not available; *Toxicity by Ingestion*: Grade 3; LD$_{50}$ 250 mg/kg (rat); *Late Toxicity*: Delayed liver, kidney, and lung damage has followed respiratory exposure to cadmium salts in industry; *Vapor (Gas) Irritant Characteristics*: Data not available; *Liquid or Solid Irritant Characteristics*: Data not available; *Odor Threshold*: Not pertinent.

Fire Hazards - *Flash Point:* Not flammable; *Flammable Limits in Air (%)*: Not flammable; *Fire Extinguishing Agents*: Not pertinent; *Fire Extinguishing Agents Not To Be Used*: Not pertinent; *Special Hazards of Combustion Products*: Toxic hydrogen fluoride and cadmium oxide fumes can form; *Behavior in Fire*: Not pertinent; *Ignition Temperature*: Not pertinent; *Electrical Hazard*: Not pertinent; *Burning Rate*: Not pertinent.

Chemical Reactivity - *Reactivity with Water*: No reaction; *Reactivity with Common Materials*: No reactions; *Stability During Transport*: Stable; *Neutralizing Agents for Acids and Caustics*: Not pertinent; *Polymerization*: Not pertinent; *Inhibitor of Polymerization*: Not pertinent.

CADMIUM NITRATE

Chemical Designations - *Synonyms*: Cadmium Nitrate Tetrahydrate; *Chemical Formula*: Cd(NO$_3$)$_2$ 4H$_2$O.

Observable Characteristics - *Physical State (as normally shipped)*: Solid; *Color*: White; *Odor*: None.

Physical and Chemical Properties - *Physical State at 15 °C and 1 atm.*: Solid; *Molecular Weight*: 308.47; *Boiling Point at 1 atm.*: Not pertinent; *Freezing Point*: 138; 59; 332; *Critical Temperature*: Not pertinent; *Critical Pressure*: Not pertinent; *Specific Gravity*: 2.45 at 20°C (solid); *Vapor (Gas) Density*: Not pertinent; *Ratio of Specific Heats of Vapor (Gas)*: Not pertinent; *Latent Heat of Vaporization*: Not pertinent; *Heat of Combustion*: Not pertinent; *Heat of Decomposition*: Not pertinent.

Health Hazards Information - *Recommended Personal Protective Equipment*: Rubber gloves, safety goggles, dust mask; *Symptoms Following Exposure*: Inhalation of fumes can produce coughing, chest constriction, headache, nausea, vomiting, pneumonitis. Chronic poisoning is characterized by emphysema and kidney injury. Ingestion causes gasrtrointestinal disturbances and sever toxic symptoms; both kidney and liver injures may occur. Contact with eyes causes irritation; *General Treatment for Exposure*: INHALATION: remove patient to fresh air; seek medical attention. INGESTION: give large amounts of water and induce vomiting, give milk or egg whites; seek medical attention. EYES: flush with copious amounts of water for 15 min; consult a physician. SKIN: Wash with soap and water; *Toxicity by Inhalation (Threshold Limit Value)*: 0.2 mg/m^3 (as cadmium); *Short-Term Exposure Limits*: Data not available; *Toxicity by Ingestion*: Grade 3; oral mouse LD_{50} = 100 mg/kg; *Late Toxicity*:; *Vapor (Gas) Irritant Characteristics*: Data not available; *Liquid or Solid Irritant Characteristics*: Data not available; *Odor Threshold*: Data not available.

Fire Hazards - *Flash Point:* Not flammable; *Flammable Limits in Air (%)*: Not flammable; *Fire Extinguishing Agents*: Not pertinent; *Fire Extinguishing Agents Not To Be Used*: Not pertinent; *Special Hazards of Combustion Products*: Toxic oxides of nitrogen and cadmium oxide fumes can form; *Behavior in Fire*: Can increase the intensity of fires when in contact with combustible materials; *Ignition Temperature*: Not pertinent; *Electrical Hazard*: Not pertinent; *Burning Rate*: Not pertinent.

Chemical Reactivity - *Reactivity with Water*: No reaction; *Reactivity with Common Materials*: Mixtures with wood and other combustibles are readily ignited; *Stability During Transport*: Stable; *Neutralizing Agents for Acids and Caustics*: Not pertinent; *Polymerization*: Not pertinent; *Inhibitor of Polymerization*: Not pertinent.

CADMIUM OXIDE

Chemical Designations - *Synonyms*: Cadmium Fume; *Chemical Formula*: CdO.

Observable Characteristics - *Physical State (as normally shipped)*: Solid; *Color*: Yellow-brown to brown; *Odor*: None.

Physical and Chemical Properties - *Physical State at 15 °C and 1 atm.*: Solid; *Molecular Weight*: 128.4; *Boiling Point at 1 atm.*: Not pertinent; *Freezing Point*: Not pertinent; *Critical Temperature*: Not pertinent; *Critical Pressure*: Not pertinent; *Specific Gravity*: 6.95 at 20°C (solid); *Vapor (Gas) Density*: Not pertinent; *Ratio of Specific Heats of Vapor (Gas)*: Not pertinent; *Latent Heat of Vaporization*: Not pertinent; *Heat of Combustion*: Not pertinent; *Heat of Decomposition*: Not pertinent.

Health Hazards Information - *Recommended Personal Protective Equipment*:; *Symptoms Following Exposure*: A single exposure to cadmium oxide fumes can cause severe or fatal lung irritation; chronic poisoning is characterized by lung injury (emphysema) and kidney disfunction. Ingestion produces severe toxic effects, both kidney and liver injures may occur. Contact with eyes causes irritation; *General Treatment for Exposure*: INHALATION; if there has been known exposure to dense cadmium oxide fume or if cough, chest tightness, or respiratory distress occur after possible exposure, place patient at bed rest and call a physician. INGESTION: induce vomiting, stop irritation by giving milk or egg whites at frequent intervals; perform gastric lavage; seek medical attention. EYES: flush with water for at least 15 min; *Toxicity by Inhalation (Threshold Limit Value)*: 0.1 mg/m^3; *Short-Term Exposure Limits*: 0.1 mg/m^3, 30 min; *Toxicity by Ingestion*: Grade 3; oral rat LD_{50} =72 mg/kg; *Late Toxicity*:; *Vapor (Gas) Irritant Characteristics*: Data not available; *Liquid or Solid Irritant Characteristics*: Data not available; *Odor Threshold*: Data not available.

Fire Hazards - *Flash Point:* Not flammable; *Flammable Limits in Air (%)*: Not flammable; *Fire Extinguishing Agents*: Not pertinent; *Fire Extinguishing Agents Not To Be Used*: Not pertinent; *Special Hazards of Combustion Products*: Toxic cadmium oxide fumes may form; *Behavior in Fire*: No data; *Ignition Temperature*: Not pertinent; *Electrical Hazard*: Not pertinent; *Burning Rate*: Not pertinent.

Chemical Reactivity - *Reactivity with Water*: No reaction; *Reactivity with Common Materials*: No reactions; *Stability During Transport*: Stable; *Neutralizing Agents for Acids and Caustics*: Not pertinent; *Polymerization*: Not pertinent; *Inhibitor of Polymerization*: Not pertinent.

CADMIUM SULFATE
Chemical Designations - *Synonyms*: No common synonyms; *Chemical Formula*: $CdSO_4$.
Observable Characteristics - *Physical State (as normally shipped)*: Solid; *Color*: White; *Odor*: None.
Physical and Chemical Properties - *Physical State at 15 °C and 1 atm.*: Solid; *Molecular Weight*: 208.46; *Boiling Point at 1 atm.*: Not pertinent; *Freezing Point*: Not pertinent; *Critical Temperature*: Not pertinent; *Critical Pressure*: Not pertinent; *Specific Gravity*: 4.7 at 20°C (solid); *Vapor (Gas) Density*: Not pertinent; *Ratio of Specific Heats of Vapor (Gas)*: Not pertinent; *Latent Heat of Vaporization*: Not pertinent; *Heat of Combustion*: Not pertinent; *Heat of Decomposition*: Not pertinent.
Health Hazards Information - *Recommended Personal Protective Equipment*: Respirator; goggles, rubber gloves; *Symptoms Following Exposure*: Inhalation may cause dryness of throat, coughing, constriction in chest, and headache. Ingestion may cause salvation, vomiting, abdominal pains or diarrhea. Contact with eyes causes irritation; *General Treatment for Exposure*: INHALATION: remove victim from exposure and call a physician. INGESTION; induce vomiting, then allay irritation with milk or egg whites given at frequent intervals; perform gastric lavage; seek medical attention, EYES: flush with water for at least 10 in; consult a physician. SKIN: wash with soap and water; *Toxicity by Inhalation (Threshold Limit Value)*: 0.2 mg/m^3 (as cadmium); *Short-Term Exposure Limits*: Data not available; *Toxicity by Ingestion*:; *Late Toxicity*: Delayed liver, kidney, and lung damage has followed respiratory exposure to cadmium salts in industry; *Vapor (Gas) Irritant Characteristics*: Data not available; *Liquid or Solid Irritant Characteristics*: Data not available; *Odor Threshold*: Data not available.
Fire Hazards - *Flash Point:* Not flammable; *Flammable Limits in Air (%)*: Not flammable; *Fire Extinguishing Agents*: Not pertinent; *Fire Extinguishing Agents Not To Be Used*: Not pertinent; *Special Hazards of Combustion Products*: Toxic cadmium oxide fumes may form; *Behavior in Fire*: No data; *Ignition Temperature*: Not pertinent; *Electrical Hazard*: Not pertinent; *Burning Rate*: Not pertinent.
Chemical Reactivity - *Reactivity with Water*: No reaction; *Reactivity with Common Materials*: No reactions; *Stability During Transport*: Stable; *Neutralizing Agents for Acids and Caustics*: Not pertinent; *Polymerization*: Not pertinent; *Inhibitor of Polymerization*: Not pertinent.

CALCIUM ARSENATE
Chemical Designations - *Synonyms*: Cucumber Dust; Tricalcium Arsenate; Tricalcium Orthoarsenate; *Chemical Formula*: $Ca_3(AsO_4)_2$.
Observable Characteristics - *Physical State (as normally shipped)*: Solid; *Color*: White; *Odor*: None.
Physical and Chemical Properties - *Physical State at 15 °C and 1 atm.*: Solid; *Molecular Weight*: 398; *Boiling Point at 1 atm.*: Not pertinent; *Freezing Point*: Not pertinent; *Critical Temperature*: Not pertinent; *Critical Pressure*: Not pertinent; *Specific Gravity*: 3.62 at 20°C (solid); *Vapor (Gas) Density*: Not pertinent; *Ratio of Specific Heats of Vapor (Gas)*: Not pertinent; *Latent Heat of Vaporization*: Not pertinent; *Heat of Combustion*: Not pertinent; *Heat of Decomposition*: Not pertinent.
Health Hazards Information - *Recommended Personal Protective Equipment*: Dust mask; goggles or face shield; protective gloves; *Symptoms Following Exposure*: Inhalation causes respiratory irritation. Ingestion causes irritation of mouth and stomach. Contact with eyes causes irritation; *General Treatment for Exposure*: INHALATION: move to fresh air. INGESTION: give victim one tablespoonful of salt in glass of water; repeat until vomit is clear; the give two tablespoonfuls of Epsom salt or milk of magnesia and force fluids; call a physician in all cases of suspected poisoning. EYES: flush with water for at least 15 min. SKIN: flush with water, wash with soap and water; *Toxicity by Inhalation (Threshold Limit Value)*: 1 mg/m^3; *Short-Term Exposure Limits*: Data not available; *Toxicity by Ingestion*: Grade 4; oral rat LD_{50} = 20 mg/kg; *Late Toxicity*: Arsenic compounds may cause skin and lung cancer; *Vapor (Gas) Irritant Characteristics*: Data not available; *Liquid or Solid Irritant Characteristics*: Data not available; *Odor Threshold*: Data not available.

Fire Hazards - *Flash Point:* Not flammable; *Flammable Limits in Air (%):* Not flammable; *Fire Extinguishing Agents:* Not pertinent; *Fire Extinguishing Agents Not To Be Used:* Not pertinent; *Special Hazards of Combustion Products:* Toxic arsenic fumes may form; *Behavior in Fire:* No data; *Ignition Temperature:* Not pertinent; *Electrical Hazard:* Not pertinent; *Burning Rate:* Not pertinent.

Chemical Reactivity - *Reactivity with Water:* No reaction; *Reactivity with Common Materials:* No reactions; *Stability During Transport:* Stable; *Neutralizing Agents for Acids and Caustics:* Not pertinent; *Polymerization:* Not pertinent; *Inhibitor of Polymerization:* Not pertinent.

CALCIUM CARBIDE

Chemical Designations - *Synonyms:* Acetylenogen; Carbide; *Chemical Formula:* CaC_2.

Observable Characteristics - *Physical State (as normally shipped):* Solid; *Color:* Gray to bluish black; *Odor:* Garlic-like.

Physical and Chemical Properties - *Physical State at 15 °C and 1 atm.:* Solid; *Molecular Weight:* 64.10; *Boiling Point at 1 atm.:* Not pertinent; *Freezing Point:* Not pertinent; *Critical Temperature:* Not pertinent; *Critical Pressure:* Not pertinent; *Specific Gravity:* 2.22 at 18°C (solid); *Vapor (Gas) Density:* Not pertinent; *Ratio of Specific Heats of Vapor (Gas):* Not pertinent; *Latent Heat of Vaporization:* Not pertinent; *Heat of Combustion:* Not pertinent; *Heat of Decomposition:* Not pertinent.

Health Hazards Information - *Recommended Personal Protective Equipment:* Chemical safety goggles and (for those exposed to unusually dusty operations) a respirator such as those approved by the U.S. Bureau of Mines for "nuisance dusts"; *Symptoms Following Exposure:* Eye and skin irritation; *General Treatment for Exposure:* INHALATION OF DUST: remove from further exposure and call a doctor. SKIN: wash with plenty of water. EYES: flush with clean running water an eye wash fountain for at least 15 min. And get medical attention; *Toxicity by Inhalation (Threshold Limit Value):* Not pertinent; *Short-Term Exposure Limits:* Not pertinent; *Toxicity by Ingestion:* Data not available; *Late Toxicity:* None; *Vapor (Gas) Irritant Characteristics:* None; *Liquid or Solid Irritant Characteristics:* Minimum hazard. If spilled on clothing and allowed to remain, may cause smarting and reddening of the skin; *Odor Threshold:* Not pertinent.

Fire Hazards - *Flash Point:* Not flammable; *Flammable Limits in Air (%):* Not flammable; *Fire Extinguishing Agents:* Dry powder; preferably allow fire to burn out; *Fire Extinguishing Agents Not To Be Used:* Water, vaporizing liquid or foam, carbon dioxide; *Special Hazards of Combustion Products:* Not pertinent; *Behavior in Fire:* When contacted with water, generates highly flammable acetylene gas; *Ignition Temperature:* Not flammable; *Electrical Hazard:* Not pertinent; *Burning Rate:* Not flammable.

Chemical Reactivity - *Reactivity with Water:* Reacts vigorously with water to form highly flammable acetylene gas which can spontaneously ignite; *Reactivity with Common Materials:* Reacts with copper and brass to form an explosive formulation; *Stability During Transport:* Stable but in absence of water; *Neutralizing Agents for Acids and Caustics:* Not pertinent; *Polymerization:* Not pertinent; *Inhibitor of Polymerization:* Not pertinent.

CALCIUM CHLORATE

Chemical Designations - *Synonyms:* No common synonyms; *Chemical Formula:* $Ca(ClO_3)_2$.

Observable Characteristics - *Physical State (as normally shipped):* Solid; *Color:* White; *Odor:* None.

Physical and Chemical Properties - *Physical State at 15 °C and 1 atm.:* Solid; *Molecular Weight:* 207; *Boiling Point at 1 atm.:* Decomposes; *Freezing Point:* 644, 340, 613; *Critical Temperature:* Not pertinent; *Critical Pressure:* Not pertinent; *Specific Gravity:* 2.710 at 0°C (solid); *Vapor (Gas) Density:* Not pertinent; *Ratio of Specific Heats of Vapor (Gas):* Not pertinent; *Latent Heat of Vaporization:* Not pertinent; *Heat of Combustion:* Not pertinent; *Heat of Decomposition:* Not pertinent.

Health Hazards Information - *Recommended Personal Protective Equipment:* Goggles or face shield; dust respirator; coveralls or other protective clothing; *Symptoms Following Exposure:* Inhalation of dust causes irritation of upper respiratory system. Dust irritates eyes and skin. Ingestion causes abdominal pain, nausea, vomiting, diarrhea, pallor, shortness of breath, unconsciousness; *General Treatment for Exposure:* INHALATION: remove to fresh air. EYES: flush with water for 15 min. SKIN: flush with water. INGESTION: induce vomiting and get medical attention; *Toxicity by Inhalation (Threshold Limit*

Value): Data not available; *Short-Term Exposure Limits*: Data not available; *Toxicity by Ingestion*: Grade 2; oral LD$_{50}$=4,500 mg/kg (rat); *Late Toxicity*: Data not available; *Vapor (Gas) Irritant Characteristics*: Not pertinent; *Liquid or Solid Irritant Characteristics*: Data not available; *Odor Threshold*: Not pertinent.

Fire Hazards - *Flash Point* Not flammable but may cause fire with other materials; *Flammable Limits in Air (%)*: Not pertinent; *Fire Extinguishing Agents*: Flood with water; *Fire Extinguishing Agents Not To Be Used*: Not pertinent; *Special Hazards of Combustion Products*: Not pertinent; *Behavior in Fire*: May cause an explosion. Irritating gases may also form upon exposure to heat; *Ignition Temperature*: Not pertinent; *Electrical Hazard*: Not pertinent; *Burning Rate*: Not pertinent.

Chemical Reactivity - *Reactivity with Water*: No reaction; *Reactivity with Common Materials*: Can form an explosive mixture with finely divided combustible materials. The mixture can ignite with application of friction; *Stability During Transport*: Stable; *Neutralizing Agents for Acids and Caustics*: Not pertinent; **Polymerization**: Not pertinent; *Inhibitor of Polymerization*: Not pertinent.

CALCIUM CHLORIDE

Chemical Designations - *Synonyms*: Calcium Chloride, Anhydrous; Calcium Chloride Hydrates; *Chemical Formula*: $CaCl_2 \cdot xH_2O$ where x=o to 6.

Observable Characteristics - *Physical State (as normally shipped)*: Solid or water solution; *Color*: White to off-white; *Odor*: None.

Physical and Chemical Properties - *Physical State at 15 °C and 1 atm.*: Solid; *Molecular Weight*: 110.99 (solute); *Boiling Point at 1 atm.*: Not pertinent; *Freezing Point*: Not pertinent; *Critical Temperature*: Not pertinent; *Critical Pressure*: Not pertinent; *Specific Gravity*: 2.15 at 20°C (solid); *Vapor (Gas) Density*: Not pertinent; *Ratio of Specific Heats of Vapor (Gas)*: Not pertinent; *Latent Heat of Vaporization*: Not pertinent; *Heat of Combustion*: Not pertinent; *Heat of Decomposition*: Not pertinent.

Health Hazards Information - *Recommended Personal Protective Equipment*: Safety glasses or face shield, dust-type respirator, rubber gloves; *Symptoms Following Exposure*: Inhalation causes irritation of nose and throat. Ingestion causes irritation of mouth and stomach. Contact with eyes (particularly by dust) causes irritation and possible transient corneal injury. Contact of solid with dry skin causes mild irritation, even a superficial burn; *General Treatment for Exposure*: INHALATION; move to fresh air, if discomfort persists, get medical attention. INGESTION; give large amounts of water; *Toxicity by Inhalation (Threshold Limit Value)*: EYES: promptly flood with water and continue washing for at least 15 min.; consult an ophthalmologist. SKIN: flush with water; *Short-Term Exposure Limits*: Data not available; *Toxicity by Ingestion*: Grade 2; oral rat LD$_{50}$=1,000 mg/kg (rat); *Late Toxicity*: Data not available; *Vapor (Gas) Irritant Characteristics*: Data not available; *Liquid or Solid Irritant Characteristics*: Data not available; *Odor Threshold*: Data not available.

Fire Hazards - *Flash Point:* Not flammable; *Flammable Limits in Air (%)*: Not flammable; *Fire Extinguishing Agents*: Not pertinent; *Fire Extinguishing Agents Not To Be Used*: Not pertinent; *Special Hazards of Combustion Products*: Not pertinent; *Behavior in Fire*: Not pertinent; *Ignition Temperature*: Not pertinent; *Electrical Hazard*: Not pertinent; *Burning Rate*: Not pertinent.

Chemical Reactivity - *Reactivity with Water*: Anhydrous grade dissolves with evolution of some heat; *Reactivity with Common Materials*: Metals slowly corrode in aqueous solutions; *Stability During Transport*: Stable; *Neutralizing Agents for Acids and Caustics*: Not pertinent; *Polymerization*: Not pertinent; *Inhibitor of Polymerization*: Not pertinent.

CALCIUM CHROMATE

Chemical Designations - *Synonyms*: Calcium Chromate (VI); Calcium Chromate Dihydrate; Gelbin Yellow Ultramarine; Steinbuhl Yellow; *Chemical Formula*: $CaCrO_4 \cdot 2H_2O$.

Observable Characteristics - *Physical State (as normally shipped)*: Solid; *Color*: Yellow; *Odor*: None.

Physical and Chemical Properties - *Physical State at 15 °C and 1 atm.*: Solid; *Molecular Weight*: 192.1; *Boiling Point at 1 atm.*: Not pertinent; *Freezing Point*: Not pertinent; *Critical Temperature*: Not pertinent; *Critical Pressure*: Not pertinent; *Specific Gravity*: > 1 at 20 °C (solid); *Vapor (Gas) Density*: Not pertinent; *Ratio of Specific Heats of Vapor (Gas)*: Not pertinent; *Latent Heat of Vaporization*: Not pertinent; *Heat of Combustion*: Not pertinent; *Heat of Decomposition*: Not pertinent.

Health Hazards Information - *Recommended Personal Protective Equipment*: Dust mask; goggles or face shield; protective gloves; *Symptoms Following Exposure*: Inhalation causes irritation of nose and throat. Ingestion causes severe circulatory collapse and chronic nephritis. Contact with eyes causes irritation. Contact with skin may cause dermatitis and ulcers; *General Treatment for Exposure*: INHALATION: remove to fresh air. INGESTION: give large amounts of water; induce vomiting. EYES: flush with water for at least 15 min. SKIN: treat local injuries like acid burns; scrub with dilute (2%) sodium hyposulfite solution; *Toxicity by Inhalation (Threshold Limit Value)*: 0.1 mg/m^3; *Short-Term Exposure Limits*: Data not available; *Toxicity by Ingestion*: Grade 3: LD_{50} 50 - 50 mg/kg; *Late Toxicity*: Lung cancer may develop; *Vapor (Gas) Irritant Characteristics*: Data not available; *Liquid or Solid Irritant Characteristics*: Data not available; *Odor Threshold*: Data not available.

Fire Hazards - *Flash Point:* Not flammable; *Flammable Limits in Air (%)*: Not flammable; *Fire Extinguishing Agents*: Not pertinent; *Fire Extinguishing Agents Not To Be Used*: Not pertinent; *Special Hazards of Combustion Products*: Toxic chromium fumes are formed during fires; *Behavior in Fire*: The hydrated salt loses water when hot and changes color, however there is no increase in hazard; *Ignition Temperature*: Not pertinent; *Electrical Hazard*: Not pertinent; *Burning Rate*: Not pertinent.

Chemical Reactivity - *Reactivity with Water*: No reaction; *Reactivity with Common Materials*: No reactions; *Stability During Transport*: Stable; *Neutralizing Agents for Acids and Caustics*: Not pertinent; *Polymerization*: Not pertinent; *Inhibitor of Polymerization*: Not pertinent.

CALCIUM FLUORIDE

Chemical Designations - *Synonyms*: Fluospar, Fluorspar; *Chemical Formula*: CaF_2.

Observable Characteristics - *Physical State (as normally shipped)*: Solid; *Color*: Gray; *Odor*: Odorless.

Physical and Chemical Properties - *Physical State at 15 °C and 1 atm.*: Solid; *Molecular Weight*: 78.08; *Boiling Point at 1 atm.*: Not pertinent; *Freezing Point*: Not pertinent; *Critical Temperature*: Not pertinent; *Critical Pressure*: Not pertinent; *Specific Gravity*: 3.18 at 20 °C (solid); *Vapor (Gas) Density*: Not pertinent; *Ratio of Specific Heats of Vapor (Gas)*: Not pertinent; *Latent Heat of Vaporization*: Not pertinent; *Heat of Combustion*: Not pertinent; *Heat of Decomposition*: Not pertinent.

Health Hazards Information - *Recommended Personal Protective Equipment*: For dust only; *Symptoms Following Exposure*: Little acute toxicity; *General Treatment for Exposure*: Usually no treatment needed; *Toxicity by Inhalation (Threshold Limit Value)*: Not pertinent; *Short-Term Exposure Limits*: Not pertinent; *Toxicity by Ingestion*:; *Late Toxicity*: Data not available; *Vapor (Gas) Irritant Characteristics*: Not pertinent; *Liquid or Solid Irritant Characteristics*: No appreciable hazard. Practically harmless to the skin; *Odor Threshold*: Not pertinent.

Fire Hazards - *Flash Point:* Not flammable; *Flammable Limits in Air (%)*: Not flammable; *Fire Extinguishing Agents*: Not pertinent; *Fire Extinguishing Agents Not To Be Used*: Not pertinent; *Special Hazards of Combustion Products*: Not pertinent; *Behavior in Fire*: Not pertinent; *Ignition Temperature*: Not pertinent; *Electrical Hazard*: Not pertinent; *Burning Rate*: Not pertinent.

Chemical Reactivity - *Reactivity with Water*: No reaction; *Reactivity with Common Materials*: No reactions; *Stability During Transport*: Stable; *Neutralizing Agents for Acids and Caustics*: Not pertinent; *Polymerization*: Not pertinent; *Inhibitor of Polymerization*: Not pertinent.

CALCIUM HYDROXIDE

Chemical Designations - *Synonyms*: Slaked Lime; *Chemical Formula*: $Ca(OH)_2$.

Observable Characteristics - *Physical State (as normally shipped)*: Solid; *Color*: White; *Odor*: None.

Physical and Chemical Properties - *Physical State at 15 °C and 1 atm.*: Solid; *Molecular Weight*: 74.09; *Boiling Point at 1 atm.*: Not pertinent; *Freezing Point*: Not pertinent; *Critical Temperature*: Not pertinent; *Critical Pressure*: Not pertinent; *Specific Gravity*: 2.24 at 20°C (solid); *Vapor (Gas) Density*: Not pertinent; *Ratio of Specific Heats of Vapor (Gas)*: Not pertinent; *Latent Heat of Vaporization*: Not pertinent; *Heat of Combustion*: Not pertinent; *Heat of Decomposition*: Not pertinent.

Health Hazards Information - *Recommended Personal Protective Equipment*: Dust-proof goggles and mask; *Symptoms Following Exposure*: Dust irritates eyes, nose and throat; *General Treatment for*

Exposure: INGESTION: have victim drink milk and water. Do NOT induce vomiting. EYES: flush with a gentle stream of water for at least 10 min. And consult an ophthalmologist for further treatment without delay. SKIN: wash off the lime and consult a physician; *Toxicity by Inhalation (Threshold Limit Value)*: Not pertinent; *Short-Term Exposure Limits*: Not pertinent; *Toxicity by Ingestion*: Grade 1; LD_{50} 5 to 15 g/kg (rat); *Late Toxicity*: None; *Vapor (Gas) Irritant Characteristics*: Not pertinent; *Liquid or Solid Irritant Characteristics*: None; *Odor Threshold*: Not pertinent.

Fire Hazards - *Flash Point:* Not flammable; *Flammable Limits in Air* (%): Not flammable; *Fire Extinguishing Agents*: Not pertinent; *Fire Extinguishing Agents Not To Be Used*: Not pertinent; *Special Hazards of Combustion Products*: Not pertinent; *Behavior in Fire*: Not pertinent; *Ignition Temperature*: Not pertinent; *Electrical Hazard*: Not pertinent; *Burning Rate*: Not pertinent.

Chemical Reactivity - *Reactivity with Water*: No reaction; *Reactivity with Common Materials*: No reactions; *Stability During Transport*: Stable; *Neutralizing Agents for Acids and Caustics*: Not pertinent; *Polymerization*: Not pertinent; *Inhibitor of Polymerization*: Not pertinent.

CALCIUM HYPOCHLORITE

Chemical Designations - *Synonyms*: HTH; HTH Dry Chlorine; Neutral Anhydrous Calcium Hypochlorite; Sentry; *Chemical Formula*: $Ca(OCl)_2$.

Observable Characteristics - *Physical State (as normally shipped)*: Solid; *Color*: White; *Odor*: Like bleaching powder.

Physical and Chemical Properties - *Physical State at 15 °C and 1 atm.*: Solid; *Molecular Weight*: 174.98; *Boiling Point at 1 atm.*: Not pertinent; *Freezing Point*: Not pertinent; *Critical Temperature*: Not pertinent; *Critical Pressure*: Not pertinent; *Specific Gravity*: 2.35 at 20°C (solid); *Vapor (Gas) Density*: Not pertinent; *Ratio of Specific Heats of Vapor (Gas)*: Not pertinent; *Latent Heat of Vaporization*: Not pertinent; *Heat of Combustion*: Not pertinent; *Heat of Decomposition*: Not pertinent.

Health Hazards Information - *Recommended Personal Protective Equipment*: Protective goggles, dust mask; *Symptoms Following Exposure*: INHALATION: hypochlorous acid fumes cause sever respiratory tract irritation and pulmonary edema. INGESTION: pain and inflamation of mouth, pharynx, esophagus, and stomach, erosion of mucous membranes, chiefly of the stomach; vomiting (hemorrhaging may cause vomitus to resemble coffee ground); circulatory collapse with cold and clammy skin, cyanosis and shallow respirations; confusion, delirum, coma; edema of pharynx, glottis and larynx, with stridor and obstruction; perforation of esophagus of stomach, with mediastinitis or peritonitis. SKIN CONTACT: may cause vesicular eruptions and eczematoic dermatitis; *General Treatment for Exposure*: INGESTION: swallow immediately milk, egg white, starch paste, milk of magnesia, aluminum hydroxide gel. Avoid sodium bicarbonate because of the release of carbon dioxide. Do not use acidic antidotes; cautious gastric lavage with tap water or a 1% solution of sodium thiosulfate; milk of magnesia (1 oz.) left in the stomach is useful as a mild antacid, adsorbent, demulcent, and catharic; demulcents, such as starch, egg white, milk, gruel; opiates for the control of pain. Treat shock vigorously with intravenous fluids. Prompt surgical intervention when indicated, e.g. tracheotomy, gastrectomy. SKIN: wash with liberal quantities of water and apply a paste of baking soda; *Toxicity by Inhalation (Threshold Limit Value)*: Not pertinent; *Short-Term Exposure Limits*: Not pertinent; *Toxicity by Ingestion*: Grade 0; LD_{50} above 15 g/kg; *Late Toxicity*: None; *Vapor (Gas) Irritant Characteristics*: Not pertinent; *Liquid or Solid Irritant Characteristics*: Irritates eyes, skin and mucous membranes; *Odor Threshold*: Not pertinent.

Fire Hazards - *Flash Point:* Not flammable; *Flammable Limits in Air* (%): Not flammable; *Fire Extinguishing Agents*: Not pertinent; *Fire Extinguishing Agents Not To Be Used*: Not pertinent; *Special Hazards of Combustion Products*: Not pertinent; *Behavior in Fire*: Poisonous gases released upon exposure to heat; *Ignition Temperature*: Not flammable; *Electrical Hazard*: Not pertinent; *Burning Rate*: Not pertinent.

Chemical Reactivity - *Reactivity with Water*: No reaction; *Reactivity with Common Materials*: Can cause fire on contact with wood or straw, and is corrosive to most metals; *Stability During Transport*: The 70 % grade decomposes violently when exposed to heat or direct sunlight. Gives off chlorine and chlorine monoxide gases above 350 of, which are poisonous gases; *Neutralizing Agents for Acids and*

Caustics: Dilute with water; *Polymerization*: Not pertinent; *Inhibitor of Polymerization*: Not pertinent.

CALCIUM NITRATE
Chemical Designations - *Synonyms*: Calcium Nitrate Tetrahydrate; *Chemical Formula*: $Ca(NO_3)_2 \cdot 4H_2O$.
Observable Characteristics - *Physical State (as normally shipped)*: Solid; *Color*: White; *Odor*: None.
Physical and Chemical Properties - *Physical State at 15 ℃ and 1 atm.*: Solid; *Molecular Weight*: 164; *Boiling Point at 1 atm.*: Decomposes; *Freezing Point*: 1.042, 561, 934; *Critical Temperature*: Not pertinent; *Critical Pressure*: Not pertinent; *Specific Gravity*: 2.50 at 18 °C (solid); *Vapor (Gas) Density*: Not pertinent; *Ratio of Specific Heats of Vapor (Gas)*: Not pertinent; *Latent Heat of Vaporization*: Not pertinent; *Heat of Combustion*: Not pertinent; *Heat of Decomposition*: Not pertinent.
Health Hazards Information - *Recommended Personal Protective Equipment*: Dust respirator and rubber gloves; *Symptoms Following Exposure*: Dust causes mild irritation of eyes; *General Treatment for Exposure*: EYES or SKIN: flush with water; *Toxicity by Inhalation (Threshold Limit Value)*: Data not available; *Short-Term Exposure Limits*: Data not available; *Toxicity by Ingestion*: Data not available; *Late Toxicity*: None; *Vapor (Gas) Irritant Characteristics*: Not pertinent; *Liquid or Solid Irritant Characteristics*: Data not available; *Odor Threshold*: Not pertinent.
Fire Hazards - *Flash Point:* Not flammable however may cause fires when in contact with flammables; *Flammable Limits in Air (%)*: Not flammable; *Fire Extinguishing Agents*: Flood with water; *Fire Extinguishing Agents Not To Be Used*: Not pertinent; *Special Hazards of Combustion Products*: Produces toxic oxides of nitrogen when involved in fires; *Behavior in Fire*: Can greatly intensify the burning of all combustible materials; *Ignition Temperature*: Not pertinent; *Electrical Hazard*: Not pertinent; *Burning Rate*: Not pertinent.
Chemical Reactivity - *Reactivity with Water*: No reaction; *Reactivity with Common Materials*: Contact with combustible materials can result in fires; *Stability During Transport*: Stable; *Neutralizing Agents for Acids and Caustics*: Not pertinent; *Polymerization*: Not pertinent; *Inhibitor of Polymerization*: Not pertinent.

CALCIUM OXIDE
Chemical Designations - *Synonyms*: Quicklime; Unslaked Lime; *Chemical Formula*: CaO.
Observable Characteristics - *Physical State (as normally shipped)*: Solid; *Color*: White yo grey; *Odor*: Odorless.
Physical and Chemical Properties - *Physical State at 15 ℃ and 1 atm.*: Solid; *Molecular Weight*: 56.08; *Boiling Point at 1 atm.*: Not pertinent; *Freezing Point;* Not pertinent; *Critical Temperature*: Not pertinent; *Critical Pressure*: Not pertinent; *Specific Gravity*: 3.3 at 20°C (solid); *Vapor (Gas) Density*: Not pertinent; *Ratio of Specific Heats of Vapor (Gas)*: Not pertinent; *Latent Heat of Vaporization*: Not pertinent; *Heat of Combustion*: Not pertinent; *Heat of Decomposition*: Not pertinent.
Health Hazards Information - *Recommended Personal Protective Equipment*: Protective gloves, goggles, and any type of respirator prescribed for fine dust; *Symptoms Following Exposure*: Cause burns on mucous membrane and skin. Inhalation of dust causes sneezing; *General Treatment for Exposure*: INGESTION: if victim is conscious, have him drink water or milk. Do NOT induce vomiting. SKIN and EYES: flush with water and seek medical help; *Toxicity by Inhalation (Threshold Limit Value)*: 5 mg/m^3; *Short-Term Exposure Limits*: 10 mg/m^3 for 30 min; *Toxicity by Ingestion*: Data not available; *Late Toxicity*: None; *Vapor (Gas) Irritant Characteristics*: Not pertinent; *Liquid or Solid Irritant Characteristics*: Causes smarting of the skin and first-degree burns on short exposure and may cause secondary burns on long exposure; *Odor Threshold*: Not pertinent.
Fire Hazards - *Flash Point:* Not flammable; *Flammable Limits in Air (%)*: Not flammable; *Fire Extinguishing Agents*: Not pertinent; *Fire Extinguishing Agents Not To Be Used*: Do not use water on adjacent fires; *Special Hazards of Combustion Products*: Not pertinent; *Behavior in Fire*: Not pertinent; *Ignition Temperature*: Not flammable; *Electrical Hazard*: Not pertinent; *Burning Rate*: Not pertinent.
Chemical Reactivity - *Reactivity with Water*: Heat causes ignition of combustible materials. The material swells during the reaction; *Reactivity with Common Materials*: No reactions unless water is

present; the principle effect is heat is liberated; *Stability During Transport*: Stable; *Neutralizing Agents for Acids and Caustics*: Not pertinent; *Polymerization*: Not pertinent; *Inhibitor of Polymerization*: Not pertinent.

CALCIUM PEROXIDE

Chemical Designations - *Synonyms*: Calcium Dioxide; *Chemical Formula*: CaO_2.

Observable Characteristics - *Physical State (as normally shipped)*: Solid; *Color*: Yellow-white; *Odor*: None.

Physical and Chemical Properties - *Physical State at 15 °C and 1 atm.*: Solid; *Molecular Weight*: 72.1; *Boiling Point at 1 atm.*: Decomposes; *Freezing Point*: Not pertinent; *Critical Temperature*: Not pertinent; *Critical Pressure*: Not pertinent; *Specific Gravity*: 2.92 at 25°C (solid); *Vapor (Gas) Density*: Not pertinent; *Ratio of Specific Heats of Vapor (Gas)*: Not pertinent; *Latent Heat of Vaporization*: Not pertinent; *Heat of Combustion*: Not pertinent; *Heat of Decomposition*: -135, -75, -3.1.

Health Hazards Information - *Recommended Personal Protective Equipment*: Toxic dust respirator; general-purpose gloves; chemical safety goggles; full cover clothing; *Symptoms Following Exposure*: Inhalation of dust irritates nose and throat. Dust also irritates eyes and skin on contact and irritates mouth and stomach if ingested; *General Treatment for Exposure*: INHALATION: remove to fresh air. EYES: flush with water for 15 min. And consult a physician. SKIN: flush with water. INGESTION: dive large amounts of water; *Toxicity by Inhalation (Threshold Limit Value)*: Data not available; *Short-Term Exposure Limits*: Data not available; *Toxicity by Ingestion*: Data not available; *Late Toxicity*: Data not available; *Vapor (Gas) Irritant Characteristics*: Data not available; *Liquid or Solid Irritant Characteristics*: Data not available; *Odor Threshold*: Not pertinent.

Fire Hazards - *Flash Point:* Not flammable but may cause fires upon contact with combustible materials; *Flammable Limits in Air (%)*: Not pertinent; *Fire Extinguishing Agents*: Flood with water or use dry powder such as graphite or powdered limestone; *Fire Extinguishing Agents Not To Be Used*: Not pertinent; *Special Hazards of Combustion Products*: Not pertinent; *Behavior in Fire*: Can increase the intensity and severity of fires; containers may explode; *Ignition Temperature*: Not pertinent; *Electrical Hazard*: Not pertinent; *Burning Rate*: Not pertinent.

Chemical Reactivity - *Reactivity with Water*: Reacts slowly with water at room temperature to form limewater and oxygen gas; *Reactivity with Common Materials*: Heavy metals and dirt can accelerate decomposition to lime and oxygen. The reaction is not explosive; *Stability During Transport*: Stable; *Neutralizing Agents for Acids and Caustics*: Flush with water; *Polymerization*: Not pertinent; *Inhibitor of Polymerization*: Not pertinent.

CALCIUM PHOSPHIDE

Chemical Designations - *Synonyms*: Photophor; *Chemical Formula*: Ca_3P_2.

Observable Characteristics - *Physical State (as normally shipped)*: Solid; *Color*: Grey; *Odor*: Musty, like acetylene.

Physical and Chemical Properties - *Physical State at 15 °C and 1 atm.*: Solid; *Molecular Weight*: 182.2; *Boiling Point at 1 atm.*: Decomposes; *Freezing Point*: (approx.) 2,910; 1,600; 1,810; *Critical Temperature*: Not pertinent; *Critical Pressure*: Not pertinent; *Specific Gravity*: 2.51 at 20°C (solid); *Vapor (Gas) Density*: Not pertinent; *Ratio of Specific Heats of Vapor (Gas)*: Not pertinent; *Latent Heat of Vaporization*: Not pertinent.

Health Hazards Information - *Recommended Personal Protective Equipment*: Dust respirator; protective gloves and clothing; goggles; *Symptoms Following Exposure*: Inhalation or ingestion causes faintness, weakness, nausea, vomiting. External contact with eyes causes irritation of eyes and skin.; *General Treatment for Exposure*: INHALATION: remove to fresh air, call a physician and alert to possibility of phosphine poisoning. EYES or SKIN: flush with water, call a physician to alert to possibility of phosphine poisoning; *Toxicity by Inhalation (Threshold Limit Value)*: Data not available; *Short-Term Exposure Limits*: Data not available; *Toxicity by Ingestion*: Data not available; *Late Toxicity*: Data not available; *Vapor (Gas) Irritant Characteristics*: Data not available; *Liquid or Solid Irritant Characteristics*: Data not available; *Odor Threshold*: 1-100 mg/m^3.

Fire Hazards - *Flash Point:* Not flammable but can spontaneously ignite if in contact with water; *Flammable Limits in Air (%):* Not flammable; *Fire Extinguishing Agents:* Extinguish adjacent fires with dry chemical or carbon dioxide; *Fire Extinguishing Agents Not To Be Used:* Water, foam; *Special Hazards of Combustion Products:* Not pertinent; *Behavior in Fire:* Can cause spontaneous ignition if wetted. Generates dense smoke of phosphoric acid; *Ignition Temperature:* Not pertinent; *Electrical Hazard:* Not pertinent; *Burning Rate:* Not pertinent.

Chemical Reactivity - *Reactivity with Water:* Reacts vigorously with water, generating phosphine, which is a poisonous and spontaneously flammable gas; *Reactivity with Common Materials:* Can react with surface moisture to generate phosphine, which is toxic and spontaneously flammable; *Stability During Transport:* Stable if kept dry; *Neutralizing Agents for Acids and Caustics:* Not pertinent; *Polymerization:* Not pertinent; *Inhibitor of Polymerization:* Not pertinent.

CAMPHENE

Chemical Designations - *Synonyms:* 2,2-Dimethyl-3-Methyl-enenorbornane; 3,3-Dimethyl-2-Methylenenorcamphane; *Chemical Formula:* $Ca_{10}H_{16}$.

Observable Characteristics - *Physical State (as normally shipped):* Solid; *Color:* White; *Odor:* Camphoraceous.

Physical and Chemical Properties - *Physical State at 15 °C and 1 atm.:* Solid; *Molecular Weight:* 136; *Boiling Point at 1 atm.:* 310; 154; 427; *Freezing Point:* 122; 50; 323; *Critical Temperature:* Not pertinent; *Critical Pressure:* Not pertinent; *Specific Gravity:* 0.87 at 15°C (solid); *Vapor (Gas) Density:* Not pertinent; *Ratio of Specific Heats of Vapor (Gas):* Not pertinent; *Latent Heat of Vaporization:* Not pertinent; *Heat of Combustion:* -19,400, -10,800, -452; *Heat of Decomposition:* Not pertinent.

Health Hazards Information - *Recommended Personal Protective Equipment:* Gloves and face shield; *Symptoms Following Exposure:* Inhalation causes irritation of nose and throat. Contact with eyes and skin causes irritation; *General Treatment for Exposure:* INHALATION: move to fresh air; call a physician immediately. EYES: flush immediately with clean, cool water; SKIN: wash with alcohol, follow with soap and water wash; *Toxicity by Inhalation (Threshold Limit Value):* Data not available; *Short-Term Exposure Limits:* Data not available; *Toxicity by Ingestion:* Data not available; *Late Toxicity:* Data not available; *Vapor (Gas) Irritant Characteristics:* Data not available; *Liquid or Solid Irritant Characteristics:* Data not available; *Odor Threshold:* Data not available.

Fire Hazards - *Flash Point (deg. F):* 108 OC, 92 CC; *Flammable Limits in Air (%):* No data; *Fire Extinguishing Agents:* Foam, dry chemical, carbon dioxide; *Fire Extinguishing Agents Not To Be Used:* Water; *Special Hazards of Combustion Products:* No data; *Behavior in Fire:* No data; *Ignition Temperature:* No data; *Electrical Hazard:* No data; *Burning Rate:* Not pertinent.

Chemical Reactivity - *Reactivity with Water:* No reaction; *Reactivity with Common Materials:* No data; *Stability During Transport:* Stable; *Neutralizing Agents for Acids and Caustics:* Not pertinent; *Polymerization:* Not pertinent; *Inhibitor of Polymerization:* Not pertinent.

CARBOLIC OIL

Chemical Designations - *Synonyms:* Middle Oil; Liquefied Phenol; *Chemical Formula:* C_6H_5OH.

Observable Characteristics - *Physical State (as normally shipped):* Liquid; *Color:* Colorless, darkens on exposure to light; *Odor:* Sweet, tar-like.

Physical and Chemical Properties - *Physical State at 15 °C and 1 atm.:* Liquid; *Molecular Weight:* 84.11; *Boiling Point at 1 atm.:* 358.2; 181.8; 455.0; *Freezing Point:* < 105.6; < 40.9; < 314.1; *Critical Temperature:* 790.0, 421.1, 694.3, *Critical Pressure:* 889, 60.5, 6.13; *Specific Gravity:* 1.04 at 41°C (liquid); *Vapor (Gas) Density:* Not pertinent; *Ratio of Specific Heats of Vapor (Gas):* 1.089; *Latent Heat of Vaporization:* 129.6; 72.0; 3.014; *Heat of Combustion:* -13,401, -7,445, -311.707; *Heat of Decomposition:* Not pertinent.

Health Hazards Information - *Recommended Personal Protective Equipment:* Fresh air mask for confined areas; rubber gloves; protective clothing; full face shield; *Symptoms Following Exposure:* Will burn eyes and skin. The analgesic action may cause loss of pain sensation. Readily absorbed through skin, causing increased heart rate, convulsions, and death; *General Treatment for Exposure:*

INHALATION: remove victim to fresh air, keep quiet and warm. If breathing stops, start artificial respiration. INGESTION: do NOT induce vomiting. Give milk, egg whites, or large amounts of water. Get medical assistance. No known antidote. EYES and SKIN: remove contaminated clothing. Flush eyes with water for 15 min. or until physician arrives. Wash skin with soap and water; *Toxicity by Inhalation (Threshold Limit Value)*: 5 ppm; *Short-Term Exposure Limits*: Data not available; *Toxicity by Ingestion*: Grade 2; LD_{50} 0.5 to 5 g/kg (rat); *Late Toxicity*: Causes cancer in experimental animals; *Vapor (Gas) Irritant Characteristics*: Vapors cause moderate irritation such that personnel will find high concentration unpleasant. The effect is temporary; *Liquid or Solid Irritant Characteristics*: Fairly severe skin irritant. May cause pain and second-degree burns after a few minutes' contact; *Odor Threshold*: 0.05 ppm.

Fire Hazards - *Flash Point (deg. F):* 117 CC; *Flammable Limits in Air* (%): No data; *Fire Extinguishing Agents*: Foam, carbon dioxide, dry chemical; *Fire Extinguishing Agents Not To Be Used*: Not pertinent; *Special Hazards of Combustion Products*: Not pertinent; *Behavior in Fire*: The solid often evaporates without first melting; *Ignition Temperature (deg. F)*: 466; *Electrical Hazard*: Not pertinent; *Burning Rate*: No data.

Chemical Reactivity - *Reactivity with Water*: No reaction; *Reactivity with Common Materials*: No reactions; *Stability During Transport*: Stable; *Neutralizing Agents for Acids and Caustics*: Not pertinent; *Polymerization*: Not pertinent; *Inhibitor of Polymerization*: Not pertinent.

CARBON DIOXIDE
Chemical Designations - *Synonyms*: Carbonic Acid Gas; Carbonic Anhydride; *Chemical Formula*: CO_2.

Observable Characteristics - *Physical State (as normally shipped)*: Liquefied compessed gas or solid ("Dry Ice"); *Color*: Colorless; *Odor*: None.

Physical and Chemical Properties - *Physical State at 15 °C and 1 atm.*: Gas; *Molecular Weight*: 44.0; *Boiling Point at 1 atm.*: Not pertinent; *Freezing Point*: -109.3, -78.5, 194.7; *Critical Temperature*: 88, 31, 304; *Critical Pressure*: 1.07; 72.9; 7.40; *Specific Gravity*: 1.56 at -79° (solid); *Vapor (Gas) Density*: 1.53; *Ratio of Specific Heats of Vapor (Gas)*: 1.0474; *Latent Heat of Vaporization*: 150, 83, 3.5; *Heat of Combustion*: Not pertinent; *Heat of Decomposition*: Not pertinent.

Health Hazards Information - *Recommended Personal Protective Equipment*: Self-contained breathing apparatus in excessively high CO_2 concentration areas. For handling liquid or solid, wear safety goggles or face shield, insulated gloves, long-sleeved shirt, and trousers worn outside boots or over high-top shoes to shed spilled liquid; *Symptoms Following Exposure*: Inhalation causes increased respiration rate, headache, subtle physiological changes for up to 5% concentration and prolonged exposure. Higher concentrations can cause unconsciousness and death. Solid can cause cold contact burns. Liquid or cold gas can cause freezing injury to skin or eyes similar to a burn; *General Treatment for Exposure*: INHALATION: move victim to fresh air. skin: treat burns from contact with solid in same way as frostbite; *Toxicity by Inhalation (Threshold Limit Value)*: 5000 ppm; *Short-Term Exposure Limits*: 30,000 ppm for 60 min.; *Toxicity by Ingestion*: Not pertinent (gas with low boiling point); *Late Toxicity*: None; *Vapor (Gas) Irritant Characteristics*: Data not available; *Liquid or Solid Irritant Characteristics*: Data not available; *Odor Threshold*: Not pertinent

Fire Hazards - *Flash Point:* Not flammable; *Flammable Limits in Air* (%): Not flammable; *Fire Extinguishing Agents*: Not pertinent; *Fire Extinguishing Agents Not To Be Used*: Not pertinent; *Special Hazards of Combustion Products*: Not pertinent; *Behavior in Fire*: Containers may explode when exposed to heat; *Ignition Temperature*: Not pertinent; *Electrical Hazard*: Not pertinent; *Burning Rate*: Not pertinent.

Chemical Reactivity - *Reactivity with Water*: No reaction; *Reactivity with Common Materials*: No reactions; *Stability During Transport*: Stable; *Neutralizing Agents for Acids and Caustics*: Not pertinent; *Polymerization*: Not pertinent; *Inhibitor of Polymerization*: Not pertinent.

CARBON MONOXIDE

Chemical Designations - *Synonyms*: Monoxide; *Chemical Formula*: CO.

Observable Characteristics - *Physical State (as normally shipped)*: Compressed gas or liquefied gas; *Color*: Colorless; *Odor*: None.

Physical and Chemical Properties - *Physical State at 15 °C and 1 atm.*: Gas; *Molecular Weight*: 28.0; *Boiling Point at 1 atm.*: -312.7, -191.5, 81.7; *Freezing Point*: -326, -199, 74; *Critical Temperature*: -220, -140, 133; *Critical Pressure*: 507.5, 34.51, 3.502; *Specific Gravity*: 0.791 at -191.5°C (liquid); *Vapor (Gas) Density*: 0.97; *Ratio of Specific Heats of Vapor (Gas)*: 1.3962; *Latent Heat of Vaporization*: 92.8, 51.6, 2.16; *Heat of Combustion*: -4,343, -2,412, -101; *Heat of Decomposition*: Not pertinent.

Health Hazards Information - *Recommended Personal Protective Equipment*: Self-contained breathing apparatus; safety glasses and safety shoes; Type D or Type N canister mask; *Symptoms Following Exposure*: Inhalation causes headache, dizziness, weakness of limbs, confusion, nausea, unconsciousness, and finally death. 0.04% conc., 2-3 hr. or 0.06% conc., 1 hr. - headache and discomfort; with moderate exercise, 0.1-0.2% will produce throbbing in the head in about ½ hr., and confusion of the mind, headache, and nausea in about 2 hrs. 0.20 - 0.25% usually produces unconsciousness in about ½ hr. Inhalation of a 0.4% conc. can prove fatal in less than 1 hr. Inhalation of high concentrations can cause sudden, unexpected collapse. Contact of liquid with skin will cause frostbite; *General Treatment for Exposure*: INHALATION: remove from exposure; give oxygen if available; support respiration; call a doctor. SKIN: if burned by liquid, treat as frostbite; *Toxicity by Inhalation (Threshold Limit Value)*: 50 ppm; *Short-Term Exposure Limits*: 400 ppm, 15 min.; *Toxicity by Ingestion*: Not pertinent (gas with low boiling point); *Late Toxicity*: Toxicity from overexposure persists for many days; *Vapor (Gas) Irritant Characteristics*: Data not available; *Liquid or Solid Irritant Characteristics*: Data not available; *Odor Threshold*: Not pertinent.

Fire Hazards - *Flash Point:* Not pertinent; *Flammable Limits in Air (%)*: 12 - 75; *Fire Extinguishing Agents*: Allow fire to burn out; shut off the flow of gas and cool adjacent exposures with water. Extinguish (only if wearing a SCBA) with dry chemicals or carbon dioxide; *Fire Extinguishing Agents Not To Be Used*: Not pertinent; *Special Hazards of Combustion Products*: Asphyxiation due to carbon dioxide production is a major concern; *Behavior in Fire*: Flame has very little color. Containers may explode in fires; *Ignition Temperature (deg. F)*: 1,128; *Electrical Hazard*: No data; *Burning Rate*: Not pertinent.

Chemical Reactivity - *Reactivity with Water*: No reaction; *Reactivity with Common Materials*: No reactions; *Stability During Transport*: Stable; *Neutralizing Agents for Acids and Caustics*: Not pertinent; *Polymerization*: Not pertinent; *Inhibitor of Polymerization*: Not pertinent.

CARBON TETRACHLORIDE

Chemical Designations - *Synonyms*: Benzinoform; Necatorina; Perchloromethane; Tetrachloromethane; *Chemical Formula*: CCl_4.

Observable Characteristics - *Physical State (as normally shipped)*: Liquid; *Color*: Colorless; *Odor*: Sweetish, aromatic; moderately strong ethereal; somewhat resembling that of chloroform.

Physical and Chemical Properties - *Physical State at 15 °C and 1 atm.*: Liquid; *Molecular Weight*: 153.83; *Boiling Point at 1 atm.*: 170, 76.5, 349.7; *Freezing Point*: -9.4, -23, 250.2; *Critical Temperature*: 541, 283, 556; *Critical Pressure*: 660, 45, 4.6; *Specific Gravity*: 1.59 at 20°C (liquid); *Vapor (Gas) Density*: 5.3; *Ratio of Specific Heats of Vapor (Gas)*: 1.111; *Latent Heat of Vaporization*: 84.2, 46.8, 1.959; *Heat of Combustion*: Not pertinent; *Heat of Decomposition*: Not pertinent.

Health Hazards Information - *Recommended Personal Protective Equipment*: Organic vapor canister with full face mask; protective clothing; rubber gloves; *Symptoms Following Exposure*: Dizziness, incoordination, anesthesia; may be accompanied by nausea and liver damage. Kidney damage also occurs, often producing decrease or stopping of urinary output; *General Treatment for Exposure*: EYES OR SKIN: flush with plenty of water; for eyes, get medical attention. Remove contaminated clothing and wash before reuse. INHALATION: immediately remove to fresh air, keep patient warm and quiet and get medical attention promptly. Start artificial respiration if breathing stops. INGESTION: induce vomiting and get medical attention promptly. No specific antidote known; *Toxicity by Inhalation (Threshold Limit Value)*: 10 ppm; *Short-Term Exposure Limits*: 25 ppm for 30 min.; *Toxicity by Ingestion*: Grade 2; LD_{50} 0.5 to 5 g/kg (rat); *Late Toxicity*: Causes severe liver damage and death if ingested; *Vapor (Gas) Irritant Characteristics*: Vapors cause moderate irritation such that personnel will find high concentrations unpleasant. The effect is temporary; *Liquid or Solid Irritant Characteristics*: Minimum hazard. If spilled on clothing and allowed to remain, may cause smarting and reddening of the skin; *Odor Threshold*: Greater than 10 ppm.

Fire Hazards - *Flash Point:* Not flammable; *Flammable Limits in Air (%)*: Not flammable; *Fire Extinguishing Agents*: Not pertinent; *Fire Extinguishing Agents Not To Be Used*: Not pertinent; *Special Hazards of Combustion Products*: Forms poisonous phosgene gas when exposed to open flames; *Behavior in Fire*: Decomposes to chloride and phosgene; *Ignition Temperature*: Not flammable; *Electrical Hazard*: Not pertinent; *Burning Rate*: Not flammable.

Chemical Reactivity - *Reactivity with Water*: No reaction; *Reactivity with Common Materials*: No reactions; *Stability During Transport*: Stable; *Neutralizing Agents for Acids and Caustics*: Not pertinent; *Polymerization*: Not pertinent; *Inhibitor of Polymerization*: Not pertinent.

CAUSTIC POTASH SOLUTION

Chemical Designations - *Synonyms*: Potassium Hydroxide Solution; Lye; *Chemical Formula*: $KOH-H_2O$.

Observable Characteristics - *Physical State (as normally shipped)*: Liquid; *Color*: Colorless; *Odor*: None.

Physical and Chemical Properties - *Physical State at 15 °C and 1 atm.*: Liquid; *Molecular Weight*: Not pertinent; *Boiling Point at 1 atm.*: >266, >130, >403; *Freezing Point*: Not pertinent; *Critical Temperature*: Not pertinent; *Critical Pressure*: Not pertinent; *Specific Gravity*: 1.45 - 1.50 at 20°C (liquid); *Vapor (Gas) Density*: Not pertinent; *Ratio of Specific Heats of Vapor (Gas)*: Not pertinent; *Latent Heat of Vaporization*: Not pertinent; *Heat of Combustion*: Not pertinent; *Heat of Decomposition*: Not pertinent.

Health Hazards Information - *Recommended Personal Protective Equipment*: Wide-brimmed hat and close-fitting safety goggles equipped with rubber side shields; long sleeved cotton shirt or jacket with buttoned collar and buttoned sleeves; rubber or rubber-coated canvas gloves. (Shirt sleeves should be buttoned over the gloves so that any spilled material will run down the outside). Rubber safety-toe shoes or boots and cotton overalls. (Trouser cuffs should be worn outside of boots). Rubber apron; *Symptoms Following Exposure*: Causes severe burns of eyes, skin, and mucous membranes; *General Treatment for Exposure*: (Act quickly!) EYES: flush with water for at least 15 min. SKIN: flush with water, than rinse with dilute vinegar (acetic acid). INGESTION: give water and milk. Do not induce vomiting. Call physician immediately, even if injury seems minor; *Toxicity by Inhalation (Threshold Limit Value)*: Not pertinent; *Toxicity by Ingestion*: oral rat LD_{50} = 365 mg/kg; *Late Toxicity*: None; *Vapor (Gas) Irritant Characteristics*: Not pertinent; *Liquid or Solid Irritant Characteristics*: This chemical is a sever skin irritant. Can cause second- and third-degree chemical burns on short contact and is very injurious to the eyes.; *Odor Threshold*: Not pertinent.

Fire Hazards - *Flash Point:* Not flammable; *Flammable Limits in Air (%)*: Not flammable; *Fire Extinguishing Agents*: Not pertinent; *Fire Extinguishing Agents Not To Be Used*: Not pertinent; *Special Hazards of Combustion Products*: Not pertinent; *Behavior in Fire*: Not pertinent; *Ignition Temperature*: Not pertinent; *Electrical Hazard*: Not pertinent; *Burning Rate*: Nor flammable.

Chemical Reactivity - *Reactivity with Water*: No reaction; *Reactivity with Common Materials*: Attacks wool, leather and some metals such as aluminum, tin, lead and zinc to produce flammable hydrogen gas. This product should be separated from easily ignitible materials; *Stability During Transport*: Stable; *Neutralizing Agents for Acids and Caustics*: Dilute with water and rinse with dilute acid such as acetic acid; *Polymerization*: Not pertinent; *Inhibitor of Polymerization*: Not pertinent.

CAUSTIC SODA SOLUTION

Chemical Designations - *Synonyms*: Sodium Hydroxide Solution, Lye; *Chemical Formula*: NaOH-H_2O.

Observable Characteristics - *Physical State (as normally shipped)*: Liquid; *Color*: Colorless; *Odor*: None.

Physical and Chemical Properties - *Physical State at 15 °C and 1 atm.*: Liquid; *Molecular Weight*: Not pertinent; *Boiling Point at 1 atm.*: >266, >130, >403; *Freezing Point*: Not pertinent; *Critical Temperature*: Not pertinent; *Critical Pressure*: Not pertinent; *Specific Gravity*: 1.5 at 20 °C (liquid); *Vapor (Gas) Density*: Not pertinent; *Ratio of Specific Heats of Vapor (Gas)*: Not pertinent; *Latent Heat of Vaporization*: Not pertinent; *Heat of Combustion*: Not pertinent; *Heat of Decomposition*: Not pertinent.

Health Hazards Information - *Recommended Personal Protective Equipment*: Wide-brimmed hat; safety goggles with rubber side shields; tight fitting cotton clothing; rubber gloves under shirt cuffs; rubber boots and apron; *Symptoms Following Exposure*: Causes severe burns to eyes, skin, and mucous membranes; *General Treatment for Exposure*: Fast response is important. EYES: flush with water for at least 15 minutes. SKIN: flush with water thoroughly and then rinse with dilute vinegar (acetic acid). INGESTION: Give victim water and milk. Do not induce vomiting. Call physician immediately, even when injury seems slight; *Toxicity by Inhalation (Threshold Limit Value)*: Not pertinent; *Short-Term Exposure Limits*: Not pertinent; *Toxicity by Ingestion*: oral rabbit LD_{50} = 500 mg/kg; *Late Toxicity*: None; *Vapor (Gas) Irritant Characteristics*: Not pertinent; *Liquid or Solid Irritant Characteristics*: Severe skin irritant. Causes second- and third-degree chemical burns on short contact and is very injurious to eyes; *Odor Threshold*: Not pertinent.

Fire Hazards - *Flash Point:* Not flammable; *Flammable Limits in Air (%)*: Not flammable; *Fire Extinguishing Agents*: Not pertinent; *Fire Extinguishing Agents Not To Be Used*: Not pertinent; *Special Hazards of Combustion Products*: Not pertinent; *Behavior in Fire*: Not pertinent; *Ignition Temperature*: Not flammable; *Electrical Hazard*: Not pertinent; *Burning Rate*: Not flammable.

Chemical Reactivity - *Reactivity with Water*: No reaction; *Reactivity with Common Materials*: Corrosive to aluminum, zinc and tin. Contact with some metals can generate flammable hydrogen gas; *Stability During Transport*: Stable; *Neutralizing Agents for Acids and Caustics*: Dilute with water and rinse with dilute acetic acid; *Polymerization*: Not pertinent; *Inhibitor of Polymerization*: Not pertinent.

CHLORDANE

Chemical Designations - *Synonyms*: Chlordan, 1,2,4,5,6,7,8,8-Octachloro-2,3,3a,4,7,7a-Hexahydro-4,7-Methanoindene, Texichlor; *Chemical Formula*: $C_{10}H_6Cl_8$.

Observable Characteristics - *Physical State (as normally shipped)*: Liquid; *Color*: Brown; *Odor*: Penetrating; aromatic; slightly pungent, like chlorine.

Physical and Chemical Properties - *Physical State at 15 °C and 1 atm.*: Liquid; *Molecular Weight*: 409.8; *Boiling Point at 1 atm.*: Decomposes; *Freezing Point*: Not pertinent; *Critical Temperature*: Not pertinent; *Critical Pressure*: Not pertinent; *Specific Gravity*: 1.6 at 25 °C (liquid); *Vapor (Gas) Density*: Not pertinent; *Ratio of Specific Heats of Vapor (Gas)*: Not pertinent; *Latent Heat of Vaporization*: Not pertinent; *Heat of Combustion*: -4,000, -2,200, -93; *Heat of Decomposition*: Not pertinent.

Health Hazards Information - *Recommended Personal Protective Equipment*: Use respirators for spray, fogs, mists or dust; goggles; rubber gloves; *Symptoms Following Exposure*: Moderately irritating to eyes and skin. Ingestion, absorption through skin, or inhalation of mist or dust may cause excitability, convulsions, nausea, vomiting, diarrhea, and local irritation of the gastrointestinal tract; *General Treatment for Exposure*: INHALATION: Administer victim oxygen and give fluid therapy; do not give epinephrine, since it may induce ventricular fibrillation; enforce complete rest. EYES: flush with water for at least 15 minutes. SKIN: wash off skin with large amounts of fresh running water and wash thoroughly with soap and water. Do not scrub infected area of skin. INGESTION: induce vomiting and follow with gastric lavage and administration of saline cathartics; ether and barbiturates may control convulsions; oxygen and fluid therapy are also recommended. Do not give epinephrine.

Since no specific antidotes are known, symptomatic therapy must be accompanied by complete rest.; *Toxicity by Inhalation (Threshold Limit Value)*: 0.5 mg/m^3; *Short-Term Exposure Limits*: 2 mg/m^3 for 30 min; *Toxicity by Ingestion*: oral LD$_{50}$ = 283 mg/kg (rat); *Late Toxicity*: Possible liver damage; loss of appetite or weight.; *Vapor (Gas) Irritant Characteristics*: No data; *Liquid or Solid Irritant Characteristics*: No data; *Odor Threshold*: No data.

Fire Hazards - *Flash Point (deg. F):* 225 OC, 132 CC. In solid form the product is not flammable; *Flammable Limits in Air (%)*: 0.7 - 5 (kerosene solution); *Fire Extinguishing Agents*: Dry chemical, foam, carbon dioxide; *Fire Extinguishing Agents Not To Be Used*: Water may be ineffective on solution fires; *Special Hazards of Combustion Products*: Produces irritating and toxic hydrogen chloride and phosgene gases when the kerosene solution of the compound burns; *Behavior in Fire*: Not pertinent; *Ignition Temperature (deg. F)*: 419 (kerosene solution); *Electrical Hazard*: No data; *Burning Rate*: Not pertinent.

Chemical Reactivity - *Reactivity with Water*: No reaction; *Reactivity with Common Materials*: No reactions; *Stability During Transport*: Product is stable below 160 of; *Neutralizing Agents for Acids and Caustics*: Not pertinent; *Polymerization*: Not pertinent; *Inhibitor of Polymerization*: Not pertinent.

CHLORINE

Chemical Designations - *Synonyms*: No common synonyms; *Chemical Formula*: Cl$_2$.

Observable Characteristics - *Physical State (as normally shipped)*: Liquefied compressed gas; *Color*: Greenish yellow; *Odor*: Characteristic choking.

Physical and Chemical Properties - *Physical State at 15 °C and 1 atm.*: Gas; *Molecular Weight*: 70.91; *Boiling Point at 1 atm.*: -29.4, -34.1, 239.1; *Freezing Point*: -150, -101, 172; *Critical Temperature*: 291, 144, 417; *Critical Pressure*: 1118, 76.05, 7.704; *Specific Gravity*: 1.424 at 15°C (liquid); *Vapor (Gas) Density*: 2.4; *Ratio of Specific Heats of Vapor (Gas)*: 1.325; *Latent Heat of Vaporization*: 124, 68.7, 2.87; *Heat of Combustion*: Not pertinent; *Heat of Decomposition*: Not pertinent.

Health Hazards Information - *Recommended Personal Protective Equipment*: Quick-opening safety shower and eye fountain; respiratory equipment approved fro chlorine service. Wear safety goggles at all times when in vicinity of liquid chlorine; *Symptoms Following Exposure*: Eye irritation, sneezing, copious salvation, general excitement and restlessness. Irritation may persists for several days. High concentrations cause respiratory distress and violent coughing, often with retching. Death may result from suffocation; *General Treatment for Exposure*: INHALATION: remove victim from source of exposure; call a doctor; support respiration; administer oxygen. EYES: flush with copious amounts of water for at least 15 min.; *Toxicity by Inhalation (Threshold Limit Value)*: 1 ppm; *Short-Term Exposure Limits*: 3 ppm for 5 min.; *Toxicity by Ingestion*: Not pertinent; ingestion unlikely (chlorine is gas above -34.5°C); *Late Toxicity*: None; *Vapor (Gas) Irritant Characteristics*: Vapors cause severe irritation of eyes and throat and can cause eye and lung injury. They cannot be tolerated even at low concentrations; *Liquid or Solid Irritant Characteristics*: Causes smarting of the skin and first-degree burns on short exposure; may cause secondary burns on long exposure; *Odor Threshold*: 3.5 ppm.

Fire Hazards - *Flash Point:* Not flammable; *Flammable Limits in Air (%)*: Not flammable; *Fire Extinguishing Agents*: Not pertinent; *Fire Extinguishing Agents Not To Be Used*: Not pertinent; *Special Hazards of Combustion Products*: Toxic products are generated when combustibles burn in the presence of chlorine; *Behavior in Fire*: Most combustible materials will burn in the presence of chlorine even though chlorine itself is not flammable; *Ignition Temperature*: Not flammable; *Electrical Hazard*: No data; *Burning Rate*: Not flammable.

Chemical Reactivity - *Reactivity with Water*: Forms a corrosive solution; *Reactivity with Common Materials*: Reacts vigorously with most metals especially at high temperatures. Copper may burn spontaneously; *Stability During Transport*: Stable; *Neutralizing Agents for Acids and Caustics*: Not pertinent; *Polymerization*: Not pertinent; *Inhibitor of Polymerization*: Not pertinent.

CHLORINE TRIFLUORIDE

Chemical Designations - *Synonyms*: CTF; *Chemical Formula*: ClF_3.

Observable Characteristics - *Physical State (as normally shipped)*: Liquefied compressed gas; *Color*: Gas: colorless. Liquid: greenish-yellow; *Odor*: Acrid; strong, pungent, sweetish; sweet and irritating.

Physical and Chemical Properties - *Physical State at 15 °C and 1 atm.*: Gas; *Molecular Weight*: 92.5; *Boiling Point at 1 atm.*: 53, 11.6, 284.8; *Freezing Point*: -105, -76.1, 197.1; *Critical Temperature*: 307, 153, 426; *Critical Pressure*: 837, 56.9, 5.77; *Specific Gravity*: 1.85 at 11°C (liquid); *Vapor (Gas) Density*: 3.2; *Ratio of Specific Heats of Vapor (Gas)*: 1.2832; *Latent Heat of Vaporization*: 128, 71.2, 2.98; *Heat of Combustion*: Not pertinent; *Heat of Decomposition*: Not pertinent.

Health Hazards Information - *Recommended Personal Protective Equipment*: Neoprene gloves and protective clothing made of glass fiber and Teflon, including full hood; self-contained breathing apparatus with full face mask; *Symptoms Following Exposure*: Inhalation causes extreme irritation of respiratory tract; pulmonary edema may result. Vapors are very irritating to eyes and skin; liquid causes severe burns; *General Treatment for Exposure*: Call physician at once after any exposure to this compound. INHALATION: remove victim to fresh air and keep him quiet; give artificial respiration if breathing has stopped; give oxygen; enforce rest for 24 hours. EYES: flush with water for at least 15 min.; get medical attention, but do not interrupt flushing for at least 10 min. SKIN: flush with water, then with 2-3% aqueous ammonia, then again with water; apply ice-cold pack of saturated Epsom salt or 70% ethyl alcohol; *Toxicity by Inhalation (Threshold Limit Value)*: 0.1 ppm (ceiling point); *Short-Term Exposure Limits*: 0.1 ppm for 5 min.; *Toxicity by Ingestion*: Grade 4; $LD_{50} < 50$ mg/kg; *Late Toxicity*: Data not available; *Vapor (Gas) Irritant Characteristics*: Vapors cause severe irritation of eyes and throat and can cause eye and lung injury. They cannot be tolerated even at low concentrations; *Liquid or Solid Irritant Characteristics*: Severe skin irritant, causes second- and third-degree burns on short contact and is very injurious to the eyes; *Odor Threshold*: Data not available.

Fire Hazards - *Flash Point:* Not flammable, but can cause fire when mixed or in contact with some materials; *Flammable Limits in Air* (%): Not pertinent; *Fire Extinguishing Agents*: Dry chemical; *Fire Extinguishing Agents Not To Be Used*: Do not use water on adjacent fires unless well protected against hydrogen fluoride gas; *Special Hazards of Combustion Products*: Fumes are highly toxic and irritating; *Behavior in Fire*: Can greatly increase the intensity of fires. Containers or vessels may explode; *Ignition Temperature*: Not pertinent; *Electrical Hazard*: Not pertinent; *Burning Rate*: Not pertinent.

Chemical Reactivity - *Reactivity with Water*: Reacts explosively with water, producing hydrogen fluoride (hydrofluoric acid) and chlorine; *Reactivity with Common Materials*: Causes ignition of all combustible materials and some inerts such as sand and concrete. The chemical is very similar to fluorine gas; *Stability During Transport*: Stable; *Neutralizing Agents for Acids and Caustics*: Flood with water; *Polymerization*: Not pertinent; *Inhibitor of Polymerization*: Not pertinent.

CHLOROACETOPHENONE

Chemical Designations - *Synonyms*: Phenacyl Chloride; omega-Chloroacetophenone; alpha-Chloroacetophenone; Phenyl Chloromethyl Ketone; Tear Gas; Chloromethyl Phenyl Ketone; *Chemical Formula*: $C_6H_5COCH_2Cl$.

Observable Characteristics - *Physical State (as normally shipped)*: Solid; *Color*: White to pale yellow; *Odor*: Pungent.

Physical and Chemical Properties - *Physical State at 15 °C and 1 atm.*: Solid; *Molecular Weight*: 154.6; *Boiling Point at 1 atm.*: 477, 247, 520; *Freezing Point*: 68 - 138, 20 - 59, 293 - 332; *Critical Temperature*: Not pertinent; *Critical Pressure*: Not pertinent; *Specific Gravity*: 1.32 at 15°C (solid); *Vapor (Gas) Density*: Not pertinent; *Ratio of Specific Heats of Vapor (Gas)*: Not pertinent; *Latent Heat of Vaporization*: Not pertinent; *Heat of Combustion*: -9,340, -5,190, -217; *Heat of Decomposition*: Not pertinent.

Health Hazards Information - *Recommended Personal Protective Equipment*: Full-face organic canister mask; self-contained breathing apparatus; rubber gloves; protective clothing; *Symptoms Following Exposure*: Inhalation causes tearing, burning of the eyes and difficulty in breathing; high concentrations may lead to development of acute pulmonary edema after latencies of 8 hrs. to several days; possible systemic manifestations include agitation, coma, contraction of pupils of eyes, loss of reflexes. External contact causes irritation of the skin and eyes. Ingestion causes agitation, coma, contraction of pupils of eye, loss of reflexes; *General Treatment for Exposure*: INHALATION: remove victim from contaminated atmosphere at once; give artificial respiration and oxygen, if necessary; watch for pulmonary edema for several days. EYES: flush with water; do not rub. SKIN: flush with water. INGESTION: get medical attention; watch for development of pulmonary edema for several days; *Toxicity by Inhalation (Threshold Limit Value)*: 0.05 ppm; *Short-Term Exposure Limits*: Data not available; *Toxicity by Ingestion*: Grade 3; oral LD_{50} 52 mg/kg (rat); *Late Toxicity*: Fatty infiltration of liver; *Vapor (Gas) Irritant Characteristics*: Data not available; *Liquid or Solid Irritant Characteristics*: Data not available; *Odor Threshold*: 0.016 ppm.

Fire Hazards - *Flash Point:* This is a combustible solid, but in solutions it has a flash point of 244 CC; *Flammable Limits in Air* (%): Not pertinent; *Fire Extinguishing Agents*: Water; *Fire Extinguishing Agents Not To Be Used*: Not pertinent; *Special Hazards of Combustion Products*: Irritating hydrogen chloride may form; *Behavior in Fire*: Unburned material may become volatile and cause severe skin and eye irritation; *Ignition Temperature*: No data; *Electrical Hazard*: No data; *Burning Rate*: Not pertinent.

Chemical Reactivity - *Reactivity with Water*: Reacts slowly, producing hydrogen chloride. The reaction is not hazardous; *Reactivity with Common Materials*: Reacts slowly with metals, causing mild corrosion; *Stability During Transport*: Stable; *Neutralizing Agents for Acids and Caustics*: Not pertinent; *Polymerization*: Not pertinent; *Inhibitor of Polymerization*: Not pertinent.

CHLOROFORM

Chemical Designations - *Synonyms*: Trichloromethane; *Chemical Formula*: $CHCl_3$.

Observable Characteristics - *Physical State (as normally shipped)*: Liquid; *Color*: Colorless; *Odor*: Pleasant, sweet; ethereal.

Physical and Chemical Properties - *Physical State at 15 °C and 1 atm.*: Liquid; *Molecular Weight*: 119.39; *Boiling Point at 1 atm.*: 142, 61.2, 334.4; *Freezing Point*: -82.3, -63.5, 209.7; *Critical Temperature*: 506, 263.2, 536.4; *Critical Pressure*: 790, 54, 5.5; *Specific Gravity*: 1.49 at 20°C (liquid); *Vapor (Gas) Density*: 4.1; *Ratio of Specific Heats of Vapor (Gas)*: 1.146; *Latent Heat of Vaporization*: 106.7, 59.3, 2.483; *Heat of Combustion*: Not pertinent; *Heat of Decomposition*: Not pertinent.

Health Hazards Information - *Recommended Personal Protective Equipment*: Chemical goggles. 50 ppm to 2%: suitable full-face gas mask. Above 2%: suitable self-contained system; *Symptoms Following Exposure*: Headache, nausea, dizziness, drunkenness, narcosis; *General Treatment for Exposure*: INHALATION: if ill effects develop, get victim to fresh air, keep him warm and quiet, and get medical attention. If breathing stops, start artificial respiration. INGESTION: induce vomiting and get medical attention. No known antidote; treat symptoms. EYES: flush with plenty of water for at least 15 min. and get medical attention. SKIN: wash with soap and water, remove contaminated clothing and free of chemical; *Toxicity by Inhalation (Threshold Limit Value)*: 10 ppm for 10-hour work day; *Short-Term Exposure Limits*: 50 ppm for 10 min.; *Toxicity by Ingestion*: Grade 2; LD_{50} 0.5 to 5 g/kg; *Late Toxicity*: None; *Vapor (Gas) Irritant Characteristics*: Vapors cause moderate irritation such that personnel will find high concentrations unpleasant. The effect is temporary; *Liquid or Solid Irritant Characteristics*: Minimum hazard. If spilled on clothing and allowed to remain, may cause smarting and reddening of the skin; *Odor Threshold*: 205-307 ppm.

Fire Hazards - *Flash Point:* Not flammable; *Flammable Limits in Air* (%): Not flammable; *Fire Extinguishing Agents*: Not pertinent; *Fire Extinguishing Agents Not To Be Used*: Not pertinent; *Special Hazards of Combustion Products*: Poisonous and irritating gases are generated upon heating; *Behavior in Fire*: Decomposes resulting in toxic vapors; *Ignition Temperature*: Not flammable; *Electrical Hazard*: Not pertinent; *Burning Rate*: Not flammable.

Chemical Reactivity - *Reactivity with Water*: No reaction; *Reactivity with Common Materials*: No reactions; *Stability During Transport*: Stable; *Neutralizing Agents for Acids and Caustics*: Not pertinent; *Polymerization*: Not pertinent; *Inhibitor of Polymerization*: Not pertinent.

CHROMIC ANHYDRIDE

Chemical Designations - *Synonyms*: Chromic Acid; Chromic Oxide; Chromium Trioxide; *Chemical Formula*: CrO_3.

Observable Characteristics - *Physical State (as normally shipped)*: Solid; *Color*: Dark red; *Odor*: None.

Physical and Chemical Properties - *Physical State at 15 °C and 1 atm.*: Solid; *Molecular Weight*: 100.01; *Boiling Point at 1 atm.*: Not pertinent; *Freezing Point*: Not pertinent; *Critical Temperature*: Not pertinent; *Critical Pressure*: Not pertinent; *Specific Gravity*: 2.70 at 20°C (solid); *Vapor (Gas) Density*: Not pertinent; *Ratio of Specific Heats of Vapor (Gas)*: Not pertinent; *Latent Heat of Vaporization*: Not pertinent; *Heat of Combustion*: Not pertinent; *Heat of Decomposition*: Not pertinent.

Health Hazards Information - *Recommended Personal Protective Equipment*: Goggles and respirator. (Special chromic acid filters are available for respirators to prevent inhalation of dust or mist); *Symptoms Following Exposure*: Very irritating to eyes and respiratory tract. Ingestion causes severe gastrointestinal symptoms. Contact with eyes or skin causes burns; prolonged contact produces dermatitis ("chrome sores"); *General Treatment for Exposure*: INGESTION: call a physician; do NOT induce vomiting. SKIN OR EYES: wash eyes throughly for at least 15 min.; flush contacted skin areas with water; remove contaminated clothing and wash before reuse; *Toxicity by Inhalation (Threshold Limit Value)*: Not pertinent; *Short-Term Exposure Limits*: Not pertinent; *Toxicity by Ingestion*: Grade 3; LD_{50} 50 to 500 mg/kg; *Late Toxicity*: Lung cancer; *Vapor (Gas) Irritant Characteristics*: Not pertinent; *Liquid or Solid Irritant Characteristics*: Severe skin irritant. Causes second- and third-degree burns on short contact; very injurious to the eyes; *Odor Threshold*: Not pertinent.

Fire Hazards - *Flash Point:* Not flammable; *Flammable Limits in Air (%)*: Not flammable; *Fire Extinguishing Agents*: Water; *Fire Extinguishing Agents Not To Be Used*: Not pertinent; *Special Hazards of Combustion Products*: Not pertinent; *Behavior in Fire*: Containers may explode. Water should be applied to cool container surfaces exposed to adjacent fires; *Ignition Temperature*: This product may ignite organic materials on contact; *Electrical Hazard*: Not pertinent; *Burning Rate*: Not flammable.

Chemical Reactivity - *Reactivity with Water*: No reaction; *Reactivity with Common Materials*: Reacts with organic materials rapidly, generating sufficient heat to cause ignition. Prolonged contact on wood floors can result in a fire hazard; *Stability During Transport*: Stable; *Neutralizing Agents for Acids and Caustics*: Flood with water and rinse with sodium bicarbonate solution; *Polymerization*: Not pertinent; *Inhibitor of Polymerization*: Not pertinent.

CHROMYL CHLORIDE

Chemical Designations - *Synonyms*: Chromium (VI) Dioxychloride; Chromium Oxychloride; *Chemical Formula*: CrO_2Cl_2.

Observable Characteristics - *Physical State (as normally shipped)*: Liquid; *Color*: Dark red; *Odor*: Acrid.

Physical and Chemical Properties - *Physical State at 15 °C and 1 atm.*: Liquid; *Molecular Weight*: 154.9; *Boiling Point at 1 atm.*: 241, 116, 389; *Freezing Point*: -141.7, -96.5, 176.7; *Critical Temperature*: Not pertinent; *Critical Pressure*: Not pertinent; *Specific Gravity*: 1.96 at 20°C (liquid); *Vapor (Gas) Density*: 5.3; *Ratio of Specific Heats of Vapor (Gas)*: 1.2832; *Latent Heat of Vaporization*: 113, 62.6, 2.62; *Heat of Combustion*: Not pertinent; *Heat of Decomposition*: Not pertinent.

Health Hazards Information - *Recommended Personal Protective Equipment*: SCBA (full face); rubber gloves; protective clothing; *Symptoms Following Exposure*: Inhalation causes severe irritation of upper respiratory system. Contact with eyes or skin causes irritation and burning. Ingestion causes burning of mouth and stomach; *General Treatment for Exposure*: Get medical attention following all exposures to this compound. INHALATION: remove from exposure; support respiration. EYES: flush with copious quantities of water for 15 min. SKIN: flush with water for 15 min. INGESTION: do NOT induce vomiting; give large amounts of water; *Toxicity by Inhalation (Threshold Limit Value)*: Data not available; *Short-Term Exposure Limits*: Data not available; *Toxicity by Ingestion*: Grade 4; $LD_{50} < 50$ mg/kg; *Late Toxicity*: Data not available; *Vapor (Gas) Irritant Characteristics*: Vapors cause severe irritation of eyes and throat and can cause eye and lung injury. Vapors cannot be tolerated even at low concentrations; *Liquid or Solid Irritant Characteristics*: Severe skin irritant. Causes second- and third-degree burns on short contact and is very injurious to the eyes; *Odor Threshold*: No data. **Fire Hazards - *Flash Point:*** Not flammable, but may cause fire on contact with combustibles; *Flammable Limits in Air (%)*: Not flammable; *Fire Extinguishing Agents*: Dry chemical or carbon dioxide; *Fire Extinguishing Agents Not To Be Used*: Do not apply water on adjacent fires unless SCBA is used to protect against toxic vapors; *Special Hazards of Combustion Products*: Not pertinent; *Behavior in Fire*: Vapors are extremely irritating to the eyes and mucous membranes. This product may increase the intensity of fires; *Ignition Temperature*: Not pertinent; *Electrical Hazard*: Not pertinent; *Burning Rate*: Not pertinent.

Chemical Reactivity - *Reactivity with Water*: Reacts violently with water forming hydrogen chloride (hydrochloric acid), chlorine gases, and chromic acid; *Reactivity with Common Materials*: Causes severe corrosion of common metals; *Stability During Transport*: Stable; *Neutralizing Agents for Acids and Caustics*: Flood with water and rinse with sodium bicarbonate; *Polymerization*: Not pertinent; *Inhibitor of Polymerization*: Not pertinent.

CITRIC ACID

Chemical Designations - *Synonyms*: 2-Hydroxy-1,2,3-Propane-Tricarboxylic Acid; beta-Hydroxytricarballylic Acid; beta-Hydroxytricarboxylic Acid; *Chemical Formula*: $HOC(CH_2CO_2H)_2CO_2H$.

Observable Characteristics - *Physical State (as normally shipped)*: Solid; *Color*: White; *Odor*: None.

Physical and Chemical Properties - *Physical State at 15 °C and 1 atm.*: Solid; *Molecular Weight*: 192.1; *Boiling Point at 1 atm.*: Not pertinent (decomposes); *Freezing Point*: 307, 153, 426; *Critical Temperature*: Not pertinent; *Critical Pressure*: Not pertinent; *Specific Gravity*: 1.54 at 20°C (solid); *Vapor (Gas) Density*: Not pertinent; *Ratio of Specific Heats of Vapor (Gas)*: Not pertinent; *Latent Heat of Vaporization*: Not pertinent; *Heat of Combustion*: -4,000, -2,220, -93; *Heat of Decomposition*: Not pertinent.

Health Hazards Information - *Recommended Personal Protective Equipment*: Dust mask; goggles or face shield; protective gloves; *Symptoms Following Exposure*: Inhalation of dust irritates nose and throat. Contact with eyes causes irritation; *General Treatment for Exposure*: INHALATION: move to fresh air. EYES: flush immediately with physiological saline or water; get medical care if irritation persists. SKIN: flush with water; *Toxicity by Inhalation (Threshold Limit Value)*: Data not available; *Short-Term Exposure Limits*: Data not available; *Toxicity by Ingestion*: Grade 1; oral LD_{50} 11.7 g/kg (rat); *Late Toxicity*: Chronic effects in humans are unknown; *Vapor (Gas) Irritant Characteristics*: Not pertinent; *Liquid or Solid Irritant Characteristics*: Data not available; *Odor Threshold*: Data not available.

Fire Hazards - *Flash Point:* Not pertinent. This is a combustible solid; *Flammable Limits in Air*: 0.28 - 2.29 kg/m^3 as dust; *Fire Extinguishing Agents*: Water, foam, dry chemical, or carbon dioxide; *Fire Extinguishing Agents Not To Be Used*: Not pertinent; *Special Hazards of Combustion Products*: No data; *Behavior in Fire*: This product melts and decomposes as a hazardous reaction; *Ignition Temperature (deg. F)*: 1,850 as powder; *Electrical Hazard*: Not pertinent; *Burning Rate*: Not pertinent.

Chemical Reactivity - *Reactivity with Water*: No reaction; *Reactivity with Common Materials*: Corrodes copper, zinc, aluminum, and alloys of these metals; *Stability During Transport*: Stable; *Neutralizing Agents for Acids and Caustics*: Not pertinent; *Polymerization*: Not pertinent; *Inhibitor of Polymerization*: Not pertinent.

COBALT ACETATE

Chemical Designations - *Synonyms*: Cobalt (II) Acetate; Cobalt Acetate Tetrahydrate; Cobaltous Acetate; *Chemical Formula*: $Co(C_2H_3O_2)_2 \cdot 4H_2O$.

Observable Characteristics - *Physical State (as normally shipped)*: Solid; *Color*: Pink; *Odor*: Slight acetic acid odor; vinegar-like.

Physical and Chemical Properties - *Physical State at 15 °C and 1 atm.*: Solid; *Molecular Weight*: 249.1; *Boiling Point at 1 atm.*: Not pertinent (decomposes); *Freezing Point*: 284, 140, 413; *Critical*

Temperature: Not pertinent; *Critical Pressure*: Not pertinent; *Specific Gravity*: 1.71 at 20°C (solid); *Vapor (Gas) Density*: Not pertinent; *Ratio of Specific Heats of Vapor (Gas)*: Not pertinent; *Latent Heat of Vaporization*: Not pertinent; *Heat of Combustion*: Not pertinent; *Heat of Decomposition*: Not pertinent.

Health Hazards Information - *Recommended Personal Protective Equipment*: Dust respirator; rubber gloves; goggles or face shield; protective clothing; *Symptoms Following Exposure*: Inhalation causes shortness of breath and coughing; permanent disability may occur. Ingestion causes pain and vomiting. Contact with eyes causes irritation. Contact with skin may cause dermatitis; *General Treatment for Exposure*: INHALATION: move to fresh air; if breathing has stopped, begin artificial respiration. INGESTION: give large amounts of water; induce vomiting. EYES: flush with water for at least 15 min. SKIN: wash with soap and water; *Toxicity by Inhalation (Threshold Limit Value)*: 0.1 mg/m^3 (as cobalt); *Short-Term Exposure Limits*: Data not available; *Toxicity by Ingestion*: Grade 3; LD$_{50}$ 50 - 500 mg/kg; *Late Toxicity*: Data not available; *Vapor (Gas) Irritant Characteristics*: Data not available; *Liquid or Solid Irritant Characteristics*: Data not available; *Odor Threshold*: Data not available.

Fire Hazards - *Flash Point:* Not flammable; *Flammable Limits in Air* (%): Not flammable; *Fire Extinguishing Agents*: Not pertinent; *Fire Extinguishing Agents Not To Be Used*: Not pertinent; *Special Hazards of Combustion Products*: Toxic cobalt oxide fumes form during fires; *Behavior in Fire*: No data; *Ignition Temperature*: Not pertinent; *Electrical Hazard*: Not pertinent; *Burning Rate*: Not pertinent.

Chemical Reactivity - *Reactivity with Water*: No reaction; *Reactivity with Common Materials*: No reactions reported; *Stability During Transport*: Stable; *Neutralizing Agents for Acids and Caustics*: Not pertinent; *Polymerization*: Not pertinent; *Inhibitor of Polymerization*: Not pertinent.

COBALT CHLORIDE

Chemical Designations - *Synonyms*: Cobalt (II) Chloride; Cobaltous Chloride; Cobaltous Chloride Dihydrate; Cobaltous Chloride Hexahydrate; *Chemical Formula*: CoCl$_2$.

Observable Characteristics - *Physical State (as normally shipped)*: Solid; *Color*: Pink to red; *Odor*: Very slight acrid.

Physical and Chemical Properties - *Physical State at 15 °C and 1 atm.*: Solid; *Molecular Weight*: 237.9; *Boiling Point at 1 atm.*: Not pertinent (decomposes); *Freezing Point*: 187, 86, 359; *Critical Temperature*: Not pertinent; *Critical Pressure*: Not pertinent; *Specific Gravity*: 1.924 at 20°C (solid); *Vapor (Gas) Density*: Not pertinent; *Ratio of Specific Heats of Vapor (Gas)*: Not pertinent; *Latent Heat of Vaporization*: Not pertinent; *Heat of Combustion*: Not pertinent; *Heat of Decomposition*: Not pertinent.

Health Hazards Information - *Recommended Personal Protective Equipment*: Rubber gloves; side-shield goggles; Bu. of Mines respirator; protective clothing; *Symptoms Following Exposure*: Inhalation causes respiratory disease, shortness of breath, and coughing; permanent disability may occur. Ingestion causes pain, vomiting, and diarrhea. Contact causes irritation of eyes and may cause skin rash; *General Treatment for Exposure*: INHALATION: move victim to fresh air; if breathing has stopped, begin artificial respiration and call a doctor. INGESTION: give large amount of water; induce vomiting. EYES: flush with water at least 15 min.; consult physician if irritation persists. SKIN: flush with water; *Toxicity by Inhalation (Threshold Limit Value)*: 0.1 mg/m^3 (as cobalt); *Short-Term Exposure Limits*: Data not available; *Toxicity by Ingestion*: Grade 3; LD$_{50}$ 50-500 mg/kg; *Late Toxicity*: Data not available; *Vapor (Gas) Irritant Characteristics*: Data not available; *Liquid or Solid Irritant Characteristics*: Data not available; *Odor Threshold*: Data not available.

Fire Hazards - *Flash Point:* Not flammable; *Flammable Limits in Air* (%): Not flammable; *Fire Extinguishing Agents*: Not pertinent; *Fire Extinguishing Agents Not To Be Used*: Not pertinent; *Special Hazards of Combustion Products*: Toxic cobalt oxide fumes can form in fire situations; *Behavior in Fire*: Not pertinent; *Ignition Temperature*: Not pertinent; *Burning Rate*: Not pertinent.

Chemical Reactivity - *Reactivity with Water*: No reaction; *Reactivity with Common Materials*: No data; *Stability During Transport*: Stable; *Neutralizing Agents for Acids and Caustics*: Not pertinent; *Polymerization*: Not pertinent; *Inhibitor of Polymerization*: Not pertinent.

COBALT NITRATE

Chemical Designations - *Synonyms*: Cobalt (II) Nitrate; Cobaltous Nitrate; Cobaltous Nitrate Hexahydrate; *Chemical Formula*: $Co(NO_3)_2 \cdot 6H_2O$.

Observable Characteristics - *Physical State (as normally shipped)*: Solid; *Color*: Red; *Odor*: None.

Physical and Chemical Properties - *Physical State at 15 °C and 1 atm.*: Solid; *Molecular Weight*: 291.04; *Boiling Point at 1 atm.*: Not pertinent (decomposes); *Freezing Point*: 131, 55, 328; *Critical Temperature*: Not pertinent; *Critical Pressure*: Not pertinent; *Specific Gravity*: 1.54 at 20°C (solid); *Vapor (Gas) Density*: Not pertinent; *Ratio of Specific Heats of Vapor (Gas)*: Not pertinent; *Latent Heat of Vaporization*: Not pertinent; *Heat of Combustion*: Not pertinent; *Heat of Decomposition*: Not pertinent.

Health Hazards Information - *Recommended Personal Protective Equipment*: Bu. Mines approved respirator; rubber gloves; safety goggles; protective clothing; *Symptoms Following Exposure*: Inhalation causes shortness of breath and coughing; permanent disability may occur. Ingestion causes pain and vomiting. Contact with eyes or skin causes irritation; *General Treatment for Exposure*: INHALATION: move to fresh air; if breathing has stopped, begin artificial respiration and call a doctor. INGESTION: give large amounts of water; induce vomiting; call a doctor. EYES: flush with water for at least 15 min. SKIN: flush with water; *Toxicity by Inhalation (Threshold Limit Value)*: Data not available; *Short-Term Exposure Limits*: Data not available; *Toxicity by Ingestion*: Grade 3; LD_{50} ~400 mg/kg (rabbit); *Late Toxicity*: Causes malignant tumors in rabbits; *Vapor (Gas) Irritant Characteristics*: Data not available; *Liquid or Solid Irritant Characteristics*: Data not available; *Odor Threshold*: Not pertinent.

Fire Hazards - *Flash Point:* Not flammable; *Flammable Limits in Air (%)*: Not flammable; *Fire Extinguishing Agents*: Not pertinent; *Fire Extinguishing Agents Not To Be Used*: Not pertinent; *Special Hazards of Combustion Products*: Toxic oxides of nitrogen form in fires; *Behavior in Fire*: Can increase fire intensity; *Ignition Temperature*: Not pertinent; *Electrical Hazard*: Not pertinent; *Burning Rate*: Not pertinent.

Chemical Reactivity - *Reactivity with Water*: No reaction; *Reactivity with Common Materials*: Contact with wood or paper may result in fire; *Stability During Transport*: Stable; *Neutralizing Agents for Acids and Caustics*: Not pertinent; *Polymerization*: Not pertinent; *Inhibitor of Polymerization*: Not pertinent.

COPPER ACETATE

Chemical Designations - *Synonyms*: Acetic Acid; Cupric Salt; Crystallized Verdigris; Cupric Acetate Monohydrate; Neutral Verdigris; *Chemical Formula*: $Cu(C_2H_3O_2)_2 \cdot H_2O$.

Observable Characteristics - *Physical State (as normally shipped)*: Solid; *Color*: Bluish green; *Odor*: None.

Physical and Chemical Properties - *Physical State at 15 °C and 1 atm.*: Solid; *Molecular Weight*: 199.65; *Boiling Point at 1 atm.*: Not pertinent (decomposes); *Freezing Point*: 239, 115, 388; *Critical Temperature*: Not pertinent; *Critical Pressure*: Not pertinent; *Specific Gravity*: 1.9 at 20°C (solid); *Vapor (Gas) Density*: Not pertinent; *Ratio of Specific Heats of Vapor (Gas)*: Not pertinent; *Latent Heat of Vaporization*: Not pertinent; *Heat of Combustion*: Not pertinent; *Heat of Decomposition*: Not pertinent.

Health Hazards Information - *Recommended Personal Protective Equipment*: Dust mask; goggles or face shield; protective gloves; *Symptoms Following Exposure*: Inhalation of dusts causes irritation of throat and lungs. Ingestion of large amounts causes violent vomiting and purging, intense pain, collapse, coma, convulsions, and paralysis. Contact with solutions irritates eyes; contact with solid causes severe eye surface injury and irritation of skin; *General Treatment for Exposure*: INHALATION: move to fresh air. INGESTION: give large amount of water; induce vomiting; get medical attention. EYES: flush with water for at least 15 min.; get medical attention if injury was caused by solid. SKIN: flush with water; *Toxicity by Inhalation (Threshold Limit Value)*: Data not available; *Short-Term Exposure Limits*: Data not available; *Toxicity by Ingestion*: Grade 2; LD_{50} 0.5 - 5 g/kg (rat); *Late Toxicity*: Causes degeneration of liver in dogs; *Vapor (Gas) Irritant Characteristics*: Data not available; *Liquid or Solid Irritant Characteristics*: Data not available; *Odor Threshold*: Data not available.

Fire Hazards - *Flash Point :* Not flammable; *Flammable Limits in Air* (%): Not flammable; *Fire Extinguishing Agents*: Not pertinent; *Fire Extinguishing Agents Not To Be Used*: Not pertinent; *Special Hazards of Combustion Products*: Irritating vapors of acetic acid form in fire situations; *Behavior in Fire*: No data; *Ignition Temperature :* Not pertinent; *Electrical Hazard*: Not pertinent; *Burning Rate*: Not pertinent.

Chemical Reactivity - *Reactivity with Water*: No reaction; *Reactivity with Common Materials*: No data; *Stability During Transport*: Stable; *Neutralizing Agents for Acids and Caustics*: Not pertinent; *Polymerization*: Not pertinent; *Inhibitor of Polymerization*: Not pertinent.

COPPER ARSENITE

Chemical Designations - *Synonyms*: Cupric Arsenite; Swedish Green; Scheele's Green; Cupric Green; Copper Orthoarsenite; *Chemical Formula*: $CuHAsO_3$.

Observable Characteristics - *Physical State (as normally shipped)*: Solid; *Color*: Green; yellowish green; *Odor*: None.

Physical and Chemical Properties - *Physical State at 15 °C and 1 atm.*: Solid; *Molecular Weight*: 277.4; *Boiling Point at 1 atm.*: Decomposes; *Freezing Point*: Not pertinent; *Critical Temperature*: Not pertinent; *Critical Pressure*: Not pertinent; *Specific Gravity*: >1.1 at 20°C (solid); *Vapor (Gas) Density*: Not pertinent; *Ratio of Specific Heats of Vapor (Gas)*: Not pertinent; *Latent Heat of Vaporization*: Not pertinent; *Heat of Combustion*: Not pertinent; *Heat of Decomposition*: Not pertinent;

Health Hazards Information - *Recommended Personal Protective Equipment*: Dust respirator; rubber gloves; goggles or face shield; *Symptoms Following Exposure*: Dust irritates eyes. Ingestion causes gastric disturbance, tremors, muscular cramps, and nervous collapse that may cause death; *General Treatment for Exposure*: Following ingestion or unusually severe exposure to dust, get medical attention. Alert doctor to possibility of arsenic poisoning. EYES: flush with water for 15 min. SKIN: wash with soap and water. INGESTION: give large amounts of water; induce vomiting; give cathartic, such as 2 oz. of Epsom salt in water; *Toxicity by Inhalation (Threshold Limit Value)*: 0.5 mg/m^3 (as arsenic); *Short-Term Exposure Limits*: Data not available; *Toxicity by Ingestion*: Grade 3; LD_{50} 50 to 500 mg/kg; *Late Toxicity*: Arsenic poisoning; *Vapor (Gas) Irritant Characteristics*: Not pertinent; *Liquid or Solid Irritant Characteristics*: Data not available; *Odor Threshold*: Not pertinent.

Fire Hazards - *Flash Point :* Not flammable; *Flammable Limits in Air* (%): Not flammable; *Fire Extinguishing Agents*: Not pertinent; *Fire Extinguishing Agents Not To Be Used*: Not pertinent; *Special Hazards of Combustion Products*: Poisonous, volatile arsenic oxides may be formed in fires; *Behavior in Fire*: Not pertinent; *Ignition Temperature :* Not pertinent; *Electrical Hazard*: Not pertinent; *Burning Rate*: Not pertinent.

Chemical Reactivity - *Reactivity with Water*: No reaction; *Reactivity with Common Materials*: No reaction; *Stability During Transport*: Stable; *Neutralizing Agents for Acids and Caustics*: Not pertinent; *Polymerization*: Not pertinent; *Inhibitor of Polymerization*: Not pertinent.

COPPER BROMIDE

Chemical Designations - *Synonyms*: Cupric Bromide, Anhydrous; *Chemical Formula*: $CuBr_2$.

Observable Characteristics - *Physical State (as normally shipped)*: Solid; *Color*: Black; *Odor*: None.

Physical and Chemical Properties - *Physical State at 15 °C and 1 atm.*: Solid; *Molecular Weight*: 223.35; *Boiling Point at 1 atm.*: Not pertinent (decomposes); *Freezing Point*: 928, 498, 771; *Critical Temperature*: Not pertinent; *Critical Pressure*: Not pertinent; *Specific Gravity*: 4.77 at 20°C (solid); *Vapor (Gas) Density*: Not pertinent; *Ratio of Specific Heats of Vapor (Gas)*: Not pertinent; *Latent Heat of Vaporization*: Not pertinent; *Heat of Combustion*: Not pertinent; *Heat of Decomposition*: Not pertinent.

Health Hazards Information - *Recommended Personal Protective Equipment*: Dust mask; goggles or face shield; protective gloves; *Symptoms Following Exposure*: Inhalation of dust causes irritation of throat and lungs. Ingestion of large amounts causes violent vomiting and purging, intense pain, collapse, coma, convulsions, and paralysis. Contact with solutions causes eye irritation; contact with solid causes severe eye surface injury and skin irritation; *General Treatment for Exposure*:

INHALATION: move to fresh air. INGESTION: give large amount of water; induce vomiting; get medical attention. EYES: flush with water for at least 15 min.; get medical attention if injury was caused by solid. SKIN: flush with water; *Toxicity by Inhalation (Threshold Limit Value)*: Data not available; *Short-Term Exposure Limits*: Data not available; *Toxicity by Ingestion*: Grade 3; LD_{50} 50 to 500 mg/kg; *Late Toxicity*: Data not available; *Vapor (Gas) Irritant Characteristics*: Data not available; *Liquid or Solid Irritant Characteristics*: Data not available; *Odor Threshold*: Data not available.

Fire Hazards - *Flash Point :* Not flammable; *Flammable Limits in Air (%)*: Not flammable; *Fire Extinguishing Agents*: Not pertinent; *Fire Extinguishing Agents Not To Be Used*: Not pertinent; *Special Hazards of Combustion Products*: Irritating hydrogen bromide gas may form in fire; *Behavior in Fire*: Not pertinent; *Ignition Temperature :* Not pertinent; *Electrical Hazard*: Not pertinent; *Burning Rate*: Not pertinent.

Chemical Reactivity - *Reactivity with Water*: No reaction; *Reactivity with Common Materials*: No reaction; *Stability During Transport*: Stable; *Neutralizing Agents for Acids and Caustics*: Not pertinent; *Polymerization*: Not pertinent; *Inhibitor of Polymerization*: Not pertinent.

COPPER CHLORIDE

Chemical Designations - *Synonyms*: Cupric Chloride Dehydrate; Eriochalcite (anhydrous); *Chemical Formula*: $CuCl_2H_2O$.

Observable Characteristics - *Physical State (as normally shipped)*: Solid; *Color*: Green; Blue-green; blue; *Odor*: None.

Physical and Chemical Properties - *Physical State at 15 °C and 1 atm.*: Solid; *Molecular Weight*: 170.48 (dihydrate); *Boiling Point at 1 atm*: Not pertinent; *Freezing Point*: not pertinent; *Critical Temperature*: not pertinent; *Critical Pressure*: Not pertinent; *Specific Gravity*: 2.54 at 20°C (solid); *Vapor (Gas) Density*: Not pertinent; *Ratio of Specific Heats of Vapor (Gas)*: Not pertinent; *Latent Heat of Vaporization*: Not pertinent; *Heat of Combustion*: Not pertinent; *Heat of Decomposition*: Not pertinent.

Health Hazards Information - *Recommended Personal Protective Equipment*: Bu. Mines approved respirator; rubber gloves; safety goggles; *Symptoms Following Exposure*: Inhalation causes coughing and sneezing. Ingestion causes pain and vomiting. Contact with solution irritates eyes: contact with solid causes severe eye surface injury and skin irritation; *General Treatment for Exposure*: INHALATION: move to fresh air. INGESTION: give large amounts of water; induce vomiting; get medical attention. EYES: flush with water for 15 min.; consult with physician if the injury was caused by solid. SKIN: flush with water; *Toxicity by Inhalation (Threshold Limit Value)*: Data not available; *Short-Term Exposure Limits*: Data not available; *Toxicity by Ingestion*: Grade 3; LD_{50}50-500mg/kg; *Late Toxicity*: causes liver damage in rabbits; *Vapor (Gas) Irritant Characteristics*: Data not available; *Liquid or Solid Irritant Characteristics*: Data not available; *Odor Threshold*: Data not available.

Fire Hazards - *Flash Point :* Not flammable; *Flammable Limits in Air (%)*: Not flammable; *Fire Extinguishing Agents*: Not pertinent; *Fire Extinguishing Agents Not To Be Used*: Not pertinent; *Special Hazards of Combustion Products*: Irritating hydrogen chloride gas may form in fire; *Behavior in Fire*: Not pertinent; *Ignition Temperature :* Not pertinent; *Electrical Hazard*: Not pertinent; *Burning Rate*: Not pertinent.

Chemical Reactivity - *Reactivity with Water*: No reaction; *Reactivity with Common Materials*: In presence of moisture may corrode metals; the reaction is not hazardous; *Stability During Transport*: Stable; *Neutralizing Agents for Acids and Caustics*: Flush with water, rinse with dilute solution of sodium bicarbonate or soda ash; *Polymerization*: Not pertinent; *Inhibitor of Polymerization*: Not pertinent.

COPPER CYANIDE

Chemical Designations - *Synonyms*: Cupricin; Cuprous Cyanide; *Chemical Formula*: CuCN.

Observable Characteristics - *Physical State (as normally shipped)*: powder; *Color*: white; *Odor*: Data not available.

Physical and Chemical Properties - *Physical State at 15 °C and 1 atm*: Solid; *Molecular Weight*:

89.56; *Boiling Point at 1 atm*: Not pertinent; *Freezing Point*: Not pertinent; *Critical* Temperature: Not pertinent; *Critical Pressure*: Not pertinent; *Specific Gravity*: 2.92 at 20°C (solid); *Vapor (Gas) Density*: Not pertinent; *Ratio of Specific Heats of Vapor (Gas)*: Not pertinent; *Latent Heat of Vaporization*: Not pertinent; *Heat of Combustion*: Not pertinent; *Heat of Decomposition*: Not pertinent.

Health Hazards Information - *Recommended Personal Protective Equipment*: Dust respirator; protective goggles or face mask; protective clothing; *Symptoms Following Exposure*: Following severe exposure to dust, symptoms of cyanide poisoning may develop (see ingestion). Ingestion causes anxiety, confusion, dizziness, sudden convulsion, and paralysis. Contact with eyes causes irritation; *General Treatment for Exposure*: Get medical attention after all exposures to this substance. INHALATION: remove victim to fresh air. INGESTION: if breathing has stopped, begin artificial respiration immediately; administer by inhalation amyl nitrite pearls for 15 - 30 seconds of every minute, while a sodium nitrite solution is being prepared; discontinue amyl nitrite and immediately inject intravenously 10 ml of a 3% sol. Of sodium nitrite (nonsterile if necessary) over a period of 2 to 4 min.; do not remove needle; through same needle infuse 50 ml of a 25% aqueous soln. Of sodium thiosulfate; injection should take about 10 min. (Concentrations of 5-50% are permissible if total dose is approx. 12 grams). Oxygen therapy may be of value in combination with the above. If symptoms recur, repeat injections of nitrite and thiosulfate at half the above doses. EYES: flush with water for at least 15 min. SKIN: flush with water wash with soap and water; *Toxicity by Inhalation (Threshold Limit Value)*: 5 mg/m^3 (as cyanide); *Short-Term Exposure Limits*: Data not available; *Toxicity by Ingestion*: Grade 4; LD_{50} <50 mg/kg; *Late* Toxicity: Data not available; *Vapor (Gas) Irritant Characteristics*: Data not available; *Liquid or Solid Irritant Characteristics*: Data not available; *Odor Threshold*: Data not available.

Fire Hazards - *Flash Point* : Not flammable; *Flammable Limits in Air* (%): Not flammable; *Fire Extinguishing Agents*: Not pertinent; *Fire Extinguishing Agents Not To Be Used*: Not pertinent; *Special Hazards of Combustion Products*: Toxic hydrogen cyanide gas may form in fires; *Behavior in Fire*: Not pertinent; *Ignition Temperature* : Not pertinent; *Electrical Hazard*: Not pertinent; *Burning Rate*: Not pertinent.

Chemical Reactivity - *Reactivity with Water*: No reaction; *Reactivity with Common Materials*: No reaction; *Stability During Transport*: Stable, in presence of moisture, toxic hydrogen cyanide gas may collect in enclosed spaces; *Neutralizing Agents for Acids and Caustics*: Not pertinent; *Polymerization*: Not pertinent; *Inhibitor of Polymerization*: Not pertinent.

COPPER FLUOROBORATE

Chemical Designations - *Synonyms*: Copper Borofluride Solution; Copper (II) Fluoborate Solution; Cupric Fluoborate Solution; *Chemical Formula*: $Cu(BF_4)_2 \cdot H_2O$.

Observable Characteristics - *Physical State (as normally shipped)*: Liquid; *Color*: Clear, dark blue; *Odor*: None.

Physical and Chemical Properties - *Physical State at 15 °C and 1 atm.*: Liquid; *Molecular Weight*: 237.16 (solute only); *Boiling Point at 1 atm.*: 121/100/373; *Freezing Point*: Data not available; *Critical Temperature*: Not pertinent; *Critical Pressure*: Not pertinent; *Specific Gravity*: 1.54 at 20°C (liquid); *Vapor (Gas) Density*: Not pertinent; *Ratio of Specific Heats of Vapor (Gas)*: Not pertinent; *Latent Heat of Vaporization*: Not pertinent; *Heat of Combustion*: Not pertinent; *Heat of Decomposition*: Not pertinent.

Health Hazards Information - *Recommended Personal Protective Equipment*: Goggles or face shield; rubber apron and gloves; *Symptoms Following Exposure*: Inhalation of mist irritates nose and throat. Ingestion causes pain and vomiting. Contact causes severe irritation of skin; *General Treatment for Exposure*: INHALATION: move to fresh air. INGESTION: give large amounts of water and induce vomiting if required. EYES: flush with water for at least 15 min.; get medical attention if irritation persists. SKIN: flush with water; *Toxicity by Inhalation (Threshold Limit Value)*: Data not available; *Short-Term Exposure Limits*: Data not available; *Toxicity by Ingestion*: Grade 3; LD_{50} 50 - 500 mg/kg; *Late Toxicity*: Data not available; *Vapor (Gas) Irritant Characteristics*: Data not available; *Liquid or Solid Irritant Characteristics*: Data not available; *Odor Threshold*: Data not available.

Fire Hazards - *Flash Point :* Not flammable; *Flammable Limits in Air* (%): Not flammable; *Fire Extinguishing Agents*: Not pertinent; *Fire Extinguishing Agents Not To Be Used*: Not pertinent; *Special Hazards of Combustion Products*: Irritating hydrogen fluoride gas may form in fires; *Behavior in Fire*: Not pertinent; *Ignition Temperature* : Not pertinent; *Electrical Hazard*: Not pertinent; *Burning Rate*: Not pertinent.

Chemical Reactivity - *Reactivity with Water*: No reaction; *Reactivity with Common Materials*: May corrode some metals; *Stability During Transport*: Stable; *Neutralizing Agents for Acids and Caustics*: Flush with water, rinse with dilute solution of sodium bicarbonate or soda ash; *Polymerization*: Not pertinent; *Inhibitor of Polymerization*: Not pertinent.

COPPER IODIDE

Chemical Designations - *Synonyms*: Cuprous Iodine; Marshite; *Chemical Formula*: CuI.

Observable Characteristics - *Physical State (as normally shipped)*: Solid; *Color*: Beige; *Odor*: None.

Physical and Chemical Properties - *Physical State at 15 °C and 1 atm.*: Solid; *Molecular Weight*: 190.4; *Boiling Point at 1 atm.*: 2,354, 1,290, 1,563; *Freezing Point*: 1,121, 605, 878; *Critical Temperature*: Not pertinent; *Critical Pressure*: Not pertinent; *Specific Gravity*: 5.62 at 20°C (solid); *Vapor (Gas) Density*: Not pertinent; *Ratio of Specific Heats of Vapor (Gas)*: Not pertinent; *Latent Heat of Vaporization*: Not pertinent; *Heat of Combustion*: Not pertinent; *Heat of Decomposition*: Not pertinent.

Health Hazards Information - *Recommended Personal Protective Equipment*: Dust mask; goggles or face shield; protective gloves; *Symptoms Following Exposure*: Inhalation causes irritation of nose and throat. Ingestion of copper salts produces violent vomiting and purging, intense pain, collapse, coma, convulsion, and paralysis. Contact with eyes or skin causes irritation.; *General Treatment for Exposure*: INHALATION: move to fresh air. INGESTION: give large amounts of water: induce vomiting; get medical attention. EYES: flush with water for at least 15 min. SKIN: flush with water; wash with soap and water; *Toxicity by Inhalation (Threshold Limit Value)*: Data not available; *Short-Term Exposure Limits*: Data not available; *Toxicity by Ingestion*: Grade 3; LD_{50} 50 - 500 mg/kg; *Late Toxicity*: Data not available; *Vapor (Gas) Irritant Characteristics*: Data not available; *Liquid or Solid Irritant Characteristics*: Data not available; *Odor Threshold*: Data not available.

Fire Hazards - *Flash Point :* Not flammable; *Flammable Limits in Air* (%): Not flammable; *Fire Extinguishing Agents*: Not pertinent; *Fire Extinguishing Agents Not To Be Used*: Not pertinent; *Special Hazards of Combustion Products*: Irritating hydrogen iodide or iodine vapors may form in fire; *Behavior in Fire*: Not pertinent; *Ignition Temperature* : Not pertinent; *Electrical Hazard*: Not pertinent; *Burning Rate*: Not pertinent.

Chemical Reactivity - *Reactivity with Water*: No reaction; *Reactivity with Common Materials*: No reaction; *Stability During Transport*: Stable; *Neutralizing Agents for Acids and Caustics*: Not pertinent; *Polymerization*: Not pertinent; *Inhibitor of Polymerization*: Not pertinent.

COPPER NAPHTHENATE

Chemical Designations - *Synonyms*: Paint Drier; *Chemical Formula*: Mixture.

Observable Characteristics - *Physical State (as normally shipped)*: Liquid; *Color*: Dark green; *Odor*: Like gasoline: slight aromatic.

Physical and Chemical Properties - *Physical State at 15 °C and 1 atm.*: Liquid; *Molecular Weight*: Mixture; *Boiling Point at 1 atm.*: 310 - 395, 154 - 202, 427 - 475; *Freezing Point*: Not pertinent; *Critical* Temperature: Not pertinent; *Critical Pressure*: Not pertinent; *Specific Gravity*: 0.93 ~ 1.05 at 25°C (liquid); *Vapor (Gas) Density*: Not pertinent; *Ratio of Specific Heats of Vapor (Gas)*: Not pertinent; *Latent Heat of Vaporization*: Not pertinent; *Heat of Combustion*: (est.) -17,600, -9,800, -410; *Heat of Decomposition*: Not pertinent.

Health Hazards Information - *Recommended Personal Protective Equipment*: Goggles or face shield; plastic gloves (as for gasoline); *Symptoms Following Exposure*: Vapor causes mild irritation of eyes and mild irritation of respiratory tract if inhaled. Ingestion causes irritation to stomach. Aspiration causes severe lung irritation and rapidly developing pulmonary edema; central nervous system excitement

followed by depression; *General Treatment for Exposure*: INHALATION: remove victim to fresh air. EYES: wash with copious amounts of water for at least 15 min. SKIN: wipe off and wash with soap and water. INGESTION: do NOT induce vomiting; guard against aspiration to lungs. ASPIRATION: enforce bed rest; give oxygen: call a doctor; *Toxicity by Inhalation (Threshold Limit Value)*: 500 ppm; *Short-Term Exposure* Limits: Data not available; *Toxicity by Ingestion*: Grade 1; oral rat LD_{50}=4 - 6 g/kg; *Late Toxicity*: Data not available; *Vapor (Gas) Irritant Characteristics*: Vapors are non-irritating to the eyes and throat; *Liquid or Solid Irritant Characteristics*: Minimum hazard. If spilled on clothing and allowed to remain, may cause smarting and reddening of skin; *Odor Threshold*: Data not available.
Fire Hazards - *Flash Point (deg. F)*: 100 CC (typical); *Flammable Limits in Air (%)*: 0.8 - 5.0 (mineral spirits); *Fire Extinguishing Agents*: Dry chemical, foam, carbon dioxide; *Fire Extinguishing Agents Not To Be Used*: Water may be ineffective; *Special Hazards of Combustion Products*: Not pertinent; *Behavior in Fire*: Not pertinent; *Ignition Temperature (deg. F)*: 540 (mineral spirits); *Electrical Hazard*: Not pertinent; *Burning Rate*: 4 mm/min.
Chemical Reactivity - *Reactivity with Water*: No reaction; *Reactivity with Common Materials*: No reaction; *Stability During Transport*: Stable; *Neutralizing Agents for Acids and Caustics*: Not pertinent; *Polymerization*: Not pertinent; *Inhibitor of Polymerization*: Not pertinent.

COPPER NITRATE

Chemical Designations - *Synonyms*: Cupric Nitrate Trihydrate; Gerhardite; *Chemical Formula*: $Cu(NO_3)\cdot3H_2O$.
Observable Characteristics - *Physical State (as normally shipped)*: Solid; *Color*: Blue; *Odor*: None.
Physical and Chemical Properties - *Physical State at 15 °C and 1 atm.*: Solid; *Molecular Weight*: 241.60; *Boiling Point at 1 atm.*: Not pertinent (decomposes); *Freezing Point*: 237.1, 114.5, 387.7; *Critical Temperature*: Non pertinent; *Critical Pressure*: Not pertinent; *Specific Gravity*: 2.32 at 20°C (solid); *Vapor (Gas) Density*: Not pertinent; *Ratio of Specific Heats of Vapor (Gas)*: Not pertinent; *Latent Heat of Vaporization*: Not pertinent; *Heat of Combustion*: Not pertinent; *Heat of Decomposition*: Not pertinent.
Health Hazards Information - *Recommended Personal Protective Equipment*: Dust mask; goggles or face shield; protective gloves; *Symptoms Following Exposure*: Inhalation causes irritation of throat and lungs. Ingestion of large amounts causes violent vomiting and purging, intense pain, collapse, coma, convulsion, and paralysis. Solution irritates eyes: contact with solid causes severe eye surface injury and skin irritation; *General Treatment for Exposure*: INHALATION: move to fresh air. INGESTION: give large amounts of water; induce vomiting; get medical attention. EYES: flush with water for at least 15 minutes; get medical attention if injury was caused by solid. SKIN: flush with water. *Toxicity by Inhalation (Threshold Limit Value)*: Data not available; *Short-Term Exposure Limits*: Data not available; *Toxicity by Ingestion*: Grade 2; LD_{50} 0.5 - 5 g/kg; *Late Toxicity*: Data not available; *Vapor (Gas) Irritant Characteristics*: Data not available; *Liquid or Solid Irritant Characteristics*: Data not available; *Odor Threshold*: Data not available.
Fire Hazards - *Flash Point :* Not flammable; *Flammable Limits in Air (%)*: Not flammable; *Fire Extinguishing Agents*: Not pertinent; *Fire Extinguishing Agents Not To Be Used*: Not pertinent; *Special Hazards of Combustion Products*: Toxic and irritating oxides of nitrogen may form in fire; *Behavior in Fire*: Can increase intensity of fire if in contact with combustible material; *Ignition Temperature :* Not pertinent; *Electrical Hazard*: Not pertinent; *Burning Rate*: Not pertinent.
Chemical Reactivity - *Reactivity with Water*: No reaction; *Reactivity with Common Materials*: Mixtures with wood, paper, and other combustibles may catch fire; *Stability During Transport*: Stable; *Neutralizing Agents for Acids and Caustics*: Not pertinent; *Polymerization*: Not pertinent; *Inhibitor of Polymerization*: Not pertinent.

COPPER OXALATE

Chemical Designations - *Synonyms*: Cupric Oxalate Hemihydrate; *Chemical Formula*: $CuC_2O_4\cdot\frac{1}{2}H_2O$.
Observable Characteristics - *Physical State (as normally shipped)*: Solid; *Color*: Bluish white; *Odor*: None.

Physical and Chemical Properties - *Physical State at 15 °C and 1 atm.*: Solid; *Molecular Weight*: 160.6; *Boiling Point at 1 atm.*: Not pertinent; *Freezing Point*: Not pertinent; *Critical Temperature*: Not pertinent; *Critical Pressure*: Not pertinent; *Specific Gravity*: > 1 at 20°C (solid); *Vapor (Gas) Density*: Not pertinent; *Ratio of Specific Heats of Vapor (Gas)*: Not pertinent; *Latent Heat of Vaporization*: Not pertinent; *Heat of Combustion*: Not pertinent; *Heat of Decomposition*: Not pertinent.

Health Hazards Information - *Recommended Personal Protective Equipment*: Dust mask; goggles or face shield; protective gloves; *Symptoms Following Exposure*: Inhalation causes irritation of nose and throat. Ingestion of very large amounts may produce symptoms of oxalate poisoning; watch for edema of the glottis and delayed constriction of esophagus. Contact with eyes causes irritation; *General Treatment for Exposure*: INHALATION; remove to fresh air; if exposure has been prolonged watch for symptoms of oxalate poisoning (nausea, shock, collapse, and convulsions). INGESTION: give large amounts of water; induce vomiting; get medical attention. EYES: flush with water foe at least 15 min. SKIN: flush with water; *Toxicity by Inhalation (Threshold Limit Value)*: Data not available; *Short-Term Exposure Limits*: Data not available; *Toxicity by Ingestion*: Data not available; *Late Toxicity*: Data not available; *Vapor (Gas) Irritant Characteristics*: Data not available; *Liquid or Solid Irritant Characteristics*: Data not available; *Odor Threshold*: Data not available.

Fire Hazards - *Flash Point :* Not flammable; *Flammable Limits in Air (%)*: Not flammable; *Fire Extinguishing Agents*: Not pertinent; *Fire Extinguishing Agents Not To Be Used*: Not pertinent; *Special Hazards of Combustion Products*: Toxic carbon monoxide gas may form in fire; *Behavior in Fire*: Not pertinent; *Ignition Temperature :* Not pertinent; *Electrical Hazard*: Not pertinent; *Burning Rate*: Not pertinent.

Chemical Reactivity - *Reactivity with Water*: No reaction; *Reactivity with Common Materials*: No reaction; *Stability During Transport*: Stable; *Neutralizing Agents for Acids and Caustics*: Not pertinent; *Polymerization*: Not pertinent; *Inhibitor of Polymerization*: Not pertinent.

COPPER SULFATE

Chemical Designations - *Synonyms*: Blue Vitriol; Copper Sulfate Pentahydrate; Cupric Sulfate; Sulfate of Copper; *Chemical Formula*: $CuSO_4$-$5H_2O$.

Observable Characteristics - *Physical State (as normally shipped)*: Solid; *Color*: Blue; *Odor*: None.

Physical and Chemical Properties - *Physical State at 15 °C and 1 atm.*: Solid; *Molecular Weight*: 249.7; *Boiling Point at 1 atm.*: Not pertinent; *Freezing Point*: Not pertinent; *Critical Temperature*: Not pertinent; *Critical Pressure*: Not pertinent; *Specific Gravity*: 2.29 at 15°C (solid); *Vapor (Gas) Density*: Not pertinent; *Ratio of Specific Heats of Vapor (Gas)*: Not pertinent; *Latent Heat of Vaporization*: Not pertinent; *Heat of Combustion*: Not pertinent; *Heat of Decomposition*: Not pertinent.

Health Hazards Information - *Recommended Personal Protective Equipment*: Filtering masks to minimize inhalation of dust; *Symptoms Following Exposure*: INGESTION: copper sulfate may induce severe gastroenteric anuria, hematuria, anemia, increase in white blood cells, icterus, coma, respiratory difficulties, and circulatory failure; *General Treatment for Exposure*: INGESTION: induce vomiting and administer gastric leverage; give a saline cathartic, fluid therapy, and transfusions if required; calcium disodium EDTA has been found moderately effective. SKIN AND EYES: wash affected tissues with water; *Toxicity by Inhalation (Threshold Limit Value)*: Not pertinent; *Short-Term Exposure Limits*: Not pertinent; *Toxicity by Ingestion*: Grade 3; LD_{50} 50 to 500 mg/kg (rat); *Late Toxicity*: Causes liver, kidney and testicular damage in rats; *Vapor (Gas) Irritant Characteristics*: Not pertinent; *Liquid or Solid Irritant Characteristics*: Causes smarting of the skin and first degree burns on short exposure; may cause second degree burns on long exposure; *Odor Threshold*: Not pertinent.

Fire Hazards - *Flash Point :* Not flammable; *Flammable Limits in Air (%)*: Not flammable; *Fire Extinguishing Agents*: Not pertinent; *Fire Extinguishing Agents Not To Be Used*: Not pertinent; *Special Hazards of Combustion Products*: Not pertinent; *Behavior in Fire*: Not pertinent; *Ignition Temperature :* Not flammable; *Electrical Hazard*: Not pertinent; *Burning Rate*: Not flammable.

Chemical Reactivity - *Reactivity with Water*: No reaction; *Reactivity with Common Materials*: No reaction; *Stability During Transport*: Stable; *Neutralizing Agents for Acids and Caustics*: Not pertinent; *Polymerization*: Not pertinent; *Inhibitor of Polymerization*: Not pertinent.

CREOSOTE, COAL TAR
Chemical Designations - *Synonyms*: Cresote Oil; Dead Oil; *Chemical Formula*: Mixture.
Observable Characteristics - *Physical State (as normally shipped)*: Liquid; *Color*: Yellow to brown to black; *Odor*: Creosote or tarry; aromatic.
Physical and Chemical Properties - *Physical State at 15 ℃ and 1 atm.*: liquid; *Molecular Weight*: Mixture; *Boiling Point at 1 atm.*: >356, >180, >353; *Freezing Point*: Not pertinent; *Critical Temperature*: Not pertinent; *Critical Pressure*: Not pertinent; *Specific Gravity*: 1.05-1.09 at 15°C (liquid); *Vapor (Gas) Density*: Not pertinent; *Ratio of Specific Heats of Vapor (Gas)*: Not pertinent; *Latent Heat of Vaporization*: Not pertinent; *Heat of Combustion*: (est). - 12,500, -6,900, -290; *Heat of Decomposition*: Not pertinent.
Health Hazards Information - *Recommended Personal Protective Equipment*: All-service canister mask; rubber gloves; chemical safety goggles and/or shield; overalls or a neoprene apron; barrier creams; *Symptoms Following Exposure*: Vapors cause moderate irritation of nose and throat. Liquid cause severe burns of eyes and reddening and itching of skin. Prolonged contact with skin can cause burns. Ingestion causes salivation, vomiting, respiratory difficulties, thready pulse, vertigo, headache, loss of pupillary reflexes, hypothermia, cyanosis, mild convulsions; *General Treatment for Exposure*: INHALATION: remove victim to fresh air; if he is not breathing, give artificial respiration, preferably mouth-to-mouth; if breathing is difficult, give oxygen; call a physician. EYES: flush immediately with plenty of water for at least 15 min. And call a physician. SKIN: wipe with vegetable oil or margarine; then wash with soap and water. INGESTION: have victim drink water or milk; do NOT induce vomiting; *Toxicity by Inhalation (Threshold Limit Value)*: Data not available; *Short-Term Exposure Limits*: Data not available; *Toxicity by Ingestion*: Grade 2; LD_{50} 0.5 to 5 g/kg; *Late Toxicity*: Repeated exposures may cause cancer of skin; *Vapor (Gas) Irritant Characteristics*: Vapor cause moderate irritation such that personnel will find high concentrations unpleasant. The effect is temporary; *Liquid or Solid Irritant Characteristics*: Fairy severe skin irritant. May cause pain and second-degree burns after a few minutes' contact; *Odor Threshold*: Data not available.
Fire Hazards - *Flash Point (deg. F)*: > 160 CC; *Flammable Limits in Air (%)*: Not pertinent; *Fire Extinguishing Agents*: Dry chemical, carbon dioxide, or foam; *Fire Extinguishing Agents Not To Be Used*: Water may be ineffective; *Special Hazards of Combustion Products*: Data not available; *Behavior in Fire*: Heavy, irritating black smoke is formed; *Ignition Temperature (deg. F)*: 637; *Electrical Hazard*: Not pertinent; *Burning Rate*: Data not available.
Chemical Reactivity - *Reactivity with Water*: No reaction; *Reactivity with Common Materials*: No reaction; *Stability During Transport*: Stable; *Neutralizing Agents for Acids and Caustics*: Not pertinent; *Polymerization*: Not pertinent; *Inhibitor of Polymerization*: Not pertinent.

CRESOLS
Chemical Designations - *Synonyms*: Cresylic Acids; Hydroxytoluenes; Methylphenols; Oxytoluenes; Tar Acids; *Chemical Formula*: $CH_3C_6H_4OH$.
Observable Characteristics - *Physical State (as normally shipped)*: Liquid or solid; *Color*: Colorless or dark to yellow; *Odor*: Sweet, tarry.
Physical and Chemical Properties - *Physical State at 15 ℃ and 1 atm.*: Liquid; *Molecular Weight*: 108.13; *Boiling Point at 1 atm.*: >350, >177, >450; *Freezing Point*: Not pertinent; *Critical Temperature*: Not pertinent; *Critical Pressure*: Not pertinent; *Specific Gravity*: 1.03 - 1.07 at 20°C (liquid); *Vapor (Gas) Density*: Not pertinent; *Ratio of Specific Heats of Vapor (Gas)*: 1.073; *Latent Heat of Vaporization*: (est.) 200, 110, 4.6; *Heat of Combustion*: -14,720 to -14,740, -8,180 to -8,190, -342.5 to -342.9; *Heat of Decomposition*: Not pertinent.
Health Hazards Information - *Recommended Personal Protective Equipment*: Organic vapor canister unit; rubber gloves; chemical safety goggles; face shield; coveralls and/or rubber apron; rubber shoes or boots; *Symptoms Following Exposure*: Vapors cause irritation of eyes, nose, and throat. Contact with skin or eyes causes severe burns. Chemical is rapidly absorbed through skin; *General Treatment for Exposure*: Call a physician. INHALATION: remove to fresh air. INGESTION: have victim drink water or milk; do not induce vomiting. SKIN OR EYES: flush immediately with plenty of water for

at least 15 min.; remove contaminated clothing immediately and wash before reuse; discard contaminated shoes; *Toxicity by Inhalation (Threshold Limit Value)*: 5 ppm; *Short-Term Exposure Limits*: Data not available; *Toxicity by Ingestion*: Grade 2; LD_{50} 0.5 to 5 g/kg (rat, rabbit); *Late Toxicity*: Data not available; *Vapor (Gas) Irritant Characteristics*: Vapors cause moderate irritation such that personnel will find high concentrations unpleasant. The effect is temporary; *Liquid or Solid Irritant Characteristics*: Fairly severe skin irritant; may cause pain and second-degree burns after a few minutes' contact; *Odor Threshold*: 5 ppm.

Fire Hazards - *Flash Point (deg. F):* 175 - 185 OC; 178 CC; *Flammable Limits in Air (%)*: LEL: 1.4 (ortho); 1.1 (meta or para); *Fire Extinguishing Agents*: Water, dry chemical, carbon dioxide, and foam; *Fire Extinguishing Agents Not To Be Used*: Not pertinent; *Special Hazards of Combustion Products*: Flammable toxic vapors given off in a fire; *Behavior in Fire*: Sealed closed containers can build up pressure if exposed to heat (fire); *Ignition Temperature (deg. F)*: 1110 (o-cresol); 1038 (m-or p-cresol); *Electrical Hazard*: Data not available; *Burning Rate*: Data not available.

Chemical Reactivity - *Reactivity with Water*: No reaction; *Reactivity with Common Materials*: No reaction; *Stability During Transport*: Stable; *Neutralizing Agents for Acids and Caustics*: Not pertinent; *Polymerization*: Not pertinent; *Inhibitor of Polymerization*: Not pertinent.

CUMENE

Chemical Designations - *Synonyms*: Cumol; Isopropylbenzene; *Chemical Formula*: $C_6H_5CH(CH_3)_2$.
Observable Characteristics - *Physical State (as normally shipped)*: Liquid; *Color*: Colorless; *Odor*: Strong, slightly irritant; fragrant; aromatic.
Physical and Chemical Properties - *Physical State at 15 °C and 1 atm.*: Liquid; *Molecular Weight*: 120.19; *Boiling Point at 1 atm.*: 306.3, 152.4, 425.6; *Freezing Point*: -140.9, -96.1, 177.1; *Critical Temperature*: 676.2, 357.9, 631.1; *Critical Pressure*: 465.5, 31.67, 3.208; *Specific Gravity*: 0.866 at 15°C (liquid); *Vapor (Gas) Density*: Not pertinent; *Ratio of Specific Heats of Vapor (Gas)*: 1.059; *Latent Heat of Vaporization*: 134, 74.6, 3.12; *Heat of Combustion*: -17,710, -9,840, -412.0; *Heat of Decomposition*: Not pertinent.
Health Hazards Information - *Recommended Personal Protective Equipment*: As necessary to avoid skin exposure. If concentration in air is greater than 1000 ppm, use self-contained breathing apparatus; *Symptoms Following Exposure*: narcotic action with long-lasting effects; depressant to central nervous system; *General Treatment for Exposure*: INHALATION: move patient immediately to fresh air; administer artificial respiration or oxygen if necessary; seek medical attention. SKIN OR EYES: wash exposed skin surfaces thoroughly; flush eyes thoroughly with water for 15 min; *Toxicity by Inhalation (Threshold Limit Value)*: 50 ppm; *Short-Term Exposure Limits*: Data not available; *Toxicity by Ingestion*: Grade 3; LD_{50}50 to 500 mg/kg; *Late Toxicity*: None reported; *Vapor (Gas) Irritant Characteristics*: Vapors cause a slight smarting of the eyes or respiratory system if present in high concentration. The effect is temporary; *Liquid or Solid Irritant Characteristics*: Minimum hazard. If spilled on clothing and allowed to remain, may cause smarting and reddening of the skin; *Odor Threshold*: 1.2 ppm.
Fire Hazards - *Flash Point (deg. F):* 111 CC; *Flammable Limits in Air (%)*: 0.9 - 6.5; *Fire Extinguishing Agents*: Foam, carbon dioxide, or dry chemical; *Fire Extinguishing Agents Not To Be Used*: Not pertinent; *Special Hazards of Combustion Products*: Not pertinent; *Behavior in Fire*: Not pertinent; *Ignition Temperature (deg. F)*: 797; *Electrical Hazard*: Data not available; *Burning Rate*: 50 mm/min.
Chemical Reactivity - *Reactivity with Water*: No reaction; *Reactivity with Common Materials*: No reaction; *Stability During Transport*: Stable; *Neutralizing Agents for Acids and Caustics*: Not pertinent; *Polymerization*: Not pertinent; *Inhibitor of Polymerization*: Not pertinent.

CYANOGEN

Chemical Designations - *Synonyms*: Ethanedenitrite; Dicyan; Oxalic Acid Dinitrile; Oxalonitrile; Dicyanogen; *Chemical Formula*: $(CN)_2$.
Observable Characteristics - *Physical State (as normally shipped)*: Liquefied compressed gas; *Color*:

Colorless; *Odor*: Characteristic almond-like; pungent, penetrating; may not be sufficiently strong to provide an adequate warning.

Physical and Chemical Properties - *Physical State at 15 °C and 1 atm.*: Gas; *Molecular Weight*: 52.0; *Boiling Point at 1 atm.*: -6.1, -21.1, 252.1; *Freezing Point*: -18.2, -27.9, 245.3; *Critical Temperature*: 259.9, 126.6, 399.8; *Critical Pressure*: 857, 58.2, 5.91; *Specific Gravity*: 0.954 at -21°C (liquid); *Vapor (Gas) Density*: 1.8; *Ratio of Specific Heats of Vapor (Gas)*: 1.205 at 25°C; *Latent Heat of Vaporization*: 200, 111, 4.65; *Heat of Combustion*: -9.059, -5.033, -210.6; *Heat of Decomposition*: Not pertinent.

Health Hazards Information - *Recommended Personal Protective Equipment*: Self-contained breathing apparatus; rubber gloves; rubber protective clothing; rubber-soled shoes; *Symptoms Following Exposure*: Vapor irritates eyes and causes giddiness, headache, fatigue, and nausea if inhaled; *General Treatment for Exposure*: In general, treatment is similar to that used following exposure to hydrogen cyanide. INHALATION: move victim to fresh air and let him lie down; do not permit him to exert himself; remove contaminated clothing but keep patient covered and comfortably warm; summon a physician; break an amyl nitrite pearl in a cloth and hold it lightly under the victim's nose for 15 seconds; repeat five times at about 15 sec. Intervals; use artificial respiration if breathing has stopped. EYES: flush with water for at least 15 min; *Toxicity by Inhalation (Threshold Limit Value)*: 10 ppm; *Short-Term Exposure Limits*: 5 mg/m$_3$ for 30 sec; *Toxicity by Ingestion*: Data not available; *Late Toxicity*: Data not available; *Vapor (Gas) Irritant Characteristics*: Data not available; *Liquid or Solid Irritant Characteristics*: Data not available; *Odor Threshold*: Data not available.

Fire Hazards - *Flash Point :* Flammable gas; *Flammable Limits in Air (%)*: 6.6 - 43; *Fire Extinguishing Agents*: Let fire burn, shut off flow of gas, cool exposed areas with water; *Fire Extinguishing Agents Not To Be Used*: Not pertinent; *Special Hazards of Combustion Products*: Unburned vapors are highly toxic; *Behavior in Fire*: Vapor is heavier than air and may travel considerable distance to a source of ignition and flash back; *Ignition Temperature :* Data not available; *Electrical Hazard*: data not available; *Burning Rate*: No data.

Chemical Reactivity - *Reactivity with Water*: No reaction, but water, provides heat to vaporize liquid cyanogen; *Reactivity with Common Materials*: No reaction; *Stability During Transport*: Stable; *Neutralizing Agents for Acids and Caustics*: Not pertinent; *Polymerization*: Not pertinent; *Inhibitor of Polymerization*: Not pertinent.

CYANOGEN BROMIDE

Chemical Designations - *Synonyms*: No common synonyms; *Chemical Formula*: BrCN.

Observable Characteristics - *Physical State (as normally shipped)*: Solid; *Color*: Colorless; *Odor*: Penetrating.

Physical and Chemical Properties - *Physical State at 15 °C and 1 atm.*: Solid; *Molecular Weight*: 105.93; *Boiling Point at 1 atm.*: Not pertinent; *Freezing Point*: 120 to 124, 49 to 51, 322 to 324; *Critical Temperature*: Not pertinent; *Critical Pressure*: Not pertinent; *Specific Gravity*: 2.015 at 20°C; *Vapor (Gas) Density*: 3.6; *Ratio of Specific Heats of Vapor (Gas)*: Not pertinent; *Latent Heat of Vaporization*: Not pertinent; *Heat of Combustion*: Not pertinent; *Heat of Decomposition*: Not pertinent.

Health Hazards Information - *Recommended Personal Protective Equipment*: Chemical cartridges respirator, goggles, protective clothing, rubber gloves; *Symptoms Following Exposure*: Same symptoms as hydrogen cyanide. Because it irritates the eyes, throat, and lungs severely, it is unlikely that anyone would voluntarily remain in areas with a high enough concentration to exert a cyanide effect; *General Treatment for Exposure*: Call a physician. INHALATION: remove victim to fresh air; if he is not breathing, give artificial respiration, preferable mouth-to-mouth; if symptoms of cyanide poisoning are observed, administer amyl nitrite as instructed for HCN. INGESTION: have victim drink water or milk; do not induce vomiting; *Toxicity by Inhalation (Threshold Limit Value)*: 0.5 ppm (suggested); *Short-Term Exposure Limits*: Data not available; *Toxicity by Ingestion*: Data not available; *Late Toxicity*: Workers exposed to solutions may develop dermatitis; *Vapor (Gas) Irritant Characteristics*: Vapors cause severe irritation of eyes and throat and can cause eye and lung injury. They cannot be tolerated even at low concentrations; *Liquid or Solid Irritant Characteristics*: Severe skin irritant.

Causes second- and third degree burns on short contact; very injurious to the eyes; *Odor Threshold*: Data not available.

Fire Hazards - *Flash Point :* Not flammable; *Flammable Limits in Air* (%): Not flammable; *Fire Extinguishing Agents*: Not pertinent; *Fire Extinguishing Agents Not To Be Used*: Not pertinent; *Special Hazards of Combustion Products*: Poison gases are produced in fire; *Behavior in Fire*: Not pertinent; *Ignition Temperature* : Not flammable; *Electrical Hazard*: Not pertinent; *Burning Rate*: Not flammable.

Chemical Reactivity - *Reactivity with Water*: No reaction; *Reactivity with Common Materials*: No reaction; *Stability During Transport*: Stable; *Neutralizing Agents for Acids and Caustics*: Strong bleaching powder solution; let stand 24 hr; *Polymerization*: Does not occur; *Inhibitor of Polymerization*: Not pertinent.

CYCLOHEXANE

Chemical Designations - *Synonyms*: Hexahydrobenzene; Hexamethylene; Hexanaphthene; *Chemical Formula*: $(CH_2)_6$.

Observable Characteristics - *Physical State (as normally shipped)*: Liquid; *Color*: Colorless; *Odor*: resembling benzene; mild, sweet, resembling chloroform.

Physical and Chemical Properties - *Physical State at 15 °C and 1 atm.*: Liquid; *Molecular Weight*: 84.16; *Boiling Point at 1 atm.*: 177.3, 80.7, 353.9; *Freezing Point* 43.8, 6.6, 279.8; *Critical Temperature*: 536.5, 280.3, 553.5; *Critical Pressure*: 591, 40.2, 4.07; *Specific Gravity*: 0.779 at 20°C (liquid); *Vapor (Gas) Density*: 2.9; *Ratio of Specific Heats of Vapor (Gas)*: 1.087; *Latent Heat of Vaporization*: 150, 85, 3.6; *Heat of Combustion*: -18,684, -10,380, -434.59; *Heat of Decomposition*: Not pertinent.

Health Hazards Information - *Recommended Personal Protective Equipment*: Hydrocarbon vapor canister, supplied-air or hose mask, hydrocarbon-insoluble rubber or plastic gloves, chemical goggles or face splash shield, hydrocarbon-insoluble rubber or plastic apron; *Symptoms Following Exposure*: Dizziness, with nausea and vomiting. Concentrated vapor may cause unconsciousness and collapse; *General Treatment for Exposure*: INHALATION: remove victim to fresh air; if breathing stops, apply artificial respiration and administer oxygen. SKIN OR EYE CONTACT: remove contaminated clothing and gently flush affected areas with water for 15 min. Calla physician; *Toxicity by Inhalation (Threshold Limit Value)*: 300 ppm; *Short-Term Exposure Limits*: 300 ppm for 60 min; *Toxicity by Ingestion*: Grade 2; LD_{50} 0.5 to 5 g/kg; *Late Toxicity*: None; *Vapor (Gas) Irritant Characteristics*: Vapors cause a slight smarting of the eyes or respiratory system if present in high concentrations. The effect is temporary; *Liquid or Solid Irritant Characteristics*: Minimum hazard. If spilled on clothing and allowed to remain, may cause smarting and reddening of the skin; *Odor Threshold*: Data not available.

Fire Hazards - *Flash Point :* - 4 CC; *Flammable Limits in Air* (%): 1.33 - 8.35; *Fire Extinguishing Agents*: Foam, carbon dioxide, dry chemical; *Fire Extinguishing Agents Not To Be Used*: Not pertinent; *Special Hazards of Combustion Products*: Not pertinent; *Behavior in Fire*: Not pertinent; *Ignition Temperature (deg. F)*: 518; *Electrical Hazard*: Data not available; *Burning Rate*: 6.8 mm/min.

Chemical Reactivity - *Reactivity with Water*: No reaction; *Reactivity with Common Materials*: No reaction; *Stability During Transport*: Stable; *Neutralizing Agents for Acids and Caustics*: Not pertinent; *Polymerization*: Not pertinent; *Inhibitor of Polymerization*: Not pertinent.

CYCLOHEXANOL

**Chemical Designations - *Synonyms*: Adronal; Anol; Cyclohexyl Alcohol; Hexalin; Hexahydrophenol; Hydroxycyclohexane; *Chemical Formula*: $(CH_2)_5CHOH$.

Observable Characteristics - *Physical State (as normally shipped)*: Solid or liquid; *Color*: Colorless to faintly yellow; *Odor*: Like camphor.

Physical and Chemical Properties - *Physical State at 15 °C and 1 atm.*: Solid; *Molecular Weight*: 100.16; *Boiling Point at 1 atm.*: 322, 161, 434; *Freezing Point*: 74.5, 23.6, 296.8; *Critical Temperature*: 666, 352, 625; *Critical Pressure*: 540, 37, 3.7; *Specific Gravity*: 0.947 at 20°C (liquid); *Vapor (Gas) Density*: Not pertinent; *Ratio of Specific Heats of Vapor (Gas)*: 1.071; *Latent Heat of*

Vaporization: 196, 109, 4.56; *Heat of Combustion*: -16,000, -8910, -373; *Heat of Decomposition)*: Not pertinent;

Health Hazards Information - *Recommended Personal Protective Equipment*: Goggles or face shield; *Symptoms Following Exposure*: Narcosis - depression of the central nervous system tending to produce sleep or unconsciousness; *General Treatment for Exposure*: Eye contact is more hazardous than inhalation, skin irritation, or ingestion. Flush eyes with water and remove victim to fresh air; *Toxicity by Inhalation (Threshold Limit Value)*: 50 ppm; *Short-Term Exposure Limits*: Data not available; *Toxicity by Ingestion*: Grade 2; LD_{50} 0.5 to 5 g/kg; *Late Toxicity*: Data not available; *Vapor (Gas) Irritant Characteristics*: Vapors cause a slight smarting of the eyes or respiratory system if present in high concentrations. The effect is temporary; *Liquid or Solid Irritant Characteristics*: Causes smarting of the skin an first-degree burns on short exposure and may cause secondary burns on long exposure; *Odor Threshold*: Data not available.

Fire Hazards - *Flash Point (deg. F):* 160 OC; 154 CC; *Flammable Limits in Air (%)*: Data not available; *Fire Extinguishing Agents*: Water, foam, carbon dioxide, or dry chemical; *Fire Extinguishing Agents Not To Be Used*: Not pertinent; *Special Hazards of Combustion Products*: Not pertinent; *Behavior in Fire*: Not pertinent; *Ignition Temperature (deg. F)*: 572; *Electrical Hazard*: Data not available; *Burning Rate*: 3.9 mm/min.

Chemical Reactivity - *Reactivity with Water*: No reaction; *Reactivity with Common Materials*: No reaction; *Stability During Transport*: Stable; *Neutralizing Agents for Acids and Caustics*: Not pertinent; *Polymerization*: Not pertinent; *Inhibitor of Polymerization*: Not pertinent.

CYCLOHEXANONE

Chemical Designations - *Synonyms*: Anone; Hytrol O; Nadone; Pimelic Ketone; Sextone; *Chemical Formula*: $(CH_2)_5CO$.

Observable Characteristics - *Physical State (as normally shipped)*: Liquid; *Color*: Colorless to slightly yellow; *Odor*: Like peppermint and acetone.

Physical and Chemical Properties - *Physical State at 15 °C and 1 atm.*: Liquid; *Molecular Weight*: 98.15; *Boiling Point at 1 atm.*: 312.4, 155.8, 429.0; *Freezing Point*: -24.2, 31.2, 242.0; *Critical Temperature*: 673, 356, 629; *Critical Pressure*: 560, 38, 3.8; *Specific Gravity*: 0.945 at 20°C (liquid); *Vapor (Gas) Density*: Not pertinent; *Ratio of Specific Heats of Vapor (Gas)*: 1.084; *Latent Heat of Vaporization*: 160, 91, 3.8; *Heat of Combustion*: -15,430, -8570, -358.8; *Heat of Decomposition*: Not pertinent.

Health Hazards Information - *Recommended Personal Protective Equipment*: Chemical goggles; *Symptoms Following Exposure*: Inhalation of vapors from hot material can cause narcosis. The liquid may cause dermatitis; *General Treatment for Exposure*: Immediately flush eyes with plenty of water; call a physician; *Toxicity by Inhalation (Threshold Limit Value)*: 50 ppm; *Short-Term Exposure Limits*: Data not available; *Toxicity by Ingestion*: Grade 2; LD_{50} 50 to 5 g/kg; *Late Toxicity*: Data not available; *Vapor (Gas) Irritant Characteristics*: Vapor is moderately irritating such that personnel will not usually tolerate moderate or high vapor concentrations; *Liquid or Solid Irritant Characteristics*: Causes smarting of the skin and first-degree burns on short exposure and may cause secondary burns on long exposure; *Odor Threshold*: 0.12 ppm.

Fire Hazards - *Flash Point (deg. F):* 129 OC; 111 CC; *Flammable Limits in Air (%)*: 1.1 LEL; *Fire Extinguishing Agents*: Water, dry chemical, or carbon dioxide; *Fire Extinguishing Agents Not To Be Used*: Not pertinent; *Special Hazards of Combustion Products*: Not pertinent; *Behavior in Fire*: Not pertinent; *Ignition Temperature (deg. F)*: 788; *Electrical Hazard*: Data not available; *Burning Rate*: 4.2 mm/min.

Chemical Reactivity - *Reactivity with Water*: No reaction; *Reactivity with Common Materials*: No reaction; *Stability During Transport*: Stable; *Neutralizing Agents for Acids and Caustics*: Not pertinent; *Polymerization*: Not pertinent; *Inhibitor of Polymerization*: Not pertinent.

CYCLOPENTANE

Chemical Designations - *Synonyms*: Pentamethylene; *Chemical Formula*: C_5H_{10}.

Observable Characteristics - *Physical State (as normally shipped)*: Liquid; *Color*: Colorless; *Odor*: Like gasoline; mild, sweet.

Physical and Chemical Properties - *Physical State at 15 °C and 1 atm.*: Liquid; *Molecular Weight*: 70.1; *Boiling Point at 1 atm.*: 120.7, 49.3, 322.5; *Freezing Point*: -137.0, -93.9, -179.3; *Critical Temperature*: 461.5, 238.6, 511.8; *Critical Pressure*: 654, 44.4, 4.51; *Specific Gravity*: 0.74 at 20°C (liquid); *Vapor (Gas) Density*: 2.4; *Ratio of Specific Heats of Vapor (Gas)*: 1.1217; *Latent Heat of Vaporization*: 179, 94, 3.9; *Heat of Combustion*: -19,990, -11,110, -465; *Heat of Decomposition*: Not pertinent.

Health Hazards Information - *Recommended Personal Protective Equipment*: Hydrobarbon canister, supplied air, or hose mask; rubber or plastic gloves; chemical goggles or face shield; *Symptoms Following Exposure*: Inhalation causes dizziness, nausea, and vomiting; concentrated vapor may cause unconsciousness and collapse. Vapor causes slight smarting of eyes. Contact with liquid causes irritation of eyes and may irritate skin if allowed to remain. Ingestion causes irritation of stomach. Aspiration produces severe lung irritation and rapidly developing pulmonary edema; central nervous excitement followed by depression; *General Treatment for Exposure*: INHALATION: remove to fresh air; if breathing stops, apply artificial respiration and administer oxygen. EYES: flush with water for at least 15 min.; call a physician. SKIN: flush well with water, then wash with soap and water. INGESTION: do NOT induce vomiting; guard against aspiration into lungs. ASPIRATION: enforce bed rest; give oxygen; get medical attention; *Toxicity by Inhalation (Threshold Limit Value)*: Data not available; *Short-Term Exposure Limits*: 300 ppm for 60 min.; *Toxicity by Ingestion*: Grade 2; LD_{50} 0.5 to 5 g/kg; *Late Toxicity*: None; *Vapor (Gas) Irritant Characteristics*: Vapors cause slight smarting of the eyes or respiratory system if present in high concentrations. The effect is temporary; *Liquid or Solid Irritant Characteristics*: Minimum hazard. If spilled on clothing and allowed to remain, may cause smarting and reddening of the skin; *Odor Threshold*: Data not available.

Fire Hazards - *Flash Point (deg. F):* <20 CC; *Flammable Limits in Air (%)*: (approx.) 1.1 - 8.7; *Fire Extinguishing Agents*: Dry chemical, foam, carbon dioxide; *Fire Extinguishing Agents Not To Be Used*: Water may be ineffective; *Special Hazards of Combustion Products*: Not pertinent; *Behavior in Fire*: Containers may explode; *Ignition Temperature (deg. F)*: 716; *Electrical Hazard*: Not pertinent; *Burning Rate*: 7.9 mm/min.

Chemical Reactivity - *Reactivity with Water*: No reaction; *Reactivity with Common Materials*: No reaction; *Stability During Transport*: Stable; *Neutralizing Agents for Acids and Caustics*: Not pertinent; *Polymerization*: Not pertinent; *Inhibitor of Polymerization*: Not pertinent.

P-CUMENE

Chemical Designations - *Synonyms*: Cymol; p-Isopropyltoluene; Isopropyltoluol 1-Methyl-4-Isopropyl-benzene; Methyl Propyl Benzene; *Chemical Formula*: $p-CH_3C_6H_4CH(CH_3)_2$.

Observable Characteristics - *Physical State (as normally shipped)*: Liquid; *Color*: Colorless; *Odor*: Mild, pleasant; aromatic, solvent-type.

Physical and Chemical Properties - *Physical State at 15 °C and 1 atm.*: Liquid; *Molecular Weight*: 134.2; *Boiling Point at 1 atm.*: 351, 177, 450; *Freezing Point*: -90.2, -67.9, 205.3; *Critical Temperature*: Not pertinent; *Critical Pressure*: Not pertinent; *Specific Gravity*: 0.857 at 20°C (liquid); *Vapor (Gas) Density*: Not pertinent; *Ratio of Specific Heats of Vapor (Gas)*: Not pertinent; *Latent Heat of Vaporization*: 122, 67.8, 2.84; *Heat of Combustion*: -18,800, -10,400, -437; *Heat of Decomposition*: Not pertinent.

Health Hazards Information - *Recommended Personal Protective Equipment*: Self-contained or air-line breathing apparatus; solvent-resistant rubber gloves; chemical splash goggles; *Symptoms Following Exposure*: Inhalation causes impairment of coordination, headache. Contact with liquid causes mild irritation of eyes and skin. Ingestion causes irritation of mouth and stomach; *General Treatment for Exposure*: INHALATION: remove victim from contaminated area; administer artificial respiration if necessary; call physician. SKIN: wipe off liquid; wash well with soap and water. INGESTION: induce vomiting; get medical attention *Toxicity by Inhalation (Threshold Limit Value)*: Data not available; *Short-Term Exposure Limits*: Data not available; *Toxicity by Ingestion*: Grade 2; oral rat LD_{50}=4,750 mg/kg. Oral human TD_{LO} (lowest toxic dose)=86 mg/kg (affects central nervous system); *Late Toxicity*: Data not available; *Vapor (Gas) Irritant Characteristics*: Vapors are nonirritating to eyes and throat; *Liquid or Solid Irritant Characteristics*: Minimum hazard. If spilled on clothing and allowed to remain, may cause smarting and reddening of the skin; *Odor Threshold*: Data not available.

Fire Hazards - *Flash Point (deg. F):* 140 OC; 117 CC; *Flammable Limits in Air (%)*: 0.7 - 5.6; *Fire Extinguishing Agents*: Foam, dry chemical, carbon dioxide; *Fire Extinguishing Agents Not To Be Used*: Water may be ineffective; *Special Hazards of Combustion Products*: Not pertinent; *Behavior in Fire*: Not pertinent; *Ignition Temperature (deg. F)*: 817; *Electrical Hazard*: Data not available; *Burning Rate*: 6.1 mm/min.

Chemical Reactivity - *Reactivity with Water*: No reaction; *Reactivity with Common Materials*: No reaction; *Stability During Transport*: Stable; *Neutralizing Agents for Acids and Caustics*: Not pertinent; *Polymerization*: Not pertinent; *Inhibitor of Polymerization*: Not pertinent.

D

DDD

Chemical Designations - *Synonyms*: 1,1-Dichloro-2,2-bis(p-Chlorophenyl) Ethane; Dichlorodiphenyldichloroethane; TDE; *Chemical Formula*: $(4-ClC_6H4)_2CH-Cl_2$.

Observable Characteristics - *Physical State (as normally shipped)*: Solid; *Color*: White; *Odor*: Data not available.

Physical and Chemical Properties - *Physical State at 15 °C and 1 atm.*: Solid; *Molecular Weight*: 320; *Boiling Point at 1 atm.*:Not pertinent (decomposes); *Freezing Point*: 234, 112, 385; *Critical Temperature*: Not pertinent; *Critical Pressure*: Not pertinent; *Specific Gravity*: 1.476 at 20°C (solid); *Vapor (Gas) Density*: Not pertinent; *Ratio of Specific Heats of Vapor (Gas)*: Not pertinent; *Latent Heat of Vaporization*: Not pertinent; *Heat of Combustion*: Data not available; *Heat of Decomposition*: Not pertinent.

Health Hazards Information - *Recommended Personal Protective Equipment*: Dust mask; goggles or face shield; rubber gloves; *Symptoms Following Exposure*: Ingestion causes vomiting and delayed symptoms similar to those caused by DDT. Contact with eyes causes irritation; *General Treatment for Exposure*: INGESTION: treatment should be given by a physician and is similar to that given following ingestion of DDT. EYES: flush with water; *Toxicity by Inhalation (Threshold Limit Value)*: Data nor available; *Short-Term Exposure Limits*: Data not available; *Toxicity by Ingestion*: Grade 2; oral LD_{50}=1.2 g/kg (mouse), 3.4 g/kg (rat); *Late Toxicity*: Data not available; *Vapor (Gas) Irritant Characteristics*: Data not available; *Liquid or Solid Irritant Characteristics*: Data not available; *Odor Threshold*: Data not available.

Fire Hazards - *Flash Point :* Not pertinent; *Flammable Limits in Air (%):* Not pertinent; *Fire Extinguishing Agents:* Water, foam, dry chemical, carbon dioxide; *Fire Extinguishing Agents Not to be Used:* Data not available; *Special Hazards of Combustion Products:* Irritating hydrogen chloride fumes may form in fire; *Behavior in Fire:* Data not available; *Ignition Temperature :* Data not available; *Electrical Hazard:* Not pertinent; *Burning Rate:* Not pertinent.

Chemical Reactivity - *Reactivity with Water :* No reaction; *Reactivity with Common Materials:* No reaction; *Stability During Transport:* Stable; *Neutralizing Agents for Acids and Caustics:* Not pertinent; *Polymerization:* Not pertinent; *Inhibitor of Polymerization:* Not pertinent.

DDT

Chemical Designations - *Synonyms*: Dichlorophenyltrichloroethane; p, p'-DDT; 1, 1, 1-Trichloro-2, 2-bis(p-Chlorophenyl) Ethane; *Chemical Formula*: $(p-ClC_6H_4)_2CHCCl_3$.

Observable Characteristics - *Physical State (as normally shipped)*: Solid; *Color*: White; *Odor*: None.
Physical and Chemical Properties - *Physical State at 15 °C and 1 atm.*: Solid; *Molecular Weight*: 354.5; *Boiling Point at 1 atm.*: Not pertinent; *Freezing Point*: 226, 108, 381; *Critical Temperature*: Not pertinent; *Critical Pressure*: Not pertinent; *Specific Gravity*: 1.56 at 15°C (solid); *Vapor (Gas) Density*: Not pertinent; *Ratio of Specific Heats of Vapor (Gas)*: Not pertinent; *Latent Heat of Vaporization*: Not pertinent; *Heat of Combustion*: Not pertinent; *Heat of Decomposition*: Not pertinent.
Health Hazards Information - *Recommended Personal Protective Equipment*: Data not available; *Symptoms Following Exposure*: Very large doses are followed promptly by vomiting, due to local gastric irritation; delayed emesis of diarrhea may occur. With smaller doses, symptoms usually appear 2-3 hours after ingestion. These include tingling of lips, tongue and face; malaise, head ache, sore throat, fatigue, coarse tremors of neck, head and eye lids; apprehension, ataxia and confusion. Convulsions may alternate with periods of coma and partial paralysis. Vital signs are essentially normal, but in severe poisoning the pulse may be irregular and abnormally slow; ventricular fibrillation and sudden death may occur at any time during acute phase . Pulmonary edema usually indicates solvent intoxication; *General Treatment for Exposure*: INGESTION; treatment should be done by a physician. In usually includes gastric lavage and administration of saline cathartic; phenobarbital, and parenteral fluids. Patient should be kept quiet and under observation for at least 24 hours; *Toxicity by Inhalation (Threshold Limit Value)*: Not pertinent; *Short-Term Exposure Limits*: Not pertinent; *Toxicity by Ingestion*: Grade 3; LD_{50} 50 to 500 mg/kg (rat); *Late Toxicity*: Data not available; *Vapor (Gas) Irritant Characteristics*: Not pertinent; *Liquid or Solid Irritant Characteristics*: Minimum hazard. If spilled on clothing and allowed to remain, may cause smarting and reddening of the skin; *Odor Threshold*: Not pertinent.
Fire Hazards - *Flash Point (deg. F):* 162 - 171 CC; *Flammable Limits in Air (%):* Not pertinent; *Fire Extinguishing Agents:* Water, foam, dry chemical or carbon dioxide; *Fire Extinguishing Agents Not to be Used:* Not pertinent; *Special Hazards of Combustion Products:* Toxic and irritating gases may be generated; *Behavior in Fire:* Melts and burns; *Ignition Temperature :* Data not available; *Electrical Hazard:* Not pertinent; *Burning Rate:* Data not available.
Chemical Reactivity - *Reactivity with Water :* No reaction; *Reactivity with Common Materials:* No reaction; *Stability During Transport:* Stable; *Neutralizing Agents for Acids and Caustics:* Not pertinent; *Polymerization:* Not pertinent; *Inhibitor of Polymerization:* Not pertinent.

DECABORANE
Chemical Designations - *Synonyms*: No common synonyms; *Chemical Formula*: $B_{10}H_{14}$.
Observable Characteristics - *Physical State (as normally shipped)*: Solid; *Color*: White; *Odor*: Pungent.
Physical and Chemical Properties - *Physical State at 15 °C and 1 atm.*: Solid; *Molecular Weight*: 122.3; *Boiling Point at 1 atm.*: 415, 213, 486; *Freezing Point*: 210, 99, 372; *Critical Temperature*: Not pertinent; *Critical Pressure*: Not pertinent; *Specific Gravity*: 0.94 at 25°C (solid); *Vapor (Gas) Density*: Not pertinent; *Ratio of Specific Heats of Vapor (Gas)*: Not pertinent; *Latent Heat of Vaporization*: Not pertinent; *Heat of Combustion*: -28,669, -15,944, -667.10; *Heat of Decomposition*: -279, -155, -6.49.
Health Hazards Information - *Recommended Personal Protective Equipment*: Self-contained breathing apparatus or positive-pressure hose mask; rubber-boots or overshoes; clothing made of material resistant to decaborane; rubber gloves; chemical-type goggles or face shield; *Symptoms Following Exposure*: (The onset of symptoms is frequently delayed until one or two days after exposure.) Inhalation or ingestion causes headache, nausea, light heatedness, drowsiness, nervousness, lack of coordination, and tremor; muscle spasms and generalized convulsions may occur. Dust irritates eyes and skin and may give same systemic symptoms as for inhalation if left on skin; *General Treatment for Exposure*: Get medical attention after all exposures to this compound. Symptoms may be delayed for 48 hours. INHALATION: move patient to fresh air; keep him warm and quiet. EYES: flush with water for at least 15 min. SKIN: immediately wash with soap and plenty of water. INGESTION: if victim is conscious, give a tablespoonful of salt in a glass of warm water and repeat until vomit is clear. *Note to physician:* Treat symptomatically; administration of methacarbamol or other muscle relaxant may

be helpful immediately following exposure and in the absence of symptoms; *Toxicity by Inhalation (Threshold Limit Value)*: 0.05 ppm; *Short-Term Exposure Limits*: Data not available; *Toxicity by Ingestion*: Grade 4; LD_{50} = 40 mg/kg (mouse); *Late Toxicity*: Data not available; *Vapor (Gas) Irritant Characteristics*: Data not available; *Liquid or Solid Irritant Characteristics*: Data not available; *Odor Threshold*: 0.05 ppm.

Fire Hazards - *Flash Point :* (Flammable solid); *Flammable Limits in Air (%):* Not pertinent; *Fire Extinguishing Agents:* Water, foam, dry chemical, and carbon dioxide; *Fire Extinguishing Agents Not to be Used:* Halogenated extinguishing agents; *Special Hazards of Combustion Products:* May give toxic fumes of unburned material; *Behavior in Fire:* May explode when hot. Burns with green-colored flame; *Ignition Temperature (deg. F):* 300; *Electrical Hazard:* Not pertinent; *Burning Rate:* Not pertinent.

Chemical Reactivity - *Reactivity with Water:* Reacts slowly to form flammable hydrogen gas, which can accumulate in closed area; *Reactivity with Common Materials:* Corrosive to natural rubber, some synthetic rubbers, some greases and some lubricants; *Stability During Transport:* Stable; *Neutralizing Agents for Acids and Caustics:* Flush with 3% aqueous ammonia solution, then with water. Methyl alcohol may also be used; *Polymerization:* Not pertinent; *Inhibitor of Polymerization:* Not pertinent.

DECAHYDRONAPHTHALENE

Chemical Designations - *Synonyms*: Bicyclo [4.4.0] Decane; Naphthalane; Perhydronaphthlene; Dec; Decalin; De Kalin; Naphthane; *sic-* or *trans*-Decahydronaphthalene; *Chemical Formula*: $C_{10}H_{18}$.

Observable Characteristics - *Physical State (as normally shipped)*: Liquid; *Color*: Colorless; *Odor*: Aromatic, like turpentine; mild, characteristic.

Physical and Chemical Properties - *Physical State at 15 °C and 1 atm.*: Liquid; *Molecular Weight*: 138.2; *Boiling Point at 1 atm.*: 383, 195, 468; *Freezing Point*: -44, -42, 231; *Critical Temperature*: Not pertinent; *Critical Pressure*: Not pertinent; *Specific Gravity*: 0.89 at 20°C (liquid); *Vapor (Gas) Density*: Not pertinent; *Ratio of Specific Heats of Vapor (Gas)*: Not pertinent; *Latent Heat of Vaporization*: 130, 71, 3.0; *Heat of Combustion*: -19,200, -10,700, -447; *Heat of Decomposition*: Not pertinent.

Health Hazards Information - *Recommended Personal Protective Equipment*: Air mask or self-contained breathing apparatus if in enclosed tank; rubber gloves or protective cream; goggles or face shield; *Symptoms Following Exposure*: Inhalation or ingestion irritates nose and throat, causes numbness, headache, vomiting; urine may become blue. Irritates eyes. Liquid de-fats skin and causes cracking and secondary infection; eczema may develop; *General Treatment for Exposure*: INHALATION: remove to fresh air. EYES: flush with water for at least 15 min. SKIN: wash with water and mild soap. INGESTION: give emetic such as warm salt water, followed by a mild catharic; direct physician to conserve liver and kidney function; *Toxicity by Inhalation (Threshold Limit Value)*: 25 ppm (suggested); *Short-Term Exposure Limits*: Data not available; *Toxicity by Ingestion*: Grade 2; LD_{50} = 4,170 mg/kg (rat); *Late Toxicity*: Data not available; *Vapor (Gas) Irritant Characteristics*: Data not available; *Liquid or Solid Irritant Characteristics*: Data not available; *Odor Threshold*: Data not available.

Fire Hazards - *Flash Point (deg. F):* 134 OC; *Flammable Limits in Air (%):* 0.7 - 5.4; *Fire Extinguishing Agents:* Water, foam, dry chemical, carbon dioxide; *Fire Extinguishing Agents Not to be Used:* Water; *Special Hazards of Combustion Products:* Not pertinent; *Behavior in Fire:* Not pertinent; *Ignition Temperature (deg. F):* 482; *Electrical Hazard:* Data not available; *Burning Rate:* 5.9 mm/min.

Chemical Reactivity - *Reactivity with Water :* No reaction; *Reactivity with Common Materials:* No reaction; *Stability During Transport:* Stable; *Neutralizing Agents for Acids and Caustics:* Not pertinent; *Polymerization:* Not pertinent; *Inhibitor of Polymerization:* Not pertinent.

DECALDEHYDE

Chemical Designations - *Synonyms*: Aldehyde C-10; Capraldehyde; Capric Aldehyde; Decanal; n-Decyl Aldehyde; *Chemical Formula*: $CH_3(CH_2)_8CHO$.

Observable Characteristics - *Physical State (as normally shipped)*: Liquid; *Color*: Colorless to pale yellow; *Odor*: Pleasant.

Physical and Chemical Properties - *Physical State at 15 °C and 1 atm.*: Liquid; *Molecular Weight*: 145.3; *Boiling Point at 1 atm.*: 404 to 410, 207 to 210, 480 to 483; *Freezing Point*: 64, 18, 291; *Critical Temperature*: Not pertinent; *Critical Pressure*: Not pertinent; *Specific Gravity*: 0.830 at 15°C (liquid); *Vapor (Gas) Density*: Not pertinent; *Ratio of Specific Heats of Vapor (Gas)*: 1.036; *Latent Heat of Vaporization*: Not pertinent; *Heat of Combustion*: (est.) -18,000, -10,000, -424; *Heat of Decomposition*: Not pertinent.

Health Hazards Information - *Recommended Personal Protective Equipment*: Protective clothing and chemical goggles; *Symptoms Following Exposure*: On direct contact can produce eye and skin irritation; *General Treatment for Exposure*: CONTACT WITH EYES AND SKIN: wash with water for 15 min; *Toxicity by Inhalation*:No data; *Short-Term Exposure Limits*: No data; *Toxicity by Ingestion*: Grade 0; $LD_{50} > 33.3$ g/kg (rat); *Late Toxicity*: Data not available; *Vapor (Gas) Irritant Characteristics*: Data not available; *Liquid or Solid Irritant Characteristics*: Skin effects are minor; *Odor Threshold*: 0.168 ppm.

Fire Hazards - *Flash Point (deg. F):* 185 OC; *Flammable Limits in Air (%):* Data not available; *Fire Extinguishing Agents:* Foam, dry chemical, carbon dioxide; *Fire Extinguishing Agents Not to be Used:* Not pertinent; *Special Hazards of Combustion Products:* Not pertinent; *Behavior in Fire:* Data not available; *Ignition Temperature :* No data; *Electrical Hazard:*No data; *Burning Rate:* No data.

Chemical Reactivity - *Reactivity with Water :* No reaction; *Reactivity with Common Materials:* No reaction; *Stability During Transport:* Stable; *Neutralizing Agents for Acids and Caustics:* Not pertinent; *Polymerization:* Not pertinent; *Inhibitor of Polymerization:* Not pertinent.

1-DECENE

Chemical Designations - *Synonyms*: alpha-Decene; *Chemical Formula*: $CH_2 = CH(CH_2)_7CH_3$.

Observable Characteristics - *Physical State (as normally shipped)*: Liquid; *Color*: Colorless; *Odor*: Mild, pleasant.

Physical and Chemical Properties - *Physical State at 15 °C and 1 atm.*: Liquid; *Molecular Weight*: 140.2; *Boiling Point at 1 atm.*: 339.1, 170.6, 443.8; *Freezing Point*: -87.3, -66.3, 206.9; *Critical Temperature*: Not pertinent; *Critical Pressure*: Not pertinent; *Specific Gravity*: 0.741 at 20°C (liquid); *Vapor (Gas) Density*: 4.8; *Ratio of Specific Heats of Vapor (Gas)*: 1.039; *Latent Heat of Vaporization*: 119, 65.9, 2.76; *Heat of Combustion*: -19,107, -10,615, -444.43; *Heat of Decomposition*: Not pertinent.

Health Hazards Information - *Recommended Personal Protective Equipment*: Organic canister or air-supplied mask, goggles or face shield; *Symptoms Following Exposure*: Vapors may produce slight irritation of eyes and respiratory tract if present in high concentration. May also act as a slight anesthetic at high concentrations; *General Treatment for Exposure*: CONTACT WITH EYES OR SKIN: splashes in the eye should be removed by thorough flushing with water. Skin areas should be washed with soap and water. Contaminated clothing should be laundered before reuse; *Toxicity by Inhalation (Threshold Limit Value)*: Data not available; *Short-Term Exposure Limits*: Data not available; *Toxicity by Ingestion*: Data not available; *Late Toxicity*: Data not available; *Vapor (Gas) Irritant Characteristics*: Slight smarting of eyes and respiratory system at high concentrations. The effect is temporary; *Liquid or Solid Irritant Characteristics*: Minimum hazard. If spilled on clothing and allowed to remain, may cause smarting and reddening of the skin; *Odor Threshold*: Data not available.

Fire Hazards - *Flash Point (deg. F):* 128 OC; *Flammable Limits in Air (%):* Not pertinent; *Fire Extinguishing Agents:* Foam, dry chemical or carbon dioxide; *Fire Extinguishing Agents Not to be Used:* Not pertinent; *Special Hazards of Combustion Products:* Not pertinent; *Behavior in Fire:* Not pertinent; *Ignition Temperature (deg. F):* 455; *Electrical Hazard:* Data not available; *Burning Rate:* 6.0 mm/min.

Chemical Reactivity - *Reactivity with Water :* No reaction; *Reactivity with Common Materials:* No reaction; *Stability During Transport:* Stable; *Neutralizing Agents for Acids and Caustics:* Not pertinent; *Polymerization:* Not pertinent; *Inhibitor of Polymerization:* Not pertinent.

N-DECYL ALCOHOL

Chemical Designations - *Synonyms*: Alcohol C-10; Capric Alcohol; 1-Decanol; Dytol S-91; Lorol-22; Nonylcarbinol; *Chemical Formula*: $CH_3(CH_2)_8CH_2OH$.

Observable Characteristics - *Physical State (as normally shipped)*: Liquid; *Color*: Colorless to light yellow; *Odor*: Faint alcoholic.

Physical and Chemical Properties - *Physical State at 15 °C and 1 atm.*: Liquid; *Molecular Weight*: 158.29; *Boiling Point at 1 atm.*: 446, 230, 503; *Freezing Point*: 44, 6.9, 280.1; *Critical Temperature*: 801, 427, 700; *Critical Pressure*: 320, 22, 2.2; *Specific Gravity*: 0.840 at 20°C (liquid); *Vapor (Gas) Density*: Not pertinent; *Ratio of Specific Heats of Vapor (Gas)*: 1.035; *Latent Heat of Vaporization*: (est.) 130, 74, 3.1; *Heat of Combustion*: -18,000, -9980, 418; *Heat of Decomposition*: Not pertinent.

Health Hazards Information - *Recommended Personal Protective Equipment*: Goggles or face shield; *Symptoms Following Exposure*: Direct contact can produce eye irritation; low general toxicity; *General Treatment for Exposure*: CONTACT WITH EYES: flush with water for 15 min.; *Toxicity by Inhalation (Threshold Limit Value)*: Not pertinent; *Short-Term Exposure Limits*: Not pertinent; *Toxicity by Ingestion*: Grade 1; LD_{50} 5 to 15 g/kg (rat); *Late Toxicity*: Data not available; *Vapor (Gas) Irritant Characteristics*: None; *Liquid or Solid Irritant Characteristics*: No appreciable hazard. Practically harmless to the skin; *Odor Threshold*: Data not available.

Fire Hazards - *Flash Point (deg. F)*: 180 OC; *Flammable Limits in Air (%)*: Data not available; *Fire Extinguishing Agents*: Dry chemical; *Fire Extinguishing Agents Not to be Used*: Not pertinent; *Special Hazards of Combustion Products*: Not pertinent; *Behavior in Fire*: Not pertinent; *Ignition Temperature*: Data not available; *Electrical Hazard*: Data not available; *Burning Rate*: Data not available.

Chemical Reactivity - *Reactivity with Water*: No reaction; *Reactivity with Common Materials*: No reaction; *Stability During Transport*: Stable; *Neutralizing Agents for Acids and Caustics*: Not pertinent; *Polymerization*: Not pertinent; *Inhibitor of Polymerization*: Not pertinent.

N-DECYLBENZENE

Chemical Designations - *Synonyms*: Decylbenzene; 1-Phenyldecane; *Chemical Formula*: $C_6H_5(CH_2)_9CH_3$.

Observable Characteristics - *Physical State (as normally shipped)*: Liquid; *Color*: Colorless; *Odor*: Data not available.

Physical and Chemical Properties - *Physical State at 15 °C and 1 atm.*: Liquid; *Molecular Weight*: 218; *Boiling Point at 1 atm.*: 572, 300, 573; *Freezing Point*: Not pertinent; *Critical Temperature*: Not pertinent; *Critical Pressure*: Not pertinent; *Specific Gravity*: 0.855 at 20°C (liquid); *Vapor (Gas) Density*: Not pertinent; *Ratio of Specific Heats of Vapor (Gas)*: Not pertinent; *Latent Heat of Vaporization*: 103.8, 57.67, 2.413; *Heat of Combustion*: -18,400, -10,200, -427; *Heat of Decomposition*: Not pertinent.

Health Hazards Information - *Recommended Personal Protective Equipment*: Goggles or face shield; rubber gloves; *Symptoms Following Exposure*: Inhalation of vapor causes slight irritation of nose and throat. Aspiration of liquid into lungs causes coughing, distress, and pulmonary edema. Ingestion causes irritation of stomach. Contact with vapor or liquid causes irritation of eyes and mild irritation of skin; *General Treatment for Exposure*: INHALATION: move to fresh air. INGESTION: do NOT induce vomiting; call a doctor. EYES: flush with water. SKIN: wipe off; flush with water; wash with soap and water. ASPIRATION: enforce bed rest; administer oxygen; call a doctor; *Toxicity by Inhalation (Threshold Limit Value)*: Data not available; *Short-Term Exposure Limits*: Data not available; *Toxicity by Ingestion*: Data not available; *Late Toxicity*: Data not available; *Vapor (Gas) Irritant Characteristics*: Data not available; *Liquid or Solid Irritant Characteristics*: Data not available; *Odor Threshold*: Data not available.

Fire Hazards - *Flash Point (deg. F)*: 225 CC; *Flammable Limits in Air (%)*: Data not available; *Fire Extinguishing Agents*: Dry chemical, carbon dioxide; *Fire Extinguishing Agents Not to be Used*: water or foam may cause frothing; *Special Hazards of Combustion Products*: Data not available; *Behavior in Fire*: Data not available; *Ignition Temperature*: Data not available; *Electrical Hazard*: Data not available; *Burning Rate*: 5.04 mm/min.

Chemical Reactivity - *Reactivity with Water* : No reaction; *Reactivity with Common Materials:* May attack some forms of plastics; *Stability During Transport:* Stable; *Neutralizing Agents for Acids and Caustics:* Not pertinent; *Polymerization:* Not pertinent; *Inhibitor of Polymerization:* Not pertinent.

2,4-D ESTERS

Chemical Designations - *Synonyms*: Butoxyethyl 2,4-Dichlorophenoxyacetate; Butyl 2,4-Dichlorophenoxyacetate; 2,4-Dichlorophenoxyacetic Acid, Butoxyethyl Ester; 2,4-Dichlorophenoxyacetic Acid, Isopropyl Ester; Isopropyl 2,4-Dichlorophenoxy Acetate; *Chemical Formula*: $2,4\text{-}Cl_2C_6H_3OCH_2COOR$, where $R=C_4H_9$, C_3H_7, or $CH_2CH_2OC_4H_9$.

Observable Characteristics - *Physical State (as normally shipped)*: Liquid; *Color*: Brown; amber; *Odor*: May have odor of fuel oil.

Physical and Chemical Properties - *Physical State at 15 ℃ and 1 atm.*: Liquid; *Molecular Weight*: 234 ~ 291; *Boiling Point at 1 atm.*: Very high; *Freezing Point*: Not pertinent; *Critical Temperature*: Not pertinent; *Critical Pressure*: Not pertinent; *Specific Gravity*: 1.088 ~ 1.237 at 20°C (liquid); *Vapor (Gas) Density*: Not pertinent; *Ratio of Specific Heats of Vapor (Gas)*: Not pertinent; *Latent Heat of Vaporization*: Data not available; *Heat of Combustion*: Data not available; *Heat of Decomposition*: Not pertinent.

Health Hazards Information - *Recommended Personal Protective Equipment*: Face shield or goggles; rubber gloves; *Symptoms Following Exposure*: Contact with eyes may cause mild irritation; *General Treatment for Exposure*: INGESTION: if large amounts are swallowed, induce vomiting and get medical help. EYES: flush with plenty of water and see a doctor. SKIN: flush with water, wash with soap and water; *Toxicity by Inhalation (Threshold Limit Value)*: Data not available; *Short-Term Exposure Limits*: Data not available; *Toxicity by Ingestion*: Grade 2 or 3: LD_{50} 320-617 mg/kg; *Late Toxicity*: Data not available; *Vapor (Gas) Irritant Characteristics*: Data not available; *Liquid or Solid Irritant Characteristics*: Data not available; *Odor Threshold*: Data not available.

Fire Hazards - *Flash Point (deg. F):* > 175 OC; *Flammable Limits in Air (%):* Data not available; *Fire Extinguishing Agents:* Foam, dry chemical , carbon dioxide; *Fire Extinguishing Agents Not to be Used:* Water may be ineffective; *Special Hazards of Combustion Products:* Irritating hydrogen chloride vapor may form in fire; *Behavior in Fire:* Data not available; *Ignition Temperature :* Data not available; *Electrical Hazard:* Data not available; *Burning Rate:* Data not available.

Chemical Reactivity - *Reactivity with Water* : No reaction; *Reactivity with Common Materials:* May attack some forms of plastics; *Stability During Transport:* Stable; *Neutralizing Agents for Acids and Caustics:* Not pertinent; *Polymerization:* Not pertinent; *Inhibitor of Polymerization:* Not pertinent.

DEXTROSE SOLUTION

Chemical Designations - *Synonyms*: Corn Sugar Solution; Glucose Solution; Grape Sugar Solution; *Chemical Formula*: $C_6H_{12}O_6-H_2O$.

Observable Characteristics - *Physical State (as normally shipped)*: Liquid; *Color*: Clear, colorless; *Odor*: None.

Physical and Chemical Properties - *Physical State at 15 ℃ and 1 atm.*: Liquid; *Molecular Weight*: Not pertinent; *Boiling Point at 1 atm*: >212, >100, >373; *Freezing Point*: <32, <0, <273; *Critical Temperature*: Not pertinent; *Critical Pressure*: Not pertinent; *Specific Gravity*: (est.) 1.20 at 20°C (liquid); *Vapor (Gas) Density*: Not pertinent; *Ratio of Specific Heats of Vapor (Gas)*: Not pertinent; *Latent Heat of Vaporization*: Not pertinent; *Heat of Combustion*: Not pertinent; *Heat of Decomposition*: Not pertinent.

Health Hazards Information - *Recommended Personal Protective Equipment*: None needed; *Symptoms Following Exposure*: No toxicity; *General Treatment for Exposure*: None needed; *Toxicity by Inhalation (Threshold Limit Value)*: Not pertinent; *Short-Term Exposure Limits*: Not pertinent; *Toxicity by Ingestion*: None; *Late Toxicity*: None; *Vapor (Gas) Irritant Characteristics*: Not pertinent; *Liquid or Solid Irritant Characteristics*: None; *Odor Threshold*: Not pertinent.

Fire Hazards - *Flash Point :* Not flammable; *Flammable Limits in Air (%):* Not flammable; *Fire Extinguishing Agents:* Not pertinent; *Fire Extinguishing Agents Not to be Used:* Not pertinent; *Special*

Hazards of Combustion Products: Not pertinent; *Behavior in Fire:* Not pertinent; *Ignition Temperature :* Not flammable; *Electrical Hazard:* Not pertinent; *Burning Rate:* Not pertinent.

Chemical Reactivity - *Reactivity with Water :* No reaction; *Reactivity with Common Materials:* No reaction; *Stability During Transport:* Stable; *Neutralizing Agents for Acids and Caustics:* Not pertinent; *Polymerization:* Not pertinent; *Inhibitor of Polymerization:* Not pertinent.

DIACETONE ALCOHOL

Chemical Designations - *Synonyms*: Diacetone; 4-Hydroxy-4-Metrhyl-2-Pentanone; Tyranton; *Chemical Formula*: $CH_3C(OH)(CH_3)CH_2COCH_3$.

Observable Characteristics - *Physical State (as normally shipped)*: Liquid; *Color*: Colorless to yellow; *Odor*: Mild, pleasant.

Physical and Chemical Properties - *Physical State at 15 °C and 1 atm.*: Liquid; *Molecular Weight*: 116.16; *Boiling Point at 1 atm.*: 336.6, 169.2, 442.4; *Freezing Point*: -45.0, -42.8, 230.4; *Critical Temperature*: 633, 334, 607; *Critical Pressure*: 380, 36, 3.6; *Specific Gravity*: 0.938 at 20°C (liquid); *Vapor (Gas) Density*: Not pertinent; *Ratio of Specific Heats of Vapor (Gas)*: 1.052; *Latent Heat of Vaporization*: 150, 85, 3.6; *Heat of Combustion*: (est.) -13,000, -7,250, -303; *Heat of Decomposition*: Not pertinent.

Health Hazards Information - *Recommended Personal Protective Equipment*: Air pack or organic canister, rubber gloves, goggles; *Symptoms Following Exposure*: Vapor is irritant to the mucous membrane of the eye and respiratory tract. Inhalation can cause dizziness, nausea, some anesthesia. Very high concentration have a narcotic effect. The liquid is not highly irritating to the skin but can cause dermatitis; *General Treatment for Exposure*: INHALATION: remove victim to fresh air. Give artificial respiration if breathing has stopped. EYES OR SKIN: wash affected skin areas with water; flush eyes with water and get medical care if discomfort persists; *Toxicity by Inhalation (Threshold Limit Value)*: 50 ppm; *Short-Term Exposure Limits*: 150 ppm for 30 min.; *Toxicity by Ingestion*: Grade 2; LD_{50} 0.5 to 5 g/kg (rat); *Late Toxicity*: Data not available; *Vapor (Gas) Irritant Characteristics*: Vapors cause moderate irritation such that personnel will find high concentrations unpleasant. The effect is temporary; *Liquid or Solid Irritant Characteristics*: Minimum hazard. If spilled on clothing and allowed to remain, may cause smarting and reddening of the skin; *Odor Threshold*: Data not available.

Fire Hazards - *Flash Point (deg. F):* 142 OC; 125 CC; *Flammable Limits in Air (%):* 1.8 - 6.9; *Fire Extinguishing Agents:* Dry chemical, alcohol foam, carbon dioxide; *Fire Extinguishing Agents Not to be Used:* Not pertinent; *Special Hazards of Combustion Products:* Not pertinent; *Behavior in Fire:* Not pertinent; *Ignition Temperature (deg. F):* 1118; *Electrical Hazard:* Not pertinent; *Burning Rate:* Data not available.

Chemical Reactivity - *Reactivity with Water :* No reaction; *Reactivity with Common Materials:* No reaction; *Stability During Transport:* Stable; *Neutralizing Agents for Acids and Caustics:* Not pertinent; *Polymerization:* Not pertinent; *Inhibitor of Polymerization:* Not pertinent.

DI-N-AMYL PHTHALATE

Chemical Designations - *Synonyms*: Diamyl Phthalate; Dipentyl Phthalate; Phthalic Acid, Diamyl Ester; Phthalic Acid, Dipentyl Ester; *Chemical Formula*: $C_6H_4(COOC_5H_{11})_2$.

Observable Characteristics - *Physical State (as normally shipped)*: Liquid; *Color*: Colorless; *Odor*: None.

Physical and Chemical Properties - *Physical State at 15 °C and 1 atm.*: Liquid; *Molecular Weight*: 306; *Boiling Point at 1 atm.*: Very high; *Freezing Point*: Not pertinent; *Critical Temperature*: Not pertinent; *Critical Pressure*: Not pertinent; *Specific Gravity*: 0.82 at 20°C (liquid); *Vapor (Gas) Density*: Not pertinent; *Ratio of Specific Heats of Vapor (Gas)*: Not pertinent; *Latent Heat of Vaporization*: 140, 76, 3.2; *Heat of Combustion*: -13,900, -7,720, -323; *Heat of Decomposition*: Not pertinent.

Health Hazards Information - *Recommended Personal Protective Equipment*: Goggles or face shield; rubber gloves; *Symptoms Following Exposure*: Inhalation of vapors from very hot material mazy cause headache, drowsiness, and convulsions. Hot vapors may irritate eyes; *General Treatment for Exposure*: INHALATION: move to fresh air. EYES: flush with water. SKIN: wipe off; flush with water; wash with soap and water; *Toxicity by Inhalation (Threshold Limit Value)*: Data not available; *Short-Term Exposure Limits*: Data not available; *Toxicity by Ingestion*: Data not available; *Late Toxicity*: Causes birth defects in rats (skeletal and gross abnormalities); *Vapor (Gas) Irritant Characteristics*: Data not available; *Liquid or Solid Irritant Characteristics*: Data not available; *Odor Threshold*: Data not available.

Fire Hazards - *Flash Point (deg. F):* 245 CC; *Flammable Limits in Air (%):* Data not available; *Fire Extinguishing Agents:* Foam, dry chemical, carbon dioxide; *Fire Extinguishing Agents Not to be Used:* Water or foam may cause frothing; *Special Hazards of Combustion Products:* Data not available; *Behavior in Fire:* Data not available; *Ignition Temperature :* Data not available; *Electrical Hazard:* Data not available; *Burning Rate:* Data not available.

Chemical Reactivity - *Reactivity with Water :* No reaction; *Reactivity with Common Materials:* May attack some forms of plastics; *Stability During Transport:* Stable; *Neutralizing Agents for Acids and Caustics:* Not pertinent; *Polymerization:* Not pertinent; *Inhibitor of Polymerization:* Not pertinent.

DIAZINON

Chemical Designations - *Synonyms*: O,O-Diethyl O-(-Isopropyl-6-Methyl-4-Pyrimidinyl) Phosphorothioate; O,O-DiethylO-2-Isopropyl-4-Methyl-6-Pyrimidyl Thiophosphate; Diethyl 2-Isopropyl-4-Methyl-6-Pyrimidyl Thionophosphate; Alpha-Tox; Saralex; Spectracide; *Chemical Formula*: $C_{12}H_{21}N_2O_3PS$.

Observable Characteristics - *Physical State (as normally shipped)*: Solid, or liquid solution; *Color*: Amber to dark brown; *Odor*: Data not available.

Physical and Chemical Properties - *Physical State at 15 °C and 1 atm.*: Liquid; *Molecular Weight*: 304.4; *Boiling Point at 1 atm.*: Very high, decomposes; *Freezing Point*: Not pertinent; *Critical Temperature*: Not pertinent; *Critical Pressure:* Not pertinent; *Specific Gravity*: 1.117 at 20°C (liquid); *Vapor (Gas) Density*: Not pertinent; *Ratio of Specific Heats of Vapor (Gas)*: Not pertinent; *Latent Heat of Vaporization*: Not pertinent; *Heat of Combustion*: (est.)-12,000, -6,500, -270; *Heat of Decomposition*: Not pertinent.

Health Hazards Information - *Recommended Personal Protective Equipment*: Goggles or face shield; rubber gloves; protective clothing; *Symptoms Following Exposure*: Ingestion or prolonged inhalation of mist causes headache, giddiness, blurred vision, nervousness, weakness, cramps, diarrhea, discomfort in the chest, sweating, miosis, tearing, salivation and other excessive respiratory tract secretion, vomiting, cyanosis, papilledema, uncontrollable muscle twitches, convulsions, coma, loss of reflexes, and loss of spincter control. Liquid irritates eyes and skin; *General Treatment for Exposure*: INHALATION: remove to fresh air; keep warm; get medical attention at once. EYES: flush with plenty of water for at lest 15 min. and get medical attention. SKIN: wash contaminated area with soap and water. INGESTION: get medical attention at once; give water slurry of charcoal; Do NOT give milk or alcohol; *Toxicity by Inhalation (Threshold Limit Value)*: Data not available; *Short-Term Exposure Limits*: Not pertinent; *Toxicity by Ingestion*: Grade 3; oral LD_{50}=76 mg/kg (rat); *Late Toxicity*: May be mutagenic; *Vapor (Gas) Irritant Characteristics*: Data not available; *Liquid or Solid Irritant Characteristics*: Data not available; *Odor Threshold*: Data not available.

Fire Hazards - *Flash Point (deg. F):* 82 - 105 CC (solutions only; pure liquid difficult to burn); *Flammable Limits in Air (%):* Not pertinent; *Fire Extinguishing Agents:* (for solutions) Foam, dry chemical, or carbon dioxide; *Fire Extinguishing Agents Not to be Used:* Water may be ineffective; *Special Hazards of Combustion Products:* Oxides of sulfur and phosphorus are generated in fires; *Behavior in Fire:* Not pertinent; *Ignition Temperature :* Not pertinent; *Electrical Hazard:* Data not available; *Burning Rate:* (for solutions) 4 mm/min.

Chemical Reactivity - *Reactivity with Water :* No reaction; *Reactivity with Common Materials:* No reaction; *Stability During Transport:* Stable; *Neutralizing Agents for Acids and Caustics:* Not pertinent; *Polymerization:* Not pertinent; *Inhibitor of Polymerization:* Not pertinent.

DIBENZOYL PEROXIDE

Chemical Designations - *Synonyms*: Benzoyl Peroxide; Benzoyl Superoxide; BP; BPO; Lucidol-70, Oxylite; *Chemical Formula*: $C_6H_5CO\cdot O\cdot O\cdot COC_6H_5$.

Observable Characteristics - *Physical State (as normally shipped)*: Solid; *Color*: White; *Odor*: None.

Physical and Chemical Properties - *Physical State at 15 °C and 1 atm.*: Solid; *Molecular Weight*: 242.22; *Boiling Point at 1 atm.*: Not pertinent; *Freezing Point*: 217, 103, 376; *Critical Temperature*: Not pertinent; *Critical Pressure*: Not pertinent; *Specific Gravity*: 1.334 at 15 °C (solid); *Vapor (Gas) Density*: Not pertinent; *Ratio of Specific Heats of Vapor (Gas)*: Not pertinent; *Latent Heat of Vaporization*: Not pertinent; *Heat of Combustion*: Not pertinent; *Heat of Decomposition*: Not pertinent.

Health Hazards Information - *Recommended Personal Protective Equipment*: Safety goggles, face shield, rubber gloves; *Symptoms Following Exposure*: CONTACT WITH EYES OR SKIN: irritates eyes. Prolonged contact may irritate skin; *General Treatment for Exposure*: INGESTION: administer an emetic to induce vomiting and call a physician. CONTACT WITH EYES OR SKIN; do not use oils or ointments; flush eyes with plenty of water, get medical attention; wash skin with plenty of soap and water; *Toxicity by Inhalation (Threshold Limit Value)*: 5 mg/m³; *Short-Term Exposure Limits*: Not pertinent; *Toxicity by Ingestion*: Grade 2; LD_{50} 0.5 to 5 g/kg; *Late Toxicity*: No data; *Vapor (Gas) Irritant Characteristics*: Not pertinent; *Liquid or Solid Irritant Characteristics*: Minimum hazard. If spilled on clothing and allowed to remain, may cause smarting and reddening of the skin; *Odor Threshold*: Not pertinent.

Fire Hazards - *Flash Point :* Highly flammable solid; explosion-sensitive to shock, heat and friction; *Flammable Limits in Air (%):* Not pertinent; *Fire Extinguishing Agents:* Difficult to extinguish once ignited. Use water spray to cool surrounding area; *Fire Extinguishing Agents Not to be Used:* Do not use hand extinguishers; *Special Hazards of Combustion Products:* Suffocating smoke evolved; *Behavior in Fire:* May explode; *Ignition Temperature :* Data not available; *Electrical Hazard:* Not pertinent; *Burning Rate:* Not pertinent.

Chemical Reactivity - *Reactivity with Water :* No reaction; *Reactivity with Common Materials:* Avoid contamination with combustible materials, various inorganic and organic acids, alkalies, alcohols, amines, easily oxidizable materials such as ethers, or materials used as accelerators in polymerizations reactions; *Stability During Transport:* Extremely explosion-sensitive to shock, heat and friction. Self-reactive; *Neutralizing Agents for Acids and Caustics:* Not pertinent; *Polymerization:* Not pertinent; *Inhibitor of Polymerization:* Not pertinent.

DI-N-BUTYLAMINE

Chemical Designations - *Synonyms*: 1-Butanamine, –butyl; Dibutylamine; *Chemical Formula*: $(C_4H_9)_2NH$.

Observable Characteristics - *Physical State (as normally shipped)*: Liquid; *Color*: Colorless; *Odor*: Weak ammonia.

Physical and Chemical Properties - *Physical State at 15 °C and 1 atm.*: Liquid; *Molecular Weight*: 129.25; *Boiling Point at 1 atm.*: 319.3, 159.6, 432.8; *Freezing Point*: -80, -62, 211; *Critical Temperature*: Not pertinent; *Critical Pressure*: Not pertinent; *Specific Gravity*: 0.759 at 20°C (liquid); *Vapor (Gas) Density*: 4.5; *Ratio of Specific Heats of Vapor (Gas)*: Not pertinent; *Latent Heat of Vaporization*: 130, 72.3, 3.03; *Heat of Combustion*: -18,800, -10,440, -436.8; *Heat of Decomposition*: Not pertinent.

Health Hazards Information - *Recommended Personal Protective Equipment*: Goggles or face shield; rubber gloves; *Symptoms Following Exposure*: Inhalation causes irritation of nose, throat, and lungs; coughing; nausea; headache. Ingestion causes irritation of mouth and stomach. Contact with eyes causes irritation. Contact with skin causes irritation and dermatitis; *General Treatment for Exposure*: INHALATION: move from exposure; if breathing has stopped, start artificial respiration. INGESTION: give large amounts of water. EYES: irrigate with water for 15 min. get medical attention for possible eye damage. SKIN: wash with large amounts of water for 15 min.; *Toxicity by Inhalation (Threshold Limit Value)*: Data not available; *Short-Term Exposure Limits*: Data not available; *Toxicity by Ingestion*: Grade 3; oral LD_{50} 360 mg/kg (rat); *Late Toxicity*: Data not available; *Vapor (Gas)*

Irritant Characteristics: Vapors cause moderate irritation such that personnel will find high concentrations unpleasant. The effect is temporary; *Liquid or Solid Irritant Characteristics*: Severe skin irritant. Causes second- and third-degree burns on short contact and is very injurious to the eyes; *Odor Threshold*: Data not available.

Fire Hazards - *Flash Point (deg. F)*: 125 OC; *Flammable Limits in Air (%)*: 1.1 (LFL); *Fire Extinguishing Agents*: "Alcohol" foam, dry chemical, carbon dioxide; *Fire Extinguishing Agents Not to be Used*: Water may be ineffective; *Special Hazards of Combustion Products*: Toxic oxides of nitrogen may form in fires; *Behavior in Fire*: Data not available; *Ignition Temperature*: Data not available; *Electrical Hazard*: Data not available; *Burning Rate*: 5.84 mm/min.

Chemical Reactivity - *Reactivity with Water*: No reaction; *Reactivity with Common Materials*: May corrode some metals and attack some forms of plastics; *Stability During Transport*: Stable; *Neutralizing Agents for Acids and Caustics*: Not pertinent; *Polymerization*: Not pertinent.

DI-N-BUTYL ETHER

Chemical Designations - *Synonyms*: n-Dibutyl Ether; n-Butyl Ether; Butyl Ether; 1-Butoxybutane; Dibutyl Ether; Dibutyl oxide; *Chemical Formula*: $C_4H_9OC_4H_9$

Observable Characteristics - *Physical State (as normally shipped)*: Liquid; *Color*: Colorless; *Odor*: Mild, ether-like.

Physical and Chemical Properties - *Physical State at 15 °C and 1 atm.*: Liquid; *Molecular Weight*: 130.2; *Boiling Point at 1 atm.*: 288, 142, 415; *Freezing Point*: -139.7, -95.4, 177.8; *Critical Temperature*: Not pertinent; *Critical Pressure*: Not pertinent; *Specific Gravity*: 0.767 at 20°C (liquid); *Vapor (Gas) Density*: 4.5; *Ratio of Specific Heats of Vapor (Gas)*: 1.0434; *Latent Heat of Vaporization*: 120, 68, 2.8; *Heat of Combustion*: -17,670, -9,820, -411; *Heat of Decomposition*: Not pertinent.

Health Hazards Information - *Recommended Personal Protective Equipment*: Goggles or face shield; rubber gloves; *Symptoms Following Exposure*: Inhalation causes irritation of nose and throat. Liquid irritates eyes and may irritate skin on prolonged contact. Ingestion causes irritation of mouth and stomach; *General Treatment for Exposure*: INHALATION: remove to fresh air. EYES: after contact with liquid, flush with water for at least 15 min. SKIN: wipe off, wash well with soap and water; *Toxicity by Inhalation (Threshold Limit Value)*: Data not available; *Short-Term Exposure Limits*: Data not available; *Toxicity by Ingestion*: Grade 1; oral LD_{50} 7,400 mg/kg; *Late Toxicity*: Data not available; *Vapor (Gas) Irritant Characteristics*: Data not available; *Liquid or Solid Irritant Characteristics*: Data not available; *Odor Threshold*: Data not available.

Fire Hazards - *Flash Point (deg. F)*: 92 OC; *Flammable Limits in Air (%)*: 1.5 - 7.6; *Fire Extinguishing Agents*: Dry chemical, "alcohol" foam, or carbon dioxide; *Fire Extinguishing Agents Not to be Used*: Water may be ineffective; *Special Hazards of Combustion Products*: Not pertinent; *Behavior in Fire*: Vapor is heavier than air and may travel a considerable distance to source of ignition and flash back; *Ignition Temperature (deg. F)*: 382; *Electrical Hazard*: Data not available; *Burning Rate*: 5.7 mm/min.

Chemical Reactivity - *Reactivity with Water No reaction.*; *Reactivity with Common Materials*: No reaction; *Stability During Transport*: Stable; *Neutralizing Agents for Acids and Caustics*: Not pertinent; *Polymerization*: Not pertinent; *Inhibitor of Polymerization*: Not pertinent.

DI-N-BUTYL KETONE

Chemical Designations - *Synonyms*: 5-Nonanone; *Chemical Formula*: $CH_3(CH_2)_3CO(CH_2)_3CH_3$.

Observable Characteristics - *Physical State (as normally shipped)*: Liquid; *Color*: White to light yellow; *Odor*: Data not available.

Physical and Chemical Properties - *Physical State at 15 °C and 1 atm.*: Liquid; *Molecular Weight*: 142; *Boiling Point at 1 atm.*: 370, 188, 461; *Freezing Point*: 21, -6, 267; *Critical Temperature*: Not pertinent; *Critical Pressure*: Not pertinent; *Specific Gravity*: 0.822 at 20°C (liquid); *Vapor (Gas) Density*: Not pertinent; *Ratio of Specific Heats of Vapor (Gas)*: Not pertinent; *Latent Heat of Vaporization*: 161, 89.6, 3.75; *Heat of Combustion*: -16,080, -8,930, -374; *Heat of Decomposition*: Not pertinent.

Health Hazards Information - *Recommended Personal Protective Equipment*: Rubber gloves; goggles or face shield; *Symptoms Following Exposure*: Inhalation causes irritation of nose and throat. Ingestion causes irritation of mouth and stomach. Contact with eyes or skin causes irritation; *General Treatment for Exposure*: INHALATION: remove to fresh air; administer artificial respiration if needed. EYES: flush with water for at least 15 min. SKIN: flush with water; *Toxicity by Inhalation (Threshold Limit Value)*: Data not available; *Short-Term Exposure Limits*: Data not available; *Toxicity by Ingestion*: Data not available; *Late Toxicity*: Data not available; *Vapor (Gas) Irritant Characteristics*: Data not available; *Liquid or Solid Irritant Characteristics*: Data not available; *Odor Threshold*: Data not available.

Fire Hazards - *Flash Point :* Data not available; *Flammable Limits in Air (%):* Data not available; *Fire Extinguishing Agents:* Foam, dry chemical, carbon dioxide; *Fire Extinguishing Agents Not to be Used:* Water may be ineffective; *Special Hazards of Combustion Products:* Data not available; *Behavior in Fire:* Data not available; *Ignition Temperature :* Data not available; *Electrical Hazard:* Data not available; *Burning Rate:* Data not available.

Chemical Reactivity - *Reactivity with Water :* No reaction; *Reactivity with Common Materials:* May attack some forms of plastics; *Stability During Transport:* Stable; *Neutralizing Agents for Acids and Caustics:* Not pertinent; *Polymerization:* Not pertinent; *Inhibitor of Polymerization:* Not pertinent.

DIBUTYLPHENOL

Chemical Designations - *Synonyms*: 2,6-Di-tert-butylphenol; *Chemical Formula*: 2,6-(t-$C_4H_9)_2C_6H_3OH$.

Observable Characteristics - *Physical State (as normally shipped)*: Solid or liquid; *Color*: Colorless; light straw; *Odor*: None.

Physical and Chemical Properties - *Physical State at 15 °C and 1 atm.*: Solid; *Molecular Weight*: 206.3; *Boiling Point at 1 atm.*: 487, 253, 526; *Freezing Point*: 97, 36, 309; *Critical Temperature*: Not pertinent; *Critical Pressure*: Not pertinent; *Specific Gravity*: 0.914 at 20°C (solid); *Vapor (Gas) Density*: Not pertinent; *Ratio of Specific Heats of Vapor (Gas)*: Not pertinent; *Latent Heat of Vaporization*: Not pertinent; *Heat of Combustion*: -18,000, -9,800, -410; *Heat of Decomposition*: Not pertinent.

Health Hazards Information - *Recommended Personal Protective Equipment*: Goggles or face shield; *Symptoms Following Exposure*: Irritates eyes and (on prolonged contact) skin. Ingestion causes irritation of mouth and stomach; *General Treatment for Exposure*: EYES: flush with water for at least 15 min. SKIN: wipe off, wash well with soap and water. INGESTION: induce vomiting; get medical attention; *Toxicity by Inhalation (Threshold Limit Value)*: Data not available; *Short-Term Exposure Limits*: Not pertinent; *Toxicity by Ingestion*: Grade 2; oral LD_{50} (2,6 Di-sec-butyl phenol); *Late Toxicity*: Data not available; *Vapor (Gas) Irritant Characteristics*: Data not available; *Liquid or Solid Irritant Characteristics*: Data not available; *Odor Threshold*: Data not available.

Fire Hazards - *Flash Point (deg. F):* >200 OC; *Flammable Limits in Air (%):* Not pertinent; *Fire Extinguishing Agents:* Dry chemical, carbon dioxide, foam; *Fire Extinguishing Agents Not to be Used:* Water may be ineffective; *Special Hazards of Combustion Products:* Not pertinent; *Behavior in Fire:* Not pertinent; *Ignition Temperature :* Data not available; *Electrical Hazard:* Data not available; *Burning Rate:* Data not available.

Chemical Reactivity - *Reactivity with Water:* No reaction; *Reactivity with Common Materials:* No reaction; *Stability During Transport:* Stable; *Neutralizing Agents for Acids and Caustics:* Not pertinent; *Polymerization:* Not pertinent; *Inhibitor of Polymerization:* Not pertinent.

DIBUTYL PHTHALATE

Chemical Designations - *Synonyms*: Butyl Phthalate; DBP; Phthalic Acid, Dibutyl Ester; RC Plasticizer DBP; Witcizer 300; *Chemical Formula*: $O-C_6H_4[COO(CH_2)_3CH_3]_2$.

Observable Characteristics - *Physical State (as normally shipped)*: Liquid; *Color*: Colorless; *Odor*: Slights characteristic ester odor; mild; practically none; slightly aromatic.

Physical and Chemical Properties - *Physical State at 15 °C and 1 atm.*: Liquid; *Molecular Weight*:

278.35; *Boiling Point at 1 atm.*: 635, 335, 608; *Freezing Point*: -31, -35, 238; *Critical Temperature*: 932, 500, 773; *Critical Pressure*: 250, 17, 1.7; *Specific Gravity*: 1.049 at 20°C (liquid); *Vapor (Gas) Density*: Not pertinent; *Ratio of Specific Heats of Vapor (Gas)*: Not pertinent; *Latent Heat of Vaporization*: Not pertinent; *Heat of Combustion (Btu/lb, cal/g, $\times 10^5$ J/kg)*: -13,300, -7,400, -310; *Heat of Decomposition*: Not pertinent.

Health Hazards Information   - *Recommended Personal Protective Equipment*: Eye protection; *Symptoms Following Exposure*: Vapors from very hot material may irritate eyes and produce headache, drowsiness, and convulsions; *General Treatment for Exposure*: Remove fresh air. Wash affected skin areas with water. Flush eyes with water; *Toxicity by Inhalation (Threshold Limit Value)*: 5 mg/m^3; *Short-Term Exposure Limits*: Not pertinent; *Toxicity by Ingestion*: Grade 1; LD$_{50}$ 5 to 15 g/kg; *Late Toxicity*: Birth defects in rats; polyneuritis in humans; *Vapor (Gas) Irritant Characteristics*: Not pertinent; *Liquid or Solid Irritant Characteristics*: No appreciable hazard. Practically harmless to the skin; *Odor Threshold*: Data not available.

Fire Hazards - *Flash Point (deg. F):* 355 OC; 315 CC; *Flammable Limits in Air (%):* 0.5 - 2.5 (calculated); *Fire Extinguishing Agents:* Dry powder, carbon dioxide, foam; *Fire Extinguishing Agents Not to be Used:* Water or foam may cause frothing; *Special Hazards of Combustion Products:* Not pertinent; *Behavior in Fire:* Not pertinent; *Ignition Temperature (deg. F):* 757; *Electrical Hazard:* Not pertinent; *Burning Rate:* Data not available.

Chemical Reactivity - *Reactivity with Water:* No reaction; *Reactivity with Common Materials:* No reaction; *Stability During Transport:* Stable; *Neutralizing Agents for Acids and Caustics:* Not pertinent; *Polymerization:* Not pertinent; *Inhibitor of Polymerization:* Not pertinent.

O-DICHLOROBENZENE

Chemical Designations - *Synonyms*: 1,2-Dichlorobenzene; Downtherm E; Orthodichlorobenzene; *Chemical Formula*: o-C$_6$H$_4$Cl$_2$.

Observable Characteristics - *Physical State (as normally shipped)*: Liquid; *Color*: Colorless; *Odor*: Aromatic; characteristic aromatic.

Physical and Chemical Properties - *Physical State at 15 °C and 1 atm.*: Liquid; *Molecular Weight*: 147.01; *Boiling Point at 1 atm.*: 356.9, 180.5, 453.7; *Freezing Point*: 0.3, -17.6, 255.6; *Critical Temperature*: Not pertinent; *Critical Pressure*: Not pertinent; *Specific Gravity*: 1.306 at 20°C (liquid); *Vapor (Gas) Density*: Not pertinent; *Ratio of Specific Heats of Vapor (Gas)*: 1.080; *Latent Heat of Vaporization*: 115, 63.9, 2.68; *Heat of Combustion*: -7,969, -4,427, -185.4; *Heat of Decomposition*: Not pertinent.

Health Hazards Information - *Recommended Personal Protective Equipment*: Organic vapor-acid gas respirator; neoprene or vinyl gloves; chemical safety spectacles, face shield, rubber footwear, apron, protective clothing; *Symptoms Following Exposure*: Chronic inhalation of mist vapors may result in damage to lungs, liver, and kidneys. Acute vapor exposure can cause symptoms ranging from coughing to central nervous system depression and transient anesthesia. Irritating to skin, eyes, and mucous membranes. May cause dermatitis; *General Treatment for Exposure*: INHALATION: remove victim to fresh air, keep him quiet and warm, and call a physician promptly. INGESTION: no known antidote; treat symptomatically; induce vomiting and get medical attention promptly. EYES AND SKIN: flush with plenty of water; get medical attention for eyes; remove contaminated clothing and wash before reuse; *Toxicity by Inhalation (Threshold Limit Value)*: 50 ppm; *Short-Term Exposure Limits*: 50 ppm for 15 min.; *Toxicity by Ingestion*: Grade 2; LD$_{50}$ 0.5 to 5 g/kg; *Late Toxicity*: Causes liver and kidney damage in rats. Effects unknown in humans; *Vapor (Gas) Irritant Characteristics*: Vapors cause moderate irritation such that personnel will find high concentrations unpleasant. The effect is temporary; *Liquid or Solid Irritant Characteristics*: Minimum hazard. If spilled on clothing and allowed to remain, may cause smarting and reddening of the skin; *Odor Threshold*: 4.0 ppm; 50 ppm.

Fire Hazards - *Flash Point (deg. F):* 165 OC; 155 CC; *Flammable Limits in Air (%):* 2.2 - 9.2; *Fire Extinguishing Agents:* Water, foam, dry chemical or carbon dioxide; *Fire Extinguishing Agents Not to be Used:* Not pertinent; *Special Hazards of Combustion Products:* Irritating vapors including hydrogen chloride gas, chlorocarbones, chlorine; *Behavior in Fire:* Not pertinent; *Ignition Temperature (deg. F):*

1198; *Electrical Hazard:* Not pertinent; *Burning Rate:* 1.3 mm/min.
Chemical Reactivity - *Reactivity with Water :* No reaction; *Reactivity with Common Materials:* No reaction; *Stability During Transport:* Stable; *Neutralizing Agents for Acids and Caustics:* Not pertinent; *Polymerization:* Not pertinent; *Inhibitor of Polymerization:* Not pertinent.

P-DICHLOROBENZENE
Chemical Designations - *Synonyms:* Dichloricide; Paradow; Paradi; Paramoth; Paradichlorobenzene; Santochlor; *Chemical Formula:* p-$C_6H_4Cl_2$.
Observable Characteristics - *Physical State (as normally shipped):* Solid; *Color:* White; *Odor:* Aromatic.
Physical and Chemical Properties - *Physical State at 15 °C and 1 atm.:* Solid; *Molecular Weight:* 147.01; *Boiling Point at 1 atm.:* 345.6, 174.2, 447.4; *Freezing Point:* 130, 53, 326; *Critical Temperature:* Not pertinent; *Critical Pressure:* Not pertinent; *Specific Gravity:* 1.458 at 20°C (solid); *Vapor (Gas) Density:* Not pertinent; *Ratio of Specific Heats of Vapor (Gas):* Not pertinent; *Latent Heat of Vaporization:* Not pertinent; *Heat of Combustion:* Not pertinent; *Heat of Decomposition:* Not pertinent.
Health Hazards Information - *Recommended Personal Protective Equipment:* Full face mask fitted with organic vapor canister for concentrations over 75 ppm; clean protective clothing; eye protection; *Symptoms Following Exposure:* INHALATION: irritation of upper respiratory tract; over-exposure may cause depression and injury to liver and kidney. EYES: pain and mild irritation; *General Treatment for Exposure:* INHALATION: if any ill effects develop, remove patient to fresh air and get medical attention. If breathing stops, give artificial respiration. EYES: flush with plenty of water and get medical attention if ill effects develop. SKIN AND INGESTION: likely no problems; *Toxicity by Inhalation (Threshold Limit Value):* 75 ppm; *Short-Term Exposure Limits:* 50 ppm for 60 min.; *Toxicity by Ingestion:* Grade 2; LD_{50} 0.5 to 5 g/kg; *Late Toxicity:* Data not available; *Vapor (Gas) Irritant Characteristics:* Vapors cause moderate irritation such that personnel will find high concentrations unpleasant. The effect is temporary; *Liquid or Solid Irritant Characteristics:* Minimum hazard. If spilled on clothing and allowed to remain, may cause smarting and reddening of the skin; *Odor Threshold:* 15 - 30 ppm.
Fire Hazards - *Flash Point (deg. F):* 165 OC; 150 CC; *Flammable Limits in Air (%):* Data not available; *Fire Extinguishing Agents:* Water, foam, carbon dioxide, or dry chemical; *Fire Extinguishing Agents Not to be Used:* Not pertinent; *Special Hazards of Combustion Products:* Vapors are irritating. Toxic chlorine, hydrogen chloride and phosgene gases may be generated in fires; *Behavior in Fire:* Not pertinent; *Ignition Temperature:* No data; *Electrical Hazard:* Not pertinent; *Burning Rate:* 1.3 mm/min (approx.).
Chemical Reactivity - *Reactivity with Water :* No reaction; *Reactivity with Common Materials:* No reaction; *Stability During Transport:* Stable; *Neutralizing Agents for Acids and Caustics:* Not pertinent; *Polymerization:* Not pertinent; *Inhibitor of Polymerization:* Not pertinent.

DI-(P-CHLOROBENZOYL) PEROXIDE
Chemical Designations - *Synonyms:* Bis-(p-chlorobenzoyl)peroxide; p-Chlorobenzoyl Peroxide; p,p`-Dichlorobenzoyl Peroxide; Di-(4-chlorobenzoyl)peroxide; Cadox PS; *Chemical Formula:* (p-$ClC_6H_4COO)_2$.
Observable Characteristics - *Physical State (as normally shipped):* Solid; or paste in silicone fluid and dibutyl phthalate; *Color:* White; *Odor:* None.
Physical and Chemical Properties - *Physical State at 15 °C and 1 atm.:* Solid; *Molecular Weight:* 311.1; *Boiling Point at 1 atm.:* Decomposes; *Freezing Point:* Not pertinent; *Critical Temperature:* Not pertinent; *Critical Pressure:* Not pertinent; *Specific Gravity:* >1.1 at 20°C (solid); *Vapor (Gas) Density:* Not pertinent; *Ratio of Specific Heats of Vapor (Gas):* Not pertinent; *Latent Heat of Vaporization:* Not pertinent; *Heat of Combustion:* -9,000, -5,000, -210; *Heat of Decomposition:* Not pertinent.
Health Hazards Information - *Recommended Personal Protective Equipment:* Goggles or face shield;

rubber gloves; protective clothing; *Symptoms Following Exposure*: Irritates eyes and (on prolonged contact) skin. Ingestion causes irritation of mouth and stomach; *General Treatment for Exposure*: EYES: wash with water for at least 15 min.; consult a doctor. SKIN: wash with soap and water. INGESTION: induce vomiting and call a doctor; *Toxicity by Inhalation (Threshold Limit Value)*: Data not available; *Short-Term Exposure Limits*: Not pertinent; *Toxicity by Ingestion*: Data not available; *Late Toxicity*: Data not available; *Vapor (Gas) Irritant Characteristics*: Data not available; *Liquid or Solid Irritant Characteristics*: Data not available; *Odor Threshold*: Not pertinent.

Fire Hazards - *Flash Point* : Not pertinent; *Flammable Limits in Air (%)*: Not pertinent; *Fire Extinguishing Agents*: Flood with water, or use dry chemical, foam, carbon dioxide; *Fire Extinguishing Agents Not to be Used*: Not pertinent; *Special Hazards of Combustion Products*: Toxic chlorinated biphenyls are formed in fires; *Behavior in Fire*: Solid may explode. Burns very rapidly when ignited. Smoke is unusually heavy when paste form is involved; *Ignition Temperature* : Data not available; *Electrical Hazard*: Data not available; *Burning Rate*: Not pertinent.

Chemical Reactivity - *Reactivity with Water* : No reaction; *Reactivity with Common Materials*: May react vigorously with combustible materials; *Stability During Transport*: Stable (below 80° F); *Neutralizing Agents for Acids and Caustics*: Not pertinent; *Polymerization*: Not pertinent; *Inhibitor of Polymerization*: Not pertinent.

DICHLOROBUTENE

Chemical Designations - *Synonyms*: 1,4-Dichloro-2-butene; 2-Butylene Dichloride; 1,4-Dichloro-2-butylene; cis-1,4-Dichloro-2-butene; trans-1,4-Dichloro-2-butene; *Chemical Formula*: $ClCH_2CH=CHCH_2Cl$.

Observable Characteristics - *Physical State (as normally shipped)*: Liquid; *Color*: Colorless; *Odor*: Characteristic; sweet, pungent.

Physical and Chemical Properties - *Physical State at 15 °C and 1 atm.*: Liquid; *Molecular Weight*: 125.0; *Boiling Point at 1 atm.*: 313, 156, 429; *Freezing Point*: cis: -54, -48, 225; trans: 37, 3, 276; *Critical Temperature*: Not pertinent; *Critical Pressure*: Not pertinent; *Specific Gravity*: 1.112 at 20°C (liquid); *Vapor (Gas) Density*: 4; *Ratio of Specific Heats of Vapor (Gas)*: 1.0874; *Latent Heat of Vaporization*: 130, 73, 3.1; *Heat of Combustion*: -17,500, -9,720, -407; *Heat of Decomposition*: Not pertinent.

Health Hazards Information - *Recommended Personal Protective Equipment*: Rubber gloves; chemical splash goggles; rubber boots and apron; barrier cream; organic canister mask; *Symptoms Following Exposure*: Inhalation of vapor irritates nose and throat. Contact with eyes causes intense irritation and tears. Contact of liquid with skin causes severe blistering and dermatitis. Ingestion causes severe irritation of mouth and stomach; *General Treatment for Exposure*: INHALATION: remove from exposure; provide low-pressure oxygen if required; keep under observation until edema is ruled out. EYES: irrigate immediately for 15 min.; call physician. SKIN: wash immediately and thoroughly with soap and water; treat as a chemical burn. INGESTION: induce vomiting; call physician; *Toxicity by Inhalation (Threshold Limit Value)*: Data not available; *Short-Term Exposure Limits*: Data not available; *Toxicity by Ingestion*: Grade 3; oral LD_{50} (1,4-dichloro-2-butene) 89 mg/kg (rat); *Late Toxicity*: Data not available; *Vapor (Gas) Irritant Characteristics*: Data not available; *Liquid or Solid Irritant Characteristics*: Data not available; *Odor Threshold*: Data not available.

Fire Hazards - *Flash Point* : Data not available; *Flammable Limits in Air (%)*: 1.5 - 4; *Fire Extinguishing Agents*: Water, foam, dry chemical ,or carbon dioxide; *Fire Extinguishing Agents Not to be Used*: Not pertinent; *Special Hazards of Combustion Products*: Decomposition vapors contain phosgene and hydrogen chloride gases; both are toxic and irritating; *Behavior in Fire*: Not pertinent; *Ignition Temperature*: Data not available; *Electrical Hazard*: Data not available; *Burning Rate*: 2.6 mm/min.

Chemical Reactivity - *Reactivity with Water* : Reacts slowly to form hydrochloric acid; *Reactivity with Common Materials*: Corrodes metal when wet; *Stability During Transport*: Stable; *Neutralizing Agents for Acids and Caustics*: Not pertinent; *Polymerization*: Not pertinent; *Inhibitor of Polymerization*: Not pertinent.

DICHLORODIFLUOROMETHANE

Chemical Designations - *Synonyms*: Eskimon 12; Genetron 12; F-12; Halon 122; Freon 12; Isotron 12; Frigen 12; Ucon 12; *Chemical Formula*: CCl_2F_2.

Observable Characteristics - *Physical State (as normally shipped)*: Liquefied compressed gas; *Color*: Colorless; *Odor*: Odorless; slight; characteristic.

Physical and Chemical Properties - *Physical State at 15 °C and 1 atm.*: Gas; *Molecular Weight*: 120.91; *Boiling Point at 1 atm.*: -21.6, -29.8, 243.4; *Freezing Point*: -251.9, -157.7, 115.5; *Critical Temperature*: 233.2, 111.8, 385; *Critical Pressure*: 598, 40.7, 4.12; *Specific Gravity*: 1.35 at 15°C (liquid); *Vapor (Gas) Density*: 4.2; *Ratio of Specific Heats of Vapor (Gas)*: 1.129; *Latent Heat of Vaporization*: 140, 77.9, 3.26; *Heat of Combustion*: Not pertinent; *Heat of Decomposition*: Not pertinent.

Health Hazards Information - *Recommended Personal Protective Equipment*: Rubber gloves, goggles; *Symptoms Following Exposure*: INHALATION: some narcosis when 10% in air is breathed; *General Treatment for Exposure*: Remove patient to non-contaminated area and apply artificial respiration if breathing has stopped; call a physician immediately; oxygen may be given; *Toxicity by Inhalation (Threshold Limit Value)*: 1000 ppm; *Short-Term Exposure Limits*: 5000 ppm for 60 min.; *Toxicity by Ingestion*: Not pertinent; *Late Toxicity*: None; *Vapor (Gas) Irritant Characteristics*: None, except at very high concentrations, which may irritates lungs; *Liquid or Solid Irritant Characteristics*: No appreciable hazard. Practically harmless to the skin because it is very volatile and evaporates quickly; *Odor Threshold*: Data not available.

Fire Hazards - *Flash Point :* Not flammable; *Flammable Limits in Air (%):* Not flammable; *Fire Extinguishing Agents:* Not pertinent; *Fire Extinguishing Agents Not to be Used:* Not pertinent; *Special Hazards of Combustion Products:* Although nonflammable, dissociation products generated in a fire may be irritating or toxic; *Behavior in Fire:* Helps extinguish fire; *Ignition Temperature :* Not flammable; *Electrical Hazard:* Not pertinent; *Burning Rate:* Not flammable.

Chemical Reactivity - *Reactivity with Water :* No reaction; *Reactivity with Common Materials:* No reaction; *Stability During Transport:* Stable; *Neutralizing Agents for Acids and Caustics:* Not pertinent; *Polymerization:* Not pertinent; *Inhibitor of Polymerization:* Not pertinent.

1,2-DICHLOROETHYLENE

Chemical Designations - *Synonyms*: Acetylene Dichloride; sym-Dichloroethylene; Dioform; cis- or trans-1,2-Dichloroethylene; *Chemical Formula*: ClCH=CHCl.

Observable Characteristics - *Physical State (as normally shipped)*: Liquid; *Color*: Colorless; *Odor*: Ethereal, slightly acrid; pleasant, chloroform-like.

Physical and Chemical Properties - *Physical State at 15 °C and 1 atm.*: Liquid; *Molecular Weight*: 97.0; *Boiling Point at 1 atm.:* cis: 140, 60, 333; trans: 118, 48, 321; *Freezing Point*: cis: -114, -81, 192; trans: -58, -50, 223; *Critical Temperature*: Not pertinent; *Critical Pressure*: Not pertinent; *Specific Gravity*: 1.27 at 25°C (liquid); *Vapor (Gas) Density*: 3.34; *Ratio of Specific Heats of Vapor (Gas)*: 1.1468; *Latent Heat of Vaporization*: 130, 72, 3.0; *Heat of Combustion*: -4,847.2, -2,692.9, -112.67; *Heat of Decomposition*: Not pertinent.

Health Hazards Information - *Recommended Personal Protective Equipment*: Rubber gloves; safety goggles; air supply mask or self-contained breathing apparatus; *Symptoms Following Exposure*: Inhalation causes nausea, vomiting, weakness, tremor, epigastric cramps, central nervous depression. Contact with liquid causes irritation of eyes and (on prolonged contact) skin. Ingestion causes slight depression to deep narcosis; *General Treatment for Exposure*: INHALATION: remove from further exposure; if breathing is difficult, give oxygen; if victim is not breathing, give artificial respiration, preferably mouth-to-mouth; give oxygen when breathing is resumed; call a physician. EYES: flush with water for at least 15 min. SKIN: wash well with soap and water. INGESTION: give gastric lavage and cathartics; *Toxicity by Inhalation (Threshold Limit Value)*: 200 ppm; *Short-Term Exposure Limits*: Data not available; *Toxicity by Ingestion*: Grade 2; oral LD_{50} 770 mg/kg (rat); *Late Toxicity*: Produces liver and kidney injury in experimental animals; *Vapor (Gas) Irritant Characteristics*: Data not available; *Liquid or Solid Irritant Characteristics*: Data not available; *Odor Threshold*: Data not available.

Fire Hazards - *Flash Point (deg. F):* 37CC; *Flammable Limits in Air (%):* 9.7 - 12.8; *Fire Extinguishing Agents:* Dry chemical, foam, carbon dioxide; *Fire Extinguishing Agents Not to be Used:* Water may be ineffective; *Special Hazards of Combustion Products:* Phosgene and hydrogen chloride fumes may form in fires; *Behavior in Fire:* Vapor is heavier than air and may travel a considerable distance to a source of ignition and flash back; *Ignition Temperature :* Data not available; *Electrical Hazard:* Data not available; *Burning Rate:* 2.6 mm/min.

Chemical Reactivity - *Reactivity with Water:* No reaction; *Reactivity with Common Materials:* No reaction; *Stability During Transport:* Stable; *Neutralizing Agents for Acids and Caustics:* Not pertinent; *Polymerization:* Will not occur under ordinary conditions of shipment. The reaction is not vigorous; *Inhibitor of Polymerization:* None used.

DICHLOROETHYL ETHER

Chemical Designations - *Synonyms*: Bis(2-chloroethyl) Ether; 2,2'-Dichloroethyl Ether; Dichlorodiethyl Ether; Di-(2-chloroethyl) Ether; Chlorex; DCEE; beta,beta'-Dichloroethyl Ether; *Chemical Formula*: $(ClCH_2CH_2)_2O$.

Observable Characteristics - *Physical State (as normally shipped)*: Liquid; *Color*: Colorless; *Odor*: Sweet, like chloroform.

Physical and Chemical Properties - *Physical State at 15 °C and 1 atm.*: Liquid; *Molecular Weight*: 143.0; *Boiling Point at 1 atm.*: 353, 178, 451; *Freezing Point*: -62, -52, 221; *Critical Temperature*: Not pertinent; *Critical Pressure*: Not pertinent; *Specific Gravity*: 1.22 at 20°C (liquid); *Vapor (Gas) Density*: 4.93; *Ratio of Specific Heats of Vapor (Gas)*: 1.0743; *Latent Heat of Vaporization*: 143, 79.5, 3.33; *Heat of Combustion*: -7,530, -4,180, -175; *Heat of Decomposition*: Not pertinent.

Health Hazards Information - *Recommended Personal Protective Equipment*: Goggles or face shield; rubber gloves; protective clothing; *Symptoms Following Exposure*: Inhalation of vapor causes irritation of nose, coughing, nausea. Liquid irritates eyes and causes mild irritation of skin. (Can be absorbed in toxic amounts through the skin). Ingestion causes irritation of mouth and stomach, symptoms of systemic poisoning; *General Treatment for Exposure*: INHALATION: remove from exposure; support respiration; call physician if needed. EYES: irrigate with copious quantities of water for 15 min.; call physician. SKIN: wipe off, wash well with soap and water. INGESTION: induce vomiting; get medical attention; *Toxicity by Inhalation (Threshold Limit Value)*: 5 ppm; *Short-Term Exposure Limits*: 35 ppm for 30 min.; *Toxicity by Ingestion*: Grade 3; oral LD_{50} 75 mg/kg (rat); *Late Toxicity*: Said to be carcinogenic; *Vapor (Gas) Irritant Characteristics*: Vapors are moderately irritating such that personnel will not usually tolerate moderate or high vapor concentrations; *Liquid or Solid Irritant Characteristics*: Causes smarting of the skin and first-degree burns on short exposure; may cause second-degree burns on long exposure; *Odor Threshold*: Data not available.

Fire Hazards - *Flash Point (deg. F):* 180 OC; 131 CC; *Flammable Limits in Air (%):* Data not available; *Fire Extinguishing Agents:* Water, foam, dry chemical, carbon dioxide; *Fire Extinguishing Agents Not to be Used:* Not pertinent; *Special Hazards of Combustion Products:* May form phosgene or hydrogen chloride in fires; *Behavior in Fire:* Not pertinent; *Ignition Temperature (deg. F):* 696; *Electrical Hazard:* Data not available; *Burning Rate:* 2.4 mm/min.

Chemical Reactivity - *Reactivity with Water :* No reaction; *Reactivity with Common Materials:* No reaction; *Stability During Transport:* Stable; *Neutralizing Agents for Acids and Caustics:* Not pertinent; *Polymerization:* V; *Inhibitor of Polymerization:* Not pertinent.

DICHLOROMETHANE

Chemical Designations - *Synonyms*: Methylene Chloride; Methylene Dichloride; *Chemical Formula*: CH_2Cl_2.

Observable Characteristics - *Physical State (as normally shipped)*: Liquid; *Color*: Colorless; *Odor*: Pleasant, aromatic; like chloroform; sweet, ethereal.

Physical and Chemical Properties - *Physical State at 15 °C and 1 atm.*: Liquid; *Molecular Weight*: 84.93; *Boiling Point at 1 atm.*: 104, 39.8, 313; *Freezing Point*: -142, -96.7, 176.5; *Critical Temperature*: 473, 245, 518; *Critical Pressure*: 895, 60.9, 6.17; *Specific Gravity*: 1.322 at 20°C (liquid); *Vapor (Gas) Density*: 2.9; *Ratio of Specific Heats of Vapor (Gas)*: 1.199; *Latent Heat of Vaporization*: 142, 78.7, 3.30; *Heat of Combustion*: Not pertinent; *Heat of Decomposition*: Not pertinent.

Health Hazards Information - *Recommended Personal Protective Equipment*: Organic vapor canister mask, safety glasses, protective clothing; *Symptoms Following Exposure*: INHALATION: anesthetic effects, nausea and drunkenness. SKIN AND EYES: skin irritation, irritation of eyes and nose; *General Treatment for Exposure*: INHALATION: remove from exposure. Give oxygen if needed. INGESTION: no specific antidote. SKIN AND EYES: remove contaminated clothing; wash skin or eyes if affected; *Toxicity by Inhalation (Threshold Limit Value)*: 500 ppm; *Short-Term Exposure Limits*: 100 ppm for 60 min.; *Toxicity by Ingestion*: Grade 2; LD_{50} 0.5 to 5 g/kg; *Late Toxicity*: None; *Vapor (Gas) Irritant Characteristics*: Vapors cause moderate irritation such that personnel will find high concentrations unpleasant. The effect is temporary; *Liquid or Solid Irritant Characteristics*: Minimum hazard. If spilled on clothing and allowed to remain, may cause smarting and reddening of the skin; *Odor Threshold*: 205-307 ppm.

Fire Hazards - *Flash Point :* Not flammable under conditions likely to be encountered; *Flammable Limits in Air (%):* 12 - 19; *Fire Extinguishing Agents:* Not pertinent; *Fire Extinguishing Agents Not to be Used:* Not pertinent; *Special Hazards of Combustion Products:* Dissociation products generated in a fire may be irritating or toxic; *Behavior in Fire:* Not pertinent; *Ignition Temperature (deg. F):* 1184; *Electrical Hazard:* Not pertinent; *Burning Rate:* Not pertinent.

Chemical Reactivity - *Reactivity with Water:* No reaction; *Reactivity with Common Materials:* No reaction; *Stability During Transport:* Stable; *Neutralizing Agents for Acids and Caustics:* Not pertinent; *Polymerization:* Not pertinent; *Inhibitor of Polymerization:* Not pertinent.

2,4-DICHLOROPHENOL

Chemical Designations - *Synonyms*: No common synonyms; *Chemical Formula*: $HOC_6H_3Cl_2$-2,4.

Observable Characteristics - *Physical State (as normally shipped)*: Solid; *Color*: White; *Odor*: None.

Physical and Chemical Properties - *Physical State at 15 °C and 1 atm.*: Solid; *Molecular Weight*: 163.01; *Boiling Point at 1 atm.*: 421, 216, 489; *Freezing Point*: 110, 45, 318; *Critical Temperature*: Not pertinent; *Critical Pressure*: Not pertinent; *Specific Gravity*: 1.40 at 15°C (solid); *Vapor (Gas) Density*: Not pertinent; *Ratio of Specific Heats of Vapor (Gas)*: Not pertinent; *Latent Heat of Vaporization*: Not pertinent; *Heat of Combustion*: Not pertinent; *Heat of Decomposition*: Not pertinent.

Health Hazards Information - *Recommended Personal Protective Equipment*: Bu. Mines approved respirator, rubber gloves, chemical goggles; *Symptoms Following Exposure*: Tremors, convulsions, shortness of breath, inhibition of respiratory system; *General Treatment for Exposure*: Inhalation - rest. Ingestion - drink water, epsom salt solution; *Toxicity by Inhalation (Threshold Limit Value)*: Not pertinent; *Short-Term Exposure Limits*: Data not available; *Toxicity by Ingestion*: Grade 2; LD_{50} 0.5 to 5 g/kg (rat); *Late Toxicity*: Data not available; *Vapor (Gas) Irritant Characteristics*: Not pertinent; *Liquid or Solid Irritant Characteristics*: Fairly severe skin irritant. May cause pain and second-degree burns after a few minutes' contact; *Odor Threshold*: Data not available.

Fire Hazards - *Flash Point (deg. F):* 200 OC, 237 CC; *Flammable Limits in Air (%):* Data not available; *Fire Extinguishing Agents:* Water, foam, carbon dioxide, dry chemical; *Fire Extinguishing Agents Not to be Used:* Water or foam may cause frothing; *Special Hazards of Combustion Products:* Toxic gases can be evolved; *Behavior in Fire:* Solid melts and burns; *Ignition Temperature :* Data not available; *Electrical Hazard:* Not pertinent; *Burning Rate:* Not pertinent.

Chemical Reactivity - *Reactivity with Water :* No reaction; *Reactivity with Common Materials:* May react vigorously with oxidizing material; *Stability During Transport:* Stable; *Neutralizing Agents for Acids and Caustics:* Not pertinent; *Polymerization:* Not pertinent.

2,4-DICHLOROPHENOXYACETIC ACID

Chemical Designations - *Synonyms*: 2,4-D; *Chemical Formula*: 2,4-$Cl_2C_6H_3OCH_2COOH$;

Observable Characteristics - *Physical State (as normally shipped)*: Solid; *Color*: White to tan; *Odor*: None.

Physical and Chemical Properties - *Physical State at 15 °C and 1 atm.*: Solid; *Molecular Weight*: 221.0; *Boiling Point at 1 atm.*: Very high; *Freezing Point*: 286, 141, 314; *Critical Temperature*: Not pertinent; *Critical Pressure*: Not pertinent; *Specific Gravity*: 1.563 at 20°C (solid); *Vapor (Gas) Density*: Not pertinent; *Ratio of Specific Heats of Vapor (Gas)*: Not pertinent; *Latent Heat of Vaporization*: Not pertinent; *Heat of Combustion*: -7,700, -4,300, -180; *Heat of Decomposition*: Not pertinent.

Health Hazards Information - *Recommended Personal Protective Equipment*: Protective dust mask; rubber gloves; chemical gloves; *Symptoms Following Exposure*: Dust may irritate eyes. Ingestion causes gastroentric distress, diarrhea, mild central nervous system depression, dysphagia, and possible transient liver and kidney injury; *General Treatment for Exposure*: EYES: flush with water for at least 15 min. SKIN: wash well with soap and water. INGESTION: induce vomiting and follow with gastric lavage and supportive therapy; *Toxicity by Inhalation (Threshold Limit Value)*: Data not available; *Short-Term Exposure Limits*: Data not available; *Toxicity by Ingestion*: Grade 3; oral rat LD_{50} 375 mg/kg (rat), 80 mg/kg (human); *Late Toxicity*: Causes birth defects in some laboratory animals; *Vapor (Gas) Irritant Characteristics*: Not pertinent; *Liquid or Solid Irritant Characteristics*: Data not available; *Odor Threshold*: Not pertinent.

Fire Hazards - *Flash Point :* Not pertinent (combustible solid); *Flammable Limits in Air (%):* Not pertinent; *Fire Extinguishing Agents:* Water, foam; *Fire Extinguishing Agents Not to be Used:* Not pertinent; *Special Hazards of Combustion Products:* Toxic and irritating hydrogen chloride or phosgene gases may form; *Behavior in Fire:* Not pertinent; *Ignition Temperature:* Not pertinent; *Electrical Hazard:* Data not available; *Burning Rate:* Not pertinent.

Chemical Reactivity - *Reactivity with Water :* No reaction; *Reactivity with Common Materials:* No reaction; *Stability During Transport:* Stable; *Neutralizing Agents for Acids and Caustics:* Flush with water, rinse with sodium bicarbonate; *Polymerization:* Not pertinent; *Inhibitor of Polymerization:* Not pertinent.

DICHLOROPROPANE

Chemical Designations - *Synonyms*: 1,2-Dichloropropane; Propylene Dichloride; *Chemical Formula*: $CH_3CHClCH_2Cl$.

Observable Characteristics - *Physical State (as normally shipped)*: Liquid; *Color*: Colorless; *Odor*: Sweet.

Physical and Chemical Properties - *Physical State at 15 °C and 1 atm.*: Liquid; *Molecular Weight*: 102.9; *Boiling Point at 1 atm.*: 206, 96.4, 369.6; *Freezing Point*: -148, -100, 173; *Critical Temperature*: Not pertinent; *Critical Pressure*: Not pertinent; *Specific Gravity*: 1.158 at 20°C (liquid); *Vapor (Gas) Density*: 3.5; *Ratio of Specific Heats of Vapor (Gas)*: 1.094; *Latent Heat of Vaporization*: 122, 67.7, 2.83; *Heat of Combustion*: 7,300, 4,100, 170; *Heat of Decomposition*: Not pertinent.

Health Hazards Information - *Recommended Personal Protective Equipment*: Air supply in confined area, rubber gloves, chemical goggles, protective coveralls and rubber footwear; *Symptoms Following Exposure*: Contact with skin or eyes may cause irritation; *General Treatment for Exposure*: INHALATION: remove to fresh air. SKIN OR EYES: wash skin thoroughly with soap and water. Flush eyes with water for 15 min. Call a doctor; *Toxicity by Inhalation (Threshold Limit Value)*: 75 ppm; *Short-Term Exposure Limits*: Data not available; *Toxicity by Ingestion*: Grade 2; LD_{50} 0.5 to 5 g/kg (guinea pig); *Late Toxicity*: Data not available; *Vapor (Gas) Irritant Characteristics*: Vapors cause moderate irritation such that personnel will find high concentrations unpleasant. The effect is temporary; *Liquid or Solid Irritant Characteristics*: Minimum hazard. If spilled on clothing and allowed to remain, may cause smarting and reddening of the skin; *Odor Threshold*: Data not available.

Fire Hazards - *Flash Point (deg. F):* 70 OC; 60 CC; *Flammable Limits in Air (%):* 3.4 - 14.5; *Fire Extinguishing Agents:* Foam, carbon dioxide, dry chemical; *Fire Extinguishing Agents Not to be Used:* Not pertinent; *Special Hazards of Combustion Products:* Toxic and irritating gases may be generated; *Behavior in Fire:* Not pertinent; *Ignition Temperature (deg. F):* 1035; *Electrical Hazard:* Not pertinent; *Burning Rate:* (est.) 3.2 mm/min.

Chemical Reactivity - *Reactivity with Water:* No reaction; *Reactivity with Common Materials:* No

reaction; *Stability During Transport:* Stable; *Neutralizing Agents for Acids and Caustics:* Not pertinent; *Polymerization:* Not pertinent; *Inhibitor of Polymerization:* Not pertinent.

DICHLOROPROPENE
Chemical Designations - *Synonyms*: 1,3-Dichloropropene; Telone; *Chemical Formula*: $ClCH_2CH=CHCl$.
Observable Characteristics - *Physical State (as normally shipped)*: Liquid; *Color*: Colorless; *Odor*: Sweet; like chloroform.
Physical and Chemical Properties - *Physical State at 15 °C and 1 atm.*: Liquid; *Molecular Weight*: 110.98; *Boiling Point at 1 atm.*: 170, 77, 350; *Freezing Point*: Not pertinent; *Critical Temperature*: Not pertinent; *Critical Pressure*: Not pertinent; *Specific Gravity*: 1.2 at 20°C (liquid); *Vapor (Gas) Density*: Not pertinent; *Ratio of Specific Heats of Vapor (Gas)*: 1.116; *Latent Heat of Vaporization*: 113, 62.8, 2.63; *Heat of Combustion*: 6,900, 3,900, 160; *Heat of Decomposition*: Not pertinent.
Health Hazards Information - *Recommended Personal Protective Equipment*: An approved full face mask equipped with a fresh black canister meeting specifications of the U.S. Bureau of Mines for organic vapors, a full face self-contained breathing apparatus, or full face air-supplied respirator; *Symptoms Following Exposure*: Smarting of skin and eyes. Prolonged contact of liquid with skin may cause second-degree burns; *General Treatment for Exposure*: INHALATION: remove patient to fresh air, keep warm and quiet; call physician immediately; give artificial respiration if breathing has stopped. INGESTION: call physician immediately. Induce vomiting by giving an emetic, e.g., 2 tablespoons table salt in glass of warm water. SKIN OR EYES: immediately remove contaminated clothing and shoes. Wash skin with soap and plenty of water. For eyes, flush immediately with plenty of water for at least 15 min. Call physician; *Toxicity by Inhalation (Threshold Limit Value)*: Data not available; *Short-Term Exposure Limits*: Data not available; *Toxicity by Ingestion*: Grade 3; LD_{50} 50 to 500 mg/kg; *Late Toxicity*: Data not available; *Vapor (Gas) Irritant Characteristics*: Vapors cause moderate irritation such that personnel will find high concentrations unpleasant. The effect is temporary; *Liquid or Solid Irritant Characteristics*: Causes smarting of the skin and first degree burns on short exposure and may cause secondary burns on long exposure; *Odor Threshold*: Data not available.
Fire Hazards - *Flash Point (deg. F):* 95 CC; *Flammable Limits in Air (%):* Data not available; *Fire Extinguishing Agents:* water, dry chemical, foam, carbon dioxide; *Fire Extinguishing Agents Not to be Used:* Not pertinent; *Special Hazards of Combustion Products:* Toxic and irritating gases may be generated; *Behavior in Fire:* Not pertinent; *Ignition Temperature :* Data not available; *Electrical Hazard:* Data not available; *Burning Rate:* (est.) 3.4 mm/min.
Chemical Reactivity - *Reactivity with Water:* No reaction; *Reactivity with Common Materials:* No reaction; *Stability During Transport:* Stable; *Neutralizing Agents for Acids and Caustics:* Not pertinent; *Polymerization:* Not pertinent; *Inhibitor of Polymerization:* Not pertinent.

DICYCLOPENTADIENE
Chemical Designations - *Synonyms*: Dicy; 3a,4,7,7a-Tetrahydro-4,7-methanoindene; *Chemical Formula*: $C_{10}H_{12}$.
Observable Characteristics - *Physical State (as normally shipped)*: Liquid; *Color*: Colorless; *Odor*: Camphor-like.
Physical and Chemical Properties - *Physical State at 15 °C and 1 atm.*: Liquid; *Molecular Weight*: 132.31; *Boiling Point at 1 atm.*: 338, 170, 443; *Freezing Point*: 41, 5, 278; *Critical Temperature*: Not pertinent; *Critical Pressure*: Not pertinent; *Specific Gravity*: 0.978 at 20°C (liquid); *Vapor (Gas) Density*: Not pertinent; *Ratio of Specific Heats of Vapor (Gas)*: Not pertinent; *Latent Heat of Vaporization*: Not pertinent; *Heat of Combustion*: -18,800, -10,400, -437; *Heat of Decomposition*: Not pertinent.
Health Hazards Information - *Recommended Personal Protective Equipment*: Air-supplied mask in confined areas, rubber gloves, safety glasses; *Symptoms Following Exposure*: Vapor irritates mucous membranes and respiratory tract, causes nausea, vomiting, headache, and dizziness. Direct contact

irritates eyes; *General Treatment for Exposure*: INHALATION: remove victim from contaminated area and call physician if unconscious; if breathing is irregular or stopped, give oxygen and start resuscitation. EYES OR SKIN: flush with plenty of water for 15 min.; *Toxicity by Inhalation (Threshold Limit Value)*: 75-100 ppm; *Short-Term Exposure Limits*: Data not available; *Toxicity by Ingestion*: Grade 2; oral rat LD_{50} 0.82 g/kg; *Late Toxicity*: Data not available; *Vapor (Gas) Irritant Characteristics*: Vapors cause moderate irritation such that personnel will find high concentrations unpleasant. The effect is temporary; *Liquid or Solid Irritant Characteristics*: Minimum hazard. If spilled on clothing and allowed to remain, may cause smarting and reddening of the skin; *Odor Threshold*: <0.003 ppm.

Fire Hazards - *Flash Point (deg. F)*: 90 OC; *Flammable Limits in Air (%)*: 0.8 - 6.3; *Fire Extinguishing Agents:* Foam, carbon dioxide, dry chemical, or water spray; *Fire Extinguishing Agents Not to be Used:* Not pertinent; *Special Hazards of Combustion Products:* Not pertinent; *Behavior in Fire:* Not pertinent; *Ignition Temperature (deg. F):* 941; *Electrical Hazard:* Data not available; *Burning Rate:* Data not available.

Chemical Reactivity - *Reactivity with Water :* No reaction; *Reactivity with Common Materials:* No reaction; *Stability During Transport:* Stable; *Neutralizing Agents for Acids and Caustics:* Not pertinent; *Polymerization:* May occur in presence of acids, but not hazardous; *Inhibitor of Polymerization:* Not pertinent.

DIELDRIN

Chemical Designations - *Synonyms*: HEOD; endo,exo-1,2,3,4,10,10-Hexachloro-6,7-epoxy-1,4,4a,5,6,7,8,8a-octahydro-1,4:5,8-dimethanonaphthalene; *Chemical Formula*: $C_{12}H_8Cl_6O$.

Observable Characteristics - *Physical State (as normally shipped)*: Solid; *Color*: Buff to light brown; *Odor*: Mild chemical.

Physical and Chemical Properties - *Physical State at 15 °C and 1 atm.*: Solid; *Molecular Weight*: 380.93; *Boiling Point at 1 atm.:* Not pertinent (decomposes); *Freezing Point*: 349, 176, 449; *Critical Temperature*: Not pertinent; *Critical Pressure*: Not pertinent; *Specific Gravity*: 1.75 at 20°C (solid); *Vapor (Gas) Density*: Not pertinent; *Ratio of Specific Heats of Vapor (Gas)*: Not pertinent; *Latent Heat of Vaporization*: Not pertinent; *Heat of Combustion*: Data not pertinent; *Heat of Decomposition*: Not pertinent.

Health Hazards Information - *Recommended Personal Protective Equipment*: U.S. Bu. Mines approved respirator; clean rubber gloves; goggles or face shield; *Symptoms Following Exposure*: Inhalation, ingestion, or skin contact causes irritability, convulsions and/or coma, nausea, vomiting, headache, fainting, tremors. Contact with eyes causes irritation; *General Treatment for Exposure*: INHALATION: move to fresh air; give oxygen and artificial respiration as required. INGESTION: induce vomiting and get medical attention. EYES: flush with plenty of water; *Toxicity by Inhalation (Threshold Limit Value)*: 0.25 mg/m³; *Short-Term Exposure Limits*: 1 mg/m³ for 30 min.; *Toxicity by Ingestion*: Grade 4; oral LD_{50} 46 mg/kg (rat), 65 mg/kg (dog); *Late Toxicity*: Banned by EPA in October 1974 because of alleged "imminent hazard to human health" as a potential carcinogen in man; *Vapor (Gas) Irritant Characteristics*: Data not pertinent; *Liquid or Solid Irritant Characteristics*: Minimum hazard. If spilled on clothing and allowed to remain, may cause smarting and reddening of the skin; *Odor Threshold*: 0.041 ppm.

Fire Hazards - *Flash Point :* Not flammable; *Flammable Limits in Air (%):* Not flammable; *Fire Extinguishing Agents:* Not pertinent; *Fire Extinguishing Agents Not to be Used:* Data not available; *Special Hazards of Combustion Products:* Toxic and irritating hydrogen chloride fumes may form in fire; *Behavior in Fire:* Not pertinent; *Ignition Temperature :* Not pertinent; *Electrical Hazard:* Not pertinent; *Burning Rate:* Not pertinent.

Chemical Reactivity - *Reactivity with Water:* No reaction; *Reactivity with Common Materials:* No reactions; *Stability During Transport:* Stable; *Neutralizing Agents for Acids and Caustics:* Not pertinent; *Polymerization:* Not pertinent; *Inhibitor of Polymerization:* Not pertinent.

DIETHANOLAMINE

Chemical Designations - *Synonyms*: Bist(2-hydroxyethyl)amine; DEA; 2,2'-Dihydroxydiethyl Amine; Di(2-hydroxyethyl)amine; 2,2'-Iminodiethanol; *Chemical Formula*: $(HOCH_2CH_2)_2NH$.

Observable Characteristics - *Physical State (as normally shipped)*: Liquid; *Color*: Colorless; *Odor*: Mild ammoniacal; faint, fishy; characteristic.

Physical and Chemical Properties - *Physical State at 15 °C and 1 atm.*: Solid; *Molecular Weight*: 105.14; *Boiling Point at 1 atm.*: 515.1, 268.4, 541.6; *Freezing Point*: 82, 28, 301; *Critical Temperature*: 828, 442, 715; *Critical Pressure*: 470, 32, 3.2; *Specific Gravity*: 1.095 at 28°C (liquid); *Vapor (Gas) Density*: Not pertinent; *Ratio of Specific Heats of Vapor (Gas)*: 1.053; *Latent Heat of Vaporization*: 266, 148, 6.20; *Heat of Combustion*: -10,790, -6,000, -251; *Heat of Decomposition*: Not pertinent.

Health Hazards Information - *Recommended Personal Protective Equipment*: Full face mask or amine vapor mask only, if required; clean body covering clothing, chemical goggles; *Symptoms Following Exposure*: Irritation of eyes and skin. Breathing vapors may cause coughing, a smothering sensation, nausea, headache; *General Treatment for Exposure*: INHALATION: no problem likely. Get medical attention if ill effects develop. INGESTION: induce vomiting if large amounts are swallowed and call a physician. Treat symptomatically. No known antidote. EYES: flush with plenty of water for at least 15 min. and get medical attention promptly. SKIN: flush with water . Wash contaminated clothing before reuse; *Toxicity by Inhalation (Threshold Limit Value)*: Not pertinent; *Short-Term Exposure Limits*: Not pertinent; *Toxicity by Ingestion*: Grade 2; LD_{50} 0.5 to 5 g/kg (rat); *Late Toxicity*: Data not available; *Vapor (Gas) Irritant Characteristics*: Vapors cause moderate irritation such that personnel will find high concentrations unpleasant. The effect is temporary; *Liquid or Solid Irritant Characteristics*: Causes smarting of the skin and first-degree burns on short exposure and may cause secondary burns on long exposure; *Odor Threshold*: Data not available.

Fire Hazards - *Flash Point (deg. F)*: 305 OC; *Flammable Limits in Air (%)*: 1.6 (calc.)- 9.8 (est); *Fire Extinguishing Agents*: Water, alcohol foam, carbon dioxide, dry chemical; *Fire Extinguishing Agents Not to be Used*: Addition of water may cause frothing; *Special Hazards of Combustion Products*: Irritating vapors are generated when heated; *Behavior in Fire*: Not pertinent; *Ignition Temperature (deg. F)*: 1224; *Electrical Hazard*: Not pertinent; *Burning Rate*: 0.74 mm/min.

Chemical Reactivity - *Reactivity with Water* : No reaction; *Reactivity with Common Materials*: No reaction; *Stability During Transport*: Stable; *Neutralizing Agents for Acids and Caustics*: Flush with water; *Polymerization*: Not pertinent; *Inhibitor of Polymerization*: Not pertinent.

DIETHYLAMINE

Chemical Designations - *Synonyms*: DEN; *Chemical Formula*: $(CH_3CH_2)_2NH$

Observable Characteristics - *Physical State (as normally shipped)*: Liquid; *Color*: Colorless; *Odor*: Ammoniacal; sharp, fishy.

Physical and Chemical Properties - *Physical State at 15 °C and 1 atm.*: Liquid; *Molecular Weight*: 73.14; *Boiling Point at 1 atm.*: 132, 55.5, 328.7; *Freezing Point*: -57.6, -49.8, 223.4; *Critical Temperature*: 434.3, 223.5, 496.7; *Critical Pressure*: 538, 36.6, 3.71; *Specific Gravity*: 0.708 at 20°C (liquid); *Vapor (Gas) Density*: 2.5; *Ratio of Specific Heats of Vapor (Gas)*: 1.079; *Latent Heat of Vaporization*: 170, 93, 3.9; *Heat of Combustion*: -17,990, -9,994, -418.4; *Heat of Decomposition*: Not pertinent.

Health Hazards Information - *Recommended Personal Protective Equipment*: Chemical safety goggles, rubber gloves, and apron; *Symptoms Following Exposure*: Irritation and burning of eyes, skin and respiratory system. High concentration of vapor can cause asphyxiation; *General Treatment for Exposure*: In case of contact, flush skin or eyes with plenty of water for at least 15 min.; for eyes, get medical attention; *Toxicity by Inhalation (Threshold Limit Value)*: 25 ppm; *Short-Term Exposure Limits*: 100 ppm for 30 min.; *Toxicity by Ingestion*: Grade 2; 0.5 to 5 g/kg (rat); *Late Toxicity*: None; *Vapor (Gas) Irritant Characteristics*: Vapor is moderately irritating such that personnel will not usually tolerate moderate or high vapor concentrations; *Liquid or Solid Irritant Characteristics*: Minimum hazard. If spilled on clothing and allowed to remain, may cause smarting and reddening of the skin;

Odor Threshold: 0.14 ppm.

Fire Hazards - *Flash Point (deg. F):* 5 OC; *Flammable Limits in Air (%):* 1.8 - 9.1; *Fire Extinguishing Agents:* Dry chemical, carbon dioxide, or alcohol foam; *Fire Extinguishing Agents Not to be Used:* Data not available; *Special Hazards of Combustion Products:* Vapors are irritating; *Behavior in Fire:* Vapors are heavier than air and may travel considerable distance to a source of ignition and flash back; *Ignition Temperature (deg. F):* 594; *Electrical Hazard:* Data not available; *Burning Rate:* 6.7 mm/min.

Chemical Reactivity - *Reactivity with Water:* No reaction; *Reactivity with Common Materials:* No hazardous reaction; *Stability During Transport:* Stable; *Neutralizing Agents for Acids and Caustics:* Flush with water; *Polymerization:* Not pertinent; *Inhibitor of Polymerization:* Not pertinent.

DIETHYLBENZENE

Chemical Designations - *Synonyms*: No common synonyms; *Chemical Formula*: $C_6H_4(C_2H_5)_2$.

Observable Characteristics - *Physical State (as normally shipped)*: Liquid; *Color*: Colorless; *Odor*: Characteristic aromatic; like benzene; like toluene,

Physical and Chemical Properties - *Physical State at 15 °C and 1 atm.*: Liquid; *Molecular Weight*: 134.21; *Boiling Point at 1 atm.*: 356, 180, 453; *Freezing Point*: < 160, < 70, < 343; *Critical Temperature*: Not pertinent; *Critical Pressure*: Not pertinent; *Specific Gravity*: 0.86 at 20 °C (liquid); *Vapor (Gas) Density*: Not pertinent; *Ratio of Specific Heats of Vapor (Gas)*: Not pertinent; *Latent Heat of Vaporization*: 140, 77, 3.2; *Heat of Combustion*: -17,800, -9,890, -414; *Heat of Decomposition*: Not pertinent.

Health Hazards Information - *Recommended Personal Protective Equipment*: Self-contained breathing apparatus, safety goggles; *Symptoms Following Exposure*: High vapor concentrations produce eye and respiratory tract irritation, dizziness, depression. Liquid irritates and may blister skin; can cause corneal injury to eye; *General Treatment for Exposure*: INHALATION: remove to fresh air and start artificial respiration. INGESTION: do NOT induce vomiting; call a doctor. EYES AND SKIN: flush with water for at least 15 min. Wash skin with soap and water; *Toxicity by Inhalation (Threshold Limit Value)*: Data not available; *Short-Term Exposure Limits*: Data not available; *Toxicity by Ingestion*: Grade 2; oral rat LD_{50} 1.2 g/kg; *Late Toxicity*: Data not available; *Vapor (Gas) Irritant Characteristics*: Vapors cause a slight smarting of the eyes and respiratory system if present in high concentrations. The effect is temporary; *Liquid or Solid Irritant Characteristics*: Minimum hazard. If spilled on clothing and allowed to remain, may cause smarting and reddening of the skin; *Odor Threshold*: Data not available.

Fire Hazards - *Flash Point (deg. F):* 135 CC; *Flammable Limits in Air (%):* Data not available; *Fire Extinguishing Agents:* Foam, water, carbon dioxide, or dry chemical; *Fire Extinguishing Agents Not to be Used:* Not pertinent; *Special Hazards of Combustion Products:* Not pertinent; *Behavior in Fire:* Not pertinent; *Ignition Temperature (deg. F):* 743 (ortho); *Electrical Hazard:* Not pertinent; *Burning Rate:* Data not available.

Chemical Reactivity - *Reactivity with Water :* No reaction; *Reactivity with Common Materials:* No reaction; *Stability During Transport:* Stable; *Neutralizing Agents for Acids and Caustics:* Not pertinent; *Polymerization:* Not pertinent; *Inhibitor of Polymerization:* Not pertinent.

DIETHYL CARBONATE

Chemical Designations - *Synonyms*: Carbonic Acid Diethyl Ester; Ethyl Carbonate; Eufin; *Chemical Formula*: $(CH_3CH_2)_2CO_3$.

Observable Characteristics - *Physical State (as normally shipped)*: Liquid; *Color*: Colorless; *Odor*: Pleasant, etheral; mild and nonresidual.

Physical and Chemical Properties - *Physical State at 15 °C and 1 atm.*: Liquid; *Molecular Weight*: 118.13; *Boiling Point at 1 atm.*: 260.2, 126.8, 400; *Freezing Point*: -45, -43, 230; *Critical Temperature:* Not pertinent; *Critical Pressure*: Not pertinent; *Specific Gravity*: 0.975 at 20°C (liquid); *Vapor (Gas) Density*: Not pertinent; *Ratio of Specific Heats of Vapor (Gas)*: 1.110; *Latent Heat of Vaporization*: 130, 73, 3.1; *Heat of Combustion*: -9,760, -5,420, -227; *Heat of Decomposition*: Not pertinent.

Health Hazards Information - *Recommended Personal Protective Equipment*: Protective clothing; rubber gloves and goggles; organic vapor canister of air mask; *Symptoms Following Exposure*: High vapor concentrations can cause headache, irritation of eyes and respiratory tract, dizziness, nausea, weakness, loss of consciousness; *General Treatment for Exposure*: INHALATION: remove from exposure; administer artificial respiration and oxygen if needed. EYES: flush with water for at least 15 min.; *Toxicity by Inhalation (Threshold Limit Value)*: Data not available; *Short-Term Exposure Limits*: Data not available; *Toxicity by Ingestion*: Data not available; *Late Toxicity*: None; *Vapor (Gas) Irritant Characteristics*: Vapors may cause slight smarting of eyes; *Liquid or Solid Irritant Characteristics*: Minimum hazard; *Odor Threshold*: Data not available.

Fire Hazards - *Flash Point (deg. F)*: 115 OC; 77 CC; *Flammable Limits in Air (%)*: Data not available; *Fire Extinguishing Agents*: Foam, carbon dioxide, dry chemical; *Fire Extinguishing Agents Not to be Used*: Water; *Special Hazards of Combustion Products*: Not pertinent; *Behavior in Fire*: Not pertinent; *Ignition Temperature* : No data; *Burning Rate*: 3.4 mm/min.

Chemical Reactivity - *Reactivity with Water* : Too slow to be hazardous; *Reactivity with Common Materials*: No reaction; *Stability During Transport*: Stable; *Neutralizing Agents for Acids and Caustics*: Not pertinent; *Polymerization*: Not pertinent; *Inhibitor of Polymerization*: Not pertinent.

DIETHYLENE GLYCOL

Chemical Designations - *Synonyms*: Bis(2-hydroxyethyl)ether; Diglycol; β,β-Dihydroxydiethyl Ether; 3-Oxa-1,5-pentanediol; 2,2'-Oxybisethanol; *Chemical Formula*: $(HOCH_2CH_2)_2O$.

Observable Characteristics - *Physical State (as normally shipped)*: Liquid; *Color*: Colorless; *Odor*: Practically odorless.

Physical and Chemical Properties - *Physical State at 15 °C and 1 atm.*: Liquid; *Molecular Weight*: 106.12; *Boiling Point at 1 atm.*: 473, 245, 518; *Freezing Point*: 20, -8, 265; *Critical Temperature*: 766, 408, 681; *Critical Pressure*: 680, 46, 4.7; *Specific Gravity*: 1.118 at 20°C (liquid); *Vapor (Gas) Density*: Not pertinent; *Ratio of Specific Heats of Vapor (Gas)*: Not pertinent; *Latent Heat of Vaporization*: 270, 150, 6.28; *Heat of Combustion*: -9,617, -5,343, -223.7; *Heat of Decomposition*: Not pertinent.

Health Hazards Information - *Recommended Personal Protective Equipment*: Full face mask with canister for short exposures to high vapors levels; rubber gloves; goggles; *Symptoms Following Exposure*: Ingestion of large amounts may cause degeneration of kidney and liver and cause death. Liquid may cause slight skin irritation; *General Treatment for Exposure*: INHALATION: no problem likely. If any ill defects do develop, get medical attention. INGESTION: induce vomiting if ingested. No known antidote; treat symptomatically. EYES AND SKIN: flush with water. If any ill defects occur, get medical attention; *Toxicity by Inhalation (Threshold Limit Value)*: 100 ppm (suggested); *Short-Term Exposure Limits*: Not pertinent; *Toxicity by Ingestion*: Grade 0; LD_{50} above 15 g/kg (rat); *Late Toxicity*: Kidney and liver damage; *Vapor (Gas) Irritant Characteristics*: None; *Liquid or Solid Irritant Characteristics*: None; *Odor Threshold*: Not pertinent.

Fire Hazards - *Flash Point (deg. F)*: 255 CC; *Flammable Limits in Air (%)*: 1.6 - 10.8; *Fire Extinguishing Agents*: Alcohol foam, carbon dioxide, dry chemical; *Fire Extinguishing Agents Not to be Used*: Water or foam may cause frothing; *Special Hazards of Combustion Products*: Not pertinent; *Behavior in Fire*: Not pertinent; *Ignition Temperature (deg. F)*: 444; *Electrical Hazard*: Not pertinent; *Burning Rate*: 1.5 mm/min.

Chemical Reactivity - *Reactivity with Water* : No reaction; *Reactivity with Common Materials*: No reaction; *Stability During Transport*: Stable; *Neutralizing Agents for Acids and Caustics*: Not pertinent; *Polymerization*: Not pertinent; *Inhibitor of Polymerization*: Not pertinent.

DIETHYLENE GLYCOL DIMETHYL ETHER

Chemical Designations - *Synonyms*: Bis(2-methoxyethyl)ether; Poly-Solv; *Chemical Formula*: $(CH_3OCH_2CH_2)_2O$.

Observable Characteristics - *Physical State (as normally shipped)*: Liquid; *Color*: Colorless; *Odor*: Mild ethereal.

Physical and Chemical Properties - *Physical State at 15 °C and 1 atm.*: Liquid; *Molecular Weight*: 134.12; *Boiling Point at 1 atm.*: 324, 162, 435; *Freezing Point*: -94, -70, 203; *Critical Temperature*: Not pertinent; *Critical Pressure*: Not pertinent; *Specific Gravity*: 0.945 at 20°C (liquid); *Vapor (Gas) Density*: Not pertinent; *Ratio of Specific Heats of Vapor (Gas)*: Not pertinent; *Latent Heat of Vaporization*: 130, 74, 3.1; *Heat of Combustion*: -11,300, -6,260, -262; *Heat of Decomposition*: Not pertinent.

Health Hazards Information - *Recommended Personal Protective Equipment*: Vinyl (not rubber) gloves; safety goggles; *Symptoms Following Exposure*: INGESTION (severe cases): nausea, vomiting, abdominal cramps, weakness progressing to coma; *General Treatment for Exposure*: INGESTION: give water and induce vomiting; oxygen and artificial respiration as needed; *Toxicity by Inhalation (Threshold Limit Value)*: Not pertinent; *Short-Term Exposure Limits*: Data not available; *Toxicity by Ingestion*: Data not available; *Late Toxicity*: Data not available; *Vapor (Gas) Irritant Characteristics*: None; *Liquid or Solid Irritant Characteristics*: None; *Odor Threshold*: Data not available.

Fire Hazards - *Flash Point (deg. F)*: 158 OC; *Flammable Limits in Air (%)*: Data not available; *Fire Extinguishing Agents*: Dry chemical, foam, carbon dioxide; *Fire Extinguishing Agents Not to be Used*: Not pertinent; *Special Hazards of Combustion Products*: Not pertinent; *Behavior in Fire*: Not pertinent; *Ignition Temperature* : Data not available; *Electrical Hazard*: Not pertinent; *Burning Rate*: Data not available.

Chemical Reactivity - *Reactivity with Water* : No reaction; *Reactivity with Common Materials*: No reaction; *Stability During Transport*: Stable; *Neutralizing Agents for Acids and Caustics*: Not pertinent; *Polymerization*: Not pertinent; *Inhibitor of Polymerization*: Not pertinent.

DIETHYLENEGLYCOL MONOBUTYL ETHER

Chemical Designations - *Synonyms*: Butoxydiethylene Glycol; Butoxydiglycol; Diglycol Monobutyl Ether; Butyl "Carbitol"; Dowanol DB; Poly-Solv DB; *Chemical Formula*: $C_4H_9OCH_2CH_2OCH_2CH_2OH$.

Observable Characteristics - *Physical State (as normally shipped)*: Liquid; *Color*: Colorless; *Odor*: Mild, characteristic; pleasant.

Physical and Chemical Properties - *Physical State at 15 °C and 1 atm.*: Liquid; *Molecular Weight*: 162.2; *Boiling Point at 1 atm.*: 448, 231, 504; *Freezing Point*: -90, -68, 205; *Critical Temperature*: Not pertinent; *Critical Pressure*: Not pertinent; *Specific Gravity*: 0.954 at 20°C (liquid); *Vapor (Gas) Density*: Not pertinent; *Ratio of Specific Heats of Vapor (Gas)*: Not pertinent; *Latent Heat of Vaporization*: 130, 74, 3.1; *Heat of Combustion*: -14,000, -7,900, -330; *Heat of Decomposition*: Not pertinent.

Health Hazards Information - *Recommended Personal Protective Equipment*: Safety goggles or face shield; *Symptoms Following Exposure*: Inhalation for brief periods has no significant effect. Contact with liquid causes moderate irritation of eyes and corneal injury. Prolonged contact with skin causes only minor irritation; *General Treatment for Exposure*: INHALATION: remove to fresh air; if ill effects are observed, call a doctor. EYES: immediately flush with plenty of water for at least 15 min. SKIN: wash well with soap and water. INGESTION: give large amounts of water; *Toxicity by Inhalation (Threshold Limit Value)*: Data not available; *Short-Term Exposure Limits*: Data not available; *Toxicity by Ingestion*: Grade 2; oral LD_{50} 2 g/kg (guinea pig); *Late Toxicity*: Data not available; *Vapor (Gas) Irritant Characteristics*: Vapors cause a slight smarting of the eyes or respiratory system if present in high concentrations. The effect is temporary; *Liquid or Solid Irritant Characteristics*: No appreciable hazard. Practically harmless to the skin; *Odor Threshold*: Data not available.

Fire Hazards - *Flash Point (deg. F)*: 230 OC; 172 CC; *Flammable Limits in Air (%)*: Not pertinent; *Fire Extinguishing Agents*: Water, "alcohol" foam, carbon dioxide, dry chemical; *Fire Extinguishing Agents Not to be Used*: Not pertinent; *Special Hazards of Combustion Products*: Not pertinent; *Behavior in Fire*: Not pertinent; *Ignition Temperature (deg. F)*: 442; *Electrical Hazard*: Not pertinent; *Burning Rate*: 3.3 mm/min.

Chemical Reactivity - *Reactivity with Water* : No reaction; *Reactivity with Common Materials*: No reaction; *Stability During Transport*: Stable; *Neutralizing Agents for Acids and Caustics*: Not pertinent;

Polymerization: Not pertinent; *Inhibitor of Polymerization:* Not pertinent.

DIETHYLENEGLYCOL MONOBUTYL ETHER ACETATE

Chemical Designations - *Synonyms*: 2-(2-Butoxyethoxyl)ethyl Acetate; Diglycol Monobutyl Ether Acetate; Butyl "Carbitol" Acetate; Ektasolve DB Acetate; *Chemical Formula*: $C_4H_9OCH_2CH_2OCH_2CH_2OCOCH_3$.

Observable Characteristics - *Physical State (as normally shipped)*: Liquid; *Color*: Colorless; *Odor*: Mild, non-residual.

Physical and Chemical Properties - *Physical State at 15 °C and 1 atm.*: Liquid; *Molecular Weight*: 204.3; *Boiling Point at 1 atm.*: 475, 246, 519; *Freezing Point*: -27, -33, 240; *Critical Temperature*: Not pertinent; *Critical Pressure*: Not pertinent; *Specific Gravity*: 0.985 at 20°C (liquid); *Vapor (Gas) Density*: Not pertinent; *Ratio of Specific Heats of Vapor (Gas)*: Not pertinent; *Latent Heat of Vaporization*: 106, 59, 2.5; *Heat of Combustion*:-13,000, -7,400, -310; *Heat of Decomposition*: Not pertinent.

Health Hazards Information - *Recommended Personal Protective Equipment*: Face shield or safety glasses; protective gloves; air mask for prolonged exposure to vapor; *Symptoms Following Exposure*: Prolonged breathing of vapor may cause irritation and nausea. Contact with liquid any cause mild irritation of eyes and skin. Can be absorbed through skin in toxic amounts; *General Treatment for Exposure*: INHALATION: move victim to fresh air; if breathing has stopped, administer artificial respiration. EYES: flush with water for at least 15 min. SKIN: wash skin with large amounts of water for 15 min.; call physician if needed. INGESTION: induce vomiting; get medical attention; *Toxicity by Inhalation (Threshold Limit Value)*: Data not available; *Short-Term Exposure Limits*: Because of high boiling point (246°C), hazards from inhalation are minimal; *Toxicity by Ingestion*: Grade 2; oral LD_{50} 2.34 g/kg; *Late Toxicity*: Kidney damage noted in animals following repeated contact with skin; *Vapor (Gas) Irritant Characteristics*: Data not available; *Liquid or Solid Irritant Characteristics*: Data not available; *Odor Threshold*: Data not available.

Fire Hazards - *Flash Point (deg. F):* 240 OC; *Flammable Limits in Air (%):* 0.8 - 5.0; *Fire Extinguishing Agents:* Water, alcohol foam, dry chemical, carbon dioxide; *Fire Extinguishing Agents Not to be Used:* Not pertinent; *Special Hazards of Combustion Products:* Not pertinent; *Behavior in Fire:* Not pertinent; *Ignition Temperature (deg. F):* 563; *Electrical Hazard:* Data not available; *Burning Rate:* No data.

Chemical Reactivity - *Reactivity with Water :* No reaction; *Reactivity with Common Materials:* No reaction; *Stability During Transport:* Stable; *Neutralizing Agents for Acids and Caustics:* Not pertinent; *Polymerization:* Not pertinent; *Inhibitor of Polymerization:* Not pertinent.

DIETHYLENE GLYCOL MONOETHYL ETHER

Chemical Designations - *Synonyms*: Carbitol; Diethylene Glycol Ethyl Ether; Dowanol DE; 2-(2-Ethoxyethoxy)ethanol; Ethoxy Diglycol; Poly-Solv DE; *Chemical Formula*: $CH_3CH_2OCH_2CH_2OCH_2CH_2OH$.

Observable Characteristics - *Physical State (as normally shipped)*: Liquid; *Color*: Colorless; *Odor*: Weakly fruity; mild and characteristic.

Physical and Chemical Properties - *Physical State at 15 °C and 1 atm.*: Liquid; *Molecular Weight*: 134.17; *Boiling Point at 1 atm.*: 396, 202, 475; *Freezing Point*: -105, -76, 197; *Critical Temperature*: Not pertinent; *Critical Pressure*: Not pertinent; *Specific Gravity*: 0.99 at 20°C (liquid); *Vapor (Gas) Density*: Not pertinent; *Ratio of Specific Heats of Vapor (Gas)*: Not pertinent; *Latent Heat of Vaporization*: 150, 85, 3.6; *Heat of Combustion*: -11,390, -6,330, -265; *Heat of Decomposition*: Not pertinent.

Health Hazards Information - *Recommended Personal Protective Equipment*: Goggles; *Symptoms Following Exposure*: None expected; *General Treatment for Exposure*: SKIN AND EYES: flush with water; *Toxicity by Inhalation (Threshold Limit Value)*: Not pertinent; *Short-Term Exposure Limits*: Data not available; *Toxicity by Ingestion*: Grade 2; LD_{50} 0.5 to 5 g/kg; *Late Toxicity*: Data not available; *Vapor (Gas) Irritant Characteristics*: None; *Liquid or Solid Irritant Characteristics*: No appreciable hazard. Practically harmless to the skin; *Odor Threshold*: Data not available.

Fire Hazards - *Flash Point (deg. F):* 201 CC; 205 OC; *Flammable Limits in Air (%):* 1.2 - 8.5 (est.); *Fire Extinguishing Agents:* alcohol foam, dry liquid, or carbon dioxide; *Fire Extinguishing Agents Not to be Used:* Not pertinent; *Special Hazards of Combustion Products:* Not pertinent; *Behavior in Fire:* Not pertinent; *Ignition Temperature (deg. F):* 400; *Electrical Hazard:* Not pertinent; *Burning Rate:* 2.5 mm/min.

Chemical Reactivity - *Reactivity with Water :* No reaction; *Reactivity with Common Materials:* No reaction; *Stability During Transport:* Stable; *Neutralizing Agents for Acids and Caustics:* Not pertinent; *Polymerization:* Not pertinent; *Inhibitor of Polymerization:* Not pertinent.

DIETHYLENETRIAMINE

Chemical Designations - *Synonyms*: Bis(2-aminoethyl)amine; 2,2'-Diaminodiethylamine; *Chemical Formula*: $NH_2(CH_2)_2NH(CH_2)_2NH_2$.

Observable Characteristics - *Physical State (as normally shipped)*: Liquid; *Color*: Colorless to light amber; yellow; *Odor*: Strong ammoniacal; mildly ammoniacal.

Physical and Chemical Properties - *Physical State at 15 °C and 1 atm.*: Liquid; *Molecular Weight*: 103.17; *Boiling Point at 1 atm.*: 405, 207, 480; *Freezing Point*: -38, -39, 234; *Critical Temperature*: Not pertinent; *Critical Pressure*: Not pertinent; *Specific Gravity*: 0.954 at 20°C (liquid); *Vapor (Gas) Density*: Not pertinent; *Ratio of Specific Heats of Vapor (Gas)*: Not pertinent; *Latent Heat of Vaporization*: Not pertinent; *Heat of Combustion*: -13,300, -7,390, -309; *Heat of Decomposition*: Not pertinent.

Health Hazards Information - *Recommended Personal Protective Equipment*: Amine respiratory cartridge mask; rubber gloves; splash-proof goggles; *Symptoms Following Exposure*: Prolonged breathing of vapors may cause asthma. Liquid burns skin and eyes. A skin rash can form; *General Treatment for Exposure*: INHALATION: remove victim to fresh air. INGESTION: do NOT induce vomiting; give large quantities of water; give at least one ounce of vinegar in an equal amount of water; get medical attention. SKIN: flush with plenty of water for at least 15 min. and get medical attention; *Toxicity by Inhalation (Threshold Limit Value)*: 1 ppm; *Short-Term Exposure Limits*: Data not available; *Toxicity by Ingestion*: Grade 2; LD_{50} 0.5 to 5 g/kg (rat); *Late Toxicity*: Data not available; *Vapor (Gas) Irritant Characteristics*: Vapors cause moderate irritation such that personnel will find high concentrations unpleasant; *Liquid or Solid Irritant Characteristics*: Causes smarting of the skin and first-degree burns on short exposure and may cause secondary burns on long exposure; *Odor Threshold*: 10 ppm.

Fire Hazards - *Flash Point (deg. F):* 200 OC; *Flammable Limits in Air (%):* 1 - 10; *Fire Extinguishing Agents:* Water spray, alcohol foam, carbon dioxide, or dry chemical; *Fire Extinguishing Agents Not to be Used:* Water or foam may cause frothing; *Special Hazards of Combustion Products:* Irritating vapors are generated when heated; *Behavior in Fire:* Not pertinent; *Ignition Temperature (deg. F):* 676; *Electrical Hazard:* Not pertinent; *Burning Rate:* Data not available.

Chemical Reactivity - *Reactivity with Water :* No reaction; *Reactivity with Common Materials:* No hazardous reaction; *Stability During Transport :*Stable; *Neutralizing Agents for Acids and Caustics:* Flush with water; *Polymerization:* Not pertinent; *Inhibitor of Polymerization:* Not pertinent.

DI(2ETHYLHEXYL) PHOSPHORIC ACID

Chemical Designations - *Synonyms*: Bis-(2-ethylhexyl) Hydrogen Phosphate; Di-(2-ethylhexyl) Phosphate; Di-(2-ethylhexyl) Acid Phosphate; *Chemical Formula*: $[CH_3CH_2CH_2CH_2CH(C_2H_5)CH_2O]_2POOH$.

Observable Characteristics - *Physical State (as normally shipped)*: Liquid; *Color*: Amber; *Odor*: None.

Physical and Chemical Properties - *Physical State at 15 °C and 1 atm.*: Liquid; *Molecular Weight*: 322.4; *Boiling Point at 1 atm.*: Decomposes; *Freezing Point*: < -76, < -60, < 213; *Critical Temperature*: Not pertinent; *Critical Pressure*: Not pertinent; *Specific Gravity*: 0.977 at 20°C (liquid); *Vapor (Gas) Density*: Not pertinent; *Ratio of Specific Heats of Vapor (Gas)*: Not pertinent; *Latent Heat* *of Vaporization*: Not pertinent; *Heat of Combustion*: -13,970, -7,760, -325; *Heat of Decomposition*: Not pertinent.

Health Hazards Information - *Recommended Personal Protective Equipment*: Goggles or face shield; rubber gloves; protective clothing; *Symptoms Following Exposure*: Contact with liquid irritates eyes and may cause serious injury; consult an eye specialist. Causes skin irritation on contact. Ingestion produces irritation, call a physician; *General Treatment for Exposure*: EYES: immediately flush with plenty of water for at least 15 min.; see a physician. SKIN: immediately flush with plenty of water for at least 15 min.; see a physician. SKIN: immediately flush with plenty of water for at least 15 min. INGESTION: induce vomiting and call a physician; *Toxicity by Inhalation (Threshold Limit Value)*: Data not available; *Short-Term Exposure Limits*: Data not available; *Toxicity by Ingestion*: Grade 2; LD_{50} 0.5 to 5 g/kg; *Late Toxicity*: No data; *Vapor (Gas) Irritant Characteristics*: Vapors are nonirritating to the eyes and throat; *Liquid or Solid Irritant Characteristics*: Causes smarting of the skin and first-degree burns on short exposure and may cause second-degree burns on long exposure; *Odor Threshold*: No data.

Fire Hazards - *Flash Point (deg. F):* 385 OC; *Flammable Limits in Air:* Not pertinent; *Fire Extinguishing Agents:* Dry chemical, alcohol foam, carbon dioxide; *Fire Extinguishing Agents Not to be Used:* Water or foam may cause frothing; *Special Hazards of Combustion Products:* Irritating phosphorus oxides may be released; *Behavior in Fire:* Not pertinent; *Ignition Temperature :* No data; *Electrical Hazard:* Not pertinent; *Burning Rate:* Data not available.

Chemical Reactivity - *Reactivity with Water:* No reaction; *Reactivity with Common Materials:* Mildly corrosive to most metals; may form flammable hydrogen gas; *Stability During Transport:* Stable; *Neutralizing Agents for Acids and Caustics:* Sodium bicarbonate or lime solution; *Polymerization:* Not pertinent; *Inhibitor of Polymerization:* Not pertinent.

DIETHYL PHTHALATE

Chemical Designations - *Synonyms*: 1,2-Benzenedicarboxylic Acid, Diethyl Ester; Ethyl Phthalate; Phthalic Acid, Diethyl Ester; *Chemical Formula*: $C_6H_4(COOC_2H_5)_2$.

Observable Characteristics - *Physical State (as normally shipped)*: Liquid; *Color*: Colorless; *Odor*: Slight characteristic ester odor.

Physical and Chemical Properties - *Physical State at 15 °C and 1 atm.*: Liquid; *Molecular Weight*: 222; *Boiling Point at 1 atm.*: 569.3, 298.5, 571.7; *Freezing Point*: 27, -3, 270; *Critical Temperature*: Not pertinent; *Critical Pressure*: Not pertinent; *Specific Gravity*: 1.12 at 20°C (liquid); *Vapor (Gas) Density*: Not pertinent; *Ratio of Specific Heats of Vapor (Gas)*: Not pertinent; *Latent Heat of Vaporization*: 170, 96, 4.0; *Heat of Combustion*: -10,920, -6,070, -254; *Heat of Decomposition*: Not pertinent.

Health Hazards Information - *Recommended Personal Protective Equipment*: Rubber gloves; goggles or face shield; *Symptoms Following Exposure*: Symptoms unlikely from any form of exposure; *General Treatment for Exposure*: INHALATION: remove to fresh air. EYES: flush with water. SKIN: flush with water, wash well with soap and water; *Toxicity by Inhalation (Threshold Limit Value)*: Data not available; *Short-Term Exposure Limits*: Data not available; *Toxicity by Ingestion*: Grade 2; oral LD_{50} 1,000 mg/kg (rabbit); *Late Toxicity*: Prolonged inhalation of heated vapor produces irritation of upper respiratory tract in humans; *Vapor (Gas) Irritant Characteristics*: Odorless; *Liquid or Solid Irritant Characteristics*: Data not available; *Odor Threshold*: Data not available.

Fire Hazards - *Flash Point (deg. F):* 305 OC; *Flammable Limits in Air (%):* LFL 0.75 (at 368° F); *Fire Extinguishing Agents:* Dry chemical, foam, carbon dioxide; *Fire Extinguishing Agents Not to be Used:* Water or foam may cause frothing; *Special Hazards of Combustion Products:* Irritating vapors of unburned chemical may form in fire; *Behavior in Fire:* Data not available; *Ignition Temperature (deg. F):* 855; *Electrical Hazard:* Data not available; *Burning Rate:* Data not available.

Chemical Reactivity - *Reactivity with Water :* No reaction; *Reactivity with Common Materials:* May attack some form of plastics; *Stability During Transport:* Stable; *Neutralizing Agents for Acids and Caustics:* Not pertinent; *Polymerization:* Not pertinent; *Inhibitor of Polymerization:* Not pertinent.

DIETHYLZINC

Chemical Designations - *Synonyms*: Zinc Diethyl; Ethyl Zinc; Zinc Ethyl; *Chemical Formula*: $(C_2H_5)_2Zn$.

Observable Characteristics - *Physical State (as normally shipped)*: Liquid; *Color*: Colorless; *Odor*: Not pertinent.

Physical and Chemical Properties - *Physical State at 15 °C and 1 atm.*: Liquid; *Molecular Weight*: 123.5; *Boiling Point at 1 atm.*: 255, 124, 397; *Freezing Point*: -18, -28, 245; *Critical Temperature*: Not pertinent; *Critical Pressure*: Not pertinent; *Specific Gravity*: 1.207 at 20°C (liquid); *Vapor (Gas) Density*: Not pertinent; *Ratio of Specific Heats of Vapor (Gas)*: Not pertinent; *Latent Heat of Vaporization*: 120, 68, 2.8; *Heat of Combustion*: -11,700, -6,495, -272; *Heat of Decomposition*: Not pertinent.

Health Hazards Information - *Recommended Personal Protective Equipment*: Cartridge-type or fresh air mask for fumes or smoke; PVC fire-retardant or asbestos gloves; full face shield, safety glasses, or goggles; fire-retardant coveralls as standard wear; for special cases, use asbestos coat or rain suit; *Symptoms Following Exposure*: Inhalation of mist or vapor causes immediate irritation of nose and throat; excessive or prolonged inhalation of fumes from ignition or decomposition may cause "metal fume fever" (sore throat, headache, fever, chills, nausea, vomiting, muscular aches, perspiration, constricting sensation in lungs, weakness, sometimes prostration); symptoms usually last 12-24 hrs., with complete recovery in 24 - 48 hrs. EYES are immediately and severely irritated on contact with liquid, vapor, or dilute solution; without thorough irrigation, cornea may be permanently damaged. Moisture in skin combines with chemical to cause thermal and acid burns; tissue may be scarred without prompt treatment. Ingestion is unlikely but would cause immediate burns at site of contact; pain, nausea, vomiting, cramps, and diarrhea may follow; if untreated, tissue may be scarred without prompt treatment. Ingestion is unlikely but would cause immediate burns at site of contact; pain, nausea, vomiting, cramps, and diarrhea may follow; if untreated, tissue may become ulcerated; *General Treatment for Exposure*: INHALATION: move victim to fresh air and call doctor immediately; give mouth-to-mouth resuscitation if needed; keep victim warm and comfortable; oxygen should be given only by experienced person, and only on doctor's instructions. EYES: flush with large amounts of running water for at least 15 min., holding eyelids apart to insure thorough washing; get medical attention as soon as possible; do not use chemical neutralizers, and avoid oils or ointments unless prescribed by doctor. SKIN: flush affected area with large amounts of water; do not use chemical neutralizers; get medical attention if irritation persists. INGESTION: do NOT induce vomiting; have victim drink large amounts of water or milk immediately; if vomiting occurs, give more fluids; get medical attention; *Toxicity by Inhalation (Threshold Limit Value)*: Not pertinent; *Short-Term Exposure Limits*: Not pertinent; *Toxicity by Ingestion*: Not pertinent; *Late Toxicity*: Not pertinent; *Vapor (Gas) Irritant Characteristics*: Data not available; *Liquid or Solid Irritant Characteristics*: Data not available; *Odor Threshold*: Not pertinent.

Fire Hazards - *Flash Point :* Not pertinent (ignites spontaneously); *Flammable Limits in Air (%):* Not pertinent; *Flammable Limits in Air (%):* Not pertinent; *Fire Extinguishing Agents:* Dry chemical, sand, or powdered limestone; *Fire Extinguishing Agents Not to be Used:* Water, foam, halogenated agents, carbon dioxide; *Special Hazards of Combustion Products:* Yields zinc fumes when burning; can cause " metal fume fever" (see 5.2); *Behavior in Fire:* Reacts spontaneously with air or oxygen, and violently with water, evolving flammable ethane gas. Contact with water applied to adjacent fires will intensify the fire; *Ignition Temperature (deg. F):* Below 0; *Electrical Hazard:* Not pertinent; *Burning Rate:* Not pertinent.

Chemical Reactivity - *Reactivity with Water* :Reacts violently to form flammable ethane gas; *Reactivity with Common Materials:* Will react with surface moisture, generating flammable ethane gas; *Stability During Transport:* Stable; *Neutralizing Agents for Acids and Caustics:* Not pertinent; *Polymerization:* Not pertinent; *Inhibitor of Polymerization:* Not pertinent.

1,1 DIFLUOROETHANE

Chemical Designations - *Synonyms*: Ethylidene Difluoride; Ethylidene Fluoride; Refrigerant 152a; *Chemical Formula*: CH_3CHF_2.

Observable Characteristics - *Physical State (as normally shipped)*: Liquefied compressed gas; *Color*: Colorless; *Odor*: Faint.

Physical and Chemical Properties - *Physical State at 15 °C and 1 atm.*: Gas; *Molecular Weight*: 66.05; *Boiling Point at 1 atm.*: 52.3, 11.3, 248.5; *Freezing Point*: -179, -117, 156; *Critical Temperature*: 236.3, 113.5. 386.6; *Critical Pressure*: 652, 44.37, 4.50; *Specific Gravity*: 0.95 at 20°C (liquid); *Vapor (Gas) Density*: 2.3; *Ratio of Specific Heats of Vapor (Gas)*: 1.141; *Latent Heat of Vaporization*: 140.5, 78.03, 3.265; *Heat of Combustion*: -7,950, -4,420, -185; *Heat of Decomposition*: Not pertinent.

Health Hazards Information - *Recommended Personal Protective Equipment*: Individual breathing devices with air supply; neoprene gloves; protective clothing; eye protection; *Symptoms Following Exposure*: Inhalation of concentrated gas will cause suffocation. Contact will liquid can damage eyes because of low temperature. Frostbite may result from contact with liquid; *General Treatment for Exposure*: INHALATION: remove to fresh air; use artificial respiration if necessary. EYES: get medical attention promptly if liquid has entered eyes. SKIN: soak in lukewarm water (for frostbite); *Toxicity by Inhalation (Threshold Limit Value)*: Data not available; *Short-Term Exposure Limits*: Data not available; *Toxicity by Ingestion*: Not pertinent (boils at -24.7°C); *Late Toxicity*: Data not available; *Vapor (Gas) Irritant Characteristics*: Data not available; *Liquid or Solid Irritant Characteristics*: Data not available; *Odor Threshold*: Data not available.

Fire Hazards - *Flash Point :* Not pertinent; *Flammable Limits in Air (%):* 3.7 - 18; *Fire Extinguishing Agents:* Shut off gas source; use water to cool adjacent combustibles; *Fire Extinguishing Agents Not to be Used:* Data not available; *Special Hazards of Combustion Products:* Irritating hydrogen fluoride fumes may form in fire; *Behavior in Fire:* Containers may explode. Vapors are heavier than air And may travel a considerable distance; *Ignition Temperature :* Data not available; *Electrical Hazard:* Data not available; *Burning Rate:* Not pertinent.

Chemical Reactivity - *Reactivity with Water :* No reaction; *Reactivity with Common Materials:* No reaction; *Stability During Transport:* Stable; *Neutralizing Agents for Acids and Caustics:* Not pertinent; *Polymerization:* Not pertinent; *Inhibitor of Polymerization:* Not pertinent.

DIFLUOROPHOSPHORIC ACID, ANHYDROUS

Chemical Designations - *Synonyms*: Difluorophosphorus Acid; *Chemical Formula*: $HOPOF_2$.

Observable Characteristics - *Physical State (as normally shipped)*: Liquid; *Color*: Colorless; *Odor*: Sharp, very irritating.

Physical and Chemical Properties - *Physical State at 15 °C and 1 atm.*: Liquid; *Molecular Weight*: 103.0; *Boiling Point at 1 atm.*: 241, 116, 389; *Freezing Point*: -139, -95, 178; *Critical Temperature*: Not pertinent; *Critical Pressure*: Not pertinent; *Specific Gravity*: 1.583 at 25°C (liquid); *Vapor (Gas) Density*: Not pertinent; *Ratio of Specific Heats of Vapor (Gas)*: Not pertinent; *Latent Heat of Vaporization*: 140, 77, 3.2; *Heat of Combustion*: Not pertinent; *Heat of Decomposition*: Not pertinent.

Health Hazards Information - *Recommended Personal Protective Equipment*: Air line mask or self-contained breathing apparatus; full protective clothing; *Symptoms Following Exposure*: Inhalation causes severe irritation of upper respiratory tract. Contact with liquid causes severe irritation of eyes and skin. Ingestion causes severe burns of mouth and stomach; *General Treatment for Exposure*: Get medical exposure attention as soon as possible following exposures to this compound. INHALATION: remove victim from exposure and support respiration. EYES: wash with copious volumes of water for at least 15 min. INGESTION: if victim is conscious, have him drink large amounts of water followed by milk or milk of magnesia; *Toxicity by Inhalation (Threshold Limit Value)*: Data not available; *Short-Term Exposure Limits*: Data not available; *Toxicity by Ingestion*: Data not available; *Late Toxicity*: Data not available; *Vapor (Gas) Irritant Characteristics*: Vapors cause moderate irritation such that personnel will find high concentrations unpleasant. The effect is temporary; *Liquid or Solid Irritant Characteristics*: Severe skin irritant. Causes second- and third-degree burns on short contact and is very injurious to the eyes; *Odor Threshold*: Data not available.

Fire Hazards - *Flash Point :* Not flammable; *Flammable Limits in Air (%):* Not flammable; *Fire Extinguishing Agents:* Not pertinent; *Fire Extinguishing Agents Not to be Used:* Do not use water on adjacent fires; *Special Hazards of Combustion Products:* Irritating and toxic fumes of hydrogen fluoride and phosphoric acid may be formed in fires; *Behavior in Fire:* Not pertinent; *Ignition Temperature :* Not pertinent; *Electrical Hazard:* Not pertinent; *Burning Rate:* Not pertinent.

Chemical Reactivity - *Reactivity with Water:* Reacts vigorously to form corrosive and toxic hydrofluoric acid; *Reactivity with Common Materials:* In the presence of moisture, is corrosive to glass, other siliceous materials, and most metals; *Stability During Transport:* Stable; *Neutralizing Agents for Acids and Caustics:* Flush with water, rinse with sodium bicarbonate or lime solution; *Polymerization:* Not pertinent; *Inhibitor of Polymerization:* Not pertinent.

DIHEPTYL PHTHALATE

Chemical Designations - *Synonyms*: Phthalic Acid, Diheptyl Ester; *Chemical Formula*: $C_6H_4(COOC_7H_{15})_2$.

Observable Characteristics - *Physical State (as normally shipped)*: Liquid; *Color*: Colorless; *Odor*: None.

Physical and Chemical Properties - *Physical State at 15 °C and 1 atm.*: Liquid; *Molecular Weight*: 362; *Boiling Point at 1 atm.*: Not pertinent (decomposes); *Freezing Point*: Not pertinent; *Critical Temperature*: Not pertinent; *Critical Pressure*: Not pertinent; *Specific Gravity*: 1.0 at 20°C (liquid); *Vapor (Gas) Density*: Not pertinent; *Ratio of Specific Heats of Vapor (Gas)*: Not pertinent; *Latent Heat of Vaporization*: Data not available; *Heat of Combustion*: -16,850, -9,370, -392; *Heat of Decomposition*: Not pertinent.

Health Hazards Information - *Recommended Personal Protective Equipment*: Goggles or face shield; rubber gloves; *Symptoms Following Exposure*: Inhalation of vapors from very hot material may cause headache, drowsiness, and convulsions. Contact with eyes may cause irritation; *General Treatment for Exposure*: INHALATION: move to fresh air. EYES: flush with water. SKIN: wipe off; flush with water; wash with soap and water; *Toxicity by Inhalation (Threshold Limit Value)*: Data not available; *Short-Term Exposure Limits*: Data not available; *Toxicity by Ingestion*: Data not available; *Late Toxicity*: Data not available; *Vapor (Gas) Irritant Characteristics*: Data not available; *Liquid or Solid Irritant Characteristics*: Data not available; *Odor Threshold*: Data not available.

Fire Hazards - *Flash Point :* Data not available; *Flammable Limits in Air (%):* Data not available; *Fire Extinguishing Agents:* Foam, dry chemical, carbon dioxide; *Fire Extinguishing Agents Not to be Used:* Water may be ineffective; *Special Hazards of Combustion Products:* Data not available; *Behavior in Fire:* Data not available; *Ignition Temperature :* Data not available; *Electrical Hazard:* Data not available; *Burning Rate:* Data not available.

Chemical Reactivity - *Reactivity with Water :* No reaction; *Reactivity with Common Materials:* May attack some form of plastics; *Stability During Transport:* Stable; *Neutralizing Agents for Acids and Caustics:* Not pertinent; *Polymerization:* Not pertinent; *Inhibitor of Polymerization:* Not pertinent.

DIISOBUTYLCARBINOL

Chemical Designations - *Synonyms*: 2,6 Dimethyl-4-heptanol; *Chemical Formula*: $[(CH_3)_2CHCH_2]_2CHOH$.

Observable Characteristics - *Physical State (as normally shipped)*: Liquid; *Color*: Colorless; *Odor*: Characteristic.

Physical and Chemical Properties - *Physical State at 15 °C and 1 atm.*: Liquid; *Molecular Weight*: 144.26; *Boiling Point at 1 atm.*: 352, 178, 451; *Freezing Point*: -85, -65, 208; *Critical Temperature:* Not pertinent; *Critical Pressure*: Not pertinent; *Specific Gravity*: 0.812 at 20°C (liquid); *Vapor (Gas) Density*: Not pertinent; *Ratio of Specific Heats of Vapor (Gas)*: Not pertinent; *Latent Heat of Vaporization*: 140, 76, 3.2; *Heat of Combustion*: -17,400, -9,680, -405; *Heat of Decomposition*: Not pertinent.

Health Hazards Information - *Recommended Personal Protective Equipment*: Air-supplied mask for prolonged exposure; plastic gloves; goggles; *Symptoms Following Exposure*: None expected; *General Treatment for Exposure*: SKIN AND EYES: flush with water; *Toxicity by Inhalation (Threshold Limit Value)*: Not pertinent; *Short-Term Exposure Limits*: Not pertinent; *Toxicity by Ingestion*: Grade 2; LD$_{50}$ 0.5 to 5 g/kg (rat); *Late Toxicity*: Data not available; *Vapor (Gas) Irritant Characteristics*: None; *Liquid or Solid Irritant Characteristics*: None; *Odor Threshold*: Data not available.

Fire Hazards - *Flash Point (deg. F)*: 162 OC; 165 CC; *Flammable Limits in Air (%)*: 0.8 - 6.1; *Fire Extinguishing Agents:* Carbon dioxide, dry chemical, alcohol foam; *Fire Extinguishing Agents Not to be Used:* Not pertinent; *Special Hazards of Combustion Products:* Not pertinent; *Behavior in Fire:* Not pertinent; *Ignition Temperature (deg. F):* 494 (calc.); *Electrical Hazard:* Not pertinent; *Burning Rate:* Data not available.

Chemical Reactivity - *Reactivity with Water :* No reaction; *Reactivity with Common Materials:* No reaction; *Stability During Transport:* Stable; *Neutralizing Agents for Acids and Caustics:* Not pertinent; *Polymerization:* Not pertinent; *Inhibitor of Polymerization:* Not pertinent.

DIISOBUTYLENE

Chemical Designations - *Synonyms*: 2,4,4-Trimethyl-1-pentene; *Chemical Formula*: $(CH_3)_3CCH_2C(CH_3)=CH_2$.

Observable Characteristics - *Physical State (as normally shipped)*: Liquid; *Color*: Colorless; *Odor*: Like gasoline.

Physical and Chemical Properties - *Physical State at 15 °C and 1 atm.*: Liquid; *Molecular Weight*: 112.22; *Boiling Point at 1 atm.*: 214.7, 101.5, 374.7; *Freezing Point*: -136.3, -93.5, 179.7; *Critical Temperature*: 548, 286.7, 559.9; *Critical Pressure*: 380, 25.85, 2.619; *Specific Gravity*: 0.715 at 20°C (liquid); *Vapor (Gas) Density*: Not pertinent; *Ratio of Specific Heats of Vapor (Gas)*: 1.049; *Latent Heat of Vaporization*: 110, 60, 2.5; *Heat of Combustion*: -18,900, -10,500, -440; *Heat of Decomposition*: Not pertinent.

Health Hazards Information - *Recommended Personal Protective Equipment*: Protective goggles; *Symptoms Following Exposure*: Low general toxicity; may act as simple asphyxiate in high vapor concentrations; *General Treatment for Exposure*: INHALATION: remove from exposure; support respiration; *Toxicity by Inhalation (Threshold Limit Value)*: Data not available; *Short-Term Exposure Limits*: Data not available; *Toxicity by Ingestion*: Data not available; *Late Toxicity*: Liver and kidney damage in exp. animals; *Vapor (Gas) Irritant Characteristics*: Vapors are nonirritating to the eyes and throat; *Liquid or Solid Irritant Characteristics*: Minimum hazard. If spilled on clothing and allowed to remain, may cause smarting and reddening of the skin; *Odor Threshold*: Data not available.

Fire Hazards - *Flash Point (deg. F)*: 35 (est.); *Flammable Limits in Air (%)*: 0.9 LEL (est.); *Fire Extinguishing Agents:* Dry chemical, foam, or carbon dioxide; *Fire Extinguishing Agents Not to be Used:* Water may be ineffective; *Special Hazards of Combustion Products:* Not pertinent; *Behavior in Fire:* Not pertinent; *Ignition Temperature (deg. F):* 788; *Electrical Hazard:* Not pertinent; *Burning Rate:* 7.9 mm/min.

Chemical Reactivity - *Reactivity with Water:* No reaction; *Reactivity with Common Materials:* No reaction; *Stability During Transport:* Stable; *Neutralizing Agents for Acids and Caustics:* Not pertinent; *Polymerization:* Not pertinent; *Inhibitor of Polymerization:* Not pertinent.

DIISOBUTYL KETONE

Chemical Designations - *Synonyms*: DIBK; sym-Diisopropylacetone; 2,6-Dimethyl-4-heptanone; Isovalerone; *Chemical Formula*: $(CH_3)_2CHCH_2COCH_2CH(CH_3)_2$ or $C_9H_{18}O$.

Observable Characteristics - *Physical State (as normally shipped)*: Liquid; *Color*: Colorless; *Odor*: Mild; characteristic ketonic.

Physical and Chemical Properties - *Physical State at 15 °C and 1 atm.*: Liquid; *Molecular Weight*: 142.23; *Boiling Point at 1 atm.*: 325, 163, 436; *Freezing Point*: -43, -42, 231; *Critical Temperature*: Not pertinent; *Critical Pressure*: Not pertinent; *Specific Gravity*: 0.806 at 20°C (liquid); *Vapor (Gas) Density*: 4.9; *Ratio of Specific Heats of Vapor (Gas)*: Not pertinent; *Latent Heat of Vaporization*: 121, 67, 2.8; *Heat of Combustion*: -16, 040, -8,910, -373; *Heat of Decomposition*: Not pertinent.

Health Hazards Information - *Recommended Personal Protective Equipment*: Air-supplied mask in confined areas; plastic gloves; face shield and safety glasses; *Symptoms Following Exposure*: Inhalation of vapors causes irritation of nose and throat. Ingestion causes irritation of mouth and stomach. Vapor irritates eyes. Contact with liquid irritates skin; *General Treatment for Exposure*: INHALATION: move to fresh air; give oxygen if breathing is difficult; call a physician. EYES: flush with plenty of water. SKIN: wipe off; flush with plenty of water; wash with soap and water; *Toxicity by Inhalation (Threshold Limit Value)*: 25 ppm; *Short-Term Exposure Limits*: 50 ppm, 30 min.; *Toxicity by Ingestion*: Grade 2; oral LD_{50} 1.4 g/kg (mouse), 5.75 g/kg (rat); *Late Toxicity*: Causes increased liver and kidney weights in rats, decreased liver weights in guinea pigs; *Vapor (Gas) Irritant Characteristics*: Vapors cause moderate irritation such that personnel will find high concentrations unpleasant. The effect is temporary; *Liquid or Solid Irritant Characteristics*: Minimum hazard. If spilled on clothing and allowed to remain, may cause smarting and reddening of the skin; *Odor Threshold*: Data not available.
Fire Hazards - *Flash Point (deg. F):* 131 OC; 120 CC; *Flammable Limits in Air (%):* 0.81 - 7.1 at 200 ° F; *Fire Extinguishing Agents:* Foam, dry chemical, carbon dioxide; *Fire Extinguishing Agents Not to be Used:* Water may be ineffective; *Special Hazards of Combustion Products:* Data not available; *Behavior in Fire:* Data not available; *Ignition Temperature (deg. F):* 745; *Electrical Hazard:* Data not available; *Burning Rate:* Data not available.
Chemical Reactivity - *Reactivity with Water :* No reaction; *Reactivity with Common Materials:* May attack some forms of plastics; *Stability During Transport:* Stable; *Neutralizing Agents for Acids and Caustics:* Not pertinent; *Polymerization:* Not pertinent; *Inhibitor of Polymerization:* Not pertinent.

DIISODECYL PHTHALATE

Chemical Designations - *Synonyms*: Phthalic Acid, Bis(8-methylonyl)ester; Phthalic Acid, Diisodecyl Ester; Plasticizer DDP; *Chemical Formula*: $C_{28}H_{46}O_4$.
Observable Characteristics - *Physical State (as normally shipped)*: Liquid; *Color*: Colorless; *Odor*: Faint.
Physical and Chemical Properties - *Physical State at 15 °C and 1 atm.*: Liquid; *Molecular Weight*: 446.7 (theor.); *Boiling Point at 1 atm.*: Very high; *Freezing Point*: -58, -50, 223; *Critical Temperature*: Not pertinent; *Critical Pressure*: Not pertinent; *Specific Gravity*: 0.967 at 20°C (liquid); *Vapor (Gas) Density*: Not pertinent; *Ratio of Specific Heats of Vapor (Gas)*: Not pertinent; *Latent Heat of Vaporization*: Data not available; *Heat of Combustion*: -16,600, -9,220, -386; *Heat of Decomposition*: Not pertinent.
Health Hazards Information - *Recommended Personal Protective Equipment*: Goggles or face shield; rubber gloves; *Symptoms Following Exposure*: No symptoms reported for any rate of exposure; *General Treatment for Exposure*: INGESTION: call physician. EYES: flush with water; call physician. SKIN: wipe off; wash with soap and water; *Toxicity by Inhalation (Threshold Limit Value)*: Data not available; *Short-Term Exposure Limits*: Data not available; *Toxicity by Ingestion*: Data not available; *Late Toxicity*: Data not available; *Vapor (Gas) Irritant Characteristics*: Data not available; *Liquid or Solid Irritant Characteristics*: Data not available; *Odor Threshold*: Data not available.
Fire Hazards - *Flash Point (deg. F):* 450 OC; *Flammable Limits in Air (%):* LFL 0.27 at 508° F; *Fire Extinguishing Agents:* Dry chemical, foam, carbon dioxide; *Fire Extinguishing Agents Not to be Used:* water may be ineffective; *Special Hazards of Combustion Products:* Data not available; *Behavior in Fire:* Data not available; *Ignition Temperature (deg. F):* 755; *Electrical Hazard:* Data not available; *Burning Rate:* Data not available.
Chemical Reactivity - *Reactivity with Water :* No reaction; *Reactivity with Common Materials:* May attack some forms of plastics; *Stability During Transport:* Stable; *Neutralizing Agents for Acids and Caustics:* Not pertinent; *Polymerization:* Not pertinent; *Inhibitor of Polymerization:* Not pertinent.

DIISOPROPANOLAMINE

Chemical Designations - *Synonyms*: 2.2'-Dihydroxydipropylamine; 1,1'-Iminodi-2-propanol; *Chemical Formula*: $[CH_3CH(OH)CH_2]NH$.
Observable Characteristics - *Physical State (as normally shipped)*: Liquid or solid; *Color*: Colorless;

Odor: Fishy; ammoniacal.

Physical and Chemical Properties - *Physical State at 15 °C and 1 atm.*: Liquid; *Molecular Weight*: 133.19; *Boiling Point at 1 atm.*: 479.7, 248.7, 521.9; *Freezing Point*: 108, 42, 315; *Critical Temperature*: 750, 399, 672; *Critical Pressure*: 529, 36, 3.6; *Specific Gravity*: 0.99 at 42°C (liquid); *Vapor (Gas) Density*: Not pertinent; *Ratio of Specific Heats of Vapor (Gas)*: Not pertinent; *Latent Heat of Vaporization*: 185, 103, 4.31; *Heat of Combustion*: -12,300, -6860, -287; *Heat of Decomposition*: Not pertinent.

Health Hazards Information - *Recommended Personal Protective Equipment*: Full face mask or amine vapor mask only if required; clean, body-covering clothing, rubber gloves, apron, boots and face shield; *Symptoms Following Exposure*: Vapor concentrations too low to irritate unless exposure is prolonged. Liquid will burn eyes and skin; *General Treatment for Exposure*: INHALATION: if ill effects occur, remove person to fresh air and get medical help. INGESTION: if swallowed and patient is conscious and not convulsing, promptly give milk or water, then induce vomiting; get medical help. No specific antidote known. EYES AND SKIN: immediately flush with plenty of water for at least 15 min. For eyes, get medical help promptly. Remove and wash contaminated clothing before reuse; *Toxicity by Inhalation (Threshold Limit Value)*: Not pertinent; *Short-Term Exposure Limits*: Not pertinent; *Toxicity by Ingestion*: Grade 2; LD_{50} 0.5 to 5 g/kg (rat); *Late Toxicity*: Data not available; *Vapor (Gas) Irritant Characteristics*: Vapors cause moderate irritation such that personnel will find high concentrations unpleasant. The effect is temporary; *Liquid or Solid Irritant Characteristics*: Causes smarting of the skin and first-degree burns on short exposure and may cause secondary burns on long exposure; *Odor Threshold*: Data not available.

Fire Hazards - *Flash Point (deg. F)*: 200 OC; *Flammable Limits in Air (%)*: 1.1 (calc.) - 5.4 (est.); *Fire Extinguishing Agents*: Water, alcohol foam, dry chemical, or carbon dioxide; *Fire Extinguishing Agents Not to be Used*: Water or foam may cause frothing; *Special Hazards of Combustion Products*: Not pertinent; *Behavior in Fire*: Not pertinent; *Ignition Temperature (deg. F)*: 580 (calc.); *Electrical Hazard*: Not pertinent; *Burning Rate*: Not pertinent.

Chemical Reactivity - *Reactivity with Water* : No reaction; *Reactivity with Common Materials*: No reaction; *Stability During Transport*: Stable; *Neutralizing Agents for Acids and Caustics*: Flush with water; *Polymerization*: Not pertinent; *Inhibitor of Polymerization*: Not pertinent.

DIISOPROPYLAMINE

Chemical Designations - *Synonyms*: No common synonyms; *Chemical Formula*: $[(CH_3)_2CH]_2NH$ or $C_6H_{15}N$.

Observable Characteristics - *Physical State (as normally shipped)*: Liquid; *Color*: Colorless; *Odor*: Amine.

Physical and Chemical Properties - *Physical State at 15 °C and 1 atm.*: Liquid; *Molecular Weight*: 101.19; *Boiling Point at 1 atm.*: 183, 83.9, 357.1; *Freezing Point*: -141.3, -96.3, 176.9; *Critical Temperature*: 480.2, 249, 522.2; *Critical Pressure*: 400, 30, 3; *Specific Gravity*: 0.717 at 20°C (liquid); *Vapor (Gas) Density*: 3.5; *Ratio of Specific Heats of Vapor (Gas)*: 1.064; *Latent Heat of Vaporization*: 121, 67.5, 2.82; *Heat of Combustion*: -19,800, -11,000, -460; *Heat of Decomposition*: Not pertinent.

Health Hazards Information - *Recommended Personal Protective Equipment*: Air-supplied mask; plastic gloves; goggles; rubber apron; *Symptoms Following Exposure*: Inhalation of vapors causes irritation, sometimes with nausea and vomiting; can also cause burns to the respiratory system. Ingestion causes irritation of mouth and stomach. Vapor irritates eyes; liquid causes severe burn, like caustic. Contact with skin causes irritation; *General Treatment for Exposure*: INHALATION: move victim to fresh air and keep him quiet and comfortably warm; give oxygen if breathing is difficult; call a physician. INGESTION: induce vomiting by giving a large volume of warm salt water; consult a physician. EYES: immediately flush eyes with plenty of water for at least 15 min., then get medical care. SKIN: flush with water; remove contaminated clothing and wash skin; if there is any redness or evidence of burning; *Toxicity by Inhalation (Threshold Limit Value)*: 5 ppm; *Short-Term Exposure Limits*: Mouse LCL_0 5,000 ppm for 20 min.; *Toxicity by Ingestion*: Grade 2; oral LD_{50} 0.7 g/kg (rat);

Late Toxicity: Data not available; *Vapor (Gas) Irritant Characteristics*: Vapors are moderately irritating such that personnel will not usually tolerate moderate or high concentrations; *Liquid or Solid Irritant Characteristics*: Causes smarting of the skin and first-degree burns on short exposure; may cause second-degree burns on long exposure; *Odor Threshold*: Data not available.

Fire Hazards - *Flash Point (deg. F):* 20 OC; 35 CC; *Flammable Limits in Air (%):* 0.8 - 7.1; *Fire Extinguishing Agents:* "Alcohol" foam, dry chemical, carbon dioxide; *Fire Extinguishing Agents Not to be Used:* water may be ineffective; *Special Hazards of Combustion Products:* Toxic oxides of nitrogen may form in fires; *Behavior in Fire:* Vapor is heavier than air and may travel to a source of ignition and flash back; *Ignition Temperature (deg. F):* 600; *Electrical Hazard:* Class I; *Burning Rate:* Data not available.

Chemical Reactivity - *Reactivity with Water :* No reaction; *Reactivity with Common Materials:* May attack some forms of plastics; *Stability During Transport:* Stable; *Neutralizing Agents for Acids and Caustics:* Not pertinent; *Polymerization:* Not pertinent; *Inhibitor of Polymerization:* Not pertinent.

DIISOPROPYLBENZENE HYDROPEROXIDE

Chemical Designations - *Synonyms*: Isopropylcumyl Hydroperoxide; *Chemical Formula*: $(CH_3)_2CHC_6H_4C(CH_3)_2OOH + (CH_3)_2CHC_6H_4CH(CH_3)_2$.

Observable Characteristics - *Physical State (as normally shipped)*: Liquid; *Color*: Colorless to pale yellow; *Odor*: Sharp, disagreeable.

Physical and Chemical Properties - *Physical State at 15 °C and 1 atm.*: Liquid; *Molecular Weight*: 194.26; *Boiling Point at 1 atm.*: Not pertinent (decomposes); *Freezing Point*: <15, <-9, <264; *Critical Temperature:* Not pertinent; *Critical Pressure*: Not pertinent; *Specific Gravity*: 0.956 at 15°C (liquid); *Vapor (Gas) Density*: Not pertinent; *Ratio of Specific Heats of Vapor (Gas)*: Not pertinent; *Latent Heat of Vaporization*: Not pertinent; *Heat of Combustion*: Data not available; *Heat of Decomposition*: Data not available.

Health Hazards Information - *Recommended Personal Protective Equipment*: Solvent-resistant gloves; chemical-resistant apron; chemical goggles or face shield; self-contained breathing apparatus; *Symptoms Following Exposure*: Inhalation causes irritation of nose and throat. Contact with eyes or skin causes throbbing sensation and irritation; *General Treatment for Exposure*: INHALATION: move to fresh air; call a doctor. EYES: flush with water for 15 min., holding eyelids open; call physician. SKIN: wash several times with soap and water; *Toxicity by Inhalation (Threshold Limit Value)*: Data not available; *Short-Term Exposure Limits*: Data not available; *Toxicity by Ingestion*: Data not available; *Late Toxicity*: Data not available; *Vapor (Gas) Irritant Characteristics*: Data not available; *Liquid or Solid Irritant Characteristics*: Data not available; *Odor Threshold*: Data not available.

Fire Hazards - *Flash Point (deg. F):* 175; *Flammable Limits in Air (%):* Data not available; *Fire Extinguishing Agents:* Foam, dry chemical, carbon dioxide; *Fire Extinguishing Agents Not to be Used:* Water may be ineffective; *Special Hazards of Combustion Products:* Flammable alcohol and ketone gases are formed in fires; *Behavior in Fire:* Burns with a flare effect. Containers may explode; *Ignition Temperature :* Data not available; *Electrical Hazard:* Data not available; *Burning Rate:* Data not available.

Chemical Reactivity - *Reactivity with Water :* No reaction; *Reactivity with Common Materials:* Aluminum, copper, brass, lead, zinc salts, mineral acids, oxidizing or reducing agents all can cause rapid decomposition; *Stability During Transport:* Unstable, slowly evolves oxygen; *Inhibitor of Polymerization:* Not pertinent..

DIMETHYLACETAMIDE

Chemical Designations - *Synonyms*: N,N-Dimethylacetamide, Acetic Acid, Dimethylamide; *Chemical Formula*: $CH_3CON(CH_3)_2$.

Observable Characteristics - *Physical State (as normally shipped)*: Liquid; *Color*: Colorless; *Odor*: Weak, fishy.

Physical and Chemical Properties - *Physical State at 15 °C and 1 atm.*: Liquid; *Molecular Weight*: 87.1; *Boiling Point at 1 atm.*: 331, 166, 439; *Freezing Point*: -4, -20, 253; *Critical Temperature*: Not pertinent; *Critical Pressure*: Not pertinent; *Specific Gravity*: 0.943 at 20°C (liquid); *Vapor (Gas) Density*: Not pertinent; *Ratio of Specific Heats of Vapor (Gas)*: Not pertinent; *Latent Heat of Vaporization*: 214, 119, 4.98; *Heat of Combustion*: -12,560, -6,980, -292; *Heat of Decomposition*: Not pertinent.

Health Hazards Information - *Recommended Personal Protective Equipment*: Goggles or face shield; rubber gloves; *Symptoms Following Exposure*: Liquid causes mild irritation of eyes and skin. Ingestion causes depression, lethargy, confusion and disorientation, visual and auditory hallucinations, perceptual distortions, delusions, emotional detachment, and affective blunting; *General Treatment for Exposure*: EYES: flush with plenty of water for 15 min.; get medical attention. SKIN: flush with plenty of water for 15 min. INGESTION: induce vomiting and follow with gastric lavage and saline cathartics; treatment for liver and kidney injury is supportive and symptomatic; *Toxicity by Inhalation (Threshold Limit Value)*: 10 ppm; *Short-Term Exposure Limits*: Data not available; *Toxicity by Ingestion*: Grade 1; oral LD_{50} 5.63 g/kg (rat); *Late Toxicity*: May produce chronic liver and kidney damage; *Vapor (Gas) Irritant Characteristics*: Data not available; *Liquid or Solid Irritant Characteristics*: Data not available; *Odor Threshold*: 46.8 ppm.

Fire Hazards - *Flash Point (deg. F):* 158 OC; *Flammable Limits in Air (%):* 1.5 - 11.5; *Fire Extinguishing Agents:* Water, dry chemical, alcohol foam; *Fire Extinguishing Agents Not to be Used:* Not pertinent; *Special Hazards of Combustion Products:* Not pertinent; *Behavior in Fire:* Not pertinent; *Ignition Temperature (deg. F):* 914; *Electrical Hazard:* Not pertinent; *Burning Rate:* 2.8 mm/min.

Chemical Reactivity - *Reactivity with Water :* No reaction; *Reactivity with Common Materials:* No reaction; *Stability During Transport:* Stable; *Neutralizing Agents for Acids and Caustics:* Not pertinent; *Polymerization:* Not pertinent; *Inhibitor of Polymerization:* Not pertinent.

DIMETHYLAMINE

Chemical Designations - *Synonyms*: No common synonyms; *Chemical Formula*: $(CH_3)_2NH$.

Observable Characteristics - *Physical State (as normally shipped)*: Compressed gas; *Color*: Colorless; *Odor*: Fishy; strongly ammoniacal.

Physical and Chemical Properties - *Physical State at 15 °C and 1 atm.*: Gas; *Molecular Weight*: 45.08; *Boiling Point at 1 atm.*: 44.42, 6.9, 280.1; *Freezing Point*: -134, -92.2, 181; *Critical Temperature*: 328.3, 164.6, 437.8; *Critical Pressure*: 770, 52.4, 5.31; *Specific Gravity*: 0.671 at 6.9°C (liquid); *Vapor (Gas) Density*: 1.6; *Ratio of Specific Heats of Vapor (Gas)*: 1.139; *Latent Heat of Vaporization*: 252.9, 140.5, 5.882; *Heat of Combustion*:-16,800, -9340, -391; *Heat of Decomposition*: Not pertinent.

Health Hazards Information - *Recommended Personal Protective Equipment*: Chemical goggles and full face shield; molded rubber acid gloves; self-contained breathing apparatus; *Symptoms Following Exposure*: Inhalation at high concentration (> 100 ppm) causes nose and throat irritation progressing all the way to pulmonary edema. Eye and skin irritation; *General Treatment for Exposure*: INHALATION: remove victim to fresh air and calla physician; if breathing has stopped, administer artificial respiration and oxygen; keep victim warm and quiet; do not give stimulants. EYES: flush continuously and thoroughly with water for at least 15 min. SKIN: remove contaminated clothing immediately; flush affected area with large amounts of water and then wash with soap and water; *Toxicity by Inhalation (Threshold Limit Value)*: 10 ppm; *Short-Term Exposure Limits*: 20 ppm for 5 min.; *Toxicity by Ingestion*: Not pertinent; *Late Toxicity*: None; *Vapor (Gas) Irritant Characteristics*: Vapors cause moderate irritation such that personnel will find high concentrations unpleasant. The effect is temporary; *Liquid or Solid Irritant Characteristics*: Causes smarting of the skin and first-degree burns on short exposure and may cause secondary burns on long exposure; *Odor Threshold*: 0.047.

Fire Hazards - *Flash Point (deg. F):* 20 CC; *Flammable Limits in Air (%):* 2.8 - 14.4; *Fire Extinguishing Agents:* Stop flow of gas. Use water spray, carbon dioxide, or dry chemical for fires in water solutions; *Fire Extinguishing Agents Not to be Used:* Do not use foam; *Special Hazards of Combustion Products:* Vapors are eye, skin and respiratory irritants; *Behavior in Fire:* Not pertinent; *Ignition Temperature (deg. F):* 756; *Electrical Hazard:* Data not available; *Burning Rate:* 4.5 mm/min.

Chemical Reactivity - *Reactivity with Water :* No reaction; *Reactivity with Common Materials:* No hazardous reaction; *Stability During Transport:* Stable; *Neutralizing Agents for Acids and Caustics:* Not pertinent; *Polymerization:* Not pertinent; *Inhibitor of Polymerization:* Not pertinent.

DIMETHYL ETHER
Chemical Designations - *Synonyms*: Methyl Ether; Wood Ether; *Chemical Formula*: CH_3OCH_3.
Observable Characteristics - *Physical State (as normally shipped)*: Liquid under pressure; *Color*: Colorless; *Odor*: Chloroform-like; sweet.
Physical and Chemical Properties - *Physical State at 15 °C and 1 atm.*: Gas; *Molecular Weight*: 46.1; *Boiling Point at 1 atm.*: -12.5, -24.7, 248.5; *Freezing Point*: -222.7, -141.5, 131.7; *Critical Temperature*: 260.4, 126.9, 400.1; *Critical Pressure*: 780, 53, 5.4; *Specific Gravity*: 0.724 at -24.7°C (liquid); *Vapor (Gas) Density*: 1.6; *Ratio of Specific Heats of Vapor (Gas)*: 1.1456; *Latent Heat of Vaporization*: 200, 111, 4.65; *Heat of Combustion*: -13,450, -7,480, -313; *Heat of Decomposition*: Not pertinent.
Health Hazards Information - *Recommended Personal Protective Equipment*: Mask for organic vapors; plastic or rubber gloves; safety glasses; *Symptoms Following Exposure*: Inhalation produces some anesthesia (but less than that of ethyl ether), blurring of vision, headache, intoxication, loss of consciousness. Liquid or concentrated vapor irritates eyes. Contact of liquid with skin may cause frostbite; *General Treatment for Exposure*: INHALATION: remove from exposure and support respiration; call physician. EYES: wash with water for at least 15 min.; consult an eye specialist. SKIN: treat frostbite by use of warm water or by wrapping the affected part blanket; *Toxicity by Inhalation (Threshold Limit Value)*: Data not available; *Short-Term Exposure Limits*: Data not available; *Toxicity by Ingestion*: Not pertinent; *Late Toxicity*: Data not available; *Vapor (Gas) Irritant Characteristics*: Data not available; *Liquid or Solid Irritant Characteristics*: Data not available; *Odor Threshold*: Data not available.
Fire Hazards - *Flash Point :* Not pertinent (flammable gas); *Flammable Limits in Air (%):* 2 - 50; *Fire Extinguishing Agents:* Let fire burn; shut off gas flow; cool exposed surroundings with water; *Fire Extinguishing Agents Not to be Used:* Not pertinent; *Special Hazards of Combustion Products:* Not pertinent; *Behavior in Fire:* Containers may explode. Vapors are heavier than air and may travel long distance to source of ignition and flash back; *Ignition Temperature (deg. F):* 662; *Electrical Hazard:* Data not available; *Burning Rate:* 6.6 mm/min.
Chemical Reactivity - *Reactivity with Water :* No reaction; *Reactivity with Common Materials:* No reaction; *Stability During Transport:* Stable; *Neutralizing Agents for Acids and Caustics:* Not pertinent; *Polymerization:* Not pertinent; *Inhibitor of Polymerization:* Not pertinent.

DIMETHYL SULFATE
Chemical Designations - *Synonyms*: No common synonyms; *Chemical Formula*: $(CH_3)_2SO_4$.
Observable Characteristics - *Physical State (as normally shipped)*: Liquid; *Color*: Colorless; *Odor*: No characteristic odor; slight, not distinctive; weak onion.
Physical and Chemical Properties - *Physical State at 15 °C and 1 atm.*: Liquid; *Molecular Weight*: 126.13; *Boiling Point at 1 atm.*: 371.8, 188.8, 462; *Freezing Point*: -25.2, -31.8, 241.4; *Critical Temperature*: Not pertinent; *Critical Pressure*: Not pertinent; *Specific Gravity*: 1.33 at 15°C (liquid); *Vapor (Gas) Density*: Not pertinent; *Ratio of Specific Heats of Vapor (Gas)*: Not pertinent; *Latent Heat of Vaporization*: Not pertinent; *Heat of Combustion*: Not pertinent; *Heat of Decomposition*: Not pertinent.
Health Hazards Information - *Recommended Personal Protective Equipment*: Chemical goggles; self-contained breathing apparatus; safety hat; rubber suit; rubber shoes; rubber gloves; safety shower and eye wash fountain; *Symptoms Following Exposure*: Severe irritation of eyes, eyelids, respiratory tract and skin. Dry, painful cough; foamy, white sputum; difficulty in breathing; malaise and fever; inflammation and edema of lungs; *General Treatment for Exposure*: Contact with dimethyl sulfate liquid or vapor (> 1 ppm) requires immediate treatment. Call a physician, even if there is no evidence of injury, as symptoms may not appear for several hours. INHALATION: get victim to fresh air

immediately; administer 100% oxygen, even if no injury is apparent, and continue for 30 min. each hour for 6 hours; give artificial respiration if breathing is weak or fails, but do not interrupt oxygen therapy; if victim's coughing prevents use of a mask, use oxygen tent under atmospheric pressure. INGESTION: do NOT induce vomiting. SKIN: wash thoroughly. EYE: flush with running water for at least 15 min.; *Toxicity by Inhalation (Threshold Limit Value)*: 1 ppm; *Short-Term Exposure Limits*: Data not available; *Toxicity by Ingestion*: Grade 3; 50 to 500 mg/kg (rat); *Late Toxicity*: Causes birth defects in rats (malignant tumors in nervous system); *Vapor (Gas) Irritant Characteristics*: Vapors cause severe irritation of eye and throat and can cause eye and lung injury. They cannot be tolerated even at low concentrations; *Liquid or Solid Irritant Characteristics*: Severe skin irritant. Causes second- and third-degree burns on short contact; very injurious to the eyes; *Odor Threshold*: Data not available.

Fire Hazards - *Flash Point (deg. F)*: 240 OC; 182 CC; *Flammable Limits in Air (%)*: Data not available; *Fire Extinguishing Agents*: Water, foam, carbon dioxide, or dry chemical; *Fire Extinguishing Agents Not to be Used*: Not pertinent; *Special Hazards of Combustion Products*: Flammable, toxic vapors generated; *Behavior in Fire*: Not pertinent; *Ignition Temperature (deg. F)*: 370; *Electrical Hazard*: Not pertinent; *Burning Rate*: Data not available.

Chemical Reactivity - *Reactivity with Water*: Slow, non-hazardous reaction; *Reactivity with Common Materials*: Corrodes metal when wet; *Stability During Transport*: Stable; *Neutralizing Agents for Acids and Caustics*: Sodium bicarbonate or lime; *Polymerization*: Not pertinent; *Inhibitor of Polymerization*: Not pertinent.

DIMETHYL SULFIDE

Chemical Designations - *Synonyms*: DMS; Methanethiomethane; Methyl Sulfide; 2-Thiapropane; *Chemical Formula*: $(CH_3)_2S$.

Observable Characteristics - *Physical State (as normally shipped)*: Liquid; *Color*: Colorless to straw; *Odor*: Ethereal, permeating; disagreeable; offensive.

Physical and Chemical Properties - *Physical State at 15 °C and 1 atm.*: Liquid; *Molecular Weight*: 62.1; *Boiling Point at 1 atm.*: 99, 37, 310; *Freezing Point*: -144, -98, 175; *Critical Temperature*: 444, 229, 502; *Critical Pressure*: 826, 56.1, 5.69; *Specific Gravity*: 0.85 at 20°C (liquid); *Vapor (Gas) Density*: 2.14; *Ratio of Specific Heats of Vapor (Gas)*: 1.1277 at 16°C; *Latent Heat of Vaporization*: 194, 108, 4.52; *Heat of Combustion*: -13,200, -7,340, -307; *Heat of Decomposition*: Not pertinent.

Health Hazards Information - *Recommended Personal Protective Equipment*: Respirator with organic vapor canister; rubber or plastic gloves; goggles or face shield; *Symptoms Following Exposure*: Inhalation causes moderate irritation of upper respiratory system. Contact of liquid with eyes causes moderate irritation. Repeated contact with skin may extract oils and result in irritation. Ingestion causes nausea and irritation of mouth and stomach; *General Treatment for Exposure*: INHALATION: remove victim to fresh air at once; enforce rest, and keep warm; get medical attention immediately. EYES: flush with water for at least 15 min.; if irritation persists, get medical attention. SKIN: flush with plenty of water and wash thoroughly; get treatment for any lasting irritation. INGESTION: if large amounts are swallowed, induce vomiting by ticking the back of the throat with the finger or by giving an emetic such as two tablespoons of common salt in a glass of warm water; get medical attention; *Toxicity by Inhalation (Threshold Limit Value)*: Data not available; *Short-Term Exposure Limits*: Data not available; *Toxicity by Ingestion*: Grade 2; oral LD_{50} 535 mg/kg (rat); *Late Toxicity*: Data not available; *Vapor (Gas) Irritant Characteristics*: Vapors cause severe irritation of eye and throat and can cause eye and lung injury. They cannot be tolerated even at low concentrations; *Liquid or Solid Irritant Characteristics*: Causes smarting of the skin and first-degree burns on short exposure and may cause second-degree burns on long exposure; *Odor Threshold*: 0.001 ppm.

Fire Hazards - *Flash Point (deg. F)*: - 36 CC; *Flammable Limits in Air (%)*: 2.2 - 19.7; *Fire Extinguishing Agents*: January 27, 1998ry chemical, foam, carbon dioxide, or alcohol foam; *Fire Extinguishing Agents Not to be Used*: Water may be ineffective; *Special Hazards of Combustion Products*: Toxic and irritating sulfur dioxide is formed; *Behavior in Fire*: Vapor is heavier than air and may travel considerable distance to source of ignition and flash back; *Ignition Temperature (deg. F)*: 403; *Electrical Hazard*: Data not available; *Burning Rate*: 4.8 mm/min.

Chemical Reactivity - *Reactivity with Water :* No reaction; *Reactivity with Common Materials:* No reaction; *Stability During Transport:* Stable; *Neutralizing Agents for Acids and Caustics:* Not pertinent; *Polymerization:* Not pertinent; *Inhibitor of Polymerization:* Not pertinent.

DIMETHYL SULFOXIDE

Chemical Designations - *Synonyms*: DMSO; Methyl Sulfoxide; *Chemical Formula*: CH_3SOCH_3.
Observable Characteristics - *Physical State (as normally shipped)*: Liquid; *Color*: Colorless; *Odor*: Slight; almost odorless.
Physical and Chemical Properties - *Physical State at 15 °C and 1 atm.*: Liquid; *Molecular Weight*: 78.13; *Boiling Point at 1 atm.*: 372, 189, 462; *Freezing Point*: 65.5, 18.6, 291.8; *Critical Temperature:* Not pertinent; *Critical Pressure*: Not pertinent; *Specific Gravity*: 1.101 at 20°C (liquid); *Vapor (Gas) Density*: Not pertinent; *Ratio of Specific Heats of Vapor (Gas)*: Not pertinent; *Latent Heat of Vaporization*: 259, 144, 6.03; *Heat of Combustion*: -10,890, -6,050, -253.3; *Heat of Decomposition*: Not pertinent.
Health Hazards Information - *Recommended Personal Protective Equipment*: Butyl rubber gloves, safety goggles. Respiratory filter if airborne sprays or drops are present; *Symptoms Following Exposure*: Slight eye irritation; *General Treatment for Exposure*: Wash eyes and skin with water; *Toxicity by Inhalation (Threshold Limit Value)*: Data not available; *Short-Term Exposure Limits*: Data not available; *Toxicity by Ingestion*: Grade 0; above 15 g/kg; *Late Toxicity*: Causes damage to eye in dogs, pigs, rats, and rabbits; *Vapor (Gas) Irritant Characteristics*: Vapors cause moderate irritation such that personnel will find high concentrations unpleasant. The effect is temporary; *Liquid or Solid Irritant Characteristics*: Minimum hazard. If spilled on clothing and allowed to remain, may cause smarting and reddening of the skin; *Odor Threshold*: Data not available.
Fire Hazards - *Flash Point (deg. F):* 203 OC; *Flammable Limits in Air (%):* 3 - 6.3; *Fire Extinguishing Agents:* Water, foam, dry chemical, or carbon dioxide; *Fire Extinguishing Agents Not to be Used:* Not pertinent; *Special Hazards of Combustion Products:* Sulfur dioxide, formaldehyde, and methyl mercaptan may form; *Behavior in Fire:* Not pertinent; *Ignition Temperature (deg. F):* 572; *Electrical Hazard:* Not pertinent; *Burning Rate:* 2.0 mm/min.
Chemical Reactivity - *Reactivity with Water :* No reaction; *Reactivity with Common Materials:* No reaction; *Stability During Transport:* Stable; *Neutralizing Agents for Acids and Caustics:* Not pertinent; *Polymerization:* Not pertinent; *Inhibitor of Polymerization:* Not pertinent.

DIMETHYL TEREPHTHALATE

Chemical Designations - *Synonyms*: Terephthalic Acid, Dimethyl Ester; *Chemical Formula*: 1,4-$CH_3OOCC_6H_4COOCH_3$.
Observable Characteristics - *Physical State (as normally shipped)*: Solid or liquid; *Color*: Colorless; *Odor*: None; weak aromatic.
Physical and Chemical Properties - *Physical State at 15 °C and 1 atm.*: Solid; *Molecular Weight*: 194.2; *Boiling Point at 1 atm.*: 540, 282, 555; *Freezing Point*: 284, 140, 413; *Critical Temperature*: Not pertinent; *Critical Pressure*: Not pertinent; *Specific Gravity*: 1.2 at 20°C (solid); *Vapor (Gas) Density*: Not pertinent; *Ratio of Specific Heats of Vapor (Gas)*: Not pertinent; *Latent Heat of Vaporization*: 121, 67.2, 2.81; *Heat of Combustion*: -10,310, -5,727, -239,6; *Heat of Decomposition*: Not pertinent.
Health Hazards Information - *Recommended Personal Protective Equipment*: Molten DMT: goggles, face shield, gauntlets, and protective clothing. Solid: dust mask, goggles; *Symptoms Following Exposure*: Molten DMT will cause severe burns on skin on contact; *General Treatment for Exposure*: EYES: flush dust from eyes with water. SKIN: wash with soap and water. If burned by molten DMT, flush area immediately with cold water for at least 15 min.; apply ice pack for at least 30 min.; do not try to rub DMT off a burn or remove clothing that DMT has penetrated, because this will remove underlying skin; seek prompt medical treatment for significant burns; *Toxicity by Inhalation (Threshold Limit Value)*: Data not available; *Short-Term Exposure Limits*: Not pertinent; *Toxicity by Ingestion*: Grade 2; oral LD_{50} 4,390 mg/kg (rat); *Late Toxicity*: Data not available; *Vapor (Gas) Irritant*

Characteristics: Not pertinent; *Liquid or Solid Irritant Characteristics*: Data not available; *Odor Threshold*: Not pertinent.

Fire Hazards - *Flash Point (deg. F):* 298 OC (molten); *Flammable Limits in Air (%):* Not pertinent; *Fire Extinguishing Agents:* Water, dry chemical, foam, carbon dioxide; *Fire Extinguishing Agents Not to be Used:* Not pertinent; *Special Hazards of Combustion Products:* Not pertinent; *Behavior in Fire:* Not pertinent; *Ignition Temperature (deg. F):* 1.058 (dust); *Electrical Hazard:* Not pertinent; *Burning Rate:* Not pertinent.

Chemical Reactivity - *Reactivity with Water :* No reaction; *Reactivity with Common Materials:* No reaction; *Stability During Transport:* Stable; *Neutralizing Agents for Acids and Caustics:* Not pertinent; *Polymerization:* Not pertinent; *Inhibitor of Polymerization:* Not pertinent.

DIMETHYLZINC

Chemical Designations - *Synonyms*: Zinc Dimethyl; Zinc Methyl; Methyl Zinc; *Chemical Formula*: $(CH_3)_2Zn$.

Observable Characteristics - *Physical State (as normally shipped)*: Liquid; *Color*: Colorless; *Odor*: Not pertinent.

Physical and Chemical Properties - *Physical State at 15 °C and 1 atm.*: Liquid; *Molecular Weight*: 95.4; *Boiling Point at 1 atm.*: 113, 45, 318; *Freezing Point*: -44, -42, 231; *Critical Temperature*: Not pertinent; *Critical Pressure*: Not pertinent; *Specific Gravity*: 1.39 at 10.5°C (liquid); *Vapor (Gas) Density*: Not pertinent; *Ratio of Specific Heats of Vapor (Gas)*: Not pertinent; *Latent Heat of Vaporization*: 134.9, 74.95, 3.138; *Heat of Combustion*: Data not available; *Heat of Decomposition*: Not pertinent.

Health Hazards Information - *Recommended Personal Protective Equipment*: Cartridge-type or fresh air mask for fumes or smoke; PVC fire-retardant or asbestos gloves; full face shield, safety glasses, or goggles; fire-retardant coveralls as standard wear; for special cases, use asbestos coat or rain suit; *Symptoms Following Exposure*: Inhalation of mists or vapor causes immediate irritation of upper respiratory tract. Excessive or prolonged inhalation of fumes from ignition or decomposition may cause "metal fume fever" (sore throat, headache, fever, chills, nausea, vomiting, muscular aches, perspiration, constricting sensation in lungs, weakness, sometimes prostration). Symptoms usually last 12-24 hrs. Eyes are immediately and severely irritated by liquid, vapor, or dilute solutions. If not removed by thorough flushing with water, chemical may permanently damage cornea. Skin will undergo thermal and acid burns when chemical reacts with moisture in skin. Unless washed quickly, skin may be scarred. Treat dilute solutions with same precautions as concentrated liquid. Ingestion, while unlikely, would cause immediate burns at site of contact. Nausea, vomiting, cramps, and diarrhea may follow. Tissues may ulcerate if not treated; *General Treatment for Exposure*: INHALATION: highly unlikely, as liquid or vapor either ignites spontaneously or reacts with moisture to form methane and zinc oxide. Move victim to clean air and administer mouth-to-mouth resuscitation if breathing has ceased; give oxygen only when authorized by physician; keep victim warm and comfortable; call physician immediately. EYES: immediately flush with large amounts of water for at least 15 min., holding eyelids apart to insure thorough irritation; use oils or ointments only when directed by physician, and do not attempt to neutralize with chemicals; get medical attention as soon as possible. SKIN: immediately flush affected area with large volumes of water; do not attempt to neutralize with chemicals; get medical attention if irritation persists. INGESTION: highly unlikely, as liquid or vapor either ignites spontaneously or reacts with moisture to form methane and zinc oxide. Do NOT induce vomiting; immediately dilute material by giving large amounts of water or milk; if vomiting occurs, give more fluids; when vomiting ceases, milk or olive oil may be given for their soothing effect; get medical attention; *Toxicity by Inhalation (Threshold Limit Value)*: Not pertinent; *Short-Term Exposure Limits*: Not pertinent; *Toxicity by Ingestion*: Not pertinent; *Late Toxicity*: Not pertinent; *Vapor (Gas) Irritant Characteristics*: Data not available; *Liquid or Solid Irritant Characteristics*: Data not available; *Odor Threshold*: Not pertinent.

Fire Hazards - *Flash Point :* Not pertinent (ignites spontaneous); *Flammable Limits in Air (%):* Not pertinent; *Fire Extinguishing Agents:* Dry chemical, sand, powdered limestone; *Fire Extinguishing*

Agents Not to be Used: Water, foam, halogenated agents, or carbon dioxide; *Special Hazards of Combustion Products:* Smoke contains zinc oxide, which can irritate lungs and cause metal fume fever; *Behavior in Fire:* Reacts spontaneously with air or oxygen and violently with water, evolving methane. Contact with water applied to adjacent fires will intensify fire; *Ignition Temperature (deg. F):* Below 0; *Electrical Hazard:* Not pertinent; *Burning Rate:* Not pertinent.

Chemical Reactivity - *Reactivity with Water :* Reacts vigorously, generating flammable methane gas; *Reactivity with Common Materials:* Will react with surface moisture to generate flammable methane; *Stability During Transport:* Stable; *Neutralizing Agents for Acids and Caustics:* Not pertinent; *Polymerization:* Not pertinent; *Inhibitor of Polymerization:* Not pertinent.

2,4-DINITROANILINE

Chemical Designations - *Synonyms*: 2,4-Dinitraniline; *Chemical Formula*: $NH_2C_6H_3(NO_2)_2$-2,4.

Observable Characteristics - *Physical State (as normally shipped)*: Solid; *Color*: Yellow; *Odor*: Slight musty.

Physical and Chemical Properties - *Physical State at 15 °C and 1 atm.*: Solid; *Molecular Weight*: 183.12; *Boiling Point at 1 atm.*: Not pertinent; *Freezing Point*: 368, 187, 460; *Critical Temperature*: Not pertinent; *Critical Pressure*: Not pertinent; *Specific Gravity*: 1.615 at 15°C (solid); *Vapor (Gas) Density*: Not pertinent; *Ratio of Specific Heats of Vapor (Gas)*: Not pertinent; *Latent Heat of Vaporization*: Not pertinent; *Heat of Combustion*: Not pertinent; *Heat of Decomposition*: Not pertinent.

Health Hazards Information - *Recommended Personal Protective Equipment*: Self-contained breathing apparatus; butyl rubber gloves; eye goggles; plastic lab coat; protective shoes; *Symptoms Following Exposure*: May cause headache, nausea, stupor. Irritating to skin and mucous membrane; *General Treatment for Exposure*: INHALATION: artificial respiration if necessary. INGESTION: induce vomiting; give universal antidote; get prompt medical care. SKIN AND EYES: remove victim from exposure; wash exposed skin with warm water and soap; flush eyes with water; *Toxicity by Inhalation (Threshold Limit Value)*: Not pertinent; *Short-Term Exposure Limits*: Not pertinent; *Toxicity by Ingestion*: Grade 3; oral rate LD_{50} 418 mg/kg; *Late Toxicity*: Data not available; *Vapor (Gas) Irritant Characteristics*: Not pertinent; *Liquid or Solid Irritant Characteristics*: Causes smarting of the skin and first degree burns on short exposure; may cause second degree burns on long exposure; *Odor Threshold*: Not pertinent.

Fire Hazards - *Flash Point (deg. F):* 435 CC; *Flammable Limits in Air (%):* Data not available; *Fire Extinguishing Agents:* For small fires, use water, dry chemical, foam, or carbon dioxide; *Fire Extinguishing Agents Not to be Used:* Water or foam may cause frothing; *Special Hazards of Combustion Products:* Vapors and combustion gases are irritating; *Behavior in Fire:* May explode; *Ignition Temperature :* Data not available; *Electrical Hazard:* Not pertinent; *Burning Rate:* Not pertinent.

Chemical Reactivity - *Reactivity with Water :* No reaction; *Reactivity with Common Materials:* Reacts with oxidizing materials; *Stability During Transport:* May detonate when heated under confinement; *Neutralizing Agents for Acids and Caustics:* Not pertinent; *Polymerization:* Not pertinent; *Inhibitor of Polymerization:* Not pertinent.

M-DINITROBENZENE

Chemical Designations - *Synonyms*: 1,3-Dinitrobenzene; 1,3-Dinitrobenzol; —DNB; meta-Dinitrobenzene; Dinitrobenzol; *Chemical Formula*: 1,3-$C_6H_4(NO_2)_2$.

Observable Characteristics - *Physical State (as normally shipped)*: Solid; *Color*: Yellow; *Odor*: Weak.

Physical and Chemical Properties - *Physical State at 15 °C and 1 atm.*: Solid; *Molecular Weight*: 168.1; *Boiling Point at 1 atm.*: 556, 291, 564; *Freezing Point*: 194, 90, 363; *Critical Temperature*: Not pertinent; *Critical Pressure*: Not pertinent; *Specific Gravity*: 1.58 at 18°C (solid); *Vapor (Gas) Density*: Not pertinent; *Ratio of Specific Heats of Vapor (Gas)*: Not pertinent; *Latent Heat of Vaporization*: Not pertinent; *Heat of Combustion*: -7,378, -4,099, -171.5; *Heat of Decomposition*: Not pertinent.

Health Hazards Information - *Recommended Personal Protective Equipment*: Dust respirator; rubber gloves; protective clothing; *Symptoms Following Exposure*: Inhalation or ingestion causes loss of color, nausea, headache, dizziness, drowsiness, and collapse. Eyes are irritated by liquid. Stains skin yellow; if contact is prolonged, can be absorbed into blood and cause same symptoms as for inhalation; *General Treatment for Exposure*: INHALATION: remove from exposure; get medical attention for methemoglobinemia. EYES: flush with water for at least 15 min. SKIN: wash well with soap and water. INGESTION: induce vomiting, if conscious; give gastric lavage and saline cathartic; get medical attention; *Toxicity by Inhalation (Threshold Limit Value)*: 1 mg/m^3; *Short-Term Exposure Limits*: Data not available; *Toxicity by Ingestion*: Grade 4; oral LD$_{50}$ 42 mg/kg; *Late Toxicity*: May cause liver damage, anemia, neuritis; *Vapor (Gas) Irritant Characteristics*: Data not available; *Liquid or Solid Irritant Characteristics*: Data not available; *Odor Threshold*: Data not available.

Fire Hazards - *Flash Point :* Not pertinent (combustible solid); *Flammable Limits in Air (%):* Not pertinent; *Fire Extinguishing Agents:* Water from protected location; *Fire Extinguishing Agents Not to be Used:* Not pertinent; *Special Hazards of Combustion Products:* Not pertinent; *Behavior in Fire:* May explode; *Ignition Temperature :* Data not available; *Electrical Hazard:* Not pertinent; *Burning Rate:* Not pertinent.

Chemical Reactivity - *Reactivity with Water :* No reaction; *Reactivity with Common Materials:* No reaction; *Stability During Transport:* Stable; *Neutralizing Agents for Acids and Caustics:* Not pertinent; *Polymerization:* Not pertinent; *Inhibitor of Polymerization:* Not pertinent.

DINITROCRESOLS

Chemical Designations - *Synonyms*: 2,6-Dinitro-p-cresol; 3,5-Dinitro-o-cresol; 4,6-Dinitro-o-cresol; *Chemical Formula*: $CH_3-C_6H_2(NO_2)_2(OH)$.

Observable Characteristics - *Physical State (as normally shipped)*: Solid; *Color*: Yellow; *Odor*: Data not available.

Physical and Chemical Properties - *Physical State at 15 °C and 1 atm.*: Solid; *Molecular Weight*: 198; *Boiling Point at 1 atm.*: Not pertinent (decomposes); *Freezing Point*: 176-187, 80-86, 353-359; *Critical Temperature*: Not pertinent; *Critical Pressure*: Not pertinent; *Specific Gravity*: > 1.1 at 20°C (solid); *Vapor (Gas) Density*: Not pertinent; *Ratio of Specific Heats of Vapor (Gas)*: Not pertinent; *Latent Heat of Vaporization*: Not pertinent; *Heat of Combustion*: -7,050, -3,920, -164; *Heat of Decomposition*: Data not available.

Health Hazards Information - *Recommended Personal Protective Equipment*: Dust mask; goggles or face shield; protective clothing; rubber gloves; *Symptoms Following Exposure*: Very high fever is prominent sign of intoxication following absorption of a toxic dose of dinitro-o-cresol. Inhalation of dust may cause same symptoms as ingestion. Ingestion causes a feeling of well-being, profuse sweating, yellow urine, increased basal metabolism, marked thirst, vomiting, convulsions, coma, and death. Contact with eyes causes irritation. Contact with skin causes local necrosis and dangerous systemic effects. Note: Some authorities recommend that all exposed workers have blood tests regularly to determine the level of this substance. Further contact should be avoided if the level exceeds 20 micrograms per gram; *General Treatment for Exposure*: INHALATION: apply ice packs to promote heat loss; replace fluids and electrolytes; allay anxiety. INGESTION: same as for inhalation; also, give large amounts of water and induce vomiting; get medical attention. EYES: flush with water for at least 15 min. SKIN: wash thoroughly with soap and water; *Toxicity by Inhalation (Threshold Limit Value)*: 0.2 mg/m$_3$; *Short-Term Exposure Limits*: 1 mg/m^3, 30 min.; *Toxicity by Ingestion*: Grade 4; LD$_{50}$ < 50 mg/kg (rat); *Late Toxicity*: Data not available; *Vapor (Gas) Irritant Characteristics*: Data not available; *Liquid or Solid Irritant Characteristics*: Data not available; *Odor Threshold*: Data not available.

Fire Hazards - *Flash Point :* Not pertinent; *Flammable Limits in Air (%):* Not pertinent; *Fire Extinguishing Agents:* Water, foam, dry chemical, carbon dioxide; *Fire Extinguishing Agents Not to be Used:* Data not available; *Special Hazards of Combustion Products:* Toxic oxides of nitrogen may form in fire; *Behavior in Fire:* Containers may explode; *Ignition Temperature :* Not pertinent; *Electrical Hazard:* Not pertinent; *Burning Rate:* Not pertinent.

Chemical Reactivity - *Reactivity with Water :* No reaction; *Reactivity with Common Materials:* No reaction; *Stability During Transport:* Stable; *Neutralizing Agents for Acids and Caustics:* Not pertinent; *Polymerization:* Not pertinent; *Inhibitor of Polymerization:* Not pertinent.

2,4-DINITROPHENOL

Chemical Designations - *Synonyms*: Aldifen; alpha-Dinitrophenol; 1-Hydroxy-2,4-dinitrobenzene; *Chemical Formula*: $HOC_6H_3(NO_2)_2$-2,4.

Observable Characteristics - *Physical State (as normally shipped)*: Solid; *Color*: Yellow; *Odor*: Musty; sweet.

Physical and Chemical Properties - *Physical State at 15 °C and 1 atm.*: Solid; *Molecular Weight*: 184.1; *Boiling Point at 1 atm.*: Not pertinent; *Freezing Point*: 235, 113, 386; *Critical Temperature*: Not pertinent; *Critical Pressure*: Not pertinent; *Specific Gravity*: 1.68 at 20°C (solid); *Vapor (Gas) Density*: Not pertinent; *Ratio of Specific Heats of Vapor (Gas)*: Not pertinent; *Latent Heat of Vaporization*: Not pertinent; *Heat of Combustion*: Not pertinent; *Heat of Decomposition*: Not pertinent.

Health Hazards Information - *Recommended Personal Protective Equipment*: Self-contained breathing apparatus; butyl rubber gloves; goggles; lab coat; protective shoes; *Symptoms Following Exposure*: Liver damage, metabolic stimulant, dermatitis, dilation of pupils; *General Treatment for Exposure*: Remove victim from contaminated area and wash exposed skin with soap and water. Administer oxygen if respiratory problems develop. Refer to a doctor; *Toxicity by Inhalation (Threshold Limit Value)*: 0.2 mg/m³; *Short-Term Exposure Limits*: Data not available; *Toxicity by Ingestion*: Grade 4; LD_{50} below 50 mg/kg; *Late Toxicity*: Produces clouding of lens of eye (cataracts) in animals and humans, birth defects in chick embryos; *Vapor (Gas) Irritant Characteristics*: Not pertinent; *Liquid or Solid Irritant Characteristics*: Causes smarting of the skin and first-degree burns on short exposure; may cause second-degree burns on long exposure; *Odor Threshold*: Data not available.

Fire Hazards - *Flash Point :* Data not available; *Flammable Limits in Air (%):* Not pertinent; *Fire Extinguishing Agents:* Water, dry chemical, carbon dioxide, foam; *Fire Extinguishing Agents Not to be Used:* Not pertinent; *Special Hazards of Combustion Products:* Vapors are toxic; *Behavior in Fire:* Can detonate or explode when heated under confinement; *Ignition Temperature :* Data not available; *Electrical Hazard:* Not pertinent; *Burning Rate:* Not pertinent.

Chemical Reactivity - *Reactivity with Water :* No reaction; *Reactivity with Common Materials:* Reacts with oxidizing materials and combustibles; *Stability During Transport:* May detonate when heated under confinement; *Neutralizing Agents for Acids and Caustics:* Not pertinent; *Polymerization:* Not pertinent; *Inhibitor of Polymerization:* Not pertinent.

DIOCTYL ADIPATE

Chemical Designations - *Synonyms*: Adipic Acid, Bis(2-ethylhexyl) Ester; Adipol 2EH; Di(2-ethylhexyl) Adipage; DOA; *Chemical Formula*: $C_8H_{17}OOC(CH_2)_4COOC_8H_{17}$.

Observable Characteristics - *Physical State (as normally shipped)*: Liquid; *Color*: Colorless; *Odor*: Mild characteristic.

Physical and Chemical Properties - *Physical State at 15 °C and 1 atm.*: Liquid; *Molecular Weight*: 371; *Boiling Point at 1 atm.*: Very high; *Freezing Point*: Not pertinent; *Critical Temperature*: Not pertinent; *Critical Pressure*: Not pertinent; *Specific Gravity*: 0.928 at 20°C (liquid); *Vapor (Gas) Density*: Not pertinent; *Ratio of Specific Heats of Vapor (Gas)*: Not pertinent; *Latent Heat of Vaporization*: Not pertinent; *Heat of Combustion*: -15,430, -8,580, -359; *Heat of Decomposition*: Not pertinent.

Health Hazards Information - *Recommended Personal Protective Equipment*: None required; *Symptoms Following Exposure*: Low toxicity; no reports of injury in industrial handling; *General Treatment for Exposure*: SKIN AND EYES: wipe off and wash skin with soap and water. Treat like lubricating oil. Flush eyes with water. Remove to fresh air; *Toxicity by Inhalation (Threshold Limit Value)*: Not pertinent; *Short-Term Exposure Limits*: Not pertinent; *Toxicity by Ingestion*: Grade 1; LD_{50} 5 to 15 g/kg; *Late Toxicity*: None; *Vapor (Gas) Irritant Characteristics*: Vapors are nonirritating to the eyes and throat; *Liquid or Solid Irritant Characteristics*: Minimum hazard. If spilled on clothing and allowed to remain, may cause smarting and reddening of the skin; *Odor Threshold*: Not pertinent.

Fire Hazards - *Flash Point (deg. F):* 390 OC; *Flammable Limits in Air (%):* Data not available; *Fire Extinguishing Agents:* Data not available; *Fire Extinguishing Agents Not to be Used:* Data not available; *Special Hazards of Combustion Products:* None; *Behavior in Fire:* Not pertinent; *Ignition Temperature:* Data not available; *Electrical Hazard:* Data not available; *Burning Rate:* Data not available.

Chemical Reactivity - *Reactivity with Water :* No reaction; *Reactivity with Common Materials:* No reaction; *Stability During Transport:* Stable; *Neutralizing Agents for Acids and Caustics:* Not pertinent; *Polymerization:* Not pertinent; *Inhibitor of Polymerization:* Not pertinent.

DIOCTYL PHTHALATE

Chemical Designations - *Synonyms*: Bis(2-ethylhexyl) Phthalate; Di(2-ethylhexyl) Phthalate; DOP; Octoil; Phthalic Acid, Bis(2-ethylhexyl) Ester; *Chemical Formula*: $o\text{-}C_6H_4[COOCH_2CH\text{-}(C_2H_5)(CH_2)_3CH_3]_2$.

Observable Characteristics - *Physical State (as normally shipped)*: Liquid; *Color*: Colorless; *Odor*: Very slight, characteristic.

Physical and Chemical Properties - *Physical State at 15 °C and 1 atm.*: Liquid; *Molecular Weight*: 390.6; *Boiling Point at 1 atm.*: 727, 386, 659; *Freezing Point*: Not pertinent; *Critical Temperature*: Not pertinent; *Critical Pressure*: Not pertinent; *Specific Gravity*: 0.980 at 25°C (liquid); *Vapor (Gas) Density*: Not pertinent; *Ratio of Specific Heats of Vapor (Gas)*: Not pertinent; *Latent Heat of Vaporization*: Not pertinent; *Heat of Combustion*: -15,130, -8410, -352; *Heat of Decomposition*: Not pertinent.

Health Hazards Information - *Recommended Personal Protective Equipment*: Not required; *Symptoms Following Exposure*: Produces no ill effects at normal temperatures but may give off irritating vapor at high temperature; *General Treatment for Exposure*: Leave contaminated area; wash skin with soap and water; flush eyes with water; *Toxicity by Inhalation (Threshold Limit Value)*: Not pertinent; *Short-Term Exposure Limits*: Not pertinent; *Toxicity by Ingestion*: Grade 0; LD_{50} above 15 g/kg (rat); *Late Toxicity*: Not established; *Vapor (Gas) Irritant Characteristics*: Nonirritating to the eyes and throat; *Liquid or Solid Irritant Characteristics*: No appreciable hazard. Practically harmless to the skin; *Odor Threshold*: Not pertinent.

Fire Hazards - *Flash Point (deg. F):* 425 OC; *Flammable Limits in Air (%):* Not pertinent; *Fire Extinguishing Agents:* Dry powder, carbon dioxide, foam; *Fire Extinguishing Agents Not to be Used:* Water or foam may cause frothing; *Special Hazards of Combustion Products:* None; *Behavior in Fire:* Not pertinent; *Ignition Temperature :* Data not available; *Electrical Hazard:* Not pertinent; *Burning Rate:* Not pertinent.

Chemical Reactivity - *Reactivity with Water :* No reaction; *Reactivity with Common Materials:* No reaction; *Stability During Transport:* Stable; *Neutralizing Agents for Acids and Caustics:* Not pertinent; *Polymerization:* Not pertinent; *Inhibitor of Polymerization:* Not pertinent.

DIOCTYL SODIUM SULFOSUCCINATE

Chemical Designations - *Synonyms*: Aerosol Surfactant; Alrowet D65; Bis(2-ethylhexyl)sodium Sulfosuccinate; Di(2-ethylhexyl) Sulfosuccinate, Sodium Salt; Sodium Dioctyl Sulfosuccinate; *Chemical Formula*: $C_8H_{17}OOCCH_2CH\text{-}(SO_3Na)COOC_8H_{17}$.

Observable Characteristics - *Physical State (as normally shipped)*: Waxy solid or water solution; *Color*: Colorless or off-white; *Odor*: None.

Physical and Chemical Properties - *Physical State at 15 °C and 1 atm.*: Solid or liquid; *Molecular Weight*: 444; *Boiling Point at 1 atm.*: Not pertinent (decomposes); *Freezing Point*: (solid form) 311, 155, 428; *Critical Temperature*: Not pertinent; *Critical Pressure*: Not pertinent; *Specific Gravity*: 1.1 at 20°C (solid or liquid); *Vapor (Gas) Density*: Not pertinent; *Ratio of Specific Heats of Vapor (Gas)*: Not pertinent; *Latent Heat of Vaporization*: Not pertinent; *Heat of Combustion*: Not pertinent; *Heat of Decomposition*: Not pertinent.

Health Hazards Information - *Recommended Personal Protective Equipment*: Chemical goggles; rubber gloves; dust respirator; *Symptoms Following Exposure*: Liquid is strong irritant to eye and may irritate to eye and may irritate skin by removing natural oils. Ingestion causes diarrhea and intestinal bloating; *General Treatment for Exposure*: EYES: irrigate with copious volumes of water for at least 15 min.; call physician. SKIN: rinse off with water. INGESTION: drink large amounts of water; *Toxicity by Inhalation (Threshold Limit Value)*: Data not available; *Short-Term Exposure Limits*: Not pertinent; *Toxicity by Ingestion*: Grade 2; oral LD_{50} 1,900 mg/kg (rat); *Late Toxicity*: Data not available; *Vapor (Gas) Irritant Characteristics*: Not pertinent; *Liquid or Solid Irritant Characteristics*: Data not available; *Odor Threshold*: Not pertinent.

Fire Hazards - *Flash Point :* Not flammable; *Flammable Limits in Air (%):* Not flammable; *Fire Extinguishing Agents:* Not pertinent; *Fire Extinguishing Agents Not to be Used:* Not pertinent; *Special Hazards of Combustion Products:* Not pertinent; *Behavior in Fire:* Cause foaming and spreading of water; *Ignition Temperature :* Not pertinent; *Electrical Hazard:* Not pertinent; *Burning Rate:* Not pertinent.

Chemical Reactivity - *Reactivity with Water :* No reaction; *Reactivity with Common Materials:* No reaction; *Stability During Transport:* Stable; *Neutralizing Agents for Acids and Caustics:* Not pertinent; *Polymerization:* Not pertinent; *Inhibitor of Polymerization:* Not pertinent.

1,4 DIOXANE

Chemical Designations - *Synonyms*: Di (Ethylene Oxide); Dioxan; p-Dioxane; *Chemical Formula*: $CH_2CH_2OCH_2CH_2O$

Observable Characteristics - *Physical State (as normally shipped)*: Liquid; *Color*: Colorless; *Odor*: Mild; somewhat alcoholic; like butyl alcohol.

Physical and Chemical Properties - *Physical State at 15 °C and 1 atm.*: Liquid; *Molecular Weight*: 88.11; *Boiling Point at 1 atm.*: 214.3, 101.3, 374.5; *Freezing Point*: 53.2, 11.8, 285.2; *Critical Temperature*: 597, 314, 587; *Critical Pressure*: 755, 51.4, 5.21; *Specific Gravity*: 1.036 at 20°C (liquid); *Vapor (Gas) Density*: Not pertinent; *Ratio of Specific Heats of Vapor (Gas)*: 1.1; *Latent Heat of Vaporization*: 178, 98.6, 4.13; *Heat of Combustion*: -11,590, -6,440, -269.6; *Heat of Decomposition*: Not pertinent.

Health Hazards Information - *Recommended Personal Protective Equipment*: Fresh air mask; rubber gloves; goggles; safety shower and eye bath; *Symptoms Following Exposure*: No significant irritation from brief exposure of skin; prolonged or repeated exposure may cause a rash or burn and absorption of toxic amounts leading to serious injury of liver and kidney. Chemical has poor warning properties; illness may be delayed. Moderately irritating to eyes; overexposure may cause corneal injury; *General Treatment for Exposure*: INHALATION: promptly remove victim to fresh air, keep him quiet and warm, and call physician; start artificial respiration if breathing stops. INGESTION: if large amounts are swallowed, quickly induce vomiting and get medical attention; no specific antidote known. SKIN AND EYES: flush with plenty of water for 15 min.; remove contaminated clothing and wash before reuse; get medical attention for eyes and if ill effects occur from skin contact; *Toxicity by Inhalation (Threshold Limit Value)*: 100 ppm; *Short-Term Exposure Limits*: 100 ppm for 60 min.; *Toxicity by Ingestion*: Grade 2; LD_{50} 0.5 to 5 g/kg (guinea pig: 3.90 g/kg); *Late Toxicity*: Causes cancer in rats; *Vapor (Gas) Irritant Characteristics*: Vapors cause a slight of the eyes or respiratory system if present in high concentrations. The effect is temporary; *Liquid or Solid Irritant Characteristics*: Minimum hazard. If spilled on clothing and allowed to remain, may cause smarting and reddening of the skin; *Odor Threshold*: 620 mg/m^3.

Fire Hazards - *Flash Point (deg. F):* 54 CC; 74 OC; *Flammable Limits in Air (%):* 1.9 - 22.5 by vol; *Fire Extinguishing Agents:* Alcohol foam, carbon dioxide, dry chemical; *Fire Extinguishing Agents Not to be Used:* Not pertinent; *Special Hazards of Combustion Products:* Toxic vapors are generated when heated*; Behavior in Fire:* Vapor is heavier than air and may travel to a source of ignition and flash back; *Ignition Temperature (deg. F):* 356; *Electrical Hazard:* Not pertinent; *Burning Rate:* Data not available.

Chemical Reactivity - *Reactivity with Water :* No reaction; *Reactivity with Common Materials:* No reaction; *Stability During Transport:* Stable; *Neutralizing Agents for Acids and Caustics:* Not pertinent; *Polymerization:* Not pertinent; *Inhibitor of Polymerization:* Not pertinent.

DIPENTENE

Chemical Designations - *Synonyms*: Limonene; para-Mentha-1,8-diene; Phellandrene; Terpinene; delta-1,8-Terpodiene; *Chemical Formula*: $C_{10}H_{16}$.

Observable Characteristics - *Physical State (as normally shipped)*: Liquid; *Color*: Colorless; *Odor*:

Pleasant, pine-like; lemon-like.

Physical and Chemical Properties - *Physical State at 15 °C and 1 atm.*: Liquid; *Molecular Weight*: 136.2; *Boiling Point at 1 atm.*: 352, 178, 451; *Freezing Point*: -40, -40, 233; *Critical Temperature*: Not pertinent; *Critical Pressure*: Not pertinent; *Specific Gravity*: 0.842 at 21°C (liquid); *Vapor (Gas) Density*: 4.9; *Ratio of Specific Heats of Vapor (Gas)*: Not pertinent; *Latent Heat of Vaporization*: 140, 77, 3.2; *Heat of Combustion*: -19,520, -10,840, -454; *Heat of Decomposition*: Not pertinent.

Health Hazards Information - *Recommended Personal Protective Equipment*: Solvent-resistant gloves; safety glasses or face shield; self-contained breathing apparatus for high vapor concentrations; *Symptoms Following Exposure*: Liquid irritates eyes; prolonged contact with skin causes irritation. Ingestion causes irritation of gastrointestinal tract; *General Treatment for Exposure*: INHALATION: remove victim from contaminated area; administer artificial respiration if necessary; call physician. EYES: flush with water for 15 min.; call physician. SKIN: wash with soap and water. INGESTION: induce vomiting; call physician; *Toxicity by Inhalation (Threshold Limit Value)*: Data not available; *Short-Term Exposure Limits*: Data not available; *Toxicity by Ingestion*: Grade 2; oral LD$_{50}$ 4,600 mg/kg (rat); *Late Toxicity*: Data not available; *Vapor (Gas) Irritant Characteristics*: Vapors cause a slight of the eyes or respiratory system if present in high concentrations. The effect is temporary; *Liquid or Solid Irritant Characteristics*: Minimum hazard. If spilled on clothing and allowed to remain, may cause smarting and reddening of the skin; *Odor Threshold*: Data not available.

Fire Hazards - *Flash Point (deg. F):* 115 CC; *Flammable Limits in Air (%):* 0.7 - 6.1; *Fire Extinguishing Agents:* Foam, dry chemical, carbon dioxide; *Fire Extinguishing Agents Not to be Used:* Water may be ineffective; *Special Hazards of Combustion Products:* Not pertinent; *Behavior in Fire:* Containers may explode; *Ignition Temperature (deg. F):* 458; *Electrical Hazard:* Data not available; *Burning Rate:* 5.5 mm/min.

Chemical Reactivity - *Reactivity with Water :* No reaction; *Reactivity with Common Materials:* No reaction; *Stability During Transport:* Stable; *Neutralizing Agents for Acids and Caustics:* Not pertinent; *Polymerization:* Not pertinent; *Inhibitor of Polymerization:* Not pertinent.

DIPHENYLAMINE

Chemical Designations - *Synonyms*: Anilinobenzene; N-Phenylaniline; *Chemical Formula*: $(C_6H_5)_2NH$.

Observable Characteristics - *Physical State (as normally shipped)*: Solid or liquid; *Color*: Very pale tan; amber to brown; *Odor*: Characteristic, pleasant.

Physical and Chemical Properties - *Physical State at 15 °C and 1 atm.*: Solid or liquid; *Molecular Weight*: 169.2; *Boiling Point at 1 atm.*: 576, 302, 575; *Freezing Point*: 127, 53, 326; *Critical Temperature*: Not pertinent; *Critical Pressure*: Not pertinent; *Specific Gravity*: 1.068 at 61°C (liquid); *Vapor (Gas) Density*: Not pertinent; *Ratio of Specific Heats of Vapor (Gas)*: Not pertinent; *Latent Heat of Vaporization*: Not pertinent; *Heat of Combustion*: -16,300, -9,060, -379; *Heat of Decomposition*: Not pertinent.

Health Hazards Information - *Recommended Personal Protective Equipment*: Respirator; safety goggles or face shield; rubber gloves; *Symptoms Following Exposure*: Inhalation may irritates mucous membranes. Overexposure, including ingestion of solid or skin contact, may cause fast pulse, hypertension, and bladder trouble. Contact with dust irritates eyes; *General Treatment for Exposure*: INHALATION: move victim to fresh air. INGESTION: get medical attention; observe for methemoglobinemia. EYES: flush with plenty of water and see physician. SKIN: wash with soap and water; *Toxicity by Inhalation (Threshold Limit Value)*: 10 mg/m^3; *Short-Term Exposure Limits*: Data not available; *Toxicity by Ingestion*: Grade 2; oral LD$_{50}$ 2,000 mg/kg (rat); *Late Toxicity*: Causes birth defects in rats (polycystic kidneys); *Vapor (Gas) Irritant Characteristics*: Data not available; *Liquid or Solid Irritant Characteristics*: Data not available; *Odor Threshold*: Data not available.

Fire Hazards - *Flash Point (deg. F):* (liquid) 307 OC; *Flammable Limits in Air (%):* Not pertinent; *Fire Extinguishing Agents:* Foam, dry chemical, carbon dioxide; *Fire Extinguishing Agents Not to be Used:* Water or foam may cause frothing; *Special Hazards of Combustion Products:* Toxic oxides of nitrogen may form in fire; *Behavior in Fire:* Dust may be explosive if mixed with air in critical proportions and in the presence of a source of ignition; *Ignition Temperature (deg. F):* 1.175; *Electrical*

Hazard: Not pertinent; *Burning Rate:* Not pertinent.
Chemical Reactivity - *Reactivity with Water :* No reaction; *Reactivity with Common Materials:* No reaction; *Stability During Transport:* Stable; *Neutralizing Agents for Acids and Caustics:* Not pertinent; *Polymerization:* Not pertinent; *Inhibitor of Polymerization:* Not pertinent.

DIPHENYL ETHER
Chemical Designations - *Synonyms*: Phenyl Ether; Diphenyl Oxide; Phenoxybenzene; *Chemical Formula*: $C_6H_5OC_6H_5$.
Observable Characteristics - *Physical State (as normally shipped)*: Liquid; *Color*: Colorless; *Odor*: Weak geranium.
Physical and Chemical Properties - *Physical State at 15 °C and 1 atm.*: Solid; *Molecular Weight*: 170.2; *Boiling Point at 1 atm.*: 495, 257, 530; *Freezing Point*: 81, 27, 300; *Critical Temperature*: 921, 494, 767; *Critical Pressure*: 478, 32.5, 3.30; *Specific Gravity*: 1.07 at 27°C (liquid); *Vapor (Gas) Density*: Not pertinent; *Ratio of Specific Heats of Vapor (Gas)*: Not pertinent; *Latent Heat of Vaporization*: 130, 72, 3.0; *Heat of Combustion*: -15,520, -8,620, -361; *Heat of Decomposition*: Not pertinent.
Health Hazards Information - *Recommended Personal Protective Equipment*: Goggles or face shield; rubber gloves; *Symptoms Following Exposure*: Inhalation may cause nausea because of disagreeable odor. Contact of liquid with eyes causes mild irritation. Prolonged exposure of skin to liquid causes reddening and irritation. Ingestion produces nausea; *General Treatment for Exposure*: EYES: flush with water for at least 15 min. SKIN: wipe off, wash with soap and water. INGESTION: induce vomiting and get medical attention; *Toxicity by Inhalation (Threshold Limit Value)*: 1 ppm; *Short-Term Exposure Limits*: Data not available; *Toxicity by Ingestion*: Grade 2; oral LD_{50} 3,370 mg/kg (rat); *Late Toxicity*: Data not available; *Vapor (Gas) Irritant Characteristics*: Data not available; *Liquid or Solid Irritant Characteristics*: Data not available; *Odor Threshold*: 0.1 ppm.
Fire Hazards - *Flash Point (deg. F):* 239 CC; *Flammable Limits in Air (%):* 0.8 - 1.5; *Fire Extinguishing Agents:* Dry chemical, carbon dioxide; *Fire Extinguishing Agents Not to be Used:* Water or foam may cause frothing; *Special Hazards of Combustion Products:* Not pertinent; *Behavior in Fire:* Not pertinent; *Ignition Temperature (deg. F):* 1.148; *Electrical Hazard:* Data not available; *Burning Rate:* 3.2 mm/min.
Chemical Reactivity - *Reactivity with Water :* No reaction; *Reactivity with Common Materials:* No reaction; *Stability During Transport:* Stable; *Neutralizing Agents for Acids and Caustics:* Not pertinent; *Polymerization:* Not pertinent; *Inhibitor of Polymerization:* Not pertinent.

DIPHENYLMETHANE DIISOCYANATE
Chemical Designations - *Synonyms*: Carwinate 125 M; Diphenylmethane-4,4`-diisocyanate; Hylene M50; MDI; Methylenebis(4-phenylisocyanate); Multrathane M; Nacconate 300; Vilrathane 4300; *Chemical Formula*: $(p-OCNC_6H_4)_2CH_2$.
Observable Characteristics - *Physical State (as normally shipped)*: Solid; *Color*: Colorless to light yellow; *Odor*: Data not available.
Physical and Chemical Properties - *Physical State at 15 °C and 1 atm.*: Solid; *Molecular Weight*: 250.3; *Boiling Point at 1 atm.*: 738, 392, 665; *Freezing Point*: 100, 37.7, 311; *Critical Temperature*: Not pertinent; *Critical Pressure*: Not pertinent; *Specific Gravity*: 1.2 at 20°C (solid); *Vapor (Gas) Density*: Not pertinent; *Ratio of Specific Heats of Vapor (Gas)*: Not pertinent; *Latent Heat of Vaporization*: Not pertinent; *Heat of Combustion*: Not pertinent; *Heat of Decomposition*: Not pertinent.
Health Hazards Information - *Recommended Personal Protective Equipment*: Mask or respirator of type approved by U.S. Bu. Mines (above 135°C); clean rubber gloves; chemical goggles; clean waterproof or freshly laundered protective clothing (coveralls, rubber boots, caps, etc.); *Symptoms Following Exposure*: Breathlessness, chest discomfort, and reduced pulmonary function; *General Treatment for Exposure*: INHALATION: treat symptomatically; give oxygen. Call a physician. SKIN: wash with soap and water. Rubbing alcohol helpful. EYE: flush with water at least 15 min. Call a physician; *Toxicity by Inhalation (Threshold Limit Value)*: 0.02 ppm; *Short-Term Exposure Limits*: Data

not available; *Toxicity by Ingestion*: Data not available; *Late Toxicity*: Data not available; *Vapor (Gas) Irritant Characteristics*: Severe irritation of eyes and throat; can cause eye and lung injury. Cannot be tolerated even at low concentrations; *Liquid or Solid Irritant Characteristics*: Minimum hazard. If spilled on clothing and allowed to remain, may cause smarting and reddening of the skin; *Odor Threshold*: Data not available.

Fire Hazards - *Flash Point (deg. F)*: 425 OC; *Flammable Limits in Air (%)*: Not pertinent; *Fire Extinguishing Agents:* Carbon dioxide or dry chemical; *Fire Extinguishing Agents Not to be Used:* Not pertinent; *Special Hazards of Combustion Products:* Toxic vapors are generated when heated; *Behavior in Fire:* Solid melts and burns; *Ignition Temperature :* Data not available; *Electrical Hazard:* Not pertinent; *Burning Rate:* Not pertinent.

Chemical Reactivity - *Reactivity with Water :* Slow, non-hazardous. Form carbon dioxide gas; *Reactivity with Common Materials:* data not available; *Stability During Transport:* Stable; *Neutralizing Agents for Acids and Caustics:* Not pertinent; *Polymerization:* May occur slowly. Is not hazardous; *Inhibitor of Polymerization:* Not pertinent.

DIPROPYLENE GLYCOL

Chemical Designations - *Synonyms*: No common synonyms; *Chemical Formula*: $(CH_3CHOHCH_2)_2O$.

Observable Characteristics - *Physical State (as normally shipped)*: liquid; *Color*: Colorless; *Odor*: Practically none.

Physical and Chemical Properties - *Physical State at 15 °C and 1 atm.*: Liquid; *Molecular Weight*: 134.17; *Boiling Point at 1 atm.*: 420, 232, 505; *Freezing Point*: >-40, >-40, >233; *Critical Temperature*: 720, 382, 655; *Critical Pressure*: 529, 36, 3.6; *Specific Gravity*: 1.023 at 20°C (liquid); *Vapor (Gas) Density*: Not pertinent; *Ratio of Specific Heats of Vapor (Gas)*: 1.0; *Latent Heat of Vaporization*: 170, 96, 4.0; *Heat of Combustion*: -11,650, -6470, -271; *Heat of Decomposition*: Not pertinent.

Health Hazards Information - *Recommended Personal Protective Equipment*: Safety glasses with side shields or goggles; shower and eye bath; *Symptoms Following Exposure*: Minor eye irritation; *General Treatment for Exposure*: EYES: irrigate briefly with water; if any ill defects, get medical attention. SKIN AND INGESTION: if any ill defects develop, get medical attention; *Toxicity by Inhalation (Threshold Limit Value)*: Not pertinent; *Short-Term Exposure Limits*: Not pertinent; *Toxicity by Ingestion*: Grade 1; LD_{50} 5 to 15 g/kg (rat); *Late Toxicity*: Data not available; *Vapor (Gas) Irritant Characteristics*: Nonirritating to the eyes and throat; *Liquid or Solid Irritant Characteristics*: No appreciable hazard. Practically harmless to the skin; *Odor Threshold*: Not pertinent.

Fire Hazards - *Flash Point (deg. F):* 280 OC; *Flammable Limits in Air (%):* LFL=2.2% (approx.); *Fire Extinguishing Agents:* Water fog, alcohol foam, carbon dioxide, dry chemical; *Fire Extinguishing Agents Not to be Used:* Water or foam may cause frothing; *Special Hazards of Combustion Products:* Not pertinent; *Behavior in Fire:* Not pertinent; *Ignition Temperature:* Data not available; *Electrical Hazard:* Not pertinent; *Burning Rate:* 2.0 mm/min.

Chemical Reactivity - *Reactivity with Water :* No reaction; *Reactivity with Common Materials:* No reaction; *Stability During Transport:* Stable; *Neutralizing Agents for Acids and Caustics:* Not pertinent; *Polymerization:* Not pertinent; *Inhibitor of Polymerization:* Not pertinent.

DISTILLATES: FLASHED FEED STOCKS

Chemical Designations - *Synonyms*: Petroleum Distillate; *Chemical Formula*: Not pertinent.

Observable Characteristics - *Physical State (as normally shipped)*: Liquid; *Color*: Colorless; *Odor*: Gasoline.

Physical and Chemical Properties - *Physical State at 15 °C and 1 atm.*: Liquid; *Molecular Weight*: Not pertinent; *Boiling Point at 1 atm.*: 58-275, 14-135, 287-408; *Freezing Point*: Not pertinent; *Critical Temperature*: Not pertinent; *Critical Pressure*: Not pertinent; *Specific Gravity*: 0.71-0.75 at 15°C (liquid); *Vapor (Gas) Density*: 3.4; *Ratio of Specific Heats of Vapor (Gas)*: 1.054; *Latent Heat of Vaporization*: 130 - 150, 71 - 81, 3.0 - 3.4; *Heat of Combustion*: -18,720, -10,400, -435.4; *Heat of Decomposition*: Not pertinent.

Health Hazards Information - *Recommended Personal Protective Equipment*: Data not available; *Symptoms Following Exposure*: INHALATION: irritation of upper respiratory tract; dizziness, headache, coma, respiratory arrest; cardiac arrhythmias may occur. ASPIRATION: severe lung irritation, coughing, pulmonary edema, signs of bronchopneumonia; acute central nervous system excitation, followed by depression. INGESTION: irritation of mouth and stomach, other symptoms as above; *General Treatment for Exposure*: Seek medical attention. INHALATION: maintain respiration, administer oxygen. ASPIRATION: enforce bed rest; administer oxygen. INGESTION: do NOT induce vomiting; lavage carefully if appreciable quantity was swallowed; guard against aspiration into lungs. EYES: wash with copious amounts of water. SKIN: wipe off and wash with soap and water; *Toxicity by Inhalation (Threshold Limit Value)*: No single TLV applicable; *Short-Term Exposure Limits*: 500 ppm for 30 min.; *Toxicity by Ingestion*: Grade 2; LD_{50} 0.5 - 5 g/kg; *Late Toxicity*: None; *Vapor (Gas) Irritant Characteristics*: Vapors cause a slight smarting of the eyes or respiratory system if present in high concentrations. The effect is temporary; *Liquid or Solid Irritant Characteristics*: Minimum hazard. If spilled on clothing and allowed to remain, may cause smarting and reddening of the skin; *Odor Threshold*: 0.25 ppm.

Fire Hazards - *Flash Point (deg. F):* (a) <0 CC; (b) 0 - 73 CC; (3) 73 - 141 CC; *Flammable Limits in Air (%):* Data not available; *Fire Extinguishing Agents:* Foam, carbon dioxide, dry chemical; *Fire Extinguishing Agents Not to be Used:* Water may be ineffective; *Special Hazards of Combustion Products:* Not pertinent; *Behavior in Fire:* Not pertinent; *Ignition Temperature:* Data not available; *Electrical Hazard:* Class I, Group D; *Burning Rate:* Approx. 4 mm/min.

Chemical Reactivity - *Reactivity with Water :* No reaction; *Reactivity with Common Materials:* No reaction; *Stability During Transport:* Stable; *Neutralizing Agents for Acids and Caustics:* Not pertinent; *Polymerization:* Not pertinent; *Inhibitor of Polymerization:* Not pertinent.

DODECANOL

Chemical Designations - *Synonyms*: Dodecyl Alcohol; Lauryl Alcohol; *Chemical Formula*: $CH_3(CH_2)_{10}CH_2OH$.

Observable Characteristics - *Physical State (as normally shipped)*: Liquid; *Color*: Colorless; *Odor*: Typical fatty alcohol odor; sweet.

Physical and Chemical Properties - *Physical State at 15 °C and 1 atm.*: Liquid; *Molecular Weight*: 186.33; *Boiling Point at 1 atm.*: 498, 259, 532; *Freezing Point*: 75, 24, 297; *Critical Temperature*: 763, 406, 679; *Critical Pressure*: 280, 19, 1.9; *Specific Gravity*: 0.831 at 24°C (liquid); *Vapor (Gas) Density*: Not pertinent; *Ratio of Specific Heats of Vapor (Gas)*: 1.030; *Latent Heat of Vaporization*: 110, 62, 2.6; *Heat of Combustion*: -18,000, -10,000, -420; *Heat of Decomposition*: Not pertinent.

Health Hazards Information - *Recommended Personal Protective Equipment*: Chemical gloves; chemical goggles; *Symptoms Following Exposure*: Liquid will cause burning of the eyes and may irritate skin; *General Treatment for Exposure*: SKIN AND EYES: wash exposed areas with water; *Toxicity by Inhalation (Threshold Limit Value)*: Not pertinent; *Short-Term Exposure Limits*: Not pertinent; *Toxicity by Ingestion*: Grade 1; LD_{50} 5 to 15 g/kg (humans); *Late Toxicity*: Data not available; *Vapor (Gas) Irritant Characteristics*: None; *Liquid or Solid Irritant Characteristics*: No appreciable hazard. Practically harmless to the skin; *Odor Threshold*: Data not available.

Fire Hazards - *Flash Point (deg. F):* 260 CC; *Flammable Limits in Air (%):* Data not available; *Fire Extinguishing Agents:* Alcohol foam, dry chemical, carbon dioxide; *Fire Extinguishing Agents Not to be Used:* Water or foam may cause frothing; *Special Hazards of Combustion Products:* Not pertinent; *Behavior in Fire:* Not pertinent; *Ignition Temperature (deg. F):* 527; *Electrical Hazard:* Not pertinent; *Burning Rate:* Data not available.

Chemical Reactivity - *Reactivity with Water :* No reaction; *Reactivity with Common Materials:* No reaction; *Stability During Transport:* Stable; *Neutralizing Agents for Acids and Caustics:* Not pertinent; *Polymerization:* Not pertinent; *Inhibitor of Polymerization:* Not pertinent.

DODECENE
Chemical Designations - *Synonyms*: Dodecene(non-linear); Propylene Tetramer; Tetrapropylene; *Chemical Formula*: $C_{12}H_{24}$.
Observable Characteristics - *Physical State (as normally shipped)*: Liquid; *Color*: Colorless; *Odor*: Characteristic.
Physical and Chemical Properties - *Physical State at 15 °C and 1 atm.*: Liquid; *Molecular Weight*: 168.31; *Boiling Point at 1 atm.*: 365-385, 185-196, 458-469; *Freezing Point*: Not pertinent; *Critical Temperature*: Not pertinent; *Critical Pressure*: Not pertinent; *Specific Gravity*: 0.77 at 20°C (liquid); *Vapor (Gas) Density*: Not pertinent; *Ratio of Specific Heats of Vapor (Gas)*: Not pertinent; *Latent Heat of Vaporization*: Not pertinent; *Heat of Combustion*: -19,100, -10,600, -444; *Heat of Decomposition*: Not pertinent.
Health Hazards Information - *Recommended Personal Protective Equipment*: Protective gloves; no respiratory protection needed if ventilation is adequate; *Symptoms Following Exposure*: No inhalation hazard expected. Aspiration hazard if ingested. Minor skin and eye irritation; *General Treatment for Exposure*: INHALATION: Remove victim to fresh air. INGESTION: do NOT induce vomiting! Do NOT lavage! Give vegetable oil and demulcents; call physician. EYE: flush with water for 15 min. SKIN: wash with soap and water; *Toxicity by Inhalation (Threshold Limit Value)*: 200 ppm; *Short-Term Exposure Limits*: Data not available; *Toxicity by Ingestion*: Grade 0; LD_{50} above 15 g/kg; *Late Toxicity*: Data not available; *Vapor (Gas) Irritant Characteristics*: Vapors cause a slight smarting of the eyes or respiratory system if present in high concentrations. The effect is temporary; *Liquid or Solid Irritant Characteristics*: Minimum hazard. If spilled on clothing and allowed to remain, may cause smarting and reddening of the skin; *Odor Threshold*: Data not available.
Fire Hazards - *Flash Point (deg. F)*: 120 CC; 134 OC; *Flammable Limits in Air (%)*: Data not available; *Fire Extinguishing Agents*: Water fog, foam, carbon dioxide, dry chemical; *Fire Extinguishing Agents Not to be Used*: Water may be ineffective; *Special Hazards of Combustion Products*: Not pertinent; *Behavior in Fire*: Not pertinent; *Ignition Temperature (deg. F)*: 400 (est.); *Electrical Hazard*: Not pertinent; *Burning Rate*: Data not available.
Chemical Reactivity - *Reactivity with Water* : No reaction; *Reactivity with Common Materials*: No reaction; *Stability During Transport*: Stable; *Neutralizing Agents for Acids and Caustics*: Not pertinent; *Polymerization*: Not pertinent; *Inhibitor of Polymerization*: Not pertinent.

1-DODECENE
Chemical Designations - *Synonyms*: Adacene-12; alpha-Dodecylene; *Chemical Formula*: $CH_3(CH_2)_9CH = CH_2$.
Observable Characteristics - *Physical State (as normally shipped)*: Liquid; *Color*: Colorless; *Odor*: Mild, pleasant.
Physical and Chemical Properties - *Physical State at 15 °C and 1 atm.*: Liquid; *Molecular Weight*: 168.31; *Boiling Point at 1 atm.*: 415, 213, 486; *Freezing Point*: -31, -35, 238; *Critical Temperature*: Not pertinent; *Critical Pressure*: Not pertinent; *Specific Gravity*: 0.758 at 20°C (liquid); *Vapor (Gas) Density*: Not pertinent; *Ratio of Specific Heats of Vapor (Gas)*: 1.032; *Latent Heat of Vaporization*: 110, 61.0, 2.55; *Heat of Combustion*: -18,911, -10,506, -439.87; *Heat of Decomposition*: Not pertinent.
Health Hazards Information - *Recommended Personal Protective Equipment*: Protective gloves; goggles or face shield; *Symptoms Following Exposure*: No inhalation hazard expected. Aspiration hazard if ingested. Minor skin and eye irritation; *General Treatment for Exposure*: INHALATION: remove victim to fresh air. INGESTION: do NOT induce vomiting! Give vegetable oil and demulcents; call physician. EYE: flush wight water for 15 min. SKIN: wash with soap and water; *Toxicity by Inhalation (Threshold Limit Value)*: Data not available; *Short-Term Exposure Limits*: Data not available; *Toxicity by Ingestion*: Data not available; *Late Toxicity*: Data not available; *Vapor (Gas) Irritant Characteristics*: Vapors cause a slight smarting of the eyes or respiratory system if present in high concentrations. The effect is temporary; *Liquid or Solid Irritant Characteristics*: Minimum hazard. If spilled on clothing and allowed to remain, may cause smarting and reddening of the skin; *Odor

Threshold: Data not available.

Fire Hazards - *Flash Point (deg. F):* 174; *Flammable Limits in Air (%):* Data not available; *Fire Extinguishing Agents:* Foam, dry chemical, carbon dioxide; *Fire Extinguishing Agents Not to be Used:* Not pertinent; *Special Hazards of Combustion Products:* Not pertinent; *Behavior in Fire:* Not pertinent; *Ignition Temperature (deg. F):* 491; *Electrical Hazard:* Not pertinent; *Burning Rate:* 5.8 mm/min.

Chemical Reactivity - *Reactivity with Water :* No reaction; *Reactivity with Common Materials:* No reaction; *Stability During Transport:* Stable; *Neutralizing Agents for Acids and Caustics:* Not pertinent; *Polymerization:* Not pertinent; *Inhibitor of Polymerization:* Not pertinent.

DODECYLTRICHLOROSILANE

Chemical Designations - *Synonyms*: No common synonyms; *Chemical Formula*: $CH_3(CH_2)_{11}SiCl_3$.

Observable Characteristics - *Physical State (as normally shipped)*: Liquid; *Color*: Colorless; *Odor*: Sharp, like hydrochloric acid; pungent and irritating.

Physical and Chemical Properties - *Physical State at 15 °C and 1 atm.*: Liquid; *Molecular Weight*: 303.7; *Boiling Point at 1 atm.*: >300, >149, >422; *Freezing Point*: Not pertinent; *Critical Temperature*: Not pertinent; *Critical Pressure*: Not pertinent; *Specific Gravity*: 1.03 at 20°C (liquid); *Vapor (Gas) Density*: Not pertinent; *Ratio of Specific Heats of Vapor (Gas)*: Not pertinent; *Latent Heat of Vaporization*: Not pertinent; *Heat of Combustion*: -11,000, -6,200, -260; *Heat of Decomposition*: Not pertinent.

Health Hazards Information - *Recommended Personal Protective Equipment*: Acid-vapor type respiratory protection; rubber gloves; chemical worker's goggles; other protective equipment as necessary to protect eyes and skin; *Symptoms Following Exposure*: Inhalation irritates mucous membrane. Contact with liquid causes severe burns of eyes and skin. Ingestion causes severe burns of mouth and stomach; *General Treatment for Exposure*: INHALATION: remove from exposure; support respiration; call physician if needed. EYES: flush with water for 15 min.; obtain medical attention immediately. SKIN: flush with water; obtain medical attention if skin is burned. INGESTION: if victim is conscious, give large amounts of water, then milk or milk of magnesia; *Toxicity by Inhalation (Threshold Limit Value)*: Data not available; *Short-Term Exposure Limits*: Data not available; *Toxicity by Ingestion*: Grade 3; LD_{50} 50 to 500 mg/kg; *Late Toxicity*: Data not available; *Vapor (Gas) Irritant Characteristics*: Vapors cause moderate irritation such that personnel will find high concentrations unpleasant. The effect is temporary; *Liquid or Solid Irritant Characteristics*: Severe skin irritant. Causes second- and third-degree burns on short contact and is very injurious to the eyes; *Odor Threshold*: Data not available.

Fire Hazards - *Flash Point (deg. F):* >150 OC; *Flammable Limits in Air (%):* Data not available; *Fire Extinguishing Agents:* Dry chemical, carbon dioxide; *Fire Extinguishing Agents Not to be Used:* Water, foam; *Special Hazards of Combustion Products:* Hydrochloric acid and phosgene fumes may form in fires; *Behavior in Fire:* Difficult to extinguish; re-ignition may occur. Contact with water applied to adjacent fires produces irritating hydrogen chloride fumes; *Ignition Temperature:* Data not available; *Electrical Hazard:* Data not available; *Burning Rate:* Data not available.

Chemical Reactivity - *Reactivity with Water :* Generates hydrogen chloride (hydrochloric acid); *Reactivity with Common Materials:* Reacts with surface moisture to generate hydrogen chloride, which is corrosive to most metals; *Stability During Transport:* Stable; *Neutralizing Agents for Acids and Caustics:* Flush with water, rinse with sodium bicarbonate or lime solution; *Polymerization:* Not pertinent; *Inhibitor of Polymerization:* Not pertinent.

E

ENDRIN

Chemical Designations - *Synonyms*: 1,2,3,4,10,10-Hexachloro-6,7-epoxy-1,4,4a,5,6,7,8,8a-octahydro-endo,endo-1,4,5,8-dimethanonaphthalene; Hexadrin; Mendrin; *Chemical Formula*: Not applicable.

Observable Characteristics - *Physical State (as shipped)*: Solid (Sometimes shipped as an emulsifiable concentrate in xylene solution); *Color*: Colorless to tan; *Odor*: None.

Physical and Chemical Properties - *Physical State at 15 °C and 1 atm.*: Solid; *Molecular Weight*: 380.92; *Boiling Point at 1 atm.*: Not pertinent (decomposes); *Freezing Point*: 392, 200, 573; *Critical Temperature*: Not pertinent; *Critical Pressure*: Not pertinent; *Specific Gravity*: 1.65 at 25 °C (solid); *Vapor (Gas) Specific Gravity*: Not pertinent; *Ratio of Specific Heats of Vapor (Gas)*: Not pertinent; *Latent Heat of Vaporization*: Not pertinent; *Heat of Combustion*: Not pertinent; *Heat of Decomposition*: Not pertinent.

Health Hazards Information - *Recommended Personal Protective Equipment*: Respirator for spray, fog, or dust; rubber gloves and boots; *Symptoms Following Exposure*: Inhalation causes moderate irritation of nose and throat; prolonged breathing may cause same toxic symptoms as for ingestion. Contact with liquid causes moderate irritation of eyes and skin. Prolonged contact with skin may cause same toxic symptoms as for ingestion. Ingestion causes frothing of the mouth, facial congestion, convulsions, violent muscular contractions, dizziness, weakness, nausea; *General Treatment for Exposure*: Get medical attention after all exposures to this compound. INHALATION: remove from exposure. EYES: flush with water for at least 15 min. SKIN: wash with plenty of soap and water, but do not scrub. INGESTION: remove from the gastrointestinal tract, either by inducing vomiting (unless hydrocarbon solvents are involved and the amount of insecticide is well below the toxic amount) or by gastric lavage with saline solution; saline cathartic may also be beneficial; fats and oils should be avoided; sedation with barbiturates is indicated if signs of CNS irritation are present; patient should have absolute quiet, expert nursing care, and a minimum of external stimuli to reduce danger of convulsions; epinephrine is contraindicated in view of the danger of precipitating ventricular fibrillation; if material ingested was dissolved in a hydrocarbon solvent, observe patient for possible development of hydrocarbon pneumonitis; *Toxicity by Inhalation (Threshold Limit Value)*: 0.1 mg/m^3; *Short-Term Inhalation Limits*: 0.5 mg/m^3 for 30 min.; *Toxicity by Ingestion*: Grade 4, oral LD_{50} = 3 mg/kg (rat); *Late Toxicity*: None known; *Vapor (Gas) Irritant Characteristics*: Data not available; *Liquid or Solid Irritant Characteristics*: Data not available; *Odor Threshold*: Not pertinent (solid).

Fire Hazards - *Flash Point* : Non flammable sold or combustible solution > 80 OC (xylene); *Flammable Limits in Air (%):* 1.1 - 7 (xylene); *Fire Extinguishing Agents:* (Solution) Dry chemical, foam, carbon dioxide; *Fire Extinguishing Agents Not To Be Used:* Water may be ineffective on solution fire; *Special Hazards of Combustion Products:* Toxic hydrogen chloride and phosgene may be generated when solution burns; *Behavior in Fire:* Not pertinent; *Ignition Temperature :* Not pertinent; *Electrical Hazard:* Not pertinent; *Burning Rate:* 4 mm/min.

Chemical Reactivity - *Reactivity with Water:* No reaction; *Reactivity with Common Materials:* No reaction; *Stability During Transport:* Stable; *Neutralizing Agents for Acids and Caustics:* Not pertinent; *Polymerization:* Not pertinent; *Inhibitor of Polymerization* Not pertinent.

EPICHLOROHYDRIN

Chemical Designations - *Synonyms*: 1,Chloro-2,3-epoxypropane; Chloromethyloxirane; gamma-Chloropropylene oxide; 3-Chloro-1,2-propylene oxide; *Chemical Formula*: O·CH$_2$·CH·CH$_2$Cl.

Observable Characteristics - *Physical State (as shipped)*: Liquid; *Color*: Colorless; *Odor*: Pungent, garlic; sweet, pungent; like chloroform.

Physical and Chemical Properties - *Physical State at 15 °C and 1 atm.*: Liquid; *Molecular Weight*: 92.53; *Boiling Point at 1 atm.*: 239.4, 115.2, 388.4; *Freezing Point*: -72.6, -58.1, 215.1; *Critical Temperature*: Not pertinent; *Critical Pressure*: Not pertinent; *Specific Gravity*: 1.18 at 20 °C (liquid);

Vapor (Gas) Specific Gravity: Not pertinent; *Ratio of Specific Heats of Vapor (Gas)*: 1.155; *Latent Heat of Vaporization*: 176, 97.9, 4.10; *Heat of Combustion*: –8143, –4524, –189.4; *Heat of Decomposition*: Not pertinent.

Health Hazards Information - *Recommended Personal Protective Equipment*: Air pack or organic canister mask; protective gloves and goggles; *Symptoms Following Exposure*: Vapor is irritating to eyes, nose, and throat; may cause headache, nausea, vomiting, and central nervous system depression. Rapidly fatal if swallowed, i.e. nausea, vomiting, and collapse. Skin contact is irritating; *General Treatment for Exposure*: INHALATION: remove victim to fresh air, keep him warm and quiet, and get medical attention immediately; if breathing stops, start artificial respiration. INGESTION: induce vomiting (but only if victim is conscious and without convulsions) and call a physician promptly; no specific antidote known. EYES OR SKIN: immediately flush with water for at least 15 min. and get medical attention; remove contaminated clothing and wash before reuse; *Toxicity by Inhalation (Threshold Limit Value)*: 5 ppm; *Short-Term Inhalation Limits*: 10 ppm for 30 min.; *Toxicity by Ingestion*: Grade 3, LD_{50} 50 to 500 mg/kg; *Late Toxicity*: Causes cancer in experimental animals; *Vapor (Gas) Irritant Characteristics*: Vapor is moderately irritating such that personnel will not usually tolerate moderate or high vapor concentrations; *Liquid or Solid Irritant Characteristics*: Fairly severe skin irritant. May cause pain and second-degree burns after a few minutes contact; *Odor Threshold*: 10 ppm.

Fire Hazards - *Flash Point (deg. F)*: 92 OC; 100 CC; *Flammable Limits in Air (%)*: 3.8 - 21.0; *Fire Extinguishing Agents:* Alcohol foam, dry chemical, carbon dioxide, water spray; *Fire Extinguishing Agents Not To Be Used:* Avoid use of dry chemical if fire occurs in container with confined vent; *Special Hazards of Combustion Products:* Toxic irritating vapors are generated when heated; *Behavior in Fire:* Containers may explode in fire because of polymerization; *Ignition Temperature (deg. F):* 804; *Electrical Hazard:* Not pertinent; *Burning Rate:* 2.6 mm/min.

Chemical Reactivity - *Reactivity with Water:* Mild reaction; not likely to be hazardous; *Reactivity with Common Materials:* No reaction; *Stability During Transport:* Stable; *Neutralizing Agents for Acids and Caustics:* Not pertinent; *Polymerization:* Can polymerize in presence of strong acids and bases, particularly when hot; *Inhibitor of Polymerization:* None used.

ETHANE

Chemical Designations - *Synonyms*: Methylmethane; *Chemical Formula*: C_2H_6.

Observable Characteristics - *Physical State (as normally shipped)*: Liquid or compressed gas; *Color*: Colorless; *Odor*: Weak, sweetish.

Physical and Chemical Properties - *Physical State at 15 °C and 1 atm.*: Gas; *Molecular Weight*: 30.07; *Boiling Point at 1 atm.*: -127.5, -88.6, 264.6; *Freezing Point*: -279.9, -183.3, 89.9; *Critical Temperature*: 90.1, 32.3, 305.5; *Critical Pressure*: 708.0, 48.16, 4.879; *Specific Gravity*: 0.546 at -88.6°C (liquid); *Vapor (Gas) Density*: 1.1; *Ratio of Specific Heats of Vapor (Gas)*: 1.191; *Latent Heat of Vaporization*: 211, 117, 4.90; *Heat of Combustion*: -20.293, -11.274, -472.02; *Heat of Decomposition*: Not pertinent.

Health Hazards Information - *Recommended Personal Protective Equipment*: Self-contained breathing apparatus for high vapor concentrations; *Symptoms Following Exposure*: In high vapor concentrations, can act as simple asphyxiant. Liquid causes severe frostbite; *General Treatment for Exposure*: Remove from Exposure, support respiration; *Toxicity by Inhalation (Threshold Limit Value)*: Not pertinent; *Short-Term Exposure Limits*: Not pertinent; *Toxicity by Ingestion*: Not pertinent; *Late Toxicity*: None; *Vapor (Gas) Irritant Characteristics*: Vapors are nonirritating to the eyes and throat; *Liquid or Solid Irritant Characteristics*: Not pertinent; appreciable hazard. Practically harmless to the skin because is very volatile and evaporates quickly; *Odor Threshold*: 899 ppm.

Fire Hazards - *Flash Point (deg. F)*: -211; *Flammable Limits in Air (%):* 2.9 - 13.0; *Fire Extinguishing Agents:* Stop flow of gas; *Fire Extinguishing Agents Not To Be Used:* Data not available; *Special Hazards of Combustion Products:* Not pertinent; *Behavior in Fire:* Not pertinent; *Ignition Temperature (deg. F):* 940; *Electrical Hazard:* Class I, Group D; *Burning Rate:* 7.3 mm/min.

Chemical Reactivity - *Reactivity with Water:* No reaction; *Reactivity with Common Materials:* No

reaction; *Stability During Transport:* Stable; *Neutralizing Agents for Acids and Caustics:* Not pertinent; *Polymerization:* Not pertinent; *Inhibitor of Polymerization:* Not pertinent.

ETHOXYLATED DODECANOL

Chemical Designations - *Synonyms*: Ethoxylated dodecyl alcohol, Ethoxylated lauryl alcohol, Poly(oxyethyl) dodecyl ether, Poly(oxyethyl) lauryl ether, Tergitol Nonionic TMN; *Chemical Formula*: $C_{12}H_{25}O(CH_2CH_2O)_nCH_2\,CH_2OH$ n=6-10 (average).

Observable Characteristics - *Physical State (as normally shipped)*: Liquid; *Color*: Colorless; *Odor*: Mild and pleasant.

Physical and Chemical Properties - *Physical State at 15 °C and 1 atm.*: Liquid; *Molecular Weight*: 450-626; *Boiling Point at 1 atm.*: Very high; *Freezing Point*: 61, 16, 289; *Critical Temperature*: Not pertinent; *Critical Pressure*: Not pertinent; *Specific Gravity*: 1.02 at 20°C (liquid); *Vapor (Gas) Density*: Not pertinent; *Ratio of Specific Heats of Vapor (Gas)*: Not pertinent; *Latent Heat of Vaporization*: Not pertinent; *Heat of Combustion*:(est.) -11,200, -6200, -260; *Heat of Decomposition*: Not pertinent.

Health Hazards Information - *Recommended Personal Protective Equipment*: Plastic gloves, goggles; *Symptoms Following Exposure*: Liquid causes eye injury and de-fats the skin, causing irritation; *General Treatment for Exposure*: Flush eyes with water for at least 15 min. Wash skin well with water. Get medical attention; *Toxicity by Inhalation (Threshold Limit Value)*: Not pertinent; *Short-Term Exposure Limits*: Not pertinent; *Toxicity by Ingestion*: Grade 1; 5 - 15 g/J/kg (rat); *Late Toxicity*: Data not available; *Vapor (Gas) Irritant Characteristics*: none; *Liquid or Solid Irritant Characteristics*: Liquid causes injury. Contact with skin may cause irritation; *Odor Threshold*: Not pertinent.

Fire Hazards - *Flash Point (deg. F)*: 470 OC; *Flammable Limits in Air (%):* Not pertinent; *Fire Extinguishing Agents:* Dry chemical, carbon dioxide, or alcohol foam; *Fire Extinguishing Agents Not To Be Used:* Not pertinent; *Special Hazards of Combustion Products:* Not pertinent; *Behavior in Fire:* Not pertinent; *Ignition Temperature :* Data not available; *Electrical Hazard:* Not pertinent; *Burning Rate:* Data not available.

Chemical Reactivity - *Reactivity with Water:* No reaction; *Reactivity with Common Materials:* No reaction; *Stability During Transport:* Stable; *Neutralizing Agents for Acids and Caustics:* Not pertinent; *Polymerization:* Not pertinent; *Inhibitor of polymerization:* Not pertinent.

ETHOXYLATED NONYLPHENOL

Chemical Designations - *Synonyms*: No common synonyms; *Chemical Formula*: $C_9H_{19}C_6H_4O(C_2H_4O)_nH$.

Observable Characteristics - *Physical State (as normally shipped)*: Liquid or solid; *Color*: White; *Odor*: Mild aromatic.

Physical and Chemical Properties - *Physical State at 15 °C and 1 atm.*: Solid or liquid; *Molecular Weight*: >500; *Boiling Point at 1 atm.*: Not pertinent; (decomposes); *Freezing Point*: Not pertinent; *Critical Temperature*: Not pertinent; *Critical Pressure*: Not pertinent; *Specific Gravity*: 0.99 - 1.07 at 25°C (liquid); *Vapor (Gas) Density*: Not pertinent; *Ratio of Specific Heats of Vapor (Gas)*: Not pertinent; *Latent Heat of Vaporization*: Not pertinent; *Heat of Combustion*: Data not available; *Heat of Decomposition*: Not pertinent.

Health Hazards Information - *Recommended Personal Protective Equipment*: Gloves and safety glasses; *Symptoms Following Exposure*: Contact with eyes causes irritation. Prolonged contact with skin causes irritation; *General Treatment for Exposure*: Not pertinent; treatment required for inhalation or ingestion. EYES: flush with copious quantities of tap water for 15 win. and seek appropriate medical attention. SKIN: wash affected areas with soap and water; *Toxicity by Inhalation (Threshold Limit Value)*: Data not available; *Short-Term Exposure Limits*: Data not available; *Toxicity by Ingestion*: Grade 2; oral LD_{50}=11.310 mg/kg (rat); *Late Toxicity*: Data not available; *Vapor (Gas) Irritant Characteristics*: Data not available; *Liquid or Solid Irritant Characteristics*: Data not available; *Odor Threshold*: Data not available.

Fire Hazards - *Flash Point (deg. F)*: (burns with difficulty) 338 - 600 OC; > 140 CC; *Flammable*

Limits in Air (%): Not pertinent; *Fire Extinguishing Agents:* Water, foam, carbon dioxide; *Fire Extinguishing Agents Not To Be Used:* Data not available; *Special Hazards of Combustion Products:* Data not available; *Behavior in Fire:* data not available; *Ignition Temperature :* Data not available; *Electrical Hazard:* Not pertinent; *Burning Rate:* Not pertinent.

Chemical Reactivity - *Reactivity with Water:* No reaction; *Reactivity with Common Materials:* No reaction; *Stability During Transport:* Stable; *Neutralizing Agents for Acids and Caustics:* Not pertinent; *Polymerization:* Not pertinent; *Inhibitor of polymerization:* Not pertinent.

ETHOXYLATED PENTADECANOL

Chemical Designations - *Synonyms*: Ethoxylated pentadecyl alcohol, Poly(oxyethyl) pentadecyl ether, Terrgitol nonionic 45-S-10; *Chemical Formula*: $C_{13}H_{31}O(CH_2CH_2O)_nCH_2CH_2OH$ n=10 (average).

Observable Characteristics - *Physical State (as normally shipped)*: Liquid; *Color*: Colorless; *Odor*: Mild, pleasant.

Physical and Chemical Properties - *Physical State at 15 °C and 1 atm.*: Liquid; *Molecular Weight*: 660; *Boiling Point at 1 atm.*: Very high; *Freezing Point*: 59, 15, 288; *Critical Temperature*: Not pertinent; *Critical Pressure*: Not pertinent; *Specific Gravity*: 1.007 at 15°C (liquid); *Vapor (Gas) Density*: Not pertinent; *Ratio of Specific Heats of Vapor (Gas)*: Not pertinent; *Latent Heat of Vaporization*: Not pertinent; *Heat of Combustion*: (est.) -11,000, -6200, -260; *Heat of Decomposition*: Not pertinent.

Health Hazards Information - *Recommended Personal Protective Equipment*: Plastic gloves, goggles; *Symptoms Following Exposure*: Liquid causes eye injury and de-fats the skin, causing irritation; *General Treatment for Exposure*: Flush eyes with water for at least 15 min. Wash the skin well with water. Get medical attention; *Toxicity by Inhalation (Threshold Limit Value)*: Not pertinent; *Short-Term Exposure Limits*: Not pertinent; *Toxicity by Ingestion*: Data not available; *Late Toxicity*: Data not available; *Vapor (Gas) Irritant Characteristics*: None; *Liquid or Solid Irritant Characteristics*: Liquid causes eye injury. Contact with skin may cause irritation; *Odor Threshold*: Not pertinent.

Fire Hazards - *Flash Point (deg. F)*: 470 OC; *Flammable Limits in Air (%):* Not pertinent; *Fire Extinguishing Agents:* Carbon dioxide or dry chemical for small fires; alcohol foam and water for large fires; *Fire Extinguishing Agents Not To Be Used:* Not pertinent; *Special Hazards of Combustion Products:* Not pertinent; *Behavior in Fire:* Not pertinent; *Ignition Temperature :* data not available; *Electrical Hazard:* Not pertinent; *Burning Rate:* Data not available.

Chemical Reactivity - *Reactivity with Water:* No reaction; *Reactivity with Common Materials:* No reaction; *Stability During Transport:* Stable; *Neutralizing Agents for Acids and Caustics:* Not pertinent; *Polymerization:* Not pertinent; *Inhibitor of Polymerization:* Not pertinent.

ETHOXYLATED TETRADECANOL

Chemical Designations - *Synonyms*: Ethoxylated myristyl alcohol, Ethoxylated tetradecyl alcohol, Poly(oxyethil) myristyl ether, Poly(oxyethil) tetradecyl ether, tergtol Nonionic 45-S-10; *Chemical Formula*: $C_{14}H_{29}O (CH2CH2O)_n CH2CH2OH$; n=5 (average).

Observable Characteristics - *Physical State (as normally shipped)*: Liquid; *Color*: Colorless; *Odor*: Mild, pleasant.

Physical and Chemical Properties - *Physical State at 15 °C and 1 atm.*: Liquid; *Molecular Weight*: 660; *Boiling Point at 1 atm.*: Very high; *Freezing Point*: 59, 15, 288; *Critical Temperature*: Not pertinent; *Critical Pressure*: Not pertinent; *Specific Gravity*: 1.007 at 15°C (liquid); *Vapor (Gas) Density*: Not pertinent; *Ratio of Specific Heats of Vapor (Gas)*: Not pertinent; *Latent Heat of Vaporization*: Not pertinent; *Heat of Combustion*: (est.) -11,000, -6200, -260; *Heat of Decomposition*: Not pertinent.

Health Hazards Information - *Recommended Personal Protective Equipment*: Plastic gloves, goggles; *Symptoms Following Exposure*: Liquid causes eye injury and de-fats the skin, causing irritation; *General Treatment for Exposure*: Flush eyes with water for at least 15 min. Wash the skin well with water. Get medical attention; *Toxicity by Inhalation (Threshold Limit Value)*: Not pertinent; *Short-Term Exposure Limits*: Not pertinent; *Toxicity by Ingestion*: Data not available; *Late Toxicity*: Data not available; *Vapor (Gas) Irritant Characteristics*: None; *Liquid or Solid Irritant Characteristics*: Liquid causes eye injury. Contact with skin may cause irritation; *Odor Threshold*: Not pertinent.

Fire Hazards - *Flash Point (deg. F)*: 470 OC; *Flammable Limits in Air (%):* Not pertinent; *Fire Extinguishing Agents:* Carbon dioxide or dry chemical for small fires; alcohol foam and water for large fires; *Fire Extinguishing Agents Not To Be Used:* Not pertinent; *Special Hazards of Combustion Products:* Not pertinent; *Behavior in Fire:* Not pertinent; *Ignition Temperature:* Data not available; *Electrical Hazard:* Not pertinent; *Burning Rate:* Data not available.

Chemical Reactivity - *Reactivity with Water:* No reaction; *Reactivity with Common Materials:* No reaction; *Stability During Transport:* Stable; *Neutralizing Agents for Acids and Caustics:* Not pertinent; *Polymerization:* Not pertinent; *Inhibitor of Polymerization:* Not pertinent.

ETHYL ACETATE

Chemical Designations - *Synonyms*: Acetic acid, ethyl ester, Acetic ester, Acetic ether, Ethyl ethanoate; *Chemical Formula*: $CH_3COOCH_2CH_3$

Observable Characteristics - *Physical State (as normally shipped)*: Liquid; *Color*: Colorless; *Odor*: Pleasant, fruity.

Physical and Chemical Properties - *Physical State at 15 °C and 1 atm.*: Liquid; *Molecular Weight*: 88.11; *Boiling Point at 1 atm.*: 171, 77, 350; *Freezing Point*: -117, -83, 190; *Critical Temperature*: 482, 250, 523; *Critical Pressure*: 558, 38, 3.8; *Specific Gravity*: 0.902 at 20°C (liquid); *Vapor (Gas) Density*: 3.0; *Ratio of Specific Heats of Vapor (Gas)*: 1.080; *Latent Heat of Vaporization*: 158, 87.6, 3.67; *Heat of Combustion*: -10,110, -5,616, -235,1; *Heat of Decomposition*: Not pertinent.

Health Hazards Information - *Recommended Personal Protective Equipment*: Organic vapor canister, or air mask; goggles or face shield; *Symptoms Following Exposure*: Head ache, irritation of respiratory passages and eyes, dizziness and nausea, weakness, loss of consciousness; *General Treatment for Exposure*: INHALATION: if victim is overcome, move him to fresh air immediately and call a physician; if breathing is irregular or stopped, star resuscitation and administer oxygen. EYES: flush with water for at least 15 min.; *Toxicity by Inhalation (Threshold Limit Value)*: 400 ppm; *Short-Term Exposure Limits*: 1000 ppm for 15 min.; *Toxicity by Ingestion*: Grade 2; LD_{50} 0.5 to 5 g/kg; *Late Toxicity*: Data not available; *Vapor (Gas) Irritant Characteristics*: Vapor cause a slight smarting of the eyes or respiratory system if present in high concentration. The effect is temporary; *Liquid or Solid Irritant Characteristics*: Minimum hazard. If spilled on clothing and allowed to remain, may cause smarting and reddening of the skin; *Odor Threshold*: 1 ppm.

Fire Hazards - *Flash Point (deg. F)*: 24 CC; 55 OC; *Flammable Limits in Air (%):* 2.2 - 9.0; *Fire Extinguishing Agents:* Alcohol foam, carbon dioxide, or dry chemical; *Fire Extinguishing Agents Not To Be Used:* Not pertinent; *Special Hazards of Combustion Products:* Not pertinent; *Behavior in Fire:* Not pertinent; *Ignition Temperature (deg. F):* 800; *Electrical Hazard:* Class I, Group D; *Burning Rate:* 3.7 mm/min.

Chemical Reactivity - *Reactivity with Water:* No reaction; *Reactivity with Common Materials:* No reaction; *Stability During Transport:* Stable; *Neutralizing Agents for Acids and Caustics:* Not pertinent; *Polymerization:* Not pertinent; *Inhibitor of Polymerization:* Not pertinent.

ETHYL ACETOACETATE

Chemical Designations - *Synonyms*: Acetoecetic acid, ethyl ester, Acetoacetic ester; Diacetic ether; *Chemical Formula*: $CH_3COCH_2COOC_2H_5$.

Observable Characteristics - *Physical State (as normally shipped)*: Liquid; *Color*: Colorless; *Odor*: Agreeable, fruity.

Physical and Chemical Properties - *Physical State at 15 °C and 1 atm.*: Liquid; *Molecular Weight*: 130.1; *Boiling Point at 1 atm.*: 363, 184, 457; *Freezing Point*: <-112, <-80, <193; *Critical Temperature*: Not pertinent; *Critical Pressure*: Not pertinent; *Specific Gravity*: 1.028 at 20°C (liquid); *Vapor (Gas) Density*: 4.48; *Ratio of Specific Heats of Vapor (Gas)*: Not pertinent; *Latent Heat of Vaporization*: 160, 91, 3.8; *Heat of Combustion*: -9.349, -5.194, -217.3; *Heat of Decomposition*: Not pertinent.

Health Hazards Information - *Recommended Personal Protective Equipment*: Goggles or face shield; rubber gloves; *Symptoms Following Exposure*: Liquid may cause mild irritation of eyes; *General Treatment for Exposure*: EYES: flush with water for 15 min.; *Toxicity by Inhalation (Threshold Limit Value)*: Data not available; *Short-Term Exposure Limits*: Data not available; *Toxicity by Ingestion*: Grade 2; oral LD_{50} = 3.980 mg/kg (rat); *Late Toxicity*: Data not available; *Vapor (Gas) Irritant Characteristics*: Data not available; *Liquid or Solid Irritant Characteristics*: Data not available; *Odor Threshold*: Data not available.

Fire Hazards - *Flash Point (deg. F)*: 176 OC; 135 CC; *Flammable Limits in Air (%):* 1.4 - 9.5; *Fire Extinguishing Agents:* Dry chemical, alcohol foam, carbon dioxide; *Fire Extinguishing Agents Not To Be Used:* Water may be ineffective; *Special Hazards of Combustion Products:* Not pertinent; *Behavior in Fire:* Not pertinent; *Ignition Temperature (deg. F):* 563; *Electrical Hazard:* Data not available; *Burning Rate:* 2.4 mm/min.

Chemical Reactivity - *Reactivity with Water:* No reaction; *Reactivity with Common Materials:* No reaction; *Stability During Transport:* Stable; *Neutralizing Agents for Acids and Caustics:* Not pertinent; *Polymerization:* Not pertinent; *Inhibitor of Polymerization:* Not pertinent.

ETHYL ACRYLATE

Chemical Designations - *Synonyms*: Acrylic acid, Ethyl ester; Ethyl 2-propenoate; *Chemical Formula*: $CH_2 = CHCOOCH_2CH_3$.

Observable Characteristics - *Physical State (as normally shipped)*: Liquid; *Color*: Colorless; *Odor*: Characteristic acrylic odor; sharp, fragrant; acrid; slightly nauseating; sharp, ester type.

Physical and Chemical Properties - *Physical State at 15 °C and 1 atm.*: Liquid; *Molecular Weight*: 100.12; *Boiling Point at 1 atm.*: 211.3, 99.6, 372.8; *Freezing Point*: -98, -72, 201; *Critical Temperature*: 534, 279, 552; *Critical Pressure*: 534, 37, 3.7; *Specific Gravity*: 0.923 at 20°C (liquid); *Vapor (Gas) Density*: Not pertinent; *Ratio of Specific Heats of Vapor (Gas)*: 1.080; *Latent Heat of Vaporization*: 149, 82.9, 3.47; *Heat of Combustion*: -11,880, -6,600, -276.3; *Heat of Decomposition*: Not pertinent.

Health Hazards Information - *Recommended Personal Protective Equipment*: Organic canister or air-supplied mask; acid goggles; impervious gloves; *Symptoms Following Exposure*: May cause irritation and burns of eyes and skin. Exposure to excessive vapor concentration can also cause drowsiness accompanied by nausea, headache, or extreme irritation of respiratory tract; *General Treatment for Exposure*: INHALATION: remove victim to fresh air and administer artificial respiration, if necessary. SKIN AND EYES: wash for 15 min. with copious quantities of water. Call a physician; *Toxicity by Inhalation (Threshold Limit Value)*: 25 ppm; *Short-Term Exposure Limits*: 50 ppm for 15 min.; *Toxicity by Ingestion*: 25 ppm; *Late Toxicity*: Repeated Exposure may develop sensitivity; *Vapor (Gas) Irritant Characteristics*: Vapor is moderately irritating such that personnel will not usually tolerate moderate or high vapor concentrations; *Liquid or Solid Irritant Characteristics*: causes smarting of the skin and first-degree burns on short Exposure and may cause secondary burns on long Exposure; *Odor Threshold*: 0.00024 ppm.

Fire Hazards - *Flash Point (deg. F)*: 44 OC; *Flammable Limits in Air (%)*: 1.8 - 9.5; *Fire Extinguishing Agents:* Dry chemical, foam, or carbon dioxide; *Fire Extinguishing Agents Not To Be Used:* Not pertinent; *Special Hazards of Combustion Products:* Toxic and irritating vapors are generated when heated; *Behavior in Fire:* Vapor is heavier than air and may travel considerable distance to a source of ignition and flash back. May polymerize and cause container to explode; *Ignition Temperature (deg. F):* 721; *Electrical Hazard:* Data not available; *Burning Rate:* 4.3 mm/min.

Chemical Reactivity - *Reactivity with Water:* No reaction; *Reactivity with Common Materials:* No reaction; *Stability During Transport:* Stable; *Neutralizing Agents for Acids and Caustics:* Not pertinent; *Polymerization:* May occur; exclude moisture, light; avoid exposure to high temperatures; store in presence of air; *Inhibitor of Polymerization:* 13 - 17 ppm monomethyl ether of hydroquinone.

ETHYL ALCOHOL

Chemical Designations - *Synonyms*: Alcohol; Cologne spirit; Denatured alcohol; Ethanol; Fermentation alcohol; Grain alcohol; Spirit; Spirits of wine; *Chemical Formula*: C_2H_5OH.

Observable Characteristics - *Physical State (as normally shipped)*: Liquid; *Color*: Colorless; *Odor*: Mild, pleasant; Like wine or whiskey. (Denatured alcohol may be unpleasant).

Physical and Chemical Properties - *Physical State at 15 °C and 1 atm.*: Liquid; *Molecular Weight*: 46.07; *Boiling Point at 1 atm.*: 172.9, 78.3, 351.5; *Freezing Point*: -173, -114, 159; *Critical Temperature*: 469.6, 243.1, 516.3; *Critical Pressure*: 926, 63.0, 6.38; *Specific Gravity*: 0.790 at 20°C (liquid); *Vapor (Gas) Density*: 1.6; *Ratio of Specific Heats of Vapor (Gas)*: 1.128; *Latent Heat of Vaporization*: 360, 200, 8.37; *Heat of Combustion*: -11,570, 6,425, -268.8; *Heat of Decomposition*: Not pertinent.

Health Hazards Information - *Recommended Personal Protective Equipment*: All-purpose canister; safety goggles. Avoid contact with liquid and irritation of vapors; *Symptoms Following Exposure*: Irritation of eyes, nose and throat. Headache and drowsiness may occur. Liquid causes intoxication; *General Treatment for Exposure*: INHALATION: if breathing is affected, remove victim to the fresh air; call physician; administer oxygen. Speed is of primary importance. EYES OR SKIN: flush with water; *Toxicity by Inhalation (Threshold Limit Value)*: 1000 ppm; *Short-Term Exposure Limits*: 5000 ppm for 15 min.; *Toxicity by Ingestion*: Grade 1; LD_{50} 5 to 15 g/kg; *Late Toxicity*: None; *Vapor (Gas) Irritant Characteristics*: Vapors cause a slight smarting of the eyes or respiratory system if present in high concentrations. The effect is temporary; *Liquid or Solid Irritant Characteristics*: No appreciable hazard. Practically harmless to the skin; *Odor Threshold*: 10 ppm.

Fire Hazards - *Flash Point (deg. F)*: 55 CC; 64 OC; *Flammable Limits in Air (%)*: 3.3 - 19; *Fire Extinguishing Agents:* Carbon dioxide, dry chemical, water spray, alcohol foam; *Fire Extinguishing Agents Not To Be Used:* None; *Special Hazards of Combustion Products:* None; *Behavior in Fire:* Not pertinent; *Ignition Temperature (deg. F):* 689; *Electrical Hazard:* Class I, Group D; *Burning Rate:* 3.9 mm/min.

Chemical Reactivity - *Reactivity with Water:* No reaction; *Reactivity with Common Materials:* No reaction; *Stability During Transport:* Stable; *Neutralizing Agents for Acids and Caustics:* Not pertinent; *Polymerization:* Not pertinent; *Inhibitor of Polymerization:* Not pertinent.

ETHYLALUMINUM DICHLORIDE

Chemical Designations - *Synonyms*: Aluminum ethyl dichloride; EADC; *Chemical Formula*: $C_2H_5AICI_2$.

Observable Characteristics - *Physical State (as normally shipped)*: Liquid; *Color*: Colorless to light amber; yellow; *Odor*: Not pertinent.

Physical and Chemical Properties - *Physical State at 15 °C and 1 atm.*: Solid; *Molecular Weight*: 130.0; *Boiling Point at 1 atm.*: 381, 194, 467; *Freezing Point*: 90, 32, 305; *Critical Temperature*: Not pertinent; *Critical Pressure*: Not pertinent; *Specific Gravity*: 1.227 at 35°C (liquid); *Vapor (Gas) Density*: Not pertinent; *Ratio of Specific Heats of Vapor (Gas)*: Not pertinent; *Latent Heat of Vaporization*: Not pertinent; *Heat of Combustion*: (est.) -5,600, -3,100, -130; *Heat of Decomposition*: Not Pertinent.

Health Hazards Information - *Recommended Personal Protective Equipment*: Full protective clothing, preferably of aluminized glass cloth; goggles, face shield, gloves; in case of fire, all-purpose canister or SCBA; *Symptoms Following Exposure*: Inhalation of smoke from fire causes metal-fume fever (flu like symptoms); acid fumes irritate nose and throat. Contact with liquid (which is not spontaneously flammable) causes severe burns of eyes and skin; *General Treatment for Exposure*: INHALATION: only fumes from fire need be considered; metal-fume fever is not critical and lasts less then 36 hrs; irritation of nose and throat by acid vapors may require treatment by a physician. EYES: flush gently with water for 15 min.; treat burns if fire occurred; get medical attention. SKIN: wash with water; treat burns caused by fire; get medical attention; *Toxicity by Inhalation (Threshold Limit Value)*: No data; *Short-Term Exposure Limits*: Not pertinent; *Toxicity by Ingestion*: Data not available; *Late Toxicity*: Metal-fume fever may develop after breathing smoke from fire; *Vapor (Gas) Irritant Characteristics*:

No data; *Liquid or Solid Irritant Characteristics*: Severe skin irritant. Causes second- and third-degree burns on short contact and is very injurious to the eyes; *Odor Threshold*: Not pertinent.

Fire Hazards - *Flash Point*: Ignites spontaneously; *Flammable Limits in Air (%)*: Not pertinent; *Fire Extinguishing Agents*: Dry chemical, inert dry powders such as sand, limestone; *Fire Extinguishing Agents Not To Be Used*: Water, foam, halogenated agents, or carbon dioxide; *Special Hazards of Combustion Products*: Intense smoke may cause metal-fume fever. Irritating hydrogen chloride also formed; *Behavior in Fire*: Contact with water applied to adjacent fires will cause formation of irritating smoke containing aluminum oxide and hydrogen chloride; *Ignition Temperature*: Ignites spontaneously in air at ambient temperature; *Electrical Hazard*: Not pertinent; *Burning Rate*: Not pertinent.

Chemical Reactivity - *Reactivity with Water*: Reacts violently to form hydrogen chloride fumes and flammable ethane gas; *Reactivity with Common Materials*: Reacts with surface moisture to generate hydrogen chloride, which is corrosive to most metals; *Stability During Transport*: Stable; *Neutralizing Agents for Acids and Caustics*: Rinse with sodium bicarbonate or lime solution; *Polymerization*: Not pertinent; *Inhibitor of Polymerization*: Not pertinent.

ETHYLAMINE

Chemical Designations - *Synonyms*: Aminoethane, Monoethylamine; *Chemical Formula*: $C_2H_5NH_2$.

Observable Characteristics - *Physical State (as normally shipped)*: Liquid; *Color*: Colorless; *Odor*: Pungent; strong ammoniacal.

Physical and Chemical Properties - *Physical State at 15 °C and 1 atm.*: Liquid; *Molecular Weight*: 45.1; *Boiling Point at 1 atm.*: 61.7, 16.5; 289.7; *Freezing Point*: -114, -81, 192; *Critical Temperature*: 361, 183, 456; *Critical Pressure*: 827, 56.2, 5.70; *Specific Gravity*: 0.687 at 15°C (liquid); *Vapor (Gas) Density*: 1.5; *Ratio of Specific Heats of Vapor (Gas)*: 1.1181; *Latent Heat of Vaporization*: 253, 146, 6.11; *Heat of Combustion*: -16,180, -8,990, -376; *Heat of Decomposition*: Not pertinent.

Health Hazards Information - *Recommended Personal Protective Equipment*: Amine-type or ammonia type mask; plastic gloves; face shield and goggles; *Symptoms Following Exposure*: Inhalation causes irritation of respiratory tract and lungs; pulmonary edema may result, liquid causes severe irritation and burn of eyes and skin, and can permanently injure eyes after 15 seconds contact. Ingestion causes severe burns of mouth and stomach; can be fatal; *General Treatment for Exposure*: Get prompt medical attention for anyone overcome or injured by Exposure to this compound. INHALATION: remove victim to fresh air, keep him warm, and administer oxygen until medical help arrives. EYES: wash for 15 min. with water; avoid pressure on eyelids. SKIN: wash with soap and water; do not use ointments for at least 24 hrs; do not cover burned area with dry clothing; keep moist with physiological saline solution. INGESTION: if victim is conscious, give large amount of water, then induce vomiting; *Toxicity by Inhalation (Threshold Limit Value)*: 10 ppm; *Short-Term Exposure Limits*: 25 ppm for 30 min.; *Toxicity by Ingestion*: Grade 3; LD_{50} = 400 mg/kg (rat); *Late Toxicity*: None; *Vapor (Gas) Irritant Characteristics*: Vapors are moderately irritating such that personnel will not usually tolerate moderate or high concentrations; *Liquid or Solid Irritant Characteristics*: Causes smarting of the skin and first-degree burns on short Exposure and may cause second- degree burns on long Exposure; *Odor Threshold*: Data not available.

Fire Hazards - *Flash Point (deg. F)*: 0 OC; *Flammable Limits in Air (%)*: 3.5 - 14; *Fire Extinguishing Agents*: Dry chemical, carbon dioxide, alcohol foam; *Fire Extinguishing Agents Not To Be Used*: Water may be ineffective; *Special Hazards of Combustion Products*: Irritating and toxic oxides of nitrogen may be formed; *Behavior in Fire*: Vapor is heavier than fire and may travel a considerable distance to a source of ignition and flash back. Containers may explode when heated; *Ignition Temperature (deg. F)*: 724; *Electrical Hazard*: Data not available; *Burning Rate*: 5.0 mm/min.

Chemical Reactivity - *Reactivity with Water*: No reaction; *Reactivity with Common Materials*: Will strip and dissolve paint; dissolves most plastic materials; can cause swelling of rubber by absorption. The reactions are not hazardous; *Stability During Transport*: Stable; *Neutralizing Agents for Acids and Caustics*: Flush with water; *Polymerization*: Not pertinent; *Inhibitor of Polymerization*: Not pertinent.

ETHYLBENZENE

Chemical Designations - *Synonyms*: EB; Phenylethane; *Chemical Formula*: $C_6H_5CH_2CH_3$.

Observable Characteristics - *Physical State (as normally shipped)*: Liquid; *Color*: Colorless; *Odor*: Aromatic.

Physical and Chemical Properties - *Physical State at 15 °C and 1 atm.*: Liquid; *Molecular Weight*: 106.17; *Boiling Point at 1 atm.*: 277.2, 136.2, 409.4; *Freezing Point*: -139, -95, 178; *Critical Temperature*: 651.0, 343.9, 617.1; *Critical Pressure*: 523, 35.6, 3.61; *Specific Gravity*: 0.867 at 20°C (liquid); *Vapor (Gas) Density*: Not pertinent; *Ratio of Specific Heats of Vapor (Gas)*: 1.071; *Latent Heat of Vaporization*: 144, 80.1, 3.35; *Heat of Combustion*: -17,780, -9,877, -413.5; *Heat of Decomposition*: Not pertinent.

Health Hazards Information - *Recommended Personal Protective Equipment*: Self-contained breathing apparatus; safety goggles; *Symptoms Following Exposure*: Inhalation may cause irritation of nose, dizziness, depression. Moderate irritation of eye with corneal injury possible. Irritates skin and may cause blisters; *General Treatment for Exposure*: INHALATION: remove victim to fresh air, keep him warm, and get medical help promptly; if breathing stops give artificial respiration. INGESTION: induce vomiting only upon physician's approval; material in lung may cause chemical pneumonitis. SKIN AND EYES: wash for 15 min. with water and get medical attention; remove and wash contaminating clothing before use; *Toxicity by Inhalation (Threshold Limit Value)*: 100 ppm; *Short-Term Exposure Limits*: 200 ppm for 30 min.; *Toxicity by Ingestion*: Grade 2; LD_{50} 0.5 to 5 g/kg (rat); *Late Toxicity*: Data not available; *Vapor (Gas) Irritant Characteristics*: Vapors are moderately irritating such that personnel; *Liquid or Solid Irritant Characteristics*: Causes smarting of the skin and first-degree burns on short Exposure and may cause secondary burns on long Exposure; *Odor Threshold*: 140 ppm.

Fire Hazards - *Flash Point (deg. F)*: 80 OC; 59 CC; *Flammable Limits in Air (%):* 1.0 - 6.7; *Fire Extinguishing Agents:* Foam(most effective), water fog, carbon dioxide or dry chemical; *Fire Extinguishing Agents Not To Be Used:* Not pertinent; *Special Hazards of Combustion Products:* Irritating vapors are generated when heated; *Behavior in Fire:* Vapor is heavier than air and may travel considerable distance to a source of ignition and flash back; *Ignition Temperature (deg. F):* 860; *Electrical Hazard:* Not pertinent; *Burning Rate:* 5.8 mm/min.

Chemical Reactivity - *Reactivity with Water:* No reaction; *Reactivity with Common Materials:* No reaction; *Stability During Transport:* Stable; *Neutralizing Agents for Acids and Caustics:* Not pertinent; *Polymerization:* Not pertinent; *Inhibitor of Polymerization:* Not pertinent.

ETHYL BUTANOL

Chemical Designations - *Synonyms*: 2-Ethyl-1-butanol; 2-Ethylbutil alcohol; sec-Hexyl alcohol; sec-Pentyl carbinol; Pseudohexyl alcohol; *Chemical Formula*: $(C_2H_5)_2CHCH_2OH$.

Observable Characteristics - *Physical State (as normally shipped)*: Liquid; *Color*: Colorless; *Odor*: Mild and non-residual.

Physical and Chemical Properties - *Physical State at 15 °C and 1 atm.*: Liquid; *Molecular Weight*: 102.17; *Boiling Point at 1 atm.*: 293, 146, 419; *Freezing Point*: -173, -114, 159; *Critical Temperature*: Not pertinent; *Critical Pressure*: Not pertinent; *Specific Gravity*: 0.843 at 20°C (liquid); *Vapor (Gas) Density*: Not pertinent; *Ratio of Specific Heats of Vapor (Gas)*: Not pertinent; *Latent Heat of Vaporization*: 196.0, 108.9, 4.559; *Heat of Combustion*: (est.) -16,660, -9,250, -387; *Heat of Decomposition*: Not pertinent.

Health Hazards Information - *Recommended Personal Protective Equipment*: Fresh-air mask; plastic gloves; coverall goggles; safety shower and eye bath; *Symptoms Following Exposure*: Liquid causes eye burn. Vapors may be mildly irritating to nose and throat; *General Treatment for Exposure*: Remove to fresh air. Remove and wash contaminated clothing. Wash affected skin areas with water. Flush eyes with water for 15 min and get medical care; *Toxicity by Inhalation (Threshold Limit Value)*: Data not available; *Short-Term Exposure Limits*: Data not available; *Toxicity by Ingestion*: Grade 2; LD_{50} 0.5 to 5 g/kg (rat); *Late Toxicity*: Data not available; *Vapor (Gas) Irritant Characteristics*: Vapors cause a slight smarting of the eyes or respiratory system if present in high concentration. The effect is temporary; *Liquid or Solid Irritant Characteristics*: Irritates eyes; moderate irritation of skin; *Odor Threshold*: Data not available.

Fire Hazards - *Flash Point (deg. F)*: 128 OC; *Flammable Limits in Air (%)*: 1.9 - 8.8; *Fire Extinguishing Agents*: Carbon dioxide or dry chemical for small fires, alcohol foam for large fires; *Fire Extinguishing Agents Not To Be Used*: Not pertinent; *Special Hazards of Combustion Products*: Not pertinent; *Behavior in Fire*: Not pertinent; *Ignition Temperature (deg. F)*: 580 (calc.); *Electrical Hazard*: Data not available; *Burning Rate*: Data not available.

Chemical Reactivity - *Reactivity with Water*: No reaction; *Reactivity with Common Materials*: No reaction; *Stability During Transport*: Stable; *Neutralizing Agents for Acids and Caustics*: Not pertinent; *Polymerization*: Not pertinent; *Inhibitor of Polymerization*: Not pertinent.

ETHYL BUTYRATE

Chemical Designations - *Synonyms*: Butyric acid, Ethyl ester; Butyric ether; Ethyl butanoate; *Chemical Formula*: $CH_3CH_2CH_2COOC_2H_5$.

Observable Characteristics - *Physical State (as normally shipped)*: Liquid; *Color*: Colorless; *Odor*: Like apple or pineapple.

Physical and Chemical Properties - *Physical State at 15 °C and 1 atm.*: Liquid; *Molecular Weight*: 116.16; *Boiling Point at 1 atm.*: 250, 121, 394; *Freezing Point*: -135, -93, 180; *Critical Temperature*: 559, 293, 566; *Critical Pressure*: 460, 31, 3.2; *Specific Gravity*: 0.879 at 20°C (liquid); *Vapor (Gas) Density*: 4.0; *Ratio of Specific Heats of Vapor (Gas)*: Not pertinent; *Latent Heat of Vaporization*: 128, 71, 3.0; *Heat of Combustion*: -13,200, -7,330, -306; *Heat of Decomposition*: Not pertinent.

Health Hazards Information - *Recommended Personal Protective Equipment*: All-purpose canister mask or chemical cartridge respirator; glass or face shield; rubber gloves; *Symptoms Following Exposure*: Inhalation or ingestion causes headache, dizziness, nausea, vomiting, and narcosis. Contact with liquid irritates eyes; *General Treatment for Exposure*: INHALATION: remove victim to fresh air and call a physician; if breathing stops give artificial respiration. INGESTION: induce vomiting and call a physician. SKIN: wash with water flush with soap and water; *Toxicity by Inhalation (Threshold Limit Value)*: Data not available; *Short-Term Exposure Limits*: Data not available; *Toxicity by Ingestion*: Grade 1; LD_{50} = 13 g/kg (rat); *Late Toxicity*: Data not available; *Vapor (Gas) Irritant Characteristics*: Data not available; *Liquid or Solid Irritant Characteristics*: Data not available; *Odor Threshold*: 0.015 ppm.

Fire Hazards - *Flash Point (deg. F)*: 85 OC; 75 CC; *Flammable Limits in Air (%)*: Data not available; *Fire Extinguishing Agents*: Dry chemical, alcohol foam, carbon dioxide; *Fire Extinguishing Agents Not To Be Used*: Water may be ineffective; *Special Hazards of Combustion Products*: Data not available; *Behavior in Fire*: Vapor is heavier than fire and may travel considerable distance to a source of ignition and flash back. Containers may explode in fire; *Ignition Temperature (deg. F)*: 865; *Electrical Hazard*: Data not available; *Burning Rate*: 4.72 mm/min.

Chemical Reactivity - *Reactivity with Water*: No reaction; *Reactivity with Common Materials*: May attack some forms of plastics; *Stability During Transport*: Stable; *Neutralizing Agents for Acids and Caustics*: Not pertinent; *Polymerization*: Not pertinent; *Inhibitor of Polymerization*: Not pertinent.

ETHYL CHLORIDE

Chemical Designations - *Synonyms*: Chloretane; Monochloretane; *Chemical Formula*: C_2H_5CL.

Observable Characteristics - *Physical State (as normally shipped)*: Liquid; *Color*: Colorless; *Odor*: Ethereal; pungent, ethereal; ether-like.

Physical and Chemical Properties - *Physical State at 15 °C and 1 atm.*: Gas; *Molecular Weight*: 64.52; *Boiling Point at 1 atm.*: 54.0, 12.2, 285.4; *Freezing Point*: -213, -136; 137; *Critical Temperature*: 369, 187.2, 460.4; *Critical Pressure*: 758, 51.6, 5.23; *Specific Gravity*: 0.906 at 12.2°C (liquid); *Vapor (Gas) Density*: 2.2; *Ratio of Specific Heats of Vapor (Gas)*: 1.155; *Latent Heat of Vaporization*: 163, 90.6, 3.79; *Heat of Combustion*: -8,100, -4,500, -188.4; *Heat of Decomposition*: Not pertinent.

Health Hazards Information - *Recommended Personal Protective Equipment*: Neoprene rubber clothing where liquid contacts is likely; chemical worker's goggles. RESPIRATORY PROTECTION: for 1000 ppm to 2% for ½ hr or less, full face mask and organic vapor canister; for greater levels, self-contained breathing apparatus or equivalent; *Symptoms Following Exposure*: Vapor causes drunkenness, anesthesia, possible lung injury. Liquid may cause frostbite on eyes and skin; *General Treatment for Exposure*: INHALATION: remove victim to fresh air. Get medical attention. SKIN: treat frostbite; *Toxicity by Inhalation (Threshold Limit Value)*: 1000 ppm; *Short-Term Exposure Limits*: Data not available; *Toxicity by Ingestion*: Not pertinent; *Late Toxicity*: Data not available; *Vapor (Gas) Irritant Characteristics*: Vapors cause a slight smarting of the eyes or respiratory system if present in high concentrations. The effect is temporary; *Liquid or Solid Irritant Characteristics*: Minimum hazard. If spilled on clothing and allowed to remain, may cause smarting and reddening of the skin; *Odor Threshold*: Data not available.

Fire Hazards - *Flash Point (deg. F)*: -58 CC; -45 OC; *Flammable Limits in Air (%):* 3.6 - 12; *Fire Extinguishing Agents:* Water fog, carbon dioxide, dry chemical. For large fires it is best to allow material to burn while cooling surrounding equipment. Stop flow of ethyl chloride; *Fire Extinguishing Agents Not To Be Used:* Not pertinent; *Special Hazards of Combustion Products:* Toxic ant irritating gases are generated in fires; *Behavior in Fire:* Containers may explode; *Ignition Temperature (deg. F):* 966; *Electrical Hazard:* Not pertinent; *Burning Rate:* 3.8 mm/min.

Chemical Reactivity - *Reactivity with Water:* No reaction; *Reactivity with Common Materials:* No reaction; *Stability During Transport:* Stable; *Neutralizing Agents for Acids and Caustics:* Not pertinent; *Polymerization:* Not pertinent; *Inhibitor of Polymerization:* Not pertinent.

ETHYL CHLOROACETATE

Chemical Designations - *Synonyms*: Chloracetic acid, ethyl ester, Ethyl chloracetate; Ethyl chloroethanoate; Monochloracetic acid, ethyl ester; Monochlorethanoic acid, ethyl ester; *Chemical Formula*: $ClCH_2COOC_2H_5$.

Observable Characteristics - *Physical State (as normally shipped)*: Liquid; *Color*: Colorless; light straw to tan; *Odor*: Extremely irritating; fruity; pungent.

Physical and Chemical Properties - *Physical State at 15 °C and 1 atm.*: Liquid; *Molecular Weight*: 122.6; *Boiling Point at 1 atm.*: 289, 143, 416; *Freezing Point*: -15, -26, 247; *Critical Temperature*: Not pertinent; *Critical Pressure*: Not pertinent; *Specific Gravity*: 1.15 at 20°C (liquid); *Vapor (Gas) Density*: 4.3; *Ratio of Specific Heats of Vapor (Gas)*: Data not available; *Latent Heat of Vaporization*: 155, 86, 3.6; *Heat of Combustion*: -7,250, -4,028, 168; *Heat of Decomposition*: Not pertinent.

Health Hazards Information - *Recommended Personal Protective Equipment*: Organic canister mask; rubber gloves; chemical goggles; *Symptoms Following Exposure*: Inhalation causes irritation of mucous membrane, headache, and nausea. Contact with liquid causes extreme eye irritation and conjunctivitis; irritates skin if nit removed at once. Ingestion causes irritation of mouth and stomach; *General Treatment for Exposure*: INHALATION: remove patient to fresh air; get medical attention. EYES: flush with copious quantities of water for at least 15 min.; get medical attention if irritation persists. SKIN: wash with soap and water. INGESTION: give large amount of water and induce vomiting; get medical attention; *Toxicity by Inhalation (Threshold Limit Value)*: Data not available; *Short-Term Exposure Limits*: Data not available; *Toxicity by Ingestion*: Grade 4; LD_{50} <50 mg/kg; *Late Toxicity*: Data not available; *Vapor (Gas) Irritant Characteristics*: Data not available; *Liquid or Solid Irritant Characteristics*: Data not available; *Odor Threshold*: Data not available.

Fire Hazards - *Flash Point (deg. F)*: 129 OC; 100 CC; *Flammable Limits in Air (%):* Data not available; *Fire Extinguishing Agents:* Water fog, foam, dry chemical, carbon dioxide; *Fire Extinguishing Agents Not To Be Used:* Not pertinent; *Special Hazards of Combustion Products:* Irritating, toxic hydrogen chloride gas may be generated in fires; *Behavior in Fire:* Not pertinent; *Ignition Temperature :* Data not available; *Electrical Hazard:* Data not available; *Burning Rate:* 2.3 mm/min.

Chemical Reactivity - *Reactivity with Water:* Very slow, not hazardous; *Reactivity with Common Materials:* Slow hydrolysis to acidic products; *Stability During Transport:* Stable; *Neutralizing Agents for Acids and Caustics:* Not pertinent; *Polymerization:* Not pertinent; *Inhibitor of Polymerization:* Not pertinent.

ETHYL CHLOROFORMATE

Chemical Designations - *Synonyms*: Chloroformic acid, ethyl ester; Ethyl chlorocarbonate; *Chemical Formula*: CICOOC$_2$H$_5$.

Observable Characteristics - *Physical State (as normally shipped)*: Liquid; *Color*: Colorless to pale yellow; *Odor*: Irritating; sharp, like hydrochloric acid.

Physical and Chemical Properties - *Physical State at 15 °C and 1 atm.*: Liquid; *Molecular Weight*: 108.5; *Boiling Point at 1 atm.*: 201, 94, 367; *Freezing Point*: -114, -81, 192; *Critical Temperature*: Not pertinent; *Critical Pressure*: Not pertinent; *Specific Gravity*: 1.135 at 20°C (liquid); *Vapor (Gas) Density*: 3.7; *Ratio of Specific Heats of Vapor (Gas)*: 1.1044; *Latent Heat of Vaporization*: (est.) 140, 79, 3.3; *Heat of Combustion*: (est.) -6,900, -3,800, -160; *Heat of Decomposition*: Not pertinent.

Health Hazards Information - *Recommended Personal Protective Equipment*: Air-line mask, self-contained breathing apparatus, or organic and canister mask; full protective clothing; *Symptoms Following Exposure*: Inhalation causes mucous membrane irritation, coughing, and sneezing. Vapor causes acid-type burns of mouth and stomach; *General Treatment for Exposure*: INHALATION: remove patient to fresh air; if breathing stops give artificial respiration. Call a doctor, keep victim quiet and administer oxygen if needed. EYES: flush with water for at least 15 min.; see a doctor. SKIN: wash liberally with water for at least 15 min., then apply dilute solution of sodium bicarbonate or commercially prepared neutralizer. INGESTION: do NOT induce vomiting; give large amount of water; get medical attention; *Toxicity by Inhalation (Threshold Limit Value)*: Data not available; *Short-Term Exposure Limits*: Data not available; *Toxicity by Ingestion*: Grade 4; oral LD$_{50}$ < 50 mg/kg (rat); *Late Toxicity*: Data not available; *Vapor (Gas) Irritant Characteristics*: Vapors are moderately irritating such the personnel will not usually tolerate moderate or high concentration; *Liquid or Solid Irritant Characteristics*: Causes smarting of the skin and first-degree burns on short Exposure and may cause second-degree burns on long Exposure; *Odor Threshold*: Data not available.

Fire Hazards - *Flash Point (deg. F)*: 82 OC; 61 CC; *Flammable Limits in Air (%)*: Data not available; *Fire Extinguishing Agents*: Water, dry chemical, carbon dioxide; *Fire Extinguishing Agents Not To Be Used*: Not pertinent; *Special Hazards of Combustion Products*: Toxic chlorine and phosgene gas may be formed in fires; *Behavior in Fire*: Not pertinent; *Ignition Temperature (deg. F)*: 932; *Electrical Hazard*: Data not available; *Burning Rate*: No data.

Chemical Reactivity - *Reactivity with Water*: Slow reaction with water, evolving hydrogen chloride (hydrochloric acid); *Reactivity with Common Materials*: Slow evolution of hydrogen chloride from surface moisture reaction can cause slow corrosion; *Stability During Transport*: Stable; *Neutralizing Agents for Acids and Caustics*: Flush with water, rinse with sodium bicarbonate or lime solution; *Polymerization*: Not pertinent; *Inhibitor of Polymerization*: Not pertinent.

ETHYLDICHLOROSILANE

Chemical Designations - *Synonyms*: No common synonyms; *Chemical Formula*: C$_2$H$_5$SiHCL$_2$.

Observable Characteristics - *Physical State (as normally shipped)*: Liquid; *Color*: Colorless; *Odor*: Sharp, hydrochloric acid-like; acrid.

Physical and Chemical Properties - *Physical State at 15 °C and 1 atm.*: Liquid; *Molecular Weight*: 129.1; *Boiling Point at 1 atm.*: 165, 74, 347; *Freezing Point*: Not pertinent; *Critical Temperature*: Not pertinent; *Critical Pressure*: Not pertinent; *Specific Gravity*: 1.092 at 20°C (liquid); *Vapor (Gas) Density*: 4.5; *Ratio of Specific Heats of Vapor (Gas)*: Data not available; *Latent Heat of Vaporization*: (est.) 104, 57.8, 2.42; *Heat of Combustion*: (est.) -6,500, -3,600, -150; *Heat of Decomposition*: Not pertinent.

Health Hazards Information - *Recommended Personal Protective Equipment*: Acid-vapor-type respiratory protection; rubber gloves; chemical worker's goggles; other equipment as necessary to protect skin and eyes; *Symptoms Following Exposure*: Inhalation irritates mucous membranes. Contact with liquid causes severe burns of eyes and skin. Ingestion causes severe burns of mouth and stomach; *General Treatment for Exposure*: Get medical attention following all Exposures to this compound.

INHALATION: remove to fresh air; give artificial respiration if required. EYES: flush with water for 15 min. SKIN: flush with water. INGESTION: do NOT induce vomiting; give large amounts of water, followed by milk or milk of magnesia; *Toxicity by Inhalation (Threshold Limit Value)*: Data not available; *Short-Term Exposure Limits*: Data not available; *Toxicity by Ingestion*: Grade 3; LD$_{50}$ 50 to 500 mg/kg; *Late Toxicity*: Data not available; *Vapor (Gas) Irritant Characteristics*: Vapors causes severe irritation of eyes and throat and canister cause eye and lung injury. They cannot be tolerated even at low concentration; *Liquid or Solid Irritant Characteristics*: Severe skin irritant. Causes second- and third-degree burns on short contact and is very injurious to the eyes; *Odor Threshold*: Data not available.

Fire Hazards - *Flash Point (deg. F)*: 30 OC; *Flammable Limits in Air (%)*: 2.9; *Fire Extinguishing Agents:* Dry chemical; *Fire Extinguishing Agents Not To Be Used:* Water, foam; *Special Hazards of Combustion Products:* Toxic hydrogen chloride and phosgene gases may be formed; *Behavior in Fire:* Difficult to extinguish; re-ignition may occur. Contact with water applied to adjacent fire, produces irritating hydrogen chloride fumes and flammable hydrogen gas; *Ignition Temperature (deg. F):* Data not available; *Electrical Hazard:* Data not available; *Burning Rate:* 3.2 mm/min.

Chemical Reactivity - *Reactivity with Water:* Reacts vigorously, evolving hydrogen chloride (hydrochloric acid); *Reactivity with Common Materials:* Reaction with surface moisture will generate hydrogen chloride, which corrodes common metals; *Stability During Transport:* Stable; *Neutralizing Agents for Acids and Caustics:* Flood with water, rinse with sodium bicarbonate or lime solution; *Polymerization:* Not pertinent; *Inhibitor of Polymerization:* Not pertinent.

ETHYLENE

Chemical Designations - *Synonyms*: Ethene; *Chemical Formula*: C_2H_4.

Observable Characteristics - *Physical State (as normally shipped)*: Compressed gas or liquefied gas; *Color*: Colorless; *Odor*: Slightly sweet; faint sweet; slight ethereal.

Physical and Chemical Properties - *Physical State at 15 °C and 1 atm.*: Gas; *Molecular Weight*: 28.05; *Boiling Point at 1 atm.*: -154.7, -103.7, 169.5; *Freezing Point*: -272.4, -169.1, 104.1; *Critical Temperature*: 49.8, 9.9, 283.1; *Critical Pressure*: 742, 50.5, 5.11; *Specific Gravity*: 0.569 at -103.8°C (liquid); *Vapor (Gas) Density*: 1.0; *Ratio of Specific Heats of Vapor (Gas)*: 1.240; *Latent Heat of Vaporization*: 207.7, 115.4, 4.823; *Heat of Combustion*: -20,290, -11,272, -471.94; *Heat of Decomposition*: Not pertinent.

Health Hazards Information - *Recommended Personal Protective Equipment*: Organic vapor canister or air-supplied mask; *Symptoms Following Exposure*: Moderate concentration inhalation causes drowsiness, dizziness, and unconsciousness. Overexposure causes headache, drowsiness, muscular weakness; *General Treatment for Exposure*: Remove victim to fresh air, give artificial respiration and oxygen if breathing has stopped, and call a physician; *Toxicity by Inhalation*: Simple asphyxiate; *Short-Term Exposure Limits*: Not pertinent; *Toxicity by Ingestion*: Not pertinent; *Late Toxicity*: Not pertinent; *Vapor (Gas) Irritant Characteristics*: Vapors are nonirritating to the eyes and throat; *Liquid or Solid Irritant Characteristics*: No appreciable hazard. Practically harmless to the skin, but may cause frostbite; *Odor Threshold*: Data not available.

Fire Hazards - *Flash Point (deg. F)*: -213 (approx.) CC; *Flammable Limits in Air (%):* 2.75 - 28.6; *Fire Extinguishing Agents:* Stop flow of gas if possible. Use carbon dioxide, dry chemical, water fog; *Fire Extinguishing Agents Not To Be Used:* Not pertinent; *Special Hazards of Combustion Products:* Vapors are anesthetic; *Behavior in Fire:* Container may explode; *Ignition Temperature (deg. F):* 842; *Electrical Hazard:* Class I, Group D; *Burning Rate:* 7.4 mm/min.

Chemical Reactivity - *Reactivity with Water:* No reaction; *Reactivity with Common Materials:* No reaction; *Stability During Transport:* Stable; *Neutralizing Agents for Acids and Caustics:* Not pertinent; *Polymerization:* Not pertinent; *Inhibitor of Polymerization:* Not pertinent.

ETHYLENE CHLOROHYDRIN

Chemical Designations - *Synonyms*: 2-Chloroethanol; 2-Chlorethanol; 2-Chloroethyl alcohol; Ethylene chlorhydryn; Glycol chlorohydrin; *Chemical Formula*: $CICH_2CH_2OH$.

Observable Characteristics - *Physical State (as normally shipped)*: Liquid; *Color*: Colorless; *Odor*: Faint, ethereal.

Physical and Chemical Properties - *Physical State at 15 °C and 1 atm.*: Liquid; *Molecular Weight*: 80.51; *Boiling Point at 1 atm.*: 263.7, 128.7, 401.9; *Freezing Point*: -80.7, -62.6, 210.6; *Critical Temperature*: Not pertinent; *Critical Pressure*: Not pertinent; *Specific Gravity*: 1.197 at 20°C (liquid); *Vapor (Gas) Density*: 2.8; *Ratio of Specific Heats of Vapor (Gas)*: Data not available; *Latent Heat of Vaporization*: 221, 123, 5.15; *Heat of Combustion*: -6,487, -3,604, -150.8; *Heat of Decomposition*: Not pertinent.

Health Hazards Information - *Recommended Personal Protective Equipment*: Organic mask or self-contained breathing apparatus; goggles or face shield; rubber gloves; *Symptoms Following Exposure*: Inhalation causes irritation of upper respiratory system, nausea, headache, delirium, coma, collapse. Liquid causes irritation of eyes and skin; prolonged contact with skin may allow penetration into body and cause same symptoms as following ingestion or inhalation. Ingestion causes nausea, headache, delirium, coma, collapse; *General Treatment for Exposure*: INHALATION: remove from Exposure; give artificial respiration if necessary; call physician. EYES: flush with copious amounts of water; call physician if contact has been prolonged. IRRITATION: give large amounts of water; get medical attention; *Toxicity by Inhalation (Threshold Limit Value)*: 5 ppm; *Short-Term Exposure Limits*: Data not available; *Toxicity by Ingestion*: Grade 3; oral LD_{50} = 71 mg/kg (rat); *Late Toxicity*: Damage to central nervous system and liver Inhalation humans; *Vapor (Gas) Irritant Characteristics*: Data not available; *Liquid or Solid Irritant Characteristics*: Data not available; *Odor Threshold*: Odorless.

Fire Hazards - *Flash Point (deg. F)*: 139 OC; *Flammable Limits in Air (%):* 4.9 - 15.9; *Fire Extinguishing Agents:* Water, alcohol foam, dry chemical, or carbon dioxide; *Fire Extinguishing Agents Not To Be Used:* Not pertinent; *Special Hazards of Combustion Products:* Toxic hydrogen chloride and phosgene fumes may be formed; *Behavior in Fire:* Vapors are heavier than air and may flash back to a source of ignition; *Ignition Temperature (deg. F):* 797; *Electrical Hazard:* Data not available; *Burning Rate:* 1.7 mm/min.

Chemical Reactivity - *Reactivity with Water:* No reaction; *Reactivity with Common Materials:* No reaction; *Stability During Transport:* Stable; *Neutralizing Agents for Acids and Caustics:* Not pertinent; *Polymerization:* Not pertinent; *Inhibitor of Polymerization:* Not pertinent.

ETHYLENE CYANOHYDRIN

Chemical Designations - *Synonyms*: 2-Cyanoethanol; Glycol cyanohydrin; 1-Hydroxy-2-cyanoethane; Hydracrylonitrile; 3-Hydroxypropanenitrile; *Chemical Formula*: $HOCH_2CH_2CN$.

Observable Characteristics - *Physical State (as normally shipped)*: Liquid; *Color*: Colorless or straw-colored; *Odor*: Practically odorless; characteristic.

Physical and Chemical Properties - *Physical State at 15 °C and 1 atm.*: Liquid; *Molecular Weight*: 71.08; *Boiling Point at 1 atm.*: 445.5, 229.7, 502.9; *Freezing Point*: -51.2, -46.2, 227.0; *Critical Temperature*: 804, 429; 702; *Critical Pressure*: 720, 49, 4.9; *Specific Gravity*: 1.047 at 20°C (liquid); *Vapor (Gas) Density*: Not pertinent; *Ratio of Specific Heats of Vapor (Gas)*: Not pertinent; *Latent Heat of Vaporization*: Not pertinent; *Heat of Combustion*: Data not available; *Heat of Decomposition*: Not pertinent.

Health Hazards Information - *Recommended Personal Protective Equipment*: Air-supplied mask; plastic gloves; rubber clothing; vapor-proof goggles; *Symptoms Following Exposure*: Liquid causes eye irritation. If swallowed, may cause severe kidney injury; *General Treatment for Exposure*: INGESTION: induce vomiting at once and call a physician. EYES: wash with flowing water for 15 min. SKIN: flush exposed area with plenty of water; *Toxicity by Inhalation (Threshold Limit Value)*: Not pertinent; *Short-Term Exposure Limits*: Not pertinent; *Toxicity by Ingestion*: Grade 2; LD_{50} 0.5 to 5 g/kg; *Late Toxicity*: Ingestion of liquid may cause severe kidney damage; *Vapor (Gas) Irritant Characteristics*: Vapors are nonirritating to the eyes and throat; *Liquid or Solid Irritant Characteristics*: No appreciable hazard. Practically harmless to the skin; *Odor Threshold*: Not pertinent.

Fire Hazards - *Flash Point (deg. F)*: 265 OC; *Flammable Limits in Air (%):* 2.3 (calc.) - 12.1 (est.); *Fire Extinguishing Agents:* Carbon dioxide or dry chemical for small fires; alcohol-type foam for large fires; *Fire Extinguishing Agents Not To Be Used:* Water or foam may cause frothing; *Special Hazards of Combustion Products:* Toxic gases are generated when heated; *Behavior in Fire:* Decomposes, generating toxic fires; *Ignition Temperature (deg. F):* 922; *Electrical Hazard:* Not pertinent; *Burning Rate:* Data not available.

Chemical Reactivity - *Reactivity with Water:* No reaction; *Reactivity with Common Materials:* No reaction; *Stability During Transport:* Stable; *Neutralizing Agents for Acids and Caustics:* Not pertinent; *Polymerization:* Not pertinent; *Inhibitor of Polymerization:* Not pertinent.

ETHYLENEDIAMINE

Chemical Designations - *Synonyms*: 1,2-Diaminoethane; 1,2-Ethanediamine; *Chemical Formula*: $NH_2CH_2CH_2NH_2$.

Observable Characteristics - *Physical State (as normally shipped)*: Liquid; *Color*: Colorless; *Odor*: Strong ammoniacal odor; ammonia-like mild and ammoniacal odor.

Physical and Chemical Properties - *Physical State at 15 °C and 1 atm.*: Liquid; *Molecular Weight*: 60.10; *Boiling Point at 1 atm.*: 243, 117, 390; *Freezing Point*: 51.8, 11.0, 284.2; *Critical Temperature*: 608, 320, 593; *Critical Pressure*: 941, 64, 6.4; *Specific Gravity*: 0.909 at 20°C (liquid); *Vapor (Gas) Density*: Not pertinent; *Ratio of Specific Heats of Vapor (Gas)*: 1.087; *Latent Heat of Vaporization*: 288, 160, 6.70; *Heat of Combustion*: -12.290, -6830, -286.0; *Heat of Decomposition*: Not pertinent.

Health Hazards Information - *Recommended Personal Protective Equipment*: Full rubber protective clothing, inc. gloves and boots; chemical worker's goggles; face shield where contact with face is likely. If necessary to enter closed area for ½ hr or less with mist, wear full-faced gas mask with canister approved by Bureau of Standards for use with ammonia; *Symptoms Following Exposure*: High concentration of vapor burns eyes and irritates the nose and throat. Liquid burns eyes and skin; *General Treatment for Exposure*: Get medical help immediately! INGESTION: drink large amounts of water or milk quickly, induce vomiting only if instructed by physician. EYES: flush immediately and thoroughly with flowing water for at least 15 min. SKIN: remove clothing and flush affected area with copious amounts of water, then wash with soap and water; severe Exposure may require showering; *Toxicity by Inhalation (Threshold Limit Value)*: 10 ppm; *Short-Term Exposure Limits*: 20 ppm to 5 min.; *Toxicity by Ingestion*: Grade 2; LD_{50} 0.5 to 5 g/kg (female rat); *Late Toxicity*: Data not available; *Vapor (Gas) Irritant Characteristics*: Vapors is moderately irritating such that personnel will no usually tolerate moderate or high vapor concentration; *Liquid or Solid Irritant Characteristics*: Fairly severe skin irritant; may cause pain and second-degree burns after a few minutes contact; *Odor Threshold*: 10 ppm.

Fire Hazards - *Flash Point (deg. F)*: 99 OC; 150 CC; *Flammable Limits in Air (%):* 5.8 - 11.1; *Fire Extinguishing Agents:* Carbon dioxide, dry chemical, water or foam; *Fire Extinguishing Agents Not To Be Used:* Do not use water in case of drum or tank fires; *Special Hazards of Combustion Products:* Irritating vapors are generated when heated; *Behavior in Fire:* Not pertinent; *Ignition Temperature (deg. F):* 715; *Electrical Hazard:* Not pertinent; *Burning Rate:* 2.2 mm/min.

Chemical Reactivity - *Reactivity with Water:* Gives off heat, but reaction is not hazardous; *Reactivity with Common Materials:* No reaction; *Stability During Transport:* Stable; *Neutralizing Agents for Acids and Caustics:* Flush with water; *Polymerization:* Not pertinent; *Inhibitor of Polymerization:* Not pertinent.

ETHYLENE DIBROMIDE

Chemical Designations - *Synonyms*: Bromofume; 1,2-Dibromoethane; sym-Dibromoethane; Dowfume 40, W-10, W-15, W-40; Ethylene bromide; Glycol dibromide; *Chemical Formula*: $BrCH_2CH_2Br$.

Observable Characteristics - *Physical State (as normally shipped)*: Liquid; *Color*: Colorless; *Odor*: Mildly sweet; like chloroform.

Physical and Chemical Properties - *Physical State at 15 °C and 1 atm.*: Liquid; *Molecular Weight*: 187.86; *Boiling Point at 1 atm.*: 268, 131, 404; *Freezing Point*: 49.6, 9.8, 283.0; *Critical Temperature*: Not pertinent; *Critical Pressure*: Not pertinent; *Specific Gravity*: 2.180 at 20°C (liquid);

Vapor (Gas) Density: Not pertinent; *Ratio of Specific Heats of Vapor (Gas)*: 1.109; *Latent Heat of Vaporization*: 82.1, 45.6, 1.91; *Heat of Combustion*: Not pertinent; *Heat of Decomposition*: Not pertinent.

Health Hazards Information - *Recommended Personal Protective Equipment*: Canister type mask or self-contained air mask; neoprene gloves; chemical safety goggles; *Symptoms Following Exposure*: Local inflammation, blisters and ulcers on skin; irritation in lungs and organic injury of liver and kidneys; may be absorbed trough skin; *General Treatment for Exposure*: Remove from Exposure. Remove contaminated clothing. Wash skin with soap and water. Flush eyes with plenty of water. Consult physician; *Toxicity by Inhalation (Threshold Limit Value)*: 3 mg/m^3; *Short-Term Exposure Limits*: 50 ppm for 5 min.; *Toxicity by Ingestion*: Grade 3; LD$_{50}$ 50 to 500 mg/kg; *Late Toxicity*: Data not available; *Vapor (Gas) Irritant Characteristics*: Vapors cause a slight smarting of the eyes or respiratory system if present Inhalation high concentrations. The effect is temporary; *Liquid or Solid Irritant Characteristics*: Minimum hazard. If spilled on clothing and allowed to remain, may cause smarting and reddening of the skin; *Odor Threshold*: Data not available.

Fire Hazards - *Flash Point* : Not flammable; *Flammable Limits in Air (%):* Not flammable; *Fire Extinguishing Agents:* Not pertinent; *Fire Extinguishing Agents Not To Be Used:* Not pertinent; *Special Hazards of Combustion Products:* Decomposition gases are toxic and irritating; *Behavior in Fire:* Decomposes into toxic irritating gases. Reacts with hot metals such as aluminum and magnesium; *Ignition Temperature :* Not flammable; *Electrical Hazard:* Not pertinent; *Burning Rate:* Not flammable.

Chemical Reactivity - *Reactivity with Water:* No reaction; *Reactivity with Common Materials:* No reaction; *Stability During Transport:* Stable; *Neutralizing Agents for Acids and Caustics:* Not pertinent; *Polymerization:* Not pertinent; *Inhibitor of Polymerization:* Not pertinent.

ETHYLENE DICHLORIDE

Chemical Designations - *Synonyms*: Brocide; 1,2-Dichlorethane; Dutch liquid; EDC; Ethylene chloride; Glycol dichloride; *Chemical Formula*: ClCH$_2$CH$_2$Cl.

Observable Characteristics - *Physical State (as normally shipped)*: Liquid; *Color*: Colorless; *Odor*: Ethereal; chloroform-like; ether-like.

Physical and Chemical Properties - *Physical State at 15 °C and 1 atm.*: Liquid; *Molecular Weight*: 98.96; *Boiling Point at 1 atm.*: 182.3, 83.5, 356.7; *Freezing Point*: -32.3, -35.7, 237.5; *Critical Temperature*: 550, 288, 561; *Critical Pressure*: 735, 50, 5.1; *Specific Gravity*: 1.253 at 20°C (liquid); *Vapor (Gas) Density*: 3.4; *Ratio of Specific Heats of Vapor (Gas)*: 1.118; *Latent Heat of Vaporization*: 138, 76.4, 3.20; *Heat of Combustion*: (est.) 3,400, 1,900, 80; *Heat of Decomposition*: Not pertinent.

Health Hazards Information - *Recommended Personal Protective Equipment*: Clean, body-covering clothing and safety glasses with side shields. Respiratory protection: up to 50 ppm, none; 50 ppm to 2%, ½ hr or less, full face mask and canister; greater than 2%, self-contained breathing apparatus; *Symptoms Following Exposure*: Inhalation of vapors causes nausea, drunkenness, depression. Contact of liquid with eyes may produce corneal injury. Prolonged contact with skin may cause a burn; *General Treatment for Exposure*: INHALATION: if victim is overcome, remove him to fresh air, keep him quiet and warm, and get medical attention immediately; if breathing stops, give artificial respiration. INGESTION: induce vomiting; call a physician; treat the symptoms. EYES: flush immediately with copious amounts of flowing water for 15 min. SKIN: remove clothing and wash skin thoroughly with soap and water; wash contaminated clothing before reuse; *Toxicity by Inhalation (Threshold Limit Value)*: 50 ppm; *Short-Term Exposure Limits*: 200 ppm for 5 min. during any 3-hour period; *Toxicity by Ingestion*: Grade 2; LD$_{50}$ 0.5 to 5 g/kg (rat); *Late Toxicity*: Data not available; *Vapor (Gas) Irritant Characteristics*: Vapors cause moderate irritation such that personnel will find high concentrations unpleasant. The effect is temporary; *Liquid or Solid Irritant Characteristics*: Causes smarting of the skin and first-degree burns on short Exposure; may cause secondary burns on long Exposure; *Odor Threshold*: 100 ppm.

Fire Hazards - *Flash Point (deg. F)*: 60 OC; 55 CC; *Flammable Limits in Air (%)*: 6.2 - 15.6; *Fire Extinguishing Agents:* Foam, carbon dioxide, dry chemical; *Fire Extinguishing Agents Not To Be Used:* Water may be ineffective; *Special Hazards of Combustion Products:* Toxic and irritating gases (hydrogen chloride and phosgene) are generated; *Behavior in Fire:* Vapor is heavier than air and may travel considerable distance to a source of ignition and flash back; *Ignition Temperature (deg. F):* 775; *Electrical Hazard:* Class I, Group D; *Burning Rate:* 1.6 mm/min.

Chemical Reactivity - *Reactivity with Water:* No reaction; *Reactivity with Common Materials:* No reaction; *Stability During Transport:* Stable; *Neutralizing Agents for Acids and Caustics:* Not pertinent; *Polymerization:* Not pertinent; *Inhibitor of Polymerization:* Not pertinent.

ETHYLENE DICHLORIDE

Chemical Designations - *Synonyms*: 1,2-Dihydroxyethane; 1,2-Ethanediol; Ethylene dihydrate; Glycol; Monoethylene glycol; *Chemical Formula*: $HOCH_2CH_2OH$.

Observable Characteristics - *Physical State (as normally shipped)*: Liquid; *Color*: Colorless; *Odor*: Slight odor.

Physical and Chemical Properties - *Physical State at 15 °C and 1 atm.*: Liquid; *Molecular Weight*: 62.07; *Boiling Point at 1 atm.*: 387.7, 197.6, 470.8; *Freezing Point*: 8.6, -13, 260; *Critical Temperature*: Not pertinent; *Critical Pressure*: Not pertinent; *Specific Gravity*: 1.115 at 20°C (liquid); *Vapor (Gas) Density*: Not pertinent; *Ratio of Specific Heats of Vapor (Gas)*: 1.095; *Latent Heat of Vaporization*: 344, 191, 8.00; *Heat of Combustion*: -7,259, -4,033, -168.9; *Heat of Decomposition*: Not pertinent.

Health Hazards Information - *Recommended Personal Protective Equipment*: Goggles; shower and eye bath; *Symptoms Following Exposure*: Inhalation of vapors is not hazardous. Ingestion causes stupor or coma, sometimes leading to fatal kidney injury; *General Treatment for Exposure*: INGESTION: induce vomiting and call a physician. SKIN AND EYES: flush with water; *Toxicity by Inhalation (Threshold Limit Value)*: 100 ppm; *Short-Term Exposure Limits*: Not pertinent; *Toxicity by Ingestion*: Grade 1; LD_{50} 5 to 15 g/kg (rat, guinea pig, mouse); *Late Toxicity*: Fatal kidney injury may result if ingested; *Vapor (Gas) Irritant Characteristics*: Vapors are nonirritating to the eyes and throat; *Liquid or Solid Irritant Characteristics*: No appreciable hazard. Practically harmless to the skin; *Odor Threshold*: Not pertinent.

Fire Hazards - *Flash Point (deg. F)*: 240 OC; 232 CC; *Flammable Limits in Air (%):* LEL=3.2; UEL not listed; *Fire Extinguishing Agents:* Water fog, alcohol foam; *Fire Extinguishing Agents Not To Be Used:* Water or foam may cause frothing; *Special Hazards of Combustion Products:* Not pertinent; *Behavior in Fire:* Not pertinent; *Ignition Temperature (deg. F):* 775; *Electrical Hazard:* Not pertinent; *Burning Rate:* 1.0 mm/min.

Chemical Reactivity - *Reactivity with Water:* No reaction; *Reactivity with Common Materials:* No reaction; *Stability During Transport:* Stable; *Neutralizing Agents for Acids and Caustics:* Not pertinent; *Polymerization:* Not pertinent; *Inhibitor of Polymerization:* Not pertinent.

ETHYL ETHER

Chemical Designations - *Synonyms:* Anestithesia Ether; Dietyl Ether; Ethoxyethane; Dietyl Oxide; Sulfuric Ether; *Chemical Formula:* $C_2H_5OC_2H_5$.

Observable Characteristics - *Physical State (as shipped):* Liquid; *Color:* Colorless; *Odor:* Sweet, pungent.

Physical and Chemical Properties - *Physical State at 15 °C and 1 atm:* Liquid; *Molecular Weight:* 74.12; *Boiling Point at 1 atm:* 94.3, 34.6, 307.8; *Freezing Point:* -177.3, -116.3, 156.9; *Critical Temperature:* 380.3, 193.5, 466.7; *Critical Pressure:* 527, 35.9, 3.64; *Specific Gravity:* 0,714 at 20 °C (liquid); *Vapor (Gas) Specific Gravity:* 2,6; *Ratio of Specific Heats of Vapor (Gas):* 1,081; *Latent Heat of Vaporization:* 153, 84.9, 3.56; *Heat of Combustion:* -14,550, -8082, -3384; *Heat of Decomposition:* Not pertinent.

Health Hazards Information- *Personal Protective Equipment:* Approved organic vapor canister mask; chemical goggles; synthetic rubber or plastic gloves. *Symptoms Following Exposure:* Vapor inhalation may causes headache, nausea, vomiting, and loss of consciousness. Contact with eyes will be irritating. Skin contact from clothing wet with the chemical may causes burns; *Treatment for Exposure:* INHALATION: remove victim to fresh air; if breathing has stopped, apply artificial respiration; if breathing is irregular, give oxygen; call a physician. EYES: flush immediately with water for 15 min.; *Toxicity by Inhalation (Threshold Limit Value):* 400 ppm; *Short-Term Inhalation Limits:* 1000 ppm for 30 min.; *Toxicity by Ingestion:* Grade 2; LD_{50} 0,5 to 5 g/kg *Late Toxicity:* None; *Vapor (Gas) Irritant Characteristics:* Vapors cause a slight smarting of the eyes or respiratory system if present in high concentrations. The effect is temporary; *Liquid or Solid Irritant Characteristics:* No appreciable hazard. Practically harmless to the skin because it is very and evaporates quickly; *Odor Threshold:* 0.83 ppm.

Fire Hazards - *Flash Point (deg. F):* -40 OC; -49 CC; *Flammable Limits in Air (%):* 1.85 - 36.5; *Fire Extinguishing Agents:* Carbon dioxide, dry chemical or foam; *Fire Extinguishing Agents Not To Be Used:* Not pertinent; *Special Hazards of Combustion Products:* Not pertinent; *Behavior in Fire:* Vapor is heavier than air and may travel considerable distance to a source of ignition and flash back. Decomposes violently when heated; *Ignition Temperature (deg. F):* 356; *Electrical Hazard:* Class I, Group C; *Burning Rate:* 6.7 mm/min.

Chemical Reactivity - *Reactivity with Water:* No reaction; *Reactivity with Common Materials:* No reaction; *Stability During Transport:* Stable; *Neutralizing Agents for Acids and Caustics:* Not pertinent; *Polymerization:* Not pertinent; *Inhibitor of Polymerization:* Not pertinent.

2-ETHYL HEXANOL

Chemical Designations - *Synonyms:* 2-Ethyl-l-hexanol; 2-Ethylhexyl alcohol; *Chemical Formula:* $CH_3(CH_2)_3CH(C_2H_5)CH_2OH$.

Observable Characteristics - *Physical State (as shipped):* Liquid; *Color:* Colorless; *Odor:* Strong.

Physical and Chemical Properties - *Physical State at 15 °C and 1 atm:* Liquid; *Molecular Weight:* 130,23; *Boiling Point at 1 atm:* 564,5, 184,7, 457,9; *Freezing Point:* <158, <70, <343; *Critical Temperature:* 711, 377, 650; *Critical Pressure:* 512, 34.8, 3.53; *Specific Gravity:* 0.834 at 20°C (liquid); *Vapor (Gas) Specific Gravity:* Not pertinent; *Ratio of Specific Heats of Vapor (Gas):* Not pertinent; *Latent Heat of Vaporization:* 167, 92.8, 3.89; *Heat of Combustion:* -17,480, -9,710, -406.5; *Heat of Decomposition:* Not pertinent.

Health Hazards Information - *Personal Protective Equipment:* Air pack or organic canister; goggles; rubber gloves. *Symptoms Following Exposure:* Anesthesia, nausea, headache, dizziness; mildly irritating to skin and eyes. *Treatment for Exposure:* INHALATION: move victim to fresh air. SKIN: wash affected areas with water. EYES: flush with water for 15 min. Get medical care. *Toxicity by Inhalation (Threshold Limit Value):* Not pertinent; *Short-Term Inhalation Limits:* Not pertinent; *Toxicity by Ingestion:* Grade 2; LD_{50} 0,5 to 5 g/kg (lab animals); *Late Toxicity:* Increased excitability of central nervous system in rats and rabbits. *Vapor (Gas) Irritant Characteristics:* Vapors cause a slight smarting of the eyes or respiratory system if present in high concentrations. The effect is temporary. *Liquid or Solid Irritant Characteristics:* Minimum hazard. If spilled on clothing and allowed to remain, may cause smarting and reddening of the skin. *Odor Threshold:* Data not available.

Fire Hazards - *Flash Point (deg. F):* 85 OC; 175 CC; *Flammable Limits in Air (%):* Data not available; *Fire Extinguishing Agents:* Foam, carbon dioxide, dry chemical; *Fire Extinguishing Agents Not To Be Used:* Not pertinent; *Special Hazards of Combustion Products:* Not pertinent; *Behavior in Fire:* Not pertinent; *Ignition Temperature (deg. F):* 581; *Electrical Hazard:* Not pertinent; *Burning Rate:* 4.0 mm/min.

Chemical Reactivity - *Reactivity with Water:* No reaction; *Reactivity with Common Materials:* No reaction; *Stability During Transport:* Stable; *Neutralizing Agents for Acids and Caustics:* Not pertinent; *Polymerization:* Not pertinent; *Inhibitor of Polymerization:* Not pertinent.

ETHYL FORMATE

Chemical Designations - *Synonyms:* Ethyl Formic Ester; Ethyl Methanoate; Formic Acid, Ethyl Ester; Formic Ether; *Chemical Formula:* $HCOOC_2H_5$.

Observable Characteristics - *Physical State (as shipped):* Liquid; *Color:* Colorless; *Odor:* Characteristic; pleasant aromatic.

Physical and Chemical Properties - *Physical State at 15 °C and 1 atm:* Liquid; *Molecular Weight:* 74,1; *Boiling Point at 1 atm:* 129.6, 54.2, 327.4; *Freezing Point:* -110, -79, 194; *Critical Temperature:*

455, 235, 508; *Critical Pressure:* 686, 46.6, 4.73; *Specific Gravity:* 0,922 at 20°C (liquid); *Vapor (Gas) Specific Gravity:* 2,6; *Ratio of Specific Heats of Vapor (Gas):* 1.1014; *Latent Heat of Vaporization:* 176, 98, 4,1; *Heat of Combustion:* -9,500, -5,300, -220; *Heat of Decomposition:* Not pertinent.

Health Hazards Information - *Personal Protective Equipment:* Organic canister gas mask; goggles or face shield; rubber gloves. *Symptoms Following Exposure:* Inhalation of vapor causes slight irritation of the eyes and rapidly increasing irritation of the nose. High concentrations cause deep narcosis within a few minutes followed by death within a few hours. Contact with liquid causes moderate irritation of eyes and mild irritation of skin. Ingestion causes irritation of mouth and stomach; may cause deep narcosis and death if not treated. *Treatment for Exposure:* INHALATION: removes from exposure; begin artificial respiration if breathing has stopped; call physician. EYES: wash with water for 15 min.; call physician if needed. SKIN: wash with water for 15 min.; call physician if irritation persists. INGESTION: do NOT induce vomiting; get medical attention at once. *Toxicity by Inhalation (Threshold Limit Value):* 100 ppm; *Short-Term Inhalation Limits:* Data not available; *Toxicity by Ingestion:* Grade 2; oral LD_{50} = 1,850 mg/kg (rat); *Late Toxicity:* Data not available; *Vapor (Gas) Irritant Characteristics:* Vapors cause moderate irritation such that personnel will find high concentrations unpleasant. The effect is temporary. *Liquid or Solid Irritant Characteristics:* Fairly severe skin irritant. May cause pain and second-degree burns after a few minutes' contact. *Odor Threshold:* No data.

Fire Hazards - *Flash Point (deg. F):* 10 OC; -4 CC; *Flammable Limits in Air (%):* 2.8 - 16.0; *Fire Extinguishing Agents:* Dry chemical, alcohol foam, carbon dioxide; *Fire Extinguishing Agents Not To Be Used:* Water may be ineffective; *Special Hazards of Combustion Products:* Not pertinent; *Behavior in Fire:* Vapor is heavier than air and may travel considerable distance to a source of ignition and flash back; *Ignition Temperature (deg. F):* 851; *Electrical Hazard:* Data not available; *Burning Rate:* 3.6 mm/min.

Chemical Reactivity - *Reactivity with Water:* No reaction; *Reactivity with Common Materials:* No reaction; *Stability During Transport:* Stable; *Neutralizing Agents for Acids and Caustics:* Not pertinent; *Polymerization:* Not pertinent; *Inhibitor of Polymerization:* Not pertinent.

ETHYL HEXYL TALLATE

Chemical Designations - *Synonyms:* Croplas EH; *Chemical Formula:* (Mixture).

Observable Characteristics - *Physical State (as shipped):* Liquid; *Color:* Pale yellow; *Odor:* Mild characteristic.

Physical and Chemical Properties - *Physical State at 15 °C and 1 atm:* Liquid; *Molecular Weight:* Not pertinent; *Boiling Point at 1 atm:* Very high; *Freezing Point:* Not pertinent; *Critical Temperature:* Not pertinent; *Critical Pressure:* Not pertinent; *Specific Gravity:* (est.) 0.95 at 20°C (liquid); *Vapor (Gas) Specific Gravity:* Not pertinent; *Ratio of Specific Heats of Vapor (Gas):* Not pertinent; *Latent Heat of Vaporization:* Not pertinent; *Heat of Combustion:* (est.) -17,000, -10,000, -400; *Heat of Decomposition:* Not pertinent.

Health Hazards Information - *Personal Protective Equipment:* Data not available; *Symptoms Following Exposure:* Data not available; *Treatment for Exposure:* Data not available; *Toxicity by Inhalation (Threshold Limit Value):* Not pertinent; *Short-Term Inhalation Limits:* Not pertinent; *Toxicity by Ingestion:* Data not available; *Late Toxicity:* Data not available; *Vapor (Gas) Irritant Characteristics:* None; *Liquid or Solid Irritant Characteristics:* None; *Odor Threshold:* Not pertinent.

Fire Hazards - *Flash Point (deg. F):* 395 OC; *Flammable Limits in Air (%):* Data not available; *Fire Extinguishing Agents:* Data not available; *Fire Extinguishing Agents Not To Be Used:* Data not available; *Special Hazards of Combustion Products:* Not pertinent; *Behavior in Fire:* Not pertinent; *Ignition Temperature :* Data not available; *Electrical Hazard:* Not pertinent; *Burning Rate:* No data.

Chemical Reactivity - *Reactivity with Water:* No reaction; *Reactivity with Common Materials:* No reaction; *Stability During Transport:* Stable; *Neutralizing Agents for Acids and Caustics:* Not pertinent; *Polymerization:* Not pertinent; *Inhibitor of Polymerization:* Not pertinent.

ETHYL LACTATE

Chemical Designations - *Synonyms:* Ethyl Alpha-hydroxy-propionate; Ethyl 2-hydroxypropanoate; Ethyl DL-Lactate; Lactic Acid; Ethyl Ester; *Chemical Formula:* $CH_3CHOHCOOC_2H_5$.

Observable Characteristics - *Physical State (as shipped):* Liquid; *Color:* Colorless; *Odor:* Mild characteristic.

Physical and Chemical Properties - *Physical State at 15 °C and 1 atm:* Liquid; *Molecular Weight:* 118.1; *Boiling Point at 1 atm:* 309, 154, 427; *Freezing Point:* Not pertinent; *Critical Temperature:* Not pertinent; *Critical Pressure:* Not pertinent; *Specific Gravity:* 1,03 at 20°C (liquid); *Vapor (Gas) Specific Gravity:* Not pertinent; *Ratio of Specific Heats of Vapor (Gas):* Not pertinent; *Latent Heat of Vaporization:* Not pertinent; *Heat of Combustion:* (est.) -11,600, -6,500, -270; *Heat of Decomposition:* Not pertinent.

Health Hazards Information - *Personal Protective Equipment:* Goggles or face shield; rubber gloves. *Symptoms Following Exposure:* Inhalation of concentrated vapor may cause drowsiness. Contact with liquid causes mild irritation of eyes and (on prolonged contact) skin. Ingestion may cause narcosis. *Treatment for Exposure:* INHALATION: remove victim to fresh air. EYES and SKIN: flush well with water. INGESTION: induce vomiting; get medical attention. *Toxicity by Inhalation (Threshold Limit Value):* Data not available; *Short-Term Inhalation Limits:* Data not available *Toxicity by Ingestion:* Grade 2; oral LD_{50} = 2,580 mg/kg; *Late Toxicity:* Data not available; *Vapor (Gas) Irritant Characteristics:* Data not available; *Liquid or Solid Irritant Characteristics:* Data not available; *Odor Threshold:* Data not available.

Fire Hazards - *Flash Point (deg. F):* 1.58 OC; 115 CC; *Flammable Limits in Air (%):* 1.5 - 11.4; *Fire Extinguishing Agents:* Water, dry chemical, alcohol foam, carbon dioxide; *Fire Extinguishing Agents Not To Be Used:* Not pertinent; *Special Hazards of Combustion Products:* Not pertinent; *Behavior in Fire:* Not pertinent; *Ignition Temperature (deg. F):* 752; *Electrical Hazard:* Data not available; *Burning Rate:* Data not available.

Chemical Reactivity - *Reactivity with Water:* No reaction; *Reactivity with Common Materials:* No reaction; *Stability During Transport:* Stable; *Neutralizing Agents for Acids and Caustics:* Not pertinent; *Polymerization:* Not pertinent; *Inhibitor of Polymerization:* Not pertinent.

ETHYL MERCAPTAN

Chemical Designations - *Synonyms:* Ethanethiol; Ethyl Sulfhydrate; Mercaptoethane; Thioethyl alcohol; *Chemical Formula:* C_2H_5SH.

Observable Characteristics - *Physical State (as shipped):* Liquid; *Color:* Colorless to yellow; *Odor:* Strong chunk; offensive garlic.

Physical and Chemical Properties - *Physical State at 15 °C and 1 atm:* Liquid; *Molecular Weight:* 62,1; *Boiling Point at 1 atm:* 93,9, 34,4, 307,6; *Freezing Point:* -234, -147, 126; *Critical Temperature:* 439, 226, 499; *Critical Pressure (psia, atm, MN/m²):* 798, 54.2, 5.50; *Specific Gravity:* 0.826 at 20°C (liquid); *Vapor (Gas) Specific Gravity:* 2.1; *Ratio of Specific Heats of Vapor (Gas):* 1.1308 at 16°C; *Latent Heat of Vaporization:* 189, 105, 4,39; *Heat of Combustion:* -15,000, -8,300, -350; *Heat of Decomposition:* Not pertinent.

Health Hazards Information - *Personal Protective Equipment:* Plastic gloves; goggles or face shield. *Symptoms Following Exposure:* Inhalation of vapor causes muscular weakness, convulsions, respiratory paralysis. High concentrations may cause pulmonary irritation. Liquid irritates eyes and skin. Ingestion causes nausea and irritation of mouth and stomach. *Treatment for Exposure:* INHALATION: move victim to fresh air; if he is unconscious, give artificial respiration and oxygen; get medical attention. EYES: flush with water for at least 15 min. following contact with liquid; get medical attention if irritation persists. SKIN: wash well with water. INGESTION: induce vomiting and follow with gastric lavage; get medical attention. *Toxicity by Inhalation (Threshold Limit Value):* 0.5 ppm *Short-Term Inhalation Limits:* Data not available *Toxicity by Ingestion:* Grade 2; oral LD_{50} = 682 mg/kg (rat) *Late Toxicity:* May impair respiratory muckle function in warm-blooded experimental animals *Vapor (Gas) Irritant Characteristics:* Data not available *Liquid or Solid Irritant Characteristics:* Data not available *Odor Threshold:* 0.001 ppm.

Fire Hazards - *Flash Point (deg. F):* <0 OC; *Flammable Limits in Air (%):* 2.8 - 18; *Fire Extinguishing Agents:* Dry chemical, foam, carbon dioxide; *Fire Extinguishing Agents Not To Be Used:* Water may be ineffective; *Special Hazards of Combustion Products:* Irritating fumes of sulfur dioxide are generated; *Behavior in Fire:* Vapor is heavier than air and may travel long distance to a source of ignition and flash back; containers may explode in a fire; offensive fumes are released when heated; *Ignition Temperature (deg. F):* 572; *Electrical Hazard:* Data not available; *Burning Rate:* 5.7 mm/min.
Chemical Reactivity - *Reactivity with Water:* No reaction; *Reactivity with Common Materials:* No reaction; *Stability During Transport:* Stability; *Neutralizing Agents for Acids and Caustics:* Not pertinent; *Polymerization:* Not pertinent; *Inhibitor of Polymerization:* Not pertinent.

ETHYL METHACRYLATE

Chemical Designations - *Synonyms:* Ethyl 2-methacrylate; Ethyl Methacrylate-inhibited; Ethyl Alphamethylmethacrylate; Ethyl 2-methyl-2-propenoate; Methacrylic Acid, Ethyl Ester; *Chemical Formula:* $CH_2 = C(CH_3)COOC_2H_5$.
Observable Characteristics - *Physical State (as shipped):* Liquid; *Color:* Colorless; *Odor:* Acrid acrylic.
Physical and Chemical Properties - *Physical State at 15 °C and 1 atm:* Liquid; *Molecular Weight:* 114; *Boiling Point at 1 atm:* 243, 117, 390; *Freezing Point:* <-58, <-50, <223; *Critical Temperature:* Not pertinent; *Critical Pressure:* Not pertinent; *Specific Gravity:* 0,9151 at 20°C (liquid); *Vapor (Gas) Specific Gravity:* 3.9; *Ratio of Specific Heats of Vapor (Gas):* 1.064; *Latent Heat of Vaporization:* 170, 96, 4.0; *Heat of Combustion:* -12,670, -7,040, -294; *Heat of Decomposition:* Not pertinent.
Health Hazards Information - *Personal Protective Equipment:* Impervious gloves; splash goggles; self-contained breathing apparatus if exposed to vapors; coveralls. *Symptoms Following Exposure:* Inhalation may cause irritation of the mucous membrane. Ingestion causes irritation of mouth and stomach. Contact with liquid irritates eyes and skin. *Treatment for Exposure:* INHALATION: remove victim to fresh air; apply artificial respiration and oxygen if indicated. INGESTION: induce vomiting; call a physician. EYES: wash with copious quantities of water for 15 min.; call a physician. SKIN: flush with water; wash with soap and water *Toxicity by Inhalation (Threshold Limit Value):* Data not available *Short-Term Inhalation Limits:* Data not available *Toxicity by Ingestion:* Grade 2; oral LD_{50} = 4 g/kg (rabbit) *Late Toxicity:* Causes birth defects in experimental animals *Vapor (Gas) Irritant Characteristics:* Data not available *Liquid or Solid Irritant Characteristics:* Data not available *Odor Threshold:* Data not available.
Fire Hazards - *Flash Point (deg. F):* 85 OC; 80 CC; *Flammable Limits in Air (%):* 1.8 (LFL); *Fire Extinguishing Agents:* Foam, dry chemical, carbon dioxide; *Fire Extinguishing Agents Not To Be Used:* Water may be ineffective; *Special Hazards of Combustion Products:* Data not available; *Behavior in Fire:* Sealed container may rupture explosively if hot. Heat can cause a violent polymerization reaction with rapid release of energy. Vapors are heavier than air and can travel considerable distance to a source of ignition and flash back; *Ignition Temperature (deg. F):* 740; *Electrical Hazard:* Data not available; *Burning Rate:* 4.56 mm/min.
Chemical Reactivity - *Reactivity with Water:* No reaction; *Reactivity with Common Materials:* No reaction; *Stability During Transport:* Stable; *Neutralizing Agents for Acids and Caustics:* Not pertinent; *Polymerization:* If proper concentration of inhibitor is not present or when material is hot, a violent polymerization reaction may occur; *Inhibitor of Polymerization:* Oxygen in the air inhibits polymerization.

ETHYL NITRITE

Chemical Designations - *Synonyms:* Nitrous ether; Spirit of ether nitrite; Sweet spirit of nitre; *Chemical Formula:* C_2H_5ONO.
Observable Characteristics - *Physical State (as shipped):* Liquid; *Color:* Colorless to pale yellow; *Odor:* Aromatic; ethereal; characteristic.
Physical and Chemical Properties - *Physical State at 15 °C and 1 atm:* Liquid; *Molecular Weight:*

75,1; *Boiling Point at 1 atm:* 63, 17, 290; *Freezing Point:* -58, -50, 223; *Critical Temperature:* Not pertinent; *Critical Pressure:* Not pertinent; *Specific Gravity:* 0.900 at 15°C (liquid); *Vapor (Gas) Specific Gravity:* 2.6; *Ratio of Specific Heats of Vapor (Gas):* Data not available; *Latent Heat of Vaporization:* 229, 127, 5.32; *Heat of Combustion:* (est.) -7,800, -4,300, -180; *Heat of Decomposition:* Not pertinent.

Health Hazards Information - *Personal Protective Equipment:* Self-contained breathing apparatus; goggles or face shield; rubber gloves. *Symptoms Following Exposure:* Inhalation or ingestion causes headache, increased pulse rate, decreased blood pressure, and unconsciousness. Contact with liquid irritates eyes and skin; *Treatment for Exposure:* INHALATION: remove victim from exposure; if breathing has stopped, give artificial respiration; call physician. EYES: flush with water, wash with soap and water. INGESTION: do not induce vomiting; call physician. *Toxicity by Inhalation (Threshold Limit Value):* Data not available; *Short-Term Inhalation Limits:* Data not available; *Toxicity by Ingestion:* Data not available; *Late Toxicity:* Data not available; *Vapor (Gas) Irritant Characteristics:* Data not available; *Liquid or Solid Irritant Characteristics:* Data not available; *Odor Threshold:* Data not available.

Fire Hazards - *Flash Point (deg. F):* -31 CC; *Flammable Limits in Air (%):* 3 - >50; *Fire Extinguishing Agents:* Water dry chemical, carbon dioxide, water, foam; *Fire Extinguishing Agents Not To Be Used:* Not pertinent; *Special Hazards of Combustion Products:* Toxic oxides of nitrogen are generated; *Behavior in Fire:* Vapors are heavier than air and may travel a considerable distance to a source of ignition and flash back; can decompose violently above 194° F; containers may explode in a fire; *Ignition Temperature (deg. F):* 194; *Electrical Hazard:* Data not available; *Burning Rate:* 2.6 mm/min.

Chemical Reactivity - *Reactivity with Water:* No reaction; *Reactivity with Common Materials:* No reaction; *Stability During Transport:* Stable if stored in a cool place and not exposed to strong light; *Neutralizing Agents for Acids and Caustics:* Not pertinent; *Polymerization:* Not pertinent; *Inhibitor of Polymerization:* Not pertinent.

ETHYLENE GLYCOL MONOBUTYL ETHER

Chemical Designations - *Synonyms:* 2-Butoxyethanol Butyl Cellosolve Dovonol EB Glycol Butyl Ether Poly-Solv EB; *Chemical Formula:* $CH_3(CH_2)_3OCH_2CH_2OH$.

Observable Characteristics - *Physical State (as shipped):* Liquid; *Color:* Colorless; *Odor:* Mild, characteristic; slightly rancid; mild ethereal.

Physical and Chemical Properties - *Physical State at 15 °C and 1 atm:* Liquid; *Molecular Weight:* 118.18; *Boiling Point at 1 atm:* 340.2, 171.2, 444.4; *Freezing Point:* -103, -75, 198; *Critical Temperature:* 694, 368, 641; *Critical Pressure:* 470, 32, 3.2; *Specific Gravity:* 0.902 at 20°C (liquid); *Vapor (Gas) Specific Gravity:* Not pertinent; *Ratio of Specific Heats of Vapor (Gas):* 1.047 *Latent Heat of Vaporization:* 157, 87.1, 3.65; *Heat of Combustion:* -13,890, -7,720, -323; *Heat of Decomposition:* Not pertinent.

Health Hazards Information - *Personal Protective Equipment:* Air pack or organic canister respirator; rubber gloves; goggles; clothing to prevent body contact with liquid. *Symptoms Following Exposure:* Vapors irritate eyes and nose. Ingestion or skin contact causes headache, nausea, vomiting, dizziness. *Treatment for Exposure:* INHALATION: remove to fresh air and call a physician. SKIN OR EYES: immediately flush with plenty of water; get medical care for eyes. *Toxicity by Inhalation (Threshold Limit Value):* 50 ppm *Short-Term Inhalation Limits:* Data not available *Toxicity by Ingestion:* Grade 2; LD_{50} 0.5 to 5 g/kg (rat) *Late Toxicity:* Data not available *Vapor (Gas) Irritant Characteristics:* Vapors cause a slight smarting of the eyes or respiratory system if present in high concentrations. The effect is temporary. *Liquid or Solid Irritant Characteristics:* Minimum hazard. If spilled on clothing and allowed to remain, may cause smarting and reddening of the skin. *Odor Threshold:* Data not available.

Fire Hazards - *Flash Point (deg. F):* 165 OC; 155 CC; *Flammable Limits in Air (%):* 1.1 - 10.6; *Fire Extinguishing Agents:* Carbon dioxide or dry chemical for small fires; alcohol-type foam for large fires; *Fire Extinguishing Agents Not To Be Used:* Data not available; *Special Hazards of Combustion Products:* Not pertinent; *Behavior in Fire:* Not pertinent; *Ignition Temperature (deg. F):* 472; *Electrical*

Hazard: Not pertinent; *Burning Rate:* 6.7 mm/min.
Chemical Reactivity - *Reactivity with Water:* No reaction; *Reactivity with Common Materials:* No reaction; *Stability During Transport:* Stable; *Neutralizing Agents for Acids and Caustics:* Not pertinent; *Polymerization:* Not pertinent; *Inhibitor of Polymerization:* Not pertinent.

ETHYLENE GLYCOL MONOETHYL ETHER
**Chemical Designations - ** *Synonyms:* Cellosolve; Dowanol EE; 2-Ethoxyethanol; Ethylene Glycol Ethyl Ether; Glycol Monoethyl ether; Poli-Solv EE; *Chemical Formula:* $HOCH_2CH_2OCH_2CH_3$.
Observable Characteristics - *Physical State (as shipped):* Liquid; *Color:* Colorless; *Odor:* Sweetish; mild, pleasant, ethereal.
Physical and Chemical Properties - *Physical State at 15 °C and 1 atm:* Liquid; *Molecular Weight:* 90.12; *Boiling Point at 1 atm:* 275.2, 135.1, 408.3; *Freezing Point:* Data not available; *Critical Temperature:* Not pertinent; *Critical Pressure:* Not pertinent; *Specific Gravity:* 0.931 at 20°C (liquid); *Vapor (Gas) Specific Gravity:* Not pertinent; *Ratio of Specific Heats of Vapor (Gas):* 1.064; *Latent Heat of Vaporization:* 191, 106, 4.44; *Heat of Combustion:* -13,000, -7,4000, -310; *Heat of Decomposition:* Not pertinent.
Health Hazards Information - *Personal Protective Equipment:* Organic das mask; goggles or face shield; rubber gloves. *Symptoms Following Exposure:* Some eye irritation. Inhalation of vapors causes irritation of nose. *Treatment for Exposure:* Flush eyes with water for 15 min. Flush skin with large volumes of water. Call a physician. *Toxicity by Inhalation (Threshold Limit Value):* 200 ppm; *Short-Term Inhalation Limits:* Not pertinent; *Toxicity by Ingestion:* Grade 2; LD_{50} 0.5 to 5 g/kg (rat, rabbit) *Late Toxicity:* Data not available; *Vapor (Gas) Irritant Characteristics:* Vapors cause a slight smarting of the eyes or respiratory system if present in high concentrations. The effect is temporary. *Liquid or Solid Irritant Characteristics:* Minimum hazard. If spilled on chopping and allowed to remain, may cause smarting and reddening of the skin. *Odor Threshold:* Data not available.
Fire Hazards - *Flash Point (deg. F):* 120 OC; 202 CC; *Flammable Limits in Air (%):* 1.8 - 14,0; *Fire Extinguishing Agents:* Alcohol foam, carbon dioxide, or dry chemical; *Fire Extinguishing Agents Not To Be Used:* Not pertinent; *Special Hazards of Combustion Products:* Not pertinent; *Behavior in Fire:* Not pertinent; *Ignition Temperature (deg. F):* 455; *Electrical Hazard:* Not pertinent; *Burning Rate:* 2.4 mm/min.
Chemical Reactivity - *Reactivity with Water:* No reaction; *Reactivity with Common Materials:* No reaction; *Stability During Transport:* Stable; *Neutralizing Agents for Acids and Caustics:* Not pertinent; *Polymerization:* Not pertinent; *Inhibitor of Polymerization:* Not pertinent.

ETHYLENE GLYCOL MONOMETHYL ETHER
Chemical Designations - *Synonyms:* Dovanol EM; Glycol Monomethyl Ether; 2-Methoxyethanol Methyl Cellosolve; Poly-Solv EM; *Chemical Formula:* $CH_3OCH_2CH_2ON$.
Observable Characteristics - *Physical State (as shipped):* Liquid; *Color:* Colorless; *Odor:* Mild ethereal.
Physical and Chemical Properties - *Physical State at 15 °C and 1 atm:* Liquid; *Molecular Weight:* 76.10; *Boiling Point at 1 atm:* 256.1, 124.5, 397.7; *Freezing Point:* -121.2, -85.1, 188.1; *Critical Temperature:* 558, 292, 565; *Critical Pressure:* 735, 50, 5.1; *Specific Gravity:* 0.966 at 20°C (liquid); *Vapor (Gas) Specific Gravity:* Not pertinent; *Ratio of Specific Heats of Vapor (Gas):* 1.079; *Latent Heat of Vaporization:* 223, 124.5, 19; *Heat of Combustion:* No data; *Heat of Decomposition:* Not pertinent.
Health Hazards Information - *Personal Protective Equipment:* Chemical safety goggles; protective clothing; supplied-air respirator for high concentrations; safety shower and eye bath. *Symptoms Following Exposure:* Irritation of skin and eyes. Chronic exposure may also cause weakness, sleepiness, headache, gastrointestinal upset, weight loss, change of personality. *Treatment for Exposure:* SKIN OR EYES: wash affected area with water for 15 min. *Toxicity by Inhalation (Threshold Limit Value):* 25 ppm *Short-Term Inhalation Limits:* Not pertinent; *Toxicity by Ingestion:* Grade 2; LD_{50} 0.5 to 5 g/kg (rat, rabbit, guinea pig); *Late Toxicity:* Causes blood disorders and damage to central nervous system in humans. *Vapor (Gas) Irritant Characteristics:* Vapors cause a slight smarting of the eyes or

respiratory system if present in high concentrations. The effect is temporary. *Liquid or Solid Irritant Characteristics:* Minimum hazard. If spilled on clothing and allowed to remain, may cause smarting and reddening of the skin. *Odor Threshold:* 0.9 ppm.

Fire Hazards - *Flash Point (deg. F):* 120 OC; 107 CC; *Flammable Limits in Air (%):* 2.5 - 19 8; *Fire Extinguishing Agents:* Dry chemical, carbon dioxide or alcohol foam; *Fire Extinguishing Agents Not To Be Used:* Not pertinent; *Special Hazards of Combustion Products:* Not pertinent; *Behavior in Fire:* Not pertinent; *Ignition Temperature (deg. F):* 551; *Electrical Hazard:* Not pertinent; *Burning Rate:* 1.8 mm/min.

Chemical Reactivity - *Reactivity with Water:* No reaction; *Reactivity with Common Materials:* No reaction; *Stability During Transport:* Stable; *Neutralizing Agents for Acids and Caustics:* Not pertinent; *Polymerization:* Not pertinent; *Inhibitor of Polymerization:* Not pertinent .

ETHYLENEIMINE

Chemical Designations - *Synonyms:* Azirane; Aziridine; *Chemical Formula:* CH_2CH_2NH.

Observable Characteristics - *Physical State (as shipped):* Liquid; *Color:* Colorless; *Odor:* Fishy; ammoniacal.

Physical and Chemical Properties - *Physical State at 15 °C and 1 atm:* Liquid; *Molecular Weight:* 43.07; *Boiling Point at 1 atm:* 133, 56, 329; *Freezing Point:* -108, -78, 195; *Critical Temperature:* Not pertinent; *Critical Pressure:* Not pertinent; *Specific Gravity:* 0,832 at 20°C (liquid); *Vapor (Gas) Specific Gravity:* 1.5; *Ratio of Specific Heats of Vapor (Gas):* 1.192; *Latent Heat of Vaporization:* 333, 185, 7.75; *Heat of Combustion:* -15,930, -8,850, -370.5; *Heat of Decomposition:* Not pertinent.

Health Hazards Information - *Personal Protective Equipment:* If exposure is possible, wear full clothing (neoprene slicker suit, rubber boots, rubber gloves, chemical goggles). If vapors may be present, wear all purpose canister or gas mask; if vapors are known to be present, use self-contained breathing apparatus. *Symptoms Following Exposure:* Material gives inadequate warning of overexposure by respiration or skin contact. May cause vomiting and possibly death when inhaled, ingested, or absorbed through skin. Severe blistering agent; can produce third-degree chemical burns of skin. Has corrosive effect on mucous membranes and may cause scaring of esophagus if swallowed. Corrosive to eye tissue; may cause permanent corneal opacity and conjunctival scarring. Effects on eye tissue, and skin may be delayed. *Treatment for Exposure:* INHALATION: remove victim from exposure and administer oxygen; steroid therapy (by physician) is recommended. SKIN OR EYES: prompt and adequate irrigation with water (within 60 seconds of exposure) can prevent serious injury. *Toxicity by Inhalation (Threshold Limit Value):* Data not available *Short-Term Inhalation Limits:* 5 ppm for 30 min *Toxicity by Ingestion:* Grade 4; LD below 50 mg/kg (rat) *Late Toxicity:* Causes cancer in mice. Effects on man unknown. *Vapor (Gas) Irritant Characteristics:* Vapor is moderately irritating such that personnel will not usually tolerate moderate or high concentrations.

Fire Hazards - *Flash Point (deg. F):* 1 OC; 12 CC; *Flammable Limits in Air (%):* 3.3 - 54.8; *Fire Extinguishing Agents:* Dry chemical, alcohol foam, carbon dioxide; *Fire Extinguishing Agents Not To Be Used:* Not pertinent; *Special Hazards of Combustion Products:* Irritating vapor generated when heated; *Behavior in Fire:* Vapor is heavier than air and may travel considerable distance to a source of ignition and flash back; *Ignition Temperature (deg. F):* 608; *Electrical Hazard:* Not pertinent; *Burning Rate:* Data not available.

Chemical Reactivity - *Reactivity with Water:* Mild reaction, non-hazardous; *Reactivity with Common Materials:* Contact with silver or aluminum may cause polymerization; *Stability During Transport:* Stable unless heated under pressure; *Neutralizing Agents for Acids and Caustics:* Flush with water; *Polymerization:* Explosive polymerization can occur when in contact with acids; *Inhibitor of Polymerization:* None used.

ETHYLIDENENORBORNENE

Chemical Designations - *Synonyms:* 5-Ethylidenebicyclo(2,2.1)hept-2-ene; Etehylidenenorbornylene; Ethylidenenorcamphene; *Chemical Formula:* $C_9H_{12.}$

Observable Characteristics - *Physical State (as shipped):* Liquid; *Color:* Colorless; *Odor:* Like turpentine.

Physical and Chemical Properties - *Physical State at 15 ℃ and 1 atm:* Liquid; *Molecular Weight:* 120.2; *Boiling Point at 1 atm:* 297.7, 147.6, 420.8; *Freezing Point:* -112, -80, 193; *Critical Temperature:* Not pertinent; *Critical Pressure:* Not pertinent; *Specific Gravity:* 0.896 at 20°C (liquid); *Vapor (Gas) Specific Gravity:* 4.1; *Ratio of Specific Heats of Vapor (Gas):* Not pertinent; *Latent Heat of Vaporization:* Data not available; *Heat of Combustion:* (est.) -18,800, -10,450, -437; *Heat of Decomposition:* Not pertinent.

Health Hazards Information - *Personal Protective Equipment:* Organic canister or air-supplied mask; goggles or face shield; rubber gloves; *Symptoms Following Exposure:* Inhalation of vapors causes headache, confusion, and respiratory distress. Ingestion causes irritation of entire digestive system. Aspiration causes severe pneumonia. Contact with liquid causes irritation of eyes and skin. *Treatment for Exposure:* INHALATION: remove victim to fresh air; administer artificial respiration and oxygen if required; call a doctor. INGESTION: dive large amount of water and induce vomiting; get medical attention at once. EYES: flush with water for at least 15 min. SKIN: wipe off, wash with soap and water. *Toxicity by Inhalation (Threshold Limit Value):* Data not available; *Short-Term Inhalation Limits:* Data not available; *Toxicity by Ingestion:* Grade 2; oral LD$_{50}$ = 2.83 g/kg (rat); *Late Toxicity:* Causes kidney lesions and gain in kidney and liver weights in rats; *Vapor (Gas) Irritant Characteristics:* Vapors are moderately irritating such that personnel will not usually tolerate moderate or high concentrations. *Liquid or Solid Irritant Characteristics:* Minimum hazard. If spilled on clothing and allowed to remain, may cause smarting and reddening of skin. *Odor Threshold:* 0.007 ppm.

Fire Hazards - *Flash Point (deg. F):* 98 OC; *Flammable Limits in Air (%):* Data not available; *Fire Extinguishing Agents:* Dry chemical, foam, carbon dioxide; *Fire Extinguishing Agents Not To Be Used:* Water may be ineffective; *Special Hazards of Combustion Products:* Data not available; *Behavior in Fire:* Data not available; *Ignition Temperature :* Data not available; *Electrical Hazard:* Data not available; *Burning Rate:* Data not available.

Chemical Reactivity - *Reactivity with Water:* No reaction; *Reactivity with Common Materials:* No reaction; *Stability During Transport:* Stable; *Neutralizing Agents for Acids and Caustics:* Not pertinent; *Polymerization:* Not pertinent; *Inhibitor of Polymerization:* Not pertinent.

F

FERRIC AMMONIUM CITRATE

Chemical Designations - *Synonyms*: Ammonium ferric citrate, Ferric ammonium citrate (brown), Ferric ammonium citrate (green); *Chemical Formula*: Mixture of $FeC_6H_5O_7$, $(NH_4)_2HC_6H_5O_7$, and water of hydration.

Observable Characteristics - *Physical State (as shipped)*: Solid; *Color*: Red, green, or brown; *Odor*: None.

Physical and Chemical Properties - *Physical State at 15 ℃ and 1 atm.*: Solid; *Molecular Weight*: Not pertinent (mixture); *Boiling Point at 1 atm.*: Not pertinent (decomposes); *Freezing Point*: Not pertinent; *Critical Temperature*: Not pertinent; *Critical Pressure*: Not pertinent; *Specific Gravity*: 1.8 at 20, (solid); *Vapor (Gas) Specific Gravity*: Not pertinent; *Ratio of Specific Heats of Vapor (Gas)*: Not pertinent; *Latent Heat of Vaporization*: Not pertinent; *Heat of Combustion*: Not pertinent; *Heat of Decomposition*: Not pertinent.

Health Hazards Information - *Recommended Personal Protective Equipment*: Approved respirator for nuisance dust, chemical goggles or face shield; *Symptoms Following Exposure*: Inhalation of dust irritates nose and throat. Ingestion causes irritation of mouth and stomach. Dust irritates eyes and causes mild irritation of skin on prolonged contact; *General Treatment for Exposure*: INGESTION: give large amount of water. EYES or SKIN: flush with water; *Toxicity by Inhalation (Threshold Limit Value)*: 1 mg/m^3 (as iron); *Short-Term Inhalation Limits*: Data not available; *Toxicity by Ingestion*: Data not available; *Late Toxicity*: Data not available; *Vapor (Gas) Irritant Characteristics*: Data not available; *Liquid or Solid Irritant Characteristics*: Data not available; *Odor Threshold*: Data not available.

Fire Hazards - *Flash Point* : Not flammable; *Flammable Limits in Air (%):* Not flammable; *Fire Extinguishing Agents:* Not pertinent; *Fire Extinguishing Agents Not To Be Used:* Not pertinent; *Special Hazards of Combustion Products:* Toxic oxides of nitrogen or ammonia gas may be formed in fire; *Behavior in Fire:* Not pertinent; *Ignition Temperature :* Not pertinent; *Electrical Hazard:* Not pertinent; *Burning Rate:* Not pertinent.

Chemical Reactivity - *Reactivity with Water:* No reaction; *Reactivity with Common Materials:* No reaction; *Stability During Transport:* Stable; *Neutralizing Agents for Acids and Caustics:* Not pertinent; *Polymerization:* Not pertinent; *Inhibitor of Polymerization:* Not pertinent.

FERRIC AMMONIUM OXALATE

Chemical Designations - *Synonyms*: Ammonium ferric oxalate trihydrate, ammonium trioxalatoferrate trihydrate; *Chemical Formula*: $Fe(NH_4)_3(C_2O_4)_3 \cdot 3H_2O$.

Observable Characteristics - *Physical State (as shipped)*: Solid; *Color*: Yellowish-green; *Odor*: Slight burnt sugar odor.

Physical and Chemical Properties - *Physical State at 15 °C and 1 atm.*: Solid; *Molecular Weight*: 428; *Boiling Point at 1 atm.*: Not pertinent (decomposes); *Freezing Point*: Not pertinent; *Critical Temperature*: Not pertinent; *Critical Pressure*: Not pertinent; *Specific Gravity*: 1.78 at 20, solid); *Vapor (Gas) Specific Gravity*: Not pertinent; *Ratio of Specific Heats of Vapor (Gas)*: Not pertinent; *Latent Heat of Vaporization*: Not pertinent; *Heat of Combustion*: Not pertinent; *Heat of Decomposition*: Not pertinent.

Health Hazards Information - *Recommended Personal Protective Equipment*: Approved dust respirator, rubber or plastic-coated gloves, chemical goggles or face shield; *Symptoms Following Exposure*: Inhalation of dust may cause irritation of nose and throat. Ingestion causes burning pain in throat and stomach; mucous membranes become white; may also cause vomiting, weak pulse, cardiovascular collapse, and death. Contact with dust irritates eyes and skin; may cause severe skin burns; *General Treatment for Exposure*: (treat victim promptly) INHALATION: move to fresh air; get medical attention if any symptoms persist. INGESTION: give immediately a dilute solution of any soluble calcium salt such as calcium lactate, lime water, chalk, or milk; induce vomiting; get medical attention. (Watch for edema of the glottis and delayed constriction of esophagus.) EYES: flush with water and get medical attention. SKIN: flush with water; *Toxicity by Inhalation (Threshold Limit Value)*: 1 mg/m³ (as iron); *Short-Term Inhalation Limits*: Data not available; *Toxicity by Ingestion*: Data not available; *Late Toxicity*: Data not available; *Vapor (Gas) Irritant Characteristics*: Data not available; *Liquid or Solid Irritant Characteristics*: Data not available; *Odor Threshold*: Data not available.

Fire Hazards - *Flash Point* : Not flammable; *Flammable Limits in Air (%):* Not flammable; *Fire Extinguishing Agents:* Not pertinent; *Fire Extinguishing Agents Not To Be Used:* Not pertinent; *Special Hazards of Combustion Products:* Toxic oxides of nitrogen, ammonia , and carbon monoxide may form in fires; *Behavior in Fire:* Not pertinent; *Ignition Temperature :* Not pertinent; *Electrical Hazard:* Not pertinent; *Burning Rate:* Not pertinent.

Chemical Reactivity - *Reactivity with Water:* No reaction; *Reactivity with Common Materials:* No reaction; *Stability During Transport:* Stable; *Neutralizing Agents for Acids and Caustics:* Not pertinent; *Polymerization:* Not pertinent; *Inhibitor of Polymerization:* Not pertinent.

FERRIC CHLORIDE

Chemical Designations - *Synonyms*: Ferric Chloride (anhydrous), Ferric Chloride (hexahydrate), Iron (III) chloride, Iron perchloride, Iron trichloride; *Chemical Formula*: $FeCl_3$ or $FeCl_3 \cdot 6H_2O$.

Observable Characteristics - *Physical State (as shipped)*: Solid; *Color*: Anhydrous: greenish black, Hidrate: brown; *Odor*: None.

Physical and Chemical Properties - *Physical State at 15 °C and 1 atm.*: Solid; *Molecular Weight*:

162.22 (anhydrous); *Boiling Point at 1 atm.*: Not pertinent (decomposes); *Freezing Point*: Not pertinent; *Critical Temperature*: Not pertinent; *Critical Pressure*: Not pertinent; *Specific Gravity*: 2.8 at 20 °C, (anhydrous solid); *Vapor (Gas) Specific Gravity*: Not pertinent; *Ratio of Specific Heats of Vapor (Gas)*: Not pertinent; *Latent Heat of Vaporization*: Not pertinent; *Heat of Combustion*: Not pertinent; *Heat of Decomposition*: Not pertinent.

Health Hazards Information - *Recommended Personal Protective Equipment*: Dust respirator if required, rubber apron and boots, chemical worker's goggles or face shield; *Symptoms Following Exposure*: Inhalation of dust may irritation nose and throat. Ingestion causes irritation of mouth and stomach. Dust irritates eyes. Prolonged contact with skin causes irritation and burns; *General Treatment for Exposure*: INGESTION: give large amounts of water; induce vomiting if large amounts have been swallowed. EYES: immediately flush with plenty of water for at least 15 min.; get medical attention promptly. SKIN: flush with water; *Toxicity by Inhalation (Threshold Limit Value)*: 1 mg/m^3 (as iron); *Short-Term Inhalation Limits*: Data not available; *Toxicity by Ingestion*: Grade 2, LD$_{50}$ 0.5—5 g/kg (rat); *Late Toxicity*: Data not available; *Vapor (Gas) Irritant Characteristics*: Not pertinent; *Liquid or Solid Irritant Characteristics*: Data not available; *Odor Threshold*: Data not available.

Fire Hazards - *Flash Point* : Not flammable; *Flammable Limits in Air (%):* Not flammable; *Fire Extinguishing Agents:* Not pertinent; *Fire Extinguishing Agents Not To Be Used:* Not pertinent; *Special Hazards of Combustion Products:* Irritating hydrogen chloride fumes may form in fire; *Behavior in Fire:* Not pertinent; *Ignition Temperature:* Not pertinent; *Electrical Hazard:* Not pertinent; *Burning Rate:* Not pertinent.

Chemical Reactivity - *Reactivity with Water:* No reaction; *Reactivity with Common Materials:* Water solutions are acidic and corrosive to most metals; *Stability During Transport:* Stable; *Neutralizing Agents for Acids and Caustics:* Flush with water, rinse with dilute sodium bicarbonate or soda ash solution; *Polymerization:* Not pertinent; *Inhibitor of Polymerization:* Not pertinent.

FERRIC GLYCEROPHOSPHATE

Chemical Designations - *Synonyms*: No common synonyms; *Chemical Formula*: (approx.) $Fe_2[C_3H_5(OH)_2PO_4]_3 \cdot H_2O$.

Observable Characteristics - *Physical State (as shipped)*: Solid; *Color*: Greenish-brown, greenish-yellow; *Odor*: None.

Physical and Chemical Properties - *Physical State at 15 °C and 1 atm.*: Solid; *Molecular Weight*: 470 (approx.); *Boiling Point at 1 atm.*: Not pertinent (decomposes); *Freezing Point*: Not pertinent; *Critical Temperature*: Not pertinent; *Critical Pressure*: Not pertinent; *Specific Gravity*: 1.5 at 20 °C, (solid); *Vapor (Gas) Specific Gravity*: Not pertinent; *Ratio of Specific Heats of Vapor (Gas)*: Not pertinent; *Latent Heat of Vaporization*: Not pertinent; *Heat of Combustion*: Not pertinent; *Heat of Decomposition*: Not pertinent.

Health Hazards Information - *Recommended Personal Protective Equipment*: Goggles or face shield, dust mask, rubber gloves; *Symptoms Following Exposure*: Inhalation of dust may irritate nose and throat. Contact with dust irritates eyes and (on prolonged contact) skin; *General Treatment for Exposure*: INHALATION: move to fresh air. INGESTION: give large amount of water; induce vomiting if large amounts have been swallowed. EYES: flush with water for at least 15 min. SKIN: flush with water; *Toxicity by Inhalation (Threshold Limit Value)*: 1 mg/m^3 (as iron); *Short-Term Inhalation Limits*: Data not available; *Toxicity by Ingestion*: Data not available; *Late Toxicity*: Data not available; *Vapor (Gas) Irritant Characteristics*: Not pertinent; *Liquid or Solid Irritant Characteristics*: Data not available; *Odor Threshold*: Odorless.

Fire Hazards - *Flash Point* : Not flammable; *Flammable Limits in Air (%):* Not flammable; *Fire Extinguishing Agents:* Not pertinent; *Fire Extinguishing Agents Not To Be Used:* Not pertinent; *Special Hazards of Combustion Products:* Not pertinent; *Behavior in Fire:* Not pertinent; *Ignition Temperature* : Not pertinent; *Electrical Hazard:* Not pertinent; *Burning Rate:* Not pertinent.

Chemical Reactivity - *Reactivity with Water:* No reaction; *Reactivity with Common Materials:* No reaction; *Stability During Transport:* Stable; *Neutralizing Agents for Acids and Caustics:* Not pertinent; *Polymerization:* Not pertinent; *Inhibitor of Polymerization:* Not pertinent.

FERRIC NITRATE

Chemical Designations - *Synonyms*: Ferric nitrate monohydrate, Nitric acid, iron (+3) salt; *Chemical Formula*: $Fe(NO_3)_3 \bullet 9H_2O$.

Observable Characteristics - *Physical State (as shipped)*: Solid; *Color*: Green, colorless to pale violet; *Odor*: None.

Physical and Chemical Properties - *Physical State at 15 °C and 1 atm.*: Solid; *Molecular Weight*: 404.02; *Boiling Point at 1 atm.*: Not pertinent (decomposes); *Freezing Point*: 117, 47, 320; *Critical Temperature*: Not pertinent; *Critical Pressure*: Not pertinent; *Specific Gravity*: 1.7 at 20, (solid); *Vapor (Gas) Specific Gravity*: Not pertinent; *Ratio of Specific Heats of Vapor (Gas)*: Not pertinent; *Latent Heat of Vaporization*: Not pertinent; *Heat of Combustion*: Not pertinent; *Heat of Decomposition*: Not pertinent.

Health Hazards Information - *Recommended Personal Protective Equipment*: Dust mask, goggles or face shield, protective gloves; *Symptoms Following Exposure*: Inhalation of dust irritates nose and throat. Ingestion causes irritation of mouth and stomach. Dust irritates eyes and can irritate skin on prolonged contact; *General Treatment for Exposure*: INHALATION: move to fresh air. INGESTION: give large amounts of water; induce vomiting if large amounts have been swallowed. EYES: flush with water; get medical attention if irritation persist. SKIN: flush with water; *Toxicity by Inhalation (Threshold Limit Value)*: 1 mg/m^3 (as iron); *Short-Term Inhalation Limits*: Data not available; *Toxicity by Ingestion*: Grade 2, LD_{50} 0.5 - 5 g/kg; *Late Toxicity*: Data not available; *Vapor (Gas) Irritant Characteristics*: Not pertinent; *Liquid or Solid Irritant Characteristics*: Data not available; *Odor Threshold*: Data not available.

Fire Hazards - *Flash Point* : Not flammable; *Flammable Limits in Air (%)*: Not flammable; *Fire Extinguishing Agents:* Not pertinent; *Fire Extinguishing Agents Not To Be Used:* Not pertinent; *Special Hazards of Combustion Products:* Toxic oxides of nitrogen and nitric acid vapor may form in fires; *Behavior in Fire:* In contact with combustible materials, will increase the intensity of a fire; *Ignition Temperature :* Not pertinent; *Electrical Hazard:* Not pertinent; *Burning Rate:* Not pertinent.

Chemical Reactivity - *Reactivity with Water:* No reaction; *Reactivity with Common Materials:* Solutions are corrosive to most metals. Contact of solid with wood or paper may cause fire; *Stability During Transport:* Stable; *Neutralizing Agents for Acids and Caustics:* Not pertinent; *Polymerization:* Not pertinent; *Inhibitor of Polymerization:* Not pertinent.

FERRIC SULFATE

Chemical Designations - *Synonyms*: Iron sesquisulfate, Iron (III) sulfate, Iron tersulfate; *Chemical Formula*: $Fe_2(SO_4)_3$.

Observable Characteristics - *Physical State (as shipped)*: Solid; *Color*: Gray-white; *Odor*: None.

Physical and Chemical Properties - *Physical State at 15 °C and 1 atm.*: Solid; *Molecular Weight*: 399.88; *Boiling Point at 1 atm.*: Not pertinent (decomposes); *Freezing Point*: Not pertinent; *Critical Temperature*: Not pertinent; *Critical Pressure*: Not pertinent; *Specific Gravity*: 3.1 at 20 °C, (solid); *Vapor (Gas) Specific Gravity*: Not pertinent; *Ratio of Specific Heats of Vapor (Gas)*: Not pertinent; *Latent Heat of Vaporization*: Not pertinent; *Heat of Combustion*: Not pertinent; *Heat of Decomposition*: Not pertinent.

Health Hazards Information - *Recommended Personal Protective Equipment*: Dust mask, goggles or face shield, protective gloves; *Symptoms Following Exposure*: Inhalation of dust irritates nose and throat. Ingestion causes irritation of mouth and stomach. Dust irritates eyes and can irritate skin on prolonged contact; *General Treatment for Exposure*: INHALATION: move to fresh air. INGESTION: give large amount of water; induce vomiting if large amounts have been swallowed. EYES: flush with water; get medical attention if irritation persists. SKIN: flush with water; *Toxicity by Inhalation (Threshold Limit Value)*: 1 mg/m^3 (as iron); *Short-Term Inhalation Limits*: Data not available; *Toxicity by Ingestion*: Data not available; *Late Toxicity*: Data not available; *Vapor (Gas) Irritant Characteristics*: Data not available; *Liquid or Solid Irritant Characteristics*: Data not available; *Odor Threshold*: Data not available.

Fire Hazards - *Flash Point* : Not flammable; *Flammable Limits in Air (%) :* Not flammable; *Fire

Extinguishing Agents: Not pertinent; *Fire Extinguishing Agents Not To Be Used:* Not pertinent; *Special Hazards of Combustion Products:* Not pertinent; *Behavior in Fire:* Not pertinent; *Ignition Temperature : Not pertinent; Electrical Hazard:* Not pertinent; *Burning Rate:* Not pertinent.

Chemical Reactivity - *Reactivity with Water:* No reaction; *Reactivity with Common Materials:* Corrosive to copper, copper alloys, mild steel. and galvanized steel; *Stability During Transport:* Stable; *Neutralizing Agents for Acids and Caustics:* Flush with water; *Polymerization:* Not pertinent; *Inhibitor of Polymerization:* Not pertinent.

FERROUS AMMONIUM SULFATE

Chemical Designations - *Synonyms*: Ammonium ferrous sulfate, Ammonium iron sulfate, Ferrous ammonium sulfate hexahydrate, Iron ammonium sulfate, Mohr's salt; *Chemical Formula*: $Fe(NH_4)_2(SO_4)_2 \cdot 6H_2O$.

Observable Characteristics - *Physical State (as shipped)*: Solid; *Color*: Pate bluish-green; *Odor*: None.

Physical and Chemical Properties - *Physical State at 15 °C and 1 atm.*: Solid; *Molecular Weight*: 392.16; *Boiling Point at 1 atm.*: Not pertinent (decomposes); *Freezing Point*: Not pertinent; *Critical Temperature*: Not pertinent; *Critical Pressure*: Not pertinent; *Specific Gravity*: 1.86 at 20 °C, (solid); *Vapor (Gas) Specific Gravity*: Not pertinent; *Ratio of Specific Heats of Vapor (Gas)*: Not pertinent; *Latent Heat of Vaporization*: Not pertinent; *Heat of Combustion*: Not pertinent; *Heat of Decomposition*: Not pertinent.

Health Hazards Information - *Recommended Personal Protective Equipment*: Dust mask, goggles or face shield, protective gloves; *Symptoms Following Exposure*: Inhalation of dust irritates nose and throat. Ingestion causes irritation of mouth and stomach. Dust irritates eyes and can irritate skin on prolonged contact; *General Treatment for Exposure*: INGESTION: give large amount of water; induce vomiting if large amounts have been swallowed. EYES or SKIN: flush with water; *Toxicity by Inhalation (Threshold Limit Value)*: 1 mg/m^3 (as iron); *Short-Term Inhalation Limits*: Data not available; *Toxicity by Ingestion*: Grade 2, LD$_{50}$ 0.5 - 5 g/kg (rat); *Late Toxicity*: May cause eye degeneration in rabbits; *Vapor (Gas) Irritant Characteristics*: Data not available; *Liquid or Solid Irritant Characteristics*: Data not available; *Odor Threshold*: Data not available.

Fire Hazards - *Flash Point* : Not flammable; *Flammable Limits in Air (%):* Not flammable; *Fire Extinguishing Agents:* Not pertinent; *Fire Extinguishing Agents Not To Be Used:* Not pertinent; *Special Hazards of Combustion Products:* Irritating and toxic ammonia and oxides of nitrogen may form in fires; *Behavior in Fire:* Not pertinent; *Ignition Temperature : Not pertinent; Electrical Hazard:* Not pertinent; *Burning Rate:* Not pertinent.

Chemical Reactivity - *Reactivity with Water:* No reaction; *Reactivity with Common Materials:* No reaction; *Stability During Transport:* Stable; *Neutralizing Agents for Acids and Caustics:* Not pertinent; *Polymerization:* Not pertinent; *Inhibitor of Polymerization:* Not pertinent.

FERROUS CHLORIDE

Chemical Designations - *Synonyms*: Ferrous chloride tetrahydrate, Iron dichloride, Iron protochloride; *Chemical Formula*: $FeCl_2 \cdot 4H_2O$.

Observable Characteristics - *Physical State (as shipped)*: Solid; *Color*: Pale green; *Odor*: None.

Physical and Chemical Properties - *Physical State at 15 °C and 1 atm.*: Solid; *Molecular Weight*: 198; *Boiling Point at 1 atm.*: Not pertinent (decomposes); *Freezing Point*: Not pertinent; *Critical Temperature*: Not pertinent; *Critical Pressure*: Not pertinent; *Specific Gravity*: 1.93 at 20, (solid); *Vapor (Gas) Specific Gravity*: Not pertinent; *Ratio of Specific Heats of Vapor (Gas)*: Not pertinent; *Latent Heat of Vaporization*: Not pertinent; *Heat of Combustion*: Not pertinent; *Heat of Decomposition*: Not pertinent.

Health Hazards Information - *Recommended Personal Protective Equipment*: Dust mask, goggles or face shield, rubber gloves; *Symptoms Following Exposure*: Inhalation of dust irritates nose and throat. Ingestion causes irritation of mouth and stomach. Dust irritates eyes and may cause skin irritation on prolonged contact; *General Treatment for Exposure*: INHALATION: move to fresh air. INGESTION:

if large amounts are swallowed, promptly induce vomiting and get medical help. EYES: flush with plenty of water for at least 15 min.; get medical help promptly if ill effect develop. SKIN: wash with soap and water; *Toxicity by Inhalation (Threshold Limit Value)*: 1 mg/m^3 (as iron); *Short-Term Inhalation Limits*: Data not available; *Toxicity by Ingestion*: Data not available; *Late Toxicity*: Grade 2, LD$_{50}$ 0.5 - 5 g/kg (rat); *Vapor (Gas) Irritant Characteristics*: Data not available; *Liquid or Solid Irritant Characteristics*: Data not available; *Odor Threshold*: Data not available.

Fire Hazards - *Flash Point* : Not flammable; *Flammable Limits in Air (%)*: Not flammable; *Fire Extinguishing Agents:* Not pertinent; *Fire Extinguishing Agents Not To Be Used:* Not pertinent; *Special Hazards of Combustion Products:* Irritating hydrogen chloride fumes may form in fire; *Behavior in Fire:* Not pertinent; *Ignition Temperature :* Not pertinent; *Electrical Hazard:* Not pertinent; *Burning Rate:* Not pertinent.

Chemical Reactivity - *Reactivity with Water:* No reaction; *Reactivity with Common Materials:* Solution may corrode metals; *Stability During Transport:* Stable; *Neutralizing Agents for Acids and Caustics:* Flush with water, rinse with dilute solution of sodium bicarbonate or soda ash; *Polymerization:* Not pertinent; *Inhibitor of Polymerization:* Not pertinent.

FERROUS FLUOROBORATE

Chemical Designations - *Synonyms*: Ferrous borofluoride; *Chemical Formula*: Fe(BF$_4$)$_2$-H$_2$O.

Observable Characteristics - *Physical State (as shipped)*: Liquid; *Color*: Yellow-green; *Odor*: Data not available.

Physical and Chemical Properties - *Physical State at 15 °C and 1 atm.*: Liquid; *Molecular Weight*: 229.5 (solute only); *Boiling Point at 1 atm.*: Not pertinent (decomposes); *Freezing Point*: Not pertinent; *Critical Temperature*: Not pertinent; *Critical Pressure*: Not pertinent; *Specific Gravity*: (est.) > 1.1 at 20 °C, (liquid); *Vapor (Gas) Specific Gravity*: Not pertinent; *Ratio of Specific Heats of Vapor (Gas)*: Not pertinent; *Latent Heat of Vaporization*: Not pertinent; *Heat of Combustion*: Not pertinent; *Heat of Decomposition*: Not pertinent.

Health Hazards Information - *Recommended Personal Protective Equipment*: Goggles or face shield, rubber gloves; *Symptoms Following Exposure*: Ingestion causes irritation of mouth and stomach. Contact with eyes or skin causes irritation; *General Treatment for Exposure*: INGESTION: give large amount of water; induce vomiting; get medical attention. EYES: flush with water for at least 15 min. SKIN: flush with water; *Toxicity by Inhalation (Threshold Limit Value)*: 1 mg/m^3 (as iron); *Short-Term Inhalation Limits*: Data not available; *Toxicity by Ingestion*: Data not available; *Late Toxicity*: Data not available; *Vapor (Gas) Irritant Characteristics*: Data not available; *Liquid or Solid Irritant Characteristics*: Data not available; *Odor Threshold*: Data not available.

Fire Hazards - *Flash Point* : Not flammable; *Flammable Limits in Air (%)*: Not flammable; *Fire Extinguishing Agents:* Not pertinent; *Fire Extinguishing Agents Not To Be Used:* Not pertinent; *Special Hazards of Combustion Products:* Not pertinent; *Behavior in Fire:* Not pertinent; *Ignition Temperature :* Not pertinent; *Electrical Hazard:* Not pertinent; *Burning Rate:* Not pertinent.

Chemical Reactivity - *Reactivity with Water:* No reaction; *Reactivity with Common Materials:* No reaction; *Stability During Transport:* Stable; *Neutralizing Agents for Acids and Caustics:* Not pertinent; *Polymerization:* Not pertinent; *Inhibitor of Polymerization:* Not pertinent.

FERROUS OXALATE

Chemical Designations - *Synonyms*: Ferrous oxalate dihydrate, Ferrox, Iron protoxalate, Oxalic acid (ferrous salt); *Chemical Formula*: FeC$_2$O$_4$•2H$_2$O.

Observable Characteristics - *Physical State (as shipped)*: Solid; *Color*: Pate yellow; *Odor*: None.

Physical and Chemical Properties - *Physical State at 15 °C and 1 atm.*: Solid; *Molecular Weight*: 179.9; *Boiling Point at 1 atm.*: Not pertinent (decomposes); *Freezing Point*: Not pertinent; *Critical Temperature*: Not pertinent; *Critical Pressure*: Not pertinent; *Specific Gravity*: 2.3 at 20 °C, (solid); *Vapor (Gas) Specific Gravity*: Not pertinent; *Ratio of Specific Heats of Vapor (Gas)*: Not pertinent; *Latent Heat of Vaporization*: Not pertinent; *Heat of Combustion*: Not pertinent; *Heat of Decomposition*: Not pertinent.

Health Hazards Information - *Recommended Personal Protective Equipment*: Dust mask, goggles or face shield, protective gloves; *Symptoms Following Exposure*: Inhalation of dust may cause irritation of nose and throat. Ingestion causes burning pain in throat and stomach; mucous membranes turn white; can also cause vomiting, weak pulse, collapse, and death. Dust irritates eyes and may irritate skin on prolonged contact; *General Treatment for Exposure*: (must be prompt) INHALATION: move to fresh air; get medical attention if any symptoms persist. INGESTION: give immediately by mouth a dilute solution of any soluble calcium salt (calcium lactate, lime water, chalk solution, or even milk); large amounts of calcium are required; give gastric lavage with dilute lime water; consult physician. Watch for edema of the glottis and constriction of esophagus. EYES: flush with water for at least 15 min. SKIN: flush with water; *Toxicity by Inhalation (Threshold Limit Value)*: Data not available; *Short-Term Inhalation Limits*: Data not available; *Toxicity by Ingestion*: Data not available; *Late Toxicity*: Data not available; *Vapor (Gas) Irritant Characteristics*: Data not available; *Liquid or Solid Irritant Characteristics*: Data not available; *Odor Threshold*: Data not available.

Fire Hazards - *Flash Point* : Not flammable; *Flammable Limits in Air (%):* Not flammable; *Fire Extinguishing Agents:* Not pertinent; *Fire Extinguishing Agents Not To Be Used:* Not pertinent; *Special Hazards of Combustion Products:* Not pertinent; *Behavior in Fire:* Not pertinent; *Ignition Temperature : Not pertinent; *Electrical Hazard:* Not pertinent; *Burning Rate:* Not pertinent.

Chemical Reactivity - *Reactivity with Water:* No reaction; *Reactivity with Common Materials:* No reaction; *Stability During Transport:* Stable; *Neutralizing Agents for Acids and Caustics:* Not pertinent; *Polymerization:* Not pertinent; *Inhibitor of Polymerization:* Not pertinent.

FERROUS FLUOROBORATE

Chemical Designations - *Synonyms*: Copperas, Green vitriol, Iron(ous) sulfate, Iron vitriol; *Chemical Formula*: $FeSO_4 \bullet 7H_2O$.

Observable Characteristics - *Physical State (as shipped)*: Solid; *Color*: Green; *Odor*: Odorless.

Physical and Chemical Properties - *Physical State at 15 °C and 1 atm.*: Solid; *Molecular Weight*: 169.96; *Boiling Point at 1 atm.*: Not pertinent; *Freezing Point*: Not pertinent; *Critical Temperature*: Not pertinent; *Critical Pressure*: Not pertinent; *Specific Gravity*: 1.90 at 15, (solid); *Vapor (Gas) Specific Gravity*: Not pertinent; *Ratio of Specific Heats of Vapor (Gas)*: Not pertinent; *Latent Heat of Vaporization*: Not pertinent; *Heat of Combustion*: Not pertinent; *Heat of Decomposition*: Not pertinent.

Health Hazards Information - *Recommended Personal Protective Equipment*: Mask if dust is present; *Symptoms Following Exposure*: INGESTION: abdominal pain, retching, diarrhea, dehydration, shock, pallor, cyanosis, rapid or weak pulse, shallow respiration, low blood pressure; *General Treatment for Exposure*: INGESTION: give milk immediately and then induce vomiting by stroking the pharynx with a blunt object such as a spoon handle. Gastric lavage with 1 pint of 5% aqueous solution of mono- or disodium phosphate if promptly available; otherwise use water. Get medical attention; *Toxicity by Inhalation (Threshold Limit Value)*: Not pertinent; *Short-Term Inhalation Limits*: Not pertinent; *Toxicity by Ingestion*: Grade 2, LD_{50} 0.5—5 g/kg; *Late Toxicity*: Data not available; *Vapor (Gas) Irritant Characteristics*: Not pertinent; *Liquid or Solid Irritant Characteristics*: None; *Odor Threshold*: Not pertinent.

Fire Hazards - *Flash Point* : Not flammable; *Flammable Limits in Air (%):* Not flammable; *Fire Extinguishing Agents:* Not pertinent; *Fire Extinguishing Agents Not To Be Used:* Not pertinent; *Special Hazards of Combustion Products:* Not pertinent; *Behavior in Fire:* Not pertinent; *Ignition Temperature : Not pertinent; *Electrical Hazard:* Not pertinent; *Burning Rate:* Not pertinent.

Chemical Reactivity - *Reactivity with Water:* No reaction; *Reactivity with Common Materials:* No reaction; *Stability During Transport:* Stable; *Neutr* **FLUOSILICIC ACID** *alizing Agents for Acids and Caustics:* Not pertinent; *Polymerization:* Not pertinent; *Inhibitor of Polymerization:* Not pertinent.

FLUORINE

Chemical Designations - *Synonyms*: No common synonyms; *Chemical Formula*: F_2.

Observable Characteristics - *Physical State (as shipped)*: Compressed gas; *Color*: Pale yellow; *Odor*: Strong, choking, intense.

Physical and Chemical Properties - *Physical State at 15 °C and 1 atm.*: Gas; *Molecular Weight*: 37.99; *Boiling Point at 1 atm.*: -306, -188, 85; *Freezing Point*: -362, -219, 54; *Critical Temperature*: -199.5, -128.6, -144.6; *Critical Pressure*: 809.7, 55.08, 5.58; *Specific Gravity*: 1.5 at -188, (liquid); *Vapor (Gas) Specific Gravity*: Not pertinent; *Ratio of Specific Heats of Vapor (Gas)*: 1.362; *Latent Heat of Vaporization*: 71.6, 39.8, 1.67; *Heat of Combustion*: Not pertinent; *Heat of Decomposition*: Not pertinent.

Health Hazards Information - *Recommended Personal Protective Equipment*: Tight-fitting chemical goggles, special clothing, not easily ignited by fluorine gas; *Symptoms Following Exposure*: Severe burning of eyes, skin and respiratory system. The burns may develop slowly after exposure; *General Treatment for Exposure*: Flush all affected parts with water for at least 15 min. Do not use ointments. Administer artificial respiration and oxygen if required; *Toxicity by Inhalation (Threshold Limit Value)*: 1 ppm; *Short-Term Inhalation Limits*: 0.5 ppm for 5 min.; *Toxicity by Ingestion*: Not pertinent; *Late Toxicity*: Severe burns may develop slowly after exposure; *Vapor (Gas) Irritant Characteristics*: Vapors cause severe irritation of eye and throat and can cause eye and lung injury. They cannot be tolerated even at low concentrations; *Liquid or Solid Irritant Characteristics*: Severe skin irritant. Causes second- and third-degree burns on short contact and is very injurious to the eyes; *Odor Threshold*: 0.035 ppm.

Fire Hazards - *Flash Point* : Not flammable; *Flammable Limits in Air (%):* Not flammable; *Fire Extinguishing Agents:* Not pertinent; *Fire Extinguishing Agents Not To Be Used:* Do not direct water onto fluorine leaks; *Special Hazards of Combustion Products:* Toxic gases generated in fires involving fluorine; *Behavior in Fire:* Dangerously reactive gas. Ignites most combustibles; *Ignition Temperature : Not flammable; *Electrical Hazard:* Not pertinent; *Burning Rate:* Not pertinent.

Chemical Reactivity - *Reactivity with Water:* Reacts with water to form hydrogen fluoride, oxygen and oxygen difluoride; *Reactivity with Common Materials:* Reacts violently with all combustible materials, except the metal cylinders in which it is shipped; *Stability During Transport:* Stable; *Neutralizing Agents for Acids and Caustics:* Not pertinent; *Polymerization:* Not pertinent; *Inhibitor of Polymerization:* Not pertinent.

FLUOSILICIC ACID

Chemical Designations - *Synonyms*: Fluorosilicic acid, Hexafluosilicic acid, Hydrofluosilicic acid, Hydrogen hexafluorosilicate, Sand acid, Silicofluoric acid; *Chemical Formula*: H_2SiF_6-H_2O.

Observable Characteristics - *Physical State (as shipped)*: Liquid; *Color*: Transparent, straw colored, colorless; *Odor*: Acrid, sharp.

Physical and Chemical Properties - *Physical State at 15 °C and 1 atm.*: Liquid; *Molecular Weight*: 144.09 (solute only); *Boiling Point at 1 atm.*: (water) ~212, ~100, ~373; *Freezing Point*: (typical) -24 to -4, -31 to -20, 242 to 253; *Critical Temperature*: Not pertinent; *Critical Pressure*: Not pertinent; *Specific Gravity*: (approx.) 1.3 at 25 °C, (liquid); *Vapor (Gas) Specific Gravity*: Not pertinent; *Ratio of Specific Heats of Vapor (Gas)*: Not pertinent; *Latent Heat of Vaporization*: Not pertinent; *Heat of Combustion*: Not pertinent; *Heat of Decomposition*: Not pertinent.

Health Hazards Information - *Recommended Personal Protective Equipment*: Rubber gloves, safety glasses, protective clothing; *Symptoms Following Exposure*: Inhalation of vapor produces severe corrosive effect on mucous membrane. Ingestion causes severe burns of mouth and stomach. Contact with liquid or vapor causes severe burns of eyes and skin; *General Treatment for Exposure*: INHALATION: remove victim to fresh air; get medical attention. INGESTION: give large amounts of water; do NOT induce vomiting. EYES: immediately wash with water for 15 min.; call physician. SKIN: wash affected parts with water; treat as for hydrogen fluoride burn with iced benzalkonium chloride soaks; *Toxicity by Inhalation (Threshold Limit Value)*: Data not available; *Short-Term Inhalation Limits*: Data not available; *Toxicity by Ingestion*: Data not available; *Late Toxicity*: Data not available; *Vapor (Gas) Irritant Characteristics*: Data not available; *Liquid or Solid Irritant Characteristics*: Data not available; *Odor Threshold*: Data not available.

Fire Hazards - *Flash Point* : Not flammable; *Flammable Limits in Air (%):* Not flammable; *Fire Extinguishing Agents:* Not pertinent; *Fire Extinguishing Agents Not To Be Used:* Not pertinent; *Special Hazards of Combustion Products:* Irritating fumes of hydrogen fluoride may form in fires; *Behavior in Fire:* Not pertinent; *Ignition Temperature :* Not pertinent; *Electrical Hazard:* Not pertinent; *Burning*

Rate: Not pertinent.
Chemical Reactivity - *Reactivity with Water:* No reaction; *Reactivity with Common Materials:* Will corrode most metals, producing flammable hydrogen gas which can collect in confined spaces; *Stability During Transport:* Stable; *Neutralizing Agents for Acids and Caustics:* Flush with water and rinse with dilute solution of sodium carbonate or soda ash; *Polymerization:* Not pertinent; *Inhibitor of Polymerization:* Not pertinent.

FLUOSULFONIC ACID

Chemical Designations - *Synonyms*: Fluorosulfuric acid, Fluorosulfonic acid; *Chemical Formula*: FSO_3H.
Observable Characteristics - *Physical State (as shipped)*: Liquid; *Color*: Somewhat cloudy, colorless to slightly yellow; *Odor*: Choking, irritating.
Physical and Chemical Properties - *Physical State at 15 °C and 1 atm.*: Liquid; *Molecular Weight*: 100.07; *Boiling Point at 1 atm.*: 324.9, 162.7, 435.9; *Freezing Point*: -125.1, -87.3, 185.9; *Critical Temperature*: Not pertinent; *Critical Pressure*: Not pertinent; *Specific Gravity*: 1.73 at 25 °C, (liquid); *Vapor (Gas) Specific Gravity*: Not pertinent; *Ratio of Specific Heats of Vapor (Gas)*: Not pertinent; *Latent Heat of Vaporization*: 170, 94, 3.9; *Heat of Combustion*: Not pertinent; *Heat of Decomposition*: Not pertinent.
Health Hazards Information - *Recommended Personal Protective Equipment*: Rubber gloves, shoes, and clothing, goggles and face shield, acid-type canister mask or air-line mask; *Symptoms Following Exposure*: Inhalation of fumes causes severe irritation of nose and throat. Contact of liquid with eyes or skin causes very severe burns. Ingestion causes very severe burns of mouth and stomach; *General Treatment for Exposure*: Get medical attention quickly following all exposures to this compound. INHALATION: remove victim to fresh air; if he is unconscious, give artificial respiration. EYES: flush with water until medical help arrives. SKIN: flush with water until medical help arrives; soak burned area in strong Epsom salt solution; pay particular attention to area around fingernails. INGESTION: give large amounts of water; *Toxicity by Inhalation (Threshold Limit Value)*: Data not available; *Short-Term Inhalation Limits*: Data not available; *Toxicity by Ingestion*: Data not available; *Late Toxicity*: Data not available; *Vapor (Gas) Irritant Characteristics*: Vapors cause severe irritation of eyes and throat and can cause eye and lung injury. They cannot be tolerated even at low concentrations; *Liquid or Solid Irritant Characteristics*: Severe skin irritant. Causes second- and third-degree burns on short contact and very injurious to the eyes; *Odor Threshold*: Data not available.
Fire Hazards - *Flash Point* : Not flammable; *Flammable Limits in Air (%)*: Not flammable; *Fire Extinguishing Agents:* Not pertinent; *Fire Extinguishing Agents Not To Be Used:* Do not use water or foam on adjacent fires; *Special Hazards of Combustion Products:* Toxic and irritating fumes of hydrogen fluoride and sulfuric acid may form in fires; *Behavior in Fire:* Contact with water or chemical foam used to fight adjacent fires can result in the formation of toxic hydrogen fluoride gas; *Ignition Temperature* : Not pertinent; *Electrical Hazard:* Not pertinent; *Burning Rate:* Not pertinent.
Chemical Reactivity - *Reactivity with Water:* Reacts violently with water forming hydrogen fluoride and sulfuric acid mists; *Reactivity with Common Materials:* Reacts with metals forming flammable hydrogen gas; *Stability During Transport:* Stable; *Neutralizing Agents for Acids and Caustics:* Flood with water and rinse with sodium bicarbonate solution or lime solution; *Polymerization:* Not pertinent; *Inhibitor of Polymerization:* Not pertinent.

FORMALDEHYDE SOLUTION

Chemical Designations - *Synonyms*: Formalin, Fyde, Formalith, Methanal, Formic aldehyde; *Chemical Formula*: $HCHO/H_2O/CH_3OH$.
Observable Characteristics - *Physical State (as shipped)*: Liquid; *Color*: Colorless; *Odor*: Pungent, irritating; characteristic, pungent.
Physical and Chemical Properties - *Physical State at 15 °C and 1 atm.*: Liquid; *Molecular Weight*: 18-30; *Boiling Point at 1 atm.*: Not pertinent; *Freezing Point*: Not pertinent; *Critical Temperature*: Not pertinent; *Critical Pressure*: Not pertinent; *Specific Gravity*: 1.1 at 25 °C, (liquid); *Vapor (Gas)*

Specific Gravity: Not pertinent; *Ratio of Specific Heats of Vapor (Gas)*: Not pertinent; *Latent Heat of Vaporization*: Not pertinent; *Heat of Combustion*: Not pertinent; *Heat of Decomposition*: Not pertinent.

Health Hazards Information - *Recommended Personal Protective Equipment*: Self-contained breathing apparatus, chemical goggles, protective clothing, synthetic rubber or plastic gloves; *Symptoms Following Exposure*: INHALATION: vapors are irritating and will cause coughing, chest pain, nausea, and vomiting. INGESTION: causes nausea, vomiting, abdominal pain, and collapse. Contact with skin and eyes causes severe irritation; *General Treatment for Exposure*: INHALATION: remove victim to fresh air; give oxygen if breathing is difficult; call a physician. INGESTION: induce vomiting at once and repeat until vomit is clear; then give milk or raw egg and call a physician. SKIN OR EYES: flush immediately with plenty of water for at least 15 min.; remove contaminated clothing, call a physician for eyes; *Toxicity by Inhalation (Threshold Limit Value)*: 2 ppm; *Short-Term Inhalation Limits*: 5 ppm for 5 min., 3 ppm for 60 min. (tentative); *Toxicity by Ingestion*: (Formaldehyde solution) Grade 2, LD_{50} 0.5 to 5 g/kg; *Late Toxicity*: None; *Vapor (Gas) Irritant Characteristics*: Vapor is moderately irritating such that personnel with not usually tolerate moderate or high concentrations; *Liquid or Solid Irritant Characteristics*: Causes smarting of the skin and first-degree burns on short exposure. May cause secondary burns on long exposure; *Odor Threshold*: 0.8 ppm.

Fire Hazards - *Flash Point (deg. F)*: 182 CC (based on solution of 37 % formaldehyde and Methanol free), 122 CC (based on solution with 15 % Methanol); *Flammable Limits in Air (%)*: 7.0 - 73; *Fire Extinguishing Agents:* Water, dry chemical, carbon dioxide, or alcohol foam; *Fire Extinguishing Agents Not To Be Used:* No data or recommendations found; *Special Hazards of Combustion Products:* Toxic vapors form; *Behavior in Fire:* Not pertinent; *Ignition Temperature (deg. F):* 806; *Electrical Hazard:* Not pertinent; *Burning Rate:* Not pertinent.

Chemical Reactivity - *Reactivity with Water:* No reaction; *Reactivity with Common Materials:* No reactions; *Stability During Transport:* Stable; *Neutralizing Agents for Acids and Caustics:* Not pertinent; *Polymerization:* Not pertinent; *Inhibitor of Polymerization:* Not pertinent.

FUMARIC ACID

Chemical Designations - *Synonyms*: Allomaleic acid; Boletic acid; trans-Butenedioic acid; trans-1,2-Ethylenedicarboxylic acid; Lichenic acid; *Chemical Formula*: $HO_2CCH=CHCO_2H$.

Observable Characteristics - *Physical State (as shipped)*: Solid; *Color*: White; *Odor*: None.

Physical and Chemical Properties - *Physical State at 15 °C and 1 atm.*: Solid; *Molecular Weight*: 116.07; *Boiling Point at 1 atm.*: Very high; *Freezing Point*: Not pertinent; *Critical Temperature*: Not pertinent; *Critical Pressure*: Not pertinent; *Specific Gravity*: 1,635 at 20 °C, (solid); *Vapor (Gas) Specific Gravity*: Not pertinent; *Ratio of Specific Heats of Vapor (Gas)*: Not pertinent; *Latent Heat of Vaporization*: Not pertinent; *Heat of Combustion*: -4,970, -2,760, -116; *Heat of Decomposition*: Not pertinent.

Health Hazards Information - *Recommended Personal Protective Equipment*: Dust mask, gloves, safety glasses, dust cap; *Symptoms Following Exposure*: Inhalation of dust may cause respiratory irritation. Compound is non-toxic when ingested. Prolonged contact with eyes or skin may cause irritation; *General Treatment for Exposure*: INHALATION: move to fresh air. EYES: flush with water; get medical attention if irritation persists. SKIN: flush with water; *Toxicity by Inhalation (Threshold Limit Value)*: Data not available; *Short-Term Inhalation Limits*: Data not available; *Toxicity by Ingestion*: Data not available; *Late Toxicity*: Data not available; *Vapor (Gas) Irritant Characteristics*: Data not available; *Liquid or Solid Irritant Characteristics*: Data not available; *Odor Threshold*: Odorless.

Fire Hazards - *Flash Point*: Not pertinent. This is a combustible solid which can present a dust explosion problem; *Flammable Limits in Air (%):* Not pertinent; *Fire Extinguishing Agents:* Water spray, dry chemical, foam, or carbon dioxide; *Fire Extinguishing Agents Not To Be Used:* Not pertinent; *Special Hazards of Combustion Products:* Irritating fumes of maleic anhydride may form in fires; *Behavior in Fire:* Dust presents significant explosion hazard. Dust should be knocked down with water fog when fighting fires; *Ignition Temperature (deg. F):* 1,364 (powder); *Electrical Hazard :*Not pertinent; *Burning Rate:* Not pertinent.

Chemical Reactivity - *Reactivity with Water:* No reaction; *Reactivity with Common Materials:* No reactions; *Stability During Transport:* Stable; *Neutralizing Agents for Acids and Caustics:* Not pertinent; *Polymerization:*; *Inhibitor of Polymerization:* Not pertinent.

FURFURAL

Chemical Designations - *Synonyms*: 2-Fyraldehyde, Furfurole, Fural, Pyromucic aldehyde, Furfuraldehyde, Quakeral; *Chemical Formula*: O-CH$_2$CH$_2$CH$_2$CHCHO.

Observable Characteristics - *Physical State (as shipped)*: Liquid; *Color*: Colorless to reddish brown; *Odor*: Almond-like.

Physical and Chemical Properties - *Physical State at 15 °C and 1 atm.*: Liquid; *Molecular Weight*: 96.08; *Boiling Point at 1 atm.*: 323.1, 161.7, 434.9; *Freezing Point*: -33.7, -36.5, 236.7; *Critical Temperature*: 745, 397, 670; *Critical Pressure*: 798, 54.3, 5.50; *Specific Gravity*: 1.159 at 20°C, (liquid); *Vapor (Gas) Specific Gravity*: Not pertinent; *Ratio of Specific Heats of Vapor (Gas)*: Not pertinent; *Latent Heat of Vaporization*: 191, 106, 4.44; *Heat of Combustion*: -10.490, -5830, -244.1; *Heat of Decomposition*: Not pertinent.

Health Hazards Information - *Recommended Personal Protective Equipment*: Skin and eye protection; *Symptoms Following Exposure*: Vapor may irritate eyes and respiratory system. Liquid irritates skin and may cause dermatitis; *General Treatment for Exposure*: INHALATION: general treatment for overexposure to vapors of toxic chemicals; keep airway open, give respiration and oxygen if necessary; observe for premonitory signs and symptoms of pulmonary edema. INGESTION: induce vomiting, then give gastric lavage and saline cathartics. SKIN AND MUCOUS MEMBRANES: flood affected tissues with water; *Toxicity by Inhalation (Threshold Limit Value)*: 5 ppm; *Short-Term Inhalation Limits*: 15 ppm for 15 min.; *Toxicity by Ingestion*: Grade 3, LD$_{50}$ 50 to 500 mg/kg; *Late Toxicity*: Causes liver damage in rats; *Vapor (Gas) Irritant Characteristics*: Vapors cause moderate irritation such that personnel will find high concentrations unpleasant. The effect is temporary; *Liquid or Solid Irritant Characteristics*: Causes smarting of the skin and first-degree burns on short exposure; may cause secondary burns on long exposure; *Odor Threshold*: Data not available.

Fire Hazards - *Flash Point (deg. F)*: 153 OC, 140 CC; *Flammable Limits in Air (%)*: 2.1 - 19.3; *Fire Extinguishing Agents:* Water, foam, carbon dioxide, dry chemical, or alcohol foam; *Fire Extinguishing Agents Not To Be Used:* Not pertinent; *Special Hazards of Combustion Products:* Irritating vapors are generated when exposed to heat; *Behavior in Fire:* Not pertinent; *Ignition Temperature (deg. F):* 739; *Electrical Hazard:* Not pertinent; *Burning Rate:* 2.6 mm/min.

Chemical Reactivity - *Reactivity with Water:* No reaction; *Reactivity with Common* Materials: No reactions; *Stability During Transport:* Stable; *Neutralizing Agents for Acids and Caustics:* Not pertinent; *Polymerization:* Not pertinent; *Inhibitor of Polymerization:* Not pertinent.

FURFURYL ALCOHOL

Chemical Designations - *Synonyms*: 2-Furancarbinol, Furfuralcohol, alpha-Furylcarbinol, 2-Furylcarbinol, 2-Hydroxymethylfuran; *Chemical Formula*: C$_5$H$_6$O$_2$.

Observable Characteristics - *Physical State (as shipped)*: Liquid; *Color*: Colorless or amber; *Odor*: Mildly irritating.

Physical and Chemical Properties - *Physical State at 15 °C and 1 atm.*: Liquid; *Molecular Weight*: 98.1; *Boiling Point at 1 atm.*: 338, 170, 443; *Freezing Point*: 5, -15, 258; *Critical Temperature*: Not pertinent; *Critical Pressure*: Not pertinent; *Specific Gravity*: 1.13 at 20 °C, (liquid); *Vapor (Gas) Specific Gravity*: 3.4; *Ratio of Specific Heats of Vapor (Gas)*: Not pertinent; *Latent Heat of Vaporization*: 230, 130, 5.4; *Heat of Combustion*: -11,200, -6,200, -260; *Heat of Decomposition*: Not pertinent.

Health Hazards Information - *Recommended Personal Protective Equipment*: Goggles or face shield, rubber gloves; *Symptoms Following Exposure*: Inhalation causes headache, nausea, and irritation of nose and throat. Vapor irritates eyes; liquid causes inflammation and corneal opacity. Contact of skin with liquid causes dryness and irritation. Ingestion causes headache, nausea, and irritation of mouth and stomach; *General Treatment for Exposure*: INHALATION: remove victim to fresh air; if breathing is difficult, call a physician. EYES: immediately flush with water for 15 min.; get medical attention. SKIN: wash promptly with soap and water. INGESTION: give large amount of water, and induce vomiting; follow with gastric lavage and saline cathartics; get medical attention; *Toxicity by Inhalation (Threshold Limit Value)*: 5 ppm; *Short-Term Inhalation Limits*: 50 ppm/30 min.; *Toxicity by Ingestion*: Grade 3; oral LD_{50} = 132 mg/kg (rat); *Late Toxicity*: Data not available; *Vapor (Gas) Irritant Characteristics*: Vapors cause moderate irritation such that personnel will find high concentrations unpleasant. The effect is temporary; *Liquid or Solid Irritant Characteristics*: Minimum hazard. If spilled on clothing and allowed to remain, may cause smarting and reddening of the skin; *Odor Threshold*: 8 ppm.

Fire Hazards - *Flash Point (deg. F)*: 167 OC, 149 CC; *Flammable Limits in Air (%):* 1.8 - 16.3; *Fire Extinguishing Agents:* Water, dry chemical, foam, or carbon dioxide; *Fire Extinguishing Agents Not To Be Used:* Not pertinent; *Special Hazards of Combustion Products:* Not pertinent; *Behavior in Fire:* Not pertinent; *Ignition Temperature (deg. F):* 736; *Electrical Hazard:* No data; *Burning Rate:* 2.3 mm/min.

Chemical Reactivity - *Reactivity with Water:* No reaction; *Reactivity with Common Materials:* No reactions; *Stability During Transport:* The product darkens and forms water insoluble material on exposure to air or acids. This reaction is accelerated at elevated temperatures; *Neutralizing Agents for Acids and* Caustics: Not pertinent; *Polymerization:* Not pertinent; *Inhibitor of Polymerization:* Not pertinent.

G

GALLIC ACID

Chemical Designations - *Synonyms*: Gallic acid monohydrate; 3,4,5-Trihydroxybenzoic acid; *Chemical Formula*: $3,4,5-(HO)_3C_6H_2COOH \bullet H_2O$.

Observable Characteristics - *Physical State (as shipped)*: Solid; *Color*: White; *Odor*: None.

Physical and Chemical Properties - *Physical State at 15 °C and 1 atm.*: Solid; *Molecular Weight*: 188; *Boiling Point at 1 atm.*: Not pertinent (decomposes); *Freezing Point*: Not pertinent; *Critical Temperature*: Not pertinent; *Critical Pressure*: Not pertinent; *Specific Gravity*: 1.7 at 20, (solid); *Vapor (Gas) Specific Gravity*: Not pertinent; *Ratio of Specific Heats of Vapor (Gas)*: Not pertinent; *Latent Heat of Vaporization*: Not pertinent; *Heat of Combustion*: -6,060, -3,370, -141; *Heat of Decomposition*: Not pertinent.

Health Hazards Information - *Recommended Personal Protective Equipment*: Bu. Mines approved respirator, rubber gloves, safety goggles; *Symptoms Following Exposure*: Inhalation of dust may irritate nose and throat. Contact with eyes or skin causes irritation; *General Treatment for Exposure*: INGESTION: give large amount of water; induce vomiting. EYES: flush with water for at least 10 min.; consult a physician in irritation persists. SKIN: wash with soap and water; *Toxicity by Inhalation (Threshold Limit Value)*: Data not available; *Short-Term Inhalation Limits*: Data not available; *Toxicity by Ingestion*: Grade 2, LD_{50} 0.5 - 5 g/kg (rat); *Late Toxicity*: Data not available; *Vapor (Gas) Irritant Characteristics*: Data not available; *Liquid or Solid Irritant Characteristics*: Data not available; *Odor Threshold*: Data not available.

Fire Hazards - *Flash Point* : Not pertinent. This is a combustible solid; *Flammable Limits in Air (%):* Not pertinent; *Fire Extinguishing Agents:* Water, foam, dry chemical, or carbon dioxide; *Fire Extinguishing Agents Not To Be Used:* Not pertinent; *Special Hazards of Combustion Products:* No data; *Behavior in Fire:* No data; *Ignition Temperature :* Not pertinent; *Electrical Hazard:* Not pertinent; *Burning Rate:* Not pertinent .

Chemical Reactivity - *Reactivity with Water:* No reaction; *Reactivity with Common Materials:* No reactions; *Stability During Transport:* Stable; *Neutralizing Agents for Acids and Caustics:* Not pertinent;

Polymerization: Not pertinent; *Inhibitor of Polymerization:* Not pertinent.

GAS OIL: CRACKED

Chemical Designations - *Synonyms*: No common synonyms; *Chemical Formula*: Not pertinent (mixture).

Observable Characteristics - *Physical State (as shipped)*: Liquid; *Color*: Yellow to brown; *Odor*: Like gasoline and petroleum.

Physical and Chemical Properties - *Physical State at 15 °C and 1 atm.*: Liquid; *Molecular Weight*: Not pertinent; *Boiling Point at 1 atm.*: 375 - 750, 190 - 399, 463 - 672; *Freezing Point*: Not pertinent; *Critical Temperature*: Not pertinent; *Critical Pressure*: Not pertinent; *Specific Gravity*: 0.848 at 16, (liquid); *Vapor (Gas) Specific Gravity*: 3.4; *Ratio of Specific Heats of Vapor (Gas)*: Not pertinent; *Latent Heat of Vaporization*: Not pertinent; *Heat of Combustion*: -18,400, -10,200, 428; *Heat of Decomposition*: Not pertinent.

Health Hazards Information - *Recommended Personal Protective Equipment*: Protective goggles, gloves; *Symptoms Following Exposure*: INHALATION: causes irritation of upper respiratory tract; simulation, then depression; dizziness, headache, incoordination, anesthesia, coma, respiratory arrest; irregular heartbeat is a complication. ASPIRATION: enforce bed rest and administer oxygen. INGESTION: give victim water or milk; do NOT induce vomiting; guard against aspiration into lungs. EYES: wash with copious quantity of water. SKIN: remove by wiping, then wash with soap and water; *General Treatment for Exposure*: Get medical attention. INHALATION: maintain respiration; administer oxygen if needed. ASPIRATION: enforce bed rest and administer oxygen. INGESTION: give victim water or milk; do NOT induce vomiting; guard against aspiration into lungs. EYES: wash with copious quantity of water. SKIN: remove by wiping, then wash with soap and water; *Toxicity by Inhalation (Threshold Limit Value)*: No single value applicable; *Short-Term Inhalation Limits*: Not pertinent; *Toxicity by Ingestion*: Grade 2, LD_{50} 0.5 to 5 g/kg; *Late Toxicity*: None; *Vapor (Gas) Irritant Characteristics*: Vapors cause a slight smarting of the eyes or respiratory system if present in high concentrations. The effect is temporary; *Liquid or Solid Irritant Characteristics*: Minimum hazard. If spilled on clothing and allowed to remain, may cause smarting and reddening of skin; *Odor Threshold*: 0.25 ppm.

Fire Hazards - *Flash Point (deg. F)*: 150 CC; *Flammable Limits in Air (%)*: 6.0 - 13.5; *Fire Extinguishing Agents:* Water, foam, dry chemical, or carbon dioxide; *Fire Extinguishing Agents Not To Be Used:* Not pertinent; *Special Hazards of Combustion Products:* Not pertinent; *Behavior in Fire:* Not pertinent; *Ignition Temperature (deg. F):* 640; *Electrical Hazard:* Not pertinent; *Burning Rate:* 4 mm/min.

Chemical Reactivity - *Reactivity with Water:* No reaction; *Reactivity with Common Materials:* No reactions; *Stability During Transport:* Stable; *Neutralizing Agents for Acids and Caustics:* Not pertinent; *Polymerization:* Not pertinent; *Inhibitor of Polymerization:* Not pertinent.

GASOLINES: AUTOMOTIVE

Chemical Designations - *Synonyms*: Motor spirit, Petrol; *Chemical Formula*: (Mixture of hydrocarbons).

Observable Characteristics - *Physical State (as shipped)*: Liquid; *Color*: Colorless to brown; *Odor*: Gasoline.

Physical and Chemical Properties - *Physical State at 15 °C and 1 atm.*: Liquid; *Molecular Weight*: Not pertinent; *Boiling Point at 1 atm.*: 140 - 390, 60 - 199, 333 - 472; *Freezing Point*: Not pertinent; *Critical Temperature*: Not pertinent; *Critical Pressure*: Not pertinent; *Specific Gravity*: 0.7321 at 20 °C, (liquid); *Vapor (Gas) Specific Gravity*: 3.4; *Ratio of Specific Heats of Vapor (Gas)*: (est.) 1.054; *Latent Heat of Vaporization*: 130 - 150, 71 - 81, 3.0 - 3.4; *Heat of Combustion*: -18,720, -10,400, -435.1; *Heat of Decomposition*: Not pertinent.

Health Hazards Information - *Recommended Personal Protective Equipment*: Protective goggles, gloves; *Symptoms Following Exposure*: Irritation of mucous membranes and stimulation followed by depression of central nervous system. Breathing of vapor may also cause dizziness, headache, and in coordination or, in more severe cases, anesthesia, coma, and, respiratory arrest. If liquid enters lungs, it will cause severe irritation, coughing, gagging, pulmonary edema, and, later, signs of bronchopneumonia and pneumonitis. Swallowing may cause irregular heartbeat; *General Treatment for Exposure*: INHALATION: maintain respiration and administer oxygen; enforce bed rest if liquid is in lungs. INGESTION: do NOT induce vomiting; stomach should be lavaged (by doctor) if appreciable quantity is swallowed. EYES: wash with copious quantity of water. SKIN: wipe off and wash with soap and water; *Toxicity by Inhalation (Threshold Limit Value)*: No single TLV applies; *Short-Term Inhalation Limits*: 500 ppm for 30 min.; *Toxicity by Ingestion*: Grade 2, LD_{50} 0.5 to 5 g/kg; *Late Toxicity*: None; *Vapor (Gas) Irritant Characteristics*: Vapors cause a slight smarting of the eyes or respiratory system if present in high concentrations. The effect is temporary; *Liquid or Solid Irritant Characteristics*: Minimum hazard. If spilled on clothing and allowed to remain, may cause smarting and reddening of the skin; *Odor Threshold*: 0.25 ppm.

Fire Hazards - *Flash Point (deg. F)*: -36 CC; *Flammable Limits in Air (%)*: 1.4 - 7.4; *Fire Extinguishing Agents:* Foam, carbon dioxide, or dry chemical; *Fire Extinguishing Agents Not To Be Used:* Water may be ineffective; *Special Hazards of Combustion Products:* None; *Behavior in Fire:* Vapor is heavier than air and may travel considerable distance to source of ignition and flash back; *Ignition Temperature (deg. F):* 853; *Electrical Hazard:* Class I, Group D; *Burning Rate:* 4 mm/min.

Chemical Reactivity - *Reactivity with Water:* No reaction; *Reactivity with Common Materials:* No reactions; *Stability During Transport:* Stable; *Neutralizing Agents for Acids and Caustics:* Not pertinent; *Polymerization:* Not pertinent; *Inhibitor of Polymerization:* Not pertinent.

GASOLINES: AVIATION

Chemical Designations - *Synonyms*: No common synonyms; *Chemical Formula*: Not pertinent.

Observable Characteristics - *Physical State (as shipped)*: Liquid; *Color*: Red, blue, green, brown, purple; *Odor*: Gasoline.

Physical and Chemical Properties - *Physical State at 15 °C and 1 atm.*: Liquid; *Molecular Weight*: Not pertinent; *Boiling Point at 1 atm.*: 160 - 340, 71 - 171, 344 -444; *Freezing Point*: <76, <24.4, <297.6; *Critical Temperature*: Not pertinent; *Critical Pressure*: Not pertinent; *Specific Gravity*: 0.711 at 15 °C, (liquid); *Vapor (Gas) Specific Gravity*: 3.4; *Ratio of Specific Heats of Vapor (Gas)*: (est.) 1.054; *Latent Heat of Vaporization*: 130 - 150, 71 - 81, 3.0 - 3.4; *Heat of Combustion*: -18,720, -10,400, -435.4; *Heat of Decomposition*: Not pertinent.

Health Hazards Information - *Recommended Personal Protective Equipment*: Protective goggles, gloves; *Symptoms Following Exposure*: INHALATION causes irritation of upper respiratory tract; central nervous system stimulation followed by depression of varying degrees ranging from dizziness, headache, and incoordination to anesthesia, coma, and respiratory arrest; irregular heartbeat is dangerous complication. ASPIRATION causes severe lung irritation with coughing, gagging, dyspnea, substernal distress, and rapidly developing pulmonary edema; later, signs of bronchopneumonia and pneumonitis; acute onset of central nervous system excitement followed by depression. INGESTION causes irritation of mucous membranes of throat, esophagus, and stomach; stimulation followed by depression of central nervous system; irregular heartbeat; *General Treatment for Exposure*: Seek medical attention. INHALATION: maintain respiration; give oxygen if needed. ASPIRATION: enforce bed rest; administer oxygen. INGESTION: do NOT induce vomiting; lavage carefully if appreciable quantity was ingested; guard against aspiration into lungs. EYES: wash with copious quantity of water. SKIN: wipe off and wash with soap and water; *Toxicity by Inhalation (Threshold Limit Value)*: No single TLV applies; *Short-Term Inhalation Limits*: 500 ppm for 30 min.; *Toxicity by Ingestion*: Grade 2, LD_{50} 0.5 to 5 g/kg; *Late Toxicity*: None; *Vapor (Gas) Irritant Characteristics*: Vapors cause a slight smarting of the eyes or respiratory system if present in high concentrations. The effect is temporary; *Liquid or Solid Irritant Characteristics*: Minimum hazard. If spilled on clothing and allowed to remain, may cause smarting and reddening of the skin; *Odor Threshold*: 0.25 ppm.

Fire Hazards - *Flash Point (deg. F)*: -50 CC; *Flammable Limits in Air (%)*: 1.2 - 7.1; *Fire Extinguishing Agents:* Foam, carbon dioxide, or dry chemical; *Fire Extinguishing Agents Not To Be Used:* Water may be ineffective; *Special Hazards of Combustion Products:* None; *Behavior in Fire:*

Vapor is heavier than air and may travel considerable distance to source of ignition and flash back; *Ignition Temperature (deg. F):* 824; *Electrical Hazard:* Class I, Group D; *Burning Rate:* 4 mm/min. **Chemical Reactivity -** *Reactivity with Water:* No reaction; *Reactivity with Common Materials:* No reactions; *Stability During Transport:* Stable; *Neutralizing Agents for Acids and Caustics:* Not pertinent; *Polymerization:* Not pertinent; *Inhibitor of Polymerization:* Not pertinent.

GASOLINE BLENDING STOCKS: ALKYLATES
Chemical Designations - *Synonyms:* No common synonyms; *Chemical Formula:* Not pertinent.
Observable Characteristics - *Physical State (as shipped):* Liquid; *Color:* Colorless; *Odor:* Gasoline.
Physical and Chemical Properties - *Physical State at 15 °C and 1 atm.:* Liquid; *Molecular Weight:* Not pertinent; *Boiling Point at 1 atm.:* 58 - 275, 14 - 135, 287 - 408; *Freezing Point:* Not pertinent; *Critical Temperature:* Not pertinent; *Critical Pressure:* Not pertinent; *Specific Gravity:* 0.71 - 0.75 at 15°C, (liquid); *Vapor (Gas) Specific Gravity:* 3.4; *Ratio of Specific Heats of Vapor (Gas):* Not pertinent; *Latent Heat of Vaporization:* 130 - 150, 71 - 81, 3.0 - 3.4; *Heat of Combustion:* -18,720, -10,400, -435.4; *Heat of Decomposition:* Not pertinent.
Health Hazards Information - *Recommended Personal Protective Equipment:* Protective goggles; gloves; *Symptoms Following Exposure:* INHALATION causes irritation of upper respiratory tract; central nervous system stimulation followed by depression of varying degrees ranging from dizziness, headache, and incoordination. ASPIRATION causes severe lung irritation with coughing, gagging, dyspnea, substernal distress, and rapidly developing pulmonary edema; later, signs of bronchopneumonia and pneumonitis; acute onset of central nervous system excitement followed by depression. INGESTION causes irritation of mucous membranes of throat, esophagus, and stomach; simulation followed by depression of central nervous system; irregular heartbeat; *General Treatment for Exposure:* Seek medical attention. INHALATION: maintain respiration; give oxygen if needed. ASPIRATION: enforce bed rest; administer oxygen. INGESTION: do not induce vomiting; lavage carefully if appreciable quantity was ingested; guard against aspiration into lungs. EYES: wash with copious quantity of water. SKIN: wipe off and wash with soap and water; *Toxicity by Inhalation (Threshold Limit Value):* No single TLV applicable; *Short-Term Inhalation Limits:* 500 ppm for 30 min.; *Toxicity by Ingestion:* Grade 2, LD_{50} 0.5 to 5 g/kg; *Late Toxicity:* None; *Vapor (Gas) Irritant Characteristics:* Vapors cause a slight smarting of the eyes or respiratory system if present in high concentrations. The effect is temporary; *Liquid or Solid Irritant Characteristics:* Minimum hazard. If spilled on clothing and allowed to remain, may cause smarting and reddening of the skin; *Odor Threshold:* 0.25 ppm.
Fire Hazards - *Flash Point (deg. F):* < 0 CC; *Flammable Limits in Air (%):* 1.1 - 8.7; *Fire Extinguishing Agents:* Dry chemical, foam, or carbon dioxide; *Fire Extinguishing Agents Not To Be Used:* Water may be ineffective; *Special Hazards of Combustion Products:* None; *Behavior in Fire:* Vapor is heavier than air and may travel considerable distance to source of ignition and flash back; *Ignition Temperature :* No data; *Electrical Hazard:* Class I, Group D; *Burning* Rate: 4 mm/min .
Chemical Reactivity - *Reactivity with Water:* No reaction; *Reactivity with Common Materials:* No reactions; *Stability During Transport:* Stable; *Neutralizing Agents for Acids and Caustics:* Not pertinent; *Polymerization:* Not pertinent; *Inhibitor of Polymerization:* Not pertinent.

GASOLINE BLENDING STOCKS: REFORMATES
Chemical Designations - *Synonyms:* No common synonyms; *Chemical Formula:* Not pertinent.
Observable Characteristics - *Physical State (as shipped):* Liquid; *Color:* Colorless; *Odor:* Gasoline.
Physical and Chemical Properties - *Physical State at 15 °C and 1 atm.:* Liquid; *Molecular Weight:* Not pertinent; *Boiling Point at 1 atm.:* 58 - 275, 14 - 135, 287 - 408; *Freezing Point:* Not pertinent; *Critical Temperature:* Not pertinent; *Critical Pressure:* Not pertinent; *Specific Gravity:* 0.7934 at 20 °C, (liquid); *Vapor (Gas) Specific Gravity:* 3.4; *Ratio of Specific Heats of Vapor (Gas):* Not pertinent; *Latent Heat of Vaporization:* 130 - 150, 71 - 81, 3.0 - 3.4; *Heat of Combustion:* -18,720, -10,400, -435.4; *Heat of Decomposition:* Not pertinent.
Health Hazards Information - *Recommended Personal Protective Equipment:* Protective goggles,

gloves; *Symptoms Following Exposure*: INHALATION causes irritation of upper respiratory tract; central nervous system stimulation followed by depression of varying degrees ranging from dizziness, headache, and incoordination to anesthesia, coma, and respiratory arrest; irregular heartbeat is dangerous complication. ASPIRATION causes severe lung irritation with coughing, gagging, dyspnea, substernal distress, and rapidly developing pulmonary edema; later, signs of bronchopneumonia and pneumonitis; acute onset of central nervous system excitement followed by depression. INGESTION causes irritation of mucous membranes of throat, esophagus, and stomach; simulation followed by depression of central nervous system; irregular heartbeat; *General Treatment for Exposure*: Seek medical attention. INHALATION: maintain respiration; give oxygen if needed. ASPIRATION: enforce bed rest; administer oxygen. INGESTION: do NOT induce vomiting; lavage carefully if appreciable quantity was ingested; guard against aspiration into lungs. EYES: wash with copious quantity of water. SKIN: wipe off and wash with soap and water; *Toxicity by Inhalation (Threshold Limit Value)*: No single TLV applicable; *Short-Term Inhalation Limits*: 500 ppm for 30 min.; *Toxicity by Ingestion*: Grade 2, LD_{50} 0.5 to 5 g/kg; *Late Toxicity*: None; *Vapor (Gas) Irritant Characteristics*: Vapors cause a slight smarting of the eyes or respiratory system if present in high concentrations. The effect is temporary; *Liquid or Solid Irritant Characteristics*: Minimum hazard. If spilled on clothing and allowed to remain, may cause smarting and reddening of the skin; *Odor Threshold*: 0.25 ppm.

   Fire Hazards - *Flash Point (deg. F)*: < 0 CC; *Flammable Limits in Air (%): 1.1 - 8.7; *Fire Extinguishing Agents:* Dry chemical, foam, or carbon dioxide; *Fire Extinguishing Agents Not To Be Used:* Water may be ineffective; *Special Hazards of Combustion Products:* None; *Behavior in Fire:* Vapor is heavier than air and may travel considerable distance to source of ignition and flash back; *Ignition Temperature :* No data; *Electrical Hazard:* Class I, Group D; *Burning* Rate: 4 mm/min.

   Chemical Reactivity - *Reactivity with Water:* No reaction; *Reactivity with Common Materials:* No reactions; *Stability During Transport:* Stable; *Neutralizing Agents for Acids and Caustics:* Not pertinent; *Polymerization:* Not pertinent; *Inhibitor of Polymerization:* Not pertinent.

GLUTARALDEHYDE SOLUTION

Chemical Designations - *Synonyms*: 1,5-Pentanedial solution; *Chemical Formula*: OHC-$(CH_2)_3$-CHO (in water).

Observable Characteristics - *Physical State (as shipped)*: Liquid; *Color*: Pale yellow; *Odor*: Like rotten apples.

Physical and Chemical Properties - *Physical State at 15 °C and 1 atm.*: Liquid; *Molecular Weight*: Mixture; *Boiling Point at 1 atm.*: >212, >100, >373; *Freezing Point*: <20, <-7, <266; *Critical Temperature*: Not pertinent; *Critical Pressure*: Not pertinent; *Specific Gravity*: 1.062 - 1.124 at 20°C, (liquid); *Vapor (Gas) Specific Gravity*: Not pertinent; *Ratio of Specific Heats of Vapor (Gas)*: Not pertinent; *Latent Heat of Vaporization*: Not pertinent; *Heat of Combustion*: Not pertinent; *Heat of Decomposition*: Not pertinent.

Health Hazards Information - *Recommended Personal Protective Equipment*: Goggles or face shield; rubber gloves; *Symptoms Following Exposure*: Contact with liquid causes severe irritation of eyes and irritation of skin. Chemical readily penetrates skin in harmful amounts. Ingestion causes irritation of mouth and stomach; *General Treatment for Exposure*: EYES: immediately flush with plenty of water for at least 15 min.; get medical attention. SKIN: immediately flush with plenty of water for at least 15 min. INGESTION: give large amounts of water and induce vomiting; get medical attention; *Toxicity by Inhalation (Threshold Limit Value)*: Data not available; *Short-Term Inhalation Limits*: Data not available; *Toxicity by Ingestion*: Grade 2; oral rat LD_{50} = 2,380 mg/kg; *Late Toxicity*: Induces contact dermatitis in some people; *Vapor (Gas) Irritant Characteristics*: Data not available; *Liquid or Solid Irritant Characteristics*: Data not available; *Odor Threshold*: Data not available.

Fire Hazards - *Flash Point* : Non flammable solution; *Flammable Limits in Air (%):* Not pertinent; *Fire Extinguishing Agents:* Not pertinent; *Fire Extinguishing Agents Not To Be Used:* Not pertinent; *Special Hazards of Combustion Products:* Not pertinent; *Behavior in Fire:* Not pertinent; *Ignition Temperature :* Not pertinent; *Electrical Hazard:* Not pertinent; *Burning Rate:* Not pertinent.

Chemical Reactivity - *Reactivity with Water:* No reaction; *Reactivity with Common Materials:* No

reactions; *Stability During Transport:* Stable; *Neutralizing Agents for Acids and Caustics:* Not pertinent; *Polymerization:* Not pertinent; *Inhibitor of Polymerization:* Not pertinent.

GLYCERINE

Chemical Designations - *Synonyms*: Glycerol; 1,2,3-Propanetriol; 1,2,3-Trihydroxypropane; *Chemical Formula*: $HOCH_2CH(OH)CH_2OH$.

Observable Characteristics - *Physical State (as shipped)*: Liquid; *Color*: Colorless; *Odor*: Odorless.

Physical and Chemical Properties - *Physical State at 15 °C and 1 atm.*: Liquid; *Molecular Weight*: 92.10; *Boiling Point at 1 atm.*: Not pertinent (decomposes)554, 290, 563; *Freezing Point*: 64.2, 17.9, 291.1; *Critical Temperature*: Not pertinent; *Critical Pressure*: Not pertinent; *Specific Gravity*: 1.261 at 20 °C, (liquid); *Vapor (Gas) Specific Gravity*: Not pertinent; *Ratio of Specific Heats of Vapor (Gas)*: Not pertinent; *Latent Heat of Vaporization*: 288, 160, 6.70; *Heat of Combustion*: -7758, -4310, -180.5; *Heat of Decomposition*: Not pertinent.

Health Hazards Information - *Recommended Personal Protective Equipment*: Rubber gloves, goggles; *Symptoms Following Exposure*: No hazard; *General Treatment for Exposure*: No hazard; *Toxicity by Inhalation (Threshold Limit Value)*: Not pertinent; *Short-Term Inhalation Limits*: Data not available; *Toxicity by Ingestion*: Grade 0; LD_{50} above 15 g/kg; *Late Toxicity*: None; *Vapor (Gas) Irritant Characteristics*: Vapors are nonirritating to the eyes and throat; *Liquid or Solid Irritant Characteristics*: No appreciable hazard. Practically harmless to the skin; *Odor Threshold*: Not pertinent.

Fire Hazards - *Flash Point (deg. F)*: 350 OC, 320 CC; *Flammable Limits in Air (%):* Not pertinent; *Fire Extinguishing Agents:* Alcohol foam, dry chemical, carbon dioxide, water fog; *Fire Extinguishing Agents Not To Be Used:* Water or foam may cause frothing; *Special Hazards of Combustion Products:* Not pertinent; *Behavior in Fire:* Not pertinent; *Ignition Temperature (deg. F):* 698; *Electrical Hazard:* Not pertinent; *Burning Rate:* 0.9 mm/min.

Chemical Reactivity - *Reactivity with Water:* No reaction; *Reactivity with Common Materials:* No reactions; *Stability During Transport:* Stable; *Neutralizing Agents for Acids and Caustics:* Not pertinent; *Polymerization:* Not pertinent; *Inhibitor of Polymerization:* Not pertinent.

GLYCIDYL METHACRYLATE

Chemical Designations - *Synonyms*: Glycidyl alpha-methyl acrylate; Methacrylic acid, 2,3-epoxypropyl ester; *Chemical Formula*: $CH_2=CH(CH_3)COOCH_2CHCH_2O$.

Observable Characteristics - *Physical State (as shipped)*: Liquid; *Color*: Colorless; *Odor*: Data not available.

Physical and Chemical Properties - *Physical State at 15 °C and 1 atm.*: Liquid; *Molecular Weight*: 142.2; *Boiling Point at 1 atm.*: Very high; *Freezing Point*: Data not available; *Critical Temperature*: Not pertinent; *Critical Pressure*: Not pertinent; *Specific Gravity*: 1.073 at 20, (liquid); *Vapor (Gas) Specific Gravity*: Not pertinent; *Ratio of Specific Heats of Vapor (Gas)*: (est.) 1.043; *Latent Heat of Vaporization*: Not pertinent; *Heat of Combustion*: (est.) -10,800, -5,980, -250; *Heat of Decomposition*: Not pertinent.

Health Hazards Information - *Recommended Personal Protective Equipment*: Polyethylene-coated apron and gloves and close-fitting goggles; *Symptoms Following Exposure*: The liquid irritates eyes about as much as soap. Prolonged contact with skin produces irritation and dermatitis; *General Treatment for Exposure*: SKIN: wash throughly with soap and water and treat as a chemical burn. EYES: irrigate with clear water for 15 min. and get medical attention; *Toxicity by Inhalation (Threshold Limit Value)*: 25 ppm; *Short-Term Inhalation Limits*: Data not available; *Toxicity by Ingestion*: Grade 2, LD_{50} 0.5 to 5 g/kg (rat); *Late Toxicity*: Data not available; *Vapor (Gas) Irritant Characteristics*: Vapor is moderately irritating such that personnel will not usually tolerate moderate or high vapor concentrations; *Liquid or Solid Irritant Characteristics*: Causes smarting of the skin and first-degree burns on short exposure; may cause secondary burns on long exposure. In eyes the irritation is similar to that caused by ordinary soap; *Odor Threshold*: Data not available.

Fire Hazards - *Flash Point (deg. F)*: 183 OC; *Flammable Limits in Air (%):* No data; *Fire Extinguishing Agents:* No data; *Fire Extinguishing Agents Not To Be Used:* No data; *Special Hazards*

of Combustion Products: Irritating vapors generated when exposed to heat; *Behavior in Fire:* Not pertinent; *Ignition Temperature :* No data; *Electrical Hazard:* Not pertinent; *Burning Rate:* No data.
Chemical Reactivity - *Reactivity with Water:* No reaction; *Reactivity with Common Materials:* No reactions; *Stability During Transport:* Stable; *Neutralizing Agents for Acids and Caustics:* Not pertinent; *Polymerization:* Heat, peroxides, and caustics cause polymerization, however the reaction is slow and generally considered non hazardous; *Inhibitor of Polymerization:* 50 ppm of Hydroquinone Monomethyl Ether.

GLYOXAL: 40 % SOLUTION
Chemical Designations - *Synonyms*: Biformal; Biformyl; Diformyl; Ethanedial; Oxal; Oxaldehyde; *Chemical Formula*: CHO-CHO (in water).
Observable Characteristics - *Physical State (as shipped)*: Liquid; *Color*: Pale yellow; *Odor*: Weak sour.
Physical and Chemical Properties - *Physical State at 15 °C and 1 atm.*: Liquid; *Molecular Weight*: Mixture; *Boiling Point at 1 atm.*: Data not available; *Freezing Point*: 5, -15, 258; *Critical Temperature*: Not pertinent; *Critical Pressure*: Not pertinent; *Specific Gravity*: 1.29 at 20°C, (liquid); *Vapor (Gas) Specific Gravity*: Not pertinent; *Ratio of Specific Heats of Vapor (Gas)*: Not pertinent; *Latent Heat of Vaporization*: Not pertinent; *Heat of Combustion*: Not pertinent; *Heat of Decomposition*: Not pertinent.
Health Hazards Information - *Recommended Personal Protective Equipment*: Goggles or face shield; rubber gloves; *Symptoms Following Exposure*: Inhalation causes some irritation of nose and throat. Contact with liquid irritates eyes and causes mild irritation of skin; stains skin yellow. (No information available on symptoms of ingestion.); *General Treatment for Exposure*: INHALATION: remove from exposure. EYES or SKIN: flood with water for 15 min. INGESTION: no information on treatment; *Toxicity by Inhalation (Threshold Limit Value)*: Data not available; *Short-Term Inhalation Limits*: Data not available; *Toxicity by Ingestion*: Grade 2; oral rat LD_{50} = 2,020 mg/kg; *Late Toxicity*: Data not available; *Vapor (Gas) Irritant Characteristics*: Vapors cause a slight smarting of the eyes or respiratory system if present in high concentrations. The effect is temporary; *Liquid or Solid Irritant Characteristics*: Minimum hazard. If spilled on clothing and allowed to remain, may cause smarting and reddening of the skin; *Odor Threshold*: Data not available.
Fire Hazards - *Flash Point :* Non flammable solution; *Flammable Limits in Air (%):* Not pertinent; *Fire Extinguishing Agents:* Not pertinent; *Fire Extinguishing Agents Not To Be Used:* Not pertinent; *Special Hazards of Combustion Products:* Not pertinent; *Behavior in Fire:* Exposure to heat can cause polymerization to a combustible, viscous material; *Ignition Temperature :* Not pertinent; *Electrical Hazard:* Not pertinent; *Burning Rate:* Not pertinent.
Chemical Reactivity - *Reactivity with Water:* No reaction; *Reactivity with Common Materials:* Corrosive to most metals. Reactions are slow but accelerated at high temperatures; *Stability During Transport:* Stable; *Neutralizing Agents for Acids and Caustics:* Not pertinent; *Polymerization:* Not pertinent; *Inhibitor of Polymerization:* Not pertinent.

H

HEPTACHLOR
Chemical Designations - *Synonyms*: E 3314; 1,4,5,6,7,8,8a-Heptachlorodicyclopentadiene; 1,4,5,6,7,8,8a-Heptachloro-3a,4,7,7a-tetrahydro-4,7-methanoindene; Velsicol; *Chemical Formula*: $C_{10}H_5Cl_7$.
Observable Characteristics - *Physical State (as shipped)*: Solid; *Color*: White; light tan; *Odor*: Camphor-like.
Physical and Chemical Properties - *Physical State at 15 °C and 1 atm.*: Solid; *Molecular Weight*: 373.5; *Boiling Point at 1 atm.*: Not pertinent (decomposes); *Freezing Point*: 115 - 165, 46 - 74, 319 -

347; *Critical Temperature*: Not pertinent; *Critical Pressure*: Not pertinent; *Specific Gravity*: 1.66 at 20 °C, (solid); *Vapor (Gas) Specific Gravity*: Not pertinent; *Ratio of Specific Heats of Vapor (Gas)*: Not pertinent; *Latent Heat of Vaporization*: Not pertinent; *Heat of Combustion*: Not pertinent; *Heat of Decomposition*: Not pertinent.

Health Hazards Information - *Recommended Personal Protective Equipment*: Protective respirator; rubber gloves; clean clothes; *Symptoms Following Exposure*: Inhalation of dust causes irritability, tremors, and collapse. Ingestion causes nausea, vomiting, diarrhea, and irritation of the gastrointestinal tract. Contact with dust causes irritation of eyes and moderate irritation of skin; *General Treatment for Exposure*: Get medical attention following all overexposure to heptachlor. INHALATION: move to fresh air; if exposure to dust was severe, get medical attention. INGESTION: lavage stomach with warm tap water (unless convulsions are imminent); fats and oils should be avoided, as they increase the rate of absorption of all chlorinated hydrocarbons. EYES: wash repeatedly with water. SKIN: flush with water, then wash with soap and water; *Toxicity by Inhalation (Threshold Limit Value)*: 0.5 mg/m^3; *Short-Term Inhalation Limits*: 2 mg/m^3 for 30 min.; *Toxicity by Ingestion*: Grade 4; oral LD_{50} = 40 mg/kg (rat); *Late Toxicity*: Liver damage may develop; *Vapor (Gas) Irritant Characteristics*: Data not available; *Liquid or Solid Irritant Characteristics*: Data not available; *Odor Threshold*: 0.02 ppm.

Fire Hazards - *Flash Point* : Not flammable; *Flammable Limits in Air (%)*: Not flammable; *Fire Extinguishing Agents:* Not pertinent; *Fire Extinguishing Agents Not To Be Used:* Not pertinent; *Special Hazards of Combustion Products:* Irritating and toxic hydrogen chloride fumes may form in fires; *Behavior in Fire:* No data; *Ignition Temperature :* Not pertinent; *Electrical Hazard:* Not pertinent; *Burning Rate:* Not pertinent.

Chemical Reactivity - *Reactivity with Water:* No reaction; *Reactivity with Common Materials:* No reactions; *Stability During Transport:* Stable; *Neutralizing Agents for Acids and Caustics:* Not pertinent; *Polymerization:* Not pertinent; *Inhibitor of Polymerization:* Not pertinent.

HEPTANE

Chemical Designations - *Synonyms*: n-Heptane; *Chemical Formula*: $CH_3(CH_2)_5CH_3$.

Observable Characteristics - *Physical State (as shipped)*: Liquid; *Color*: Colorless; *Odor*: Gasoline.

Physical and Chemical Properties - *Physical State at 15 °C and 1 atm.*: Liquid; *Molecular Weight*: 100.21; *Boiling Point at 1 atm.*: 209.1, 98.4, 371.6; *Freezing Point*: -131, -90.6, 182.6; *Critical Temperature*: 513, 267, 540; *Critical Pressure*: 400, 27, 2.7; *Specific Gravity*: 0.6838 at 20 °C, (liquid); *Vapor (Gas) Specific Gravity*: Data not available; *Ratio of Specific Heats of Vapor (Gas)*: 1.054; *Latent Heat of Vaporization*: 136.1, 75.61, 3.166; *Heat of Combustion*: -19,170, -10,650, -445.9; *Heat of Decomposition*: Not pertinent.

Health Hazards Information - *Recommended Personal Protective Equipment*: Safety glasses; gloves; similar to gasoline; *Symptoms Following Exposure*: INHALATION: irritation of respiratory tract, coughing, depression, cardiac arrhythmia. ASPIRATION: severe lung irritation, pulmonary edema, mild excitement followed by depression. INGESTION: nausea, vomiting, swelling of abdomen, depression, headache; *General Treatment for Exposure*: INHALATION: maintain respiration; give oxygen if needed. ASPIRATION: enforce bed rest; administer oxygen. INGESTION: do not induce vomiting. SKIN OR EYES: remove contaminated clothing, wipe and wash skin area with soap and water; wash eyes with plenty of water; *Toxicity by Inhalation (Threshold Limit Value)*: 500 ppm; *Short-Term Inhalation Limits*: 500 ppm for 30 min.; *Toxicity by Ingestion*: Grade 0; LD_{50} above 15 g/kg; *Late Toxicity*: None; *Vapor (Gas) Irritant Characteristics*: Vapors are nonirritating to the eyes and throat; *Liquid or Solid Irritant Characteristics*: Minimum hazard. If spilled on clothing and allowed to remain, may cause smarting and reddening of the skin; *Odor Threshold*: 220 ppm.

Fire Hazards - *Flash Point (deg. F)*: 25 CC; *Flammable Limits in Air (%):* 1.0 - 7.0; *Fire Extinguishing Agents:* Foam, dry chemical, or carbon dioxide; *Fire Extinguishing Agents Not To Be Used:* Not pertinent; *Special Hazards of Combustion Products:* Not pertinent; *Behavior in Fire:* Not pertinent; *Ignition Temperature (deg. F):* 433; *Electrical Hazard:* Class I, Group D; *Burning Rate:* 6.8 mm/min.

Chemical Reactivity - *Reactivity with Water:* No reaction; *Reactivity with Common Materials:* No

reactions; *Stability During Transport:* Stable; *Neutralizing Agents for Acids and Caustics:* Not pertinent; *Polymerization:* Not pertinent; *Inhibitor of Polymerization:* Not pertinent.

HEPTANOL

Chemical Designations - *Synonyms*: Enanthic alcohol; 1-Heptanol; Heptyl alcohol; 1-Hydroxyheptane; *Chemical Formula*: $CH_3(CH_2)_5CH_2OH$.

Observable Characteristics - *Physical State (as shipped)*: Liquid; *Color*: Colorless; *Odor*: Weak alcoholic.

Physical and Chemical Properties - *Physical State at 15 °C and 1 atm.*: Liquid; *Molecular Weight*: 116.20; *Boiling Point at 1 atm.*: 349, 176, 449; *Freezing Point*: -29, -34, 239; *Critical Temperature*: 680, 360, 633; *Critical Pressure*: 440,30, 3.0; *Specific Gravity*: 0.822 at 20°C, (liquid); *Vapor (Gas) Specific Gravity*: Data not available; *Ratio of Specific Heats of Vapor (Gas)*: 1.049; *Latent Heat of Vaporization*: 189, 105, 4.40; *Heat of Combustion*: -18,810, -8784, -367.8; *Heat of Decomposition*: Not pertinent.

Health Hazards Information - *Recommended Personal Protective Equipment*: Chemical goggles or face shield; *Symptoms Following Exposure*: Low toxicity; liquid may irritate eyes; *General Treatment for Exposure*: Flush all affected parts with plenty of water; *Toxicity by Inhalation (Threshold Limit Value)*: Data not available; *Short-Term Inhalation Limits*: Data not available; *Toxicity by Ingestion*: Grade 2; oral rat LD_{50} = 1.87 g/kg; *Late Toxicity*: Data not available; *Vapor (Gas) Irritant Characteristics*: Nonirritating; *Liquid or Solid Irritant Characteristics*: Liquid may irritate eyes; it is not irritating to skin; *Odor Threshold*: 0.49 ppm.

Fire Hazards - *Flash Point (deg. F)*: 170 OC; *Flammable Limits in Air (%):* No data; *Fire Extinguishing Agents:* Foam, carbon dioxide, or dry chemical; *Fire Extinguishing Agents Not To Be Used:* Not pertinent; *Special Hazards of Combustion Products:* Not pertinent; *Behavior in Fire:* Not pertinent; *Ignition Temperature :* No data; *Electrical Hazard:* Not pertinent; *Burning Rate:* 3.2 mm/min.

Chemical Reactivity - *Reactivity with Water:* No reaction; *Reactivity with Common Materials:* No reactions; *Stability During Transport:* Stable; *Neutralizing Agents for Acids and Caustics:* Not pertinent; *Polymerization:* Not pertinent; *Inhibitor of Polymerization:* Not pertinent.

1-HEPTENE

Chemical Designations - *Synonyms*: Heptylene; *Chemical Formula*: $CH_3(CH_2)_4CH=CH_2$.

Observable Characteristics - *Physical State (as shipped)*: Liquid; *Color*: Colorless; *Odor*: Like gasoline.

Physical and Chemical Properties - *Physical State at 15 °C and 1 atm.*: Liquid; *Molecular Weight*: 98.18; *Boiling Point at 1 atm.*: 200.5, 93.6, 366.8; *Freezing Point*: -182, -119, 154; *Critical Temperature*: 507.4, 264.1, 537.3; *Critical Pressure*: 420, 28.57, 2.89; *Specific Gravity*: 0.697 at 20°C, (liquid); *Vapor (Gas) Specific Gravity*: 3.4; *Ratio of Specific Heats of Vapor (Gas)*: 1.057; *Latent Heat of Vaporization*: 137, 76.3, 3.20; *Heat of Combustion*: -19,377, -10,765, -450.71; *Heat of Decomposition*: Not pertinent.

Health Hazards Information - *Recommended Personal Protective Equipment*: Safety goggles or face shield; similar to gasoline; *Symptoms Following Exposure*: High concentrations may produce slight irritation of eyes and respiratory tract; may also act as simple asphyxiate and slight anesthetic; *General Treatment for Exposure*: Remove from exposure. Administer artificial respirator if needed; *Toxicity by Inhalation (Threshold Limit Value)*: Data not available; *Short-Term Inhalation Limits*: Data not available; *Toxicity by Ingestion*: Data not available; *Late Toxicity*: Data not available; *Vapor (Gas) Irritant Characteristics*: Vapors cause a slight smarting of the eyes or respiratory system if present in high concentrations. The effect is temporary; *Liquid or Solid Irritant Characteristics*: Minimum hazard. If spilled on clothing and allowed to remain, may cause smarting and reddening of the skin; *Odor Threshold*: Data not available.

Fire Hazards - *Flash Point (deg. F)*: 25 CC; *Flammable Limits in Air (%):* 1.0 (LEL); *Fire

Extinguishing Agents: Foam, carbon dioxide, or dry chemical; *Fire Extinguishing Agents Not To Be Used:* Not pertinent; *Special Hazards of Combustion Products:* Not pertinent; *Behavior in Fire:* Not pertinent; *Ignition Temperature (deg. F):* 500; *Electrical Hazard:* Not pertinent; *Burning Rate:* 3.2 mm/min.

Chemical Reactivity - *Reactivity with Water:* No reaction; *Reactivity with Common Materials:* No reactions; *Stability During Transport:* Stable; *Neutralizing Agents for Acids and Caustics:* Not pertinent; *Polymerization:* Not pertinent; *Inhibitor of Polymerization:* Not pertinent.

HEXADECYL SULFATE, SODIUM SALT

Chemical Designations - *Synonyms*: Cetyl sodium sulfate; Sodium cetyl sulfate solution; *Chemical Formula*: $CH_3(CH_2)_{14}CH_2OSO_3Na-H_2O$.

Observable Characteristics - *Physical State (as shipped)*: Pasty solid or liquid; *Color*: White; *Odor*: Mild.

Physical and Chemical Properties - *Physical State at 15 °C and 1 atm.*: Solid or liquid; *Molecular Weight*: Not pertinent (mixture); *Boiling Point at 1 atm.*: Not pertinent; *Freezing Point*: Not pertinent; *Critical Temperature*: Not pertinent; *Critical Pressure*: Not pertinent; *Specific Gravity*: 1 at 20, (liquid); *Vapor (Gas) Specific Gravity*: Not pertinent; *Ratio of Specific Heats of Vapor (Gas)*: Not pertinent; *Latent Heat of Vaporization*: Not pertinent; *Heat of Combustion*: Not pertinent; *Heat of Decomposition*: Not pertinent.

Health Hazards Information - *Recommended Personal Protective Equipment*: Plastic or rubber gloves; goggles or face shield; *Symptoms Following Exposure*: Contact with eyes causes mild irritation. May cause skin to dry out and become irritated; *General Treatment for Exposure*: EYES or SKIN: flush with water; *Toxicity by Inhalation (Threshold Limit Value)*: Data not available; *Short-Term Inhalation Limits*: Data not available; *Toxicity by Ingestion*: Data not available; *Late Toxicity*: Data not available; *Vapor (Gas) Irritant Characteristics*: Data not available; *Liquid or Solid Irritant Characteristics*: Data not available; *Odor Threshold*: Data not available.

Fire Hazards - *Flash Point* : Not flammable; *Flammable Limits in Air (%):* Not flammable; *Fire Extinguishing Agents:* Not pertinent; *Fire Extinguishing Agents Not To Be Used:* Not pertinent; *Special Hazards of Combustion Products:* Not pertinent; *Behavior in Fire:* Not pertinent; *Ignition Temperature : * Not pertinent; *Electrical Hazard:* Not pertinent; *Burning Rate:* Not pertinent.

Chemical Reactivity - *Reactivity with Water:* No reaction; *Reactivity with Common Materials:* No reactions; *Stability During Transport:* Stable; *Neutralizing Agents for Acids and Caustics:* Not pertinent; *Polymerization:* Not pertinent; *Inhibitor of Polymerization:* Not pertinent.

HEXADECYLTRIMETHYLAMMONIUM CHLORIDE

Chemical Designations - *Synonyms*: Cetyltrimethylammonium chloride solution; *Chemical Formula*: $C_{16}H_{33}(CH_3)_3NCl-H_2O-(CH_3)_2CHOH$.

Observable Characteristics - *Physical State (as shipped)*: Liquid; *Color*: Almost clear to pale yellow; *Odor*: Like rubbing alcohol.

Physical and Chemical Properties - *Physical State at 15 °C and 1 atm.*: Liquid; *Molecular Weight*: 319 (solute only); *Boiling Point at 1 atm.*: (isopropyl alcohol) 180, 82.3, 355.5; *Freezing Point*: Not pertinent; *Critical Temperature*: Not pertinent; *Critical Pressure*: Not pertinent; *Specific Gravity*: (approx.) 0.9 at 25°C, (liquid); *Vapor (Gas) Specific Gravity*: Not pertinent; *Ratio of Specific Heats of Vapor (Gas)*: Not pertinent; *Latent Heat of Vaporization*: Not pertinent; *Heat of Combustion*: Not pertinent; *Heat of Decomposition*: Not pertinent.

Health Hazards Information - *Recommended Personal Protective Equipment*: Goggles or face shield; rubber gloves; *Symptoms Following Exposure*: Ingestion may produce toxic effects. Contact with eyes or skin may cause severe damage; *General Treatment for Exposure*: INGESTION: do not induce vomiting; drink large quantities of fluid and call a physician immediately. EYES: flush with water for at least 15 min. and call a physician. SKIN: flush with water; *Toxicity by Inhalation (Threshold Limit Value)*: Data not available; *Short-Term Inhalation Limits*: Data not available; *Toxicity by Ingestion*: Grade 3; LD_{50} 250 mg/kg (rat); *Late Toxicity*: Data not available; *Vapor (Gas) Irritant Characteristics*:

Data not available; *Liquid or Solid Irritant Characteristics*: Data not available; *Odor Threshold*: Data not available.

Fire Hazards - *Flash Point (deg. F)*: 69 CC (for isopropyl alcohol solutions); *Flammable Limits in Air (%)*: 2 - 12 (isopropyl alcohol); *Fire Extinguishing Agents:* Dry chemical, alcohol foam, or carbon dioxide; *Fire Extinguishing Agents Not To Be Used:* Water may be ineffective; *Special Hazards of Combustion Products:* Irritating fumes of hydrogen chloride may form in fires; *Behavior in Fire:* Solvent vapors are heavier than air and may travel to a source of ignition and flash back; *Ignition Temperature (deg. F):* 750 (isopropyl alcohol solutions); *Electrical Hazard:* Class I, Group D; *Burning Rate:* 2.3 mm/min.

Chemical Reactivity - *Reactivity with Water:* No reaction; *Reactivity with Common Materials:* No reactions; *Stability During Transport:* Stable; *Neutralizing Agents for Acids and Caustics:* Not pertinent; *Polymerization:* Not pertinent; *Inhibitor of Polymerization:* Not pertinent.

N-HEXALDEHYDE

Chemical Designations - *Synonyms*: Caproaldehyde; Caproic aldehyde; Capronaldehyde; n-Caproylaldehyde; Hexanal; *Chemical Formula*: $CH_3(CH_2)_4CHO$.

Observable Characteristics - *Physical State (as shipped)*: Liquid; *Color*: Colorless; *Odor*: Pungent.

Physical and Chemical Properties - *Physical State at 15 °C and 1 atm.*: Liquid; *Molecular Weight*: 100; *Boiling Point at 1 atm.*: 262, 128, 401; *Freezing Point*: Not pertinent; *Critical Temperature*: Not pertinent; *Critical Pressure*: Not pertinent; *Specific Gravity*: 0.83 at 20°C, (liquid); *Vapor (Gas) Specific Gravity*: 3.5; *Ratio of Specific Heats of Vapor (Gas)*: (est.) 1.061 at 20°C; *Latent Heat of Vaporization*: (est.) 153, 85, 3.6; *Heat of Combustion*: (est.) -17,000, -9,430, -394; *Heat of Decomposition*: Not pertinent.

Health Hazards Information - *Recommended Personal Protective Equipment*: Goggles or face shield; rubber gloves; *Symptoms Following Exposure*: Ingestion causes irritation of mouth and stomach. Contact with vapor or liquid irritates eyes. Liquid irritates skin; *General Treatment for Exposure*: INGESTION: give large amount of water and induce vomiting. EYES: flush with water for at least 15 min. SKIN: wipe off; wash with soap and water; *Toxicity by Inhalation (Threshold Limit Value)*: Data not available; *Short-Term Inhalation Limits*: Data not available; *Toxicity by Ingestion*: Grade 2; oral LD_{50} = 4,890 mg/kg (rat); *Late Toxicity*: Data not available; *Vapor (Gas) Irritant Characteristics*: Data not available; *Liquid or Solid Irritant Characteristics*: Data not available; *Odor Threshold*: Data not available.

Fire Hazards - *Flash Point (deg. F)*: 90 OC; *Flammable Limits in Air (%)*: No data; *Fire Extinguishing Agents:* Dry chemical, foam, or carbon dioxide; *Fire Extinguishing Agents Not To Be Used:* Water may be ineffective; *Special Hazards of Combustion Products:* None; *Behavior in Fire:* Vapor is heavier than air and may travel considerable distances to a source of ignition and flash back; *Ignition Temperature :* No data; *Electrical Hazard:* No data; *Burning Rate:* 5.2 mm/min.

Chemical Reactivity - *Reactivity with Water:* No reaction; *Reactivity with Common Materials:* May attack some plastics; *Stability During Transport:* Stable; *Neutralizing Agents for Acids and Caustics:* Not pertinent; *Polymerization:* Not pertinent; *Inhibitor of Polymerization:* Not pertinent.

HEXAMETHYLENEDIAMINE

Chemical Designations - *Synonyms*: 1,6-Diaminoxexane; 1,6-Hexanediamine; *Chemical Formula*: $NH_2(CH_2)_6NH_2$.

Observable Characteristics - *Physical State (as shipped)*: Solid (anhydrous) or liquid (70% solution); *Color*: Glassy solid; clear liquid; *Odor*: Weak, fishy.

Physical and Chemical Properties - *Physical State at 15 °C and 1 atm.*: Solid (anhydrous); Liquid (10% solution); *Molecular Weight*: 116.21; *Boiling Point at 1 atm.*: 478, 205, 401; *Freezing Point*: (anhydrous) 104.9, 40.5, 313.7; (70% solution) 28, -2, 269; *Critical Temperature*: Not pertinent; *Critical Pressure*: Not pertinent; *Specific Gravity*: (anhydrous) 0.799 at 20 °C, (liquid); (70 % solution) 0.933 at 20°C, (liquid); *Vapor (Gas) Specific Gravity*: Not pertinent; *Ratio of Specific Heats of Vapor (Gas)*: Not pertinent; *Latent Heat of Vaporization*: 203, 113, 4.73; *Heat of Combustion*: (est.) -12,200,

-6,790, -284; *Heat of Decomposition*: Not pertinent.

Health Hazards Information - *Recommended Personal Protective Equipment*: Protective clothing; eye protection; *Symptoms Following Exposure*: Vapors cause irritation of eyes and respiratory tract. Liquid irritates eyes and skin, may cause dermatitis; *General Treatment for Exposure*: SKIN OR EYES: flush immediately with water for 15 min.; call a physician; *Toxicity by Inhalation (Threshold Limit Value)*: Data not available; *Short-Term Inhalation Limits*: Data not available; *Toxicity by Ingestion*: Data not available; *Late Toxicity*: Repeated exposure can cause anemia and damage kidney and liver; *Vapor (Gas) Irritant Characteristics*: Vapors cause a slight smarting of the eyes or respiratory system if present in high concentrations. The effect is temporary; *Liquid or Solid Irritant Characteristics*: Causes smarting of the skin and first-degree burns on short exposure; may cause secondary burns on long exposure; *Odor Threshold*: 0.0041 mg/m^3.

Fire Hazards - *Flash Point (deg. F)*: 160 OC; *Flammable Limits in Air (%)*: 0.7 - 6.3; *Fire Extinguishing Agents:* Carbon dioxide, dry chemical; *Fire Extinguishing Agents Not To Be Used:* No data; *Special Hazards of Combustion Products:* No data; *Behavior in Fire:* No data; *Ignition Temperature :* No data; *Electrical Hazard:* No data; *Burning Rate:* No data.

Chemical Reactivity - *Reactivity with Water:* No reaction; *Reactivity with Common Materials:* No reactions; *Stability During Transport:* Stable; *Neutralizing Agents for Acids and Caustics:* Flush with water; *Polymerization:* Not pertinent; *Inhibitor of Polymerization:* Not pertinent.

HEXAMETHYLENEIMINE

Chemical Designations - *Synonyms*: Azacycloheptane; Hexahydroazepine; Homopiperidine; *Chemical Formula*: $CH_2CH_2CH_2CH_2CH_2CH_2NH$

Observable Characteristics - *Physical State (as shipped)*: Liquid; *Color*: Colorless to light yellow; *Odor*: Ammonia-like

Physical and Chemical Properties - *Physical State at 15 °C and 1 atm.*: Liquid; *Molecular Weight*: 99; *Boiling Point at 1 atm.*: 270, 132, 405; *Freezing Point*: Not pertinent; *Critical Temperature*: Not pertinent; *Critical Pressure*: Not pertinent; *Specific Gravity*: 0.880 at 20 °C, (liquid); *Vapor (Gas) Specific Gravity*: Not pertinent; *Ratio of Specific Heats of Vapor (Gas)*: Not pertinent; *Latent Heat of Vaporization*: Data not available; *Heat of Combustion*: Data not available; *Heat of Decomposition*: Not pertinent

Health Hazards Information - *Recommended Personal Protective Equipment*: Self-contained breathing apparatus; impervious gloves; chemical safety goggles; impervious apron and boots; *Symptoms Following Exposure*: Inhalation of vapor irritates respiratory tract; high concentrations may cause disturbance of central nervous system. Ingestion causes burns of mouth and stomach. Cause contact with liquid causes burns of eyes and skin; *General Treatment for Exposure*: INHALATION: remove victim to uncontaminated atmosphere; get medical attention. INGESTION: give large amount of water; do NOT induce vomiting; get medical attention if large amount was swallowed. EYES: flush with water for 15 min. and get medical attention. SKIN: flush with water; wash with soap and water; *Toxicity by Inhalation (Threshold Limit Value)*: Data not available; *Short-Term Inhalation Limits*: Data not available; *Toxicity by Ingestion*: Grade 4; oral LD_{50} = 32 mg/kg (rat); *Late Toxicity*: Data not available; *Vapor (Gas) Irritant Characteristics*: Data not available; *Liquid or Solid Irritant Characteristics*: Data not available; *Odor Threshold*: Data not available.

Fire Hazards - *Flash Point (deg. F)*: 99 OC; *Flammable Limits in Air (%)*: 1.6 - 2.3; *Fire Extinguishing Agents:* Dry chemical, alcohol foam, or carbon dioxide; *Fire Extinguishing Agents Not To Be Used:* Water may be ineffective; *Special Hazards of Combustion Products:* Toxic oxides of nitrogen may form in fires; *Behavior in Fire:* Vapor is heavier than air and may travel to a source of ignition and flash back; *Ignition Temperature :* No data; *Electrical Hazard:* No data; *Burning Rate:* No data.

Chemical Reactivity - *Reactivity with Water:* No reaction; *Reactivity with Common Materials:* No reactions; *Stability During Transport:* Stable; *Neutralizing Agents for Acids and Caustics:* Not pertinent; *Polymerization:* Not pertinent; *Inhibitor of Polymerization:* Not pertinent.

HEXAMETHYLENETETRAMINE

Chemical Designations - *Synonyms*: Aminoform; Ammoform; Hexamine; Metheneamine; Urotropin; *Chemical Formula*: $C_6H_{12}N_4$.

Observable Characteristics - *Physical State (as shipped)*: Solid; *Color*: White; *Odor*: Mild ammonia-like.

Physical and Chemical Properties - *Physical State at 15 ℃ and 1 atm.*: Solid; *Molecular Weight*: 140.19; *Boiling Point at 1 atm.*: Not pertinent; *Freezing Point*: Not pertinent; *Critical Temperature*: Not pertinent; *Critical Pressure*: Not pertinent; *Specific Gravity*: 1.35 at 20°C, (solid); *Vapor (Gas) Specific Gravity*: Not pertinent; *Ratio of Specific Heats of Vapor (Gas)*: Not pertinent; *Latent Heat of Vaporization*: Not pertinent; *Heat of Combustion*: -13,300, -7400, -310; *Heat of Decomposition*: Not pertinent.

Health Hazards Information - *Recommended Personal Protective Equipment*: Gloves; for dusty or spatter conditions, use dust filter respirator and goggles; *Symptoms Following Exposure*: Prolonged and repeated contact may cause skin irritation; *General Treatment for Exposure*: Wash skin or eyes throughly with water. Call a physician; *Toxicity by Inhalation (Threshold Limit Value)*: Not pertinent; *Short-Term Inhalation Limits*: Data not available; *Toxicity by Ingestion*: Grade 2, LD_{50} 0.5 to 5 g/kg (human); *Late Toxicity*: None; *Vapor (Gas) Irritant Characteristics*: Vapors are nonirritating to the eyes and throat; *Liquid or Solid Irritant Characteristics*: Minimum hazard. If spilled on clothing and allowed to remain, may cause smarting and reddening of the skin; *Odor Threshold*: Not pertinent.

Fire Hazards - *Flash Point (deg. F)*: 482 CC; *Flammable Limits in Air (%):* Not pertinent; *Fire Extinguishing Agents:* Water, foam, carbon dioxide, or dry chemical; *Fire Extinguishing Agents Not To Be Used:* Not pertinent; *Special Hazards of Combustion Products:* Formaldehyde gas and ammonia may be given off when exposed to heat; *Behavior in Fire:* Not pertinent; *Ignition Temperature (deg. F):* > 700; *Electrical Hazard:* Not pertinent; *Burning Rate:* Not pertinent.

Chemical Reactivity - *Reactivity with Water:* No reaction; *Reactivity with Common Materials:* No reactions; *Stability During Transport:* Stable; *Neutralizing Agents for Acids and Caustics:* Not pertinent; *Polymerization:* Not pertinent; *Inhibitor of Polymerization:* Not pertinent.

HEXANE

Chemical Designations - *Synonyms*: No common synonyms; *Chemical Formula*: $CH_3(CH_2)_4CH_3$.

Observable Characteristics - *Physical State (as shipped)*: Liquid; *Color*: Colorless; *Odor*: Like gasoline.

Physical and Chemical Properties - *Physical State at 15 ℃ and 1 atm.*: Liquid; *Molecular Weight*: 86.17; *Boiling Point at 1 atm.*: 155.7, 68.7, 341.9; *Freezing Point*: -219.3, -139.6, 133.6; *Critical Temperature*: 453.6, 234.2, 507.4; *Critical Pressure*: 436.6, 29.7, 3.01; *Specific Gravity*: 0.659 at 20 °C, (liquid); *Vapor (Gas) Specific Gravity*: 3.0; *Ratio of Specific Heats of Vapor (Gas)*: 1.063; *Latent Heat of Vaporization*: 144, 80.0, 3.35; *Heat of Combustion*: -19,246, -10,692, -447.65; *Heat of Decomposition*: Not pertinent.

Health Hazards Information - *Recommended Personal Protective Equipment*: Eye protection (like gasoline); *Symptoms Following Exposure*: INHALATION causes irritation of respiratory tract, cough, mild depression, cardiac arrhythmias. ASPIRATION causes severe lung irritation, coughing, pulmonary edema; excitement followed by depression. INGESTION causes nausea, vomiting, swelling of abdomen, headache, depression; *Toxicity by Inhalation (Threshold Limit Value)*: Call a doctor. INHALATION: maintain respiration; give oxygen if needed. ASPIRATION: enforce bed rest; give oxygen if needed. INGESTION: do not induce vomiting. SKIN OR EYES: wipe off; wash skin with soap and water; wash eyes with copious amounts of water; *General Treatment for Exposure*: 500 ppm; *Short-Term Inhalation Limits*: 500 ppm for 30 min.; *Toxicity by Ingestion*: Very slight; *Late Toxicity*: None; *Vapor (Gas) Irritant Characteristics*: Vapors are nonirritating to the eyes and throat; *Liquid or Solid Irritant Characteristics*: No appreciable hazard. Practically harmless to the skin; *Odor Threshold*: Data not available.

Fire Hazards - *Flash Point (deg. F)*: -7 CC; *Flammable Limits in Air (%):* 1.2 - 7.7; *Fire Extinguishing Agents:* Foam, dry chemical, or carbon dioxide; *Fire Extinguishing Agents Not To Be*

Used: Not pertinent; *Special Hazards of Combustion Products:* Not pertinent; *Behavior in Fire:* Vapors may explode; *Ignition Temperature (deg. F):* 437; *Electrical Hazard:* Class I, Group D; *Burning Rate:* 7.3 mm/min.

Chemical Reactivity - *Reactivity with Water:* No reaction; *Reactivity with Common Materials:* No reactions; *Stability During Transport:* Stable; *Neutralizing Agents for Acids and Caustics:* Not pertinent; *Polymerization:* Not pertinent; *Inhibitor of Polymerization:* Not pertinent.

HEXANOL

Chemical Designations - *Synonyms*: Amylcarbinol; 1-Hexanol; Hexyl alcohol; 1-Hydroxyhexane; *Chemical Formula*: $CH_3(CH_2)_4CH_2OH$.

Observable Characteristics - *Physical State (as shipped)*: Liquid; *Color*: Colorless; *Odor*: Sweet; mild.

Physical and Chemical Properties - *Physical State at 15 °C and 1 atm.*: Liquid; *Molecular Weight*: 102.18; *Boiling Point at 1 atm.*: 314.8, 157.1, 430.3; *Freezing Point*: -48.3, -44.6, 228.6; *Critical Temperature*: 638.6, 337, 610.2; *Critical Pressure*: 485, 33, 3.34; *Specific Gravity*: 0.850 at 20°C, (liquid); *Vapor (Gas) Specific Gravity*: Not pertinent; *Ratio of Specific Heats of Vapor (Gas)*: 1.057; *Latent Heat of Vaporization*: 209, 116, 4.86; *Heat of Combustion*: -16,810, -9340, -391.0; *Heat of Decomposition*: Not pertinent.

Health Hazards Information - *Recommended Personal Protective Equipment*: Chemical gloves; chemical goggles; *Symptoms Following Exposure*: Liquid causes eye burns and skin irritation. Breathing vapors is not expected to cause systemic illness; *General Treatment for Exposure*: In case of contact, immediately flush skin and eyes with plenty of water. Wash eyes at least 15 min. and get medical care; *Toxicity by Inhalation (Threshold Limit Value)*: Data not available; *Short-Term Inhalation Limits*: Data not available; *Toxicity by Ingestion*: Grade 2, LD_{50} 0.5 to 5 g/kg (rat); *Late Toxicity*: Data not available; *Vapor (Gas) Irritant Characteristics*: Data not available; *Liquid or Solid Irritant Characteristics*: Causes smarting of the skin and first-degree burns on short exposure; may cause second-degree | burns on long exposure; *Odor Threshold*: Data not available.

Fire Hazards - *Flash Point (deg. F)*: 149 OC, 145 CC; *Flammable Limits in Air (%)*: 1.2 - 7.7; *Fire Extinguishing Agents:* Alcohol foam, dry chemical, or carbon dioxide; *Fire Extinguishing Agents Not To Be Used:* Not pertinent; *Special Hazards of Combustion Products:* Not pertinent; *Behavior in Fire:* Not pertinent; *Ignition Temperature (deg. F):* 580; *Electrical Hazard:* Not pertinent; *Burning Rate:* No data.

Chemical Reactivity - *Reactivity with Water:* No reaction; *Reactivity with Common Materials:* No reactions; *Stability During Transport:* Stable; *Neutralizing Agents for Acids and Caustics:* Not pertinent; *Polymerization:* Not pertinent; *Inhibitor of Polymerization:* Not pertinent.

1—HEXENE

Chemical Designations - *Synonyms*: alpha-Hexene; *Chemical Formula*: $CH_3(CH_2)_3CH=CH_2$.

Observable Characteristics - *Physical State (as shipped)*: Liquid; *Color*: Colorless; *Odor*: Mild, pleasant.

Physical and Chemical Properties - *Physical State at 15 °C and 1 atm.*: Liquid; *Molecular Weight*: 84.16; *Boiling Point at 1 atm.*: 146.3, 63.5, 336.7; *Freezing Point*: -219.6, -139.8, 133.4; *Critical Temperature*: 447.4, 230.8, 504.0; *Critical Pressure*: 460, 31.3, 3.17; *Specific Gravity*: 0.673 at 20 °C, (liquid); *Vapor (Gas) Specific Gravity*: 2.9; *Ratio of Specific Heats of Vapor (Gas)*: 1.068; *Latent Heat of Vaporization*: 140, 80, 3.3; *Heat of Combustion*: -19,134, -10,630, -445.06; *Heat of Decomposition*: Not pertinent.

Health Hazards Information - *Recommended Personal Protective Equipment*: Approved organic vapor respirator or air-line mask; protective goggles or face shield; *Symptoms Following Exposure*: Inhalation may cause giddiness or incoordination similar to that from gasoline vapor. Prolonged exposure to high concentrations may induce loss of consciousness or death; *General Treatment for Exposure*: SKIN OR EYES: wash exposed skin areas with soap and water; thoroughly flush eyes with water to remove any splashes; launder contaminated clothing before reuse; *Toxicity by Inhalation (Threshold Limit Value)*:

500 ppm (suggested); *Short-Term Inhalation Limits*: Data not available; *Toxicity by Ingestion*: Data not available; *Late Toxicity*: Data not available; *Vapor (Gas) Irritant Characteristics*: Slight smarting of the eyes or respiratory system if present in high concentrations. Effect is temporary; *Liquid or Solid Irritant Characteristics*: Data not available; *Odor Threshold*: Data not available.

Fire Hazards - *Flash Point (deg. F)*: - 15 CC; *Flammable Limits in Air (%)*: 1.2 (LEL); *Fire Extinguishing Agents:* Foam, dry chemical, or carbon dioxide; *Fire Extinguishing Agents Not To Be Used:* Water may be ineffective; *Special Hazards of Combustion Products:* Not pertinent; *Behavior in Fire:* Not pertinent; *Ignition Temperature (deg. F):* 521; *Electrical Hazard:* Not pertinent; *Burning Rate:* 8.1 mm/min.

Chemical Reactivity - *Reactivity with Water:* No reaction; *Reactivity with Common Materials:* No reactions; *Stability During Transport:* Stable; *Neutralizing Agents for Acids and Caustics:* Not pertinent; *Polymerization:* Not pertinent; *Inhibitor of Polymerization:* Not pertinent.

HEXYLENE GLYCOL

Chemical Designations - *Synonyms*: No common synonyms; *Chemical Formula*: N_2H_4.

Observable Characteristics - *Physical State (as shipped)*: Liquid; *Color*: Colorless; *Odor*: Ammonia-like.

Physical and Chemical Properties - *Physical State at 15 °C and 1 atm.*: Liquid; *Molecular Weight*: 32.05; *Boiling Point at 1 atm.*: 236.3, 113.5, 386.7; *Freezing Point*: 34.7, 1.5, 274.7; *Critical Temperature*: 716, 380, 653; *Critical Pressure*: 2130, 145, 14.7; *Specific Gravity*: 1.008 at 20°C, (liquid); *Vapor (Gas) Specific Gravity*: Not pertinent; *Ratio of Specific Heats of Vapor (Gas)*: 1.191; *Latent Heat of Vaporization*: 538, 299, 12.5; *Heat of Combustion*: -8345, -4636, -194.1; *Heat of Decomposition*: Not pertinent.

Health Hazards Information - *Recommended Personal Protective Equipment*: Ammonia-type gas mask; self-contained breathing apparatus; plastic-coated or rubber gloves, clothes, and apron; safety shower must be available; *Symptoms Following Exposure*: Vapors cause itching, swelling, and blistering of eyelids, skin, nose and throat; symptoms may be delayed for several hours. Temporary blindness may occur. Liquid causes a caustic-like burn if not washed off at once. Ingestion or absorption through skin causes nausea, dizziness, headache. Severe exposure may cause death; *General Treatment for Exposure*: Call a doctor at once. INHALATION: remove to fresh air; observe for development of delayed symptoms. Keep quiet. INGESTION: do not induce vomiting; give egg whites or other emollient. SKIN OR EYES: wash with large amounts of water for at least 15 min.; *Toxicity by Inhalation (Threshold Limit Value)*: 1 ppm; *Short-Term Inhalation Limits*: 1 ppm for 30 min.; *Toxicity by Ingestion*: Grade 3; LD_{50} 50 to 500 mg/kg (rat); *Late Toxicity*: Data not available; *Vapor (Gas) Irritant Characteristics*: Vapor is moderately irritating such that personnel will not usually tolerate moderate or high vapor concentrations; *Liquid or Solid Irritant Characteristics*: Severe skin irritant. Causes second- and third-degree burns on short contact; very injurious to the eyes; *Odor Threshold*: Data not available.

Fire Hazards - *Flash Point (deg. F)*: 126 OC; *Flammable Limits in Air (%):* 4.7 - 100; *Fire Extinguishing Agents:* Water, alcohol foam, dry chemical, or carbon dioxide; *Fire Extinguishing Agents Not To Be Used:* Not pertinent; *Special Hazards of Combustion Products:* Toxic vapors are generated upon exposure to heat; *Behavior in Fire:* May explode if vapors are confined; *Ignition Temperature :* May ignite spontaneously; *Electrical Hazard:* Not pertinent; *Burning Rate:* 1 mm/min.

Chemical Reactivity - *Reactivity with Water:* No reaction; *Reactivity with Common Materials:* Can catch fire when in contact with porous materials such as wood, asbestos, cloth, soil, or rusty metals; *Stability During Transport:* Stable at ordinary temperatures, however when heated this material can decompose to nitrogen and ammonia gases. The decomposition is not generally hazardous unless it occurs in confined spaces; *Neutralizing Agents for Acids and Caustics:* Flush with water and neutralize the resulting solution with calcium hypochlorite; *Polymerization:* Not pertinent; *Inhibitor of Polymerization:* Not pertinent.

HYDRAZINE

Chemical Designations - *Synonyms*: No common synonyms; *Chemical Formula*: N_2H_4.

Observable Characteristics - *Physical State (as shipped)*: Liquid; *Color*: Colorless; *Odor*: Ammonia-like.

Physical and Chemical Properties - *Physical State at 15 °C and 1 atm.*: Liquid; *Molecular Weight*: 32.05; *Boiling Point at 1 atm.*: 236.3, 113.5, 386.7; *Freezing Point*: 34.7, 1.5, 274.7; *Critical Temperature*: 716, 380, 653; *Critical Pressure*: 2130, 145, 14.7; *Specific Gravity*: 1.008 at 20°C, (liquid); *Vapor (Gas) Specific Gravity*: Not pertinent; *Ratio of Specific Heats of Vapor (Gas)*: 1.191; *Latent Heat of Vaporization*: 538, 299, 12.5; *Heat of Combustion*: -8345, -4636, -194.1; *Heat of Decomposition*: Not pertinent.

Health Hazards Information - *Recommended Personal Protective Equipment*: Ammonia-type gas mask; self-contained breathing apparatus; plastic-coated or rubber gloves, clothes, and apron; safety shower must be available; *Symptoms Following Exposure*: Vapors cause itching, swelling, and blistering of eyelids, skin, nose and throat; symptoms may be delayed for several hours. Temporary blindness may occur. Liquid causes a caustic-like burn if not washed off at once. Ingestion or absorption through skin causes nausea, dizziness, headache. Severe exposure may cause death; *General Treatment for Exposure*: Call a doctor at once. INHALATION: remove to fresh air; observe for development of delayed symptoms. Keep quiet. INGESTION: do NOT induce vomiting; give egg whites or other emollient. SKIN OR EYES: wash with large amounts of water for at least 15 min.; *Toxicity by Inhalation (Threshold Limit Value)*: 1 ppm; *Short-Term Inhalation Limits*: 1 ppm for 30 min.; *Toxicity by Ingestion*: Grade 3; LD_{50} 50 to 500 mg/kg (rat); *Late Toxicity*: Causes lung cancer in mice; *Vapor (Gas) Irritant Characteristics*: Vapor is moderately irritating such that personnel will not usually tolerate moderate or high vapor concentrations; *Liquid or Solid Irritant Characteristics*: Severe skin irritant. Causes second- and third-degree burns on short contact; very injurious to the eyes; *Odor Threshold*: 3 - 4 ppm.

Fire Hazards - *Flash Point (deg. F)*: 100 OC; *Flammable Limits in Air (%):* 4.7 - 100; *Fire Extinguishing Agents:* Water, alcohol foam, dry chemical, or carbon dioxide; *Fire Extinguishing Agents Not To Be Used:* Not pertinent; *Special Hazards of Combustion Products:* Toxic vapors are generated when heated; *Behavior in Fire:* May explode if confined; *Ignition Temperature :* May ignite spontaneously, 518; *Electrical Hazard:* Not pertinent; *Burning Rate:* 1 mm/min.

Chemical Reactivity - *Reactivity with Water:* No reaction; *Reactivity with Common Materials:* Can catch fire when in contact with porous materials such as wood, asbestos, cloth, soil, or rusty metals; *Stability During Transport:* Stable at ordinary temperatures, however when heated this material can decompose to nitrogen and ammonia gases. The decomposition is not generally hazardous unless it occurs in confined spaces; *Neutralizing Agents for Acids and Caustics:* Flush with water and neutralize the resulting solution with calcium hypochlorite; *Polymerization:* Not pertinent; *Inhibitor of Polymerization:* Not pertinent.

HYDROCHLORIC ACID

Chemical Designations - *Synonyms*: Muriatic acid; *Chemical Formula*: $HCl-H_2O$.

Observable Characteristics - *Physical State (as shipped)*: Liquid; *Color*: Colorless to light yellow; *Odor*: Pungent; sharp, pungent, irritating.

Physical and Chemical Properties - *Physical State at 15 °C and 1 atm.*: Liquid; *Molecular Weight*: Not pertinent; *Boiling Point at 1 atm.*: 123, 50.5, 323.8; *Freezing Point*: Not pertinent; *Critical Temperature*: Not pertinent; *Critical Pressure*: Not pertinent; *Specific Gravity*: 1.19 at 20°C, (liquid); *Vapor (Gas) Specific Gravity*: Not pertinent; *Ratio of Specific Heats of Vapor (Gas)*: Not pertinent; *Latent Heat of Vaporization*: 178, 98.6, 4.13; *Heat of Combustion*: Not pertinent; *Heat of Decomposition*: Not pertinent.

Health Hazards Information - *Recommended Personal Protective Equipment*: Self-contained breathing equipment, air-line mask, or industrial canister-type gas mask; rubber-coated gloves, apron, coat, overalls, shoes; *Symptoms Following Exposure*: Inhalation of fumes results in coughing and choking sensation, and irritation of nose and lungs. Liquid causes burns; *General Treatment for Exposure*:

INHALATION: remove person to fresh air; keep him warm and quiet and get medical attention immediately; start artificial respiration if breathing stops. INGESTION: have person drink water or milk; do NOT induce vomiting. EYES: immediately flush with plenty of water for at least 15 min. and get medical attention; continue flushing for another 15 min. if physician does not arrive promptly. SKIN: immediately flush skin while removing contaminated clothing; get medical attention promptly; use soap and wash area for at least 15 min.; *Toxicity by Inhalation (Threshold Limit Value)*: 5 ppm; *Short-Term Inhalation Limits*: 5 ppm for 5 min.; *Toxicity by Ingestion*: Data not available; *Late Toxicity*: None; *Vapor (Gas) Irritant Characteristics*: Vapor is moderately irritating such that personnel will not usually tolerate moderate or high vapor concentrations; *Liquid or Solid Irritant Characteristics*: Fairly severe skin irritant; may cause pain and second-degree burns after a few minutes contact; *Odor Threshold*: 1 - 5 ppm.

Fire Hazards - *Flash Point* : Not flammable; *Flammable Limits in Air (%):* Not flammable; *Fire Extinguishing Agents:* Not pertinent; *Fire Extinguishing Agents Not To Be Used:* Not pertinent; *Special Hazards of Combustion Products:* Toxic and irritating vapors are generated upon heating; *Behavior in Fire:* Not pertinent; *Ignition Temperature :* Not flammable; *Electrical Hazard:* Not pertinent; *Burning Rate:* Not flammable.

Chemical Reactivity - *Reactivity with Water:* No reaction; *Reactivity with Common Materials:* Corrosive to most metals with the evolution of flammable and explosive hydrogen gas; *Stability During Transport:* Stable; *Neutralizing Agents for Acids and Caustics:* Flush with water and apply powdered limestone, slaked lime, soda ash, or sodium bicarbonate; *Polymerization:* Not pertinent; *Inhibitor of Polymerization:* Not pertinent.

HYDROFLUORIC ACID

Chemical Designations - *Synonyms*: No common synonyms; *Chemical Formula*: $HF-H_2O$.

Observable Characteristics - *Physical State (as shipped)*: Liquid; *Color*: Colorless to slightly yellow; *Odor*: Pungent, irritating.

Physical and Chemical Properties - *Physical State at 15 °C and 1 atm.*: Liquid; *Molecular Weight*: Not pertinent; *Boiling Point at 1 atm.*: 152, 67, 340; *Freezing Point*: Not pertinent; *Critical Temperature*: Not pertinent; *Critical Pressure*: Not pertinent; *Specific Gravity*: 1.258 at 25°C, (liquid); *Vapor (Gas) Specific Gravity*: Not pertinent; *Ratio of Specific Heats of Vapor (Gas)*: Not pertinent; *Latent Heat of Vaporization*: 649, 361, 15.1; *Heat of Combustion*: Not pertinent; *Heat of Decomposition*: Not pertinent.

Health Hazards Information - *Recommended Personal Protective Equipment*: Proper protective clothing must be worn that encapsulates the body including the face. A shower and an eye wash must be available; *Symptoms Following Exposure*: Serious and painful burns of eyes and skin; *General Treatment for Exposure*: INGESTION: have victim drink water or milk; do NOT induce vomiting. SKIN: if victim has come in contact with liquid or vapor, put him in a shower and call a physician. EYES: flush with water for at least 15 min. and consult physician; *Toxicity by Inhalation (Threshold Limit Value)*: 3 ppm; *Short-Term Inhalation Limits*: 500 ppm for 60 min.; *Toxicity by Ingestion*: Data not available; *Late Toxicity*: Data not available; *Vapor (Gas) Irritant Characteristics*: Vapors cause severe irritation of eye and throat and can cause eye and lung injury. They cannot be tolerated even at low concentrations; *Liquid or Solid Irritant Characteristics*: Severe skin irritant. Causes second- and third-degree burns on short contact; very injurious to the eyes; *Odor Threshold*: Data not available.

Fire Hazards - *Flash Point* : Not flammable; *Flammable Limits in Air (%):* Not flammable; *Fire Extinguishing Agents:* Not pertinent; *Fire Extinguishing Agents Not To Be Used:* Not pertinent; *Special Hazards of Combustion Products:* Toxic and irritating vapors are generated upon heating; *Behavior in Fire:* Not pertinent; *Ignition Temperature :* Not flammable; *Electrical Hazard:* Not pertinent; *Burning Rate:* Not flammable.

Chemical Reactivity - *Reactivity with Water:* No reaction; *Reactivity with Common Materials:* Will attack glass, concrete and certain metals containing silica, such as cast iron. Will attack natural rubber, leather, and many organic materials. Can generate flammable and explosive hydrogen when in contact with some metals; *Stability During Transport:* Stable but requires special packaging; *Neutralizing*

Agents for Acids and Caustics: Flush with water and apply powdered limestone, slaked lime, soda ash, or sodium bicarbonate; *Polymerization:* Not pertinent; *Inhibitor of Polymerization:* Not pertinent.

HYDROGEN BROMIDE

Chemical Designations - *Synonyms:* Hydrobromic acid, anhydrous; Hydrogen bromide, anhydrous; *Chemical Formula:* HBr.

Observable Characteristics - *Physical State (as shipped):* Compressed; liquefied gas; *Color:* Colorless; *Odor:* Sharp, pungent, irritating.

Physical and Chemical Properties - *Physical State at 15 °C and 1 atm:* Gas; *Molecular Weight:* 80.92; *Boiling Point at 1 atm:* -88.2, -66.8, 206.4; *Freezing Point:* Not pertinent; *Critical Temperature:* 193.6, 89.8, 363.0; *Critical Pressure:* 1,235, 84, 8.52; *Specific Gravity:* 2.14 at -67 °C (liquid); *Vapor (Gas) Specific Gravity:* 2.71; *Ratio of Specific Heats of Vapor (Gas):* 1.38; *Latent Heat of Vaporization:* 923, 51.3, 2.15; *Heat of Combustion:* Not pertinent; *Heat of Decomposition:* Not pertinent.

Health Hazards - *Personal Protective Equipment:* Full face mask and acid gas canister, self-contained breathing apparatus: chemical goggles: rubber apron and gloves: acid-proof clothing: safety shower; *Symptoms Following Exposure:* Inhalation causes severe irritation of nose and upper respiratory tract, lung injury. Ingestion causes burns of mouth and stomach. Contact with eyes causes severe irritation and burns. Contact with skin causes irritation and burns; *Treatment for Exposure:* Get medical attention after all overexposure to this chemical. INHALATION: move victim to fresh air and keep him warm and quiet: if a qualified person is available lo give oxygen, such treatment may be helpful. INGESTION: give large amounts of water or milk: do not induce vomiting. EYES: flush with water for at least 15 min. SKIN: flush with water; treat acid burns; *Toxicity by Inhalation (Threshold Limit Value):* 3 ppm; *Short-Term Inhalation Limits:* 5 ppm for 5 min.; *Toxicity by Ingestion:* Data not available; *Late Toxicity:* Data not available; *Vapor (Gas) Irritant Characteristics:* Data not available; *Liquid or Solid Irritant Characteristics:* Data not available; *Odor Threshold:* Data not available.

Fire Hazards - *Flash Point :* Not flammable; *Flammable Limits in Air (%):* Not flammable; *Fire Extinguishing Agents:* Not pertinent; *Fire Extinguishing Agents Not To Be Used:* Not pertinent; *Special Hazards of Combustion Products:* Not pertinent; *Behavior in Fire:* Pressurized containers may explode and release toxic and irritating vapors; *Ignition Temperature :* Not pertinent; *Electrical Hazard:* Not pertinent; *Burning Rate:* Not pertinent.

Chemical Reactivity - *Reactivity with Water:* Moderate reaction with the evolution of heat; *Reactivity with Common Materials:* Rapidly absorbs moisture , forming hydrobromic acid . Highly corrosive to most metals, with the evolution of flammable and explosive hydrogen gas; *Stability During Transport:* Stable; *Neutralizing Agents for Acids and Caustics:* Flush with water and apply powdered limestone, slaked lime, soda ash, or sodium bicarbonate; *Polymerization:* Not pertinent; *Inhibitor of Polymerization:* Not pertinent.

HYDROGEN CHLORIDE

Chemical Designations - *Synonyms:* Hydrochloric acid, anhydrous; *Chemical Formula:* Hcl.

Observable Characteristics - *Physical State (as shipped):* Compressed liquefied gas; *Color:* Colorless to slightly yellow; *Odor:* Sharp, pungent, irritating.

Physical and Chemical Properties - *Physical State at 15 °C and 1 atm:* Gas; *Molecular Weight:* 36.42; *Boiling Point at 1 atm:* -121, -85.0, 188.2; *Freezing Point:* -175, -115, 158; *Critical Temperature:* 124.5, 51.4, 324.6; *Critical Pressure(psia, atm, MN/m²):* 1200, 81.6, 8.27; *Specific Gravity:* 1.191 at -85°C (liquid); *Vapor (Gas) Specific Gravity:* 1.3; *Ratio of Specific Heats of Vapor (Gas):* 1.398; *Latent Heat of Vaporization:* 185, 103, 4.31; *Heat of Combustion:* Not pertinent; *Heat of Decomposition:* Not pertinent.

Health Hazards - *Personal Protective Equipment:* Full face mask and acid gas canister; self-contained breathing apparatus; chemical goggles; rubber apron and gloves; acid-proof clothing; safety shower; *Symptoms Following Exposure:* Severely irritating to nose and upper respiratory tract; lung injury; *Treatment for Exposure:* INHALATION: immediately remove patient to fresh air, keep him warm and quiet, and call a physician immediately; if a qualified person is available to give oxygen, such treatment may be helpful. INGESTION: have victim drink water or milk; do NOT induce vomiting. EYES OR SKIN: immediately flush with plenty of water for at least 15 min.; for eyes get medical attention promptly; air contaminated clothing and wash before reuse; *Toxicity by Inhalation (Threshold Limit Value):* 5 ppm; *Short-Term Inhalation Limits:* 5 ppm for 5 min.; *Toxicity by Ingestion:* Data not available; *Late Toxicity:* None; *Vapor (Gas) Irritant Characteristics:* Vapors cause severe irritation eye and lung injury. They cannot be tolerated even at low concentrations; *Liquid or Solid Irritant Characteristics:* Fairly severe skin irritant; may cause pain and second-degree burns after few minutes contact; *Odor Threshold:* 1 - 5 ppm.

Fire Hazards - *Flash Point* : Not flammable; *Flammable Limits in Air (%):* Not flammable; *Fire Extinguishing Agents:* Not flammable; *Fire Extinguishing Agents Not To Be Used:* Not flammable; *Special Hazards of Combustion Products:* Not flammable; *Behavior in Fire:* Pressurized containers may explode releasing toxic and irritating vapors; *Ignition Temperature :* Not flammable; *Electrical Hazard:*; *Burning Rate:* Not flammable.

Chemical Reactivity - *Reactivity with Water:* Moderate reaction with the evolution of heat; *Reactivity with Common Materials:* Rapidly absorbs moisture forming hydrochloric acid. Very corrosive to metals with the evolution of flammable and explosive hydrogen gas; *Stability During Transport:* Stable; *Neutralizing Agents for Acids and Caustics:* Flush with water and apply powdered limestone, slaked lime, soda ash, or sodium bicarbonate; *Polymerization:* Not pertinent; *Inhibitor of Polymerization:* Not pertinent.

HYDROGEN CYANIDE

Chemical Designations - *Synonyms*: Hydrocyanic acid; Prussic acid; *Chemical Formula*: NCH.

Observable Characteristics - *Physical State (as shipped)*: Liquid; *Color*: Colorless to bluish white; *Odor*: Characteristic sweetish, like almond.

Physical and Chemical Properties - *Physical State at 15 °C and 1 atm.*: Liquid; *Molecular Weight*: 27.03; *Boiling Point at 1 atm.*: 78.3, 25.7, 298.9; *Freezing Point*: 8.1, −13.3, 259.9; *Critical Temperature*: 362.3, 183.5, 456.7; *Critical Pressure*: 735, 50, 5.07; *Specific Gravity*: 0.689 at 20 °C (liquid); *Vapor (Gas) Specific Gravity*: 0.9; *Ratio of Specific Heats of Vapor (Gas)*: 1.303; *Latent Heat of Vaporization*: 444, 247, 10.3; *Heat of Combustion*: −10,560, −5864, −245.3; *Heat of Decomposition*: Not pertinent.

Health Hazards Information - *Recommended Personal Protective Equipment*: Escape purposes only — air escape mask with 5-minute air cylinder. Work purposes- vapor-proof emergency suit or vinyl-coated coverall, plus air mask with clear-view facepiece, speaking diaphragm, demand regulator, and 30-minute air cylinder. Rubber gloves; chemical safety goggles; quick-opening safety shower; *Symptoms Following Exposure*: Irritation of throat, palpitation, difficult breathing, reddening of eyes, salivation, nausea, headache, weakness of arms and legs, giddiness-followed by collapse and convulsions; *General Treatment for Exposure*: Call a doctor. If breathing has stopped, give artificial respiration until doctor arrives. INHALATION: remove patient to fresh air. SKIN CONTACT: remove contaminated clothing and wash skin thoroughly with copious quantities of water for at least 15 min. If patient is unconscious, administer amyl nitrite by crushing a pearl (ampule) in a cloth and holding this under patient's nose for 15 seconds in every spent. Continue treatment until patient's condition improves or doctor arrives; *Toxicity by Inhalation (Threshold Limit Value)*: 10 ppm; *Short-Term Inhalation Limits*: 20 ppm for 30 min.; *Toxicity by Ingestion*: Grade 4, LD_{50} less than 50 mg/kg; *Late Toxicity*: Data not available; *Vapor (Gas) Irritant Characteristics*: Vapor is not very irritating but is extremely poisonous; *Liquid or Solid Irritant Characteristics*: Liquid is not irritating but is extremely if absorbed through skin or eyes; *Odor Threshold*: 1 mg/m^3.

Fire Hazards - *Flash Point (deg. F)*: 0 CC; *Flammable Limits in Air (%):* 5.6 - 40; *Fire Extinguishing Agents:* Stop flow of gas if practical; *Fire Extinguishing Agents Not To Be Used:* None; *Special Hazards of Combustion Products:* Extremely toxic vapors are generated even at ordinary temperatures; *Behavior in Fire:* Containers may explode and contents spontaneously ignite; *Ignition Temperature (deg. F):* 1,004; *Electrical Hazard:* No data; *Burning Rate:* 1.8 mm/min.

Chemical Reactivity - *Reactivity with Water:* Dissolves with a moderate reaction; *Reactivity with Common Materials:* None; *Stability During Transport:* May become unstable and subject to explosion if stored for extended periods of time or is exposed to high temperatures and pressures; *Neutralizing Agents for Acids and Caustics:* The weak acidity can be neutralized by slaked lime, however this does not destroy the hazardous properties of the material; *Polymerization:* Not pertinent; *Inhibitor of Polymerization:* Not pertinent.

HYDROGEN FLUORIDE
Chemical Designations - *Synonyms*: Hydrofluoric acid, anhydrous; *Chemical Formula*: H_2F_2
Observable Characteristics - *Physical State (as shipped)*: Liquid; *Color*: Colorless; *Odor*: pungent, irritating.
Physical and Chemical Properties - *Physical State at 15 °C and 1 atm.*: Liquid; *Molecular Weight*: 20.01; *Boiling Point at 1 atm.*: 67.1, 19.5, 292.7; *Freezing Point*: –134, –92.2, 181.0; *Critical Temperature*: 447, 230.6, 503.8; *Critical Pressure*: 1100, 74.8, 7.58; *Specific Gravity*: 0.992 at 19 °C (liquid); *Vapor (Gas) Specific Gravity*: 0.7; *Ratio of Specific Heats of Vapor (Gas)*: 1.399; *Latent Heat of Vaporization*: 145, 80.85, 3.37; *Heat of Combustion*: Not pertinent; *Heat of Decomposition*: Not pertinent.
Health Hazards Information - *Recommended Personal Protective Equipment*: Acid-resistant hat, safety goggles, face shield, jacket, overalls, gauntlet-type gloves, and boots. The goggles and face shield must have plastic lenses. There must be a shower and eye wash. Observe all precautions in the Manufacturing Chemists Association Chemical Safety Data Sheet SD-25; *Symptoms Following Exposure*: Serious and painful burns of eyes, skin and respiratory tract; pulmonary edema; *General Treatment for Exposure*: INGESTION: have victim drink water or milk; do NOT induce vomiting. SKIN: flush with water; consult physician. EYES: flush with water for at least 15 min.; consult physician; *Toxicity by Inhalation (Threshold Limit Value)*: 3 ppm; *Short-Term Inhalation Limits*: 3 ppm for 15 min.; *Toxicity by Ingestion*: Oral LD_{50} = 80 mg/kg (guinea pig); *Late Toxicity*: Data not available; *Vapor (Gas) Irritant Characteristics*: Vapors cause severe irritation of eye and throat and can cause eye and lung injury. They cannot be tolerated even at low concentrations; *Liquid or Solid Irritant Characteristics*: Severe skin irritant. Causes second- and third-degree burns on short contact; very injurious to the eyes; *Odor Threshold*: 0.03 mg/m³.
Fire Hazards - *Flash Point* : Not flammable; *Flammable Limits in Air (%):* Not flammable; *Fire Extinguishing Agents:* Not pertinent; *Fire Extinguishing Agents Not To Be Used:* Not pertinent; *Special Hazards of Combustion Products:* Toxic and irritating vapors are generated when exposed to hear; *Behavior in Fire:* Not pertinent; *Ignition Temperature :* Not flammable; *Electrical Hazard:* Not pertinent; *Burning Rate:* Not flammable.
Chemical Reactivity - *Reactivity with Water:* Dissolves with liberation of heat; *Reactivity with Common Materials:* Will attack glass, concrete and certain metals containing silica, such as cast iron. Will attack natural rubber, leather, and many organic materials. Can generate flammable and explosive hydrogen when in contact with some metals; *Stability During Transport:* Stable but requires special packaging; *Neutralizing Agents for Acids and Caustics:* Flush with water and apply powdered limestone, slaked lime, soda ash, or sodium bicarbonate; *Polymerization:* Not pertinent; *Inhibitor of Polymerization:* Not pertinent.

HYDROGEN, LIQUEFIED
Chemical Designations - *Synonyms*: Liquid hydrogen; para Hydrogen; *Chemical Formula*: H_2.
Observable Characteristics - *Physical State (as shipped)*: Liquid; *Color*: Colorless; *Odor*: None.
Physical and Chemical Properties - *Physical State at 15 °C and 1 atm.*: Gas; *Molecular Weight*: 2.0; *Boiling Point at 1 atm.*: -423, -253, 20; *Freezing Point*: -434, -259, 14; *Critical Temperature*: -400, -240, 33; *Critical Pressure*: 188, 12.8, 1.30; *Specific Gravity*: 0.071 at -253 °C (liquid); *Vapor (Gas) Specific Gravity*: 0.067; *Ratio of Specific Heats of Vapor (Gas)*: 1.3962; *Latent Heat of Vaporization*: 190.5, 105.8, 4.427; *Heat of Combustion*: -50,080, -27,823, -1164.1; *Heat of Decomposition*: Not pertinent.

Health Hazards Information - *Recommended Personal Protective Equipment*: Safety goggles or face shield; insulated gloves and long sleeves; cuffless trousers worn outside boots or over high-top shoes to shed spilled liquid; self-contained breathing apparatus containing air (never use oxygen); *Symptoms Following Exposure*: If atmosphere does not contain enough oxygen, inhalation can cause dizziness, unconsciousness, or even death. Contact of liquid with eyes or skin causes freezing similar to a burn; *General Treatment for Exposure*: The only effect of exposure to liquid hydrogen is that caused by its unusually low temperature and its action as a simple asphyxiate. INHALATION: if victim is unconscious (due to oxygen deficiency), move him to fresh air and apply resuscitation method; call physician. EYES: treat for frostbite. SKIN: treat for frostbite; soak in lukewarm water; get medical attention if burn is severe; *Toxicity by Inhalation (Threshold Limit Value)*: Gas is non-poisonous but can act as a simple asphyxiate; *Short-Term Inhalation Limits*: Not pertinent; *Toxicity by Ingestion*: Not pertinent (gas with low boiling point); *Late Toxicity*: None; *Vapor (Gas) Irritant Characteristics*: Data not available; *Liquid or Solid Irritant Characteristics*: Data not available; *Odor Threshold*: Not pertinent.

Fire Hazards - *Flash Point* : Not pertinent; *Flammable Limits in Air (%)*: 4.0 - 75.0; *Fire Extinguishing Agents:* Let fire burn; shut off gas supply; *Fire Extinguishing Agents Not To Be Used:* Carbon dioxide; *Special Hazards of Combustion Products:* Not pertinent; *Behavior in Fire:* Burns with an almost invisible flame; *Ignition Temperature (deg. F):* 1,065; *Electrical Hazard:* Class I, Group B; *Burning Rate:* 9.9 mm/min.

Chemical Reactivity - *Reactivity with Water:* Ambient temperature of water will cause vigorous vaporization of hydrogen; *Reactivity with Common Materials:* No chemical reaction, but low temperature causes most materials to become very brittle; *Stability During Transport:* Stable; *Neutralizing Agents for Acids and Caustics:* Not pertinent; *Polymerization:* Not pertinent; *Inhibitor of Polymerization:* Not pertinent.

HYDROGEN PEROXIDE

Chemical Designations - *Synonyms*: Peroxide, Albone, Superoxol; *Chemical Formula*: H_2O_2-H_2O.

Observable Characteristics - *Physical State (as shipped)*: Liquid; *Color*: Colorless; *Odor*: Slightly sharp.

Physical and Chemical Properties - *Physical State at 15 °C and 1 atm.*: Liquid; *Molecular Weight*: 34.01; *Boiling Point at 1 atm.*: 257, 125, 398; *Freezing Point*: -40.5, -40.3, 232.9; *Critical Temperature*: Not pertinent; *Critical Pressure*: Not pertinent; *Specific Gravity*: 1.29 at 20 °C (liquid); *Vapor (Gas) Specific Gravity*: Not pertinent; *Ratio of Specific Heats of Vapor (Gas)*: 1.241; *Latent Heat of Vaporization*: 542, 301, 12.6; *Heat of Combustion*: Not pertinent; *Heat of Decomposition*: -1220, -676, -28.3.

Health Hazards Information - *Recommended Personal Protective Equipment*: Protective garments, both outer and inner, made of a woven polyester fabric or of modacrylic or polyvinylidene fabrics; impermeable apron made of polyvinyl chloride or polyethylene film; neoprene gloves and boots; goggles; *Symptoms Following Exposure*: Although solutions and vapors are nontoxic, they are irritating. Vapor causes discomfort of eyes and nose. Moderately concentrated liquid causes whitening of the skin and severe stinging sensation. In most causes the stinging subsides quickly and the skin gradually returns to normal without any damage. Highly concentrated liquid can cause blistering of skin if left on for any length of time; can also cause eye damage; *General Treatment for Exposure*: Contact should be avoided, but immediate flushing with water will prevent any reaction in case of accidental contact; *Toxicity by Inhalation (Threshold Limit Value)*: 1 ppm; *Short-Term Inhalation Limits*: Data not available; *Toxicity by Ingestion*: Data not available; *Late Toxicity*: None; *Vapor (Gas) Irritant Characteristics*: Vapors cause moderate irritation, such that personnel will find high concentrations unpleasant. The effect is temporary; *Liquid or Solid Irritant Characteristics*: Fairly severe skin irritant. May cause pain and second-degree burns after a few minutes' contact; *Odor Threshold*: Not pertinent.

Fire Hazards - *Flash Point*: Not flammable but may cause fire and violent reactions on contact with combustibles and metals; *Flammable Limits in Air (%):* Not flammable; *Fire Extinguishing Agents:* Water for fires resulting from spillage; *Fire Extinguishing Agents Not To Be Used:* Not pertinent;

Special Hazards of Combustion Products: Not pertinent; *Behavior in Fire:* May explode in fires; *Ignition Temperature :* Not flammable; *Electrical Hazard:* Not pertinent; *Burning Rate:* Not flammable. **Chemical Reactivity -** *Reactivity with Water:* No reaction; *Reactivity with Common Materials:* Dirt and metals can cause rapid decomposition with the liberation of oxygen gas; *Stability During Transport:* Pure grades are stable, but contamination by dirt and metals can cause rapid or violent decomposition; *Neutralizing Agents for Acids and Caustics:* Not pertinent; *Polymerization:* Not pertinent; *Inhibitor of Polymerization:* Not pertinent.

HYDROGEN SULFIDE

Chemical Designations - *Synonyms*: Sulfuretted hydrogen; *Chemical Formula*: H_2S.
; **(ii) Observable Characteristics -** *Physical State (as shipped)*: Liquid under pressure; *Color*: Colorless; *Odor*: Offensive odor, like rotten eggs.
Physical and Chemical Properties - *Physical State at 15 °C and 1 atm.*: Gas; *Molecular Weight*: 34.08; *Boiling Point at 1 atm.*: -76.7, -60.4, 212.8; *Freezing Point*: -117, -82.8, 190.4; *Critical Temperature*: 212.7, 100.4, 373.6; *Critical Pressure*: 1300, 88.9, 9.01; *Specific Gravity*: 0.916 at -60 °C (liquid); *Vapor (Gas) Specific Gravity*: 1.2; *Ratio of Specific Heats of Vapor (Gas)*: 1.322; *Latent Heat of Vaporization*: 234, 5.44; *Heat of Combustion*: -6552, -3640, -152.4; *Heat of Decomposition*: Not pertinent.
Health Hazards Information - *Recommended Personal Protective Equipment*: Rubber-framed goggles; approved respiratory protection; *Symptoms Following Exposure*: Irritation of eyes, nose and throat. If high concentrations are inhaled, hyperpnea and respiratory paralysis may occur. Very high concentrations may produce pulmonary edema; *General Treatment for Exposure*: INHALATION: remove victim from exposure; if breathing has stopped, give artificial respiration; administer oxygen if needed; consult physician. EYES: wash with plenty of water; *Toxicity by Inhalation (Threshold Limit Value)*: 10 ppm; *Short-Term Inhalation Limits*: 200 ppm for 10 min.; 100 ppm for 30 min. and 50 ppm for 60 min.; *Toxicity by Ingestion*: Hydrogen sulfide is present as a gas at room temperature, so ingestion not likely; *Late Toxicity*: Data not available; *Vapor (Gas) Irritant Characteristics*: Vapor is moderately irritating such that personnel will not usually tolerate moderate or high vapor concentrations; *Liquid or Solid Irritant Characteristics*: Minimum hazard. If spilled on clothing and allowed to remain, may cause smarting and reddening of skin; *Odor Threshold*: 0.0047 ppm.
Fire Hazards - *Flash Point* : This is a flammable gas; *Flammable Limits in Air (%)*: 4.3 - 45; *Fire Extinguishing Agents:* Stop the flow of gas; *Fire Extinguishing Agents Not To Be Used:* Not pertinent; *Special Hazards of Combustion Products:* Toxic gases are generated in fires; *Behavior in Fire:* Vapor is heavier than air and can travel to a source of ignition and flash back; *Ignition Temperature (deg. F):* 500; *Electrical Hazard:* Not pertinent; *Burning Rate:* 2.3 mm/min.
Chemical Reactivity - *Reactivity with Water:* No reaction; *Reactivity with Common Materials:* No reactions; *Stability During Transport:* Stable; *Neutralizing Agents for Acids and Caustics:* Not pertinent; *Polymerization:* Not pertinent; *Inhibitor of Polymerization:* Not pertinent.

HYDROQUINONE

Chemical Designations - *Synonyms*: 1,4-Benzenediol; p-Dihydroxybenzene; Hydroquinol; Pyrogentisic acid; Quinol; *Chemical Formula*: $1,4\text{-}C_6H_4(OH)_2$.
Observable Characteristics - *Physical State (as shipped)*: Solid; *Color*: Light tan to light gray; white; *Odor*: None.
Physical and Chemical Properties - *Physical State at 15 °C and 1 atm.*: Solid; *Molecular Weight*: 110.11; *Boiling Point at 1 atm.*: 545, 285, 558; *Freezing Point*: 338, 170, 443; *Critical Temperature*: Not pertinent; *Critical Pressure*: Not pertinent; *Specific Gravity*: 1.33 at 20 °C (solid); *Vapor (Gas) Specific Gravity*: Not pertinent; *Ratio of Specific Heats of Vapor (Gas)*: Not pertinent; *Latent Heat of Vaporization*: Not pertinent; *Heat of Combustion*: -11,200, -6,220, -260; *Heat of Decomposition*: Not pertinent.
Health Hazards Information - *Recommended Personal Protective Equipment*: Goggles; respiratory protection if dust is present; *Symptoms Following Exposure*: Ingestion can cause ringing in the ears,

nausea, dizziness, a sense of suffocation, increased respiration rate, vomiting, pallor, muscular twitching, headache, dyspnea, cyanosis, delirium, and collapse; the urine is green or brownish-green. Lethal adult dose is 2 grams. Direct contamination of the eye with particles of hydroquinone can cause immediate irritation and may result in ulceration of the cornea. Contact with skin may cause dermatitis; *General Treatment for Exposure*: INGESTION: induce vomiting; perform gastric lavage, and follow with a saline cathartic and demulcents; get medical attention. EYES: flush immediately with plenty of water for 15 min. and get medical attention. SKIN: wash with soap and water; *Toxicity by Inhalation (Threshold Limit Value)*: 2 mg/m^3; *Short-Term Inhalation Limits*: Data not available; *Toxicity by Ingestion*: Grade 3; LD$_{50}$ 370 mg/kg (rat); *Late Toxicity*: Causes bladder cancer in mice, discoloration of eyelids and eye changes in men; *Vapor (Gas) Irritant Characteristics*: Data not available; *Liquid or Solid Irritant Characteristics*: Data not available; *Odor Threshold*: Data not available.

Fire Hazards - *Flash Point (deg. F)*: 350 OC; *Flammable Limits in Air (%):* Not pertinent; *Fire Extinguishing Agents:* Water, foam, dry chemical, or carbon dioxide; *Fire Extinguishing Agents Not To Be Used:* No data; *Special Hazards of Combustion Products:* No data; *Behavior in Fire:* Dust explosion is high probability; *Ignition Temperature :* No data; *Electrical Hazard:* No data; *Burning Rate:* Not pertinent.

Chemical Reactivity - *Reactivity with Water:* No reaction; *Reactivity with Common Materials:* No reactions; *Stability During Transport:* Stable; *Neutralizing Agents for Acids and Caustics:* Not pertinent; *Polymerization:* Not pertinent; *Inhibitor of Polymerization:* Not pertinent.

2-HYDROXYETHYL ACRYLATE, INHIBITED

Chemical Designations - *Synonyms*: beta-Hydroxyethyl acrylate; 2-Hydroxyethyl 2-propenoate; *Chemical Formula*: CH$_2$=CHCOOCH$_2$CH$_2$OH.

Observable Characteristics - *Physical State (as shipped)*: Liquid; *Color*: Colorless; *Odor*: Sweet, pleasant.

Physical and Chemical Properties - *Physical State at 15 °C and 1 atm.*: Liquid; *Molecular Weight*: 116.1; *Boiling Point at 1 atm.*: >346, >210, >583; *Freezing Point*: -76, -60, 213; *Critical Temperature*: Not pertinent; *Critical Pressure*: Not pertinent; *Specific Gravity*: 1.10 at 25 °C (liquid); *Vapor (Gas) Specific Gravity*: Not pertinent; *Ratio of Specific Heats of Vapor (Gas)*: Not pertinent; *Latent Heat of Vaporization*: Not pertinent; *Heat of Combustion*: (est.) -10,800, -6,000, -250; *Heat of Decomposition*: Not pertinent.

Health Hazards Information - *Recommended Personal Protective Equipment*: Goggles or face shield; rubber gloves; *Symptoms Following Exposure*: Inhalation causes irritation of nose and throat. Contact with liquid irritates eyes and skin; *General Treatment for Exposure*: INHALATION: remove victim from exposure; support respiration; call physician if needed. EYES: wash with large amounts of water for 15 min.; call physician. SKIN: flush with water; *Toxicity by Inhalation (Threshold Limit Value)*: Data not available; *Short-Term Inhalation Limits*: Data not available; *Toxicity by Ingestion*: Grade 2; oral LD$_{50}$ = 1,070 mg/kg (rat); *Late Toxicity*: Data not available; *Vapor (Gas) Irritant Characteristics*: Vapors cause severe irritation of eyes and throat and can cause eye and lung injury. They cannot be tolerated even at low concentrations; *Liquid or Solid Irritant Characteristics*: Severe skin irritant. Causes second- and third-degree burns on short contact and is very injurious to the eyes; *Odor Threshold*: Data not available.

Fire Hazards - *Flash Point (deg. F)*: 220 OC; *Flammable Limits in Air (%):* No data; *Fire Extinguishing Agents:* Water, dry chemical, alcohol foam, or carbon dioxide; *Fire Extinguishing Agents Not To Be Used:* Not pertinent; *Special Hazards of Combustion Products:* Not pertinent; *Behavior in Fire:* Containers may explode; *Ignition Temperature :* No data; *Electrical Hazard:* No data; *Burning Rate:* 2.0 mm/min.

Chemical Reactivity - *Reactivity with Water:* No reaction; *Reactivity with Common Materials:* No reactions; *Stability During Transport:* Stable; *Neutralizing Agents for Acids and Caustics:*; *Polymerization:* In the absence of inhibitor, polymerization will occur especially at elevated temperature; *Inhibitor of Polymerization:* Monomethyl Ether of Hydroquinone (400 ppm).

HYDROXYLAMINE SULFATE

Chemical Designations - *Synonyms*: Oxammonium sulfate; *Chemical Formula*: $(NH_2OH)_2 \cdot H_2SO_4$.

Observable Characteristics - *Physical State (as shipped)*: Crystals; *Color*: White; *Odor*: None.

Physical and Chemical Properties - *Physical State at 15 °C and 1 atm.*: Solid; *Molecular Weight*: 164.14; *Boiling Point at 1 atm.*: Not pertinent (decomposes); *Freezing Point*: Not pertinent; *Critical Temperature*: Not pertinent; *Critical Pressure*: Not pertinent; *Specific Gravity*: >1 at 20 °C (solid); *Vapor (Gas) Specific Gravity*: Not pertinent; *Ratio of Specific Heats of Vapor (Gas)*: Not pertinent; *Latent Heat of Vaporization*: Not pertinent; *Heat of Combustion*: Not pertinent; *Heat of Decomposition*: Not pertinent.

Health Hazards Information - *Recommended Personal Protective Equipment*: Acid-resistant protective clothing, including coveralls, wrist-length gloves, cap, goggles, and dust mask; *Symptoms Following Exposure*: Inhalation of dust or ingestion may cause systemic poisoning characterized by cyanosis, methemoglobinemia, convulsions, and coma. Contact with eyes or skin causes irritation; *General Treatment for Exposure*: INHALATION: remove victim to fresh air; seek immediate medical attention if symptoms occur. INGESTION: give large amounts of water; induce vomiting; get medical attention. EYES: flush with water for at least 15 min., and get medical attention. SKIN: flush immediately with plenty of water, then wash with soap and water; *Toxicity by Inhalation (Threshold Limit Value)*: Data not available; *Short-Term Inhalation Limits*: Data not available; *Toxicity by Ingestion*: Grade 3; LD_{50} 50 - 500 mg/kg; *Late Toxicity*: Data not available; *Vapor (Gas) Irritant Characteristics*: Data not available; *Liquid or Solid Irritant Characteristics*: Data not available; *Odor Threshold*: Data not available.

Fire Hazards - *Flash Point*: Not flammable; *Flammable Limits in Air (%):* Not flammable; *Fire Extinguishing Agents:* Not pertinent; *Fire Extinguishing Agents Not To Be Used:* Not pertinent; *Special Hazards of Combustion Products:* Sulfuric acid fumes may form when exposed to heat or fires; *Behavior in Fire:* Not pertinent; *Ignition Temperature :* Not pertinent; *Electrical Hazard:* Not pertinent; *Burning Rate:* Not pertinent.

Chemical Reactivity - *Reactivity with Water:* No reaction; *Reactivity with Common Materials:* Can be corrosive to metals in the presence of moisture; *Stability During Transport:* Stable; *Neutralizing Agents for Acids and Caustics:* Flush with water; *Polymerization:* Not pertinent; *Inhibitor of Polymerization:* Not pertinent.

HYDROXYPROPYL ACRYLATE

Chemical Designations - *Synonyms*: 1,2-Propanediol 1-acrylate; Propylene glycol mono-acrylate; *Chemical Formula*: $CH_3CHOHCH_2OCOCH=CH_2$.

Observable Characteristics - *Physical State (as shipped)*: Liquid; *Color*: Colorless; *Odor*: Slightly acrylic.

Physical and Chemical Properties - *Physical State at 15 °C and 1 atm.*: Liquid; *Molecular Weight*: 130; *Boiling Point at 1 atm.*: Not pertinent (decomposes); *Freezing Point*: Not pertinent; *Critical Temperature*: Not pertinent; *Critical Pressure*: Not pertinent; *Specific Gravity*: 1.06 at 25 °C (liquid); *Vapor (Gas) Specific Gravity*: 4.5; *Ratio of Specific Heats of Vapor (Gas)*: Not pertinent; *Latent Heat of Vaporization*: Not pertinent; *Heat of Combustion*: (est.) -12,300, -6,850, -287; *Heat of Decomposition*: Not pertinent.

Health Hazards Information - *Recommended Personal Protective Equipment*: Rubber gloves, apron, and boots; worker's goggles or face shield; *Symptoms Following Exposure*: Inhalation irritates nose and throat and causes coughing; lung injury may occur. Ingestion causes irritation and burning of mouth and stomach. Vapor irritates eyes. Contact with liquid causes severe burns of eyes and burns of skin; *General Treatment for Exposure*: INHALATION: if ill effect occurs, get patient to fresh air, keep him quiet and warm, and get medical attention; if breathing stops, start artificial respiration. INGESTION: force milk or water immediately; induce vomiting only at physician's recommendation. EYES: promptly flush with plenty of water and get medical attention if victim complains of severe irritation or burning; *Toxicity by Inhalation (Threshold Limit Value)*: Data not available; *Short-Term Inhalation Limits*: Data not available; *Toxicity by Ingestion*: Grade 2; oral LD_{50} = 1,230 mg/kg (rat); *Late*

Toxicity: Data not available; *Vapor (Gas) Irritant Characteristics*: Vapors are moderately irritating such that personnel will not usually tolerate moderate or high concentrations; *Liquid or Solid Irritant Characteristics*: This chemical is a severe skin irritant. Direct contact with this chemical causes second- and third-degree chemical burns - even short contact, and is very injurious to the eyes; *Odor Threshold*: Data not available.

Fire Hazards - *Flash Point (deg. F)*: 212 OC; *Flammable Limits in Air (%):* 1.8 (LEL); *Fire Extinguishing Agents:* Dry chemical, alcohol foam, or carbon dioxide; *Fire Extinguishing Agents Not To Be Used:* Water may be ineffective; *Special Hazards of Combustion Products:* No data; *Behavior in Fire:* No data; *Ignition Temperature :* No data; *Electrical Hazard:* No data; *Burning Rate:* No data.

Chemical Reactivity - *Reactivity with Water:* No reaction; *Reactivity with Common Materials:* No reactions; *Stability During Transport:* Stable; *Neutralizing Agents for Acids and Caustics:* Not pertinent; *Polymerization:* Polymerization may occur. Avoid exposure to high temperatures, ultraviolet light, and free-radical catalysts; *Inhibitor of Polymerization:* 200 ppm Hydroquinone.

HYDROXYPROPYL METHACRYLATE

Chemical Designations - *Synonyms:* 1,2-propanediol 1-methacrylate; Propylene glycol monomethacrylate; *Chemical Formula:* $CH_3CHOHCH_2OCOC(CH_3)=CH_2$.

Observable Characteristics - *Physical State (as shipped)*: Liquid; *Color*: Colorless; *Odor*: Slight acrylic.

Physical and Chemical Properties - *Physical State at 15 °C and 1 atm.*: Liquid; *Molecular Weight*: 144; *Boiling Point at 1 atm.*: Not pertinent (decomposes); *Freezing Point*: Not pertinent; *Critical Temperature*: Not pertinent; *Critical Pressure*: Not pertinent; *Specific Gravity*: 1.06 at 20 °C (liquid); *Vapor (Gas) Specific Gravity*: Not pertinent; *Ratio of Specific Heats of Vapor (Gas)*: Not pertinent; *Latent Heat of Vaporization*: Data not available; *Heat of Combustion*: No data; *Heat of Decomposition*: Not pertinent.

Health Hazards Information - *Recommended Personal Protective Equipment*: Face shield; rubber gloves; *Symptoms Following Exposure*: Inhalation causes coughing and irritation of nose and throat; lung injury may occur. Ingestion causes irritation and burning of mouth and stomach. Vapor contact with eyes causes irritation. Liquid causes severe eye burns and skin irritation; *General Treatment for Exposure*: INHALATION: immediately remove victim to fresh air; if required, start artificial respiration and call a doctor. INGESTION: force milk or water at once; get medical attention. EYES: flush with water for at least 15 min.; get medical attention if irritation persist. SKIN: flush with copious amounts of water; seek immediate medical attention for chemical burns; *Toxicity by Inhalation*: No data; *Short-Term Inhalation Limits*: No data; *Toxicity by Ingestion*: Grade 1; LD_{50} 5-15 g/kg (mouse); *Late Toxicity*: No data; *Vapor (Gas) Irritant Characteristics*: No data; *Liquid or Solid Irritant Characteristics*: Data not available; *Odor Threshold*: No data.

Fire Hazards - *Flash Point (deg. F)*: 250 OC; *Flammable Limits in Air:* No data; *Fire Extinguishing Agents:* Foam, dry chemical, or carbon dioxide; *Fire Extinguishing Agents Not To Be Used:* Water may be ineffective; *Special Hazards of Combustion Products:* No data; *Behavior in Fire:* Can polymerize when hot and burst containers; *Ignition Temperature:* No data; *Electrical Hazard:* No data; *Burning Rate:* No data.

Chemical Reactivity - *Reactivity with Water:* No reaction; *Reactivity with Common Materials:* No reactions; *Stability During Transport:* Stable; *Neutralizing Agents for Acids and Caustics:* Not pertinent; *Polymerization:* Polymerizes when exposed to heat, ultraviolet light, or free-radical catalysts; *Inhibitor of Polymerization:* 200 ppm Hydroquinone.

I

ISOAMYL ALCOHOL

Chemical Designations - *Synonyms*: Fermentation amyl alcohol; Fusel oil; Isobutylcarbinol; Isopentyl alcohol; 3-Methyl-1-butanol; Potato-spirit oil; *Chemical Formula*: $(CH_3)_2CHCH_2CH_2OH$.

Observable Characteristics - *Physical State (as shipped)*: Liquid; *Color*: Colorless; *Odor*: Mild odor; alcoholic, non-residual.

Physical and Chemical Properties - *Physical State at 15 °C and 1 atm.*: Liquid; *Molecular Weight*: 88.15; *Boiling Point at 1 atm.*: 270, 132, 405; *Freezing Point*: Not pertinent; *Critical Temperature*: 585, 307, 580; *Critical Pressure*: Not pertinent; *Specific Gravity*: 0.81 at 20°C (liquid); *Vapor (Gas) Specific Gravity*: Not pertinent; *Ratio of Specific Heats of Vapor (Gas)*: (est.) 1.062; *Latent Heat of Vaporization*: 215.6, 119.8, 5.016; *Heat of Combustion*: -16,200, -9,000, -376.8; *Heat of Decomposition*: Not pertinent.

Health Hazards Information - *Recommended Personal Protective Equipment*: Face shield to avoid splash; *Symptoms Following Exposure*: Very high vapor concentrations irritate eyes and upper respiratory tract. Continued contact with skin may cause irritation; *General Treatment for Exposure*: EYES: immediately flush with plenty of water for at least 15 min.; get medical attention. SKIN: flush with water; wash with soap and water; *Toxicity by Inhalation (Threshold Limit Value)*: 100 ppm; *Short-Term Inhalation Limits*: Data not available; *Toxicity by Ingestion*: Grade 2, LD_{50} 0.5 - 5 g/kg; *Late Toxicity*: None; *Vapor (Gas) Irritant Characteristics*: Vapor is moderately irritating such that personnel will not usually tolerate moderate or high vapor concentrations; *Liquid or Solid Irritant Characteristics*: Liquid may irritate skin; *Odor Threshold*: Data not available.

Fire Hazards - *Flash Point (deg. F)*: 114 OC; *Flammable Limits in Air (%)*: 1.2 - 9.0 (212 of); *Fire Extinguishing Agents:* Water spray, dry chemical, alcohol foam, or carbon dioxide; *Fire Extinguishing Agents Not To Be Used:* Not pertinent; *Special Hazards of Combustion Products:* Not pertinent; *Behavior in Fire:* Not pertinent; *Ignition Temperature (deg. F):* 662; *Electrical Hazard:* Class I, Group C; *Burning Rate:* 3.6 mm/min.

Chemical Reactivity - *Reactivity with Water:* No reaction; *Reactivity with Common Materials:* No reactions; *Stability During Transport:* Stable; *Neutralizing Agents for Acids and Caustics:* Not pertinent; *Polymerization:* Not pertinent; *Inhibitor of Polymerization:* Not pertinent.

ISOBUTANE

Chemical Designations - *Synonyms*: 2-Methylpropane; *Chemical Formula*: $CH_3CH(CH_3)_2$.

Observable Characteristics - *Physical State (as shipped)*: Liquid under pressure; *Color*: Colorless; *Odor*: Like gasoline.

Physical and Chemical Properties - *Physical State at 15 °C and 1 atm.*: Gas; *Molecular Weight*: 58.12; *Boiling Point at 1 atm.*: 10.8, -11.8, 261.4; *Freezing Point*: -427.5, -255.3, 17.9; *Critical Temperature*: 275, 135, 408; *Critical Pressure*: 529, 36.0, 3.65; *Specific Gravity*: 0.557 at 20°C (liquid); *Vapor (Gas) Specific Gravity*: 2.0; *Ratio of Specific Heats of Vapor (Gas)*: 1.095; *Latent Heat of Vaporization*: 158, 87.5, 3.66; *Heat of Combustion*: -19.458, -10.810, -452.59; *Heat of Decomposition*: Not pertinent.

Health Hazards Information - *Recommended Personal Protective Equipment*: Self-contained breathing apparatus; safety goggles; *Symptoms Following Exposure*: Central nervous system depression ranging from dizziness and incoordination to anesthesia and respiratory arrest, depending on concentrations and extent of inhalation. Irregular heartbeat is rare but is a dangerous complication at anesthetic levels; *General Treatment for Exposure*: INHALATION: protect victim against self-injure if he is stuporous, confused, or anesthetized; apply artificial respiration if breathing has stopped; avoid administration of epinephrine or other sympathomimetic amines; prevent aspiration of vomitus by proper positioning of head; give symptomatic and supportive treatment. INGESTION OR ASPIRATION: no treatment required; *Toxicity by Inhalation (Threshold Limit Value)*: Data not available; *Short-Term Inhalation*

Limits: Data not available; *Toxicity by Ingestion*: Not pertinent; *Late Toxicity*: None; *Vapor (Gas) Irritant Characteristics*: None; *Liquid or Solid Irritant Characteristics*: No appreciable hazard. Practically harmless to skin because it is very volatile and evaporates quickly. Some frostbite possible; *Odor Threshold*: Data not available.

Fire Hazards - *Flash Point (deg. F)*: -117 CC; *Flammable Limits in Air (%)*: 1.8 - 8.4; *Fire Extinguishing Agents:* Stop the flow of gas; *Fire Extinguishing Agents Not To Be Used:* Not pertinent; *Special Hazards of Combustion Products:* Not pertinent; *Behavior in Fire:* Not pertinent; *Ignition Temperature (deg. F):* 890; *Electrical Hazard:* Not pertinent; *Burning Rate:* 9.3 mm/min.

Chemical Reactivity - *Reactivity with Water:* No reaction; *Reactivity with Common Materials:* No reactions; *Stability During Transport:* Stable; *Neutralizing Agents for Acids and Caustics:* Not pertinent; *Polymerization:* Not pertinent; *Inhibitor of Polymerization:* Not pertinent.

ISOBUTYL ACETATE

Chemical Designations - *Synonyms*: Acetic acid, isobutyl ester; 2-Methyl-1-propyl acetate; beta-Methylpropyl ethanoate; *Chemical Formula*: $CH_3COOCH_2CH(CH_3)_2$.

Observable Characteristics - *Physical State (as shipped)*: Liquid; *Color*: Colorless; *Odor*: Agreeable fruity odor in low concentrations, disagreeable in higher concentrations; mild, characteristic ester; nonresidual.

Physical and Chemical Properties - *Physical State at 15 °C and 1 atm.*: Liquid; *Molecular Weight*: 116.16; *Boiling Point at 1 atm.*: 243.1, 117.3, 390.5; *Freezing Point*: -142.8, -97.1, -176.1; *Critical Temperature*: 565, 296, 569; *Critical Pressure*: 470 psia, 32, 3.2; *Specific Gravity*: 0.871 at 20 °C (liquid); *Vapor (Gas) Specific Gravity*: Not pertinent; *Ratio of Specific Heats of Vapor (Gas)*: Not pertinent; *Latent Heat of Vaporization*: 133, 73.7, 3.09; *Heat of Combustion*: (est.) -13,000, -7220, -302; *Heat of Decomposition*: Not pertinent.

Health Hazards Information - *Recommended Personal Protective Equipment*: Air pack or organic canister mask; chemical goggles; *Symptoms Following Exposure*: Vapors may irritate upper respiratory tract and cause nausea, vomiting, dizziness and loss of consciousness. Liquid irritates eyes and may irritate skin; *General Treatment for Exposure*: INHALATION: remove from exposure; if breathing is irregular or has stopped, start resuscitation and give oxygen; call a doctor. EYES: flush with water for at least 15 minutes; *Toxicity by Inhalation (Threshold Limit Value)*: 150 ppm; *Short-Term Inhalation Limits*: Data not available; *Toxicity by Ingestion*: Data not available; *Late Toxicity*: Data not available; *Vapor (Gas) Irritant Characteristics*: Vapors causes a slight smarting of the eyes or respiratory system if present in high concentrations. The effect is temporary; *Liquid or Solid Irritant Characteristics*: Minimum hazard. If spilled on clothing and allowed to remain, may cause smarting and reddening of skin; *Odor Threshold*: Data not available.

Fire Hazards - *Flash Point* : 62 CC, 85 OC; *Flammable Limits in Air (%)*: 2.4 - 10.5; *Fire Extinguishing Agents:* Foam, carbon dioxide, and dry chemical; *Fire Extinguishing Agents Not To Be Used:* Water may be ineffective; *Special Hazards of Combustion Products:* Not pertinent; *Behavior in Fire:* Not pertinent; *Ignition Temperature :* 793; *Electrical Hazard:* Class I, Group D; *Burning Rate:* No data.

Chemical Reactivity - *Reactivity with Water:* No reaction; *Reactivity with Common Materials:* Softens and dissolves many types of plastics; *Stability During Transport:* Stable; *Neutralizing Agents for Acids and Caustics:* Not pertinent; *Polymerization:* Not pertinent; *Inhibitor of Polymerization:* Not pertinent.

ISOBUTYL ALCOHOL

Chemical Designations - *Synonyms*: Isobutanol; Isopropylcarbinol; 2-methyl-1-propanol; Fermentation butyl alcohol; *Chemical Formula*: $(CH_3)_2CHCH_2OH$.

Observable Characteristics - *Physical State (as shipped)*: Liquid; *Color*: Colorless; *Odor*: Slightly suffocating; nonresidual alcoholic.

Physical and Chemical Properties - *Physical State at 15 °C and 1 atm.*: Liquid; *Molecular Weight*: 74.12; *Boiling Point at 1 atm.*: 226.2, 107.9, 381.1; *Freezing Point*: -162, -108, 165; *Critical Temperature*: 526.3, 274.6, 547.8; *Critical Pressure*: 623, 42.4, 4.30; *Specific Gravity*: 0.802 at 20

°C (liquid); *Vapor (Gas) Specific Gravity*: Not pertinent; *Ratio of Specific Heats of Vapor (Gas)*: Not pertinent; *Latent Heat of Vaporization*: 248, 138, 5.78; *Heat of Combustion*: -14,220, -7,900, -330.8; *Heat of Decomposition*: Not pertinent.

Health Hazards Information - *Recommended Personal Protective Equipment*: Air pack or organic canister mask; chemical goggles; *Symptoms Following Exposure*: Contact with eyes is extremely irritating and may cause burns. Breathing vapors will be irritating to the nose and throat. In high concentrations, may cause nausea, dizziness, headache, and stupor; *General Treatment for Exposure*: INHALATION: if victim is overcome by vapors, remove him from exposure immediately; call a physician; if breathing is irregular or has stopped, start resuscitation; administer oxygen. EYES: flush with water for at least 15 minutes; *Toxicity by Inhalation (Threshold Limit Value)*: 100 ppm; *Short-Term Inhalation Limits*: 200 ppm for 60 min.; *Toxicity by Ingestion*: Grade 2, LD_{50} 0.5 to 5 g/kg (rat); *Late Toxicity*: Data not available; *Vapor (Gas) Irritant Characteristics*: Vapors causes a slight smarting of the eyes or respiratory system if present in high concentrations. The effect is temporary; *Liquid or Solid Irritant Characteristics*: No appreciable hazard. Practically harmless to the skin; *Odor Threshold*: Data not available.

Fire Hazards - *Flash Point (deg. F)*: 82 CC, 90 OC; *Flammable Limits in Air (%)*: 1.6 - 10.9; *Fire Extinguishing Agents:* Alcohol foam, dry chemical, or carbon dioxide; *Fire Extinguishing Agents Not To Be Used:* Water may be ineffective; *Special Hazards of Combustion Products:* Not pertinent; *Behavior in Fire:* Not pertinent; *Ignition Temperature (deg. F):* 800; *Electrical Hazard:* Not pertinent; *Burning Rate:* 3.5 mm/min.

Chemical Reactivity - *Reactivity with Water:* No reaction; *Reactivity with Common Materials:* No reactions; *Stability During Transport:* Stable; *Neutralizing Agents for Acids and Caustics:* Not pertinent; *Polymerization:* Not pertinent; *Inhibitor of Polymerization:* Not pertinent.

ISOBUTYLAMINE

Chemical Designations - *Synonyms*: 1-Amino-2-methylpropane; iso-Butylamine; Monoisobutylamine; *Chemical Formula*: $(CH_3)_2CHCH_2NH_2$.

Observable Characteristics - *Physical State (as shipped)*: Liquid; *Color*: Colorless; *Odor*: Strong ammoniacal.

Physical and Chemical Properties - *Physical State at 15 °C and 1 atm.*: Liquid; *Molecular Weight*: 73.1; *Boiling Point at 1 atm.*: 153.3, 67.4, 340.6; *Freezing Point*: -121.9, -85.5, 187.7; *Critical Temperature*: 469.4, 243.0, 516.2; *Critical Pressure*: 620, 42, 4.3; *Specific Gravity*: 0.739 at 20 °C (liquid); *Vapor (Gas) Specific Gravity*: 2.5; *Ratio of Specific Heats of Vapor (Gas)*: 1.073 at 20 °C; *Latent Heat of Vaporization*: 182, 101, 4.23; *Heat of Combustion*: -17,550, -9,760, -408; *Heat of Decomposition*: Not pertinent.

Health Hazards Information - *Recommended Personal Protective Equipment*: Self-contained breathing apparatus; butyl rubber gloves; chemical face shield; butyl rubber apron; *Symptoms Following Exposure*: Inhalation causes severe coughing and chest pain due to irritation of air passages; can cause lung edema. Compound is sympathomimetic and is also a cardiac depressant and convulsant; ingestion causes nausea and profuse salivation. Contact with skin causes severe irritation; *General Treatment for Exposure*: INHALATION: remove victim to fresh air; if he is not breathing, give artificial respiration; if breathing is difficult, give oxygen; call a physician. INGESTION: give large amount of water followed by dilute vinegar or lemon juice; keep patient warm. EYES: flush with water for 15 min. SKIN: flush with water; *Toxicity by Inhalation (Threshold Limit Value)*: 5 ppm; *Short-Term Inhalation Limits*: Data not available; *Toxicity by Ingestion*: Grade 3; oral LD_{50} = 120 mg/kg (rabbit), 250 mg/kg (rat); *Late Toxicity*: Data not available; *Vapor (Gas) Irritant Characteristics*: Vapors are moderately irritating such that personnel will not usually tolerate moderate or high concentrations; *Liquid or Solid Irritant Characteristics*: Severe skin irritant. Causes second- and third-degree burns on short contact and is very injurious to the eyes; *Odor Threshold*: Data not available.

Fire Hazards - *Flash Point (deg. F)*: 15 CC; *Flammable Limits in Air (%)*: 3.4 - 9; *Fire Extinguishing Agents:* Dry chemical, alcohol foam, or carbon dioxide; *Fire Extinguishing Agents Not To Be Used:* Water may be ineffective; *Special Hazards of Combustion Products:* Toxic oxides of nitrogen form in fires; *Behavior in Fire:* Vapor is heavier than air and may travel to source of ignition and flash back. Containers may explode; *Ignition Temperature (deg. F):* 712; *Electrical Hazard:* No data; *Burning Rate:* 6.0 mm/min.

Chemical Reactivity - *Reactivity with Water:* No reaction; *Reactivity with Common Materials:* No reaction; *Stability During Transport:* Stable; *Neutralizing Agents for Acids and Caustics:* Not pertinent; *Polymerization:* Not pertinent; *Inhibitor of Polymerization:* Not pertinent.

ISOBUTYLENE

Chemical Designations - *Synonyms*: Isobutene; 2-Methylpropene; *Chemical Formula*: $(CH_3)_2C = CH_2$.
Observable Characteristics - *Physical State (as shipped)*: Liquid under pressure; *Color*: Colorless; *Odor*: Mild and sweet.
Physical and Chemical Properties - *Physical State at 15 °C and 1 atm.*: Gas; *Molecular Weight*: 56.10; *Boiling Point at 1 atm.*: 19.6, -6.9, 266.3; *Freezing Point*: -220, -140.3, 132.9; *Critical Temperature*: 292.5, 144.7, 417.9; *Critical Pressure*: 580, 39.48, 3.99; *Specific Gravity*: 0.59 at 20 °C (liquid); *Vapor (Gas) Specific Gravity*: 1.9; *Ratio of Specific Heats of Vapor (Gas)*: 1.061; *Latent Heat of Vaporization*: 170, 94.3, 3.95; *Heat of Combustion*: -19,359, -10,755, -450.29; *Heat of Decomposition*: Not pertinent.
Health Hazards Information - *Recommended Personal Protective Equipment*: Chemical gloves and eye protection; organic vapor canister or self-contained breathing apparatus; *Symptoms Following Exposure*: Inhalation of moderate concentrations causes dizziness, drowsiness, and unconsciousness. Contact with eyes or skin may cause irritation; the liquid may cause frostbite; *General Treatment for Exposure*: INHALATION: remove victim to fresh air and apply resuscitation; call a physician promptly if victim is unconscious. EYES: if irritated, wash with water. SKIN: if irritated, wash with soap and water; *Toxicity by Inhalation (Threshold Limit Value)*: 1000 ppm (8 hr); *Short-Term Inhalation Limits*: Data not available; *Toxicity by Ingestion*: Not pertinent; *Late Toxicity*: None; *Vapor (Gas) Irritant Characteristics*: Vapors are non-irritating to eyes and throat; *Liquid or Solid Irritant Characteristics*: No appreciable hazard. Practically harmless to the skin because it is very volatile and evaporates quickly. May cause frostbite; *Odor Threshold*: Data not available.
Fire Hazards - *Flash Point (deg. F)*: -105 CC; *Flammable Limits in Air (%)*: 1.8 - 9.6; *Fire Extinguishing Agents:* Recommended to allow fire to burn. Stop the flow of gas if feasible. Water fog, dry chemical, or carbon dioxide may be used on small firs; *Fire Extinguishing Agents Not To Be Used:* Not pertinent; *Special Hazards of Combustion Products:* Not pertinent; *Behavior in Fire:* Containers may explode in fires. Vapors are heavier than air and can travel to source of ignition and flash back; *Ignition Temperature (deg. F):* 869; *Electrical Hazard:* Not pertinent; *Burning Rate:* No data.
Chemical Reactivity - *Reactivity with Water:* No reaction; *Reactivity with Common Materials:* No reactions; *Stability During Transport:* Stable; *Neutralizing Agents for Acids and Caustics:* Not pertinent; *Polymerization:* Not pertinent; *Inhibitor of Polymerization:* Not pertinent.

ISOBUTYRIC ACID

Chemical Designation - *Synonyms*: Dimethylacetic acid; Isopropylformic acid; 2-Methylpropanoic acid; alpha-Methylpropionic acid; Propane-2-carboxylic acid; *Chemical Formula*: $(CH_3)_2CHCOOH$.
Observable Characteristics - *Physical State (as shipped)*: Liquid; *Color*: Colorless; *Odor*: Unpleasant, acrid.
Physical and Chemical Properties - *Physical State at 15 °C and 1 atm.*: Liquid; *Molecular Weight*: 88; *Boiling Point at 1 atm.*: 309, 154, 427; *Freezing Point*: -51, -46, 227; *Critical Temperature*: 637, 336, 609; *Critical Pressure*: 588, 40, 4.06; *Specific Gravity*: 0.949 at 20 °C (liquid); *Vapor (Gas) Specific Gravity*: 3.0; *Ratio of Specific Heats of Vapor (Gas)*: Not pertinent; *Latent Heat of Vaporization*: 202, 112, 4.68; *Heat of Combustion*: -10,600, -5,880, -246; *Heat of Decomposition*: Not pertinent.
Health Hazards Information - *Recommended Personal Protective Equipment*: Organic chemical respirator; goggles or face shield; rubber gloves; *Symptoms Following Exposure*: Inhalation causes irritation of nose and throat. Ingestion causes irritation of mouth and stomach. Contact with eyes or skin

causes irritation; *General Treatment for Exposure*: INHALATION: move to fresh air. INGESTION: give large amounts of water. EYES: flush with water for at least 15 min.; get medical attention if irritation persists. SKIN: flush with water; *Toxicity by Inhalation (Threshold Limit Value)*: Data not available; *Short-Term Inhalation Limits*: Data not available; *Toxicity by Ingestion*: Grade 3; oral LD_{50} =280 mg/kg (rat); *Late Toxicity*: Data not available; *Vapor (Gas) Irritant Characteristics*: Data not available; *Liquid or Solid Irritant Characteristics*: Data not available; *Odor Threshold*: Data not available.

Fire Hazards - *Flash Point (deg. F)*: 170 OC; *Flammable Limits in Air (%):* No data; *Fire Extinguishing Agents:* Dry chemical, alcohol foam, or carbon dioxide; *Fire Extinguishing Agents Not To Be Used:* Water may be ineffective; *Special Hazards of Combustion Products:* None; *Behavior in Fire:* No data; *Ignition Temperature :* 935; *Electrical Hazard:* No data; *Burning Rate:* 2.6 mm/min.

Chemical Reactivity - *Reactivity with Water:* No reaction; *Reactivity with Common Materials:* Corrosive to aluminum and other metals. Flammable hydrogen gas may accumulate in enclosed spaces; *Stability During Transport:* Stable; *Neutralizing Agents for Acids and Caustics:* Flush with water; *Polymerization:* Not pertinent; *Inhibitor of Polymerization:* Not pertinent.

ISOBUTYRONITRILE

Chemical Designations - *Synonyms*: IBN; Isopropyl cyanide; 2-Methilpropanenitrile; 2-Methilpropionitrile; *Chemical Formula*: $(CH_3)_2CHCN$.

Observable Characteristics - *Physical State (as shipped)*: Liquid; *Color*: Colorless; *Odor*: Like almonds or benzaldehyde.

Physical and Chemical Properties - *Physical State at 15 °C and 1 atm.*: Liquid; *Molecular Weight*: 69.1; *Boiling Point at 1 atm.*: 219, 104, 377; *Freezing Point*: Not pertinent; *Critical Temperature*: Data not available; *Critical Pressure*: Data not available; *Specific Gravity*: 0.774 at 20 °C (liquid); *Vapor (Gas) Specific Gravity*: 2.4; *Ratio of Specific Heats of Vapor (Gas)*: Data not available; *Latent Heat of Vaporization*: 200, 110, 4.7; *Heat of Combustion*: -14,960, -8,310, -348; *Heat of Decomposition*: Not pertinent.

Health Hazards Information - *Recommended Personal Protective Equipment*: Self-contained breathing apparatus; goggles; rubber gloves; *Symptoms Following Exposure*: Inhalation, ingestion, or skin contact causes weakness, headache, confusion, nausea, vomiting; acute cyanide poisoning may result. Contact with eyes causes irritation; *General Treatment for Exposure*: Get medical attention following all overexposures to this chemical. Watch for symptoms of cyanide poisoning. INHALATION: move patient to fresh air; apply artificial respiration if breathing stops. INGESTION: break an amyl nitrite pearl in a cloth and hold lightly under patient's nose for 15 sec.; if he is conscious, induce vomiting and repeat until vomit is clear; repeat inhalation of amyl nitrite 5 times at 15-sec. intervals. EYES: flush with water for at least 15 min. SKIN: flush with water; remove contaminated clothing; destroy contaminated shoes; *Toxicity by Inhalation (Threshold Limit Value)*: Data not available; *Short-Term Inhalation Limits*: Data not available; *Toxicity by Ingestion*: Grade 3; oral LD_{50} = 100 mg/kg (rat); *Late Toxicity*: Data not available; *Vapor (Gas) Irritant Characteristics*: Data not available; *Liquid or Solid Irritant Characteristics*: Data not available; *Odor Threshold*: Data not available.

Fire Hazards - *Flash Point (deg.F)*: 47 CC; *Flammable Limits in Air (%):* No data; *Fire Extinguishing Agents:* Dry chemical, foam, or carbon dioxide; *Fire Extinguishing Agents Not To Be Used:* Water may be ineffective; *Special Hazards of Combustion Products:* Toxic oxides of nitrogen may form; *Behavior in Fire:* No data; *Ignition Temperature :* No data; *Electrical Hazard:* No data; *Burning Rate:* No data.

Chemical Reactivity - *Reactivity with Water:* No reaction; *Reactivity with Common Materials:* No reactions; *Stability During Transport:*; *Neutralizing Agents for Acids and Caustics:* Not pertinent; *Polymerization:* Not pertinent; *Inhibitor of Polymerization:* Not pertinent.

ISOOCTALDEHYDE

Chemical Designations - *Synonyms*: Dimethylhexanals; Isooctyl aldehyde; 6-Methyl-1-heptanal; *Chemical Formula*: $(CH_3)_2CH(CH_2)_4CHO$.

Observable Characteristics - *Physical State (as shipped)*: Liquid; *Color*: Colorless; *Odor*: Mild fruity.

Physical and Chemical Properties - *Physical State at 15 °C and 1 atm.*: Liquid; *Molecular Weight*: 128.22; *Boiling Point at 1 atm.*: 307 - 352, 153 - 178, 426 - 451; *Freezing Point*: -180, -118, 155; *Critical Temperature*: Not pertinent; *Critical Pressure*: Not pertinent; *Specific Gravity*: 0.825 at 20 °C (liquid); *Vapor (Gas) Specific Gravity*: Not pertinent; *Ratio of Specific Heats of Vapor (Gas)*: (est.) 1.040; *Latent Heat of Vaporization*: 140, 77, 3.2; *Heat of Combustion*: (est.) -17,000, -9600, -400; *Heat of Decomposition*: Not pertinent.

Health Hazards Information - *Recommended Personal Protective Equipment*: Chemical goggles; *Symptoms Following Exposure*: High vapor concentrations produce eye irritation. Liquid may irritate eyes; *General Treatment for Exposure*: Remove from exposure. Wash eyes with water for 15 min.; *Toxicity by Inhalation (Threshold Limit Value)*: Data not available; *Short-Term Inhalation Limits*: Data not available; *Toxicity by Ingestion*: Data not available; *Late Toxicity*: Data not available; *Vapor (Gas) Irritant Characteristics*: Vapors causes a slight smarting of the eyes or respiratory system if present in high concentrations. The effect is temporary; *Liquid or Solid Irritant Characteristics*: No appreciable hazard. Practically harmless to the skin; *Odor Threshold*: Data not available.

Fire Hazards - *Flash Point (deg. F)*: 104 CC; *Flammable Limits in Air (%):* No data; *Fire Extinguishing Agents:* No data; *Fire Extinguishing Agents Not To Be Used:* No data; *Special Hazards of Combustion Products:* Not pertinent; *Behavior in Fire:* Not pertinent; *Ignition Temperature :* Not pertinent; *Electrical Hazard:* 320; *Burning Rate:* No data.

Chemical Reactivity - *Reactivity with Water:* No reaction; *Reactivity with Common Materials:* No reactions; *Stability During Transport:* Stable; *Neutralizing Agents for Acids and Caustics:* Not pertinent; *Polymerization:* Not pertinent; *Inhibitor of Polymerization:* Not pertinent.

ISODECALDEHYDE

Chemical Designations - *Synonyms*: Isodecaldehde, mixed isomers; Trimethylheptanals; *Chemical Formula*: $C_9H_{19}CHO$.

Observable Characteristics - *Physical State (as shipped)*: Liquid; *Color*: Colorless; *Odor*: Somewhat fruity.

Physical and Chemical Properties - *Physical State at 15 °C and 1 atm.*: Liquid; *Molecular Weight*: 156.28; *Boiling Point at 1 atm.*: Data not available; *Freezing Point*: Data not available; *Critical Temperature*: Not pertinent; *Critical Pressure*: Not pertinent; *Specific Gravity*: (est.) 0.84 at 15 °C (liquid); *Vapor (Gas) Specific Gravity*: Not pertinent; *Ratio of Specific Heats of Vapor (Gas)*: Not pertinent; *Latent Heat of Vaporization*: Data not available; *Heat of Combustion*: Data not available; *Heat of Decomposition*: Not pertinent.

Health Hazards Information - *Recommended Personal Protective Equipment*: Protective clothing; chemical goggles; *Symptoms Following Exposure*: Low general toxicity. Liquid may irritate eyes and skin; *General Treatment for Exposure*: Wash eyes and skin with plenty of water for at least 15 min.; *Toxicity by Inhalation (Threshold Limit Value)*: Data not available; *Short-Term Inhalation Limits*: Data not available; *Toxicity by Ingestion*: Data not available; *Late Toxicity*: Data not available; *Vapor (Gas) Irritant Characteristics*: Vapors causes a slight smarting of the eyes or respiratory system if present in high concentrations. The effect is temporary; *Liquid or Solid Irritant Characteristics*: Minimum hazard. If spilled on clothing and allowed to remain, may cause smarting and reddening of skin; *Odor Threshold*: Data not available.

Fire Hazards - *Flash Point (deg. F)*: 185 OC; *Flammable Limits in Air (%):* No data; *Fire Extinguishing Agents:* Foam, dry chemical, or carbon dioxide; *Fire Extinguishing Agents Not To Be Used:* Not pertinent; *Special Hazards of Combustion Products:* Not pertinent; *Behavior in Fire:* Not pertinent; *Ignition Temperature :* No data; *Electrical Hazard:* Not pertinent; *Burning Rate:* No data.

Chemical Reactivity - *Reactivity with Water:* No reaction; *Reactivity with Common Materials:* No reactions; *Stability During Transport:* Stable; *Neutralizing Agents for Acids and Caustics:* Not pertinent; *Polymerization:* Not pertinent; *Inhibitor of Polymerization:* Not pertinent.

ISODECYL ACRYLATE, INHIBITED

Chemical Designations - *Synonyms*: iso-Decyl acrylate; *Chemical Formula*: $CH_2=CHCOOC_{10}H_{21}$.

Observable Characteristics - *Physical State (as shipped)*: Liquid; *Color*: Colorless; *Odor*: Weak acrylate.

Physical and Chemical Properties - *Physical State at 15 °C and 1 atm.*: Liquid; *Molecular Weight*: 212.4; *Boiling Point at 1 atm.*: Not pertinent (polymerizes); *Freezing Point*: -148, -100, 173; *Critical Temperature*: Not pertinent; *Critical Pressure*: Not pertinent; *Specific Gravity*: 0.885 at 20 °C (liquid); *Vapor (Gas) Specific Gravity*: Not pertinent; *Ratio of Specific Heats of Vapor (Gas)*: Not pertinent; *Latent Heat of Vaporization*: 110, 61, 2.6; *Heat of Combustion*: (est.) -16,300, -9,100, -380; *Heat of Decomposition*: Not pertinent.

Health Hazards Information - *Recommended Personal Protective Equipment*: Goggles or face shield; rubber gloves; *Symptoms Following Exposure*: Inhalation causes mild irritation of nose and throat. Eyes are mildly irritated by vapor, more severely by liquid. Prolonged contact of liquid with skin may cause irritation; *General Treatment for Exposure*: INHALATION: move to fresh air. EYES: flush with water for at least 15 min. after contact with liquid. SKIN: wipe off, wash well with soap and water; *Toxicity by Inhalation (Threshold Limit Value)*: Data not available; *Short-Term Inhalation Limits*: Data not available; *Toxicity by Ingestion*: Grade 1; LD_{50} 5 to 15 g/kg; *Late Toxicity*: Data not available; *Vapor (Gas) Irritant Characteristics*: Vapors causes a slight smarting of the eyes or respiratory system if present in high concentrations. The effect is temporary; *Liquid or Solid Irritant Characteristics*: Minimum hazard. If spilled on clothing and allowed to remain, may cause smarting and reddening of skin; *Odor Threshold*: Data not available.

Fire Hazards - *Flash Point (deg. F)*: 240 OC; *Flammable Limits in Air (%):* No data; *Fire Extinguishing Agents:* Foam, dry chemical, or carbon dioxide; *Fire Extinguishing Agents Not To Be Used:* Water may be ineffective; *Special Hazards of Combustion Products:* Not pertinent; *Behavior in Fire:* May polymerize to form a gummy material, but the reaction is not violent; *Ignition Temperature : * No data; *Electrical Hazard:* No data; *Burning Rate:* Not pertinent.

Chemical Reactivity - *Reactivity with Water:* No reaction; *Reactivity with Common Materials:* No reactions; *Stability During Transport:* Stable if inhibited; *Neutralizing Agents for Acids and Caustics:* Not pertinent; *Polymerization:* In the absence of inhibitor, polymerization will occur, especially when heated; *Inhibitor of Polymerization:* Monomethyl Ether of Hydroquinone (25 ppm).

ISODECYL ALCOHOL

Chemical Designations - *Synonyms*: No common synonyms; *Chemical Formula*: $C_{10}H_{21}OH$.

Observable Characteristics - *Physical State (as shipped)*: Liquid; *Color*: Colorless; *Odor*: Weak alcoholic.

Physical and Chemical Properties - *Physical State at 15 °C and 1 atm.*: Liquid; *Molecular Weight*: 158.29; *Boiling Point at 1 atm.*: 428, 220, 493; *Freezing Point*: <140, <60, <333; *Critical Temperature*: Not pertinent; *Critical Pressure*: Not pertinent; *Specific Gravity*: 0.841 at 20 °C (liquid); *Vapor (Gas) Specific Gravity*: Not pertinent; *Ratio of Specific Heats of Vapor (Gas)*: (est.) 1.032; *Latent Heat of Vaporization*: (est.) 120, 67, 2.8; *Heat of Combustion*: Data not available; *Heat of Decomposition*: Not pertinent.

Health Hazards Information - *Recommended Personal Protective Equipment*: Chemical goggles; *Symptoms Following Exposure*: Direct contact with skin can produce irritation; *General Treatment for Exposure*: Wash affected area with water for 15 min.; *Toxicity by Inhalation (Threshold Limit Value)*: Data not available; *Short-Term Inhalation Limits*: Data not available; *Toxicity by Ingestion*: Data not available; *Late Toxicity*: Data not available; *Vapor (Gas) Irritant Characteristics*: Vapors are non-irritating to the eyes and throat; *Liquid or Solid Irritant Characteristics*: Causes smarting of the skin and first-degree burns on short exposures; may cause secondary burns on long exposure; *Odor Threshold*: Data not available.

Fire Hazards - *Flash Point (deg. F)*: 220 OC; *Flammable Limits in Air (%):* No data; *Fire Extinguishing Agents:* Alcohol foam, dry chemical, or carbon dioxide; *Fire Extinguishing Agents Not To Be Used:* Water or foam may cause frothing; *Special Hazards of Combustion Products:* Not pertinent; *Behavior in Fire:* Not pertinent; *Ignition Temperature:* No data; *Electrical Hazard:* Not pertinent; *Burning Rate:* No data.

Chemical Reactivity - *Reactivity with Water:* No reaction; *Reactivity with Common Materials:* No reactions; *Stability During Transport:* Stable; *Neutralizing Agents for Acids and Caustics:* Not pertinent; *Polymerization:* Not pertinent; *Inhibitor of Polymerization:* Not pertinent.

ISOHEXANE

Chemical Designations - *Synonyms:* 2-Methylpentane; *Chemical Formula:* $CH_3CH(CH_3)CH_2CH_2CH_3$.
Observable Characteristics - *Physical State (as shipped):* Liquid; *Color:* Colorless; *Odor:* Gasoline.
Physical and Chemical Properties - *Physical State at 15 °C and 1 atm.:* Liquid; *Molecular Weight:* 86.18; *Boiling Point at 1 atm.:* 140.5, 60.3, 333.5; *Freezing Point:* -244.6, -153.7, 119.5; *Critical Temperature:* 435.7, 224.3, 497.5; *Critical Pressure:* 437, 29.7, 3.01; *Specific Gravity:* 0.653 at 20 °C (liquid); *Vapor (Gas) Specific Gravity:* 2.9; *Ratio of Specific Heats of Vapor (Gas):* 1.062; *Latent Heat of Vaporization:* 139, 77.1, 3.23; *Heat of Combustion:* -19,147, -10,637, -445.35; *Heat of Decomposition:* Not pertinent.
Health Hazards Information - *Recommended Personal Protective Equipment:* Eye protection (as for gasoline); *Symptoms Following Exposure:* Inhalation causes irritation of respiratory tract, cough, mild depression, cardiac arrhythmia. Aspiration causes severe lung irritation, coughing, pulmonary edema; excitement followed by depression. Ingestion causes nausea, vomiting, swelling of abdomen, headache, depression; *General Treatment for Exposure:* INHALATION: maintain respiration, give oxygen if needed. ASPIRATION: enforce bed rest; give oxygen. INGESTION: do NOT induce vomiting; call a doctor. EYES: wash with copious amount of water. SKIN: wipe off, wash with soap and water; *Toxicity by Inhalation (Threshold Limit Value):* Data not available; *Short-Term Inhalation Limits:* Data not available; *Toxicity by Ingestion:* Data not available; *Late Toxicity:* None; *Vapor (Gas) Irritant Characteristics:* Vapors are nonirritating to the eyes and throat; *Liquid or Solid Irritant Characteristics:* No appreciable hazard. Practically harmless to the skin; *Odor Threshold:* Data not available.
Fire Hazards - *Flash Point (deg. F):* -20 CC; *Flammable Limits in Air (%):* 1.2 - 7.7; *Fire Extinguishing Agents:* Foam, dry chemical, or carbon dioxide; *Fire Extinguishing Agents Not To Be Used:* Water may be ineffective; *Special Hazards of Combustion Products:* Not pertinent; *Behavior in Fire:* Not pertinent; *Ignition Temperature (deg. F) :* 585; *Electrical Hazard:* Not pertinent; *Burning Rate:* 8.2 mm/min.
Chemical Reactivity - *Reactivity with Water:* No reaction; *Reactivity with Common Materials:* No reactions; *Stability During Transport:* Stable; *Neutralizing Agents for Acids and Caustics:* Not pertinent; *Polymerization:* Not pertinent; *Inhibitor of Polymerization:* Not pertinent.

ISOOCTYL ALCOHOL

Chemical Designations - *Synonyms:* Dimethyl-1-hexanols; 6-Methyl-1-heptanol; *Chemical Formula:* $(CH_3)_2CH(CH_2)_4CH_2OH$.
Observable Characteristics - *Physical State (as shipped):* Liquid; *Color:* Colorless; *Odor:* Mild; characteristic.
Physical and Chemical Properties - *Physical State at 15 °C and 1 atm.:* Liquid; *Molecular Weight:* 130.22; *Boiling Point at 1 atm.:* 367, 186, 459; *Freezing Point:* <212, <100, <373; *Critical Temperature:* Not pertinent; *Critical Pressure:* Not pertinent; *Specific Gravity:* 0.832 at 20 °C (liquid); *Vapor (Gas) Specific Gravity:* Not pertinent; *Ratio of Specific Heats of Vapor (Gas):* (est.) 1.040; *Latent Heat of Vaporization:* (est.) 140, 77, 3.2; *Heat of Combustion:* (est.) -17,400, -9650, -404,; *Heat of Decomposition:* Not pertinent.
Health Hazards Information - *Recommended Personal Protective Equipment:* Air-supplied mask in confined areas; plastic gloves; goggles; eye bath and safety shower; *Symptoms Following Exposure:* Inhalation hazard slight. Skin contact results in moderate irritation. Liquid contact with eyes causes severe irritation and possible eye damage; *General Treatment for Exposure:* Remove to fresh air. Flush skin and eye contact area at once for at least 15 min. Get medical care for eyes; *Toxicity by Inhalation (Threshold Limit Value):* Data not available; *Short-Term Inhalation Limits:* Data not available; *Toxicity by Ingestion:* Grade 2, LD_{50} 0.5 to 5 g/kg (lab animals); *Late Toxicity:* Data not available; *Vapor (Gas) Irritant Characteristics:* Vapors are nonirritating to the eyes and throat; *Liquid or Solid Irritant*

Characteristics: No appreciable hazard. Practically harmless to the skin; *Odor Threshold*: Data not available.
Fire Hazards - *Flash Point (deg. F)*: 180 OC; *Flammable Limits in Air (%):* 0.9 - 5.7; *Fire Extinguishing Agents:* Water, foam, dry chemical, or carbon dioxide; *Fire Extinguishing Agents Not To Be Used:* Not pertinent; *Special Hazards of Combustion Products:* Not pertinent; *Behavior in Fire:* Not pertinent; *Ignition Temperature (deg. F):* 530; *Electrical Hazard:* Not pertinent; *Burning Rate:* No data.
Chemical Reactivity - *Reactivity with Water:* No reaction; *Reactivity with Common Materials:* No reactions; *Stability During Transport:* Stable; *Neutralizing Agents for Acids and Caustics:* Not pertinent; *Polymerization:* Not pertinent; *Inhibitor of Polymerization:* Not pertinent.

ISOPENTANE
Chemical Designations - *Synonyms*: 2-Methylbutane; *Chemical Formula*: $(CH_3)_2CHCH_2CH_3$.
Observable Characteristics - *Physical State (as shipped)*: Liquid; *Color*: Colorless; *Odor*: Like gasoline.
Physical and Chemical Properties - *Physical State at 15 °C and 1 atm.*: Liquid; *Molecular Weight*: 72.15; *Boiling Point at 1 atm.*: 82.2, 27.9, 301.1; *Freezing Point*: -255.8, -159.9, 113.3; *Critical Temperature*: 369.0, 187.2, 460.4; *Critical Pressure*: 491.0, 33.4, 3.38; *Specific Gravity*: 0.620 at 20 °C (liquid); *Vapor (Gas) Specific Gravity*: 2.5; *Ratio of Specific Heats of Vapor (Gas)*: 1.076; *Latent Heat of Vaporization*: 146, 81.0, 3.39; *Heat of Combustion*: -19,314, -10,730, -449.24; *Heat of Decomposition*: Not pertinent.
Health Hazards Information - *Recommended Personal Protective Equipment*: Eye protection (as for gasoline); *Symptoms Following Exposure*: Inhalation causes irritation of respiratory tract, cough, mild depression, irregular heartbeat. Aspiration causes severe lung irritation, coughing, pulmonary edema; excitement followed by depression. Ingestion causes nausea, vomiting, swelling of abdomen, headache, depression; *General Treatment for Exposure*: INHALATION: maintain respiration, give oxygen if needed. ASPIRATION: enforce bed rest; give oxygen. INGESTION: do NOT induce vomiting; call a doctor. EYES: wash with copious amount of water. SKIN: wipe off, wash with soap and water; *Toxicity by Inhalation (Threshold Limit Value)*: Data not available; *Short-Term Inhalation Limits*: Data not available; *Toxicity by Ingestion*: Grade 1, LD_{50} 5 to 15 g/kg; *Late Toxicity*: None; *Vapor (Gas) Irritant Characteristics*: Vapors are nonirritating to the eyes and throat; *Liquid or Solid Irritant Characteristics*: No appreciable hazard. Practically harmless to the skin; *Odor Threshold*: Data not available.
Fire Hazards - *Flash Point (deg. F)*: -70 CC; *Flammable Limits in Air (%):* 1.4 - 8.3; *Fire Extinguishing Agents:* Dry chemical, foam, or carbon dioxide; *Fire Extinguishing Agents Not To Be Used:* Water may be ineffective; *Special Hazards of Combustion Products:* Not pertinent; *Behavior in Fire:* This is a highly volatile liquid. The vapors are explosive when mixed with air; *Ignition Temperature (deg. F):* 800; *Electrical Hazard:* Not pertinent; *Burning Rate:* 7.4 mm/min.
Chemical Reactivity - *Reactivity with Water:* No reaction; *Reactivity with Common Materials:* No reaction; *Stability During Transport:* Stable; *Neutralizing Agents for Acids and Caustics:* Not pertinent; *Polymerization:* Not pertinent; *Inhibitor of Polymerization:* Not pertinent.

ISOPHORONE
Chemical Designations - *Synonyms*: 3,5,5-Trimethyl-2-cyclohexene-1-one; *Chemical Formula*: $COCH=C(CH_3)CH_2C(CH_3)_2CH_2$.
Observable Characteristics - *Physical State (as shipped)*: Liquid; *Color*: Colorless; *Odor*: Like camphor.
Physical and Chemical Properties - *Physical State at 15 °C and 1 atm.*: Liquid; *Molecular Weight*: 138.2; *Boiling Point at 1 atm.*: 419.5, 215.3, 488.5; *Freezing Point*: 17.4, -8.1, 265.1; *Critical Temperature*: Not pertinent; *Critical Pressure*: Not pertinent; *Specific Gravity*: 0.921 at 25 °C (liquid); *Vapor (Gas) Specific Gravity*: 4.75; *Ratio of Specific Heats of Vapor (Gas)*: Not pertinent; *Latent Heat of Vaporization*: 135, 75, 3.14; *Heat of Combustion*: -16,170, -8,980, -376; *Heat of Decomposition*:

Not pertinent.

Health Hazards Information - *Recommended Personal Protective Equipment*: Self-contained breathing apparatus with full face mask; rubber gloves; *Symptoms Following Exposure*: Inhalation irritates eye, nose and throat; causes central depression and has some anesthetic effect. Contact of liquid with eyes causes severe irritation and possible tissue damage. Skin is irritated by liquid and may crack on prolonged contact. Ingestion causes irritation of mouth and stomach; *General Treatment for Exposure*: INHALATION: remove victim promptly from contaminated atmosphere; if breathing has stopped, give artificial respiration and oxygen. EYES: flood with water for at least 15 min.; consult an eye specialist as soon as possible. SKIN: flood with water. INGESTION: do not induce vomiting; call a doctor; *Toxicity by Inhalation (Threshold Limit Value)*: 10 ppm; *Short-Term Inhalation Limits*: Data not available; *Toxicity by Ingestion*: Grade 2; oral LD_{50} = 2,330 mg/kg (rat); *Late Toxicity*: Data not available; *Vapor (Gas) Irritant Characteristics*: Vapors cause moderate irritation such that personnel will find high concentrations unpleasant. The effect is temporary; *Liquid or Solid Irritant Characteristics*: Causes smarting of the skin and first-degree burns on short exposure; may cause secondary burns on long exposure; *Odor Threshold*: Data not available.

Fire Hazards - *Flash Point (deg. F)*: 205 OC, 184 CC; *Flammable Limits in Air (%)*: 0.84 - 3.8; *Fire Extinguishing Agents:* Dry chemical, foam, or carbon dioxide; *Fire Extinguishing Agents Not To Be Used:* Water may be ineffective; *Special Hazards of Combustion Products:* Not pertinent; *Behavior in Fire:* Not pertinent; *Ignition Temperature (deg. F):* 864; *Electrical Hazard:* No data; *Burning Rate:* 4.0 mm/min.

Chemical Reactivity - *Reactivity with Water:* No reaction; *Reactivity with Common Materials:* No reactions; *Stability During Transport:* Stable; *Neutralizing Agents for Acids and Caustics:* Not pertinent; *Polymerization:* Not pertinent; *Inhibitor of Polymerization:* Not pertinent.

ISOPHTHALIC ACID

Chemical Designations - *Synonyms*: Benzene-1,3-dicarboxylic acid; m-Phthalic acid; *Chemical Formula*: $1,3-C_6H_4(COOH)_2$.

Observable Characteristics - *Physical State (as shipped)*: Solid; *Color*: White; *Odor*: Slightly acrid.

Physical and Chemical Properties - *Physical State at 15 °C and 1 atm.*: Solid; *Molecular Weight*: 166; *Boiling Point at 1 atm.*: Not pertinent (sublimes); *Freezing Point*: 653, 345, 618; *Critical Temperature*: Not pertinent; *Critical Pressure*: Not pertinent; *Specific Gravity*: 1.54 at 25 °C (solid); *Vapor (Gas) Specific Gravity*: Not pertinent; *Ratio of Specific Heats of Vapor (Gas)*: Not pertinent; *Latent Heat of Vaporization*: Not pertinent; *Heat of Combustion*: -8,340, -4,630, -194; *Heat of Decomposition*: Not pertinent.

Health Hazards Information - *Recommended Personal Protective Equipment*: If without adequate ventilation, use respirator with dust filter, goggles, and gloves; *Symptoms Following Exposure*: May cause slight to moderate irritation of eyes, skin, and mucous membranes on prolonged contact. Ingestion may cause gastrointestinal irritation; *General Treatment for Exposure*: INHALATION: remove victim to uncontaminated area; get medical attention if complications arise. INGESTION: get medical attention if complications arise. EYES: flush with large amounts of water for 15 min.; get promptly medical attention. SKIN: wash with water; *Toxicity by Inhalation (Threshold Limit Value)*: Data not available; *Short-Term Inhalation Limits*: Data not available; *Toxicity by Ingestion*: Grade 1; LD_{50} 12.2 g/kg (rat); *Late Toxicity*: Data not available; *Vapor (Gas) Irritant Characteristics*: Data not available; *Liquid or Solid Irritant Characteristics*: Data not available; *Odor Threshold*: Data not available.

Fire Hazards - *Flash Point* : Not pertinent. This is a combustible solid; *Flammable Limits in Air (%):* Not pertinent; *Fire Extinguishing Agents:* Water, dry powder, foam, or carbon dioxide; *Fire Extinguishing Agents Not To Be Used:* None; *Special Hazards of Combustion Products:* None; *Behavior in Fire:* Dust forms explosive mixture in air; *Ignition Temperature:* No data; *Electrical Hazard:* Not pertinent; *Burning Rate:* Not pertinent.

Chemical Reactivity - *Reactivity with Water:* No reaction; *Reactivity with Common Materials:* No reactions; *Stability During Transport:* Stable; *Neutralizing Agents for Acids and Caustics:* Not pertinent;

Polymerization: Not pertinent; *Inhibitor of Polymerization:* Not pertinent.

ISOPRENE

Chemical Designations - *Synonyms*: beta-Methylbivinyl; 2-Methyl-1,3-butadiene; *Chemical Formula*: $CH_2=C(CH_3)CH=CH_2$.

Observable Characteristics - *Physical State (as shipped)*: Liquid; *Color*: Colorless; *Odor*: Mild, aromatic.

Physical and Chemical Properties - *Physical State at 15 °C and 1 atm.*: Liquid; *Molecular Weight*: 68.12; *Boiling Point at 1 atm.*: 93.4, 34.1, 307.3; *Freezing Point*: -230.7, -145.9, 127.3; *Critical Temperature*: 412, 211.1, 484.3; *Critical Pressure*: 550, 37.4, 3.79; *Specific Gravity*: 0.681 at 20 °C (liquid); *Vapor (Gas) Specific Gravity*: 2.3; *Ratio of Specific Heats of Vapor (Gas)*: 1.091; *Latent Heat of Vaporization*: 150, 85, 3.6; *Heat of Combustion*: -18,848, -10,471, -438.40; *Heat of Decomposition*: Not pertinent.

Health Hazards Information - *Recommended Personal Protective Equipment*: Vapor-proof goggles; self-contained breathing apparatus; leather or rubber safety shoes; rubber gloves; *Symptoms Following Exposure*: Vapor produces no effects other than slight irritation of the eyes and upper respiratory tract. Liquid may irritate eyes; like gasoline; *General Treatment for Exposure*: INHALATION: remove victim promptly from irritating or asphyxiating atmosphere; if symptoms of asphyxiation persist, administer artificial respiration and oxygen; treat symptomatically thereafter; call a physician. EYES: flush with water for at least 15 min.; *Toxicity by Inhalation (Threshold Limit Value)*: Data not available; *Short-Term Inhalation Limits*: Data not available; *Toxicity by Ingestion*: Data not available; *Late Toxicity*: None; *Vapor (Gas) Irritant Characteristics*: Vapors causes a slight smarting of the eyes or respiratory system if present in high concentrations. The effect is temporary; *Liquid or Solid Irritant Characteristics*: Minimum hazard. If spilled on clothing and allowed to remain, may cause smarting and reddening of skin; *Odor Threshold*: Data not available.

Fire Hazards - *Flash Point (deg. F)*: -65 CC; *Flammable Limits in Air (%)*: 2 - 9; *Fire Extinguishing Agents:* Dry chemical, foam, or carbon dioxide; *Fire Extinguishing Agents Not To Be Used:* Water may be ineffective; *Special Hazards of Combustion Products:* Toxic vapors are generated upon heating; *Behavior in Fire:* May polymerize in containers and explode; *Ignition Temperature (deg. F):* 428; *Electrical Hazard:* Class I, Group C; *Burning Rate:* 8.6 mm/min.

Chemical Reactivity - *Reactivity with Water:* No reaction; *Reactivity with Common Materials:* No reactions; *Stability During Transport:* Stable; *Neutralizing Agents for Acids and Caustics:* Not pertinent; *Polymerization:* Polymerization is accelerated by heat and exposure to oxygen, as well as the presence of contamination such as iron rust. Iron surfaces should be treated with an appropriate reducing agent such as sodium nitrate, before being placed into isoprene service; *Inhibitor of Polymerization:* Tertiary butyl catechol (0.06 %). Di-n-butylamine, phenyl-beta-naphthylamine and phenyl-alpha-naphthylamine are also recommended.

ISOPROPYL ACETATE

Chemical Designations - *Synonyms*: Acetic acid, isopropyl ester; 2-Propyl acetate; *Chemical Formula*: $CH_3COOCH(CH_3)_2$.

Observable Characteristics - *Physical State (as shipped)*: Liquid; *Color*: Colorless; *Odor*: Pleasant, fruity; nonresidual.

Physical and Chemical Properties - *Physical State at 15 °C and 1 atm.*: Liquid; *Molecular Weight*: 102.13; *Boiling Point at 1 atm.*: 191.3, 88.5, 361.7; *Freezing Point*: -92.7, -69.3, 203.9; *Critical Temperature*: 509, 265, 538; *Critical Pressure*: 529, 36, 3.65; *Specific Gravity*: 0.874 at 20 °C (liquid); *Vapor (Gas) Specific Gravity*: 3.5; *Ratio of Specific Heats of Vapor (Gas)*: (est.) 1.074; *Latent Heat of Vaporization*: 150, 81, 3.4; *Heat of Combustion*: -9420, -5230, -219; *Heat of Decomposition*: Not pertinent.

Health Hazards Information - *Recommended Personal Protective Equipment*: Organic vapor canister or air-supplied mask; chemical goggles or face splash shield; *Symptoms Following Exposure*: Vapors irritate eyes and respiratory tract; high concentrations can be anesthetic. Liquid irritates eyes but causes no serious injury; may cause dermatitis; no serious effects if swallowed; *General Treatment for Exposure*: INHALATION: if victim is overcome by vapors, remove from exposure immediately; call a physician; if breathing is irregular or stopped, start resuscitation and administer oxygen. EYES: flush with water for at least 15 min.; *Toxicity by Inhalation (Threshold Limit Value)*: 250 ppm; *Short-Term Inhalation Limits*: Data not available; *Toxicity by Ingestion*: Grade 2, LD_{50} 0.5 to 5 g/kg (rat); *Late Toxicity*: Data not available; *Vapor (Gas) Irritant Characteristics*: Vapors causes a slight smarting of the eyes or respiratory system if present in high concentrations. The effect is temporary; *Liquid or Solid Irritant Characteristics*: Minimum hazard. If spilled on clothing and allowed to remain, may cause smarting and reddening of skin; *Odor Threshold*: Data not available.

Fire Hazards - *Flash Point (deg. F)*: 37 CC, 60 CC; *Flammable Limits in Air (%)*: 1.8 - 8.0; *Fire Extinguishing Agents:* Alcohol foam, dry chemical, or carbon dioxide; *Fire Extinguishing Agents Not To Be Used:* Not pertinent; *Special Hazards of Combustion Products:* Not pertinent; *Behavior in Fire:* Not pertinent; *Ignition Temperature (deg. F):* 860; *Electrical Hazard:* Not pertinent; *Burning Rate:* No data.

Chemical Reactivity - *Reactivity with Water:* No reaction; *Reactivity with Common Materials:* No reactions; *Stability During Transport:* Stable; *Neutralizing Agents for Acids and Caustics:* Not pertinent; *Polymerization:* Not pertinent; *Inhibitor of Polymerization:* Not pertinent.

ISOPROPYL ALCOHOL

Chemical Designations - *Synonyms*: Dimethylcarbinol; 2-Propanol; Isopropanol; sec-Propyl alcohol; Petrohol; Rubbing alcohol; *Chemical Formula*: $CH_3CH(OH)CH_3$.

Observable Characteristics - *Physical State (as shipped)*: Liquid; *Color*: Colorless; *Odor*: Like ethyl alcohol; sharp, somewhat unpleasant; characteristic mild alcoholic; nonresidual.

Physical and Chemical Properties - *Physical State at 15 °C and 1 atm.*: Liquid; *Molecular Weight*: 60.10; *Boiling Point at 1 atm.*: 180.1, 82.3, 355.5; *Freezing Point*: -127.3, -88.5, 184.7; *Critical Temperature*: 455.4, 235.2, 508.4; *Critical Pressure*: 691, 47.0, 4.76,; *Specific Gravity*: 0.785 at 20 °C (liquid); *Vapor (Gas) Specific Gravity*: 2.1; *Ratio of Specific Heats of Vapor (Gas)*: 1.105; *Latent Heat of Vaporization*: 286, 159, 6.66; *Heat of Combustion*: -12,960, -7,201, -301.5; *Heat of Decomposition*: Not pertinent.

Health Hazards Information - *Recommended Personal Protective Equipment*: Organic vapor canister or air-supplied mask; chemical goggles or face shield; *Symptoms Following Exposure*: Vapors cause mild irritation of eyes and upper respiratory tract; high concentrations may be anesthetic. Liquid irritates eyes and may cause injury; harmless to skin; if ingested causes drunkenness and vomiting; *General Treatment for Exposure*: INHALATION: if victim is overcome by vapors, remove from exposure immediately; call a physician; if breathing is irregular or has stopped, start resuscitation and administer oxygen. EYES: flush with water for at least 15 min.; *Toxicity by Inhalation (Threshold Limit Value)*: 400 ppm; *Short-Term Inhalation Limits*: 400 ppm for 10 min.; *Toxicity by Ingestion*: Grade 1; LD_{50} 5 to 15 g/kg (rat LD_{50}: 5.84 g/kg); *Late Toxicity*: Data not available; *Vapor (Gas) Irritant Characteristics*: Vapors causes a slight smarting of the eyes or respiratory system if present in high concentrations. The effect is temporary; *Liquid or Solid Irritant Characteristics*: No appreciable hazard. Practically harmless to the skin; *Odor Threshold*: 90 mg/m^3.

Fire Hazards - *Flash Point (deg. F):* 65 OC, 53 CC; *Flammable Limits in Air (%):* 2.3 - 12.7; *Fire Extinguishing Agents:* Alcohol foam, dry chemical, or carbon dioxide; *Fire Extinguishing Agents Not To Be Used:* Water may be ineffective; *Special Hazards of Combustion Products:* Not pertinent; *Behavior in Fire:* Not pertinent; *Ignition Temperature (deg. F):* 750; *Electrical Hazard:* Class I, Group D; *Burning Rate:* 2.3 mm/min.

Chemical Reactivity - *Reactivity with Water:* No reaction; *Reactivity with Common Materials:* No reactions; *Stability During Transport:* Stable; *Neutralizing Agents for Acids and Caustics:* Not pertinent; *Polymerization:* Not pertinent; *Inhibitor of Polymerization:* Not pertinent.

ISOPROPYLAMINE

Chemical Designations - *Synonyms*: 2-Aminopropane; Monoisopropylamine; iso-Propylamine; *Chemical Formula*: $(CH_3)_2CHNH_2$.

Observable Characteristics - *Physical State (as shipped)*: Liquid; *Color*: Colorless; *Odor*: Strong ammoniacal; pungent, irritating, typical amine.

Physical and Chemical Properties - *Physical State at 15 °C and 1 atm.*: Liquid; *Molecular Weight*: 59.11; *Boiling Point at 1 atm.*: 90.3, 32.4, 305.6; *Freezing Point*: -139, -95, 178; *Critical Temperature*: 396, 202, 475; *Critical Pressure*: 740, 50, 5.1; *Specific Gravity*: (est.) 0.691 at 20°C (liquid); *Vapor (Gas) Specific Gravity*: 2.04; *Ratio of Specific Heats of Vapor (Gas)*: Not pertinent; *Latent Heat of Vaporization*: 193, 107, 4.48; *Heat of Combustion*: -16,940, -9,420, -394; *Heat of Decomposition*: Not pertinent.

Health Hazards Information - *Recommended Personal Protective Equipment*: Self-contained breathing apparatus; butyl rubber gloves and apron; chemical face shield or safety goggles; *Symptoms Following Exposure*: Inhalation causes nose and throat irritation, severe coughing, and chest pain due to irritation of air passages; can cause lung edema and loss of consciousness. Ingestion causes nausea, salivation and severe irritation of mouth and stomach. Contact with skin causes severe irritation; *General Treatment for Exposure*: INHALATION: remove victim to fresh air; if he is not breathing, give artificial respiration; give oxygen if breathing is difficult; call a physician. INGESTION: call a physician immediately; encourage the drinking of large quantities of water followed by dilute vinegar, lemon juice, cider, or other weak acids; keep patient warm. EYES: flush with water for 15 min., bolding eyelids apart; call physician as soon as possible, preferably an eye specialist. SKIN: flush with water; *Toxicity by Inhalation (Threshold Limit Value)*: 5 ppm; *Short-Term Inhalation Limits*: Data not available; *Toxicity by Ingestion*: Grade 2, oral LD_{50} = 820 mg/kg (rat), 600 mg/kg (mouse); *Late Toxicity*: Data not available; *Vapor (Gas) Irritant Characteristics*: Vapor is moderately irritating such that personnel will not usually tolerate moderate or high vapor concentrations; *Liquid or Solid Irritant Characteristics*: Causes smarting of the skin and first-degree burns on short exposure; may cause secondary burns on long exposure; *Odor Threshold*: 5 ppm.

Fire Hazards - *Flash Point (deg. F)*: -15 OC; *Flammable Limits in Air (%)*: 2.3 - 12; *Fire Extinguishing Agents:* Dry chemical, alcohol foam, or carbon dioxide; *Fire Extinguishing Agents Not To Be Used:* Water may be ineffective; *Special Hazards of Combustion Products:* Toxic oxides of nitrogen form in fires; *Behavior in Fire:* Burning product is difficult to control because of the ease of reigniting of vapors. Vapors are heavier than air and may travel to a source of ignition and flash back. There is danger of container explosion; *Ignition Temperature (deg. F):* 756; *Electrical Hazard:* No data; *Burning Rate:* 6.3 mm/min.

Chemical Reactivity - *Reactivity with Water:* No reaction; *Reactivity with Common Materials:* No reactions; *Stability During Transport:* Stable; *Neutralizing Agents for Acids and Caustics:* Not pertinent; *Polymerization:* Not pertinent; *Inhibitor of Polymerization:* Not pertinent.

ISOPROPYL ETHER

Chemical Designations - *Synonyms*: Diisopropyl ether; Diisopropyl oxide; 2-Isopropoxypropane; *Chemical Formula*: $(CH_3)_2CHOCH(CH_3)_2$.

Observable Characteristics - *Physical State (as shipped)*: Liquid; *Color*: Colorless; *Odor*: Sweet, slightly sharp; characteristic pungent; ethereal; like amphor and ethyl ether.

Physical and Chemical Properties - *Physical State at 15 °C and 1 atm.*: Liquid; *Molecular Weight*: 102.2; *Boiling Point at 1 atm.*: 156, 69, 342; *Freezing Point*: -123, -86, 187; *Critical Temperature*: 440.4, 226.9, 500.1; *Critical Pressure*: 418, 28.4, 2.88; *Specific Gravity*: 0.724 at 20 °C (liquid); *Vapor (Gas) Specific Gravity*: 3.5; *Ratio of Specific Heats of Vapor (Gas)*: 1.0590; *Latent Heat of Vaporization*: 131, 73, 3.1; *Heat of Combustion*: -16,900, -9,390, -393; *Heat of Decomposition*: Not pertinent.

Health Hazards Information - *Recommended Personal Protective Equipment*: Air pack or organic canister mask; rubber gloves; goggles; *Symptoms Following Exposure*: Inhalation causes anesthesia, nausea, dizziness, headache, and irritation of the eyes and nose. Contact of liquid with eyes causes only

minor injury; repeated contact with skin will remove natural oils and may cause dermatitis; *General Treatment for Exposure*: INHALATION: remove victim to fresh air and obtain medical attention immediately; keep him warm and at rest, and give artificial respiratory if breathing stops; maintain an open airway. EYES: flush with water for 15 min. SKIN: flush with water. INGESTION: do NOT induce vomiting; get medical attention; *Toxicity by Inhalation (Threshold Limit Value)*: 250 ppm (tentative); *Short-Term Inhalation Limits*: Grade 1; oral LD_{50} = 8,470 mg/kg (rat); *Toxicity by Ingestion*: Data not available; *Late Toxicity*: Data not available; *Vapor (Gas) Irritant Characteristics*: Vapors causes a slight smarting of the eyes or respiratory system if present in high concentrations. The effect is temporary; *Liquid or Solid Irritant Characteristics*: Minimum hazard. If spilled on clothing and allowed to remain, may cause smarting and reddening of skin; *Odor Threshold*: Data not available.
Fire Hazards - *Flash Point (deg. F)*: -15 OC, -18 CC; *Flammable Limits in Air (%):* 1.4 - 7.9; *Fire Extinguishing Agents:* Dry chemical, alcohol foam, or carbon dioxide; *Fire Extinguishing Agents Not To Be Used:* Water may be ineffective; *Special Hazards of Combustion Products:* Not pertinent; *Behavior in Fire:* Vapor is heavier than air and may travel to a source of ignition and flash back. Containers may explode; *Ignition Temperature (deg. F):* 830; *Electrical Hazard:* No data; *Burning Rate:* 5.0 mm/min.
Chemical Reactivity - *Reactivity with Water:* No reaction; *Reactivity with Common Materials:* No reactions; *Stability During Transport:* Unstable peroxides may form if the product contacts air for long time periods. These may explode spontaneously or when heated; *Neutralizing Agents for Acids and Caustics:* Not pertinent; *Polymerization:* Not pertinent; *Inhibitor of Polymerization:* Not pertinent.

ISOPROPYL MERCAPTAN
Chemical Designations - *Synonyms*: 2-Propanethiol; Propane-2-thiol; *Chemical Formula*: $(CH_3)_2CHSH$.
Observable Characteristics - *Physical State (as shipped)*: Liquid; *Color*: Colorless; *Odor*: Powerful skunk.
Physical and Chemical Properties - *Physical State at 15 °C and 1 atm.*: Liquid; *Molecular Weight*: 76.2; *Boiling Point at 1 atm.*: 126.6, 52.5, 325.8; *Freezing Point*: -202.8, -130.5, 142.7; *Critical Temperature*: Data not available; *Critical Pressure*: Data not available; *Specific Gravity*: 0.814 at 20 °C (liquid); *Vapor (Gas) Specific Gravity*: 2.6; *Ratio of Specific Heats of Vapor (Gas)*: 1.0964 at 15.6 °C; *Latent Heat of Vaporization*: 165.7, 92.1, 3.83; *Heat of Combustion*: -14,920, -8,290, -347; *Heat of Decomposition*: Not pertinent.
Health Hazards Information - *Recommended Personal Protective Equipment*: Self-contained breathing apparatus; goggles or face shield; rubber gloves; *Symptoms Following Exposure*: Inhalation causes loss of sense of smell, muscular weakness, convulsions, respiratory paralysis. Ingestion causes nausea and vomiting. Contact with eyes or skin causes irritation; *General Treatment for Exposure*: INHALATION: move victim to fresh air; start artificial respiratory and give oxygen if required; observe for signs of pulmonary edema; get medical attention. INGESTION: give large amount of water; induce vomiting. EYES or SKIN: flush with water; *Toxicity by Inhalation (Threshold Limit Value)*: Data not available; *Short-Term Inhalation Limits*: Data not available; *Toxicity by Ingestion*: Grade 2; oral LD_{50} = 1,790 mg/kg (rat); *Late Toxicity*: Data not available; *Vapor (Gas) Irritant Characteristics*: Data not available; *Liquid or Solid Irritant Characteristics*: Data not available; *Odor Threshold*: 0.25 ppm.
Fire Hazards - *Flash Point (deg. F)*: -30 OC; *Flammable Limits in Air (%):* No data; *Fire Extinguishing Agents:* Dry chemical, alcohol foam, or carbon dioxide; *Fire Extinguishing Agents Not To Be Used:* Water may be ineffective; *Special Hazards of Combustion Products:* Irritating sulfur dioxide gas is formed in fires; *Behavior in Fire:* Vapor is heavier than air and may travel to a source of ignition and flash back. Containers may explode; *Ignition Temperature :* No data; *Electrical Hazard:* No data; *Burning Rate:* No data.
Chemical Reactivity - *Reactivity with Water:* No reaction; *Reactivity with Common Materials:* No reactions; *Stability During Transport:* Stable; *Neutralizing Agents for Acids and Caustics:* Not pertinent; *Polymerization:* Not pertinent; *Inhibitor of Polymerization:* Not pertinent.

ISOPROPYL PERCARBONATE

Chemical Designations - *Synonyms*: Diisopropyl percarbonate; Diisopropyl peroxydicarbonate; Isopropyl peroxydicarbonate; Peroxydicarbonic acid, bis (1-methylethyl) ester; Peroxydicarbonic acid, diisopropyl ester; *Chemical Formula*: $C_3H_7OOCOOCOOC_3H_7$.

Observable Characteristics - *Physical State (as shipped)*: Solid (containers packed in "Dry Ice"); *Color*: White; *Odor*: Disagreeable; pungent.

Physical and Chemical Properties - *Physical State at 15 °C and 1 atm.*: Liquid; *Molecular Weight*: 206.2; *Boiling Point at 1 atm.*: Not pertinent (decomposes); *Freezing Point*: 46 - 50, 8 - 10, 281 - 283; *Critical Temperature*: Not pertinent; *Critical Pressure*: Not pertinent; *Specific Gravity*: 1.08 at 15 °C (solid); *Vapor (Gas) Specific Gravity*: Not pertinent; *Ratio of Specific Heats of Vapor (Gas)*: Not pertinent; *Latent Heat of Vaporization*: Not pertinent; *Heat of Combustion*: -8,500, -4,720, -198; *Heat of Decomposition*: -670, -370, -15.5.

Health Hazards Information - *Recommended Personal Protective Equipment*: Rubber gloves and shoes; hard hat; chemical splash goggles; plastic apron; respirator (depending on solvent used); *Symptoms Following Exposure*: Inhalation overexposure unlikely, but prolonged exposure may cause lung edema. Contact with eyes may cause irritation. Solutions are severe primary skin irritants; *General Treatment for Exposure*: INHALATION: move to uncontaminated atmosphere; if breathing is difficult, give oxygen. EYES: flush with copious amounts of water. SKIN: wash off with isopropyl alcohol and water; call a physician; *Toxicity by Inhalation (Threshold Limit Value)*: Data not available; *Short-Term Inhalation Limits*: Data not available; *Toxicity by Ingestion*: Grade 2, LD_{50} 0.5-5 g/kg; *Late Toxicity*: Data not available; *Vapor (Gas) Irritant Characteristics*: Data not available; *Liquid or Solid Irritant Characteristics*: Data not available; *Odor Threshold*: Data not available.

Fire Hazards - *Flash Point* : Not pertinent. This is a combustible solid; *Flammable Limits in Air (%)*: Not pertinent; *Fire Extinguishing Agents*: Water; *Fire Extinguishing Agents Not To Be Used*: All extinguishing agents may be ineffective; *Special Hazards of Combustion Products*: Flammable and toxic vapors are formed in fires, including acetone, isopropyl alcohol, acetaldehyde, and ethane; *Behavior in Fire*: This product undergoes auto-accelerated decomposition and can self-ignite. Fires are very difficult to extinguish because air is not needed for combustion; *Ignition Temperature* : No data; *Electrical Hazard*: Not pertinent; *Burning Rate*: Not pertinent.

Chemical Reactivity - *Reactivity with Water*: No reaction; *Reactivity with Common Materials*: May decompose with the formation of oxygen when in contact with metals; *Stability During Transport*: Unstable at temperatures above 0 of with the formation of oxygen gas; *Neutralizing Agents for Acids and Caustics*: Not pertinent; *Polymerization*: Not pertinent; *Inhibitor of Polymerization*: Not pertinent.

ISOVALERALDEHYDE

Chemical Designations - *Synonyms*: Isovaleral; Isovaleric aldehyde; 3-Methylbutanal; 3-Methylbutyraldehyde; *Chemical Formula*: $(CH_3)_2CHCH_2CHO$.

Observable Characteristics - *Physical State (as shipped)*: Liquid; *Color*: Colorless; *Odor*: Weakly suffocating.

Physical and Chemical Properties - *Physical State at 15 °C and 1 atm.*: Liquid; *Molecular Weight*: 86.1; *Boiling Point at 1 atm.*: 198.5, 92.5, 365.7; *Freezing Point*: -60, -51, 222; *Critical Temperature*: Not pertinent; *Critical Pressure*: Not pertinent; *Specific Gravity*: 0.785 at 20 °C (liquid); *Vapor (Gas) Specific Gravity*: 3; *Ratio of Specific Heats of Vapor (Gas)*: (est.) 1.0736; *Latent Heat of Vaporization*: (est.) 167, 93, 3.9; *Heat of Combustion*: -15,500, -8,620, -360; *Heat of Decomposition*: Not pertinent.

Health Hazards Information - *Recommended Personal Protective Equipment*: Goggles or face shield; rubber gloves; air mask or self-contained breathing apparatus for high vapor concentrations; *Symptoms Following Exposure*: Inhalation causes chest discomfort, nausea, vomiting, and headache. Contact with eyes or skin causes irritation. Ingestion causes irritation of mouth and stomach; *General Treatment for Exposure*: INHALATION: remove victim to fresh air; apply artificial respiration if required; get medical attention. EYES: flush with water for at least 15 min. SKIN: wipe off, wash well with soap and water. INGESTION: induce vomiting; get medical attention; *Toxicity by Inhalation (Threshold Limit Value)*: Data not available; *Short-Term Inhalation Limits*: Data not available; *Toxicity by*

Ingestion: Grade 3; oral LD_{50} =3,200 mg/kg (rat); *Late Toxicity*: Data not available; *Vapor (Gas) Irritant Characteristics*: Data not available; *Liquid or Solid Irritant Characteristics*: Data not available; *Odor Threshold*: Data not available.

Fire Hazards - *Flash Point (deg. F)*: 55 OC; *Flammable Limits in Air (%):* No data; *Fire Extinguishing Agents:* Dry chemical, foam, or carbon dioxide; *Fire Extinguishing Agents Not To Be Used:* Water may be ineffective; *Special Hazards of Combustion Products:* Not pertinent; *Behavior in Fire:* Not pertinent; *Ignition Temperature:* No data; *Electrical Hazard:* No data; *Burning Rate:* 5.3 mm/min.

Chemical Reactivity - *Reactivity with Water:* No reaction; *Reactivity with Common Materials:* No reactions; *Stability During Transport:* Stable; *Neutralizing Agents for Acids and Caustics:* Not pertinent; *Polymerization:* Not pertinent; *Inhibitor of Polymerization:* Not pertinent.

K

KEROSENE

Chemical Designations - *Synonyms*: No. 1 Fuel oil; Kerosine; Illuminating oil; Range oil; JP-1; *Chemical Formula*: C_nH_{2n+2}.

Observable Characteristics - *Physical State (as shipped)*: Liquid; *Color*: Colorless to light brown; *Odor*: Characteristic; like fuel oil.

Physical and Chemical Properties - *Physical State at 15 °C and 1 atm.*: Liquid; *Molecular Weight*: Not pertinent; *Boiling Point at 1 atm.*: 392 - 500, 200 - 260, 473 - 533; *Freezing Point*: -50, -45.6, 227.6; *Critical Temperature*: Not pertinent; *Critical Pressure*: Not pertinent; *Specific Gravity*: 0.80 at 15 °C (liquid); *Vapor (Gas) Specific Gravity*: Not pertinent; *Ratio of Specific Heats of Vapor (Gas)*: Not pertinent; *Latent Heat of Vaporization*: 110, 60, 2.5; *Heat of Combustion*: -18,540, -10,300, -431.24; *Heat of Decomposition*: Not pertinent.

Health Hazards Information - *Recommended Personal Protective Equipment*: Protective gloves; goggles or face shield; *Symptoms Following Exposure*: Vapor causes slight irritation of eyes and nose. Liquid irritates stomach; if taken into lung, causes coughing, distress, and rapidly developing pulmonary edema; *General Treatment for Exposure*: ASPIRATION: enforce bed rest; administer oxygen; call a doctor. INGESTION: do not induce vomiting; call a doctor. EYES: wash with plenty of water. SKIN: wipe off and wash with soap and water; *Toxicity by Inhalation (Threshold Limit Value)*: 200 ppm; *Short-Term Inhalation Limits*: 2500 mg/m^3 for 60 min.; *Toxicity by Ingestion*: Grade 1; LD_{50} 5 to 15 g/kg; *Late Toxicity*: Data not available; *Vapor (Gas) Irritant Characteristics*: Vapors causes a slight smarting of the eyes or respiratory system if present in high concentrations. The effect is temporary; *Liquid or Solid Irritant Characteristics*: Minimum hazard. If spilled on clothing and allowed to remain, may cause smarting and reddening of skin; *Odor Threshold*: Data not available.

Fire Hazards - *Flash Point (deg. F)*: 100 CC; *Flammable Limits in Air (%): 0.7 - 5*; *Fire Extinguishing Agents*: Foam, dry chemical, or carbon dioxide; *Fire Extinguishing Agents Not to be Used:* Water may be ineffective; *Special Hazards of Combustion Products*: Not pertinent; *Behavior in Fire*: Not pertinent; *Ignition Temperature (deg. F)*: 444; *Electrical Hazard*: Not pertinent; *Burning Rate*: 4 mm/min.

Chemical Reactivity - *Reactivity with Water*: No reaction; *Reactivity with Common Materials*: No reactions; *Stability During Transport*: Stable; *Neutralizing Agents for Acids and Caustics:* Not pertinent; *Polymerization:* Not pertinent; *Inhibitor of Polymerization*: Not pertinent.

L

LACTIC ACID

Chemical Designations - *Synonyms*: 2-Hydroxypropanoic acid; alpha-Hydroxypropionic acid; Milk acid; Racemic acid; *Chemical Formula*: $CH_3CHOHCOOH-H_2O$.

Observable Characteristics - *Physical State (as shipped)*: Syrupy liquid; *Color*: Yellow to colorless; *Odor*: None or weak acidic.

Physical and Chemical Properties - *Physical State at 15 °C and 1 atm.*: Liquid; *Molecular Weight*: 90; *Boiling Point at 1 atm.*: Not pertinent (decomposes); *Freezing Point*: Not pertinent; *Critical Temperature*: Not pertinent; *Critical Pressure*: Not pertinent; *Specific Gravity*: 1.20 at 20 °C (liquid); *Vapor (Gas) Specific Gravity*: Not pertinent; *Ratio of Specific Heats of Vapor (Gas)*: Not pertinent; *Latent Heat of Vaporization*: Not pertinent; *Heat of Combustion*: -6,520, -3,620, -152; *Heat of Decomposition*: Not pertinent.

Health Hazards Information - *Recommended Personal Protective Equipment*: Rubber gloves; goggles; Self-contained breathing apparatus where high concentrations of mist are present; *Symptoms Following Exposure*: Inhalation of mist causes coughing and irritation of mucous membranes. Ingestion, even of diluted preparations, has a corrosive effect on the esophagus and stomach. Contact with more concentrated solutions can cause severe burns of eyes or skin; *General Treatment for Exposure*: INHALATION: move to fresh air. INGESTION: give large amount of water. EYES: flush with water for 15 min. SKIN: flush with water; wash well with soap and water; *Toxicity by Inhalation (Threshold Limit Value)*: Data not available; *Short-Term Inhalation Limits*: Data not available; *Toxicity by Ingestion*: Grade 2; oral LD_{50} = 1,810 mg/kg (guinea pig); *Late Toxicity*: Data not available; *Vapor (Gas) Irritant Characteristics*: Data not available; *Liquid or Solid Irritant Characteristics*: Data not available; *Odor Threshold*: 4.0×10^{-7} ppm.

Fire Hazards - *Flash Point* : Not pertinent (not flammable); *Flammable Limits in Air (%):* Not pertinent; *Fire Extinguishing Agents:* Water, foam, dry chemical, carbon dioxide; *Fire Extinguishing Agents Not To Be Used:* Not pertinent; *Special Hazards of Combustion Products:* Not pertinent; *Behavior in Fire:* Not pertinent; *Ignition Temperature :* Not pertinent; *Electrical Hazard:* Not pertinent; *Burning Rate:* Not pertinent.

Chemical Reactivity - *Reactivity with Water:* No reaction; *Reactivity with Common Materials:* Slowly corrodes most metals; *Stability During Transport:* Stable; *Neutralizing Agents for Acids and Caustics:* Dilute with water, rinse with sodium bicarbonate or lime solution; *Polymerization:* Not pertinent; *Inhibitor of Polymerization:* Not pertinent.

LATEX, LIQUID SYNTHETIC

Chemical Designations - *Synonyms*: Plastic latex; Synthetic rubber latex; *Chemical Formula*: Not pertinent.

Observable Characteristics - *Physical State (as shipped)*: Liquid; *Color*: Milky; *Odor*: Each type has a characteristic odor.

Physical and Chemical Properties - *Physical State at 15 °C and 1 atm.*: Liquid; *Molecular Weight*: Not pertinent; *Boiling Point at 1 atm.*: Very high; *Freezing Point*: Not pertinent; *Critical Temperature*: Not pertinent; *Critical Pressure*: Not pertinent; *Specific Gravity*: 1.057 at 25 °C (liquid); *Vapor (Gas) Specific Gravity*: Not pertinent; *Ratio of Specific Heats of Vapor (Gas)*: Not pertinent; *Latent Heat of Vaporization*: Not pertinent; *Heat of Combustion*: Not pertinent; *Heat of Decomposition*: Not pertinent.

Health Hazards Information - *Recommended Personal Protective Equipment*: Chemical goggles or face shield; *Symptoms Following Exposure*: Irritation of eyes; *General Treatment for Exposure*: EYES: flush with water for at least 15 minutes; *Toxicity by Inhalation (Threshold Limit Value)*: Not pertinent; *Short-Term Inhalation Limits*: Not pertinent; *Toxicity by Ingestion*: Data not available; *Late Toxicity*: Data not available; *Vapor (Gas) Irritant Characteristics*: None; *Liquid or Solid Irritant Characteristics*: Contact with eyes can cause irritation; *Odor Threshold*: Not pertinent.

Fire Hazards - *Flash Point* : Not flammable unless coagulated; *Flammable Limits in Air (%):* Not flammable; *Fire Extinguishing Agents:* Not pertinent; *Fire Extinguishing Agents Not To Be Used:* Not pertinent; *Special Hazards of Combustion Products:* If the latex dries out and the burns, hydrochloric acid, hydrogen cyanide and styrene gases may be evolved. All are irritating and poisonous; *Behavior in Fire:* Heat may coagulate the latex and form sticky plastic lumps which may burn; *Ignition Temperature :* Not flammable; *Electrical Hazard:* Data not available; *Burning Rate:* Not flammable.
Chemical Reactivity - *Reactivity with Water:* No reaction; *Reactivity with Common Materials:* No reaction; *Stability During Transport:* Coagulated by heat and acids to gummy, flammable material; *Neutralizing Agents for Acids and Caustics:* Not pertinent; *Polymerization:* Not pertinent; *Inhibitor of Polymerization:* Not pertinent.

LAUROYL PEROXIDE
Chemical Designations - *Synonyms*: Dilauroyl peroxide; Dodecanoyl peroxyde; *Chemical Formula*: $[CH_3(CH_2)_{10}COO]_2$.
Observable Characteristics - *Physical State (as shipped)*: Solid; *Color*: White; *Odor*: Faint pungent; bland, soapy.
Physical and Chemical Properties - *Physical State at 15 °C and 1 atm.*: Solid; *Molecular Weight*: 399; *Boiling Point at 1 atm.*: Decomposes; *Freezing Point*: 129, 54, 327; *Critical Temperature*: Not pertinent; *Critical Pressure*: Not pertinent; *Specific Gravity*: 0.91 at 25 °C (solid); *Vapor (Gas) Specific Gravity*: Not pertinent; *Ratio of Specific Heats of Vapor (Gas)*: Not pertinent; *Latent Heat of Vaporization*: Not pertinent; *Heat of Combustion*: (est.) -16,300, -9,100, -380; *Heat of Decomposition*: Data not available.
Health Hazards Information - *Recommended Personal Protective Equipment*: Protective goggles, gloves; *Symptoms Following Exposure*: Contact with liquid irritates eyes and skin. Ingestion causes irritation of mouth and stomach; *General Treatment for Exposure*: EYES: flush with plenty of water for 15 min. and get medical attention. SKIN: wash with soap and water. INGESTION: administer en emetic to induce vomiting and call a physician; *Toxicity by Inhalation (Threshold Limit Value)*: Data not available; *Short-Term Inhalation Limits*: Not pertinent; *Toxicity by Ingestion*: Data not available; *Late Toxicity*: Weak carcinogen in mice; *Vapor (Gas) Irritant Characteristics*: Data not available; *Liquid or Solid Irritant Characteristics*: Data not available; *Odor Threshold*: Data not available.
Fire Hazards - *Flash Point* : Not pertinent (oxidizing combustible solid); *Flammable Limits in Air (%):* Not pertinent; *Fire Extinguishing Agents:* Water, dry chemical, foam, or carbon dioxide; *Fire Extinguishing Agents Not To Be Used:* Not pertinent; *Special Hazards of Combustion Products:* Not pertinent; *Behavior in Fire:* Can increase the severity of fire. Becomes sensitive to shock when hot. Containers may explode in a fire. May ignite or explode spontaneously if mixed with flammable materials; *Ignition Temperature :* Not pertinent; *Electrical Hazard:* Not pertinent; *Burning Rate:* Not pertinent.
Chemical Reactivity - *Reactivity with Water:* No reaction; *Reactivity with Common Materials:* May ignite or explode spontaneously when mixed with combustible materials; *Stability During Transport:* Stable if not overheated; *Neutralizing Agents for Acids and Caustics:* Not pertinent; *Polymerization:* Not pertinent; *Inhibitor of Polymerization:* Not pertinent.

LAURYL MERCAPTAN
Chemical Designations - *Synonyms*: 1-Dodecanethiol; Dodecyl mercaptan; *Chemical Formula*: $CH_3(CH_2)_{10}CH_2SH$.
Observable Characteristics - *Physical State (as shipped)*: Liquid; *Color*: Colorless; *Odor*: Mild skunk.
Physical and Chemical Properties - *Physical State at 15 °C and 1 atm.*: Liquid; *Molecular Weight*: 202; *Boiling Point at 1 atm.*: Very high; *Freezing Point*: 19.4, -7.0, 266.2; *Critical Temperature*: Not pertinent; *Critical Pressure*: Not pertinent; *Specific Gravity*: 0.85 at 15 °C (liquid); *Vapor (Gas) Specific Gravity*: Not pertinent; *Ratio of Specific Heats of Vapor (Gas)*: Not pertinent; *Latent Heat of Vaporization*: (est.) 110, 60, 2.5; *Heat of Combustion*: (est.) -18,200, -10,100, -422; *Heat of Decomposition*: Not pertinent.

Health Hazards Information - *Recommended Personal Protective Equipment*: Respirator when mist is present; rubber or vinyl gloves; chemical goggles; rubber shoes and apron; *Symptoms Following Exposure*: Liquid is irritating to skin, eyes, and mucous membranes. Ingestion may cause nausea. Repeated skin exposure can cause dermatitis and may produce a sensitizing effect; *General Treatment for Exposure*: Get medical attention for all eye exposures and may other serious overexposures. INHALATION (mist): rinse mouth repeatedly with cold water; treatment is symptomatic. INGESTION: dilute by drinking water; if vomiting occurs, drink more water; administer saline laxative. EYES: flush thoroughly with water; ventilation by electric fan is helpful in removing last traces, especially around eyes and eyelids. SKIN: remove contaminated clothing; flush skin with water; wash exposed area with soap and water; *Toxicity by Inhalation (Threshold Limit Value)*: Data not available; *Short-Term Inhalation Limits*: Data not available; *Toxicity by Ingestion*: Data not available; *Late Toxicity*: Causes decline in kidney and liver function in rats; *Vapor (Gas) Irritant Characteristics*: Irritating concentrations of vapor unlikely, but mist can cause irritation of eyes and upper respiratory tract; *Liquid or Solid Irritant Characteristics*: Minimum hazard. If spilled on clothing and allowed to remain, may cause smarting and reddening of skin; *Odor Threshold*: 4 mg/m^3.

Fire Hazards - *Flash Point (deg. F)*: 262 OC; *Flammable Limits in Air (%):* Data not available; *Fire Extinguishing Agents:* Dry chemical, foam, or carbon dioxide; *Fire Extinguishing Agents Not To Be Used:* Water or foam may cause frothing; *Special Hazards of Combustion Products:* Poisonous and irritating gases (e.g. sulfur dioxide) are generated in fires; *Behavior in Fire:* Not pertinent; *Ignition Temperature :* Data not available; *Electrical Hazard:* Not pertinent; *Burning Rate:* Data not available.

Chemical Reactivity - *Reactivity with Water:* No reaction; *Reactivity with Common Materials:* No reaction; *Stability During Transport:* Stable; *Neutralizing Agents for Acids and Caustics:* Not pertinent; *Polymerization:* Not pertinent; *Inhibitor of Polymerization:* Not pertinent.

LEAD ACETATE

Chemical Designations - *Synonyms*: Lead acetate trihydrate; Neutral lead acetate; Normal lead acetate; Salt of Saturn; Sugar of lead; *Chemical Formula*: $Pb(C_2H_3O_2)_2 \cdot 3H_2O$.

Observable Characteristics - *Physical State (as shipped)*: Solid; *Color*: White (commercial grades are frequently brown or grey lumps); *Odor*: None.

Physical and Chemical Properties - *Physical State at 15 °C and 1 atm.*: Solid; *Molecular Weight*: 379.3; *Boiling Point at 1 atm.*: Not pertinent (decomposes); *Freezing Point*: Not pertinent; *Critical Temperature*: Not pertinent; *Critical Pressure*: Not pertinent; *Specific Gravity*: 2.55 at 20 °C (solid); *Vapor (Gas) Specific Gravity*: Not pertinent; *Ratio of Specific Heats of Vapor (Gas)*: Not pertinent; *Latent Heat of Vaporization*: Not pertinent; *Heat of Combustion*: Not pertinent; *Heat of Decomposition*: Not pertinent.

Health Hazards Information - *Recommended Personal Protective Equipment*: Dust mask and protective gloves; *Symptoms Following Exposure*: Early symptoms of lead intoxication via inhalation or ingestion are most commonly gastrointestinal disorders, colic, constipation, etc; weakness, which may go on to paralysis, chiefly of the extensor muscles of the wrists and less often of the ankles, is noticeable in the most serious cases. Ingestion of a large amount causes local irritation of the alimentary tract; pain leg cramps, muscle weakness, paresthesias, depression, coma, and death may follow in 1 or 2 days. Contact with eyes causes irritation; *General Treatment for Exposure*: Remove at once all cases of lead intoxication from further exposure until the blood level is reduced to a safe value; immediately place the individual under medical care. INGESTION: give gastric lavage using 1% solution of sodium or magnesium sulfate; leave 15-30 gm magnesium sulfate in 6-8 oz. of water in the stomach as antidote and cathartic; egg white, milk, and tannin are useful demulcents; atropine sulfate and other antispasmodics may relieve abdominal pain, but morphine may be necessary. EYES or SKIN: flush with water; *Toxicity by Inhalation (Threshold Limit Value)*: 0.2 mg/m^3 (as lead); *Short-Term Inhalation Limits*: Data not available; *Toxicity by Ingestion*: Grade 2, LD_{50} 0.5-5 g/kg; *Late Toxicity*: Data not available; *Vapor (Gas) Irritant Characteristics*: Data not available; *Liquid or Solid Irritant Characteristics*: Data not available; *Odor Threshold*: Data not available.

Fire Hazards - *Flash Point :* Not flammable; *Flammable Limits in Air (%):* Not flammable; *Fire*

Extinguishing Agents: Not pertinent; *Fire Extinguishing Agents Not To Be Used:* Not pertinent; *Special Hazards of Combustion Products:* Irritating acid fumes may be formed in fires; *Behavior in Fire:* Not pertinent; *Ignition Temperature :* Not pertinent; *Electrical Hazard:* Not pertinent; *Burning Rate:* Not pertinent.

Chemical Reactivity - *Reactivity with Water:* No reaction; *Reactivity with Common Materials:* No reaction; *Stability During Transport:* Stable; *Neutralizing Agents for Acids and Caustics:* Not pertinent; *Polymerization:* Not pertinent; *Inhibitor of Polymerization:* Not pertinent.

LEAD ARSENATE

Chemical Designations - *Synonyms*: Lead arsenate, acid; Plumbous arsenate; *Chemical Formula*: $PbHASO_4$.

Observable Characteristics - *Physical State (as shipped)*: Solid; *Color*: White; *Odor*: None.

Physical and Chemical Properties - *Physical State at 15 °C and 1 atm.*: Solid; *Molecular Weight*: 347.12; *Boiling Point at 1 atm.*: Decomposes; *Freezing Point*: Not pertinent; *Critical Temperature*: Not pertinent; *Critical Pressure*: Not pertinent; *Specific Gravity*: 5.79 at 15 °C (solid); *Vapor (Gas) Specific Gravity*: Not pertinent; *Ratio of Specific Heats of Vapor (Gas)*: Not pertinent; *Latent Heat of Vaporization*: Not pertinent; *Heat of Combustion*: Not pertinent; *Heat of Decomposition*: Not pertinent.

Health Hazards Information - *Recommended Personal Protective Equipment*: Dust respirator; protective clothing to prevent accidental inhalation or ingestion of dust; *Symptoms Following Exposure*: Inhalation or ingestion causes dizziness, headache, paralysis, cramps, constipation, collapse, coma. Subacute doses cause irritability, loss of weight, anemia, constipation. Blood and urine concentrations of lead increase; *General Treatment for Exposure*: A specific medical treatment is used for exposure to this chemical; call a physician immediately! Give victim a tablespoon of salt in glass of warm water and repeat until vomit is clear. Then give two tablespoon of epsom salt or milk of magnesia in water, and plenty of milk and water. Have victim lie down and keep quiet; *Toxicity by Inhalation (Threshold Limit Value)*: (dust) 0.15 mg/m^3; *Short-Term Inhalation Limits*: Not pertinent; *Toxicity by Ingestion*: Grade 4, LD_{50} below 50 mg/kg (rabbit, rat); *Late Toxicity*: Lead poisoning; *Vapor (Gas) Irritant Characteristics*: Not pertinent; *Liquid or Solid Irritant Characteristics*: None; *Odor Threshold*: Not pertinent.

Fire Hazards - *Flash Point :* Not flammable; *Flammable Limits in Air (%):* Not flammable; *Fire Extinguishing Agents:* Not pertinent; *Fire Extinguishing Agents Not To Be Used:* Not pertinent; *Special Hazards of Combustion Products:* Not pertinent; *Behavior in Fire:* Not pertinent; *Ignition Temperature :* Not pertinent; *Electrical Hazard:* Not pertinent; *Burning Rate:* Not pertinent.

Chemical Reactivity - *Reactivity with Water:* No reaction; *Reactivity with Common Materials:* No reaction; *Stability During Transport:* Stable; *Neutralizing Agents for Acids and Caustics:* Not pertinent; *Polymerization:* Not pertinent; *Inhibitor of Polymerization:* Not pertinent.

LEAD FLUOROBORATE

Chemical Designations - *Synonyms*: Lead fluoroborate; Lead fluoroborate solution; *Chemical Formula*: $Pb(BF_4)_2$-H_2O.

Observable Characteristics — *Physical State (as shipped)*: Liquid; *Color*: Colorless; *Odor*: Faint.

Physical and Chemical Properties - *Physical State at 15 °C and 1 atm.*: Liquid; *Molecular Weight*: Mixture; *Boiling Point at 1 atm.*: Not pertinent; *Freezing Point*: Not pertinent; *Critical Temperature*: Not pertinent; *Critical Pressure*: Not pertinent; *Specific Gravity*: 1.75 at 20 °C (liquid); *Vapor (Gas) Specific Gravity*: Not pertinent; *Ratio of Specific Heats of Vapor (Gas)*: Not pertinent; *Latent Heat of Vaporization*: Not pertinent; *Heat of Combustion*: Not pertinent; *Heat of Decomposition*: Not pertinent.

Health Hazards Information - *Recommended Personal Protective Equipment*: Rubber gloves; face shield; rubber apron; *Symptoms Following Exposure*: Early symptoms of lead intoxication via inhalation or ingestion are most commonly gastrointestinal disorders, colic, constipation, etc; weakness, which may go on to paralysis, chiefly of the extensor muscles of the wrists and less often of the alimentary tract; pain, leg cramps, muscle weakness, paresthesias, coma, and death may follow in 1 or 2 days. Contact with skin or eyes may cause burns and/or irritation; *General Treatment for Exposure*: Remove at once all cases of lead intoxication from further exposure until the blood level is reduced to a safe value; immediately place the individual under medical care. INGESTION: give gastric lavage using 1% solution of sodium or magnesium sulfate; leave 15-30 gm magnesium sulfate in 6-8 oz. of water in the stomach as antidote and cathartic; egg white, milk, and tannin are useful demulcents; atropine sulfate and other antispasmodics may relieve abdominal pain, but morphine may be necessary. EYES: flush with copious quantities of water for 15 min. SKIN: wash area with soap and water; treat as an acid burn; *Toxicity by Inhalation (Threshold Limit Value)*: 0.2 mg/m^3 (as lead); *Short-Term Inhalation Limits*: Data not available; *Toxicity by Ingestion*: Grade 2, LD$_{50}$ 0.5-5 g/kg; *Late Toxicity*: Data not available; *Vapor (Gas) Irritant Characteristics*: Data not available; *Liquid or Solid Irritant Characteristics*: Data not available; *Odor Threshold*: Data not available.

Fire Hazards - *Flash Point* : Not flammable; *Flammable Limits in Air (%):* Not flammable; *Fire Extinguishing Agents:* Not pertinent; *Fire Extinguishing Agents Not To Be Used:* Not pertinent; *Special Hazards of Combustion Products:* Toxic and irritating hydrogen fluoride gas may form in fire; *Behavior in Fire:* Not pertinent; *Ignition Temperature :* Not pertinent; *Electrical Hazard:* Not pertinent; *Burning Rate:* Not pertinent.

Chemical Reactivity - *Reactivity with Water:* No reaction; *Reactivity with Common Materials:* Solution is acidic and will corrode most metals; *Stability During Transport:* Stable; *Neutralizing Agents for Acids and Caustics:* Flush with water; rinse with dilute solution of sodium bicarbonate of soda ash; *Polymerization:* Not pertinent; *Inhibitor of Polymerization:* Not pertinent.

LEAD FLUORIDE

Chemical Designations - *Synonyms*: Lead difluoride; Plumbous fluoride; *Chemical Formula*: PbF$_2$.
Observable Characteristics - *Physical State (as shipped)*: Solid; *Color*: White; *Odor*: None.
Physical and Chemical Properties - *Physical State at 15 °C and 1 atm.*: Solid; *Molecular Weight*: 245.19; *Boiling Point at 1 atm.*: Not pertinent; *Freezing Point*: Not pertinent; *Critical Temperature*: Not pertinent; *Critical Pressure*: Not pertinent; *Specific Gravity*: 8.24 at 20 °C (solid); *Vapor (Gas) Specific Gravity*: Not pertinent; *Ratio of Specific Heats of Vapor (Gas)*: Not pertinent; *Latent Heat of Vaporization*: Not pertinent; *Heat of Combustion*: Not pertinent; *Heat of Decomposition*: Not pertinent.
Health Hazards Information - *Recommended Personal Protective Equipment*: Respirator for heavy dust exposure; safety goggles; *Symptoms Following Exposure*: Not irritating to skin or mucous membranes; protect against chronic poisoning. Early symptoms of lead intoxication via inhalation or ingestion are most commonly gastrointestinal disorders, colic, constipation, etc; weakness, which may go on to paralysis, chiefly of the extensor muscles of the wrists and less often of the ankles, is noticeable in the most serious cases. Ingestion of a large amount causes local irritation of the alimentary tract; pain, leg cramps, muscle weakness, paresthesias, coma, and death may follow in 1 or 2 days. Contact with eyes causes irritation; *General Treatment for Exposure*: Remove at once all cases of lead intoxication from further exposure until the blood level is reduced to a safe value; immediately place the individual under medical care. INGESTION: give gastric lavage using 1% solution of sodium or magnesium sulfate; leave 15-30 gm magnesium sulfate in 6-8 oz. of water in the stomach as antidote and cathartic; egg white, milk, and tannin are useful demulcents; atropine sulfate and other antispasmodics may relieve abdominal pain, but morphine may be necessary. EYES or SKIN: flush with water; *Toxicity by Inhalation (Threshold Limit Value)*: 0.15 mg/m^3 (as lead); *Short-Term Inhalation Limits*: Data not available; *Toxicity by Ingestion*: Grade 2, LD$_{50}$ 0.5 - 5 g/kg; *Late Toxicity*: Data not available; *Vapor (Gas) Irritant Characteristics*: Data not available; *Liquid or Solid Irritant Characteristics*: Data not available; *Odor Threshold*: Data not available.
Fire Hazards - *Flash Point* : Not flammable; *Flammable Limits in Air (%):* Not flammable; *Fire Extinguishing Agents:* Not pertinent; *Fire Extinguishing Agents Not To Be Used:* Not pertinent; *Special Hazards of Combustion Products:* Not pertinent; *Behavior in Fire:* Not pertinent; *Ignition Temperature :* Not pertinent; *Electrical Hazard:* Not pertinent; *Burning Rate:* Not pertinent.
Chemical Reactivity - *Reactivity with Water:* No reaction; *Reactivity with Common Materials:* No reaction; *Stability During Transport:* Stable; *Neutralizing Agents for Acids and Caustics:* Not pertinent; *Polymerization:* Not pertinent; *Inhibitor of Polymerization:* Not pertinent.

LEAD IODIDE

Chemical Designations - *Synonyms*: No common synonyms; *Chemical Formula*: PbI_2.

Observable Characteristics - *Physical State (as shipped)*: Solid; *Color*: Bright yellow; *Odor*: None.

Physical and Chemical Properties - *Physical State at 15 °C and 1 atm.*: Solid; *Molecular Weight*: 461.03; *Boiling Point at 1 atm.*: Not pertinent; *Freezing Point*: Not pertinent; *Critical Temperature*: Not pertinent; *Critical Pressure*: Not pertinent; *Specific Gravity*: 6.16 at 20 °C (solid); *Vapor (Gas) Specific Gravity*: Not pertinent; *Ratio of Specific Heats of Vapor (Gas)*: Not pertinent; *Latent Heat of Vaporization*: Not pertinent; *Heat of Combustion*: Not pertinent; *Heat of Decomposition*: Not pertinent.

Health Hazards Information - *Recommended Personal Protective Equipment*: Dust mask and protective gloves; *Symptoms Following Exposure*: Early symptoms of lead intoxication via inhalation or ingestion are most commonly gastrointestinal disorders, colic, constipation, etc; weakness, which may go on to paralysis, chiefly of the extensor muscles of the wrists and less often of the ankles, is noticeable in the most serious cases. Ingestion of a large amount causes local irritation of the alimentary tract. Pain, leg cramps, muscle weakness, paresthesias, coma, and death may follow in 1 or 2 days. Contact with eyes causes irritation; *General Treatment for Exposure*: Remove at once all cases of lead intoxication from further exposure until the blood level is reduced to a safe value; immediately place the individual under medical care. INGESTION: give gastric lavage using 1% solution of sodium or magnesium sulfate; leave 15-30 gm magnesium sulfate in 6-8 oz. of water in the stomach as antidote and cathartic; egg white, milk, and tannin are useful demulcents; atropine sulfate and other antispasmodics may relieve abdominal pain, but morphine may be necessary. EYES or SKIN: flush with water; *Toxicity by Inhalation (Threshold Limit Value)*: 0.2 mg/m^3 (as lead); *Short-Term Inhalation Limits*: Data not available; *Toxicity by Ingestion*: Grade 2, LD_{50} 0.5-5 g/kg; *Late Toxicity*: Data not available; *Vapor (Gas) Irritant Characteristics*: Data not available; *Liquid or Solid Irritant Characteristics*: Data not available; *Odor Threshold*: Data not available.

Fire Hazards - *Flash Point* : Not flammable; *Flammable Limits in Air (%):* Not flammable; *Fire Extinguishing Agents:* Not pertinent; *Fire Extinguishing Agents Not To Be Used:* Not pertinent; *Special Hazards of Combustion Products:* Not pertinent; *Behavior in Fire:* Not pertinent; *Ignition Temperature : Not pertinent; *Electrical Hazard:* Not pertinent; *Burning Rate:* Not pertinent.

Chemical Reactivity - *Reactivity with Water:* No reaction; *Reactivity with Common Materials:* No reaction; *Stability During Transport:* Stable; *Neutralizing Agents for Acids and Caustics:* Not pertinent; *Polymerization:* Not pertinent; *Inhibitor of Polymerization:* Not pertinent.

LEAD NITRATE

Chemical Designations - *Synonyms*: Nitric acid, lead (2+) salt; *Chemical Formula*: $Pb(NO_3)_2$.

Observable Characteristics - *Physical State (as shipped)*: Solid; *Color*: White; *Odor*: None.

Physical and Chemical Properties - *Physical State at 15 °C and 1 atm.*: Solid; *Molecular Weight*: 331.2; *Boiling Point at 1 atm.*: Not pertinent; *Freezing Point*: Not pertinent; *Critical Temperature*: Not pertinent; *Critical Pressure*: Not pertinent; *Specific Gravity*: 4.53 at 20 °C (solid); *Vapor (Gas) Specific Gravity*: Not pertinent; *Ratio of Specific Heats of Vapor (Gas)*: Not pertinent; *Latent Heat of Vaporization*: Not pertinent; *Heat of Combustion*: Not pertinent; *Heat of Decomposition*: Not pertinent.

Health Hazards Information - *Recommended Personal Protective Equipment*: Dust mask and protective gloves; *Symptoms Following Exposure*: Early symptoms of lead intoxication via inhalation or ingestion are most commonly gastrointestinal disorders, colic, constipation, etc.; weakness, which may go on to paralysis, chiefly of the extensor muscles of the wrists and less often of the ankles, is noticeable in the most serious cases. Ingestion of a large amount causes local irritation of the alimentary tract. Pain, leg cramps, muscle weakness, paresthesias, coma, and death may follow in 1 or 2 days. Contact with eyes causes irritation; *General Treatment for Exposure*: Remove at once all cases of lead intoxication from further exposure until the blood level is reduced to a safe value; immediately place the individual under medical care. INGESTION: give gastric lavage using 1% solution of sodium or magnesium sulfate; leave 15-30 gm magnesium sulfate in 6-8 oz. of water in the stomach as antidote and cathartic; egg white, milk, and tannin are useful demulcents; atropine sulfate and other antispasmodics may relieve abdominal pain, but morphine may be necessary. EYES or SKIN: flush with water; *Toxicity by Inhalation (Threshold Limit Value)*: 0.2 mg/m^3 (as lead); *Short-Term Inhalation Limits*: Data not available; *Toxicity by Ingestion*: Grade 2, LD$_{50}$ 0.5-5 g/kg; *Late Toxicity*: Data not available; *Vapor (Gas) Irritant Characteristics*: Data not available; *Liquid or Solid Irritant Characteristics*: Data not available; *Odor Threshold*: Data not available.

Fire Hazards - *Flash Point* : Not flammable; *Flammable Limits in Air (%):* Not flammable; *Fire Extinguishing Agents:* Not pertinent; *Fire Extinguishing Agents Not To Be Used:* Not pertinent; *Special Hazards of Combustion Products:* Toxic oxides of nitrogen may form in fire; *Behavior in Fire:* Increases the intensity of a fire when in contact with burning material. Use plenty of water to cool containers or spilled material; *Ignition Temperature :* Not pertinent; *Electrical Hazard:* Not pertinent; *Burning Rate:* Not pertinent.

Chemical Reactivity - *Reactivity with Water:* No reaction; *Reactivity with Common Materials:* Contact with wood or paper may cause fire; *Stability During Transport:* Stable; *Neutralizing Agents for Acids and Caustics:* Not pertinent; *Polymerization:* Not pertinent; *Inhibitor of Polymerization:* Not pertinent.

LEAD TETRAACETATE

Chemical Designations - *Synonyms*: Lead (IV) acetate; *Chemical Formula*: Pb(C$_2$H$_3$O$_2$)$_4$-CH$_3$COOH.

Observable Characteristics - *Physical State (as shipped)*: Crystals wet with glacial acetic acid; *Color*: Faintly pink; *Odor*: Like acetic acid or vinegar.

Physical and Chemical Properties - *Physical State at 15 °C and 1 atm.*: Solid; *Molecular Weight*: 443.39; *Boiling Point at 1 atm.*: Not pertinent (decomposes); *Freezing Point*: 347, 175, 448; *Critical Temperature*: Not pertinent; *Critical Pressure*: Not pertinent; *Specific Gravity*: 2.2 at 20 °C (solid); *Vapor (Gas) Specific Gravity*: Not pertinent; *Ratio of Specific Heats of Vapor (Gas)*: Not pertinent; *Latent Heat of Vaporization*: Not pertinent; *Heat of Combustion*: Not pertinent; *Heat of Decomposition*: Not pertinent.

Health Hazards Information - *Recommended Personal Protective Equipment*: Goggles or face shield; rubber gloves; *Symptoms Following Exposure*: Early symptoms of lead intoxication via inhalation or ingestion are most commonly gastrointestinal disorders, colic, constipation, etc; weakness, which may go on to paralysis, chiefly of the extensor muscles of the wrists and less often of the ankles, is noticeable in the most serious cases. Ingestion of a large amount causes local irritation of the alimentary tract. Pain, leg cramps, muscle weakness, paresthesias, coma, and death may follow in 1 or 2 days. Contact with eyes cause severe irritation and can burn skin; *General Treatment for Exposure*: Remove at once all cases of lead intoxication from further exposure until the blood level is reduced to a safe value; immediately place the individual under medical care. INGESTION: give gastric lavage using 1% solution of sodium or magnesium sulfate; leave 15-30 gm magnesium sulfate in 6-8 oz. of water in the stomach as antidote and cathartic; egg white, milk, and tannin are useful demulcents; atropine sulfate and other antispasmodics may relieve abdominal pain, but morphine may be necessary. EYES: flush with water for at least 15 min. SKIN: wash contaminated skin with large amounts of water for 15 min.; *Toxicity by Inhalation (Threshold Limit Value)*: 0.2 mg/m^3 (as lead); *Short-Term Inhalation Limits*: Data not available; *Toxicity by Ingestion*: Grade 2, LD$_{50}$ 0.5-5 g/kg; *Late Toxicity*: Data not available; *Vapor (Gas) Irritant Characteristics*: Data not available; *Liquid or Solid Irritant Characteristics*: Data not available; *Odor Threshold*: Data not available.

Fire Hazards - *Flash Point* : Not flammable; *Flammable Limits in Air (%):* Not flammable; *Fire Extinguishing Agents:* Not pertinent; *Fire Extinguishing Agents Not To Be Used:* Not pertinent; *Special Hazards of Combustion Products:* Not pertinent; *Behavior in Fire:* Can increase the intensity of a fire when in contact with combustible material. Cool containers with plenty of water; *Ignition Temperature :* Not pertinent; *Electrical Hazard:* Not pertinent; *Burning Rate:* Not pertinent.

Chemical Reactivity - *Reactivity with Water:* Forms lead dioxide and acetic acid in a reaction that is not violent; *Reactivity with Common Materials:* May corrode metals when moist; *Stability During Transport:* Stable; *Neutralizing Agents for Acids and Caustics:* Dilute with water, rinse with dilute sodium bicarbonate or lime solution; *Polymerization:* Not pertinent; *Inhibitor of Polymerization:* Not pertinent.

LEAD THIOCYANATE

Chemical Designations -*Synonyms*: Lead sulfocyanate; *Chemical Formula*: $Pb(SCN)_2$.

Observable Characteristics - *Physical State (as shipped)*: Solid; *Color*: White; *Odor*: None.

Physical and Chemical Properties - *Physical State at 15 °C and 1 atm.*: Solid; *Molecular Weight*: 323.4; *Boiling Point at 1 atm.*: Not pertinent (decomposes); *Freezing Point*: Not pertinent; *Critical Temperature*: Not pertinent; *Critical Pressure*: Not pertinent; *Specific Gravity*: 3.82 at 20 °C (solid); *Vapor (Gas) Specific Gravity*: Not pertinent; *Ratio of Specific Heats of Vapor (Gas)*: Not pertinent; *Latent Heat of Vaporization*: Not pertinent; *Heat of Combustion*: Not pertinent; *Heat of Decomposition*: Not pertinent.

Health Hazards Information - *Recommended Personal Protective Equipment*: Dust mask; goggles or face shield; rubber gloves; *Symptoms Following Exposure*: Early symptoms of lead intoxication via inhalation or ingestion are most commonly gastrointestinal disorders, colic, constipation, etc; weakness, which may go on to paralysis, chiefly of the extensor muscles of the wrists and less often of the ankles, is noticeable in the most serious cases. Ingestion of a large amount causes local irritation of the alimentary tract. Pain, leg cramps, muscle weakness, paresthesias, coma, and death may follow in 1 or 2 days. Contact causes irritation of eyes and mild irritation of skin; *General Treatment for Exposure*: Remove at once all cases of lead intoxication from further exposure until the blood level is reduced to a safe value; immediately place the individual under medical care. INGESTION: give gastric lavage using 1% solution of sodium or magnesium sulfate; leave 15-30 gm magnesium sulfate in 6-8 oz. of water in the stomach as antidote and cathartic; egg white, milk, and tannin are useful demulcents; atropine sulfate and other antispasmodics may relieve abdominal pain, but morphine may be necessary. EYES: flush with water for at least 15 min. SKIN: wash well with soap and water; *Toxicity by Inhalation (Threshold Limit Value)*: 0.2 mg/m^3 (as lead); *Short-Term Inhalation Limits*: Data not available; *Toxicity by Ingestion*: Grade 2, LD_{50} 0.5-5 g/kg; *Late Toxicity*: Data not available; *Vapor (Gas) Irritant Characteristics*: Data not available; *Liquid or Solid Irritant Characteristics*: Data not available; *Odor Threshold*: Data not available.

Fire Hazards - *Flash Point* : Not flammable; *Flammable Limits in Air (%):* Not flammable; *Fire Extinguishing Agents:* Not pertinent; *Fire Extinguishing Agents Not To Be Used:* Not pertinent; *Special Hazards of Combustion Products:* Irritating sulfur dioxide gas may form in fire; *Behavior in Fire:* Not pertinent; *Ignition Temperature :* Not pertinent; *Electrical Hazard:* Not pertinent; *Burning Rate:* Not pertinent.

Chemical Reactivity - *Reactivity with Water:* No reaction; *Reactivity with Common Materials:* No reaction; *Stability During Transport:* Stable; *Neutralizing Agents for Acids and Caustics:* Not pertinent; *Polymerization:* Not pertinent; *Inhibitor of Polymerization:* Not pertinent.

LINEAR ALCOHOLS

Chemical Designations - *Synonyms:* Dodecanol; Tridecanol; Tetradecanol; Pentadecanol; (Could be any of the above or mixtures thereof); *Chemical Formula:* $CH_3(CH_2)_{10-13}CH_2OH$.

Observable Characteristics - *Physical State (as shipped):* Liquid; *Color:* Colorless; *Odor:* Mild.

Physical and Chemical Properties - *Physical State at 15 °C and 1 atm:* Liquid or solid; *Molecular Weight:* >186; *Boiling Point at 1 atm:* >486, >252, >525; *Freezing Point:* >66, >19, >292; *Critical Temperature:* Not pertinent; *Critical Pressure:* Not pertinent; *Specific Gravity:* 0,84 at 20°C (liquid); *Vapor (Gas) Specific Gravity:* Not pertinent; *Ratio of Specific Heats of Vapor (Gas):* Not pertinent; *Latent Heat of Vaporization:* Not pertinent; *Heat of Combustion:* (est.) -18,500, -10,300, -429; *Heat of Decomposition:* Not pertinent.

Health Hazards - *Personal Protective Equipment:* Eye protection. *Symptoms Following Exposure:* Direct contact can produce eye irritation. Low general toxicity. *Treatment for Exposure:* Wash eyes with water for at least 15 min. *Toxicity by Inhalation (Threshold Limit Value):* Not pertinent; *Short-Term Inhalation Limits:* Not pertinent; *Toxicity by Ingestion:* Grade 1; LD_{50} 5 to 15 g/kg (rat); *Late Toxicity:* Data not available; *Vapor (Gas) Irritant Characteristics:* None; *Liquid or Solid Irritant Characteristics:* No appreciable hazard. Practically harmless to skin. *Odor Threshold:* Data not available.

Fire Hazards - *Flash Point (deg. F):* 180 - 285 OC; *Flammable Limits in Air (%):* Data not available; *Fire Extinguishing Agents:* Alcohol foam, dry chemical, or carbon dioxide; *Fire Extinguishing Agents Not To Be Used:* Water or foam may cause frothing; *Special Hazards of Combustion Products:* Not pertinent; *Behavior in Fire:* Not pertinent; *Ignition Temperature :* Data not available; *Electrical Hazard:* Not pertinent; *Burning Rate:* Data not available.

Chemical Reactivity - *Reactivity with Water:* No reaction; *Reactivity with Common Materials:* No reaction; *Stability During Transport:* Stable; *Neutralizing Agents for Acids and Caustics:* Not pertinent; *Polymerization:* Not pertinent; *Inhibitor of Polymerization:* Not pertinent.

LIQUEFIED NATURAL GAS

Chemical Designations - *Synonyms:* LNG; *Chemical Formula:* $CH_4 + C_2H_6$.

Observable Characteristics - *Physical State (as shipped):* Liquefied gas; *Color:* Colorless; *Odor:* Mild, sweet.

Physical and Chemical Properties - *Physical State at 15 °C and 1 atm:* Gas; *Molecular Weight:* >16; *Boiling Point at 1 atm:* -258, -161, 112; *Freezing Point:* -296, -182,2, 91,0; *Critical Temperature:* -116, -82,2, 191,0; *Critical Pressure (psia, atm, MN/m^2):* 673, 45,78, 4,64; *Specific Gravity:* 0,415-0,45 at -162°C (liquid); *Vapor (Gas) Specific Gravity:* 0,55-1,0; *Ratio of Specific Heats of Vapor (Gas):* 1,306; *Latent Heat of Vaporization:* (est.) 220, 120, 5,1; *Heat of Combustion:* -21,600 to -23,400, -12,000 to -13,000, -502,4 to 544,3; *Heat of Decomposition:* Not pertinent.

Health Hazards - *Personal Protective Equipment:* Self-contained breathing apparatus; protective clothing if exposed to liquid. *Symptoms Following Exposure:* If concentration of gas in high enough, may cause asphyxiation. No detectable systemic effects, even at 5% concentration in air. *Treatment for Exposure:* Remove victim to open air. If the is overcome by gas, apply artificial resuscitation. *Toxicity by Inhalation (Threshold Limit Value):* Data not available; *Short-Term Inhalation Limits:* Data not available; *Toxicity by Ingestion:* Not pertinent; *Late Toxicity:* None; *Vapor (Gas) Irritant Characteristics:* Vapors are nonirritating to the eyes and throat. *Liquid or Solid Irritant Characteristics:* No appreciable hazard. Practically harmless to the skin because it is very volatile and evaporates quickly. May cause some frostbite. *Odor Threshold:* Data not available.

Fire Hazards - *Flash Point :* Flammable gas; *Flammable Limits in Air (%):* 5.3 - 14.0; *Fire Extinguishing Agents:* Do not extinguish large spill fires. Allow to burn while cooling adjacent equipment with water spray. Shut off leak if possible. Extinguish small fires with dry chemicals; *Fire Extinguishing Agents Not To Be Used:* Water; *Special Hazards of Combustion Products:* Not pertinent; *Behavior in Fire:* Not pertinent; *Ignition Temperature (deg. F):* 999; *Electrical Hazard:* Class I, Group D; *Burning Rate:* 12.5 mm/min.

Chemical Reactivity - *Reactivity with Water:* No reaction; *Reactivity with Common Materials:* No reaction; *Stability During Transport:* Stable; *Neutralizing Agents for Acids and Caustics:* Not pertinent; *Polymerization:* Not pertinent; *Inhibitor of Polymerization:* Not pertinent.

LIQUEFIED PETROLEUM GAS

Chemical Designations - *Synonyms*: Bottled gas; Propane-butane-(propylene) Pyrofax; LPG; *Chemical Formula*: C_3H_6-C_3H_8-C_4H_{10}.

Observable Characteristics - *Physical State (as shipped)*: Liquefied compressed gas; *Color*: Colorless; *Odor*: Mild. But commercial LPG has a skunk-like odorant added as a warning.

Physical and Chemical Properties - *Physical State at 15 °C and 1 atm.*: Gas; *Molecular Weight*: >44; *Boiling Point at 1 atm.*: >-40, >-40, >233; *Freezing Point*: Not pertinent; *Critical Temperature*: -142.01, -96.67, 176.53; *Critical Pressure*: 616.5, 41.94, 4.249; *Specific Gravity*: 0.51-0.58 at -50 °C (liquid); *Vapor (Gas) Specific Gravity*: 1.5; *Ratio of Specific Heats of Vapor (Gas)*: 1.130; *Latent Heat of Vaporization*: 183.2, 101.8, 4.262; *Heat of Combustion*: -19,782, -10,990, -460.13; *Heat of Decomposition*: Not pertinent.

Health Hazards Information - *Recommended Personal Protective Equipment*: Self-contained breathing apparatus for high concentrations of gas; *Symptoms Following Exposure*: Concentrations in air greater than 10% cause dizziness in a few minutes. 1% concentrations give the same symptom in 10 min. High concentrations cause asphyxiation; *General Treatment for Exposure*: Remove victim to open air. If he is overcome by gas, apply artificial respiration. Guard against self-injury if confused; *Toxicity by Inhalation (Threshold Limit Value)*: 1000 ppm; *Short-Term Inhalation Limits*: Data not available; *Toxicity by Ingestion*: Not pertinent; *Late Toxicity*: None; *Vapor (Gas) Irritant Characteristics*: Vapors are nonirritating to the eyes and throat; *Liquid or Solid Irritant Characteristics*: No appreciable hazard. Practically harmless to the skin because it is very volatile and evaporates quickly. May cause frostbite; *Odor Threshold*: 5000-20,000 ppm.

Fire Hazards - *Flash Point (deg. F)*: Propane: -156 CC; butane: -76 CC; *Flammable Limits in Air (%):* Propane: 2.2 - 9.5; butane: 1.8 - 8.4; *Fire Extinguishing Agents:* Allow to burn while cooling adjacent equipment with water spray. Extinguish small fires with dry chemical. Shut off leak if possible; *Fire Extinguishing Agents Not To Be Used:* Water (let fire burn); *Special Hazards of Combustion Products:* Not pertinent; *Behavior in Fire:* Containers may explode. Vapor is heavier than air and may travel a long distance to a source of ignition and flash back; *Ignition Temperature :* No data; *Electrical Hazard:* No data; *Burning Rate:* No data.

Chemical Reactivity - *Reactivity with Water:* No reaction; *Reactivity with Common Materials:* No reaction; *Stability During Transport:* Stable; *Neutralizing Agents for Acids and Caustics:* Not pertinent; *Polymerization:* Not pertinent; *Inhibitor of Polymerization:* Not pertinent.

LITHARGE

Chemical Designations - *Synonyms*: Lead monoxide; Lead oxide, yellow; Lead protoxide; Massicot; Plumbous oxide; *Chemical Formula*: PbO.

Observable Characteristics - *Physical State (as shipped)*: Solid; *Color*: Yellow to red; low-metal-content oxides; yellow to green to brown; high-metal-content oxides; gray to brown; *Odor*: None.

Physical and Chemical Properties - *Physical State at 15 °C and 1 atm.*: Solid; *Molecular Weight*: 223.2; *Boiling Point at 1 atm.*: Not pertinent (decomposes); *Freezing Point*: Not pertinent; *Critical Temperature*: Not pertinent; *Critical Pressure*: Not pertinent; *Specific Gravity*: 9.5 at 20 °C (solid); *Vapor (Gas) Specific Gravity*: Not pertinent; *Ratio of Specific Heats of Vapor (Gas)*: Not pertinent; *Latent Heat of Vaporization*: Not pertinent; *Heat of Combustion*: Not pertinent; *Heat of Decomposition*: Not pertinent.

Health Hazards Information - *Recommended Personal Protective Equipment*: Dust or metal fume respirator; gloves; goggles; *Symptoms Following Exposure*: General symptoms of lead poisoning (delayed). Inhalation or ingestion causes abdominal pain (lead colic), metallic taste in mouth, loss of weight, pain in muscles, and muscular weakness. Dust may irritate eyes; *General Treatment for Exposure*: Consult physician after ingestion or exposure to high concentrations of dust. INGESTION: call physician at once; as first aid, induce vomiting and give milk and magnesium sulfate (Epsom salt); *Toxicity by Inhalation (Threshold Limit Value)*: 0.2 mg/m^3; *Short-Term Inhalation Limits*: Data not available; *Toxicity by Ingestion*: Data not available; *Late Toxicity*: Impairs development of human fetal connective tissue cells; *Vapor (Gas) Irritant Characteristics*: Not pertinent; *Liquid or Solid Irritant Characteristics*: Data not available; *Odor Threshold*: Not pertinent.

Fire Hazards - *Flash Point* : Not flammable; *Flammable Limits in Air (%):* Not flammable; *Fire Extinguishing Agents:* Not pertinent; *Fire Extinguishing Agents Not To Be Used:* Not pertinent; *Special Hazards of Combustion Products:* Not pertinent; *Behavior in Fire:* Not pertinent; *Ignition Temperature :* Not pertinent; *Electrical Hazard:* Not pertinent; *Burning Rate:* Not pertinent.

Chemical Reactivity - *Reactivity with Water:* No reaction; *Reactivity with Common Materials:* No reaction; *Stability During Transport:* Stable; *Neutralizing Agents for Acids and Caustics:* Not pertinent; *Polymerization:* Not pertinent; *Inhibitor of Polymerization:* Not pertinent.

LITHIUM ALUMINUM HYDRIDE

Chemical Designations - *Synonyms*: LAH; *Chemical Formula*: LiAlH$_4$

Observable Characteristics - *Physical State (as shipped)*: Solid; *Color*: White to gray; *Odor*: None.

Physical and Chemical Properties - *Physical State at 15 °C and 1 atm.*: Solid; *Molecular Weight*: 37.94; *Boiling Point at 1 atm.*: Decomposes; *Freezing Point*: Not pertinent; *Critical Temperature*: Not pertinent; *Critical Pressure*: Not pertinent; *Specific Gravity*: 0.917 at 15 °C (solid); *Vapor (Gas) Specific Gravity*: Not pertinent; *Ratio of Specific Heats of Vapor (Gas)*: Not pertinent; *Latent Heat of Vaporization*: Not pertinent; *Heat of Combustion*: Not pertinent; *Heat of Decomposition*: Not pertinent.

Health Hazards Information - *Recommended Personal Protective Equipment*: Rubberized gloves; full face shield; *Symptoms Following Exposure*: Contact of solid with eyes and skin causes severe burns similar to those caused by caustic soda; *General Treatment for Exposure*: In case of accidental contact with the skin, wipe off excess with a dry paper towel. Wash the affected area with a large volume of water to prevent localized heating of the skin; *Toxicity by Inhalation (Threshold Limit Value)*: Not pertinent; *Short-Term Inhalation Limits*: Not pertinent; *Toxicity by Ingestion*: Data not available; *Late Toxicity*: Data not available; *Vapor (Gas) Irritant Characteristics*: Not pertinent; *Liquid or Solid Irritant Characteristics*: Moisture of skin causes caustic burns; *Odor Threshold*: Not pertinent.

Fire Hazards - *Flash Point* : Solid; *Flammable Limits in Air (%):* Not pertinent; *Fire Extinguishing Agents:* Powdered graphite, powdered salt, or powdered limestone; *Fire Extinguishing Agents Not To Be Used:* Do not use water, soda acid, carbon dioxide or dry chemical; *Special Hazards of Combustion Products:* Data not available; *Behavior in Fire:* Decomposes at 257 °F to form hydrogen gas. The heat generated may cause ignition and/or explosion; *Ignition Temperature :* Data not available; *Electrical Hazard:* Class I, Group B; *Burning Rate:* Not pertinent.

Chemical Reactivity - *Reactivity with Water:* Reacts violently with water as a dry solid or when dissolved in ether. The hydrogen produced by the reaction with water is a major hazard and necessitates adequate ventilation; *Reactivity with Common Materials:* Can burn in heated or moist air; *Stability During Transport:* Normally stable; unstable at high temperatures; *Neutralizing Agents for Acids and Caustics:* Not pertinent; *Polymerization:* Not pertinent; *Inhibitor of Polymerization:* Not pertinent.

LITHIUM HYDRIDE

Chemical Designations - *Synonyms*: No common synonyms; *Chemical Formula*: LiH.

Observable Characteristics - *Physical State (as shipped)*: Solid; *Color*: Gray-blue crystalline mass; finely ground material ranges in color from white to gray; *Odor*: None.

Physical and Chemical Properties - *Physical State at 15 °C and 1 atm.*: Solid; *Molecular Weight*: 7.95; *Boiling Point at 1 atm.*: Not pertinent (decomposes); *Freezing Point*: Not pertinent; *Critical Temperature*: Not pertinent; *Critical Pressure*: Not pertinent; *Specific Gravity*: 0.78 at 20 °C (solid); *Vapor (Gas) Specific Gravity*: Not pertinent; *Ratio of Specific Heats of Vapor (Gas)*: Not pertinent; *Latent Heat of Vaporization*: Not pertinent; *Heat of Combustion*: Not pertinent; *Heat of Decomposition*: Not pertinent.

Health Hazards Information - *Recommended Personal Protective Equipment*: Goggles or face shield; rubberized gloves; flame proof outer clothing; respirator; high boots or shoes; *Symptoms Following Exposure*: Inhalation of dust causes coughing, sneezing, and burning of nose and throat. Ingestion causes severe burns of mouth and stomach; symptoms of central nervous system damage may occur. Contact with eyes causes severe caustic burns; *General Treatment for Exposure*: Lithium hydride burns of the eyes, skin, or respiratory tract appear to be worse than those caused by an equivalent amount of sodium hydroxide. INHALATION: remove victim to fresh air; if irritation persists get medical attention at once. INGESTION: give large volumes of water and milk; gastric lavage may be indicated. EYES: flush with copious quantities of running water for at least 15 min; get medical attention. SKIN: flush with water; treat as a caustic burn; *Toxicity by Inhalation (Threshold Limit Value)*: 0.025 mg/m^3; *Short-Term Inhalation Limits*: Data not available; *Toxicity by Ingestion*: Data not available; *Late Toxicity*: Data not available; *Vapor (Gas) Irritant Characteristics*: Data not available; *Liquid or Solid Irritant Characteristics*: Data not available; *Odor Threshold*: Data not available.

Fire Hazards - *Flash Point* : Not pertinent (combustible solid); *Flammable Limits in Air (%):* Not pertinent; *Fire Extinguishing Agents:* Dry nitrogen, graphite, or lithium chloride; *Fire Extinguishing Agents Not To Be Used:* Never use water, foam, halogenated hydrocarbons, soda acid, dry chemical, or carbon dioxide; *Special Hazards of Combustion Products:* Irritating alkali fumes may form in fire; *Behavior in Fire:* May decompose when hot to form flammable hydrogen gas. Reacts violently with water to produce hydrogen, which may explode in air; *Ignition Temperature (deg. F):* 392; *Electrical*

Hazard: Not pertinent; *Burning Rate:* Not pertinent.

Chemical Reactivity - *Reactivity with Water:* Reacts violently with water to produce flammable hydrogen gas and strong caustic solution; ignition may occur, especially with powder; *Reactivity with Common Materials:* May ignite combustible materials if they are damp; *Stability During Transport:* Stable, if air and moisture are excluded; *Neutralizing Agents for Acids and Caustics:* Residues should be washed well with water, then rinsed with dilute acetic acid; *Polymerization:* Not pertinent; *Inhibitor of Polymerization:* Not pertinent.

LITHIUM, METALLIC
Chemical Designations - *Synonyms*: No common synonyms; *Chemical Formula*: Li.
Observable Characteristics - *Physical State (as shipped)*: Solid; *Color*: White; light silvery; *Odor*: None.
Physical and Chemical Properties - *Physical State at 15 °C and 1 atm.*: Solid; *Molecular Weight*: 6.939; *Boiling Point at 1 atm.*: Not pertinent; *Freezing Point*: Not pertinent; *Critical Temperature*: Not pertinent; *Critical Pressure*: Not pertinent; *Specific Gravity*: 0.53 at 20 °C (solid); *Vapor (Gas) Specific Gravity*: Not pertinent; *Ratio of Specific Heats of Vapor (Gas)*: Not pertinent; *Latent Heat of Vaporization*: Not pertinent; *Heat of Combustion*: -18,470, -10,260, -429.3; *Heat of Decomposition*: Not pertinent.
Health Hazards Information - *Recommended Personal Protective Equipment*: Rubber or plastic gloves; face shield; respirator; fire-retardant clothing; *Symptoms Following Exposure*: Contact with eyes causes caustic irritation or burn. In contact with skin lithium react with body moisture to cause chemical burns; foil, ribbon, and wire react relatively slowly; *General Treatment for Exposure*: EYES or SKIN: flush with water and treat with boric acid; *Toxicity by Inhalation (Threshold Limit Value)*: Data not available; *Short-Term Inhalation Limits*: Data not available; *Toxicity by Ingestion*: Data not available; *Late Toxicity*: Data not available; *Vapor (Gas) Irritant Characteristics*: Data not available; *Liquid or Solid Irritant Characteristics*: Data not available; *Odor Threshold*: Data not available.
Fire Hazards - *Flash Point* : Not pertinent (combustible solid); *Flammable Limits in Air (%):* Not pertinent; *Fire Extinguishing Agents:* Graphite, lithium chloride; *Fire Extinguishing Agents Not To Be Used:* Water, sand, halogenated hydrocarbons, carbon dioxide, soda-acid , or dry chemical; *Special Hazards of Combustion Products:* Strong alkali fumes are formed in fire; *Behavior in Fire:* Molten lithium is quite easily ignited and is then difficult to extinguish. Hot or burning lithium will react with all gases except those of the helium-argon group. It also reacts violently with concrete, wood, asphalt, sand, asbestos, and in fact, nearly everything except metal. Do not apply water to adjacent fires. Hydrogen explosion may result; *Ignition Temperature* : No data; *Electrical Hazard:* No data; *Burning Rate:* No data.
Chemical Reactivity - *Reactivity with Water:* Reacts violently to form flammable hydrogen gas and strong caustic solution. Ignition usually occurs; *Reactivity with Common Materials:* May ignite combustible materials if they are damp; *Stability During Transport:* Stable, if air and moisture are excluded; *Neutralizing Agents for Acids and Caustics:* Residues should be flushed with water, then rinsed with dilute acetic acid; *Polymerization:* Not pertinent; *Inhibitor of Polymerization:* Not pertinent.

M

MAGNESIUM
Chemical Designations - *Synonyms*: No common synonyms; *Chemical Formula*: Mg.
Observable Characteristics — *Physical State (as shipped)*: Solid; *Color*: Silvery; looks like aluminum; *Odor*: None.
Physical and Chemical Properties - *Physical State at 15 °C and 1 atm.*: Solid; *Molecular Weight*: 24.3; *Boiling Point at 1 atm.*: 2,012, 1,100, 1,373; *Freezing Point*: 1,202, 650, 923; *Critical*

Temperature: Not pertinent; *Critical Pressure*: Not pertinent; *Specific Gravity*: 1.74 at 20 °C (solid); *Vapor (Gas) Specific Gravity*: Not pertinent; *Ratio of Specific Heats of Vapor (Gas)*: Not pertinent; *Latent Heat of Vaporization*: Not pertinent; *Heat of Combustion*: -11,950, -6,650, -278; *Heat of Decomposition*: Not pertinent.

Health Hazards Information - *Recommended Personal Protective Equipment*: Eye protection; *Symptoms Following Exposure*: Dust irritates eyes in same way as any foreign material. Penetration of skin by fragments of metal is likely to produce local irritation, blisters, and ulcers which may become infected; *General Treatment for Exposure*: EYES: flush with water to remove dust. SKIN: treat as any puncture; *Toxicity by Inhalation (Threshold Limit Value)*: Data not available; *Short-Term Inhalation Limits*: Not pertinent; *Toxicity by Ingestion*: Oral LDL_0 (lowest lethal dose) = 230 mg/kg (dog); *Late Toxicity*: Data not available; *Vapor (Gas) Irritant Characteristics*: Not pertinent; *Liquid or Solid Irritant Characteristics*: Data not available; *Odor Threshold*: Not pertinent.

Fire Hazards - *Flash Point* : Not pertinent (solid). Flammable when in the form of turnings or powder; *Flammable Limits in Air (%)*: Not pertinent; *Fire Extinguishing Agents:* Inert dry powders (e.g. graphite, limestone, salt); *Fire Extinguishing Agents Not To Be Used:* Water, foam, halogenated agents, carbon dioxide; *Special Hazards of Combustion Products:* Not pertinent; *Behavior in Fire:* Forms dense white smoke. Flame is very bright; *Ignition Temperature (deg. F):* 883; *Electrical Hazard:* Class I, Group E; *Burning Rate:* Not pertinent.

Chemical Reactivity - *Reactivity with Water:* In finely divided form, reacts with water and acids to release flammable hydrogen gas; *Reactivity with Common Materials:* No reaction; *Stability During Transport:* Stable; *Neutralizing Agents for Acids and Caustics:* Not pertinent; *Polymerization:* Not pertinent; *Inhibitor of Polymerization:* Not pertinent.

MAGNESIUM PERCHLORATE

Chemical Designations - *Synonyms*: Anhydrone; Dehydrite; Magnesium perchlorate, anhydrous; Magnesium perchlorate, hexahydrate; *Chemical Formula*: $Mg(ClO_4)_2$.

Observable Characteristics - *Physical State (as shipped)*: Solid; *Color*: White; *Odor*: None.

Physical and Chemical Properties - *Physical State at 15 °C and 1 atm.*: Solid; *Molecular Weight*: 223.2; *Boiling Point at 1 atm.*: Decomposes above 250°C; *Freezing Point*: Not pertinent; *Critical Temperature*: Not pertinent; *Critical Pressure*: Not pertinent; *Specific Gravity*: 2.21 at 20 °C (solid); *Vapor (Gas) Specific Gravity*: Not pertinent; *Ratio of Specific Heats of Vapor (Gas)*: Not pertinent; *Latent Heat of Vaporization*: Not pertinent; *Heat of Combustion*: Not pertinent; *Heat of Decomposition*: Data not available.

Health Hazards Information - *Recommended Personal Protective Equipment*: U.S. Bu. Mines approved respirator; chemical safety goggles; face shield; *Symptoms Following Exposure*: Inhalation of dust irritation of mucous membrane. Ingestion of large amounts may be fatal; immediate symptoms include abdominal pains, nausea and vomiting, diarrhea, pallor, blueness, shortness of breath, unconsciousness. Contact with eyes or skin causes irritation; *General Treatment for Exposure*: INHALATION: remove victim to fresh air; get medical attention if irritation persists. INGESTION: give large amounts of water; induce vomiting; call a physician. EYES: flush with copious quantities of water for 15 min.; call a physician. SKIN: flush with water; *Toxicity by Inhalation (Threshold Limit Value)*: Data not available; *Short-Term Inhalation Limits*: Data not available; *Toxicity by Ingestion*: Data not available; *Late Toxicity*: Data not available; *Vapor (Gas) Irritant Characteristics*: Data not available; *Liquid or Solid Irritant Characteristics*: Data not available; *Odor Threshold*: Data not available.

Fire Hazards - *Flash Point* : Not flammable, but may cause or increase the intensity of a fire; *Flammable Limits in Air (%):* Not flammable; *Fire Extinguishing Agents:* Not pertinent; *Fire Extinguishing Agents Not To Be Used:* Not pertinent; *Special Hazards of Combustion Products:* Not pertinent; *Behavior in Fire:* Can form explosive mixture with combustible material or finely powdered metals. Increases the intensity of fires; *Ignition Temperature :* Not pertinent; *Electrical Hazard:* Not pertinent; *Burning Rate:* Not pertinent.

Chemical Reactivity - *Reactivity with Water:* Dissolves with liberation of heat. May cause spattering;

Reactivity with Common Materials: Contact with wood, paper, oils, grease, or finely divided metals may cause fires and explosions; *Stability During Transport:* Stable; *Neutralizing Agents for Acids and Caustics:* Not pertinent; *Polymerization:* Not pertinent; *Inhibitor of Polymerization:* Not pertinent.

MALATHION

Chemical Designations - *Synonyms*: CYTHION Insecticide; S-[1,2-Bis(ethoxycarbonyl) ethyl]; 0,0-dimethyl phosphorodithioate; *Chemical Formula*: $C_{10}H_{19}O_6PS_2$.

Observable Characteristics - *Physical State (as shipped)*: Liquid; *Color*: Yellow to dark brown; *Odor*: Characteristic skunk-like mercaptan.

Physical and Chemical Properties - *Physical State at 15 °C and 1 atm.*: Liquid; *Molecular Weight*: 330.36; *Boiling Point at 1 atm.*: Very high; *Freezing Point*: 37, 2.9, 276; *Critical Temperature*: Not pertinent; *Critical Pressure*: Not pertinent; *Specific Gravity*: 1.234 at 20 °C (liquid); *Vapor (Gas) Specific Gravity*: Not pertinent; *Ratio of Specific Heats of Vapor (Gas)*: Not pertinent; *Latent Heat of Vaporization*: Not pertinent; *Heat of Combustion*: Data not available; *Heat of Decomposition*: Not pertinent.

Health Hazards Information - *Recommended Personal Protective Equipment*: Wear self-contained breathing apparatus (or respirator for organophosphate pesticides) and rubber clothing while fighting fires of malathion with chlorine bleach solution. All clothing contaminated by fumes and vapors must be decontaminated; *Symptoms Following Exposure*: Exposure to fumes from a fire or to liquid causes headache, blurred vision, constricted pupils of the eyes, weakness, nausea, cramps, diarrhea, and tightness in the chest. Muscles twitch and convulsions may follow. The symptoms may develop over a period of 8 hours; *General Treatment for Exposure*: Speed is essential. INHALATION: in the non-breathing victim immediately institute artificial respiration, using the mouth-to-mouth, the mouth-to-nose, or the mouth-to-oropharyngeal method. Call physician! INGESTION: administer milk, water or salt-water and induce vomiting repeatedly. SKIN OR EYE CONTACT: flood and wash exposed skin areas thoroughly with water. Remove contaminated clothing under a shower. Administer atropine, 2 mg (1/30 gr) intramuscularly or intravenously as soon as any local or systemic signs or symptoms of an intoxication are noted; repeat the administration of atropine every 3-8 min. until signs of atropinization (mydriasis, dry mouth, rapid pulse, hot and dry skin) occur; initiate treatment in children with 1 mg of atropine. Watch respiration, and remove bronchial secretion if they appear to be obstructing the airway; incubate if necessary. Give 2-PAM (Pralidoxime; Protopan), 2.5 gm in 100 ml of sterile water or in 5% dextrose and water, intravenously, slowly, in 15-30 min.; if sufficient fluid is not available, give 1 gm of 2-PAM in 3 ml of distilled water by deep intramuscular injection; repeat this every half hour if respiration weakens or if muscle fasciculation or convulsions recur; *Toxicity by Inhalation (Threshold Limit Value)*: 10 mg/m^3; *Short-Term Inhalation Limits*: Data not available; *Toxicity by Ingestion*: Grade 2, LD_{50} 0.5 to 5 g/kg (rat); *Late Toxicity*: Data not available; *Vapor (Gas) Irritant Characteristics*: None likely; *Liquid or Solid Irritant Characteristics*: Minimum hazard. If spilled on clothing and allowed to remain, may cause smarting and reddening of skin; *Odor Threshold*: Data not available.

Fire Hazards - *Flash Point (deg. F)*: > 325; *Flammable Limits in Air (%):* Data not available; *Fire Extinguishing Agents:* Dry chemical, carbon dioxide, foam, water spray; *Fire Extinguishing Agents Not To Be Used:* Not pertinent; *Special Hazards of Combustion Products:* Vapors and fumes from fires are hazardous. They include sulfur dioxide and phosphoric acid; *Behavior in Fire:* Gives off hazardous fumes. Area surrounding fire should be diked to prevent water runoff; *Ignition Temperature :* Data not available; *Electrical Hazard:* Not pertinent; *Burning Rate:* Data not available.

Chemical Reactivity - *Reactivity with Water:* None; *Reactivity with Common Materials:* No hazardous reaction; *Stability During Transport:* Not pertinent; *Neutralizing Agents for Acids and Caustics:* Liquid bleach solution for decontamination; *Polymerization:* Not pertinent; *Inhibitor of Polymerization:* Not pertinent.

MALEIC ACID

Chemical Designations - *Synonyms*: cis-Butenedioic acid; cis-1,2-Ethylenedicarboxylic acid; Maleinic acid; Malenic acid; Toxilic acid; *Chemical Formula*: HOOC-CH=CH-COOH.

Observable Characteristics - *Physical State (as shipped)*: Solid; *Color*: White; *Odor*: None.

Physical and Chemical Properties - *Physical State at 15 °C and 1 atm.*: Solid; *Molecular Weight*: 116.1; *Boiling Point at 1 atm.*: Not pertinent (decomposes); *Freezing Point*: Not pertinent; *Critical Temperature*: Not pertinent; *Critical Pressure*: Not pertinent; *Specific Gravity*: 1.59 at 20°C (solid); *Vapor (Gas) Specific Gravity*: Not pertinent; *Ratio of Specific Heats of Vapor (Gas)*: Not pertinent; *Latent Heat of Vaporization*: Not pertinent; *Heat of Combustion*: -5,000, -2,800, -117; *Heat of Decomposition*: Not pertinent.

Health Hazards Information - *Recommended Personal Protective Equipment*: Dust mask; goggles or face shield; protective gloves; *Symptoms Following Exposure*: Inhalation causes irritation of nose and throat. Contact with eyes or skin causes irritation; *General Treatment for Exposure*: INHALATION: remove to fresh air. EYES: immediately flush with plenty of water for at least 15 min.; get medical attention if irritation persist. SKIN: wash with soap and water; *Toxicity by Inhalation (Threshold Limit Value)*: Data not available; *Short-Term Inhalation Limits*: Data not available; *Toxicity by Ingestion*: Grade 2; oral LD_{50} = 708 mg/kg (rat); *Late Toxicity*: Data not available; *Vapor (Gas) Irritant Characteristics*: Data not available; *Liquid or Solid Irritant Characteristics*: Data not available; *Odor Threshold*: Data not available.

Fire Hazards - *Flash Point* : Not pertinent (combustible solid); *Flammable Limits in Air (%):* Not pertinent; *Fire Extinguishing Agents:* Water, foam, carbon dioxide, dry chemical; *Fire Extinguishing Agents Not To Be Used:* Not pertinent; *Special Hazards of Combustion Products:* Irritating smoke containing maleic anhydride may form in fire; *Behavior in Fire:* Not pertinent; *Ignition Temperature :* Data not available; *Electrical Hazard:* Not pertinent; *Burning Rate:* Not pertinent.

Chemical Reactivity - *Reactivity with Water:* No reaction; *Reactivity with Common Materials:* May corrode metals when wet; *Stability During Transport:* Stable; *Neutralizing Agents for Acids and Caustics:* Flush with water, rinse with dilute solution of sodium bicarbonate or soda ash; *Polymerization:* Not pertinent; *Inhibitor of Polymerization:* Not pertinent.

MALEIC ANHYDRIDE

Chemical Designations - *Synonyms*: cis-Butenedioic anhydride; 2,5-Furanedione; Toxilic anhydride; *Chemical Formula*: OCOCH=CHCO.

Observable Characteristics - *Physical State (as shipped)*: Solid or liquid; *Color*: Colorless; *Odor*: Acrid; choking.

Physical and Chemical Properties - *Physical State at 15 °C and 1 atm.*: Solid; *Molecular Weight*: 98.06; *Boiling Point at 1 atm.*: 392, 200, 473; *Freezing Point*: 127, 53, 326; *Critical Temperature*: Not pertinent; *Critical Pressure*: Not pertinent; *Specific Gravity*: 1.43 at 15 °C (solid); *Vapor (Gas) Specific Gravity*: Not pertinent; *Ratio of Specific Heats of Vapor (Gas)*: Not pertinent; *Latent Heat of Vaporization*: Not pertinent; *Heat of Combustion*: -5936, -3298, -138.1; *Heat of Decomposition*: Not pertinent.

Health Hazards Information - *Recommended Personal Protective Equipment*: Approved organic vapor-acid gas canister; chemical goggles and face shield; rubber gloves and boots; coveralls or rubber apron; *Symptoms Following Exposure*: Inhalation causes coughing, sneezing, throat irritation. Skin contact causes irritation and redness. Vapors cause severe eye irritation; photophobia and double vision may occur; *General Treatment for Exposure*: INHALATION: give oxygen. EYE OR SKIN CONTACT: flush with lots of water for at least 15 min.; for eyes, call a physician. For molten maleic burns, remove crust and treat as chemical and thermal burn; *Toxicity by Inhalation (Threshold Limit Value)*: 0.25 ppm; *Short-Term Inhalation Limits*: Data not available; *Toxicity by Ingestion*: Grade 2, LD_{50} 0.5 to 5 g/kg; *Late Toxicity*: None; *Vapor (Gas) Irritant Characteristics*: Vapors cause moderate irritation such that personnel will find high concentrations unpleasant. The effect is temporary; *Liquid or Solid Irritant Characteristics*: Causes smarting of the skin and first-degree burns on short exposure; may cause secondary burns on long exposure; *Odor Threshold*: 1.3-2.0 mg/m^3.

Fire Hazards - *Flash Point (deg. F)*: (liquid) 215 CC; 230 OC; *Flammable Limits in Air (%)*: 1.4-7.1; *Fire Extinguishing Agents:* Alcohol foam, dry chemical, or carbon dioxide; *Fire Extinguishing Agents Not To Be Used:* Water or foam may cause frothing; *Special Hazards of Combustion Products:* Not pertinent; *Behavior in Fire:* When heated above 300°F in the presence of various materials may generate heat and carbon dioxide. Will explode if confined; *Ignition Temperature (deg. F):* 878; *Electrical Hazard:* Class I, Group D; *Burning Rate:* 1.4 mm/min.

Chemical Reactivity - *Reactivity with Water:* Hot water may cause frothing. Reaction with cold water is slow and non-hazardous; *Reactivity with Common Materials:* No reaction; *Stability During Transport:* Stable; *Neutralizing Agents for Acids and Caustics:* Solid spills can usually be recovered before any significant reaction with water occurs. Flush area of spill with water; *Polymerization:* Very unlikely at ordinary temperatures, even in the molten state; *Inhibitor of Polymerization:* None.

MALEIC HYDRAZIDE

Chemical Designations - *Synonyms:* 1,2-Dihydro-3,6-pyridazinedione; 6-Hydroxy-3-(2H)-pyridazinone; Maleic acid hydrazide; Malazide; MH; Regulox; *Chemical Formula:* $C_4H_4N_2O_2$.

Observable Characteristics - *Physical State (as shipped):* Solid; *Color:* White; *Odor:* None.

Physical and Chemical Properties - *Physical State at 15 °C and 1 atm.*: Solid; *Molecular Weight:* 112.1; *Boiling Point at 1 atm.*: Not pertinent (decomposes); *Freezing Point:* 558, 292, 565; *Critical Temperature:* Not pertinent; *Critical Pressure:* Not pertinent; *Specific Gravity:* 1.60 at 25 °C (solid); *Vapor (Gas) Specific Gravity:* Not pertinent; *Ratio of Specific Heats of Vapor (Gas):* Not pertinent; *Latent Heat of Vaporization:* Not pertinent; *Heat of Combustion:* (est.) -8,200, -4,500, -190; *Heat of Decomposition:* Not pertinent.

Health Hazards Information - *Recommended Personal Protective Equipment:* Goggles or face shield; dust mask; *Symptoms Following Exposure:* Inhalation of dust may causes irritation of nose and throat. Contact with eyes or skin causes irritation. Ingestion has been observed to cause tremors and muscle spasms in test animals; *General Treatment for Exposure:* INHALATION: move to fresh air. EYES: flush with water for at least 15 min. SKIN: flush with water; wash with soap and water. INGESTION: get medical attention; *Toxicity by Inhalation (Threshold Limit Value):* Data not available; *Short-Term Inhalation Limits:* Data not available; *Toxicity by Ingestion:* Grade 2; oral LD_{50} = 3,800 mg/kg (rat); *Late Toxicity:* Causes cancer in rats; *Vapor (Gas) Irritant Characteristics:* Data not available; *Liquid or Solid Irritant Characteristics:* Data not available; *Odor Threshold:* Not pertinent.

Fire Hazards - *Flash Point* : Not pertinent (combustible solid); *Flammable Limits in Air (%):* Not pertinent; *Fire Extinguishing Agents:* Water, dry chemical, carbon dioxide, foam; *Fire Extinguishing Agents Not To Be Used:* Not pertinent; *Special Hazards of Combustion Products:* Toxic nitrogen oxides are produced; *Behavior in Fire:* Not pertinent; *Ignition Temperature :* Not pertinent; *Electrical Hazard:* Not pertinent; *Burning Rate:* Not pertinent.

Chemical Reactivity - *Reactivity with Water:* No reaction; *Reactivity with Common Materials:* No reaction; *Stability During Transport:* Stable; *Neutralizing Agents for Acids and Caustics:* Not pertinent; *Polymerization:* Not pertinent; *Inhibitor of Polymerization:* Not pertinent.

MERCURIC ACETATE

Chemical Designations - *Synonyms:* No common synonyms; *Chemical Formula:* $(CH_3COO)_2Hg$.

Observable Characteristics - *Physical State (as shipped):* Solid; *Color:* White; *Odor:* Slight acetic.

Physical and Chemical Properties - *Physical State at 15 °C and 1 atm.*: Solid; *Molecular Weight:* 318.7; *Boiling Point at 1 atm.*: Not pertinent (decomposes); *Freezing Point:* Not pertinent; *Critical Temperature:* Not pertinent; *Critical Pressure:* Not pertinent; *Specific Gravity:* 3.27 at 20 °C (solid); *Vapor (Gas) Specific Gravity:* Not pertinent; *Ratio of Specific Heats of Vapor (Gas):* Not pertinent; *Latent Heat of Vaporization:* Not pertinent; *Heat of Combustion:* Not pertinent; *Heat of Decomposition:* Not pertinent.

Health Hazards Information - *Recommended Personal Protective Equipment:* Rubber gloves, dust mask, goggles; *Symptoms Following Exposure:* The general symptoms are those of mercury poisoning, developing rapidly after ingestion but more slowly after low repeated exposures. Contact with eyes causes irritation and ulceration. Skin contact may cause dermatitis. Ingestion causes pain, vomiting, ulceration of mouth and stomach, kidney failure, metallic taste, pallor, and rapid, weak pulse; *General Treatment for Exposure*: Have physician treat for mercury poisoning. EYES or SKIN: flush with water. INGESTION: call physician; poison should be removed from stomach as soon as possible; give milk or white of eggs beaten with water, than tablespoon of salt in a glass of warm water and repeat until vomit fluid is clear; repeat milk or white of eggs beaten with water; *Toxicity by Inhalation (Threshold Limit Value)*: 0.05 mg/m^3 (as mercury); *Short-Term Inhalation Limits*: No data; *Toxicity by Ingestion*: Grade 3; oral LD_{50} = 76 mg/kg (rat); *Late Toxicity*: Intestinal bleeding and kidney damage may develop; *Vapor (Gas) Irritant Characteristics*: Not pertinent; *Liquid or Solid Irritant Characteristics*: No data; *Odor Threshold*: Not pertinent.

Fire Hazards - *Flash Point* : Not flammable; *Flammable Limits in Air (%)*: Not flammable; *Fire Extinguishing Agents:* Not pertinent; *Fire Extinguishing Agents Not To Be Used:* Not pertinent; *Special Hazards of Combustion Products:* Smoke may contain toxic mercury or mercury oxide fumes; *Behavior in Fire:* Not pertinent; *Ignition Temperature :* Not pertinent; *Electrical Hazard:* Not pertinent; *Burning Rate:* Not pertinent.

Chemical Reactivity - *Reactivity with Water:* No reaction; *Reactivity with Common Materials:* No reaction; *Stability During Transport:* Stable; *Neutralizing Agents for Acids and Caustics:* Not pertinent; *Polymerization:* Not pertinent; *Inhibitor of Polymerization:* Not pertinent.

MERCURIC AMMONIUM CHLORIDE

Chemical Designations - *Synonyms*: Albus; Aminomercuric chloride; Ammoniated mercury; Mercuric chloride, ammoniated; Mercury ammonium chloride; Mercury (II) chloride ammonobasic; *Chemical Formula*: $HgNH_2Cl$.

Observable Characteristics - *Physical State (as shipped)*: Solid; *Color*: White; *Odor*: None.

Physical and Chemical Properties - *Physical State at 15 °C and 1 atm.*: Solid; *Molecular Weight*: 252.1; *Boiling Point at 1 atm.*: Not pertinent (sublimes at red heat); *Freezing Point*: Not pertinent (infusible); *Critical Temperature*: Not pertinent; *Critical Pressure*: Not pertinent; *Specific Gravity*: 5.7 at 20 °C (solid); *Vapor (Gas) Specific Gravity*: Not pertinent; *Ratio of Specific Heats of Vapor (Gas)*: Not pertinent; *Latent Heat of Vaporization*: Not pertinent; *Heat of Combustion*: Not pertinent; *Heat of Decomposition*: Not pertinent.

Health Hazards Information - *Recommended Personal Protective Equipment*: Gloves, goggles, respirator; *Symptoms Following Exposure*: The general symptoms are those of mercury poisoning, developing rapidly after ingestion but more slowly after low repeated exposures. Contact with eyes causes irritation and ulceration. Skin contact may cause dermatitis. Ingestion causes pain, vomiting, ulceration of mouth and stomach, kidney failure, metallic taste, pallor, and rapid, weak pulse; *General Treatment for Exposure*: Have physician treat for mercury poisoning. EYES or SKIN: flush with water. INGESTION: call physician; poison should be removed from stomach as soon as possible; give milk or white of eggs beaten with water, than tablespoon of salt in a glass of warm water and repeat until vomit fluid is clear; repeat milk or white of eggs; *Toxicity by Inhalation (Threshold Limit Value)*: 0.05 mg/m^3 (as mercury); *Short-Term Inhalation Limits*: No data; *Toxicity by Ingestion*: Grade 3; oral LD_{50} = 76 mg/kg (rat); *Late Toxicity*: Intestinal bleeding and kidney damage may develop; *Vapor (Gas) Irritant Characteristics*: Data not available; *Liquid or Solid Irritant Characteristics*: No data; *Odor Threshold*: Not pertinent.

Fire Hazards - *Flash Point* : Not flammable; *Flammable Limits in Air (%)*: Not flammable; *Fire Extinguishing Agents:* Not pertinent; *Fire Extinguishing Agents Not To Be Used:* Not pertinent; *Special Hazards of Combustion Products:* Smoke may contain toxic mercury compounds; *Behavior in Fire:* Not pertinent; *Ignition Temperature :* Not pertinent; *Electrical Hazard:* Not pertinent; *Burning Rate:* Not pertinent.

Chemical Reactivity - *Reactivity with Water:* No reaction; *Reactivity with Common Materials:* No reaction; *Stability During Transport:* Stable; *Neutralizing Agents for Acids and Caustics:* Not pertinent; *Polymerization:* Not pertinent; *Inhibitor of Polymerization:* Not pertinent.

MERCURIC CHLORIDE

Chemical Designations - *Synonyms*: Calochlor; Corrosive mercury chloride; Corrosive sublimate; Mercury bichloride; Mercury (II) chloride; Mercury perchloride; *Chemical Formula*: $HgCl_2$.

Observable Characteristics - *Physical State (as shipped)*: Solid; *Color*: White; colorless; *Odor*: None.

Physical and Chemical Properties - *Physical State at 15 °C and 1 atm.*: Solid; *Molecular Weight*: 271.50; *Boiling Point at 1 atm.*: 576, 302, 575; *Freezing Point*: 531, 277, 550; *Critical Temperature*: Not pertinent; *Critical Pressure*: Not pertinent; *Specific Gravity*: 5.4 at 20 °C (solid); *Vapor (Gas) Specific Gravity*: Not pertinent; *Ratio of Specific Heats of Vapor (Gas)*: Not pertinent; *Latent Heat of Vaporization*: Not pertinent; *Heat of Combustion*: Not pertinent; *Heat of Decomposition*: Not pertinent.

Health Hazards Information - *Recommended Personal Protective Equipment*: Bu. Mines approved airline respirator; impervious suit; appropriate eye protection; *Symptoms Following Exposure*: All forms of exposure to this compound are hazardous; acute systemic mercurialism may be fatal within a few minutes, but death by uremic poisoning is usually delayed 5-12 days. Acute poisoning has resulted from inhaling dust concentrations of 1.2-8.5 mg/m^3 of air; symptoms include tightness and pain in chest, coughing, and difficulty in breathing. Ingestion causes necrosis, pain, vomiting, and severe purging; as little as 0.5 gm cam be fatal. Contact with eyes causes ulceration of conjunctiva and cornea. Contact with skin causes irritation and possible dermatitis; systemic poisoning can occur by absorption through skin; *General Treatment for Exposure*: Act promptly! Alimentary absorption is very rapid, and first 10-15 minutes determine the prognosis. INHALATION: remove victim to fresh air; get medical attention. INGESTION: give egg whites, milk, or activated charcoal; induce vomiting; consult physician. EYES or SKIN: wash with water for 15 min.; *Toxicity by Inhalation (Threshold Limit Value)*: 0.05 mg/m^3 (as mercury); *Short-Term Inhalation Limits*: Data not available; *Toxicity by Ingestion*: Grade 4, oral LD_{50} = 1 mg/kg (rat); *Late Toxicity*: Data not available; *Vapor (Gas) Irritant Characteristics*: Data not available; *Liquid or Solid Irritant Characteristics*: Data not available; *Odor Threshold*: Data not available.

Fire Hazards - *Flash Point* : Not flammable; *Flammable Limits in Air (%)*: Not flammable; *Fire Extinguishing Agents*: Not pertinent; *Fire Extinguishing Agents Not To Be Used*: Not pertinent; *Special Hazards of Combustion Products*: Heat of fire may cause material to form fumes of mercuric chloride, which are toxic; *Behavior in Fire*: Not pertinent; *Ignition Temperature* : Not pertinent; *Electrical Hazard*: Not pertinent; *Burning Rate*: Not pertinent.

Chemical Reactivity - *Reactivity with Water*: No reaction; *Reactivity with Common Materials*: No reaction; *Stability During Transport*: Stable; *Neutralizing Agents for Acids and Caustics*: Not pertinent; *Polymerization*: Not pertinent; *Inhibitor of Polymerization*: Not pertinent.

MERCURIC CYANIDE

Chemical Designations - *Synonyms*: Cianurina; Mercury cyanide; Mercury (II) cyanide; *Chemical Formula*: $Hg(CN)_2$.

Observable Characteristics - *Physical State (as shipped)*: Solid; *Color*: White or colorless; *Odor*: None.

Physical and Chemical Properties - *Physical State at 15 °C and 1 atm.*: Solid; *Molecular Weight*: 252.63; *Boiling Point at 1 atm.*: Not pertinent (decomposes); *Freezing Point*: Not pertinent; *Critical Temperature*: Not pertinent; *Critical Pressure*: Not pertinent; *Specific Gravity*: 4.0 at 20 °C (solid); *Vapor (Gas) Specific Gravity*: Not pertinent; *Ratio of Specific Heats of Vapor (Gas)*: Not pertinent; *Latent Heat of Vaporization*: Not pertinent; *Heat of Combustion*: Not pertinent; *Heat of Decomposition*: Not pertinent.

Health Hazards Information - *Recommended Personal Protective Equipment*: Dust mask, goggles or face shield, rubber gloves; *Symptoms Following Exposure*: Symptoms of both cyanide and mercury intoxication can occur. Acute poisoning has resulted from inhaling dust concentrations of 1.2-8.5 mg/m^3 of air; symptoms include tightness and pain in chest, coughing, and difficulty in breathing; cyanide poisoning can cause anxiety, confusion, dizziness, and shortness of breath, with possible unconsciousness, convulsions, and paralysis; breath may smell like bitter almonds. Ingestion causes necrosis, pain, vomiting, and severe purging, plus the above symptoms. Contact with eyes causes ulceration of conjunctiva and cornea. Contact with skin causes irritation and possible dermatitis; systemic poisoning can occur by absorption through skin; *General Treatment for Exposure*: Act quickly; call physician. INHALATION: if victim has stopped breathing, start artificial respiration immediately; using amyl nitrite pearls, administer amyl nitrite by inhalation for 15-30 seconds of every minute while sodium nitrite solution is being prepared; discontinue amyl nitrite and immediately inject intravenously 10 ml of a 3% solution of sodium nitrite (nonsterile if necessary) over a period of 2-4 min.; without removing needle, infuse intravenously 50 ml of a 25% aqueous solution of sodium thiosulphate; injection should take about 10 min. (concentrations of 5-50% may be used, but keep total dose approx. 12 gm). Oxygen therapy may be helpful in combination with the above. INGESTION: Alimentary absorption is very rapid, and first 10-15 minutes determine the prognosis. Give egg whites, milk, or activated charcoal; induce vomiting; treat for cyanide poisoning as above. EYES or SKIN: wash with water for 15 min.; *Toxicity by Inhalation (Threshold Limit Value)*: 0.05 mg/m^3 (as mercury); *Short-Term Inhalation Limits*: Data not available; *Toxicity by Ingestion*: Grade 4, oral LD_{50} = 25 mg/kg (rat); *Late Toxicity*: Data not available; *Vapor (Gas) Irritant Characteristics*: Data not available; *Liquid or Solid Irritant Characteristics*: Data not available; *Odor Threshold*: Odorless.

Fire Hazards - *Flash Point* : Not flammable; *Flammable Limits in Air (%):* Not flammable; *Fire Extinguishing Agents:* Not pertinent; *Fire Extinguishing Agents Not To Be Used:* Not pertinent; *Special Hazards of Combustion Products:* Fumes from fire may contain toxic mercury and hydrogen cyanide; *Behavior in Fire:* Not pertinent; *Ignition Temperature :* Not pertinent; *Electrical Hazard:* Not pertinent; *Burning Rate:* Not pertinent.

Chemical Reactivity - *Reactivity with Water:* No reaction; *Reactivity with Common Materials:* Contact with any acidic material will form poisonous hydrogen cyanide gas, which may collect in enclosed spaces; *Stability During Transport:* Stable; *Neutralizing Agents for Acids and Caustics:* Not pertinent; *Polymerization:* Not pertinent; *Inhibitor of Polymerization:* Not pertinent.

MERCURIC IODIDE

Chemical Designations - *Synonyms*: Mercuric iodide, red; Mercury biniodide; *Chemical Formula*: HgI_2.

Observable Characteristics - *Physical State (as shipped)*: Solid; *Color*: Red; *Odor*: None.

Physical and Chemical Properties - *Physical State at 15 °C and 1 atm.*: Solid; *Molecular Weight*: 454.90; *Boiling Point at 1 atm.*: 669, 354, 627; *Freezing Point*: 495, 257, 530; *Critical Temperature*: Not pertinent; *Critical Pressure*: Not pertinent; *Specific Gravity*: 6.3 at 20 °C (solid); *Vapor (Gas) Specific Gravity*: Not pertinent; *Ratio of Specific Heats of Vapor (Gas)*: Not pertinent; *Latent Heat of Vaporization*: Not pertinent; *Heat of Combustion*: Not pertinent; *Heat of Decomposition*: Not pertinent.

Health Hazards Information - *Recommended Personal Protective Equipment*: Dust mask, goggles or face shield, protective gloves; *Symptoms Following Exposure*: All forms of exposure to this compound are hazardous; acute systemic mercurialism may be fatal within a few minutes, but death by uremic poisoning is usually delayed 5-12 days. Acute poisoning has resulted from inhaling dust concentrations of 1.2-8.5 mg/m^3 of air; symptoms include tightness and pain in chest, coughing, and difficulty in breathing. Ingestion causes necrosis, pain, vomiting, and severe purging. Contact with eyes causes ulceration of conjunctiva and cornea. Contact with skin causes irritation and possible dermatitis; systemic poisoning can occur by absorption through skin; *General Treatment for Exposure*: INHALATION: remove victim to fresh air; get medical attention. INGESTION: Alimentary absorption is very rapid, and first 10-15 minutes determine the prognosis. Give egg whites, milk, or activated charcoal; induce vomiting; consult physician. EYES: wash with water for at least 15 min. SKIN: flush with water; wash with soap and water; *Toxicity by Inhalation (Threshold Limit Value)*: 0.05 mg/m^3 (as mercury); *Short-Term Inhalation Limits*: Data not available; *Toxicity by Ingestion*: Grade 4, oral LD_{50} = 40 mg/kg (rat); *Late Toxicity*: Data not available; *Vapor (Gas) Irritant Characteristics*: Data not available; *Liquid or Solid Irritant Characteristics*: Data not available; *Odor Threshold*: Odorless.

Fire Hazards - *Flash Point* : Not flammable; *Flammable Limits in Air (%):* Not flammable; *Fire Extinguishing Agents:* Not pertinent; *Fire Extinguishing Agents Not To Be Used:* Not pertinent; *Special Hazards of Combustion Products:* Fumes from fire may contain toxic mercury vapor; *Behavior in Fire:*

Not pertinent; *Ignition Temperature :* Not pertinent; *Electrical Hazard:* Not pertinent; *Burning Rate:* Not pertinent.
Chemical Reactivity - *Reactivity with Water:* No reaction; *Reactivity with Common Materials:* No reaction; *Stability During Transport:* Stable; *Neutralizing Agents for Acids and Caustics:* Not pertinent; *Polymerization:* Not pertinent; *Inhibitor of Polymerization:* Not pertinent.

MESITYL OXIDE
Chemical Designations - *Synonyms*: Isobutenyl methyl ketone; Isopropylideneacetone; Methyl isobutenyl ketone; 4-Methyl -3-pentene-2-one; *Chemical Formula*: $CH_3COCH=C(CH_3)_2$.
Observable Characteristics - *Physical State (as shipped)*: Liquid; *Color*: Colorless to slightly yellow; *Odor*: Strong; peppermint; honeylike.
Physical and Chemical Properties - *Physical State at 15 °C and 1 atm.*: Liquid; *Molecular Weight*: 98.2; *Boiling Point at 1 atm.*: 266, 130, 403; *Freezing Point*: -51, -46, 227; *Critical Temperature*: Not pertinent; *Critical Pressure*: Not pertinent; *Specific Gravity*: 0.853 at 20 °C (liquid); *Vapor (Gas) Specific Gravity*: 3.4; *Ratio of Specific Heats of Vapor (Gas)*: Not pertinent; *Latent Heat of Vaporization*: 157, 87, 3.7; *Heat of Combustion*: -14,400, -8,000, -330; *Heat of Decomposition*: Not pertinent.
Health Hazards Information - *Recommended Personal Protective Equipment*: Air pack or organic canister mask; rubber gloves; goggles; *Symptoms Following Exposure*: Inhalation causes irritation of nose and throat, dizziness, headache, difficult breathing. Contact with liquid or concentrated vapor causes severe eye irritation. Liquid irritates skin. Ingestion causes irritation of mouth and stomach; *General Treatment for Exposure*: INHALATION: remove victim to fresh air and restore breathing; call physician. EYES: immediately flush with plenty of water for at least 15 min. SKIN: wash with water. INGESTION: give large amount of water; call physician; *Toxicity by Inhalation (Threshold Limit Value)*: 25 ppm; *Short-Term Inhalation Limits*: 1,000 ppm for 60 min.; *Toxicity by Ingestion*: Grade 2; oral LD_{50} = 1,120 mg/kg (rat); *Late Toxicity*: Data not available; *Vapor (Gas) Irritant Characteristics*: Vapors cause moderate irritation such that personnel will find high concentrations unpleasant. The effect is temporary; *Liquid or Solid Irritant Characteristics*: Causes smarting of the skin and first-degree burns on short exposure; may cause secondary burns on long exposure; *Odor Threshold*: 12 ppm.
Fire Hazards - *Flash Point (deg. F)*: 84 OC; 73 CC; *Flammable Limits in Air (%):* Data not available; *Fire Extinguishing Agents:* Alcohol foam, dry chemical, carbon dioxide; *Fire Extinguishing Agents Not To Be Used:* Water may be ineffective; *Special Hazards of Combustion Products:* Not pertinent; *Behavior in Fire:* Vapor is heavier than air and may travel a considerable distance to a source of ignition and flash back; *Ignition Temperature (deg. F):* 652; *Electrical Hazard:* Data not available; *Burning Rate:* 4.2 mm/min.
Chemical Reactivity - *Reactivity with Water:* No reaction; *Reactivity with Common Materials:* No reaction; *Stability During Transport:* Stable; *Neutralizing Agents for Acids and Caustics:* Not pertinent; *Polymerization:* Not pertinent; *Inhibitor of Polymerization:* Not pertinent.

METHALLYL CHLORIDE
Chemical Designations - *Synonyms*: gamma-Chloroisobutylene; 3-Chloro-2-methylpropene; beta-Methallyl chloride; beta-Methylallyl chloride; *Chemical Formula*: $CH_2=C(CH_3)CH_2Cl$.
Observable Characteristics - *Physical State (as shipped)*: Liquid; *Color*: Colorless to straw; *Odor*: Pungent; penetrating.
Physical and Chemical Properties - *Physical State at 15 °C and 1 atm.*: Liquid; *Molecular Weight*: 90.55; *Boiling Point at 1 atm.*: 162.0, 72.2, 345.4; *Freezing Point*: <-112, <-80, <193; *Critical Temperature*: Not pertinent; *Critical Pressure*: Not pertinent; *Specific Gravity*: 0.928 at 20 °C (liquid); *Vapor (Gas) Specific Gravity*: 3.12; *Ratio of Specific Heats of Vapor (Gas)*: 1.0893; *Latent Heat of Vaporization*: 160, 89, 3.7; *Heat of Combustion*: (est.) -11,600, -6,500, -270; *Heat of Decomposition*: Not pertinent.
Health Hazards Information - *Recommended Personal Protective Equipment*: Organic canister mask;

goggles; rubber gloves; *Symptoms Following Exposure*: Inhalation causes irritation of nose and throat. Contact with vapor or liquid causes irritates eyes. Liquid irritates skin. Ingestion causes irritation of mouth and stomach; *General Treatment for Exposure*: INHALATION: remove victim to fresh air; give oxygen if breathing stops; give artificial respiration and oxygen; subsequent treatment is symptomatic and supportive. EYES: flush with water for at least 15 min.; get medical attention if exposure has been to liquid. SKIN: flush with water; get medical attention if skin is burned. INGESTION: induce vomiting and follow with gastric lavage, demulcents, and saline cathartics; get medical attention; *Toxicity by Inhalation (Threshold Limit Value)*: Data not available; *Short-Term Inhalation Limits*: Data not available; *Toxicity by Ingestion*: Data not available; *Late Toxicity*: Data not available; *Vapor (Gas) Irritant Characteristics*: Data not available; *Liquid or Solid Irritant Characteristics*: Data not available; *Odor Threshold*: Data not available.

Fire Hazards - *Flash Point (deg. F)*: 14 OC; *Flammable Limits in Air (%)*: 2.3 - 9.3; *Fire Extinguishing Agents:* Alcohol foam, dry chemical, carbon dioxide; *Fire Extinguishing Agents Not To Be Used:* Water may be ineffective; *Special Hazards of Combustion Products:* Irritating and toxic hydrogen chloride and phosgene vapors may be formed; *Behavior in Fire:* Vapor is heavier than air and may travel a considerable distance to a source of ignition and flash back; *Ignition Temperature :* Data not available; *Electrical Hazard:* Data not available; *Burning Rate:* 4.4 mm/min.

Chemical Reactivity - *Reactivity with Water:* No reaction; *Reactivity with Common Materials:* No reaction; *Stability During Transport:* Stable; *Neutralizing Agents for Acids and Caustics:* Not pertinent; *Polymerization:* Not pertinent; *Inhibitor of Polymerization:* Not pertinent.

METHANE

Chemical Designations - *Synonyms*: Marsh gas; *Chemical Formula*: CH_4.

Observable Characteristics - *Physical State (as shipped)*: Liquefied gas; *Color*: Colorless; *Odor*: Mild; sweet.

Physical and Chemical Properties - *Physical State at 15 °C and 1 atm.*: Gas; *Molecular Weight*: 16.04; *Boiling Point at 1 atm.*: -258.7, -161.5, 111.7; *Freezing Point*: -296.5, -182.5, 90.7; *Critical Temperature*: -116.5, -82.5, 190.7; *Critical Pressure*: 668, 45.44, 4.60; *Specific Gravity*: 0.422 at =160 °C (liquid); *Vapor (Gas) Specific Gravity*: 0.55 1.0; *Ratio of Specific Heats of Vapor (Gas)*: 1.306; *Latent Heat of Vaporization*: 219.4, 121.9, 5.100; *Heat of Combustion*: -21,517, -11,954, -500.2; *Heat of Decomposition*: Not pertinent.

Health Hazards Information - *Recommended Personal Protective Equipment*: Self-contained breathing apparatus for high concentrations; protective clothing if exposed to liquid; *Symptoms Following Exposure*: High concentrations may cause asphyxiation. No systemic effects, even at 5% concentration in air; *General Treatment for Exposure*: Remove to fresh air. Support respiration; *Toxicity by Inhalation (Threshold Limit Value)*: Not pertinent(methane is an asphyxiant, and limiting factor is available oxygen); *Short-Term Inhalation Limits*: Data not available; *Toxicity by Ingestion*: Not pertinent; *Late Toxicity*: None; *Vapor (Gas) Irritant Characteristics*: Vapors are nonirritating to the eyes and throat; *Liquid or Solid Irritant Characteristics*: No appreciable hazard. Practically harmless to the skin because it is very volatile and evaporates quickly, but may cause some frostbite; *Odor Threshold*: 200 ppm.

Fire Hazards - *Flash Point :* Flammable gas; *Flammable Limits in Air (%):* 5.0 - 15.0; *Fire Extinguishing Agents:* Stop flow of gas; *Fire Extinguishing Agents Not To Be Used:* Water; *Special Hazards of Combustion Products:* None; *Behavior in Fire:* Not pertinent; *Ignition Temperature (deg. F):* 1004; *Electrical Hazard:* Class I, Group D; *Burning Rate:* 12.5 mm/min.

Chemical Reactivity - *Reactivity with Water:* No reaction; *Reactivity with Common Materials:* No reaction; *Stability During Transport:* Stable; *Neutralizing Agents for Acids and Caustics:* Not pertinent; *Polymerization:* Not pertinent; *Inhibitor of Polymerization:* Not pertinent.

METHOXYCHLOR

Chemical Designations - *Synonyms*: 2,2-Bis(p-methoxyphenyl)-1,1,1-trichloroethane; 2,2-Di-(p-anisyl)-1,1,1- trichloroethane; DMDT; Marlate 50; Methoxy-DDT; *Chemical Formula*: $C_{16}H_{15}Cl_3O_2$.

Observable Characteristics - *Physical State (as shipped)*: Solid; *Color*: Light cream; white to light yellow; *Odor*: Slightly fruity.

Physical and Chemical Properties - *Physical State at 15 °C and 1 atm.*: Solid; *Molecular Weight*: 345.7; *Boiling Point at 1 atm.*: Not pertinent (decomposes); *Freezing Point*: 171 - 192, 77 - 89, 350 - 362; *Critical Temperature*: Not pertinent; *Critical Pressure*: Not pertinent; *Specific Gravity*: 1.41 at 25 °C (solid); *Vapor (Gas) Specific Gravity*: Not pertinent; *Ratio of Specific Heats of Vapor (Gas)*: Not pertinent; *Latent Heat of Vaporization*: Not pertinent; *Heat of Combustion*: Not pertinent; *Heat of Decomposition*: Not pertinent.

Health Hazards Information - *Recommended Personal Protective Equipment*: Dust respirator if needed; gloves and goggles; *Symptoms Following Exposure*: Toxicity is relatively low. Inhalation or ingestion causes generalized depression; *General Treatment for Exposure*: EYES: flush with water if irritated. SKIN: wash well with soap and water. INGESTION: consult physician; *Toxicity by Inhalation (Threshold Limit Value)*: 10 mg/m^3; *Short-Term Inhalation Limits*: Data not available; *Toxicity by Ingestion*: Grade 1; LD$_{50}$ 5 to 15 g/kg; *Late Toxicity*: Data not available; *Vapor (Gas) Irritant Characteristics*: Data not available; *Liquid or Solid Irritant Characteristics*: Data not available; *Odor Threshold*: Not pertinent.

Fire Hazards - *Flash Point* : Burns only at high temperatures. For liquid forms, see Kerosene; *Flammable Limits in Air (%):* Not pertinent; *Fire Extinguishing Agents:* Water, foam, dry chemical, carbon dioxide; *Fire Extinguishing Agents Not To Be Used:* Not pertinent; *Special Hazards of Combustion Products:* Irritating and toxic hydrogen chloride gas may be formed in fire; *Behavior in Fire:* Not pertinent; *Ignition Temperature :* Not pertinent; *Electrical Hazard:* Not pertinent; *Burning Rate:* Not pertinent.

Chemical Reactivity - *Reactivity with Water:* No reaction; *Reactivity with Common Materials:* No reaction; *Stability During Transport:* Stable; *Neutralizing Agents for Acids and Caustics:* Not pertinent; *Polymerization:* Not pertinent; *Inhibitor of Polymerization:* Not pertinent.

METHYL ACETATE

Chemical Designations - *Synonyms*: Acetic acid, methyl ester; *Chemical Formula*: CH_3COOCH_3.

Observable Characteristics - *Physical State (as shipped)*: Liquid; *Color*: Colorless; *Odor*: Slightly acrid, sweet; fragrant.

Physical and Chemical Properties - *Physical State at 15 °C and 1 atm.*: Liquid; *Molecular Weight*: 74.1; *Boiling Point at 1 atm.*: 134.6, 57.0, 330.2; *Freezing Point*: -145.3, -98.5, 174.7; *Critical Temperature*: 452.7, 233.7, 506.9; *Critical Pressure*: 666, 45.3, 4.60; *Specific Gravity*: 0.927 at 20 °C (liquid); *Vapor (Gas) Specific Gravity*: 2.8; *Ratio of Specific Heats of Vapor (Gas)*: 1.1192; *Latent Heat of Vaporization*: 174, 97, 4.1; *Heat of Combustion*: 9,260, 5,150, 215; *Heat of Decomposition*: Not pertinent.

Health Hazards Information - *Recommended Personal Protective Equipment*: Air mask or organic canister mask; goggles or face shield; *Symptoms Following Exposure*: (Very similar to those of methyl alcohol, which constitutes 20% of commercial grade.) Inhalation causes headache, fatigue, and drowsiness; high concentrations can produce central nervous system depression and optic nerve damage. Liquid irritates eyes and may cause defatting and cracking of skin. Ingestion causes headache, dizziness, drowsiness, fatigue; may cause severe eye damage; *General Treatment for Exposure*: INHALATION: remove victim from affected area; if breathing has ceased, apply artificial respiration; call doctor. EYES: irrigate thoroughly with clear water for 15 min. and call doctor. SKIN: wash affected area with water. INGESTION: get medical attention for methyl alcohol poisoning; *Toxicity by Inhalation (Threshold Limit Value)*: 200 ppm; *Short-Term Inhalation Limits*: 400 ppm for 5 min.; *Toxicity by Ingestion*: Grade 2; oral LD$_{50}$ = 3,700 mg/kg (rabbit); *Late Toxicity*: Optic nerve may be damaged following overexposure to vapor or liquid; *Vapor (Gas) Irritant Characteristics*: Vapors cause moderate irritation such that personnel will find high concentrations unpleasant. The effect is temporary; *Liquid or Solid Irritant Characteristics*: No appreciable hazard. Practically harmless to the skin; *Odor Threshold*: Data not available.

Fire Hazards - *Flash Point (deg. F)*: 22 OC; 14 CC; *Flammable Limits in Air (%):* 3.1 - 16; *Fire

Extinguishing Agents: Dry chemical, alcohol foam, carbon dioxide; *Fire Extinguishing Agents Not To Be Used:* Water may be ineffective; *Special Hazards of Combustion Products:* Not pertinent; *Behavior in Fire:* Vapor is heavier than fire and may travel a considerable distance to a source of ignition and flash back; *Ignition Temperature (deg. F):* 935; *Electrical Hazard:* Data not available; *Burning Rate:* 3.7 mm/min.

Chemical Reactivity - *Reactivity with Water:* Reacts slowly to form acetic acid and methyl alcohol; the reaction is not violent; *Reactivity with Common Materials:* No reaction; *Stability During Transport:* Stable; *Neutralizing Agents for Acids and Caustics:* Not pertinent; *Polymerization:* Not pertinent; *Inhibitor of Polymerization:* Not pertinent.

METHYL ACETYLENE-PROPADIENE MIXTURE

Chemical Designations - *Synonyms*: Allene-methylacetylene mixture; MAPP gas; Methylacetylene-allene mixture; Propadiene-methylacetylene mixture; *Chemical Formula*: $CH_3C=CH+CH_2=C=CH_2$.

Observable Characteristics - *Physical State (as shipped)*: Liquefied compressed gas; *Color*: Colorless; *Odor*: Offensive, like acetylene.

Physical and Chemical Properties - *Physical State at 15 °C and 1 atm.*: Gas; *Molecular Weight*: 40.1; *Boiling Point at 1 atm.*: -36 to -4, -38 to -20, 235 to 253; *Freezing Point*: Not pertinent; *Critical Temperature*: Not pertinent; *Critical Pressure*: Not pertinent; *Specific Gravity*: 0.576 at 15 °C (liquid); *Vapor (Gas) Specific Gravity*: 1.48; *Ratio of Specific Heats of Vapor (Gas)*: 1.1686; *Latent Heat of Vaporization*: 227, 126, 5.28; *Heat of Combustion*: -19,800, -11,000, -460; *Heat of Decomposition*: Not pertinent.

Health Hazards Information - *Recommended Personal Protective Equipment*: Self-contained breathing apparatus for high concentrations; safety goggles; protective gloves; *Symptoms Following Exposure*: Simple asphyxiant. Toxicology of propadiene component not fully established. Contact with liquid may burn eyes and cause frostbite of skin; *General Treatment for Exposure*: INHALATION: remove to fresh air; give artificial respiration if necessary. EYES or SKIN: treat burns caused by cold liquid; *Toxicity by Inhalation (Threshold Limit Value)*: 1000 ppm; *Short-Term Inhalation Limits*: Data not available; *Toxicity by Ingestion*: Not pertinent; *Late Toxicity*: Lung irritation in rats and dogs; *Vapor (Gas) Irritant Characteristics*: Data not available; *Liquid or Solid Irritant Characteristics*: Data not available; *Odor Threshold*: 100 ppm.

Fire Hazards - *Flash Point* : Not pertinent (flammable liquefied compressed gas); *Flammable Limits in Air (%)*: 3-11; *Fire Extinguishing Agents:* Let fire burn; shut off gas supply; cool adjacent exposures; *Fire Extinguishing Agents Not To Be Used:* Not pertinent; *Special Hazards of Combustion Products:* Not pertinent; *Behavior in Fire:* Containers may explode; *Ignition Temperature (deg. F):* 850; *Electrical Hazard:* Data not available; *Burning Rate:* No data.

Chemical Reactivity - *Reactivity with Water:* No reaction; *Reactivity with Common Materials:* No reaction, except forms explosive compounds in contact with alloys containing more than 67% copper at high pressures; *Stability During Transport:* Stable; *Neutralizing Agents for Acids and Caustics:* Not pertinent; *Polymerization:* Not pertinent; *Inhibitor of Polymerization:* Not pertinent.

METHYL ACRYLATE

Chemical Designations - *Synonyms*: Acrylic acid, methyl ester; Methyl 2-propenoate; *Chemical Formula*: $CH_2=CHCOOCH_3$.

Observable Characteristics - *Physical State (as shipped)*: Liquid; *Color*: Colorless; *Odor*: Characteristic acrylic; sweet, sharp; sharp, fragrant.

Physical and Chemical Properties - *Physical State at 15 °C and 1 atm.*: Liquid; *Molecular Weight*: 86.09; *Boiling Point at 1 atm.*: 177, 80.6, 353.8; *Freezing Point*: -105.7, -76.5, 196.7; *Critical Temperature*: 505, 263, 536; *Critical Pressure*: 630, 43, 4.3; *Specific Gravity*: 0.956 at 20 °C (liquid); *Vapor (Gas) Specific Gravity*: 3.0; *Ratio of Specific Heats of Vapor (Gas)*: 1.102; *Latent Heat of Vaporization*: 160, 90, 3.8; *Heat of Combustion*: (est.) -9900, -5500, -230; *Heat of Decomposition*: Not pertinent.

Health Hazards Information - *Recommended Personal Protective Equipment*: Organic canister for high vapor concentrations; rubber gloves; chemical goggles or face shield; *Symptoms Following Exposure*: May irritate skin, eyes, respiratory system, and gastrointestinal tract. Fumes cause tears; *General Treatment for Exposure*: INHALATION: remove to fresh air; lay patient down; keep him warm; administer artificial respiration if breathing has stopped; administer oxygen. SKIN OR EYES: flush with plenty water for 15 min.; consult physician for eye exposure; *Toxicity by Inhalation (Threshold Limit Value)*: 10 ppm; *Short-Term Inhalation Limits*: 25 ppm for 30 min.; *Toxicity by Ingestion*: Grade 3; LD_{50} 50 to 500 mg/kg (rabbit); *Late Toxicity*: Data not available; *Vapor (Gas) Irritant Characteristics*: Vapors cause moderate irritation such that personnel will find high concentrations unpleasant. The effect is temporary; *Liquid or Solid Irritant Characteristics*: Causes smarting of the skin and first-degree burns on short exposure; may cause secondary burns on long exposure; *Odor Threshold*: Data not available.

  Fire Hazards - *Flash Point (deg. F)*: 2.7 CC; 44 OC; *Flammable Limits in Air (%)*: 2.8 - 25; *Fire Extinguishing Agents:* Foam, dry chemical, or carbon dioxide; *Fire Extinguishing Agents Not To Be Used:* Water may be ineffective; *Special Hazards of Combustion Products:* Irritating vapors are generated in fires; *Behavior in Fire:* May polymerize. Vapor is heavier than fire and may travel a considerable distance to a source of ignition and flash back; *Ignition Temperature:* Data not available; *Electrical Hazard:* Not pertinent; *Burning Rate:* Data not available.

Chemical Reactivity - *Reactivity with Water:* No reaction; *Reactivity with Common Materials:* No reaction; *Stability During Transport:* Stable; *Neutralizing Agents for Acids and Caustics:* Not pertinent; *Polymerization:* Heat may cause an explosive polymerization. Strong ultraviolet light can also initiate polymerization; *Inhibitor of Polymerization:* Hydroquinone and its methyl ether, in presence of air .

METHYL ALCOHOL

Chemical Designations - *Synonyms*: Colonial spirit; Wood alcohol; Columbian spirit; Wood naphtha; Methanol; Wood spirit; *Chemical Formula*: CH_3OH.

Observable Characteristics - *Physical State (as shipped)*: Liquid; *Color*: Colorless; *Odor*: Faint alcohol; like ethyl alcohol; faintly sweet; characteristic pungent.

Physical and Chemical Properties - *Physical State at 15 °C and 1 atm.*: Liquid; *Molecular Weight*: 32.04; *Boiling Point at 1 atm.*: 148.1, 64.5, 337.7; *Freezing Point*: -144.0, -97.8, 175.4; *Critical Temperature*: 464, 240, 513; *Critical Pressure*: 1142.0, 77.7, 7.87; *Specific Gravity*: 0.792 at 20 °C (liquid); *Vapor (Gas) Specific Gravity*: 1.1; *Ratio of Specific Heats of Vapor (Gas)*: 1.254; *Latent Heat of Vaporization*: 473.0, 262.8, 11.00; *Heat of Combustion*: -8419, -4677, -195.8; *Heat of Decomposition*: Not pertinent.

Health Hazards Information - *Recommended Personal Protective Equipment*: Approved canister mask for high vapor concentrations; safety goggles; rubber gloves; *Symptoms Following Exposure*: Exposure to excessive vapor causes eye irritation, headache, fatigue and drowsiness. High concentrations can produce central nervous system depression and optic nerve damage. 50,000 ppm will probably cause death in 1 to 2 hrs. Can be absorbed trough skin. Swallowing may cause death or eye damage; *General Treatment for Exposure*: Remove victim from exposure and apply artificial respiration if breathing has ceased. INGESTION: induce vomiting, then give 2 teaspoons of baking soda in glass of water; call a physician. EYES or SKIN: flush with water for 15 min.; *Toxicity by Inhalation (Threshold Limit Value)*: 200 ppm; *Short-Term Inhalation Limits*: 260 mg/m³ for 60 min.; *Toxicity by Ingestion*: Grade 1; 5 to 15 g/kg (rat); *Late Toxicity*: None; *Vapor (Gas) Irritant Characteristics*: Vapors causes a slight smarting of the eyes or respiratory system if present in high concentrations. The effect is temporary; *Liquid or Solid Irritant Characteristics*: Minimum hazard. If spilled on clothing and allowed to remain, may cause smarting and reddening of skin; *Odor Threshold*: 100 ppm.

Fire Hazards - *Flash Point (deg. F)*: 54 CC; 61 OC; *Flammable Limits in Air (%)*: 6.0 - 36; *Fire Extinguishing Agents:* Alcohol foam, dry chemical, carbon dioxide; *Fire Extinguishing Agents Not To Be Used:* Water may be ineffective; *Special Hazards of Combustion Products:* Not pertinent; *Behavior in Fire:* Containers may explode; *Ignition Temperature (deg. F):* 867; *Electrical Hazard:* Class I, Group D; *Burning Rate:* 1.7 mm/min.

Chemical Reactivity - *Reactivity with Water:* No reaction; *Reactivity with Common Materials:* No

reaction; *Stability During Transport:* Stable; *Neutralizing Agents for Acids and Caustics:* Not pertinent; *Polymerization:* Not pertinent; *Inhibitor of Polymerization:* Not pertinent.

METHYLAMINE
Chemical Designations - *Synonyms*: Aminomethane; Mercurialin; Monomethylamine, anhydrous; *Chemical Formula*: CH_3NH_2
Observable Characteristics - *Physical State (as shipped)*: Liquefied; *Color*: Colorless; *Odor*: Like ammonia, pungent, fishy, suffocating.
Physical and Chemical Properties - *Physical State at 15 °C and 1 atm.*: Gas; *Molecular Weight*: 31.1; *Boiling Point at 1 atm.*: 20.3, -6.5, 266.7; *Freezing Point*: -134.5, -92.5, 180.7; *Critical Temperature*: 318, 159, 432; *Critical Pressure*: 1,080, 73.6, 7.47; *Specific Gravity*: 0.693 at -6.5 °C (liquid); *Vapor (Gas) Specific Gravity*: 1.1; *Ratio of Specific Heats of Vapor (Gas)*: 1.1946; *Latent Heat of Vaporization*: 358, 199, 8.33; *Heat of Combustion*: -15,000, -8,340, -34.9; *Heat of Decomposition*: Not pertinent.
Health Hazards Information - *Recommended Personal Protective Equipment*: Goggles or face mask; rubber suit, apron, sleeves, and/or gloves; rubber or leather safety shoes; air-line mask, positive-pressure hose mask, self-contaminated breathing apparatus, or industrial canister-type gas mask; *Symptoms Following Exposure*: Inhalation causes irritation of nose and throat, followed by violent sneezing, burning sensation in throat, coughing, constriction of larynx and difficulty in breathing, pulmonary congestion, edema of the lung, and conjunctivitis. Contact with liquid burn eyes and skin. (Severe exposure may cause blindness.) Vapors may cause dermatitis. Ingestion causes burns of the mouth, throat, and esophagus; *General Treatment for Exposure*: Get medical attention for anyone overcome or injured by exposure to this compound. INHALATION: remove victim to fresh air at once; apply artificial respiration if breathing has stopped; administer oxygen. EYES: flush with water for at least 15 min. SKIN: flush with water; if skin is burned do not use ointments or cover for 24 hours. INGESTION: do NOT induce vomiting; give large amount of water; *Toxicity by Inhalation (Threshold Limit Value)*: 10 ppm; *Short-Term Inhalation Limits*: Data not available; *Toxicity by Ingestion*: Grade 2, LD_{50} 0.5 to 5 g/kg; *Late Toxicity*: Data not available; *Vapor (Gas) Irritant Characteristics*: Vapors are moderately irritating such that personnel will not usually tolerate moderate or high concentrations; *Liquid or Solid Irritant Characteristics*: Causes smarting of the skin and first-degree burns on short exposure; may cause secondary burns on long exposure; *Odor Threshold*: 0.021 ppm.
Fire Hazards - *Flash Point* : Not pertinent (flammable liquefied compressed gas); *Flammable Limits in Air (%)*: 4.3 - 21; *Fire Extinguishing Agents:* Let gas fire burn; stop flow of gas. Extinguish solution fires with dry chemical, alcohol foam, or carbon dioxide; *Fire Extinguishing Agents Not To Be Used:* Not pertinent; *Special Hazards of Combustion Products:* Toxic nitrogen oxides may be formed; *Behavior in Fire:* Vapor is heavier than air and may travel a considerable distance to a source of ignition and flash back; *Ignition Temperature (deg. F):* 806; *Electrical Hazard:* Data not available; *Burning Rate:* Not pertinent.
Chemical Reactivity - *Reactivity with Water:* Dissolves completely; *Reactivity with Common Materials:* Corrosive to copper, copper alloys, zinc alloys, aluminum, and galvanized surfaces; *Stability During Transport:* Stable; *Neutralizing Agents for Acids and Caustics:* Not pertinent; *Polymerization:* Not pertinent; *Inhibitor of Polymerization:* Not pertinent.

METHYL AMYL ACETATE
Chemical Designations - *Synonyms*: Hexyl acetate; MAAc; Methylisobutylcarbinyl acetate; 4-Methyl-2-pentanol, acetate; 4-Methyl-2-pentyl acetate; *Chemical Formula*: $CH_3COOCH(CH_3)CH_2CH(CH_3)_2$.
Observable Characteristics - *Physical State (as shipped)*: Liquid; *Color*: Colorless; *Odor*: Fruity; mild, pleasant; mild and nonresidual.
Physical and Chemical Properties - *Physical State at 15 °C and 1 atm.*: Liquid; *Molecular Weight*: 144.22; *Boiling Point at 1 atm.*: 295.2, 146.2, 419.4; *Freezing Point*: -82.8, -63.8, 209.4; *Critical Temperature*: 606, 319, 592; *Critical Pressure*: 382, 26, 2.6; *Specific Gravity*: 0.860 at 20 °C (liquid); *Vapor (Gas) Specific Gravity*: Not pertinent; *Ratio of Specific Heats of Vapor (Gas)*: 1.046; *Latent Heat*

of Vaporization: 225, 125, 5.23; *Heat of Combustion*: (est.) -14,400, -8000, -335; *Heat of Decomposition*: Not pertinent.

Health Hazards Information - *Recommended Personal Protective Equipment*: Organic canister or air pack; rubber gloves; goggles; *Symptoms Following Exposure*: Headache, dizziness, nausea, irritation to respiratory passages. Irritates eyes; *General Treatment for Exposure*: INHALATION: remove from exposure immediately; call a physician; if breathing is irregular or has stopped, start resuscitation and administer oxygen. EYE CONTACT: flush with water for at least 15 min.; *Toxicity by Inhalation (Threshold Limit Value)*: Data not available; *Short-Term Inhalation Limits*: Data not available; *Toxicity by Ingestion*: Grade 1; LD_{50} 5 to 15 g/kg; *Late Toxicity*: None; *Vapor (Gas) Irritant Characteristics*: Vapors cause moderate irritation such that personnel will find high concentrations unpleasant. The effect is temporary; *Liquid or Solid Irritant Characteristics*: Minimum hazard. If spilled on clothing and allowed to remain, may cause smarting and reddening of skin; *Odor Threshold*: Data not available.

Fire Hazards - *Flash Point (deg. F)*: 113 CC; 110 OC; *Flammable Limits in Air (%)*: 0.9 - 5.7 (calc.); *Fire Extinguishing Agents:* Alcohol foam, carbon dioxide, or dry chemical; *Fire Extinguishing Agents Not To Be Used:* Not pertinent; *Special Hazards of Combustion Products:* Not pertinent; *Behavior in Fire:* Not pertinent; *Ignition Temperature (deg. F):* 510 (calc.); *Electrical Hazard:* Not pertinent; *Burning Rate:* Data not available.

Chemical Reactivity - *Reactivity with Water:* No reaction; *Reactivity with Common Materials:* No reaction; *Stability During Transport:* Stable; *Neutralizing Agents for Acids and Caustics:* Not pertinent; *Polymerization:* Not pertinent; *Inhibitor of Polymerization:* Not pertinent.

METHYL AMYL ALCOHOL

Chemical Designations - *Synonyms*: Isobutylmethyl carbinol; Isobutyl methylmethanol; MAOH; Methylisobutyl carbinol; 4-Methyl-2-pentanol; MIC; *Chemical Formula*: $(CH_3)_2CHCH_2CH(OH)CH_3$.

Observable Characteristics — *Physical State (as shipped)*: Liquid; *Color*: Colorless; *Odor*: Sharp; mild and nonresidual.

Physical and Chemical Properties - *Physical State at 15 °C and 1 atm.*: Liquid; *Molecular Weight*: 102.18; *Boiling Point at 1 atm.*: 269.2, 131.8, 405; *Freezing Point*: < -130, < -90, < 183; *Critical Temperature*: 556, 291, 564; *Critical Pressure*: Not pertinent; *Specific Gravity*: 0.807 at 20 °C (liquid); *Vapor (Gas) Specific Gravity*: Not pertinent; *Ratio of Specific Heats of Vapor (Gas)*: 1.053; *Latent Heat of Vaporization*: 162, 90.1, 3.77; *Heat of Combustion*: -16,640, -9240, -387; *Heat of Decomposition*: Not pertinent.

Health Hazards Information - *Recommended Personal Protective Equipment*: Air pack or organic canister mask; rubber gloves; goggles or face shield; *Symptoms Following Exposure*: Vapor irritates eyes and nose. May cause anesthesia. Prolonged contact with liquid causes irritation and cracking of skin, and irritates eyes; *General Treatment for Exposure*: INHALATION: remove victim to fresh air; give artificial respiration if needed; call a doctor. SKIN: flush with water. EYES: flood with water for at least 15 min.; consult a doctor; *Toxicity by Inhalation (Threshold Limit Value)*: 25 ppm; *Short-Term Inhalation Limits*: Data not available; *Toxicity by Ingestion*: Grade 2, LD_{50} 0.5 to 5 g/kg (rat); *Late Toxicity*: None; *Vapor (Gas) Irritant Characteristics*: Vapors cause moderate irritation such that personnel will find high concentrations unpleasant. The effect is temporary; *Liquid or Solid Irritant Characteristics*: Minimum hazard. If spilled on clothing and allowed to remain, may cause smarting and reddening of skin; *Odor Threshold*: Data not available.

Fire Hazards - *Flash Point (deg. F)*: 120 - 130 OC; 106 CC; *Flammable Limits in Air (%):* 1.0 - 5.5; *Fire Extinguishing Agents:* Alcohol foam, dry chemical, or carbon dioxide; *Fire Extinguishing Agents Not To Be Used:* Not pertinent; *Special Hazards of Combustion Products:* Not pertinent; *Behavior in Fire:* Not pertinent; *Ignition Temperature (deg. F):* 583 (calc.); *Electrical Hazard:* Not pertinent; *Burning Rate:* 4.7 mm/min.

Chemical Reactivity - *Reactivity with Water:* No reaction; *Reactivity with Common Materials:* No reaction; *Stability During Transport:* Stable; *Neutralizing Agents for Acids and Caustics:* Not pertinent; *Polymerization:* Not pertinent; *Inhibitor of Polymerization:* Not pertinent.

N-METHYLANILINE

Chemical Designations - *Synonyms*: Anilinomethane; N-Methylaminobenzene; Methylaniline (mono); Methylphenylamine; *Chemical Formula*: $C_6H_5NHCH_3$.

Observable Characteristics - *Physical State (as shipped)*: Liquid; *Color*: Yellow to light brown; *Odor*: Moderate aniline-type.

Physical and Chemical Properties - *Physical State at 15 °C and 1 atm.*: Liquid; *Molecular Weight*: 107.2; *Boiling Point at 1 atm.*: 384.6, 195.9, 469.1; *Freezing Point*: -71, -57, 216; *Critical Temperature*: 802, 428, 701; *Critical Pressure*: 754, 51.3, 5.20; *Specific Gravity*: 0.989 at 20 °C (liquid); *Vapor (Gas) Specific Gravity*: 3.70; *Ratio of Specific Heats of Vapor (Gas)*: Not pertinent; *Latent Heat of Vaporization*: 180, 100, 4.20; *Heat of Combustion*: -16,350, -9,085, -380.1; *Heat of Decomposition*: Not pertinent.

Health Hazards Information - *Recommended Personal Protective Equipment*: Approved respirator; rubber gloves; splash proof goggles; *Symptoms Following Exposure*: Inhalation causes dizziness and headache. Ingestion causes bluish discoloration (cyanosis) of lips, ear lobes, and fingernail beds. Liquid irritates eyes. Absorption through skin produces same symptoms as for ingestion; *General Treatment for Exposure*: INHALATION: remove victim to fresh air and call a physician at once; administer oxygen until physician arrives. INGESTION: give large amount of water; get medical attention at once. EYES or SKIN: flush with plenty of water for at least 15 min.; if cyanosis is present, shower with soap and warm water, with special attention to scalp and finger nails; remove any contaminated clothing; *Toxicity by Inhalation (Threshold Limit Value)*: Data not available; *Short-Term Inhalation Limits*: Data not available; *Toxicity by Ingestion*: Data not available; *Late Toxicity*: Data not available; *Vapor (Gas) Irritant Characteristics*: Data not available; *Liquid or Solid Irritant Characteristics*: Data not available; *Odor Threshold*: Data not available.

Fire Hazards - *Flash Point (deg. F)*: 175 CC; *Flammable Limits in Air (%)*: Data not available; *Fire Extinguishing Agents:* Dry chemical, foam, carbon dioxide; *Fire Extinguishing Agents Not To Be Used:* Water may be ineffective; *Special Hazards of Combustion Products:* Toxic vapors are generated when heated; *Behavior in Fire:* Data not available; *Ignition Temperature :* Data not available; *Electrical Hazard:* Data not available; *Burning Rate:* 3.65 mm/min.

Chemical Reactivity - *Reactivity with Water:* No reaction; *Reactivity with Common Materials:* May attack some forms of plastic; *Stability During Transport:* Stable; *Neutralizing Agents for Acids and Caustics:* Not pertinent; *Polymerization:* Not pertinent; *Inhibitor of Polymerization:* Not pertinent.

METHYL BROMIDE

Chemical Designations - *Synonyms*: Bromomethane; Embafume; M-B-C Fumigant; Monobromomethane; *Chemical Formula*: CH_3Br.

Observable Characteristics - *Physical State (as shipped)*: Liquefied gas; *Color*: Colorless; *Odor*: Relatively odorless; sweet, chloroform-like.

Physical and Chemical Properties - *Physical State at 15 °C and 1 atm.*: Gas; *Molecular Weight*: 94.95; *Boiling Point at 1 atm.*: 38.5, 3.6, 276.8; *Freezing Point*: -135, -93, 180; *Critical Temperature*: 376, 191, 464; *Critical Pressure*: Not pertinent; *Specific Gravity*: 1.68 at 20 °C (liquid); *Vapor (Gas) Specific Gravity*: 3.3; *Ratio of Specific Heats of Vapor (Gas)*: 1.247; *Latent Heat of Vaporization*: 108, 59.7, 2.50; *Heat of Combustion*: -3188, -1771, -74.15; *Heat of Decomposition*: Not pertinent.

Health Hazards Information — *Recommended Personal Protective Equipment*: Self-contained breathing apparatus; goggles; *Symptoms Following Exposure*: Inhalation of vapor causes lung congestion and pulmonary edema. Higher concentrations cause rapid narcosis and death. Contact with liquid irritates eyes and burns skin; *General Treatment for Exposure*: INHALATION: remove victim to fresh air; give artificial respiration if needed. SKIN OR EYES: flush with water for at least 15 min.; *Toxicity by Inhalation (Threshold Limit Value)*: 15 ppm; *Short-Term Inhalation Limits*: 20 ppm for 5 min.; *Toxicity by Ingestion*: Data not available; *Late Toxicity*: Data not available; *Vapor (Gas) Irritant Characteristics*: Vapors are moderately irritating such that personnel will not usually tolerate moderate or high concentrations; *Liquid or Solid Irritant Characteristics*: Fairly severe skin irritant; may cause pain and second-degree burns after a few minutes contact; *Odor Threshold*: Data not available.

Fire Hazards - *Flash Point* : Practically not flammable; *Flammable Limits in Air (%):* 10 - 15; *Fire Extinguishing Agents:* Not pertinent; *Fire Extinguishing Agents Not To Be Used:* Not pertinent; *Special Hazards of Combustion Products:* Toxic and irritating gases are generated when exposed to fire or heat; *Behavior in Fire:* Containers may explode; *Ignition Temperature (deg. F):* 999; *Electrical Hazard:* Not pertinent; *Burning Rate:* Not pertinent.

Chemical Reactivity - *Reactivity with Water:* No reaction; *Reactivity with Common Materials:* No reaction; *Stability During Transport:* Stable; *Neutralizing Agents for Acids and Caustics:* Not pertinent; *Polymerization:* Not pertinent; *Inhibitor of Polymerization:* Not pertinent.

METHYL N-BUTYL KETONE

Chemical Designations - *Synonyms*: n-Butyl methyl ketone; 2-Hexanone; *Chemical Formula*: $CH_3(CH_2)_3COCH_3$

Observable Characteristics - *Physical State (as shipped)*: Liquid; *Color*: Clear; *Odor*: Characteristic; strong, disagreeable odor resembling acetone.

Physical and Chemical Properties - *Physical State at 15 °C and 1 atm.*: Liquid; *Molecular Weight*: 100.16; *Boiling Point at 1 atm.*: 261, 127, 400; *Freezing Point*: -70.4, -56.9, 216.3; *Critical Temperature*: Not pertinent; *Critical Pressure*: Not pertinent; *Specific Gravity*: 0.812 at 20 °C (liquid); *Vapor (Gas) Specific Gravity*: 3.5; *Ratio of Specific Heats of Vapor (Gas)*: Not pertinent; *Latent Heat of Vaporization*: 148, 82, 3.4; *Heat of Combustion*: -16,100, -8,940, -374; *Heat of Decomposition*: Not pertinent.

Health Hazards Information - *Recommended Personal Protective Equipment*: Protective gloves; goggles or face shield; approved respirator (for major spills); *Symptoms Following Exposure*: Inhalation of high concentrations of vapor may result in narcosis; peripheral neuropathy may develop. Ingestion of large amounts may cause some systemic injury. Contact with eyes causes mild to moderate irritation. Liquid irritates skin; prolonged or repeated contact may cause defatting of the skin with resultant dermatitis; *General Treatment for Exposure*: INHALATION: move to uncontaminated atmosphere and treat symptomatically; alert physician to possible development of peripheral neuropathy. INGESTION: give large amount of water and induce vomiting. EYES: irrigate immediately and thoroughly with water for 15 min. and get medical attention. SKIN: flush exposed areas thoroughly with water; *Toxicity by Inhalation (Threshold Limit Value)*: 100 ppm; *Short-Term Inhalation Limits*: Data not available; *Toxicity by Ingestion*: Grade 2; oral LD_{50} = 2,590 mg/kg (rat); *Late Toxicity*: Peripheral neuropathy in experimental animals and man (disease of motor and/or sensor nerves); *Vapor (Gas) Irritant Characteristics*: Data not available; *Liquid or Solid Irritant Characteristics*: Data not available; *Odor Threshold*: Data not available.

Fire Hazards - *Flash Point (deg. F)*: 83 OC; 77 CC; *Flammable Limits in Air (%):* 1.3 - 8.0; *Fire Extinguishing Agents:* Dry chemical, alcohol foam, carbon dioxide; *Fire Extinguishing Agents Not To Be Used:* Water may be ineffective; *Special Hazards of Combustion Products:* Data not available; *Behavior in Fire:* Data not available; *Ignition Temperature (deg. F):* 795; *Electrical Hazard:* Data not available; *Burning Rate:* 4.8 mm/min.

Chemical Reactivity - *Reactivity with Water:* No reaction; *Reactivity with Common Materials:* No reaction; *Stability During Transport:* Stable; *Neutralizing Agents for Acids and Caustics:* Not pertinent; *Polymerization:* Not pertinent; *Inhibitor of Polymerization:* Not pertinent.

METHYL CHLORIDE

Chemical Designations - *Synonyms*: Arctic; Chloromethane; *Chemical Formula*: CH_3Cl.

Observable Characteristics - *Physical State (as shipped)*: Liquefied gas; *Color*: Colorless; *Odor*: Faint, sweet, non-irritating; faint ether-like.

Physical and Chemical Properties - *Physical State at 15 °C and 1 atm.*: Gas; *Molecular Weight*: 50.49; *Boiling Point at 1 atm.*: -11.6, -24.2, 249; *Freezing Point*: -143.9, 97.7, 175.5; *Critical Temperature*: 290.5, 143.6, 416.8; *Critical Pressure*: 969, 65.9, 6.68; *Specific Gravity*: 0.997 at -24 °C (liquid); *Vapor (Gas) Specific Gravity*: 1.7; *Ratio of Specific Heats of Vapor (Gas)*: 1.259; *Latent Heat of Vaporization*: 182.3, 101.3, 4.241; *Heat of Combustion*: -5290, -2939, -123.1; *Heat of*

Decomposition: Not pertinent.

Health Hazards Information - *Recommended Personal Protective Equipment*: Approved canister mask; leather or vinyl gloves; goggles or face shield; *Symptoms Following Exposure*: Inhalation causes nausea, vomiting, weakness, headache, emotional disturbances; high concentrations cause mental confusion, eye disturbances, muscular tremors, cyanosis, convulsions. Contact of liquid with skin may cause frostbite; *General Treatment for Exposure*: Remove to fresh air. Call a doctor and have patient hospitalized for observation of slowly developing symptoms; *Toxicity by Inhalation (Threshold Limit Value)*: 100 ppm; *Short-Term Inhalation Limits*: 100 ppm for 5 min.; *Toxicity by Ingestion*: Not pertinent; *Late Toxicity*: None; *Vapor (Gas) Irritant Characteristics*: Vapors are nonirritating to eyes and throat; *Liquid or Solid Irritant Characteristics*: No appreciable hazard. Practically harmless to the skin because it is very volatile and evaporates quickly. May cause frostbite; *Odor Threshold*: Data not available.

Fire Hazards - *Flash Point (deg. F)*: < 32 CC; *Flammable Limits in Air (%):* 8.1 - 17.2; *Fire Extinguishing Agents:* Dry chemical or carbon dioxide. Stop flow of gas; *Fire Extinguishing Agents Not To Be Used:* Not pertinent; *Special Hazards of Combustion Products:* Toxic and irritating gases are generated in fires; *Behavior in Fire:* Containers may explode; *Ignition Temperature (deg. F):* 1170; *Electrical Hazard:* Not pertinent; *Burning Rate:* 2.2 mm/min.

Chemical Reactivity - *Reactivity with Water:* No reaction; *Reactivity with Common Materials:* Reacts with zinc, aluminum, magnesium, and their alloys; reaction is not violent; *Stability During Transport:* Stable; *Neutralizing Agents for Acids and Caustics:* Not pertinent; *Polymerization:* Not pertinent; *Inhibitor of Polymerization:* Not pertinent.

METHYL CHLOROFORMATE

Chemical Designations - *Synonyms*: Chlorocarbonic acid, methyl ester; Chloroformic acid, methyl ester; Methyl chlorocarbonate; *Chemical Formula*: $ClCOOCH_3$.

Observable Characteristics - *Physical State (as shipped)*: Liquid; *Color*: Colorless to light yellow; *Odor*: Acrid.

Physical and Chemical Properties - *Physical State at 15 °C and 1 atm.*: Liquid; *Molecular Weight*: 94.5; *Boiling Point at 1 atm.*: 160, 71, 344; *Freezing Point*: <-114, <-81, <192; *Critical Temperature*: Not pertinent; *Critical Pressure*: Not pertinent; *Specific Gravity*: 1.22 at 20 °C (liquid); *Vapor (Gas) Specific Gravity*: 3.25; *Ratio of Specific Heats of Vapor (Gas)*: 1.1544; *Latent Heat of Vaporization*: (est.) 153, 85, 3.6; *Heat of Combustion*: -4,690, -2,600, -109; *Heat of Decomposition*: Not pertinent.

Health Hazards Information - *Recommended Personal Protective Equipment*: Acid- or organic-canister mask or self-contained breathing apparatus; goggles or face shield; plastic gloves; *Symptoms Following Exposure*: Inhalation of vapor irritates nose and throat and can cause delayed pulmonary edema. Liquid irritates eyes and causes severe skin burns if allowed to remain. Ingestion causes burns of mouth and stomach; *General Treatment for Exposure*: INHALATION: remove victim from exposure; if breathing has stopped, administer artificial respiration; call physician. EYES: irrigate with copious amounts of water for at least 15 min.; call a physician if needed. SKIN: flush with water for 15 min.; get medical attention for burns. INGESTION: give large amount of water; do NOT induce vomiting; get medical attention; *Toxicity by Inhalation (Threshold Limit Value)*: Data not available; *Short-Term Inhalation Limits*: Data not available; *Toxicity by Ingestion*: Grade 4, oral LD_{50} < 50 mg/kg (rat); *Late Toxicity*: Data not available; *Vapor (Gas) Irritant Characteristics*: Data not available; *Liquid or Solid Irritant Characteristics*: Data not available; *Odor Threshold*: Data not available.

Fire Hazards - *Flash Point (deg. F)*: 76 OC; 73 CC; *Flammable Limits in Air (%):* LEL = 6.7; *Fire Extinguishing Agents:* Water, dry chemical, foam, carbon dioxide; *Fire Extinguishing Agents Not To Be Used:* Not pertinent; *Special Hazards of Combustion Products:* Irritating and toxic hydrogen chloride and phosgene may be formed; *Behavior in Fire:* Vapor is heavier than air and may travel a considerable distance to a source of ignition and flash back; *Ignition Temperature :* Data not available; *Electrical Hazard:* Data not available; *Burning Rate:* 2.0 mm/min.

Chemical Reactivity - *Reactivity with Water:* Reacts slowly, evolving hydrogen chloride (hydrochloric acid). Reaction can be hazardous if water is hot; *Reactivity with Common Materials:* Corrodes rubber; *Stability During Transport:* Stable; *Neutralizing Agents for Acids and Caustics:* Flush with water, rinse with sodium bicarbonate or lime solution; *Polymerization:* Not pertinent; *Inhibitor of Polymerization:* Not pertinent.

METHYLCYCLOPENTADIENYLMANGANESE TRICARBONYL
Chemical Designations - *Synonyms:* Combustion Improver C12; *Chemical Formula:* $C_9H_7O_3Mn$.
Observable Characteristics - *Physical State (as shipped):* Liquid; *Color:* Straw to dark orange; *Odor:* Faint, pleasant, herbaceous.
Physical and Chemical Properties - *Physical State at 15 °C and 1 atm.:* Liquid; *Molecular Weight:* 218.1; *Boiling Point at 1 atm.:* 451, 233, 506; *Freezing Point:* 34, 1, 274; *Critical Temperature:* Not pertinent; *Critical Pressure:* Not pertinent; *Specific Gravity:* 1.39 at 20 °C (liquid); *Vapor (Gas) Specific Gravity:* Not pertinent; *Ratio of Specific Heats of Vapor (Gas):* Not pertinent; *Latent Heat of Vaporization:* Not pertinent; *Heat of Combustion:* (est.) -9,900, -5,500, -230; *Heat of Decomposition:* Not pertinent.
Health Hazards Information - *Recommended Personal Protective Equipment:* Organic vapor canister mask; rubber gloves and apron; protective goggles or face shield; *Symptoms Following Exposure:* Inhalation, ingestion, or skin contact affect central nervous system, causing convulsions, respiratory depression, cyanosis, and coma. Liquid irritates eyes; *General Treatment for Exposure:* Get medical attention following all exposures to this compound. INHALATION: remove victim from exposure; give artificial respiration if necessary. EYES: flush with plenty of water for at least 15 min. SKIN: wash well with soap and water. INGESTION: induce vomiting; *Toxicity by Inhalation (Threshold Limit Value):* 0.1 ppm (as manganese); *Short-Term Inhalation Limits:* Data not available; *Toxicity by Ingestion:* Grade 4, oral LD_{50} = 23 mg/kg (rat); *Late Toxicity:* Data not available; *Vapor (Gas) Irritant Characteristics:* Data not available; *Liquid or Solid Irritant Characteristics:* Data not available; *Odor Threshold:* Data not available.
Fire Hazards - *Flash Point (deg. F):* >200 CC; *Flammable Limits in Air (%):* Data not available; *Fire Extinguishing Agents:* Dry chemical, foam, water spray, carbon dioxide; *Fire Extinguishing Agents Not To Be Used:* Not pertinent; *Special Hazards of Combustion Products:* Toxic vapors are formed in a fire; *Behavior in Fire:* Not pertinent; *Ignition Temperature :* Data not available; *Electrical Hazard:* Data not available; *Burning Rate:* Data not available.
Chemical Reactivity - *Reactivity with Water:* No reaction; *Reactivity with Common Materials:* Data not available; *Stability During Transport:* Stable; *Neutralizing Agents for Acids and Caustics:* Not pertinent; *Polymerization:* Not pertinent; *Inhibitor of Polymerization:* Not pertinent.

METHYL CYCLOPENTANE
Chemical Designations - *Synonyms:* Cyclopentane, methyl; *Chemical Formula:* C_6H_{12}.
Observable Characteristics - *Physical State (as shipped):* Liquid; *Color:* Colorless; *Odor:* Like gasoline.
Physical and Chemical Properties - *Physical State at 15 °C and 1 atm.:* Liquid; *Molecular Weight:* 84.2; *Boiling Point at 1 atm.:* 161.3, 71.8, 345.0; *Freezing Point:* -224, -142, 131; *Critical Temperature:* 499.3, 259.6, 532.8; *Critical Pressure:* 550, 37.4, 3.79; *Specific Gravity:* 0.749 at 20 °C (liquid); *Vapor (Gas) Specific Gravity:* 2.9; *Ratio of Specific Heats of Vapor (Gas):* 1.0834; *Latent Heat of Vaporization:* 162, 90, 3.8; *Heat of Combustion:* (liquid) -18,900, -10,500, -440; *Heat of Decomposition:* Not pertinent.
Health Hazards Information - *Recommended Personal Protective Equipment:* SCBA; goggles or face shield; rubber gloves; *Symptoms Following Exposure:* Inhalation causes dizziness, nausea, and vomiting; concentrated vapor may cause unconsciousness and collapse. Liquid causes irritation of eyes and mild irritation of skin if allowed to remain. Ingestion causes irritation of stomach. Aspiration causes severe lung irritation, rapidly developing pulmonary edema, and central nervous system excitement followed by depression; *General Treatment for Exposure:* INHALATION: remove victim from exposure; if breathing has stopped, administer artificial respiration; call physician. EYES: flush

with water for 15 min.; call physician. SKIN: flush with water, then wash with soap and water. INGESTION: do NOT induce vomiting; guard against aspiration into lungs. ASPIRATION: enforce bed rest; give oxygen; seek medical attention; *Toxicity by Inhalation*: No data; *Short-Term Inhalation Limits*: 300 ppm for 60 min.; *Toxicity by Ingestion*: Grade 1; LD_{50} 5 to 15 g/kg; *Late Toxicity*: No data; *Vapor (Gas) Irritant Characteristics*: Vapors are nonirritating to the eyes and throat; *Liquid or Solid Irritant Characteristics*: If spilled on clothing and allowed to remain, may cause smarting and reddening of skin; *Odor Threshold*: No data.

Fire Hazards - *Flash Point (deg. F)*: < 0 CC; *Flammable Limits in Air (%)*: 1.1 - 8.7; *Fire Extinguishing Agents:* Dry chemical, foam, carbon dioxide; *Fire Extinguishing Agents Not To Be Used:* Water may be ineffective; *Special Hazards of Combustion Products:* Not pertinent; *Behavior in Fire:* Vapor is heavier than air and may travel a considerable distance to a source of ignition and flash back; *Ignition Temperature (deg. F):* 624; *Electrical Hazard:* No data; *Burning Rate:* 7.1 mm/min.

Chemical Reactivity - *Reactivity with Water:* No reaction; *Reactivity with Common Materials:* No reaction; *Stability During Transport:* Stable; *Neutralizing Agents for Acids and Caustics:* Not pertinent; *Polymerization:* Not pertinent; *Inhibitor of Polymerization:* Not pertinent.

METHYLDICHLOROSILANE

Chemical Designations - *Synonyms*: No common synonyms; *Chemical Formula*: CH_3SiHCl_2.

Observable Characteristics - *Physical State (as shipped)*: Liquid; *Color*: Colorless; *Odor*: Acrid; sharp, hydrochloric acid-like.

Physical and Chemical Properties - *Physical State at 15 °C and 1 atm.*: Liquid; *Molecular Weight*: 115; *Boiling Point at 1 atm.*: 106.7, 41.5, 314.7; *Freezing Point*: -135, -93, 180; *Critical Temperature*: Not pertinent; *Critical Pressure*: Not pertinent; *Specific Gravity*: 1.11 at 25 °C (liquid); *Vapor (Gas) Specific Gravity*: 4; *Ratio of Specific Heats of Vapor (Gas)*: Data not available; *Latent Heat of Vaporization*: 106, 59, 2.5; *Heat of Combustion*: (est.) -4,700, -2,600, -110; *Heat of Decomposition*: Not pertinent.

Health Hazards Information - *Recommended Personal Protective Equipment*: Full protective clothing; acid-vapor-type respiratory protection; rubber gloves; chemical worker's goggles; other protective equipment as necessary to protect skin and eyes; *Symptoms Following Exposure*: Inhalation causes irritation of respiratory tract; heavy exposure can cause pulmonary edema. Contact of liquid with skin or eyes causes severe burns. Ingestion causes burns of mouth and stomach; *General Treatment for Exposure*: Get medical attention following all exposure to this compound. INHALATION: remove victim from exposure; if breathing has stopped, begin artificial respiration. EYES: flush with water for 15 min. SKIN: flush with water. INGESTION: do NOT induce vomiting; give large amounts of water; *Toxicity by Inhalation (Threshold Limit Value)*: Data not available; *Short-Term Inhalation Limits*: Data not available; *Toxicity by Ingestion*: Grade 3; LD_{50} 50 to 500 mg/kg; *Late Toxicity*: Data not available; *Vapor (Gas) Irritant Characteristics*: Vapors cause severe irritation of eye and throat and can cause eye and lung injury. They cannot be tolerated even at low concentrations; *Liquid or Solid Irritant Characteristics*: Severe skin irritant. Causes second- and third-degree burns on short contact and is very injurious to the eyes; *Odor Threshold*: Data not available.

Fire Hazards - *Flash Point (deg. F)*: - 14 OC; *Flammable Limits in Air (%):* 6 - 55; *Fire Extinguishing Agents:* Dry chemical or carbon dioxide; *Fire Extinguishing Agents Not To Be Used:* Water, foam; *Special Hazards of Combustion Products:* Toxic hydrogen chloride and phosgene gases may be formed; *Behavior in Fire:* Difficult to extinguish; re-ignition may occur. Contact with water, applied to adjacent fires will generate irritating hydrogen chloride gas; *Ignition Temperature (deg. F):* >600; *Electrical Hazard:* Data not available; *Burning Rate:* 3.0 mm/min.

Chemical Reactivity - *Reactivity with Water:* Reacts violently to form hydrogen chloride (hydrochloric acid); *Reactivity with Common Materials:* Reacts with surface moisture to evolve hydrogen chloride, which is corrosive to common metals; *Stability During Transport:* Stable; *Neutralizing Agents for Acids and Caustics:* Flood with water, rinse with sodium bicarbonate or lime solution; *Polymerization:* Not pertinent; *Inhibitor of Polymerization:* Not pertinent.

METHYL ETHYL KETONE

Chemical Designations - *Synonyms*: 2-Butanone; Ethyl methyl ketone; MEK; *Chemical Formula*: $CH_3COCH_2CH_3$.

Observable Characteristics - *Physical State (as shipped)*: Liquid; *Color*: Colorless; *Odor*: Like acetone; pleasant; pungent.

Physical and Chemical Properties - *Physical State at 15 °C and 1 atm.*: Liquid; *Molecular Weight*: 72.11; *Boiling Point at 1 atm.*: 175.3, 79.6, 352.8; *Freezing Point*: -123.3, -86.3, 186.9; *Critical Temperature*: 504.5, 262.5, 535.7; *Critical Pressure*: 603, 41.0, 4.15; *Specific Gravity*: 0.806 at 20 °C (liquid); *Vapor (Gas) Specific Gravity*: 2.5; *Ratio of Specific Heats of Vapor (Gas)*: 1.075; *Latent Heat of Vaporization*: 191, 106, 4.44; *Heat of Combustion*: -13,480, -7491, -313.6; *Heat of Decomposition*: Not pertinent.

Health Hazards Information - *Recommended Personal Protective Equipment*: Organic canister or air pack; plastic gloves; goggles or face shield; *Symptoms Following Exposure*: Liquid causes eye burn. Vapor irritates eye, nose, and throat; can cause headache, dizziness, nausea, weakness, and loss of consciousness; *General Treatment for Exposure*: INHALATION: remove victim from exposure; if breathing has stopped, start resuscitation and administer oxygen. EYES: wash with plenty of water for at least 15 min. and call physician; *Toxicity by Inhalation (Threshold Limit Value)*: 200 ppm; *Short-Term Inhalation Limits*: 290 mg/m^3 for 60 min.; *Toxicity by Ingestion*: Grade 2, LD_{50} 0.5 to 5 g/kg (rat); *Late Toxicity*: None; *Vapor (Gas) Irritant Characteristics*: Vapors causes a slight smarting of the eyes or respiratory system if present in high concentrations. The effect is temporary; *Liquid or Solid Irritant Characteristics*: Minimum hazard. If spilled on clothing and allowed to remain, may cause smarting and reddening of skin; *Odor Threshold*: 10 ppm.

Fire Hazards - *Flash Point (deg. F)*: 20 CC; 22 OC; *Flammable Limits in Air (%):* 1.8 - 11.5; *Fire Extinguishing Agents:* Alcohol foam dry chemical, carbon dioxide; *Fire Extinguishing Agents Not To Be Used:* Water may be ineffective; *Special Hazards of Combustion Products:* Not pertinent; *Behavior in Fire:* Not pertinent; *Ignition Temperature (deg. F):* 961; *Electrical Hazard:* Class I, Group D; *Burning Rate:* 4.1 mm/min.

Chemical Reactivity - *Reactivity with Water:* No reaction; *Reactivity with Common Materials:* No reaction; *Stability During Transport:* Stable; *Neutralizing Agents for Acids and Caustics:* Not pertinent; *Polymerization:* Not pertinent; *Inhibitor of Polymerization:* Not pertinent.

METHYLETHYLPYRIDINE

Chemical Designations - *Synonyms*: Aldehyde-collidine; Aldehydine; 5-Ethyl-2-Methylpyridine; 5-Ethyl-2-picoline; MEP; *Chemical Formula*: $C_8H_{11}N$.

Observable Characteristics - *Physical State (as shipped)*: Liquid; *Color*: Colorless; *Odor*: Sharp, penetrating.

Physical and Chemical Properties - *Physical State at 15 °C and 1 atm.*: Liquid; *Molecular Weight*: 121.18; *Boiling Point at 1 atm.*: 252, 178, 451; *Freezing Point*: -94.5, -70.3, 202.9; *Critical Temperature*: Not pertinent; *Critical Pressure*: Not pertinent; *Specific Gravity*: 0.922 at 20 °C (liquid); *Vapor (Gas) Specific Gravity*: Not pertinent; *Ratio of Specific Heats of Vapor (Gas)*: Not pertinent; *Latent Heat of Vaporization*: Data not available; *Heat of Combustion*: Data not available; *Heat of Decomposition*: Not pertinent.

Health Hazards Information - *Recommended Personal Protective Equipment*: Air-supplied mask for high vapor concentrations; plastic gloves; goggles or face shield; *Symptoms Following Exposure*: Breathing of vapors will cause vomiting and chest discomfort. Contact with liquid causes skin and eye burns; *General Treatment for Exposure*: INHALATION: remove victim from exposure; give oxygen if breathing is difficult; call a physician. EYES OR SKIN: immediately flush with plenty of water for at least 15 min.; get medical care for eyes; *Toxicity by Inhalation (Threshold Limit Value)*: 1700 ppm/3.7 hr/lethal (rat); *Short-Term Inhalation Limits*: Data not available; *Toxicity by Ingestion*: Grade 2, LD_{50} 0.5 to 5 g/kg (rat); *Late Toxicity*: Data not available; *Vapor (Gas) Irritant Characteristics*: Vapors cause moderate irritation such that personnel will find high concentrations unpleasant. The effect is temporary; *Liquid or Solid Irritant Characteristics*: Causes smarting of the skin and first-degree burns on short exposure; may cause secondary burns on long exposure; *Odor Threshold*: Data not available.

Fire Hazards - *Flash Point (deg. F)*: 155 OC; *Flammable Limits in Air (%):* 1.1 - 6.6; *Fire Extinguishing Agents:* Foam, carbon dioxide, dry chemical; *Fire Extinguishing Agents Not To Be Used:* Not pertinent; *Special Hazards of Combustion Products:* Irritating vapors are generated when heated; *Behavior in Fire:* Not pertinent; *Ignition Temperature (deg. F):* 939; *Electrical Hazard:* Not pertinent; *Burning Rate:* Data not available.

Chemical Reactivity - *Reactivity with Water:* No reaction; *Reactivity with Common Materials:* No reaction; *Stability During Transport:* Stable; *Neutralizing Agents for Acids and Caustics:* Flush with water, neutralize with dilute acetic acid; *Polymerization:* Not pertinent; *Inhibitor of Polymerization:* Not pertinent.

METHYL FORMAL

Chemical Designations - *Synonyms*: Dimethoxymethane; Dimethylformal; Formaldehyde dimethylacetat; Methylal; Methylene dimethyl ether; *Chemical Formula*: $CH_2(OCH_3)_2$.

Observable Characteristics — *Physical State (as shipped)*: Liquid; *Color*: Colorless; *Odor*: Mild, ethereal; chloroform-like.

Physical and Chemical Properties - *Physical State at 15 °C and 1 atm.*: Liquid; *Molecular Weight*: 76.1; *Boiling Point at 1 atm.*: 108, 42, 315; *Freezing Point*: -157, -105, 168; *Critical Temperature*: 419, 215, 488; *Critical Pressure*: Not pertinent; *Specific Gravity*: 0.861 at 20 °C (liquid); *Vapor (Gas) Specific Gravity*: 2.6; *Ratio of Specific Heats of Vapor (Gas)*: 1.0888; *Latent Heat of Vaporization*: 161.5, 89.8, 3.76; *Heat of Combustion*: -10,970, -6,100, -255; *Heat of Decomposition*: Not pertinent.

Health Hazards Information - *Recommended Personal Protective Equipment*: Self-contained breathing apparatus or all-purpose canister mask; gloves; rubber gloves; chemical safety goggles; impervious apron and boots; *Symptoms Following Exposure*: Inhalation causes irritation of respiratory system and depression of central nervous system. Liquid causes irritation of eyes and will irritate skin if allowed to remain. Ingestion causes depression of central nervous system; *General Treatment for Exposure*: INHALATION: remove victim from contaminated area and administer artificial respiration and oxygen if necessary. EYES: flush with plenty of water; get medical attention. SKIN: flush with plenty of water. INGESTION: induce vomiting, then give gastric lavage and saline cathartics; subsequent treatment is symptomatic and supportive; *Toxicity by Inhalation (Threshold Limit Value)*: 1,000 ppm; *Short-Term Inhalation Limits*: Data not available; *Toxicity by Ingestion*: Grade 1; LD_{50} 5 to 15 g/kg; *Late Toxicity*: Liver and kidney injury may follow high exposures; *Vapor (Gas) Irritant Characteristics*: Vapors causes a slight smarting of the eyes or respiratory system if present in high concentrations. The effect is temporary; *Liquid or Solid Irritant Characteristics*: Minimum hazard. If spilled on clothing and allowed to remain, may cause smarting and reddening of skin; *Odor Threshold*: Data not available.

Fire Hazards - *Flash Point (deg. F)*: 0 OC; *Flammable Limits in Air (%):* 1.6 - 17.6; *Fire Extinguishing Agents:* Dry chemical, foam, carbon dioxide; *Fire Extinguishing Agents Not To Be Used:* Water may be ineffective; *Special Hazards of Combustion Products:* Irritating formaldehyde smoke may be present in smoke; *Behavior in Fire:* Not pertinent; *Ignition Temperature (deg. F):* 459; *Electrical Hazard:* Data not available; *Burning Rate:* 5.5 mm/min.

Chemical Reactivity - *Reactivity with Water:* No reaction; *Reactivity with Common Materials:* No reaction; *Stability During Transport:* Stable; *Neutralizing Agents for Acids and Caustics:* Not pertinent; *Polymerization:* Not pertinent; *Inhibitor of Polymerization:* Not pertinent.

METHYL FORMATE

Chemical Designations - *Synonyms*: Formic acid, methyl ester; *Chemical Formula*: $HCOOH_3$.

Observable Characteristics — *Physical State (as shipped)*: Liquid; *Color*: Colorless; *Odor*: Pleasant; agreeable.

Physical and Chemical Properties - *Physical State at 15 °C and 1 atm.*: Liquid; *Molecular Weight*: 60.1; *Boiling Point at 1 atm.*: 89.2, 31.8, 305; *Freezing Point*: -147.6, -99.8, 173.4; *Critical Temperature*: 417, 214, 487; *Critical Pressure*: 870, 59.2, 6.00; *Specific Gravity*: 0.977 at 20 °C (liquid); *Vapor (Gas) Specific Gravity*: 2.07; *Ratio of Specific Heats of Vapor (Gas)*: 1.1446; *Latent Heat of Vaporization*: 202, 112, 4.696; *Heat of Combustion*: -6,980, -3,880, -162; *Heat of Decomposition*: Not pertinent.

Health Hazards Information - *Recommended Personal Protective Equipment*: Goggles or safety glasses; self-contained breathing apparatus; rubber gloves; *Symptoms Following Exposure*: Inhalation causes irritation of mucous membranes. Prolonged inhalation can produce narcosis and central nervous symptoms, including some temporary visual disturbance. Contact with liquid irritates eyes and may irritate skin if allowed to remain. Ingestion causes irritation of mouth and stomach and central nervous system depression, including visual disturbances; *General Treatment for Exposure*: INHALATION: remove to fresh air and rest; if pulmonary edema develops, administer oxygen; call physician. EYES: irrigate with water for 15 min SKIN: wash thoroughly with soap and water. INGESTION: do NOT induce vomiting; get medical attention; *Toxicity by Inhalation (Threshold Limit Value)*: 100 ppm; *Short-Term Inhalation Limits*: Data not available; *Toxicity by Ingestion*: Grade 1; LD_{50} 5 to 15 g/kg; *Late Toxicity*: Data not available; *Vapor (Gas) Irritant Characteristics*: Vapors are moderately irritating such that personnel will not usually tolerate moderate or high concentrations; *Liquid or Solid Irritant Characteristics*: Minimum hazard. If spilled on clothing and allowed to remain, may cause smarting and reddening of skin; *Odor Threshold*: Data not available.

Fire Hazards - *Flash Point (deg. F)*: - 26 CC; *Flammable Limits in Air (%):* 5 - 22.7; *Fire Extinguishing Agents:* Dry chemical, carbon dioxide, alcohol foam; *Fire Extinguishing Agents Not To Be Used:* Water may be ineffective; *Special Hazards of Combustion Products:* Not pertinent; *Behavior in Fire:* Vapor is heavier than air and may travel a considerable distance to source of ignition and flash back; *Ignition Temperature (deg. F):* 853; *Electrical Hazard:* Data not available; *Burning Rate:* 2.5 mm/min.

Chemical Reactivity - *Reactivity with Water:* Slow reaction to form formic acid and methyl alcohol; reaction is not hazardous; *Reactivity with Common Materials:* No reaction; *Stability During Transport:* Stable; *Neutralizing Agents for Acids and Caustics:* Not pertinent; *Polymerization:* Not pertinent; *Inhibitor of Polymerization:* Not pertinent.

METHYLHYDRAZINE

Chemical Designations - *Synonyms*: Monomethylhydrazine; MMH; *Chemical Formula*: CH_3NHNH_2.
Observable Characteristics - *Physical State (as shipped)*: Liquid; *Color*: Colorless; *Odor*: Like ammonia.

Physical and Chemical Properties - *Physical State at 15 °C and 1 atm.*: Liquid; *Molecular Weight*: 46.1; *Boiling Point at 1 atm.*: 189.5, 87.5, 360.7; *Freezing Point*: -62.3, -52.4, 220.8; *Critical Temperature*: 594, 312, 585; *Critical Pressure*: 1,195, 81.3, 8.25; *Specific Gravity*: 0.878 at 20 °C (liquid); *Vapor (Gas) Specific Gravity*: 1.59; *Ratio of Specific Heats of Vapor (Gas)*: 1.1326; *Latent Heat of Vaporization*: 376, 209, 8.75; *Heat of Combustion*: -12,178, -6,766, -283.1; *Heat of Decomposition*: Not pertinent.

Health Hazards Information - *Recommended Personal Protective Equipment*: Organic canister mask or self-contained breathing apparatus; goggles or face shield; rubber gloves; protective clothing; *Symptoms Following Exposure*: Tremors and convulsions follow absorption by any route. Inhalation causes local irritation of respiratory tract, respiratory distress, and systemic effects. Contact of liquid with eyes or skin causes irritation and burns. Ingestion causes irritation of mouth and stomach; *General Treatment for Exposure*: Get medical attention at once following all exposures to this compound. INHALATION: move victim to fresh air and keep him quiet; give artificial respiration if breathing stops. EYES: flush with large quantities of water for at least 15 minutes. SKIN: immediately flush with large quantities of water and treat as for alkali burn. INGESTION: give egg whites or other emollient, followed by a 5% salt solution or other mild emetic. Keep patient as quiet as possible. To control convulsions, short-acting barbiturates may be administered parenterally by a physician with due regard for depression of respiration; *Toxicity by Inhalation (Threshold Limit Value)*: 0.2 ppm; *Short-Term Inhalation Limits*: 90 ppm for 10 min.; 30 ppm for 30 min.; 15 ppm for 60 min.; *Toxicity by Ingestion*: Grade 4, oral $LD_{50} = 33$ mg/kg (rat); *Late Toxicity*: Hemolytic anemia may result from large doses by any route; *Vapor (Gas) Irritant Characteristics*: Vapors are moderately irritating such that personnel will not usually tolerate moderate or high concentrations; *Liquid or Solid Irritant Characteristics*: Severe skin irritant. Causes second- and third-degree burns on short contact and is very injurious to the eyes; *Odor Threshold*: 1-3 ppm.

Fire Hazards - *Flash Point (deg. F)*: 62 OC; *Flammable Limits in Air (%)*: 2.5 - 98; *Fire Extinguishing Agents:* Water or dry chemical; *Fire Extinguishing Agents Not To Be Used:* Not pertinent; *Special Hazards of Combustion Products:* Irritating nitrogen oxides are produced; *Behavior in Fire:* May explode; *Ignition Temperature (deg. F):* 382; *Electrical Hazard:* Data not available; *Burning Rate:* 2.0 mm/min.

Chemical Reactivity - *Reactivity with Water:* No reaction; *Reactivity with Common Materials:* Reacts slowly with air, but heat may cause ignition of rags, rust or other combustibles; *Stability During Transport:* Stable if not in contact with iron, copper or their alloys; *Neutralizing Agents for Acids and Caustics:* Flush with water; *Polymerization:* Not pertinent; *Inhibitor of Polymerization:* Not pertinent.

METHYL ISOBUTYL CARBINOL

Chemical Designations - *Synonyms*: Isobutyl methyl carbinol; Methyl Alcohol; MAOH; 4-Methyl-2-pentanol; MIBC; MIC; *Chemical Formula*: $(CH_3)_2CHCH_2CH(OH)CH_3$.

Observable Characteristics - *Physical State (as shipped)*: Liquid; *Color*: Colorless; *Odor*: Mild, sharp, non-residual.

Physical and Chemical Properties - *Physical State at 15 °C and 1 atm.*: Liquid; *Molecular Weight*: 102.18; *Boiling Point at 1 atm.*: 269.2, 131.8, 405; *Freezing Point*: < -130, < -90, < 183; *Critical Temperature*: 556, 291, 564; *Critical Pressure*: Not pertinent; *Specific Gravity*: 0.807 at 20 °C (liquid); *Vapor (Gas) Specific Gravity*: Not pertinent; *Ratio of Specific Heats of Vapor (Gas)*: 1.053; *Latent Heat of Vaporization*: 162, 90.1, 3.77; *Heat of Combustion*: (est.) -16,600, -9,300, -387; *Heat of Decomposition*: Not pertinent.

Health Hazards Information - *Recommended Personal Protective Equipment*: Air pack or organic canister mask; rubber gloves; goggles or face shield; *Symptoms Following Exposure*: Vapor irritates eyes and nose; may cause anesthesia. Prolonged contact with liquid causes irritation and cracking of skin; also irritates eyes; *General Treatment for Exposure*: INHALATION: remove to fresh air; give artificial respiration if needed; call a doctor. SKIN: flush with water. EYES: flush with water for at least 15 min.; consult a doctor; *Toxicity by Inhalation (Threshold Limit Value)*: 25 ppm; *Short-Term Inhalation Limits*: Data not available; *Toxicity by Ingestion*: Grade 2, LD_{50} 0.5 to 5 g/kg (rat); *Late Toxicity*: None; *Vapor (Gas) Irritant Characteristics*: Vapors cause moderate irritation such that personnel will find high concentrations unpleasant. The effect is temporary; *Liquid or Solid Irritant Characteristics*: Minimum hazard. If spilled on clothing and allowed to remain, may cause smarting and reddening of skin; *Odor Threshold*: Data not available.

Fire Hazards - *Flash Point (deg. F)*: 120 -130 OC; 106 CC; *Flammable Limits in Air (%):* 1.0 - 5.5; *Fire Extinguishing Agents:* Alcohol foam, dry chemical, or carbon dioxide; *Fire Extinguishing Agents Not To Be Used:* Not pertinent; *Special Hazards of Combustion Products:* Not pertinent; *Behavior in Fire:* Not pertinent; *Ignition Temperature :* Data not available; *Electrical Hazard:* Not pertinent; *Burning Rate:* Data not available.

Chemical Reactivity - *Reactivity with Water:* No reaction; *Reactivity with Common Materials:* No reaction; *Stability During Transport:* Stable; *Neutralizing Agents for Acids and Caustics:* Not pertinent; *Polymerization:* Not pertinent; *Inhibitor of Polymerization.*

METHYL ISOBUTYL KETONE

Chemical Designations - *Synonyms*: Hexone; Isobutyl methyl ketone; Isopropylacetone; 4-Methyl-2-pentanone; MIBK; MIK; *Chemical Formula*: $(CH_3)_2CHCH_2COCH_3$

Observable Characteristics - *Physical State (as shipped)*: Liquid; *Color*: Colorless; *Odor*: Pleasant; mild, characteristic; sharp; non-residual; ketonic.

Physical and Chemical Properties - *Physical State at 15 °C and 1 atm.*: Liquid; *Molecular Weight*: 100.16; *Boiling Point at 1 atm.*: 241.2, 116.2, 389.4; *Freezing Point*: -119, -84, 189; *Critical*

Temperature: 568.9, 298.3, 571.5; *Critical Pressure*: 475, 32.3, 3.27; *Specific Gravity*: 0.802 at 20 °C (liquid); *Vapor (Gas) Specific Gravity*: Not pertinent; *Ratio of Specific Heats of Vapor (Gas)*: 1.061; *Latent Heat of Vaporization*: 149, 82.5, 3.45; *Heat of Combustion*: (est.) -10,400, -5,800, -242; *Heat of Decomposition*: Not pertinent.

Health Hazards Information - *Recommended Personal Protective Equipment*: Air pack or organic canister mask; rubber gloves; goggles or face shield; *Symptoms Following Exposure*: Vapor irritates eyes and nose; high concentrations can cause anesthesia and depression. Liquid dries out skin and may cause dermatitis; generally irritates eyes but does not injure them; *General Treatment for Exposure*: INHALATION: remove victim to fresh air and give artificial respiration if needed; seek medical attention immediately. SKIN OR EYES: flush eyes thoroughly with water; wash skin with water until irritation stops; *Toxicity by Inhalation (Threshold Limit Value)*: 100 ppm; *Short-Term Inhalation Limits*: 100 ppm for 60 min.; *Toxicity by Ingestion*: Grade 2, LD_{50} 0.5 to 5 g/kg (rat); *Late Toxicity*: None; *Vapor (Gas) Irritant Characteristics*: Vapors cause moderate irritation such that personnel will find high concentrations unpleasant. The effect is temporary; *Liquid or Solid Irritant Characteristics*: Minimum hazard. If spilled on clothing and allowed to remain, may cause smarting and reddening of skin; *Odor Threshold*: 0.47 ppm.

Fire Hazards - *Flash Point (deg. F)*: 73 CC; 75 OC; *Flammable Limits in Air (%)*: 1.4 - 7.5; *Fire Extinguishing Agents:* Alcohol foam, dry chemical, or carbon dioxide; *Fire Extinguishing Agents Not To Be Used:* Water may be ineffective; *Special Hazards of Combustion Products:* Irritating vapors are generated when heated; *Behavior in Fire:* Vapors may travel a considerable distance and ignite, causing a flash back; *Ignition Temperature (deg. F):* 854; *Electrical Hazard:* Class I, Group D; *Burning Rate:* Data not available.

Chemical Reactivity - *Reactivity with Water:* No reaction; *Reactivity with Common Materials:* No reaction; *Stability During Transport:* Stable; *Neutralizing Agents for Acids and Caustics:* Not pertinent; *Polymerization:* Not pertinent; *Inhibitor of Polymerization:* Not pertinent.

METHYL ISOPROPENYL KETONE, INHIBITED

Chemical Designations - *Synonyms*: Isopropenyl methyl ketone; 2-Methyl-1-butene-3-one; *Chemical Formula*: $CH_3COC(CH_3)=CH_2$.

Observable Characteristics - *Physical State (as shipped)*: Liquid; *Color*: Colorless; *Odor*: Very pungent; pleasant; sweet.

Physical and Chemical Properties - *Physical State at 15 °C and 1 atm.*: Liquid; *Molecular Weight*: 84.1; *Boiling Point at 1 atm.*: 208, 98, 371; *Freezing Point*: -65, -54, 219; *Critical Temperature*: Not pertinent; *Critical Pressure*: Not pertinent; *Specific Gravity*: 0.85 at 20 °C (liquid); *Vapor (Gas) Specific Gravity*: 2.9; *Ratio of Specific Heats of Vapor (Gas)*: 1.0796 at 20 °C (68°F); *Latent Heat of Vaporization*: (est.) 182, 101, 4.23; *Heat of Combustion*: (est.) -15,500, -8,600, -360; *Heat of Decomposition*: Not pertinent.

Health Hazards Information - *Recommended Personal Protective Equipment*: Goggles or face shield; rubber gloves; *Symptoms Following Exposure*: Inhalation causes irritation of nose and throat. Liquid may cause severe damage to eyes, resulting possibly in some permanent impairment of vision; vapor produces tears. If not removed promptly from skin, liquid may cause delayed pain and blistering. Ingestion causes irritation of mouth and stomach; *General Treatment for Exposure*: INHALATION: remove victim from exposure; give artificial respiration if needed; call physician. EYES: immediately irrigate with copious amounts of water for 15 min.; call physician if burn has occurred. INGESTION: induce vomiting; call physician; *Toxicity by Inhalation (Threshold Limit Value)*: Data not available; *Short-Term Inhalation Limits*: Data not available; *Toxicity by Ingestion*: Grade 3; oral LD_{50} = 180 mg/kg (rat); *Late Toxicity*: Data not available; *Vapor (Gas) Irritant Characteristics*: Vapors are moderately irritating such that personnel will not usually tolerate moderate or high concentrations; *Liquid or Solid Irritant Characteristics*: Causes smarting of the skin and first-degree burns on short exposure; may cause secondary burns on long exposure; *Odor Threshold*: Data not available.

Fire Hazards - *Flash Point (deg. F)*: < 73 CC; *Flammable Limits in Air (%)*: 1.8 - 9.0; *Fire Extinguishing Agents:* Dry chemical, foam, carbon dioxide; *Fire Extinguishing Agents Not To Be Used:*

Water may be ineffective; *Special Hazards of Combustion Products:* Not pertinent; *Behavior in Fire:* May polymerize and explode; *Ignition Temperature :* Data not available; *Electrical Hazard:* Data not available; *Burning Rate:* 4.7 mm/min.

Chemical Reactivity - *Reactivity with Water:* No reaction; *Reactivity with Common Materials:* No reaction; *Stability During Transport:* Stable; *Neutralizing Agents for Acids and Caustics:* Not pertinent; *Polymerization:* Will polymerize in the absence of inhibitor, especially when heated; *Inhibitor of Polymerization:* Up to 1% hydroquinone.

METHYL MERCAPTAN
Chemical Designations - *Synonyms*: Mercaptomethane; Methanethiol; Methyl sulfidrate; Thiomethyl alcohol; *Chemical Formula*: CH_3SH.
Observable Characteristics - *Physical State (as shipped)*: Liquefied compressed gas; *Color*: Colorless; *Odor*: Garlic; foul; strong offensive.
Physical and Chemical Properties - *Physical State at 15 °C and 1 atm.*: Gas; *Molecular Weight*: 48.1; *Boiling Point at 1 atm.*: 43.2, 6.2, 279.4; *Freezing Point*: -189, -123, 150; *Critical Temperature*: 386.2, 196.8, 470.0; *Critical Pressure*: 1,050, 71.4, 7.25; *Specific Gravity*: 0.892 at 6 °C (liquid); *Vapor (Gas) Specific Gravity*: 1.66; *Ratio of Specific Heats of Vapor (Gas)*: 1.1988; *Latent Heat of Vaporization*: 220, 122, 5.10; *Heat of Combustion*: -11,054, -6,141, -257.0; *Heat of Decomposition*: Not pertinent.
Health Hazards Information - *Recommended Personal Protective Equipment*: Self-contained breathing apparatus or air-line mask; goggles or face shield; rubber gloves; *Symptoms Following Exposure*: Inhalation causes irritation of respiratory system, tremors, paralysis, unconsciousness; death may follow respiratory paralysis. Contact with liquid irritates eyes and skin. Ingestion causes irritation of mouth and stomach plus symptoms described for inhalation; *General Treatment for Exposure*: INHALATION: remove patient immediately from the contaminated area; keep him warm and at complete rest; if necessary give artificial respiration until medical assistance can be obtained; oxygen or oxygen-CO_2 inhalation is recommended, continuing after spontaneous breathing has returned. EYES: for exposure to vapor, apply hot and cold compresses to reduce pain of conjunctivitis; for exposure to liquid, wash with water and obtain medical assistance. SKIN: wash with water. INGESTION: induce vomiting and follow with gastric lavage; *Toxicity by Inhalation (Threshold Limit Value)*: 0.5 ppm; *Short-Term Inhalation Limits*: 20 ppm for 5 min.; *Toxicity by Ingestion*: Data not available; *Late Toxicity*: Data not available; *Vapor (Gas) Irritant Characteristics*: Data not available; *Liquid or Solid Irritant Characteristics*: Data not available; *Odor Threshold*: 0.0021 ppm.
Fire Hazards - *Flash Point :* Not pertinent (flammable, liquefied compressed gas); *Flammable Limits in Air (%):* 3.9 - 21.8; *Fire Extinguishing Agents:* Preferably let fire burn, stop gas flow. Fires may be extinguished with dry chemical, foam, or carbon dioxide; *Fire Extinguishing Agents Not To Be Used:* Water may be ineffective; *Special Hazards of Combustion Products:* Irritating sulfur dioxide is produced; *Behavior in Fire:* Containers may explode; *Ignition Temperature :* Data not available; *Electrical Hazard:* Data not available; *Burning Rate:* 3.8 mm/min.
Chemical Reactivity - *Reactivity with Water:* No reaction; *Reactivity with Common Materials:* No reaction; *Stability During Transport:* Stable; *Neutralizing Agents for Acids and Caustics:* Not pertinent; *Polymerization:* Not pertinent; *Inhibitor of Polymerization:* Not pertinent.

METHYL METHACRYLATE
Chemical Designations - *Synonyms*: Methacrylate monomer; Methacrylic acid, methyl ester; Methyl alpha-methylacrylate; Methyl 2-methyl-2-propenoate; *Chemical Formula*: $CH_2=C(CH_3)COOCH_3$.
Observable Characteristics - *Physical State (as shipped)*: Liquid; *Color*: Colorless; *Odor*: Sharp, fragrant; pleasant smelling; pungent ester.
Physical and Chemical Properties - *Physical State at 15 °C and 1 atm.*: Liquid; *Molecular Weight*: 100.12; *Boiling Point at 1 atm.*: 214, 101, 374; *Freezing Point*: -54, -48, 225; *Critical Temperature*: 561, 294, 567; *Critical Pressure*: 485, 33, 3.3; *Specific Gravity*: 0.945 at 20 °C (liquid); *Vapor (Gas) Specific Gravity*: Not pertinent; *Ratio of Specific Heats of Vapor (Gas)*: 1.059; *Latent Heat of*

Vaporization: 140, 77, 3.2; *Heat of Combustion*: (est.) -11,400, -6,310, -264; *Heat of Decomposition*: Not pertinent.

Health Hazards Information - *Recommended Personal Protective Equipment*: Air mask; plastic gloves; goggles; *Symptoms Following Exposure*: Irritation of eyes, nose, and throat. Nausea and vomiting. Liquid may cause skin irritation; *General Treatment for Exposure*: INHALATION: remove to fresh air; apply artificial respiration and oxygen if needed; refer to physician. SKIN OR EYES: flush with plenty of water for 15 min.; refer to physician for eye exposure; *Toxicity by Inhalation (Threshold Limit Value)*: 100 ppm; *Short-Term Inhalation Limits*: Data not available; *Toxicity by Ingestion*: Grade 1; LD_{50} 5 to 15 g/kg (rat); *Late Toxicity*: Data not available; *Vapor (Gas) Irritant Characteristics*: Vapor is moderately irritating such that personnel will not usually tolerate moderate or high vapor concentrations; *Liquid or Solid Irritant Characteristics*: Causes smarting of the skin and first-degree burns on short exposure; may cause secondary burns on long exposure; *Odor Threshold*: Data not available.

Fire Hazards - *Flash Point (deg. F)*: 50 OC; *Flammable Limits in Air (%)*: 2.1 - 12.5; *Fire Extinguishing Agents:* Foam, carbon dioxide, dry chemical; *Fire Extinguishing Agents Not To Be Used:* Water may be ineffective; *Special Hazards of Combustion Products:* Not pertinent; *Behavior in Fire:* Vapor is heavier than air and may travel a considerable distance to a source of ignition and flash back. Containers may explode in fire or when heated because of polymerization; *Ignition Temperature (deg. F):* 790; *Electrical Hazard:* Not pertinent; *Burning Rate:* 2.5 mm/min.

Chemical Reactivity - *Reactivity with Water:* No reaction; *Reactivity with Common Materials:* No reaction; *Stability During Transport:* Stable; *Neutralizing Agents for Acids and Caustics:* Not pertinent; *Polymerization:* Heat, oxidizing agents, and ultraviolet light may cause polymerization; *Inhibitor of Polymerization:* Hydroquinone, 22 - 65 ppm; hydroquinone methyl ether, 22 - 120 ppm; dimethyl tert-butylphenol, 45 - 65 ppm.

METHYL PARATHION

Chemical Designations - *Synonyms*: MPT; O,O-Dimethyl O-(p-Nitrophenyl) Phosphorothiorate; O,O-Dimethyl O-p-Nitrophenyl thiophosphate; Parathion-methyl; *Chemical Formula*: $(CH_3O)_2PSOC_6H_4NO_2$-p.

Observable Characteristics - *Physical State (as shipped)*: Solid or liquid; *Color*: Colorless solid or brownish liquid; *Odor*: Characteristic; like rotten eggs or garlic.

Physical and Chemical Properties - *Physical State at 15 °C and 1 atm.*: Liquid or solid; *Molecular Weight*: 263.2; *Boiling Point at 1 atm.*: Very high; *Freezing Point*: 65, 18, 291; *Critical Temperature*: Not pertinent; *Critical Pressure*: Not pertinent; *Specific Gravity*: 1.360 at 20 °C (liquid); *Vapor (Gas) Specific Gravity*: Not pertinent; *Ratio of Specific Heats of Vapor (Gas)*: Not pertinent; *Latent Heat of Vaporization*: Not pertinent; *Heat of Combustion*: Data not available; *Heat of Decomposition*: Not pertinent.

Health Hazards Information - *Recommended Personal Protective Equipment*: Approved mask or respirator; natural rubber gloves, overshoes; protective clothing; goggles; *Symptoms Following Exposure*: Exposure to fumes from a fire, or to the liquid, causes headache, blurred vision, constricted pupils of the eyes, weakness, nausea, cramps, diarrhea, and tightness in the chest. Muscle twitch and convulsions may follow. Symptoms may develop over a period of 8 hrs; *General Treatment for Exposure*: Speed is essential. INGESTION: call a doctor! If victim is not breathing, immediately institute artificial respiration by mouth-to-mouth, mouth-to-nose, or mouth-to-oropharyngeal method; when victim is conscious, give milk, water, or salt-water and induce vomiting repeatedly. SKIN OR EYES: flood and wash exposed areas thoroughly with water; remove contaminated clothing under a shower; *Toxicity by Inhalation (Threshold Limit Value)*: 0.2 mg/m^3 (solid); 100 ppm (liquid); *Short-Term Inhalation Limits*: Data not available; *Toxicity by Ingestion*: Grade 4; LD_{50} below 50 mg/kg (rat); *Late Toxicity*: Data not available; *Vapor (Gas) Irritant Characteristics*: Not pertinent; *Liquid or Solid Irritant Characteristics*: Poisonous when absorbed thorough skin; *Odor Threshold*: Data not available.

Fire Hazards - *Flash Point (deg. F)*: 115 OC; *Flammable Limits in Air (%):* Data not available; *Fire

Extinguishing Agents: Water; *Fire Extinguishing Agents Not To Be Used:* Not pertinent; *Special Hazards of Combustion Products:* Toxic gases are produced in fires; *Behavior in Fire:* Drums may rupture violently; *Ignition Temperature :* Data not available; *Electrical Hazard:* Not pertinent; *Burning Rate:* Data not available.

Chemical Reactivity - *Reactivity with Water:* Half decomposed in 8 days at 40° C; *Reactivity with Common Materials:* Is absorbed in wood. etc., which must be replaced to eliminate poison hazard; *Stability During Transport:* Decomposes above 50 °C with possible explosive force; *Neutralizing Agents for Acids and Caustics:* Apply caustic or soda ash slurry until yellow stains disappear; *Polymerization:* Not pertinent; *Inhibitor of Polymerization:* Not pertinent.

METHYL PHOSPHONOTHIOIC DICHLORIDE (ANHYDROUS)

Chemical Designations - *Synonyms:* MPTD; *Chemical Formula*: CH_3PSCl_2.

Observable Characteristics - *Physical State (as shipped)*: Liquid; *Color*: Clear; *Odor*: Acrid.

Physical and Chemical Properties - *Physical State at 15 °C and 1 atm.*: Liquid; *Molecular Weight*: 149; *Boiling Point at 1 atm.*: Data not available; *Freezing Point*: -14.1, -25.6, 247.6; *Critical Temperature*: Not pertinent; *Critical Pressure*: Not pertinent; *Specific Gravity*: 1.42 at 20 °C (liquid); *Vapor (Gas) Specific Gravity*: Not pertinent; *Ratio of Specific Heats of Vapor (Gas)*: Not pertinent; *Latent Heat of Vaporization*: (est.) 110, 60, 2.5; *Heat of Combustion*: Data not available; *Heat of Decomposition*: Not pertinent.

Health Hazards Information - *Recommended Personal Protective Equipment*: Rubber or neoprene gloves; respiratory protection; goggles; *Symptoms Following Exposure*: Inhalation causes irritation of nose and throat; effects are quite similar to those of phosgene. Ingestion causes irritation of mouth and stomach. Delayed, painful eye irritation may occur from exposure to vapor; liquid causes severe irritation. Contact with skin causes irritation and burns; *General Treatment for Exposure*: Get medical attention after all exposures to this compound. INHALATION: remove victim to fresh air; alert physician to delayed effects similar to those of phosgene. INGESTION: give large amount of water and induce vomiting. EYES: flush with water for at least 15 min. SKIN: flood with water; *Toxicity by Inhalation (Threshold Limit Value)*: Data not available; *Short-Term Inhalation Limits*: Data not available; *Toxicity by Ingestion*: Data not available; *Late Toxicity*: Data not available; *Vapor (Gas) Irritant Characteristics*: Data not available; *Liquid or Solid Irritant Characteristics*: Data not available; *Odor Threshold*: Data not available.

Fire Hazards - *Flash Point (deg. F)*: > 122 OC; *Flammable Limits in Air (%):* Data not available; *Fire Extinguishing Agents:* Dry chemical or carbon dioxide; *Fire Extinguishing Agents Not To Be Used:* Water or foam; *Special Hazards of Combustion Products:* Irritating hydrogen chloride, sulfur dioxide and other fumes may be formed in fire; *Behavior in Fire:* Data not available; *Ignition Temperature :* Data not available; *Electrical Hazard:* Data not available; *Burning Rate:* Data not available.

Chemical Reactivity - *Reactivity with Water:* Reacts with water to form hydrochloric acid and/or hydrogen chloride vapor. The reaction may be violent; *Reactivity with Common Materials:* Corrosive to metals because of its high acidity; *Stability During Transport:* Stable; *Neutralizing Agents for Acids and Caustics:* Flush with water, rinse with dilute sodium bicarbonate or soda ash solution; *Polymerization:* Not pertinent; *Inhibitor of Polymerization:* Not pertinent.

1-METHYLPYRROLIDONE

Chemical Designations - *Synonyms:* 1-Methyl-2-pyrrolidinone; N-Methylpyrrolidinone; –Methyl-pyrrolidone; –Methyl-alpha-pyrrolidone; *Chemical Formula:* C_5H_9NO.

Observable Characteristics - *Physical State (as shipped)*: Liquid; *Color*: Colorless; *Odor*: Amine-like.

Physical and Chemical Properties - *Physical State at 15 °C and 1 atm.*: Liquid; *Molecular Weight*: 99; *Boiling Point at 1 atm.*: 396, 202, 475; *Freezing Point*: 1, -17, 256; *Critical Temperature*: Not pertinent; *Critical Pressure*: Not pertinent; *Specific Gravity*: 1.03 at 25 °C (liquid); *Vapor (Gas) Specific Gravity*: 3.4; *Ratio of Specific Heats of Vapor (Gas)*: Not pertinent; *Latent Heat of Vaporization*: Data not available; *Heat of Combustion*: -13,000, -7,220, -302; *Heat of Decomposition*: Not pertinent.

Health Hazards Information - *Recommended Personal Protective Equipment*: Goggles or face shield; rubber gloves; *Symptoms Following Exposure*: Inhalation of hot vapors can irritate nose and throat. Ingestion causes irritation of mouth and stomach. Contact with eyes causes irritation. Repeated and prolonged skin contact produces a mild, transient irritation; *General Treatment for Exposure*: INHALATION: remove to fresh air. INGESTION: give large amounts of water and induce vomiting. EYES: flush with water for at least 15 min. SKIN: remove from skin and eyes by flooding the affected tissues with water; wash with soap and water; *Toxicity by Inhalation (Threshold Limit Value)*: Data not available; *Short-Term Inhalation Limits*: Data not available; *Toxicity by Ingestion*: Grade 2; oral LD_{50} = 3.5 mg/kg (rabbit); *Late Toxicity*: Causes blood abnormalities in rats; *Vapor (Gas) Irritant Characteristics*: Data not available; *Liquid or Solid Irritant Characteristics*: Data not available; *Odor Threshold*: Data not available.

Fire Hazards - *Flash Point (deg. F)*: 204 OC; *Flammable Limits in Air (%):* Data not available; *Fire Extinguishing Agents:* Dry chemical, "alcohol foam", or carbon dioxide; *Fire Extinguishing Agents Not To Be Used:* Water may be ineffective; *Special Hazards of Combustion Products:* Toxic oxides of nitrogen may be formed in fire; *Behavior in Fire:* Data not available; *Ignition Temperature :* Data not available; *Electrical Hazard:* Data not available; *Burning Rate:* Data not available.

Chemical Reactivity - *Reactivity with Water:* No reaction; *Reactivity with Common Materials:* No reaction; *Stability During Transport:* Stable; *Neutralizing Agents for Acids and Caustics:* Not pertinent; *Polymerization:* Not pertinent; *Inhibitor of Polymerization:* Not pertinent.

ALPHA-METHYLSTYRENE

Chemical Designations - *Synonyms*: Isopropenylbenzene; 1-Methyl-1-phenylethylene; Phenylpropylene; *Chemical Formula*: $C_6H_5C(CH_3)=CH_2$.

Observable Characteristics - *Physical State (as shipped)*: Liquid; *Color*: Colorless; *Odor*: Characteristic.

Physical and Chemical Properties - *Physical State at 15 °C and 1 atm.*: Liquid; *Molecular Weight*: 118.17; *Boiling Point at 1 atm.*: 329, 165, 438; *Freezing Point*: -9.8, -23.2, 250.0; *Critical Temperature*: 719.1, 381.7, 654.9; *Critical Pressure*: 494, 33.6, 3.41; *Specific Gravity*: 0.91 at 20 °C (liquid); *Vapor (Gas) Specific Gravity*: 4.08; *Ratio of Specific Heats of Vapor (Gas)*: 1.060 at 27 °C; *Latent Heat of Vaporization*: 140.4, 78.0, 3.26; *Heat of Combustion*: -17,690, -9,830, -411; *Heat of Decomposition*: Not pertinent.

Health Hazards Information - *Recommended Personal Protective Equipment*: Neoprene gloves; splashproof goggles or face shield; *Symptoms Following Exposure*: Inhalation causes irritation of respiratory tract, headache, dizziness, light-heatedness, and breathlessness. Ingestion causes irritation of mouth and stomach. Contact with liquid irritates eyes. Prolonged skin contact can cause severe rashes, swelling, and blistering; *General Treatment for Exposure*: INHALATION: remove victim to fresh air; if he is not breathing, give artificial respiration; contact a physician; keep victim quiet and warm. INGESTION: do NOT induce vomiting; call a physician. EYES: flush with water for at least 15 min.; get medical attention. SKIN: wash area with soap and water; *Toxicity by Inhalation (Threshold Limit Value)*: 100 ppm; *Short-Term Inhalation Limits*: 100 ppm, 30 min.; *Toxicity by Ingestion*: Grade 2, LD_{50} 0.5-5 g/kg; *Late Toxicity*: Data not available; *Vapor (Gas) Irritant Characteristics*: Data not available; *Liquid or Solid Irritant Characteristics*: Data not available; *Odor Threshold*: < 10 ppm.

Fire Hazards - *Flash Point (deg. F)*: 137 CC; *Flammable Limits in Air (%):* 1.9 - 6.1; *Fire Extinguishing Agents:* Dry chemical, foam, carbon dioxide; *Fire Extinguishing Agents Not To Be Used:* Water may be ineffective; *Special Hazards of Combustion Products:* Data not available; *Behavior in Fire:* Data not available; *Ignition Temperature (deg. F):* 1,066; *Electrical Hazard:* Data not available; *Burning Rate:* Data not available.

Chemical Reactivity - *Reactivity with Water:* No reaction; *Reactivity with Common Materials:* May attack some forms of plastics; *Stability During Transport:* Stable; *Neutralizing Agents for Acids and Caustics:* Not pertinent; *Polymerization:* Hazardous polymerization unlikely to occur except when in contact with alkali metals or metallo-organic compounds; *Inhibitor of Polymerization:* 10 -20 ppm tert-butylcatechol.

METHYLTRICHLOROSILANE

Chemical Designations - *Synonyms*: Trichloromethylsilane; *Chemical Formula*: CH_3SiCl_3.

Observable Characteristics - *Physical State (as shipped)*: Liquid; *Color*: Colorless; *Odor*: Acrid; sharp, like hydrochloric acid.

Physical and Chemical Properties - *Physical State at 15 °C and 1 atm.*: Liquid; *Molecular Weight*: 149.5; *Boiling Point at 1 atm.*: 151.5, 66.4, 339.6; *Freezing Point*: -130, -90, 183; *Critical Temperature*: Not pertinent; *Critical Pressure*: Not pertinent; *Specific Gravity*: 1.27 at 25 °C (liquid); *Vapor (Gas) Specific Gravity*: 5.16; *Ratio of Specific Heats of Vapor (Gas)*: Data not available; *Latent Heat of Vaporization*: 89.3, 49.6, 2.08; *Heat of Combustion*: (est.) -3,000, -1,700, -70; *Heat of Decomposition*: Not pertinent.

Health Hazards Information - *Recommended Personal Protective Equipment*: Full protective clothing; acrid-vapor-type respiratory protection; rubber gloves; chemical worker's goggles; other protective equipment as necessary to protect skin and eyes; *Symptoms Following Exposure*: Inhalation causes irritation of mucous membrane. Contact with liquid causes severe burns of eyes and skin. Ingestion causes severe burns of mouth and stomach; *General Treatment for Exposure*: Get medical attention at once following all exposures to this compound. INHALATION: remove victim from exposure; give artificial respiration if breathing has ceased. EYES: flush with water for 15 min. SKIN: flush with water. INGESTION: do NOT induce vomiting; give large amounts of water; *Toxicity by Inhalation (Threshold Limit Value)*: Data not available; *Short-Term Inhalation Limits*: Data not available; *Toxicity by Ingestion*: Grade 3; LD_{50} 50 to 500 mg/kg; *Late Toxicity*: Data not available; *Vapor (Gas) Irritant Characteristics*: Vapors cause severe irritation of eye and throat and can cause eye and lung injury. They cannot be tolerated even at low concentrations; *Liquid or Solid Irritant Characteristics*: Severe skin irritant. Causes second- and third-degree burns on short contact and is very injurious to the eyes; *Odor Threshold*: Decomposes in moist air, creating HCl with odor threshold of 1 ppm.

Fire Hazards - *Flash Point (deg. F)*: 45 OC; 15 CC; *Flammable Limits in Air (%)*: 5.1 - >20; *Fire Extinguishing Agents:* Dry chemical, carbon dioxide; *Fire Extinguishing Agents Not To Be Used:* Water, foam; *Special Hazards of Combustion Products:* Toxic hydrogen chloride and phosgene gases may form in fires; *Behavior in Fire:* Difficult to extinguish; re-ignition may occur. Contact with water applied to adjacent fires produces irritating hydrogen chloride; *Ignition Temperature (deg. F):* >760; *Electrical Hazard:* Data not available; *Burning Rate:* 1.9 mm/min.

Chemical Reactivity - *Reactivity with Water:* Reacts violently to form hydrogen chloride, which is corrosive to metals; *Reactivity with Common Materials:* Reacts with surface moisture to evolve hydrogen chloride, which is corrosive to metals; *Stability During Transport:* Stable; *Neutralizing Agents for Acids and Caustics:* Flush with water, rinse with sodium bicarbonate or lime solution; *Polymerization:* Not pertinent; *Inhibitor of Polymerization:* Not pertinent.

METHYL VINYL KETONE

Chemical Designations - *Synonyms*: 3-Buten-2-one; *Chemical Formula*: $CH_3COCH=CH_2$.

Observable Characteristics - *Physical State (as shipped)*: Liquid; *Color*: Colorless to light yellow; *Odor*: Powerfully irritating.

Physical and Chemical Properties - *Physical State at 15 °C and 1 atm.*: Liquid; *Molecular Weight*: 70.1; *Boiling Point at 1 atm.*: 178.5, 81.4, 354.6; *Freezing Point*: 20, -7, 266; *Critical Temperature*: Not pertinent; *Critical Pressure*: Not pertinent; *Specific Gravity*: 0.864 at 20 °C (liquid); *Vapor (Gas) Specific Gravity*: 2.4; *Ratio of Specific Heats of Vapor (Gas)*: 1.1053; *Latent Heat of Vaporization*: (est.) 203, 113, 4.73; *Heat of Combustion*: (est.) -14,600, -8,100, -340; *Heat of Decomposition*: Not pertinent.

Health Hazards Information - *Recommended Personal Protective Equipment*: Self-contained breathing apparatus with full face piece; rubber gloves; chemical goggles or face piece of breathing apparatus; *Symptoms Following Exposure*: Inhalation causes irritation of nose and throat. Vapor causes tears; contact with liquid can burn eyes. Liquid irritates skin and will cause burn if not removed at once. Ingestion causes irritation of mouth and stomach; *General Treatment for Exposure*: Get medical attention for all exposures to this compound. INHALATION: remove victim to fresh air; administer artificial respiration if necessary. EYES or SKIN: flush with copious quantities of water for 15 min. INGESTION: do NOT induce vomiting; *Toxicity by Inhalation (Threshold Limit Value)*: Data not available; *Short-Term Inhalation Limits*: Data not available; *Toxicity by Ingestion*: Grade 4, $LD_{50} < 50$ mg/kg; *Late Toxicity*: Data not available; *Vapor (Gas) Irritant Characteristics*: Vapors cause severe irritation of eye and throat and can cause eye and lung injury. They cannot be tolerated even at low concentrations; *Liquid or Solid Irritant Characteristics*: Severe skin irritant. Causes second- and third-degree burns on short contact and is very injurious to the eyes; *Odor Threshold*: 0.5 mg/m^3.

Fire Hazards - *Flash Point (deg. F)*: 30 OC; 20 CC; *Flammable Limits in Air (%):* 2.1 - 15.6; *Fire Extinguishing Agents:* Dry chemical, alcohol foam, carbon dioxide; *Fire Extinguishing Agents Not To Be Used:* Water may be ineffective; *Special Hazards of Combustion Products:* Not pertinent; *Behavior in Fire:* Vapor is heavier than air and may travel a considerable distance to a source of ignition and flash back. At elevated temperatures (fire conditions) polymerization may take place in containers, causing violent rupture. Unburned vapors are very irritating; *Ignition Temperature (deg. F):* 915; *Electrical Hazard:* Data not available; *Burning Rate:* 4.5 mm/min.

Chemical Reactivity - *Reactivity with Water:* No reaction; *Reactivity with Common Materials:* No reaction; *Stability During Transport:* Stable; *Neutralizing Agents for Acids and Caustics:* Not pertinent; *Polymerization:* Polymerize spontaneously upon exposure to heat or sunlight; *Inhibitor of Polymerization:* Up to 1% hydroquinone.

MINERAL SPIRITS

Chemical Designations - *Synonyms*: Naphtha; Petroleum spirits; *Chemical Formula*: Not applicable.

Observable Characteristics - *Physical State (as shipped)*: Liquid; *Color*: Colorless; *Odor*: Like gasoline.

Physical and Chemical Properties - *Physical State at 15 °C and 1 atm.*: Liquid; *Molecular Weight*: Not pertinent; *Boiling Point at 1 atm.*: 310-395, 154-202, 428-475; *Freezing Point*: Not pertinent; *Critical Temperature*: Not pertinent; *Critical Pressure*: Not pertinent; *Specific Gravity*: 0.78 at 20 °C (liquid); *Vapor (Gas) Specific Gravity*: Not pertinent; *Ratio of Specific Heats of Vapor (Gas)*: (est.) 1.030; *Latent Heat of Vaporization*: Data not available; *Heat of Combustion*: Data not available; *Heat of Decomposition*: Not pertinent.

Health Hazards Information - *Recommended Personal Protective Equipment*: Plastic gloves; goggles or face shield (as for gasoline); *Symptoms Following Exposure*: INHALATION: mild irritation of respiratory tract. ASPIRATION: severe lung irritation and rapidly developing pulmonary edema; central nervous system excitement followed by depression. INGESTION: irritation of stomach; *General Treatment for Exposure*: INHALATION: remove victim to fresh air. ASPIRATION: enforced bed rest; give oxygen; call a doctor. INGESTION: do NOT induce vomiting; guard against aspiration into lungs. EYES: wash with copious amounts of water. SKIN: wipe off and wash with soap and water; *Toxicity by Inhalation (Threshold Limit Value)*: 200 ppm; *Short-Term Inhalation Limits*: 4000-7000 ppm for 60 min.; *Toxicity by Ingestion*: Grade 2, LD_{50} 0.5 to 5 g/kg; *Late Toxicity*: Data not available; *Vapor (Gas) Irritant Characteristics*: Vapors are nonirritating to the eyes and throat; *Liquid or Solid Irritant Characteristics*: Minimum hazard. If spilled on clothing and allowed to remain, may cause smarting and reddening of skin; *Odor Threshold*: Data not available.

Fire Hazards - *Flash Point (deg. F)*: 105 - 140 CC, depending on grade; *Flammable Limits in Air (%):* 0.8 - 5.0; *Fire Extinguishing Agents:* Foam, carbon dioxide, dry chemical; *Fire Extinguishing Agents Not To Be Used:* Do not use straight hose water stream; *Special Hazards of Combustion Products:* Not pertinent; *Behavior in Fire:* Not pertinent; *Ignition Temperature (deg. F):* 540; *Electrical Hazard:* Not pertinent; *Burning Rate:* 4 mm/min.

Chemical Reactivity - *Reactivity with Water:* No reaction; *Reactivity with Common Materials:* No reaction; *Stability During Transport:* Stable; *Neutralizing Agents for Acids and Caustics:* Not pertinent; *Polymerization:* Not pertinent; *Inhibitor of Polymerization:* Not pertinent.

MOLYBDIC TRIOXIDE

Chemical Designations - *Synonyms*: Molybdenum trioxide; Molybdic anhydride; *Chemical Formula*: MoO_3.

Observable Characteristics - *Physical State (as shipped)*: Solid; *Color*: Colorless or white-yellow to yellow; *Odor*: None.

Physical and Chemical Properties - *Physical State at 15 ℃ and 1 atm.*: Solid; *Molecular Weight*: 143.94; *Boiling Point at 1 atm.*: Not pertinent; *Freezing Point*: Not pertinent; *Critical Temperature*: Not pertinent; *Critical Pressure*: Not pertinent; *Specific Gravity*: 4.69 at 20 °C (solid); *Vapor (Gas) Specific Gravity*: Not pertinent; *Ratio of Specific Heats of Vapor (Gas)*: Not pertinent; *Latent Heat of Vaporization*: Not pertinent; *Heat of Combustion*: Not pertinent; *Heat of Decomposition*: Not pertinent.

Health Hazards Information - *Recommended Personal Protective Equipment*: Approved by U.S. Bu. Mines respirator; safety glasses or face shield; protective gloves; *Symptoms Following Exposure*: Compound is relatively nontoxic. Dust irritates eyes; *General Treatment for Exposure*: No treatment necessary except those applicable to any nontoxic dust. EYES: flush with water; *Toxicity by Inhalation (Threshold Limit Value)*: 5 mg/m^3; *Short-Term Inhalation Limits*: Data not available; *Toxicity by Ingestion*: Grade 3; LD_{50} 50-500 mg/kg; *Late Toxicity*: Data not available; *Vapor (Gas) Irritant Characteristics*: Data not available; *Liquid or Solid Irritant Characteristics*: Data not available; *Odor Threshold*: Data not available.

Fire Hazards - *Flash Point* : Not flammable; *Flammable Limits in Air (%)*: Not flammable; *Fire Extinguishing Agents*: Not pertinent; *Fire Extinguishing Agents Not To Be Used*: Not pertinent; *Special Hazards of Combustion Products*: Not pertinent; *Behavior in Fire*: Not pertinent; *Ignition Temperature* : Not pertinent; *Electrical Hazard*: Not pertinent; *Burning Rate*: Not pertinent.

Chemical Reactivity - *Reactivity with Water*: No reaction; *Reactivity with Common Materials*: No reaction; *Stability During Transport*: Stable; *Neutralizing Agents for Acids and Caustics*: Not pertinent; *Polymerization*: Not pertinent; *Inhibitor of Polymerization*: Not pertinent.

N

2-NITROPHENOL

Chemical Designations - *Synonyms*: 2-Hydroxynitrobenzene; o-Nitrophenol; ONP; *Chemical Formula*: $1,2\text{-}HOC_6H_4NO_2$.

Observable Characteristics - *Physical State (as shipped)*: Solid; *Color*: Yellow; *Odor*: Peculiar aromatic.

Physical and Chemical Properties - *Physical State at 15 ℃ and 1 atm.*: Solid; *Molecular Weight*: 139.1; *Boiling Point at 1 atm.*: 417, 214, 487; *Freezing Point*: 111, 44, 313; *Critical Temperature*: Not pertinent; *Critical Pressure*: Not pertinent; *Specific Gravity*: 1.49 at 20 °C (solid); *Vapor (Gas) Specific Gravity*: Not pertinent; *Ratio of Specific Heats of Vapor (Gas)*: Not pertinent; *Latent Heat of Vaporization*: Not pertinent; *Heat of Combustion*: -8,910, -4,950, -207; *Heat of Decomposition*: Not pertinent.

Health Hazards Information - *Recommended Personal Protective Equipment*: Self-contained breathing apparatus for fumes; goggles; rubber gloves; *Symptoms Following Exposure*: Inhalation or ingestion causes headache, drowsiness, nausea, and blue color in lips, ears, and fingernails (cyanosis). Contact with eyes causes irritation. Can be absorbed through the intact skin to give same symptoms as for inhalation; *General Treatment for Exposure*: INHALATION or INGESTION: remove victim to fresh air; give artificial respiration; call a doctor if symptoms persist. EYES: flood with water for at least 15 min.; get medical attention. SKIN: cleanse thoroughly with soap and water; *Toxicity by Inhalation (Threshold Limit Value)*: Data not available; *Short-Term Inhalation Limits*: Data not available; *Toxicity by Ingestion*: Grade 2; oral LD_{50} = 1,297 mg/kg (rat); *Late Toxicity*: Data not available; *Vapor (Gas) Irritant Characteristics*: Data not available; *Liquid or Solid Irritant Characteristics*: Data not available;

Odor Threshold: Data not available.

Fire Hazards - *Flash Point* : Not pertinent (combustible solid); *Flammable Limits in Air (%):* Not pertinent; *Fire Extinguishing Agents:* Water, foam, dry chemical, carbon dioxide; *Fire Extinguishing Agents Not To Be Used:* Data not available; *Special Hazards of Combustion Products:* Toxic and irritating fumes of unburned materials and oxides of nitrogen can form in fire; *Behavior in Fire:* Data not available; *Ignition Temperature :* Data not available; *Electrical Hazard:* Not pertinent; *Burning Rate:* Not pertinent.

Chemical Reactivity - *Reactivity with Water:* No reaction; *Reactivity with Common Materials:* No reaction; *Stability During Transport:* Stable; *Neutralizing Agents for Acids and Caustics:* Not pertinent; *Polymerization:* Not pertinent; *Inhibitor of Polymerization:* Not pertinent.

4-NITROPHENOL

Chemical Designations - *Synonyms*: 2-Hydroxynitrobenzene; p-Nitrophenol; PNP; *Chemical Formula*: $1,4-HOC_6H_4NO_2$.

Observable Characteristics - *Physical State (as shipped)*: Solid; *Color*: Yellow to brown; *Odor*: Slight characteristic, sweet.

Physical and Chemical Properties - *Physical State at 15 °C and 1 atm.*: Solid; *Molecular Weight*: 139.1; *Boiling Point at 1 atm.*: Not pertinent (decomposes); *Freezing Point*: 235, 113, 386; *Critical Temperature*: Not pertinent; *Critical Pressure*: Not pertinent; *Specific Gravity*: 1.48 at 20 °C (solid); *Vapor (Gas) Specific Gravity*: Not pertinent; *Ratio of Specific Heats of Vapor (Gas)*: Not pertinent; *Latent Heat of Vaporization*: Not pertinent; *Heat of Combustion*: -8,870, -4,930, -206; *Heat of Decomposition*: Not pertinent.

Health Hazards Information - *Recommended Personal Protective Equipment*: Self-contained breathing apparatus or dust mask; butyl rubber gloves; side-shield safety glasses; *Symptoms Following Exposure*: Inhalation or ingestion causes headache, drowsiness, nausea, and blue color in lips, ears, and fingernails (cyanosis). Contact with eyes causes irritation; can be absorbed through the intact skin to give same symptoms as for inhalation; *General Treatment for Exposure*: INHALATION or INGESTION: remove victim to fresh air; give artificial respiration; call a doctor if symptoms persist. EYES: flood with water for at least 15 min.; get medical attention. SKIN: cleanse thoroughly with soap and water; *Toxicity by Inhalation (Threshold Limit Value)*: Data not available; *Short-Term Inhalation Limits*: Data not available; *Toxicity by Ingestion*: Grade 3; LD_{50} 50-500 mg/kg; *Late Toxicity*: Data not available; *Vapor (Gas) Irritant Characteristics*: Data not available; *Liquid or Solid Irritant Characteristics*: Data not available; *Odor Threshold*: Odorless.

Fire Hazards - *Flash Point* : Not pertinent (combustible solid); *Flammable Limits in Air (%):* Not pertinent; *Fire Extinguishing Agents:* Water, foam, dry chemical, carbon dioxide; *Fire Extinguishing Agents Not To Be Used:* No data; *Special Hazards of Combustion Products:* Toxic oxides of nitrogen and fumes of unburned material may form in fire; *Behavior in Fire:* Decomposes violently at 279° C and will burn even in absence of air; *Ignition Temperature :* Data not available; *Electrical Hazard:* Not pertinent; *Burning Rate:* Not pertinent.

Chemical Reactivity - *Reactivity with Water:* No reaction; *Reactivity with Common Materials:* NO reaction; *Stability During Transport:* Stable; *Neutralizing Agents for Acids and Caustics:* Not pertinent; *Polymerization:* Not pertinent; *Inhibitor of Polymerization:* Not pertinent.

2-NITROPROPANE

Chemical Designations - *Synonyms*: Izonitropropane; sec-Nitropropane; 2-NP; *Chemical Formula*: $CH_3CH(NO_2)CH_3$.

Observable Characteristics - *Physical State (as shipped)*: Liquid; *Color*: Colorless; *Odor*: Mild, fruity.

Physical and Chemical Properties - *Physical State at 15 °C and 1 atm.*: Liquid; *Molecular Weight*: 89.09; *Boiling Point at 1 atm.*: 245.5, 120.3, 393.5; *Freezing Point*: -132,-91, 182; *Critical Temperature*: Data not available; *Critical Pressure*: Data not available; *Specific Gravity*: 0.99 at 20 °C (liquid); *Vapor (Gas) Specific Gravity*: 3.06 at 16°C; *Ratio of Specific Heats of Vapor (Gas)*: 1.090 at

20 °C; *Latent Heat of Vaporization*: 178, 99, 4.1; *Heat of Combustion*: -9,650, -5,360, -224; *Heat of Decomposition*: Not pertinent.

Health Hazards Information - *Recommended Personal Protective Equipment*: Self-contained breathing apparatus; goggles or face shield; rubber gloves; *Symptoms Following Exposure*: Inhalation causes respiratory tract irritation, headache, dizziness, nausea, and causes mild irritation of skin; *General Treatment for Exposure*: INHALATION: in case of pulmonary symptoms, enforce bed rest and give oxygen; get medical attention at once. INGESTION: give large amount of water and induce vomiting. EYES or SKIN: flush with water; *Toxicity by Inhalation (Threshold Limit Value)*: 25 ppm; *Short-Term Inhalation Limits*: Data not available; *Toxicity by Ingestion*: Grade 2; oral rat $LD_{50} = 720$ g/kg; *Late Toxicity*: Causes liver cancer in rats; *Vapor (Gas) Irritant Characteristics*: Vapors cause a slight smarting of eyes or respiratory system if present in high concentrations. The effect is temporary; *Liquid or Solid Irritant Characteristics*: Minimum hazard. If spilled on clothing and allowed to remain, may cause smarting and reddening of skin; *Odor Threshold*: 300 ppm.

Fire Hazards - *Flash Point (deg. F)*: 100 OC; 82 CC; *Flammable Limits in Air (%)*: 2.6 (LEL); *Fire Extinguishing Agents:* Foam, dry chemical, carbon dioxide; *Fire Extinguishing Agents Not To Be Used:* "Alcohol" foam; water may be ineffective; *Special Hazards of Combustion Products:* Toxic oxides of nitrogen may form in fire; *Behavior in Fire:* No data; *Ignition Temperature (deg. F):* 802; *Electrical Hazard:* Data not available; *Burning Rate:* Data not available.

Chemical Reactivity - *Reactivity with Water:* No reaction; *Reactivity with Common Materials:* May attack some forms of plastics; *Stability During Transport:* Stable; *Neutralizing Agents for Acids and Caustics:* Not pertinent; *Polymerization:* Not pertinent; *Inhibitor of Polymerization:* Not pertinent.

NITROSYL CHLORIDE

Chemical Designations - *Synonyms*: No common synonyms; *Chemical Formula*: NOCl.

Observable Characteristics - *Physical State (as shipped)*: Liquefied gas; *Color*: Yellow to red; *Odor*: irritating, choking.

Physical and Chemical Properties - *Physical State at 15 °C and 1 atm.*: Gas; *Molecular Weight*: 65,46; *Boiling Point at 1 atm.*: 21.6, -5.8, 267.4; *Freezing Point*: -74, -59, 214; *Critical Temperature*: 334, 168, 441; *Critical Pressure*: 1300, 90, 9.1; *Specific Gravity*: 1.36 at -5.7 °C (liquid); *Vapor (Gas) Specific Gravity*: 2.3; *Ratio of Specific Heats of Vapor (Gas)*: 1.229; *Latent Heat of Vaporization*: 164, 91.0, 3.81; *Heat of Combustion*: Not pertinent; *Heat of Decomposition*: Not pertinent.

Health Hazards Information - *Recommended Personal Protective Equipment*: Self-contained breathing apparatus (approved mask may be used for short exposures only); rubberized clothing; gloves; shoes; chemical goggles; *Symptoms Following Exposure*: Gas is highly toxic. Inhalation causes severe irritation of respiratory tract and damage to mucous membranes. Delayed effects, which include severe pulmonary edema, may not be apparent for several hours; *General Treatment for Exposure*: INHALATION: remove victim to fresh air; call a doctor; enforce complete rest until doctor arrives; observe at least 24 hours for delayed effects. SKIN OR EYES: flush with water for at least 15 min.; consult physician; *Toxicity by Inhalation (Threshold Limit Value)*: 1 ppm (recommended); *Short-Term Inhalation Limits*: Data not available; *Toxicity by Ingestion*: Data not available; *Late Toxicity*: None; *Vapor (Gas) Irritant Characteristics*: Vapors cause severe irritation of eye and throat and can cause eye and lung injury. They cannot be tolerated even at low concentrations; *Liquid or Solid Irritant Characteristics*: Severe burns to eyes and skin; *Odor Threshold*: Data not available.

Fire Hazards - *Flash Point* : Not flammable; *Flammable Limits in Air (%):* Not flammable; *Fire Extinguishing Agents:* Not pertinent; *Fire Extinguishing Agents Not To Be Used:* Not pertinent; *Special Hazards of Combustion Products:* Very toxic gases are generated when heated; *Behavior in Fire:* Not pertinent; *Ignition Temperature :* Not flammable; *Electrical Hazard:* Not pertinent; *Burning Rate:* Not flammable.

Chemical Reactivity - *Reactivity with Water:* Dissolves and reacts to form acid solution and toxic red oxides of nitrogen; *Reactivity with Common Materials:* Corrosive to most metals, but reaction is not hazardous; *Stability During Transport:* Stable; *Neutralizing Agents for Acids and Caustics:* Flush with water. Residual acid may be neutralized with soda ash; *Polymerization:* Not pertinent; *Inhibitor of*

Polymerization: Not pertinent.

NITROUS OXIDE

Chemical Designations - *Synonyms*: Dinitrogen monoxide; *Chemical Formula*: N_2O.

Observable Characteristics - *Physical State (as shipped)*: Liquefied compressed gas; *Color*: Colorless; *Odor*: None; slightly sweetish.

Physical and Chemical Properties - *Physical State at 15 °C and 1 atm.*: Gas; *Molecular Weight*: 44.0; *Boiling Point at 1 atm.*: -129.1, -89.5, 183.7; *Freezing Point*: -131.5, -90.8, 182.4; *Critical Temperature*: 97.7, 36.5, 309.7; *Critical Pressure*: 1.054, 71.7, 7.28; *Specific Gravity*: 1.266 at -89 °C (liquid); *Vapor (Gas) Specific Gravity*: 1.53; *Ratio of Specific Heats of Vapor (Gas)*: 1.303 at 25 °C; *Latent Heat of Vaporization*: 161.7, 89.9, 3.76; *Heat of Combustion*: Not pertinent; *Heat of Decomposition*: Not pertinent.

Health Hazards Information - *Recommended Personal Protective Equipment*: Self-contained breathing apparatus for high vapor concentrations; *Symptoms Following Exposure*: Inhalation causes intense analgesia; concentrations of over 40-60% cause loss of consciousness preceded by hysteria. Contact of liquid with eyes or skin causes frostbite burn; *General Treatment for Exposure*: INHALATION: remove to fresh air. EYES: get medical attention for frostbite burn. SKIN: treat frostbite burn; soak in lukewarm water; *Toxicity by Inhalation (Threshold Limit Value)*: Data not available; *Short-Term Inhalation Limits*: 56.000 ppm for 60 min.; *Toxicity by Ingestion*: Grade 0, LD_{50} 0.5 > 15 g/kg; *Late Toxicity*: Causes birth defects in rats; can cause lethal effects in chick eggs; *Vapor (Gas) Irritant Characteristics*: Vapors are nonirritating to eyes and throat; *Liquid or Solid Irritant Characteristics*: No appreciable hazard; practically harmless to the skin; *Odor Threshold*: Data not available.

Fire Hazards - *Flash Point* : Not pertinent (nonflammable compressed gas); *Flammable Limits in Air (%):* Not pertinent; *Fire Extinguishing Agents:* Not pertinent; *Fire Extinguishing Agents Not To Be Used:* Not pertinent; *Special Hazards of Combustion Products:* Not pertinent; *Behavior in Fire:* Will support combustion, and may increase intensity of fire. Containers may explode when heated; *Ignition Temperature :* Not pertinent; *Electrical Hazard:* Not pertinent; *Burning Rate:* Not pertinent.

Chemical Reactivity - *Reactivity with Water:* No reaction; *Reactivity with Common Materials:* Supports combustion but does not cause spontaneous combustion; *Stability During Transport:* Stable; *Neutralizing Agents for Acids and Caustics:* Not pertinent; *Polymerization:* Not pertinent; *Inhibitor of Polymerization:* Not pertinent.

NONANE

Chemical Designations - *Synonyms*: n-Noname; *Chemical Formula*: C_9N_{20}.

Observable Characteristics - *Physical State (as shipped)*: Liquid; *Color*: Colorless; *Odor*: Like gasoline.

Physical and Chemical Properties - *Physical State at 15 °C and 1 atm.*: Liquid; *Molecular Weight*: 128.3; *Boiling Point at 1 atm.*: 304, 151, 424; *Freezing Point*: -64.3, -53.5, 219.7; *Critical Temperature*: 610.5, 321.4, 594.6; *Critical Pressure*: 335, 22.8, 2.31; *Specific Gravity*: 0.718 at 20 °C (liquid); *Vapor (Gas) Specific Gravity*: 4.4; *Ratio of Specific Heats of Vapor (Gas)*: 1.042 at 16°C; *Latent Heat of Vaporization*: 127, 2.95; *Heat of Combustion*: -19,067, -10,593, -443.21; *Heat of Decomposition*: Not pertinent.

Health Hazards Information - *Recommended Personal Protective Equipment*: Self-contained breathing apparatus for vapor concentrations; goggles or face shield; rubber gloves; *Symptoms Following Exposure*: Inhalation of concentrated vapor causes depression, irritation of respiratory tract, and pulmonary edema. Liquid can irritate eyes and (on prolonged contact) skin. Ingestion causes irritation of mouth and stomach. Aspiration causes severe lung irritation, rapidly developing pulmonary edema, and central nervous system excitement followed by depression; *General Treatment for Exposure*: INHALATION: remove victim from exposure; give artificial respiration if needed. EYES: irrigate with large amounts of water for 15 min. SKIN: flush with water; wash with soap and water. INGESTION: do NOT induce vomiting; call a physician. ASPIRATION: enforce bed rest; give oxygen; get medical attention; *Toxicity by Inhalation (Threshold Limit Value)*: Data not available; *Short-Term Inhalation*

Limits: Data not available; *Toxicity by Ingestion*: Grade 0, LD_{50} 0.5 > 15 g/kg; *Late Toxicity*: Data not available; *Vapor (Gas) Irritant Characteristics*: Vapors are nonirritating to eyes and throat; *Liquid or Solid Irritant Characteristics*: No appreciable hazard. Practically harmless to the skin; *Odor Threshold*: Data not available.

Fire Hazards - *Flash Point (deg. F)*: 88 CC; *Flammable Limits in Air (%)*: 0.87 - 2.9; *Fire Extinguishing Agents:* Dry chemical, foam, carbon dioxide; *Fire Extinguishing Agents Not To Be Used:* Water may be ineffective; *Special Hazards of Combustion Products:* Not pertinent; *Behavior in Fire:* Not pertinent; *Ignition Temperature (deg. F):* 401; *Electrical Hazard:* Class I, Group D; *Burning Rate:* 5.8 mm/min.

Chemical Reactivity - *Reactivity with Water:* No reaction; *Reactivity with Common Materials:* No reaction; *Stability During Transport:* Stable; *Neutralizing Agents for Acids and Caustics:* Not pertinent; *Polymerization:* Not pertinent; *Inhibitor of Polymerization:* Not pertinent.

NONANOL

Chemical Designations - *Synonyms*: 1-Nonanol; Octylcarbinol; Nonilalcohol; Pelargonic alcohol; *Chemical Formula*: $CH_3(CH_2)_7CH_2OH$.

Observable Characteristics - *Physical State (as shipped)*: Liquid; *Color*: Colorless; *Odor*: Rose-citrus.

Physical and Chemical Properties - *Physical State at 15 °C and 1 atm.*: Liquid; *Molecular Weight*: 144.26; *Boiling Point at 1 atm.*: 415, 213, 486; *Freezing Point*: 23, -5, 268; *Critical Temperature*: 759, 404, 677; *Critical Pressure*: 350, 24, 2.4; *Specific Gravity*: 0.827 at 20 °C (liquid); *Vapor (Gas) Specific Gravity*: Not pertinent; *Ratio of Specific Heats of Vapor (Gas)*: 1.039; *Latent Heat of Vaporization*: 131, 72.5, 3.04; *Heat of Combustion*: -17,800, -9860, -413; *Heat of Decomposition*: Not pertinent.

Health Hazards Information - *Recommended Personal Protective Equipment*: Goggles or face shield; rubber gloves; *Symptoms Following Exposure*: Liquid irritates eyes; *General Treatment for Exposure*: Flush eyes and skin with water for at least 15 min.; *Toxicity by Inhalation (Threshold Limit Value)*: Data not available; *Short-Term Inhalation Limits*: Data not available; *Toxicity by Ingestion*: Grade 2; 0.5 to 5 g/kg (rat); *Late Toxicity*: None; *Vapor (Gas) Irritant Characteristics*: Vapors are nonirritating to the eyes and throat; *Liquid or Solid Irritant Characteristics*: No appreciable hazard. Practically harmless to the skin; *Odor Threshold*: Data not available.

Fire Hazards - *Flash Point (deg. F)*: 210 OC; 165 CC; *Flammable Limits in Air (%):* 0,8 - 6.1; *Fire Extinguishing Agents:* Alcohol foam, dry chemical, or carbon dioxide; *Fire Extinguishing Agents Not To Be Used:* Water may be ineffective; *Special Hazards of Combustion Products:* Not pertinent; *Behavior in Fire:* Not pertinent; *Ignition Temperature :* Data not available; *Electrical Hazard:* Not pertinent; *Burning Rate:* Data not available.

Chemical Reactivity - *Reactivity with Water:* No reaction; *Reactivity with Common Materials:* No reaction; *Stability During Transport:* Stable; *Neutralizing Agents for Acids and Caustics:* Not pertinent; *Polymerization:* Not pertinent; *Inhibitor of Polymerization:* Not pertinent.

NONENE

Chemical Designations - *Synonyms*: Nonene (non-linear); Propylene trimer; Tripropylene; *Chemical Formula*: C_9H_{18}.

Observable Characteristics - *Physical State (as shipped)*: Liquid; *Color*: Colorless; *Odor*: Characteristic; like gasoline.

Physical and Chemical Properties - *Physical State at 15 °C and 1 atm.*: Liquid; *Molecular Weight*: 126.2; *Boiling Point at 1 atm.*: 275-284, 135-140, 408-413; *Freezing Point*: Not pertinent; *Critical Temperature*: Data not available; *Critical Pressure*: Data not available; *Specific Gravity*: 0.739 at 20 °C (liquid); *Vapor (Gas) Specific Gravity*: Not pertinent; *Ratio of Specific Heats of Vapor (Gas)*: (est.) 1.044; *Latent Heat of Vaporization*: (est.) 124, 68.9, 2.88; *Heat of Combustion*: -19,100, -10,600, -445; *Heat of Decomposition*: Not pertinent.

Health Hazards Information - *Recommended Personal Protective Equipment*: Respiratory organic vapor canister or air-supplied mask; face splash shield; *Symptoms Following Exposure*: High vapor concentrations irritate eyes and respiratory tract and act as an anesthetic; *General Treatment for Exposure*: INHALATION: remove patient to fresh air; if breathing stops, apply artificial respiration and administer oxygen; call a physician. INGESTION: do NOT induce vomiting because of aspiration hazard; *Toxicity by Inhalation (Threshold Limit Value)*: Data not available; *Short-Term Inhalation Limits*: Data not available; *Toxicity by Ingestion*: Data not available; *Late Toxicity*: Data not available; *Vapor (Gas) Irritant Characteristics*: Vapors cause a slight smarting of the eyes or respiratory system if present at high concentrations. The effect is temporary; *Liquid or Solid Irritant Characteristics*: Minimum hazard. If spilled on clothing and allowed to remain, may cause smarting and reddening of the skin; *Odor Threshold*: Data not available.

Fire Hazards - *Flash Point (deg. F)*: 78 OC; *Flammable Limits in Air (%)*: 0.7 - 3.9; *Fire Extinguishing Agents:* Foam, carbon dioxide, or dry chemical; *Fire Extinguishing Agents Not To Be Used:* Water may be ineffective; *Special Hazards of Combustion Products:* Not pertinent; *Behavior in Fire:* Not pertinent; *Ignition Temperature :* Data not available; *Electrical Hazard:* Not pertinent; *Burning Rate:* 6.0 mm/min.

Chemical Reactivity - *Reactivity with Water:* No reaction; *Reactivity with Common Materials:* No reaction; *Stability During Transport:* Stable; *Neutralizing Agents for Acids and Caustics:* Not pertinent; *Polymerization:* Not pertinent; *Inhibitor of Polymerization:* Not pertinent.

1-NONENE

Chemical Designations - *Synonyms*: n-Heptylethylene; 1-Nonylene; *Chemical Formula*: $CH_3(CH_2)_6CH=CH_2$.

Observable Characteristics - *Physical State (as shipped)*: Liquid; *Color*: Colorless; *Odor*: Hydrocarbon odor; like gasoline.

Physical and Chemical Properties - *Physical State at 15 °C and 1 atm.*: Liquid; *Molecular Weight*: 126.2; *Boiling Point at 1 atm.*: 297, 147, 420; *Freezing Point*: -115, -81.7, 191.5; *Critical Temperature*: 622, 327.8, 601.0; *Critical Pressure*: 360, 24.5, 2.98; *Specific Gravity*: 0.733 at 20 °C (liquid); *Vapor (Gas) Specific Gravity*: Not pertinent; *Ratio of Specific Heats of Vapor (Gas)*: 1.044; *Latent Heat of Vaporization*: 124, 68.8, 2.88; *Heat of Combustion*: -18,979, -10,544, -441.46; *Heat of Decomposition*: Not pertinent.

Health Hazards Information - *Recommended Personal Protective Equipment*: Respiratory organic vapor canister or air-supplied mask; face splash shield; *Symptoms Following Exposure*: High vapor concentrations irritate eyes and respiratory tract and act as an anesthetic; *General Treatment for Exposure*: INHALATION: remove to fresh air; if breathing stops, apply artificial respiration; administer oxygen; call a physician. INGESTION: if swallowed, do NOT induce vomiting because of aspiration hazard; *Toxicity by Inhalation (Threshold Limit Value)*: Data not available; *Short-Term Inhalation Limits*: Data not available; *Toxicity by Ingestion*: Data not available; *Late Toxicity*: Data not available; *Vapor (Gas) Irritant Characteristics*: Vapors cause a slight smarting of the eyes or respiratory system if present at high concentrations. The effect is temporary; *Liquid or Solid Irritant Characteristics*: Minimum hazard. If spilled on clothing and allowed to remain, may cause smarting and reddening of the skin; *Odor Threshold*: Data not available.

Fire Hazards - *Flash Point :* Data not available; *Flammable Limits in Air (%):* 0.8 (LEL); *Fire Extinguishing Agents:* Foam, dry chemical, or carbon dioxide; *Fire Extinguishing Agents Not To Be Used:* Water may be ineffective; *Special Hazards of Combustion Products:* Not pertinent; *Behavior in Fire:* Not pertinent; *Ignition Temperature :* Data not available; *Electrical Hazard:* Not pertinent; *Burning Rate:* 6.0 mm/min.

Chemical Reactivity - *Reactivity with Water:* No reaction; *Reactivity with Common Materials:* No reaction; *Stability During Transport:* Stable; *Neutralizing Agents for Acids and Caustics:* Not pertinent; *Polymerization:* Not pertinent; *Inhibitor of Polymerization:* Not pertinent.

NONYLPHENOL

Chemical Designations - *Synonyms*: No common synonyms; *Chemical Formula*: $p-HOC_6H_4(CH_2)_8CH_3$.

Observable Characteristics - *Physical State (as shipped)*: Liquid; *Color*: Pale yellow; *Odor*: Phenolic;

like disinfectant.

Physical and Chemical Properties - *Physical State at 15 °C and 1 atm.*: Liquid; *Molecular Weight*: 220.36; *Boiling Point at 1 atm.*: 579, 304, 577; *Freezing Point*: Not pertinent; *Critical Temperature*: 878, 470, 743; *Critical Pressure*: Not pertinent; *Specific Gravity*: 0.9494 at 25 °C (liquid); *Vapor (Gas) Specific Gravity*: Not pertinent; *Ratio of Specific Heats of Vapor (Gas)*: Not pertinent; *Latent Heat of Vaporization*: Not pertinent; *Heat of Combustion*: (est.) -17,500, -9730, -407; *Heat of Decomposition*: Not pertinent.

Health Hazards Information - *Recommended Personal Protective Equipment*: Rubber gloves and splash-proof goggles; *Symptoms Following Exposure*: Moderately toxic if swallowed. Severely irritating to skin and eyes; *General Treatment for Exposure*: EYES: wash with water for 15 min. and get medical attention. SKIN: wash with soap and water; *Toxicity by Inhalation (Threshold Limit Value)*: Data not available; *Short-Term Inhalation Limits*: Data not available; *Toxicity by Ingestion*: Grade 2; LD_{50} 0.5-5 g/kg; *Late Toxicity*: Data not available; *Vapor (Gas) Irritant Characteristics*: Vapors cause a slight smarting of the eyes or respiratory system if present in high concentrations. The effect is temporary; *Liquid or Solid Irritant Characteristics*: Causes smarting of the skin and first-degree burns on short exposure; may cause secondary burns on long exposure; *Odor Threshold*: Data not available.

Fire Hazards - *Flash Point (deg. F)*: 300 OC; 285 CC; *Flammable Limits in Air (%)*: Approx 1% (calc. LEL); *Fire Extinguishing Agents:* Alcohol foam, dry chemical, or carbon dioxide; *Fire Extinguishing Agents Not To Be Used:* Water or foam may cause frothing; *Special Hazards of Combustion Products:* Not pertinent; *Behavior in Fire:* Not pertinent; *Ignition Temperature :* Data not available; *Electrical Hazard:* Not pertinent; *Burning Rate:* Data not available.

Chemical Reactivity - *Reactivity with Water:* No reaction; *Reactivity with Common Materials:* No reaction; *Stability During Transport:* Stable; *Neutralizing Agents for Acids and Caustics:* Not pertinent; *Polymerization:* Not pertinent; *Inhibitor of Polymerization:* Not pertinent.

O

OCTANE

Chemical Designations - *Synonyms*: n-Octane; *Chemical Formula*: C_8H_{18}.

Observable Characteristics - *Physical State (as shipped)*: Liquid; *Color*: Colorless; *Odor*: Like gasoline.

Physical and Chemical Properties - *Physical State at 15 °C and 1 atm.*: Liquid; *Molecular Weight*: 114.2; *Boiling Point at 1 atm.*: 258.1, 125.6, 398.9; *Freezing Point*: -70.2, -56.8, 216.4; *Critical Temperature*: 563.7, 295.4, 568.6; *Critical Pressure*: 361, 24.5, 2.49; *Specific Gravity*: 0.703 at 20 °C (liquid); *Vapor (Gas) Specific Gravity*: 3.9; *Ratio of Specific Heats of Vapor (Gas)*: 1.047 at 16°C; *Latent Heat of Vaporization*: 130.4, 72.5, 3.03; *Heat of Combustion*: -19,112, -10,618, -442.26; *Heat of Decomposition*: Not pertinent.

Health Hazards Information - *Recommended Personal Protective Equipment*: Self-contained breathing apparatus for high vapor concentrations; goggles or face shield; rubber gloves; *Symptoms Following Exposure*: Inhalation of concentrated vapor may cause irritation of respiratory tract, depression, and pulmonary edema. Liquid can cause irritation of eyes and (on prolonged contact) irritation and cracking of skin. Ingestion causes irritation of mouth and stomach. Aspiration causes severe lung irritation, rapidly developing pulmonary edema, and central nervous system excitement, followed by depression; *General Treatment for Exposure*: INHALATION: remove victim from exposure; apply artificial respiration if breathing has stopped; call physician if needed. EYES: irrigate with copious quantities of water for 15 min. SKIN: flush with water; wash with soap and water. INGESTION: do NOT induce vomiting; call a physician. ASPIRATION: enforce bed rest; give oxygen; get medical attention; *Toxicity by Inhalation (Threshold Limit Value)*: 440 ppm; *Short-Term Inhalation Limits*: 500 ppm for 30 min.; *Toxicity by Ingestion*: Data not available; *Late Toxicity*: Data not available; *Vapor (Gas)*

Irritant Characteristics: Data not available; *Liquid or Solid Irritant Characteristics*: Data not available; *Odor Threshold*: 4 ppm.

Fire Hazards - *Flash Point (deg. F)*: 56 CC; *Flammable Limits in Air (%)*: 1.0 - 6.5; *Fire Extinguishing Agents*: Dry chemical, foam, carbon dioxide; *Fire Extinguishing Agents Not to be Used*: Water may be ineffective; *Special Hazards of Combustion Products*: Not pertinent; *Behavior in Fire*: Vapor is heavier than fire and may travel a considerable distance to a source of ignition and flash back; *Ignition Temperature (deg. F)*: 428; *Electrical Hazard*: Class I, Group D; *Burning Rate*: 6.3 mm/min.

Chemical Reactivity - *Reactivity with Water*: No reaction; *Reactivity with Common Materials*: No reaction; *Stability During Transport*: Stable; *Neutralizing Agents for Acids and Caustics:* Not pertinent; *Polymerization:* Not pertinent; *Inhibitor of Polymerization*: Not pertinent.

OCTANOL

Chemical Designations - *Synonyms*: Alcohol C-8; Heptylcarbinol; 1-Octanol; Octyl alcohol; *Chemical Formula*: $CH_3(CH_2)_6CH_2OH$.

Observable Characteristics - *Physical State (as shipped)*: Liquid; *Color*: Colorless; *Odor*: sweet.

Physical and Chemical Properties - *Physical State at 15 °C and 1 atm.*: Liquid; *Molecular Weight*: 130.23; *Boiling Point at 1 atm.*: 383, 195, 468; *Freezing Point*: 5, -15, 258; *Critical Temperature*: 725, 385, 658; *Critical Pressure*: 400, 27, 2.7; *Specific Gravity*: 0.829 at 20 °C (liquid); *Vapor (Gas) Specific Gravity*: Not pertinent; *Ratio of Specific Heats of Vapor (Gas)*: 1.044; *Latent Heat of Vaporization*: 176, 97.5, 4.08; *Heat of Combustion*: -16,130, -8,963, -375.3; *Heat of Decomposition*: Not pertinent.

Health Hazards Information - *Recommended Personal Protective Equipment*: Chemical gloves and chemical goggles; *Symptoms Following Exposure*: Irritates skin and eyes; *General Treatment for Exposure*: Flush with copious amounts of water; *Toxicity by Inhalation (Threshold Limit Value)*: Data not available; *Short-Term Inhalation Limits*: Data not available; *Toxicity by Ingestion*: Grade 1; oral rat LD_{50} = 3.2 g/kg; *Late Toxicity*: None; *Vapor (Gas) Irritant Characteristics*: Vapors are nonirritating to the eyes and throat; *Liquid or Solid Irritant Characteristics*: No appreciable hazard. Practically harmless to the skin; *Odor Threshold*: 0.49.

Fire Hazards - *Flash Point (deg. F)*: 178 CC; *Flammable Limits in Air (%)*: Data not available; *Fire Extinguishing Agents*: Foam, carbon dioxide, or dry chemical; *Fire Extinguishing Agents Not to be Used*: Not pertinent; *Special Hazards of Combustion Products*: Not pertinent; *Behavior in Fire*: Not pertinent; *Ignition Temperature* : Data not available; *Electrical Hazard*: Not pertinent; *Burning Rate*: 3.7 mm/min (approx.).

Chemical Reactivity - *Reactivity with Water*: No reaction; *Reactivity with Common Materials*: No reaction; *Stability During Transport*: Stable; Neutralizing Agents for Acids and Caustics: Not pertinent; *Polymerization:* Not pertinent; *Inhibitor of Polymerization*: Not pertinent.

1-OCTENE

Chemical Designations - *Synonyms*: Caprylene; alpha-Octylene; *Chemical Formula*: $CH_3(CH_2)_5CH=CH_2$.

Observable Characteristics - *Physical State (as shipped)*: Liquid; *Color*: Colorless; *Odor*: Like gasoline.

Physical and Chemical Properties - *Physical State at 15 °C and 1 atm.*: Liquid; *Molecular Weight*: 112.22; *Boiling Point at 1 atm.*: 250.3, 121.3, 194.5; *Freezing Point*: -151, -102, 172; *Critical Temperature*: 560.1, 293.4, 566.6; *Critical Pressure*: 400, 27.2, 2.76; *Specific Gravity*: 0.715 at 20 °C (liquid); *Vapor (Gas) Specific Gravity*: Not pertinent; *Ratio of Specific Heats of Vapor (Gas)*: 1.050; *Latent Heat of Vaporization*: 129, 71.9, 3.01; *Heat of Combustion*: -19,170, -10,650, -445.89; *Heat of Decomposition*: Not pertinent.

Health Hazards Information - *Recommended Personal Protective Equipment*: Organic vapor canister; goggles or face shield; *Symptoms Following Exposure*: Generally low toxicity. Mildly anesthetic at high vapor concentrations. May irritate eyes; *General Treatment for Exposure*: INHALATION: remove from exposure; support respiration. INGESTION: do NOT induce vomiting; *Toxicity by Inhalation*

(Threshold Limit Value): Data not available; *Short-Term Inhalation Limits*: Data not available; *Toxicity by Ingestion*: Data not available; *Late Toxicity*: Data not available; *Vapor (Gas) Irritant Characteristics*: Vapors cause a slight smarting of the eyes or respiratory system if present at high concentrations. The effect is temporary; *Liquid or Solid Irritant Characteristics*: Minimum hazard. If spilled on clothing and allowed to remain, may cause smarting and reddening of the skin; *Odor Threshold*: Data not available.

Fire Hazards - *Flash Point (deg. F):* 70 OC; *Flammable Limits in Air (%):* 0.9 (LEL); *Fire Extinguishing Agents*: Dry chemical, foam, or carbon dioxide; *Fire Extinguishing Agents Not to be Used:* Water may be ineffective; *Special Hazards of Combustion Products*: Not pertinent; *Behavior in Fire*: Not pertinent; *Ignition Temperature (deg. F)*: 493; *Electrical Hazard*: Not pertinent; *Burning Rate*: 6.5 mm/min.

Chemical Reactivity - *Reactivity with Water*: No reaction; *Reactivity with Common Materials*: No reaction; *Stability During Transport*: Stable; *Neutralizing Agents for Acids and Caustics:* Not pertinent; *Polymerization:* Not pertinent; *Inhibitor of Polymerization*: Not pertinent.

OCTYL EPOXY TALLATE

Chemical Designations - *Synonyms*: Epoxidized tall oil, octyl ester; *Chemical Formula*: Mixture.

Observable Characteristics - *Physical State (as shipped)*: Liquid; *Color*: Pale yellow; *Odor*: Mild.

Physical and Chemical Properties - *Physical State at 15 °C and 1 atm.*: Liquid; *Molecular Weight*: 420 (approx.); *Boiling Point at 1 atm.*: Not pertinent; *Freezing Point*: Not pertinent; *Critical Temperature*: Not pertinent; *Critical Pressure*: Not pertinent; *Specific Gravity*: (est.) 1.002 at 20 °C (liquid); *Vapor (Gas) Specific Gravity*: Not pertinent; *Ratio of Specific Heats of Vapor (Gas)*: Not pertinent; *Latent Heat of Vaporization*: Not pertinent; *Heat of Combustion*: Data not available; *Heat of Decomposition*: Not pertinent.

Health Hazards Information - *Recommended Personal Protective Equipment*: Chemical goggles; face shield; oil-resistant gloves; *Symptoms Following Exposure*: Contact with eyes causes mild inflammation. Contact with skin may produce allergic response; *General Treatment for Exposure*: EYES or SKIN: remove excess oil with cloth or absorbent paper; then wash with soapy water and flush with clear water; consult a physician; *Toxicity by Inhalation (Threshold Limit Value)*: Data not available; *Short-Term Inhalation Limits*: Data not available; *Toxicity by Ingestion*: Grade 0; $LD_{50} > 15$ g/kg; *Late Toxicity*: Data not available; *Vapor (Gas) Irritant Characteristics*: Data not available; *Liquid or Solid Irritant Characteristics*: Data not available; *Odor Threshold*: Data not available.

Fire Hazards - *Flash Point (deg. F):* 450 OC; *Flammable Limits in Air (%):* Not pertinent; *Fire Extinguishing Agents*: Foam, dry chemical, carbon dioxide; *Fire Extinguishing Agents Not to be Used:* Water may be ineffective; *Special Hazards of Combustion Products*: Data not available; *Behavior in Fire*: Data not available; *Ignition Temperature* : Data not available; *Electrical Hazard*: Data not available; *Burning Rate*: Data not available.

Chemical Reactivity - *Reactivity with Water*: No reaction; *Reactivity with Common Materials*: May attack some forms of plastics; *Stability During Transport*: Stable; *Neutralizing Agents for Acids and Caustics:* Not pertinent; *Polymerization:* Not pertinent; *Inhibitor of Polymerization*: Not pertinent.

OILS: CLARIFIED

Chemical Designations - *Synonyms*: No common synonyms; *Chemical Formula*: Not applicable.

Observable Characteristics - *Physical State (as shipped)*: Liquid; *Color*: Data not available; *Odor*: Data not available.

Physical and Chemical Properties - *Physical State at 15 °C and 1 atm.*: Liquid; *Molecular Weight*: Not pertinent; *Boiling Point at 1 atm.*: Data not available; *Freezing Point*: Not pertinent; *Critical Temperature*: Not pertinent; *Critical Pressure*: Not pertinent; *Specific Gravity*: (etc.) 0.85 at 20 °C (liquid); *Vapor (Gas) Specific Gravity*: Not pertinent; *Ratio of Specific Heats of Vapor (Gas)*: Not pertinent; *Latent Heat of Vaporization*: Not pertinent; *Heat of Combustion*: (est.) -18,000, -10,000, -420; *Heat of Decomposition*: Not pertinent.

Health Hazards Information - *Recommended Personal Protective Equipment*: Goggles or face shield;

Symptoms Following Exposure: If liquid is ingested, an increased frequency of bowel movements will occur; *General Treatment for Exposure*: INGESTION: Do not induce vomiting; SKIN: Wipe off, wash with soap and water; EYES: Wash with water for at least 15 min.; *Toxicity by Inhalation (Threshold Limit Value)*: No single TLV applicable; *Short-Term Inhalation Limits*: Data not available; *Toxicity by Ingestion*: Grade 1; LD_{50} 50 to 15 g/kg; *Late Toxicity*: Data not available; *Vapor (Gas) Irritant Characteristics*: Data not available; *Liquid or Solid Irritant Characteristics*: Data not available; *Odor Threshold*: Data not available.

Fire Hazards - *Flash Point :* Data not available; *Flammable Limits in Air (%):* Data not available; *Fire Extinguishing Agents*: Dry chemical, foam, or carbon dioxide; *Fire Extinguishing Agents Not to be Used:* Water may be ineffective; *Special Hazards of Combustion Products*: Not pertinent; *Behavior in Fire*: Not pertinent; *Ignition Temperature* : Data not available; *Electrical Hazard*: Not pertinent; *Burning Rate*: 4 mm/min.

Chemical Reactivity - *Reactivity with Water*: No reaction; *Reactivity with Common Materials*: No reaction; *Stability During Transport*: Stable; *Neutralizing Agents for Acids and Caustics:* Not pertinent; *Polymerization:* Not pertinent; *Inhibitor of Polymerization*: Not pertinent.

OILS: CRUDE

Chemical Designations - *Synonyms*: Petroleum; *Chemical Formula*: Not applicable.

Observable Characteristics - *Physical State (as shipped)*: Liquid; *Color*: Dark; *Odor*: Offensive; tarry.

Physical and Chemical Properties - *Physical State at 15 °C and 1 atm.*: Liquid; *Molecular Weight*: Not pertinent; *Boiling Point at 1 atm.*: 90- >750, 32- >400, 306- >673; *Freezing Point*: Not pertinent; *Critical Temperature*: Not pertinent; *Critical Pressure*: Not pertinent; *Specific Gravity*: 0.70 0.98 at 15 °C (liquid); *Vapor (Gas) Specific Gravity*: Not pertinent; *Ratio of Specific Heats of Vapor (Gas)*: Not pertinent; *Latent Heat of Vaporization*: 140-150, 76-86, 3.2-3.6; *Heat of Combustion*: -18,252, - 10,140, -424.54; *Heat of Decomposition*: Not pertinent.

Health Hazards Information - *Recommended Personal Protective Equipment*: Goggles or face shield; rubber gloves and boots; *Symptoms Following Exposure*: May irritate eyes and skin; *General Treatment for Exposure*: EYES: Flush with water for at least 15 min. SKIN: wipe off and wash with soap and water; *Toxicity by Inhalation (Threshold Limit Value)*: Data not available; *Short-Term Inhalation Limits*: Data not available; *Toxicity by Ingestion*: Data not available; *Late Toxicity*: Data not available; *Vapor (Gas) Irritant Characteristics*: Vapors are nonirritating to the eyes and throat; *Liquid or Solid Irritant Characteristics*: Minimum hazard. If spilled on clothing and allowed to remain, may cause smarting and reddening of the skin; *Odor Threshold*: Data not available.

Fire Hazards - *Flash Point :* Data not available; *Flammable Limits in Air (%):* Data not available; *Fire Extinguishing Agents*: Dry chemical, foam, or carbon dioxide; *Fire Extinguishing Agents Not to be Used:* Water may be ineffective; *Special Hazards of Combustion Products*: Not pertinent; *Behavior in Fire*: Not pertinent; *Ignition Temperature* : Data not available; *Electrical Hazard*: Not pertinent; *Burning Rate*: 4 mm/min.

Chemical Reactivity - *Reactivity with Water*: No reaction; *Reactivity with Common Materials*: No reaction; *Stability During Transport*: Stable; *Neutralizing Agents for Acids and Caustics:* Not pertinent; *Polymerization:* Not pertinent; *Inhibitor of Polymerization*: Not pertinent.

OILS: DIESEL

Chemical Designations - *Synonyms*: Fuel Oil 1-D; Fuel Oil 2-D; *Chemical Formula*: Not applicable.

Observable Characteristics - *Physical State (as shipped)*: Liquid; *Color*: Light brown; *Odor*: Like fuel oil.

Physical and Chemical Properties - *Physical State at 15 °C and 1 atm.*: Liquid; *Molecular Weight*: Not pertinent; *Boiling Point at 1 atm.*: 550-640, 288-338, 561-612; *Freezing Point*: 0 to -30, -18 to -34, 255 to 239; *Critical Temperature*: Not pertinent; *Critical Pressure*: Not pertinent; *Specific Gravity*: 0.841 at 16 °C (liquid); *Vapor (Gas) Specific Gravity*: Not pertinent; *Ratio of Specific Heats of Vapor (Gas)*: Not pertinent; *Latent Heat of Vaporization*: Not pertinent; *Heat of Combustion*: -18,400, -

10,200, 429; *Heat of Decomposition*: Not pertinent.
Health Hazards Information - *Recommended Personal Protective Equipment*: Goggles or face shield; *Symptoms Following Exposure*: If liquid is ingested, an increased frequency of bowel movements will occur; *General Treatment for Exposure*: INGESTION: do NOT induce vomiting. SKIN: wipe off, wash with soap and water. EYES: wash with copious amounts of water for at least 15 min.; *Toxicity by Inhalation (Threshold Limit Value)*: No single TLV applicable; *Short-Term Inhalation Limits*: Data not available; *Toxicity by Ingestion*: Grade 1; LD_{50} 5 to 15 g/kg; *Late Toxicity*: Data not available; *Vapor (Gas) Irritant Characteristics*: Vapors cause a slight smarting of the eyes or respiratory system if present in high concentrations. The effect is temporary; *Liquid or Solid Irritant Characteristics*: Minimum hazard. If spilled on clothing and allowed to remain, may cause smarting and reddening of the skin; *Odor Threshold*: Data not available.
Fire Hazards - *Flash Point (deg. F):* (1 -D) 100 CC; (2 - D) 125 CC; *Flammable Limits in Air (%):* 1.3 - 6.0 vol; *Fire Extinguishing Agents*: Dry chemical, foam, or carbon dioxide; *Fire Extinguishing Agents Not to be Used:* Water may be ineffective; *Special Hazards of Combustion Products*: Not pertinent; *Behavior in Fire*: Not pertinent; *Ignition Temperature (deg. F)*: (1 - D) 350 -625; (2 - D) 490 - 545; *Electrical Hazard*: Not pertinent; *Burning Rate*: 4 mm/min.
Chemical Reactivity - *Reactivity with Water*: No reaction; *Reactivity with Common Materials*: No reaction; *Stability During Transport*: Stable; *Neutralizing Agents for Acids and Caustics:* Not pertinent; *Polymerization:* Not pertinent; *Inhibitor of Polymerization*: Not pertinent.

OILS, EDIBLE: CASTOR

Chemical Designations - *Synonyms*: No common synonyms; *Chemical Formula*: Not applicable.
Observable Characteristics - *Physical State (as shipped)*: Liquid; *Color*: Light yellow to green; *Odor*: Characteristic; odorless.
Physical and Chemical Properties - *Physical State at 15 °C and 1 atm.*: Liquid; *Molecular Weight*: Not pertinent; *Boiling Point at 1 atm.*: Varies, depending on composition; *Freezing Point*: 10, -12, 261; *Critical Temperature*: Not pertinent; *Critical Pressure*: Not pertinent; *Specific Gravity*: 0.96 at 25 °C (liquid); *Vapor (Gas) Specific Gravity*: Not pertinent; *Ratio of Specific Heats of Vapor (Gas)*: Not pertinent; *Latent Heat of Vaporization*: Not pertinent; *Heat of Combustion*: -15,950, -8,860, -371.0; *Heat of Decomposition*: Not pertinent.
Health Hazards Information - *Recommended Personal Protective Equipment*: Goggles or face shield; *Symptoms Following Exposure*: If ingested causes severe diarrhea; *General Treatment for Exposure*: INGESTION: if more than 2 tablespoons, consult physician. EYES: flush with water for at least 15 min. SKIN: wipe off, wash with soap and water; *Toxicity by Inhalation (Threshold Limit Value)*: None; *Short-Term Inhalation Limits*: Not pertinent; *Toxicity by Ingestion*: Grade 1; LD_{50} 5 to 15 g/kg (Fatal dose unknown but presumably large); *Late Toxicity*: None; *Vapor (Gas) Irritant Characteristics*: None; *Liquid or Solid Irritant Characteristics*: None; *Odor Threshold*: None.
Fire Hazards - *Flash Point (deg. F):* 445 CC; *Flammable Limits in Air (%):* Data not available; *Fire Extinguishing Agents*: Dry chemical, foam, or carbon dioxide; *Fire Extinguishing Agents Not to be Used:* Water or foam may cause frothing; *Special Hazards of Combustion Products*: Not pertinent; *Behavior in Fire*: Not pertinent; *Ignition Temperature (deg. F)*: 840; *Electrical Hazard*: Not pertinent; *Burning Rate*: Data not available.
Chemical Reactivity - *Reactivity with Water*: No reaction; *Reactivity with Common Materials*: No reaction; *Stability During Transport*: Stable; *Neutralizing Agents for Acids and Caustics:* Not pertinent; *Polymerization:* Not pertinent; *Inhibitor of Polymerization*: Not pertinent.

OILS, EDIBLE: COCONUT

Chemical Designations - *Synonyms*: Coconut butter; Coconut oil; Copra oil; *Chemical Formula*: Not applicable.
Observable Characteristics - *Physical State (as shipped)*: Liquid or solid; *Color*: Light yellow-orange; *Odor*: Weak acid.
Physical and Chemical Properties - *Physical State at 15 °C and 1 atm.*: Solid of liquid; *Molecular

Weight: Not pertinent; *Boiling Point at 1 atm.*: Not pertinent (very high); *Freezing Point*: (approx.) 76, 24, 297; *Critical Temperature*: Not pertinent; *Critical Pressure*: Not pertinent; *Specific Gravity*: 0.922 at 25 °C (liquid); *Vapor (Gas) Specific Gravity*: Not pertinent; *Ratio of Specific Heats of Vapor (Gas)*: Not pertinent; *Latent Heat of Vaporization*: Not pertinent; *Heat of Combustion*: (est.) -15,500, -8,600, -360; *Heat of Decomposition*: Not pertinent.

Health Hazards Information - *Recommended Personal Protective Equipment*: Goggles or face shield; rubber gloves; *Symptoms Following Exposure*: Oil is essentially nontoxic, but can cause mild irritation of eyes on contact; *General Treatment for Exposure*: EYES: flush with water for at least 15 min. INGESTION: do not induce vomiting; *Toxicity by Inhalation (Threshold Limit Value)*: Data not available; *Short-Term Inhalation Limits*: Data not available; *Toxicity by Ingestion*: Data not available; *Late Toxicity*: None; *Vapor (Gas) Irritant Characteristics*: Data not available; *Liquid or Solid Irritant Characteristics*: Data not available; *Odor Threshold*: Not pertinent.

Fire Hazards - *Flash Point (deg. F):* 420 CC (crude); 580 CC (refined); *Flammable Limits in Air (%):* Not pertinent; *Fire Extinguishing Agents*: Dry chemical, foam, or carbon dioxide; *Fire Extinguishing Agents Not to be Used:* Water or foam may cause frothing; water may be ineffective; *Special Hazards of Combustion Products*: Not pertinent; *Behavior in Fire*: Not pertinent; *Ignition Temperature* : Data not available; *Electrical Hazard*: Not pertinent; *Burning Rate*: 4 mm/min.

Chemical Reactivity - *Reactivity with Water*: No reaction; *Reactivity with Common Materials*: No reaction; *Stability During Transport*: Stable; *Neutralizing Agents for Acids and Caustics:* Not pertinent; *Polymerization:* Not pertinent; *Inhibitor of Polymerization*: Not pertinent.

OILS, EDIBLE: COTTONSEED

Chemical Designations - *Synonyms*: No common synonyms; *Chemical Formula*: Not applicable.

Observable Characteristics - *Physical State (as shipped)*: Liquid; *Color*: Pale yellow; *Odor*: Odorless.

Physical and Chemical Properties - *Physical State at 15 °C and 1 atm.*: Liquid; *Molecular Weight*: Not pertinent; *Boiling Point at 1 atm.*: Very high; *Freezing Point*: 32, 0, 273; *Critical Temperature*: Not pertinent; *Critical Pressure*: Not pertinent; *Specific Gravity*: 0.922 at 20 °C (liquid); *Vapor (Gas) Specific Gravity*: Not pertinent; *Ratio of Specific Heats of Vapor (Gas)*: Not pertinent; *Latent Heat of Vaporization*: Not pertinent; *Heat of Combustion*: (est.) -16,000, -8,870, -371; *Heat of Decomposition*: Not pertinent.

Health Hazards Information - *Recommended Personal Protective Equipment*: Goggles or face shield; *Symptoms Following Exposure*: None; is used as a food; *General Treatment for Exposure*: EYES: wash with water for at least 15 min.; *Toxicity by Inhalation (Threshold Limit Value)*: None; *Short-Term Inhalation Limits*: Not pe; *Toxicity by Ingestion*: None; *Late Toxicity*: None; *Vapor (Gas) Irritant Characteristics*: None; *Liquid or Solid Irritant Characteristics*: None; *Odor Threshold*: Not pertinent.

Fire Hazards - *Flash Point (deg. F):* 486 CC (refined oil); 610 CC (cooking oil); *Flammable Limits in Air (%):* Data not available; *Fire Extinguishing Agents*: Dry chemical, foam, or carbon dioxide; *Fire Extinguishing Agents Not to be Used:* Water or foam may cause frothing; *Special Hazards of Combustion Products*: Not pertinent; *Behavior in Fire*: Not pertinent; *Ignition Temperature (deg. F)*: 650 (refined oil); *Electrical Hazard*: Not pertinent; *Burning Rate*: Data not available.

Chemical Reactivity - *Reactivity with Water*: No reaction; *Reactivity with Common Materials*: No reaction; *Stability During Transport*: Stable; *Neutralizing Agents for Acids and Caustics:* Not pertinent; *Polymerization:* Not pertinent; *Inhibitor of Polymerization*: Not pertinent.

OILS, EDIBLE: FISH

Chemical Designations - *Synonyms*: No common synonyms; *Chemical Formula*: Not applicable.

Observable Characteristics - *Physical State (as shipped)*: Liquid; *Color*: Pale yellow; *Odor*: Fishy.

Physical and Chemical Properties - *Physical State at 15 °C and 1 atm.*: Liquid; *Molecular Weight*: Not pertinent; *Boiling Point at 1 atm.*: Very high; *Freezing Point*: Not pertinent; *Critical Temperature*: Not pertinent; *Critical Pressure*: Not pertinent; *Specific Gravity*: 0.93 at 20 °C (liquid); *Vapor (Gas) Specific Gravity*: Not pertinent; *Ratio of Specific Heats of Vapor (Gas)*: Not pertinent; *Latent Heat of Vaporization*: Not pertinent; *Heat of Combustion*: (est.) -16,000, -8,870, -371; *Heat of Decomposition*:

Not pertinent.

Health Hazards Information - *Recommended Personal Protective Equipment*: Goggles or face shield; *Symptoms Following Exposure*: None — is a food; *General Treatment for Exposure*: EYES: flush with water for at least 15 min.; *Toxicity by Inhalation (Threshold Limit Value)*: Not pertinent; *Short-Term Inhalation Limits*: Not pertinent; *Toxicity by Ingestion*: None; *Late Toxicity*: None; *Vapor (Gas) Irritant Characteristics*: None; *Liquid or Solid Irritant Characteristics*: None; *Odor Threshold*: Data not available.

Fire Hazards - *Flash Point (deg. F):* 420 CC; *Flammable Limits in Air (%):* Data not available; *Fire Extinguishing Agents*: Dry chemical, foam, or carbon dioxide; *Fire Extinguishing Agents Not to be Used:* Water or foam may cause frothing; *Special Hazards of Combustion Products*: Not pertinent; *Behavior in Fire*: Not pertinent; *Ignition Temperature* : Data not available; *Electrical Hazard*: Not pertinent; *Burning Rate*: Data not available.

Chemical Reactivity - *Reactivity with Water*: No reaction; *Reactivity with Common Materials*: No reaction; *Stability During Transport*: Stable; *Neutralizing Agents for Acids and Caustics:* Not pertinent; *Polymerization:* Not pertinent; *Inhibitor of Polymerization*: Not pertinent.

OILS, EDIBLE: LARD

Chemical Designations - *Synonyms*: Kettle-rendered lard; Leaf lard; Prime steam lard; *Chemical Formula*: Not applicable.

Observable Characteristics - *Physical State (as shipped)*: Solid or liquid; *Color*: Colorless or pale yellow; *Odor*: Fatty.

Physical and Chemical Properties - *Physical State at 15 °C and 1 atm.*: Solid; *Molecular Weight*: Not pertinent; *Boiling Point at 1 atm.*: Not pertinent; *Freezing Point*: 66 - 99, 19 - 37, 292 - 310; *Critical Temperature*: Not pertinent; *Critical Pressure*: Not pertinent; *Specific Gravity*: 0.861 at 20 °C (liquid); *Vapor (Gas) Specific Gravity*: Not pertinent; *Ratio of Specific Heats of Vapor (Gas)*: Not pertinent; *Latent Heat of Vaporization*: Not pertinent; *Heat of Combustion*: -16,750, -9,320, -390; *Heat of Decomposition*.

Health Hazards Information - *Recommended Personal Protective Equipment*: Goggles or face shield; rubber gloves; *Symptoms Following Exposure*: Substance is essentially nontoxic. Prolonged contact with skin may cause dermatitis (oil acne). Hot liquid can burn eyes or skin; *General Treatment for Exposure*: EYES: flush with water for at least 15 min.; get medical attention for burn. SKIN: wipe off; get medical attention for burn. INGESTION: do NOT induce vomiting; *Toxicity by Inhalation (Threshold Limit Value)*: Not pertinent; *Short-Term Inhalation Limits*: Not pertinent; *Toxicity by Ingestion*: Not pertinent; *Late Toxicity*: None; *Vapor (Gas) Irritant Characteristics*: Data not available; *Liquid or Solid Irritant Characteristics*: Data not available; *Odor Threshold*: Not pertinent.

Fire Hazards - *Flash Point (deg. F):* 395 CC; *Flammable Limits in Air (%):* Not pertinent; *Fire Extinguishing Agents*: Dry chemical, foam, or carbon dioxide; *Fire Extinguishing Agents Not to be Used:* Water or foam may cause frothing; water may be ineffective; *Special Hazards of Combustion Products*: Not pertinent; *Behavior in Fire*: Not pertinent; *Ignition Temperature (deg. F)*: 833; *Electrical Hazard*: Not pertinent; *Burning Rate*: 4 mm/min.

Chemical Reactivity - *Reactivity with Water*: No reaction; *Reactivity with Common Materials*: No reaction; *Stability During Transport*: Stable; *Neutralizing Agents for Acids and Caustics:* Not pertinent; *Polymerization:* Not pertinent; *Inhibitor of Polymerization*: Not pertinent.

OILS, EDIBLE: OLIVE

Chemical Designations - *Synonyms*: No common synonyms; *Chemical Formula*: Not applicable.

Observable Characteristics - *Physical State (as shipped)*: Liquid; *Color*: Pale yellow-green; *Odor*: Weak, characteristic.

Physical and Chemical Properties - *Physical State at 15 °C and 1 atm.*: Liquid; *Molecular Weight*: Not pertinent; *Boiling Point at 1 atm.*: Very high; *Freezing Point*: Not pertinent; *Critical Temperature*: Not pertinent; *Critical Pressure*: Not pertinent; *Specific Gravity*: 0.915 at 20 °C (liquid); *Vapor (Gas) Specific Gravity*: Not pertinent; *Ratio of Specific Heats of Vapor (Gas)*: Not pertinent; *Latent Heat of*

Vaporization: Not pertinent; *Heat of Combustion*: (est.) = 16,000, -8,870, -371; *Heat of Decomposition*: Not pertinent.

Health Hazards Information - *Recommended Personal Protective Equipment*: Goggles or face shield; *Symptoms Following Exposure*: None - is a food; *General Treatment for Exposure*: No treatment necessary; *Toxicity by Inhalation (Threshold Limit Value)*: Not pertinent; *Short-Term Inhalation Limits*: Not pertinent; *Toxicity by Ingestion*: None; *Late Toxicity*: None; *Vapor (Gas) Irritant Characteristics*: None; *Liquid or Solid Irritant Characteristics*: None; *Odor Threshold*: Data not available.

Fire Hazards - *Flash Point (deg. F)*: 437 CC; *Flammable Limits in Air (%)*: Data not available; *Fire Extinguishing Agents*: Dry chemical, foam, or carbon dioxide; *Fire Extinguishing Agents Not to be Used*: Water or foam may cause frothing; *Special Hazards of Combustion Products*: Not pertinent; *Behavior in Fire*: Not pertinent; *Ignition Temperature (deg. F)*: 650; *Electrical Hazard*: Not pertinent; *Burning Rate*: Data not available.

Chemical Reactivity - *Reactivity with Water*: No reaction; *Reactivity with Common Materials*: No reaction; *Stability During Transport*: Stable; *Neutralizing Agents for Acids and Caustics:* Not pertinent; *Polymerization:* Not pertinent; *Inhibitor of Polymerization*: Not pertinent.

OILS, EDIBLE

Chemical Designations - *Synonyms*: Palm butter; Palm fruit oil; *Chemical Formula*: Not applicable.

Observable Characteristics - *Physical State (as shipped)*: Semi-solid to liquid; *Color*: Orange-red; *Odor*: Pleasant, characteristic.

Physical and Chemical Properties - *Physical State at 15 °C and 1 atm.*: Solid to liquid; *Molecular Weight*: Not pertinent; *Boiling Point at 1 atm.*: Not pertinent (very high); *Freezing Point*: 70 - 80, 21 - 27, 294 - 300; *Critical Temperature*: Not pertinent; *Critical Pressure*: Not pertinent; *Specific Gravity*: 0.906 at 38 °C (liquid); *Vapor (Gas) Specific Gravity*: Not pertinent; *Ratio of Specific Heats of Vapor (Gas)*: Not pertinent; *Latent Heat of Vaporization*: Not pertinent; *Heat of Combustion*: (est.) -15,500, -8,600, -360; *Heat of Decomposition*: Not pertinent.

Health Hazards Information - *Recommended Personal Protective Equipment*: Goggles or face shield; rubber gloves; *Symptoms Following Exposure*: Oil is essentially nontoxic; may cause mild irritation of eyes; *General Treatment for Exposure*: EYES: flush with water for at least 15 min. INGESTION: do NOT induce vomiting; *Toxicity by Inhalation (Threshold Limit Value)*: Data not available; *Short-Term Inhalation Limits*: Not pertinent; *Toxicity by Ingestion*: Data not available; *Late Toxicity*: None; *Vapor (Gas) Irritant Characteristics*: Data not available; *Liquid or Solid Irritant Characteristics*: Data not available; *Odor Threshold*: Not pertinent.

Fire Hazards - *Flash Point (deg. F)*: 373 CC; *Flammable Limits in Air (%)*: Not pertinent; *Fire Extinguishing Agents*: Dry chemical, foam, or carbon dioxide; *Fire Extinguishing Agents Not to be Used*: Water or foam may cause frothing; water may be ineffective; *Special Hazards of Combustion Products*: Not pertinent; *Behavior in Fire*: Not pertinent; *Ignition Temperature (deg. F)*: 600; *Electrical Hazard*: Not pertinent; *Burning Rate*: 4 mm/min.

Chemical Reactivity - *Reactivity with Water*: No reaction; *Reactivity with Common Materials*: No reaction; *Stability During Transport*: Stable; *Neutralizing Agents for Acids and Caustics:* Not pertinent; *Polymerization:* Not pertinent; *Inhibitor of Polymerization*: Not pertinent.

OILS, EDIBLE: PEANUT

Chemical Designations - *Synonyms*: No common synonyms; *Chemical Formula*: Not applicable.

Observable Characteristics - *Physical State (as shipped)*: Liquid; *Color*: Pale yellow; *Odor*: Characteristic slight nutty odor.

Physical and Chemical Properties - *Physical State at 15 °C and 1 atm.*: Liquid; *Molecular Weight*: Not pertinent; *Boiling Point at 1 atm.*: Very high; *Freezing Point*: 28, -2, 271; *Critical Temperature*: Not pertinent; *Critical Pressure*: Not pertinent; *Specific Gravity*: 0.919 at 20°C (liquid); *Vapor (Gas) Specific Gravity*: Not pertinent; *Ratio of Specific Heats of Vapor (Gas)*: Not pertinent; *Latent Heat of Vaporization*: Not pertinent; *Heat of Combustion*: (est.) -16,000, -8,870, -371; *Heat of Decomposition*: Not pertinent.

Health Hazards Information - *Recommended Personal Protective Equipment*: Goggles or face shield; *Symptoms Following Exposure*: None — is a food; *General Treatment for Exposure*: EYES: flush with water for at least 15 min.; *Toxicity by Inhalation (Threshold Limit Value)*: Not pertinent; *Short-Term Inhalation Limits*: Not pertinent; *Toxicity by Ingestion*: None; *Late Toxicity*: None; *Vapor (Gas) Irritant Characteristics*: None; *Liquid or Solid Irritant Characteristics*: None; *Odor Threshold*: Data not available.

Fire Hazards - *Flash Point (deg. F):* 640 OC; 540 CC; *Flammable Limits in Air (%):* Data not available; *Fire Extinguishing Agents*: Dry chemical, foam, or carbon dioxide; *Fire Extinguishing Agents Not to be Used:* Water or foam may cause frothing; *Special Hazards of Combustion Products*: Not pertinent; *Behavior in Fire*: Not pertinent; *Ignition Temperature (deg. F)*: 833; *Electrical Hazard*: Not pertinent; *Burning Rate*: Data not available.

Chemical Reactivity - *Reactivity with Water*: No reaction; *Reactivity with Common Materials*: No reaction; *Stability During Transport*: Stable; *Neutralizing Agents for Acids and Caustics:* Not pertinent; *Polymerization:* Not pertinent; *Inhibitor of Polymerization*: Not pertinent.

OILS, EDIBLE: SAFFLOWER

Chemical Designations - *Synonyms*: Carthamus tinctorius oil; Safflower seed oil; *Chemical Formula*: Not applicable.

Observable Characteristics - *Physical State (as shipped)*: Liquid; *Color*: Pale yellow; *Odor*: Bland, fatty.

Physical and Chemical Properties - *Physical State at 15 °C and 1 atm.*: Liquid; *Molecular Weight*: Not pertinent; *Boiling Point at 1 atm.*: Not pertinent (very high); *Freezing Point*: Not pertinent; *Critical Temperature*: Not pertinent; *Critical Pressure*: Not pertinent; *Specific Gravity*: 0.923 at 25°C (liquid); *Vapor (Gas) Specific Gravity*: Not pertinent; *Ratio of Specific Heats of Vapor (Gas)*: Not pertinent; *Latent Heat of Vaporization*: Not pertinent; *Heat of Combustion*: (est.) -15,500, -8,600, -360; *Heat of Decomposition*: Not pertinent.

Health Hazards Information - *Recommended Personal Protective Equipment*: Goggles or face shield; rubber gloves; *Symptoms Following Exposure*: Oil essentially nontoxic. Contact with eyes can cause mild vomiting; *General Treatment for Exposure*: EYES: flush with water for at least 15 min. INGESTION: do not induce vomiting; *Toxicity by Inhalation (Threshold Limit Value)*: Data not available; *Short-Term Inhalation Limits*: Not pertinent; *Toxicity by Ingestion*: Data not available; *Late Toxicity*: None; *Vapor (Gas) Irritant Characteristics*: Data not available; *Liquid or Solid Irritant Characteristics*: Data not available; *Odor Threshold*: Not pertinent.

Fire Hazards - *Flash Point :* Data not available; *Flammable Limits in Air (%):* Not pertinent; *Fire Extinguishing Agents*: Dry chemical, foam, or carbon dioxide; *Fire Extinguishing Agents Not to be Used:* Water or foam may cause frothing; water may be ineffective; *Special Hazards of Combustion Products*: Not pertinent; *Behavior in Fire*: Not pertinent; *Ignition Temperature :* Data not available; *Electrical Hazard*: Not pertinent; *Burning Rate*: 4 mm/min.

Chemical Reactivity - *Reactivity with Water*: No reaction; *Reactivity with Common Materials*: No reaction; *Stability During Transport*: Stable; *Neutralizing Agents for Acids and Caustics:* Not pertinent; *Polymerization:* Not pertinent; *Inhibitor of Polymerization*: Not pertinent.

OILS, EDIBLE: SOYA BEAN

Chemical Designations - *Synonyms*: Soybean oil; *Chemical Formula*: Not applicable.

Observable Characteristics - *Physical State (as shipped)*: Liquid; *Color*: Pale yellow; *Odor*: Weak.

Physical and Chemical Properties - *Physical State at 15 °C and 1 atm.*: Liquid; *Molecular Weight*: Not pertinent; *Boiling Point at 1 atm.*: Very high; *Freezing Point*: -4, -20, 253; *Critical Temperature*: Not pertinent; *Critical Pressure*: Not pertinent; *Specific Gravity*: 0.22 at 20 °C (liquid); *Vapor (Gas) Specific Gravity*: Not pertinent; *Ratio of Specific Heats of Vapor (Gas)*: Not pertinent; *Latent Heat of Vaporization*: Not pertinent; *Heat of Combustion*: (est.) -16,000, -8,870, -371; *Heat of Decomposition*: Not pertinent.

Health Hazards Information - *Recommended Personal Protective Equipment*: Goggles or face shield;

Symptoms Following Exposure: None — is a food; *General Treatment for Exposure*: EYES: flush with water for at least 15 min.; *Toxicity by Inhalation (Threshold Limit Value)*: Not pertinent; *Short-Term Inhalation Limits*: Not pertinent; *Toxicity by Ingestion*: None; *Late Toxicity*: None; *Vapor (Gas) Irritant Characteristics*: None; *Liquid or Solid Irritant Characteristics*: None; *Odor Threshold*: Data not available.

Fire Hazards - *Flash Point (deg. F)*: 540 CC; *Flammable Limits in Air (%)*: Data not available; *Fire Extinguishing Agents*: Dry chemical, foam, or carbon dioxide; *Fire Extinguishing Agents Not to be Used*: Water or foam may cause frothing; *Special Hazards of Combustion Products*: Not pertinent; *Behavior in Fire*: Not pertinent; *Ignition Temperature (deg. F)*: 833; *Electrical Hazard*: Not pertinent; *Burning Rate*: Data not available.

Chemical Reactivity - *Reactivity with Water*: No reaction; *Reactivity with Common Materials*: No reaction; *Stability During Transport*: Stable; *Neutralizing Agents for Acids and Caustics*: Not pertinent; *Polymerization*: Not pertinent; *Inhibitor of Polymerization*: Not pertinent.

OILS, EDIBLE: TUCUM

Chemical Designations - *Synonyms*: American palm kernel oil; Aouara oil; Palm seed oil; *Chemical Formula*: Not applicable.

Observable Characteristics - *Physical State (as shipped)*: Liquid; *Color*: Light yellow; *Odor*: Weak, acid.

Physical and Chemical Properties - *Physical State at 15 °C and 1 atm.*: Liquid; *Molecular Weight*: Not pertinent; *Boiling Point at 1 atm.*: Not pertinent (very high); *Freezing Point*: 86, 30, 303; *Critical Temperature*: Not pertinent; *Critical Pressure*: Not pertinent; *Specific Gravity*: 0.908 at 60 °C (liquid); *Vapor (Gas) Specific Gravity*: Not pertinent; *Ratio of Specific Heats of Vapor (Gas)*: Not pertinent; *Latent Heat of Vaporization*: Not pertinent; *Heat of Combustion*: (est.) -15,500, -8,600, -360; *Heat of Decomposition*: Not pertinent.

Health Hazards Information - *Recommended Personal Protective Equipment*: Goggles or face shield; rubber gloves; *Symptoms Following Exposure*: Oil essentially nontoxic. Contact with eyes cause mild irritation, and prolonged contact with skin may cause dermatitis; *General Treatment for Exposure*: EYES: flush with water for at least 15 min. INGESTION: do NOT induce vomiting; *Toxicity by Inhalation (Threshold Limit Value)*: Data not available; *Short-Term Inhalation Limits*: Not pertinent; *Toxicity by Ingestion*: Data not available; *Late Toxicity*: None known; *Vapor (Gas) Irritant Characteristics*: Data not available; *Liquid or Solid Irritant Characteristics*: Data not available; *Odor Threshold*: Not pertinent.

Fire Hazards - *Flash Point (deg. F)*: 398 CC; *Flammable Limits in Air (%)*: Not pertinent; *Fire Extinguishing Agents*: Dry chemical, foam, or carbon dioxide; *Fire Extinguishing Agents Not to be Used*: Water or foam may cause frothing; water may be ineffective; *Special Hazards of Combustion Products*: Not pertinent; *Behavior in Fire*: Not pertinent; *Ignition Temperature*: Data not available; *Electrical Hazard*: Not pertinent; *Burning Rate*: 4 mm/min.

Chemical Reactivity - *Reactivity with Water*: No reaction; *Reactivity with Common Materials*: No reaction; *Stability During Transport*: Stable; *Neutralizing Agents for Acids and Caustics*: Not pertinent; *Polymerization*: Not pertinent; *Inhibitor of Polymerization*: Not pertinent.

OILS, EDIBLE: VEGETABLE

Chemical Designations - *Synonyms*: No common synonyms; *Chemical Formula*: Not applicable.

Observable Characteristics - *Physical State (as shipped)*: Liquid; *Color*: Colorless to pale yellow; *Odor*: Weak fatty.

Physical and Chemical Properties - *Physical State at 15 °C and 1 atm.*: Liquid; *Molecular Weight*: Not pertinent; *Boiling Point at 1 atm.*: Very high; *Freezing Point*: Not pertinent; *Critical Temperature*: Not pertinent; *Critical Pressure*: Not pertinent; *Specific Gravity*: 0.923 at 25 °C (liquid); *Vapor (Gas) Specific Gravity*: Not pertinent; *Ratio of Specific Heats of Vapor (Gas)*: Not pertinent; *Latent Heat of Vaporization*: Not pertinent; *Heat of Combustion*: (est.) -16,000, -8,870, -371; *Heat of Decomposition*: Not pertinent.

Health Hazards Information - *Recommended Personal Protective Equipment*: Goggles or face shield; *Symptoms Following Exposure*: None — is a food; *General Treatment for Exposure*: EYES: flush with water for at least 15 min.; *Toxicity by Inhalation (Threshold Limit Value)*: Not pertinent; *Short-Term Inhalation Limits*: Not pertinent; *Toxicity by Ingestion*: None; *Late Toxicity*: None; *Vapor (Gas) Irritant Characteristics*: None; *Liquid or Solid Irritant Characteristics*: None; *Odor Threshold*: Data not available.

Fire Hazards - *Flash Point (deg. F):* 610 CC; *Flammable Limits in Air (%):* Data not available; *Fire Extinguishing Agents*: Dry chemical, foam, or carbon dioxide; *Fire Extinguishing Agents Not to be Used:* Water may be ineffective; *Special Hazards of Combustion Products*: Not pertinent; *Behavior in Fire*: Not pertinent; *Ignition Temperature* : Data not available; *Electrical Hazard*: Not pertinent; *Burning Rate*: Data not available.

Chemical Reactivity - *Reactivity with Water*: No reaction; *Reactivity with Common Materials*: No reaction; *Stability During Transport*: Stable; *Neutralizing Agents for Acids and Caustics:* Not pertinent; *Polymerization:* Not pertinent; *Inhibitor of Polymerization*: Not pertinent.

OILS, FUEL: 2

Chemical Designations - *Synonyms*: Home heating oil; *Chemical Formula*: Not applicable.

Observable Characteristics - *Physical State (as shipped)*: Liquid; *Color*: Light brown; *Odor*: Like kerosine; characteristic.

Physical and Chemical Properties - *Physical State at 15 °C and 1 atm.*: Liquid; *Molecular Weight*: Not pertinent; *Boiling Point at 1 atm.*: 540-640, 282-338, 555-611; *Freezing Point*: -20, -29, 244; *Critical Temperature*: Not pertinent; *Critical Pressure*: Not pertinent; *Specific Gravity*: 0.879 at 20 °C (liquid); *Vapor (Gas) Specific Gravity*: Not pertinent; *Ratio of Specific Heats of Vapor (Gas)*: Not pertinent; *Latent Heat of Vaporization*: Not pertinent; *Heat of Combustion*: -19,440, -10,800, -452.17; *Heat of Decomposition*: Not pertinent.

Health Hazards Information - *Recommended Personal Protective Equipment*: Protective gloves; goggles or face shield; *Symptoms Following Exposure*: INHALATION: causes headache and slight giddiness. INGESTION: causes nausea, vomiting, and cramping; depression of central nervous system ranging from mild headache to anesthesia, coma, and death; pulmonary irritation secondary to exhalation of solvent; signs of kidney and liver damage may be delayed. ASPIRATION: causes severe lung irritation with coughing, gagging, dyspnea, substernal distress, and rapidly developing pulmonary edema; later, signs of bronchopneumonia and pneumonitis; acute onset of central nervous system excitement followed by depression; *General Treatment for Exposure*: INGESTION: do NOT induce vomiting. ASPIRATION: enforce bed rest; administer oxygen; seek medical attention. EYES: wash with copious quantity of water. SKIN: remove solvent by wiping and wash with soap and water; *Toxicity by Inhalation (Threshold Limit Value)*: No single value applicable; *Short-Term Inhalation Limits*: Data not available; *Toxicity by Ingestion*: Grade 1; LD_{50} 5-15 g/kg; *Late Toxicity*: Data not available; *Vapor (Gas) Irritant Characteristics*: Slight smarting of eyes or respiratory system if present in high concentrations. The effects is temporary; *Liquid or Solid Irritant Characteristics*: Minimum hazard. If spilled on clothing and allowed to remain, may cause smarting and reddening of skin; *Odor Threshold*: Data not available.

Fire Hazards - *Flash Point (deg. F):* 136 CC; *Flammable Limits in Air (%):* Data not available; *Fire Extinguishing Agents*: Dry chemical, foam, or carbon dioxide; *Fire Extinguishing Agents Not to be Used:* Water may be ineffective; *Special Hazards of Combustion Products*: Not pertinent; *Behavior in Fire*: Not pertinent; *Ignition Temperature (deg. F)*: 494; *Electrical Hazard*: Not pertinent; *Burning Rate*: 4 mm/min.

Chemical Reactivity - *Reactivity with Water*: No reaction; *Reactivity with Common Materials*: No reaction; *Stability During Transport*: Stable; *Neutralizing Agents for Acids and Caustics:* Not pertinent; *Polymerization:* Not pertinent; *Inhibitor of Polymerization*: Not pertinent.

OILS, FUEL: 4

Chemical Designations - *Synonyms*: Residual fuel oil, No 4; *Chemical Formula*: Not applicable.

Observable Characteristics - *Physical State (as shipped)*: Liquid; *Color*: Brown; *Odor*: Like kerosine; characteristic.

Physical and Chemical Properties - *Physical State at 15 °C and 1 atm.*: Liquid; *Molecular Weight*: Not pertinent; *Boiling Point at 1 atm.*: 214 to 1092, 101 to > 588, 374 to 861; *Freezing Point*: -20 to +15, -29 to -9, 244 to 264; *Critical Temperature*: Not pertinent; *Critical Pressure*: Not pertinent; *Specific Gravity*: 0.904 at 15 °C (liquid); *Vapor (Gas) Specific Gravity*: Not pertinent; *Ratio of Specific Heats of Vapor (Gas)*: Not pertinent; *Latent Heat of Vaporization*: Not pertinent; *Heat of Combustion*: -17,460, -9,700, -406.17; *Heat of Decomposition*: Not pertinent.

Health Hazards Information - *Recommended Personal Protective Equipment*: Protective gloves; goggles or face shield; *Symptoms Following Exposure*: INGESTION: do NOT lavage or induce vomiting. ASPIRATION: Treatment probably not required; delayed development of pulmonary irritation can be detected by serial chest x-rays; consider prophylactic antibiotic regime if condition warrants. EYES: wash with copious quantity of water. SKIN: wipe off and wash with soap and water; *Toxicity by Inhalation (Threshold Limit Value)*: Not pertinent; *Short-Term Inhalation Limits*: Not pertinent; *Toxicity by Ingestion*: Grade 1; LD_{50} 5 to 15 g/kg; *Late Toxicity*: Data not available; *Vapor (Gas) Irritant Characteristics*: None; *Liquid or Solid Irritant Characteristics*: Minimum hazard. If spilled on clothing and allowed to remain, may cause smarting and reddening of skin; *Odor Threshold*: Data not available.

Fire Hazards - *Flash Point (deg. F)*: > 130 CC; *Flammable Limits in Air (%)*: 1 -5; *Fire Extinguishing Agents*: Dry chemical, foam, or carbon dioxide; *Fire Extinguishing Agents Not to be Used:* Water may be ineffective; *Special Hazards of Combustion Products*: Not pertinent; *Behavior in Fire*: Not pertinent; *Ignition Temperature (deg. F)*: 505; *Electrical Hazard*: Not pertinent; *Burning Rate*: 4 mm/min.

Chemical Reactivity - *Reactivity with Water*: No reaction; *Reactivity with Common Materials*: No reaction; *Stability During Transport*: Stable; *Neutralizing Agents for Acids and Caustics:* Not pertinent; *Polymerization:* Not pertinent; *Inhibitor of Polymerization*: Not pertinent.

OILS, FUEL: 5

Chemical Designations - *Synonyms*: Residual fuel oil, No 5; *Chemical Formula*: Not applicable.

Observable Characteristics - *Physical State (as shipped)*: Liquid; *Color*: Brown; *Odor*: Like kerosine; characteristic.

Physical and Chemical Properties - *Physical State at 15 °C and 1 atm.*: Liquid; *Molecular Weight*: Not pertinent; *Boiling Point at 1 atm.*: 426 - > 1062, 218 - > 570, 491 - > 843; *Freezing Point*: 0, -18, 255; *Critical Temperature*: Not pertinent; *Critical Pressure*: Not pertinent; *Specific Gravity*: 0.936 at 16 °C (liquid); *Vapor (Gas) Specific Gravity*: Not pertinent; *Ratio of Specific Heats of Vapor (Gas)*: Not pertinent; *Latent Heat of Vaporization*: Not pertinent; *Heat of Combustion*: -18,000, -10,000, -418.68; *Heat of Decomposition*: Not pertinent.

Health Hazards Information - *Recommended Personal Protective Equipment*: Protective gloves; goggles or face shield; *Symptoms Following Exposure*: INGESTION: do NOT lavage or induce vomiting. ASPIRATION: Treatment probably not required; delayed development of pulmonary irritation can be detected by serial chest x-rays; consider prophylactic antibiotic regime if condition warrants. EYES: wash with copious quantity of water. SKIN: wipe off and wash with soap and water; *Toxicity by Inhalation (Threshold Limit Value)*: Not pertinent; *Short-Term Inhalation Limits*: Not pertinent; *Toxicity by Ingestion*: Grade 1; LD_{50} 5 to 15 g/kg; *Late Toxicity*: Data not available; *Vapor (Gas) Irritant Characteristics*: None; *Liquid or Solid Irritant Characteristics*: Minimum hazard. If spilled on clothing and allowed to remain, may cause smarting and reddening of skin; *Odor Threshold*: Data not available.

Fire Hazards - *Flash Point (deg. F)*: > 130 CC; *Flammable Limits in Air (%)*: 1 -5; *Fire Extinguishing Agents*: Dry chemical, foam, or carbon dioxide; *Fire Extinguishing Agents Not to be Used:* Water may be ineffective; *Special Hazards of Combustion Products*: Not pertinent; *Behavior in Fire*: Not pertinent; *Ignition Temperature* : Data not available; *Electrical Hazard*: Not pertinent; *Burning Rate*: 4 mm/min.

Chemical Reactivity - *Reactivity with Water*: No reaction; *Reactivity with Common Materials*: No reaction; *Stability During Transport*: Stable; *Neutralizing Agents for Acids and Caustics:* Not pertinent;

Polymerization: Not pertinent; *Inhibitor of Polymerization*: Not pertinent.

OILS, FUEL: 1-D
Chemical Designations - *Synonyms*: Diesel oil, light; *Chemical Formula*: Not applicable.
Observable Characteristics - *Physical State (as shipped)*: Liquid; *Color*: Light brown; *Odor*: Characteristic.
Physical and Chemical Properties - *Physical State at 15 °C and 1 atm.*: Liquid; *Molecular Weight*: Not pertinent; *Boiling Point at 1 atm.*: 380-560, 193-293, 466-566; *Freezing Point*: -30, -34, 240; *Critical Temperature*: Not pertinent; *Critical Pressure*: Not pertinent; *Specific Gravity*: 0.81-0.85 at 15 °C (liquid); *Vapor (Gas) Specific Gravity*: Not pertinent; *Ratio of Specific Heats of Vapor (Gas)*: Not pertinent; *Latent Heat of Vaporization*: 110, 60, 2.5; *Heat of Combustion*: -18,540, -10,300, -431.24; *Heat of Decomposition*: Not pertinent.
Health Hazards Information - *Recommended Personal Protective Equipment*: Protective gloves; goggles or face shield; *Symptoms Following Exposure*: INHALATION: causes headache and slight giddiness. INGESTION: causes nausea, vomiting, and cramping; depression of central nervous system ranging from mild headache to anesthesia, coma, and death; pulmonary irritation secondary to exhalation of solvent; signs of kidney and liver damage may be delayed. ASPIRATION: causes severe lung irritation with coughing, gagging, dyspnea, substernal distress, and rapidly developing pulmonary edema; later, signs of bronchopneumonia and pneumonitis; acute onset of central nervous system excitement followed by depression; *General Treatment for Exposure*: INGESTION: do NOT induce vomiting; seek medical attention. ASPIRATION: enforce bed rest; administer oxygen; EYES: wash with copious quantity of water. SKIN: remove solvent by wiping and wash with soap and water; *Toxicity by Inhalation (Threshold Limit Value)*: No single value applicable; *Short-Term Inhalation Limits*: Data not available; *Toxicity by Ingestion*: Grade 1; LD_{50} 5-15 g/kg; *Late Toxicity*: Data not available; *Vapor (Gas) Irritant Characteristics*: Slight smarting of eyes or respiratory system if present in high concentrations. The effects is temporary; *Liquid or Solid Irritant Characteristics*: Minimum hazard. If spilled on clothing and allowed to remain, may cause smarting and reddening of skin; *Odor Threshold*: Data not available.
Fire Hazards - *Flash Point (deg. F)*: 100 CC; *Flammable Limits in Air (%)*: 1.3 - 6; *Fire Extinguishing Agents*: Dry chemical, foam, or carbon dioxide; *Fire Extinguishing Agents Not to be Used:* Water may be ineffective; *Special Hazards of Combustion Products*: Not pertinent; *Behavior in Fire*: Not pertinent; *Ignition Temperature (deg. F)*: 350 - 625; *Electrical Hazard*: Not pertinent; *Burning Rate*: 4 mm/min.
Chemical Reactivity - *Reactivity with Water*: No reaction; *Reactivity with Common Materials*: No reaction; *Stability During Transport*: Stable; *Neutralizing Agents for Acids and Caustics:* Not pertinent; *Polymerization:* Not pertinent; *Inhibitor of Polymerization*: Not pertinent.

OILS, FUEL: 2-D
Chemical Designations - *Synonyms*: Diesel oil, medium; *Chemical Formula*: Not applicable.
Observable Characteristics - *Physical State (as shipped)*: Liquid; *Color*: Light brown; *Odor*: Characteristic; like kerosine.
Physical and Chemical Properties - *Physical State at 15 °C and 1 atm.*: Liquid; *Molecular Weight*: Not pertinent; *Boiling Point at 1 atm.*: 540-640, 282-338, 555-611; *Freezing Point*: -0, -18, 255; *Critical Temperature*: Not pertinent; *Critical Pressure*: Not pertinent; *Specific Gravity*: 0.87-0.90 at 20 °C (liquid); *Vapor (Gas) Specific Gravity*: Not pertinent; *Ratio of Specific Heats of Vapor (Gas)*: Not pertinent; *Latent Heat of Vaporization*: Not pertinent; *Heat of Combustion*: -19,440, -10,800, -452.17; *Heat of Decomposition*: Not pertinent.
Health Hazards Information - *Recommended Personal Protective Equipment*: Protective gloves; goggles or face shield; *Symptoms Following Exposure*: INGESTION: causes nausea, vomiting, and cramping; depression of central nervous system ranging from mild headache to anesthesia, coma, and death; pulmonary irritation secondary to exhalation of solvent; signs of kidney and liver damage may be delayed. ASPIRATION: causes severe lung irritation with coughing, gagging, dyspnea, substernal distress, and rapidly developing pulmonary edema; later, signs of bronchopneumonia and pneumonitis;

acute onset of central nervous system excitement followed by depression; *General Treatment for Exposure*: INGESTION: do NOT induce vomiting; seek medical attention. ASPIRATION: enforce bed rest; administer oxygen; EYES: wash with copious quantity of water. SKIN: remove solvent by wiping and wash with soap and water; *Toxicity by Inhalation (Threshold Limit Value)*: No single TLV applicable; *Short-Term Inhalation Limits*: Data not available; *Toxicity by Ingestion*: Grade 1; LD_{50} 5-15 g/kg; *Late Toxicity*: Data not available; *Vapor (Gas) Irritant Characteristics*: Slight smarting of eyes or respiratory system if present in high concentrations. The effects is temporary; *Liquid or Solid Irritant Characteristics*: Minimum hazard. If spilled on clothing and allowed to remain, may cause smarting and reddening of skin; *Odor Threshold*: Data not available.

Fire Hazards - *Flash Point (deg. F):* 125 CC; *Flammable Limits in Air (%):* 1.3 - 6; *Fire Extinguishing Agents*: Dry chemical, foam, or carbon dioxide; *Fire Extinguishing Agents Not to be Used:* Water may be ineffective; *Special Hazards of Combustion Products*: Not pertinent; *Behavior in Fire*: Not pertinent; *Ignition Temperature (deg. F)*: 490 - 545; *Electrical Hazard*: Not pertinent; *Burning Rate*: 4 mm/min.

Chemical Reactivity - *Reactivity with Water*: No reaction; *Reactivity with Common Materials*: No reaction; *Stability During Transport*: Stable; *Neutralizing Agents for Acids and Caustics:* Not pertinent; *Polymerization:* Not pertinent; *Inhibitor of Polymerization*: Not pertinent.

OILS, FUEL: NO. 1

Chemical Designations - *Synonyms*: JP-1; Kerosene; Kerosine; Range oil; *Chemical Formula*: Not applicable.

Observable Characteristics - *Physical State (as shipped)*: Liquid; *Color*: Colorless to light brown; *Odor*: Characteristic.

Physical and Chemical Properties - *Physical State at 15 °C and 1 atm.*: Liquid; *Molecular Weight*: Not pertinent; *Boiling Point at 1 atm.*: 380 - 560, 193 - 293, 466 - 566; *Freezing Point*: -45 to -55, -43 to -48, 230 to 225; *Critical Temperature*: Not pertinent; *Critical Pressure*: Not pertinent; *Specific Gravity*: 0.81-0.85 at 15 °C (liquid); *Vapor (Gas) Specific Gravity*: Not pertinent; *Ratio of Specific Heats of Vapor (Gas)*: Not pertinent; *Latent Heat of Vaporization*: 110, 60, 2.5; *Heat of Combustion*: -18,540, -10,300, -421.24; *Heat of Decomposition*: Not pertinent.

Health Hazards Information - *Recommended Personal Protective Equipment*: Protective gloves; goggles or face shield; *Symptoms Following Exposure*: INGESTION: causes irritation of gastrointestinal tract; pulmonary tract irritation secondary to exhalation of vapors. ASPIRATION: causes severe lung irritation with coughing, gagging, dyspnea, substernal distress, and rapidly developing pulmonary edema; later, signs of bronchopneumonia and pneumonitis appear later; minimal central nervous system depression; *General Treatment for Exposure*: INGESTION: do NOT lavage induce vomiting; call a physician. ASPIRATION: enforce bed rest; administer oxygen; call a physician. EYES: wash with plenty of water. SKIN: wipe off and wash with soap and water; *Toxicity by Inhalation (Threshold Limit Value)*: 200 ppm (suggested); *Short-Term Inhalation Limits*: Data not available; *Toxicity by Ingestion*: Grade 1; LD_{50} 0.5-5 g/kg; *Late Toxicity*: Data not available; *Vapor (Gas) Irritant Characteristics*: Vapors causes a slight smarting of the eyes or respiratory system if present in high concentrations. The effect is temporary; *Liquid or Solid Irritant Characteristics*: Minimum hazard. If spilled on clothing and allowed to remain, may cause smarting and reddening of skin; *Odor Threshold*: 1ppm.

Fire Hazards - *Flash Point (deg. F):* 100 CC; *Flammable Limits in Air (%):* 0.7 - 5; *Fire Extinguishing Agents*: Dry chemical, foam, or carbon dioxide; *Fire Extinguishing Agents Not to be Used:* Water may be ineffective; *Special Hazards of Combustion Products*: Not pertinent; *Behavior in Fire*: Not pertinent; *Ignition Temperature (deg. F)*: 444; *Electrical Hazard*: Not pertinent; *Burning Rate*: 4 mm/min.

Chemical Reactivity - *Reactivity with Water*: No reaction; *Reactivity with Common Materials*: No reaction; *Stability During Transport*: Stable; *Neutralizing Agents for Acids and Caustics:* Not pertinent; *Polymerization:* Not pertinent; *Inhibitor of Polymerization*: Not pertinent.

OILS, MISCELLANEOUS: ABSORPTION

Chemical Designations - *Synonyms*: Absorbent oil; *Chemical Formula*: Not applicable.

Observable Characteristics - *Physical State (as shipped)*: Liquid; *Color*: Pale yellow to colorless;

Odor: Like fuel oil.

Physical and Chemical Properties - *Physical State at 15 °C and 1 atm.*: Liquid; *Molecular Weight*: Not pertinent; *Boiling Point at 1 atm.*: > 500, > 200, > 533; *Freezing Point*: Not pertinent; *Critical Temperature*: Not pertinent; *Critical Pressure*: Not pertinent; *Specific Gravity*: (est.) 0.85 at 20 °C (liquid); *Vapor (Gas) Specific Gravity*: Not pertinent; *Ratio of Specific Heats of Vapor (Gas)*: Not pertinent; *Latent Heat of Vaporization*: Not pertinent; *Heat of Combustion*: (est.) -18,000, -10,000, -420; *Heat of Decomposition*: Not pertinent.

Health Hazards Information - *Recommended Personal Protective Equipment*: Protective gloves; goggles or face shield; *Symptoms Following Exposure*: INGESTION: irritation of stomach. ASPIRATION: pulmonary irritation is normally minimal but may become more several hours after exposure. (Delayed development can be detected by serial chest x-rays.); *General Treatment for Exposure*: INGESTION: have victim drink water or milk; do NOT induce vomiting. EYES: wash with copious amounts of water. SKIN: wipe off and wash with soap and water; *Toxicity by Inhalation (Threshold Limit Value)*: Not pertinent; *Short-Term Inhalation Limits*: Not pertinent; *Toxicity by Ingestion*: Grade 1; LD_{50} 5-15 g/kg; *Late Toxicity*: Data not available; *Vapor (Gas) Irritant Characteristics*: None; *Liquid or Solid Irritant Characteristics*: Minimum hazard. If spilled on clothing and allowed to remain, may cause smarting and reddening of skin; *Odor Threshold*: Data not available.

Fire Hazards - *Flash Point (deg. F)*: 255; *Flammable Limits in Air (%)*: Data not available; *Fire Extinguishing Agents*: Dry chemical, foam, or carbon dioxide; *Fire Extinguishing Agents Not to be Used*: Water may be ineffective; *Special Hazards of Combustion Products*: Not pertinent; *Behavior in Fire*: Not pertinent; *Ignition Temperature (deg. F)*: 300; *Electrical Hazard*: Not pertinent; *Burning Rate*: 4 mm/min.

Chemical Reactivity - *Reactivity with Water*: No reaction; *Reactivity with Common Materials*: No reaction; *Stability During Transport*: Stable; *Neutralizing Agents for Acids and Caustics*: Not pertinent; *Polymerization*: Not pertinent; *Inhibitor of Polymerization*: Not pertinent.

OILS, MISCELLANEOUS: COAL TAR

Chemical Designations - *Synonyms*: Light oil; *Chemical Formula*: Not applicable.

Observable Characteristics - *Physical State (as shipped)*: Liquid; *Color*: Colorless to yellow; *Odor*: Aromatic; like benzene; pleasant.

Physical and Chemical Properties - *Physical State at 15 °C and 1 atm.*: Liquid; *Molecular Weight*: Not pertinent; *Boiling Point at 1 atm.*: 233-333, 106-167, 379-440; *Freezing Point*: Not pertinent; *Critical Temperature*: Not pertinent; *Critical Pressure*: Not pertinent; *Specific Gravity*: (est.) 0.90 at 20 °C (liquid); *Vapor (Gas) Specific Gravity*: Not pertinent; *Ratio of Specific Heats of Vapor (Gas)*: (est.) 1.071; *Latent Heat of Vaporization*: (est.) 107, 59.8, 2.5; *Heat of Combustion*: -17,440, -9,690, -405.7; *Heat of Decomposition*: Not pertinent.

Health Hazards Information - *Recommended Personal Protective Equipment*: Protective gloves; goggles or face shield; *Symptoms Following Exposure*: Vapor causes slight irritation of nose and throat, smarting of eyes. Liquid may irritate skin on prolonged contact; *General Treatment for Exposure*: INGESTION: have victim drink water or milk; do NOT induce vomiting. EYES: flush with water for at least 15 min. SKIN: wipe off and wash with soap and water; *Toxicity by Inhalation (Threshold Limit Value)*: Data not available; *Short-Term Inhalation Limits*: Data not available; *Toxicity by Ingestion*: Data not available; *Late Toxicity*: Data not available; *Vapor (Gas) Irritant Characteristics*: Vapors causes a slight smarting of the eyes or respiratory system if present in high concentrations. The effect is temporary; *Liquid or Solid Irritant Characteristics*: Minimum hazard. If spilled on clothing and allowed to remain, may cause smarting and reddening of skin; *Odor Threshold*: Data not available.

Fire Hazards - *Flash Point (deg. F)*: 60 - 77 CC; *Flammable Limits in Air (%)*: 1.3 - 8; *Fire Extinguishing Agents*: Dry chemical, foam, or carbon dioxide; *Fire Extinguishing Agents Not to be Used*: Water may be ineffective; *Special Hazards of Combustion Products*: Not pertinent; *Behavior in Fire*: Not pertinent; *Ignition Temperature*: Data not available; *Electrical Hazard*: Not pertinent; *Burning Rate*: 4 mm/min.

Chemical Reactivity - *Reactivity with Water*: No reaction; *Reactivity with Common Materials*: No reaction; *Stability During Transport*: Stable; *Neutralizing Agents for Acids and Caustics:* Not pertinent; *Polymerization:* Not pertinent; *Inhibitor of Polymerization*: Not pertinent.

OILS, MISCELLANEOUS: CROTON

Chemical Designations - *Synonyms*: Crotonoel; *Chemical Formula*: Not applicable.

Observable Characteristics - *Physical State (as shipped)*: Liquid; *Color*: Dark; *Odor*: Unpleasant, acrid.

Physical and Chemical Properties - *Physical State at 15 °C and 1 atm.*: Liquid; *Molecular Weight*: Not pertinent; *Boiling Point at 1 atm.*: Not pertinent (very high); *Freezing Point*: 0 to 18, -18 to -8, 255 to 265; *Critical Temperature*: Not pertinent; *Critical Pressure*: Not pertinent; *Specific Gravity*: (est.) 0.946 at 15 °C (liquid); *Vapor (Gas) Specific Gravity*: Not pertinent; *Ratio of Specific Heats of Vapor (Gas)*: Not pertinent; *Latent Heat of Vaporization*: Not pertinent; *Heat of Combustion*: (est.) - 16,800, -9,300, -390; *Heat of Decomposition*: Not pertinent.

Health Hazards Information - *Recommended Personal Protective Equipment*: Goggles or face shield; rubber gloves and other protective clothing to prevent contact with skin; *Symptoms Following Exposure*: Contact of liquid with eyes causes severe irritation. May induce severe skin irritation, inflammation, swelling, and pustule formation. Absorption through the skin may cause purging. Ingestion causes burning of the mouth and stomach and drastic purging, possibly leading to collapse and death. Small doses have a strong laxative effect; *General Treatment for Exposure*: EYES: flush with water; a 2.5% hydroxycortisone ointment is recommended. SKIN: remove as much liquid as possible from skin by use of a good solvent such as acetone or alcohol; wash with soap and water. INGESTION: for gastrointestinal symptoms, use demulcents; further treatment is symptomatic; do NOT induce vomiting; *Toxicity by Inhalation (Threshold Limit Value)*: Data not available; *Short-Term Inhalation Limits*: Data not available; *Toxicity by Ingestion*: Grade 4, $LD_{50} < 50$ mg/kg; *Late Toxicity*: Has been used in cancer research as a promoter for other compounds that cause skin cancer; *Vapor (Gas) Irritant Characteristics*: Data not available; *Liquid or Solid Irritant Characteristics*: Data not available; *Odor Threshold*: Data not available.

Fire Hazards - *Flash Point :* Data not available; *Flammable Limits in Air (%):* Not pertinent; *Fire Extinguishing Agents*: Dry chemical, foam, or carbon dioxide; *Fire Extinguishing Agents Not to be Used:* Water may be ineffective; *Special Hazards of Combustion Products*: Not pertinent; *Behavior in Fire*: Not pertinent; *Ignition Temperature* : Data not available; *Electrical Hazard*: Not pertinent; *Burning Rate*: 4 mm/min.

Chemical Reactivity - *Reactivity with Water*: No reaction; *Reactivity with Common Materials*: No reaction; *Stability During Transport*: Stable; *Neutralizing Agents for Acids and Caustics:* Not pertinent; *Polymerization:* Not pertinent; *Inhibitor of Polymerization*: Not pertinent.

OILS, MISCELLANEOUS: LINSEED

Chemical Designations - *Synonyms*: Flaxseed oil; Raw linseed oil; *Chemical Formula*: Not applicable.

Observable Characteristics - *Physical State (as shipped)*: Liquid; *Color*: Pale yellow to dark amber; *Odor*: Like oil-base paint.

Physical and Chemical Properties - *Physical State at 15 °C and 1 atm.*: Liquid; *Molecular Weight*: Not pertinent; *Boiling Point at 1 atm.*: Not pertinent (very high); *Freezing Point*: -2, -19, 254; *Critical Temperature*: Not pertinent; *Critical Pressure*: Not pertinent; *Specific Gravity*: 0.932 at 20 °C (liquid); *Vapor (Gas) Specific Gravity*: Not pertinent; *Ratio of Specific Heats of Vapor (Gas)*: Not pertinent; *Latent Heat of Vaporization*: Not pertinent; *Heat of Combustion*: -16,800, -9,300, -390; *Heat of Decomposition*: Not pertinent.

Health Hazards Information - *Recommended Personal Protective Equipment*: Goggles or face shield; rubber gloves; *Symptoms Following Exposure*: Contact of liquid with eyes causes mild irritation. Prolonged contact with skin can cause dermatitis. Ingestion of large doses (over 1 oz) has laxative effect; *General Treatment for Exposure*: EYES: flesh with water for at least 15 min. SKIN: wipe off; wash with soap and water. INGESTION: do NOT induce vomiting; *Toxicity by Inhalation (Threshold Limit Value)*: Data not available; *Short-Term Inhalation Limits*: Not pertinent; *Toxicity by Ingestion*:

Grade 0, LD_{50} >15 g/kg; *Late Toxicity*: Liver damage in rats (from addition of oil to diet); *Vapor (Gas) Irritant Characteristics*: Data not available; *Liquid or Solid Irritant Characteristics*: Data not available; *Odor Threshold*: Data not available.

Fire Hazards - *Flash Point (deg. F)*: 535 OC; 403 CC; *Flammable Limits in Air (%):* Not pertinent; *Fire Extinguishing Agents*: Dry chemical, foam, or carbon dioxide; *Fire Extinguishing Agents Not to be Used:* Water or foam may cause frothing; water may be ineffective; *Special Hazards of Combustion Products*: Not pertinent; *Behavior in Fire*: Not pertinent; *Ignition Temperature (deg. F)*: 650; *Electrical Hazard*: Not pertinent; *Burning Rate*: 4 mm/min.

Chemical Reactivity - *Reactivity with Water*: No reaction; *Reactivity with Common Materials*: No reaction; *Stability During Transport*: Stable; *Neutralizing Agents for Acids and Caustics:* Not pertinent; *Polymerization:* Not pertinent; *Inhibitor of Polymerization*: Not pertinent.

OILS, MISCELLANEOUS: LUBRICATING

Chemical Designations - *Synonyms*: Crankcase oil; Motor oil; Transmission oil; *Chemical Formula*: Not applicable.

Observable Characteristics - *Physical State (as shipped)*: Liquid; *Color*: Yellow fluorescent; *Odor*: Characteristic.

Physical and Chemical Properties - *Physical State at 15 °C and 1 atm.*: Liquid; *Molecular Weight*: Not pertinent; *Boiling Point at 1 atm.*: Very high; *Freezing Point*: Not pertinent; *Critical Temperature*: Not pertinent; *Critical Pressure*: Not pertinent; *Specific Gravity*: (est.) 0.902 at 20 °C (liquid); *Vapor (Gas) Specific Gravity*: Not pertinent; *Ratio of Specific Heats of Vapor (Gas)*: Not pertinent; *Latent Heat of Vaporization*: Not pertinent; *Heat of Combustion*: -18,486, -19,270, -429.98; *Heat of Decomposition*: Not pertinent.

Health Hazards Information - *Recommended Personal Protective Equipment*: Protective gloves; goggles or face shield; *Symptoms Following Exposure*: INGESTION: minimal gastrointestinal tract irritation; increased frequency of bowel passage may occur. ASPIRATION: pulmonary irritation is normally minimal but may become more severe several hours after exposure; *General Treatment for Exposure*: INGESTION: do NOT lavage or induce vomiting. ASPIRATION: treatment probably not required; delayed development of pulmonary irritation can be detected by serial chest x-rays. EYES: wash with copious quantity of water. SKIN: wipe off and wash with soap and water; *Toxicity by Inhalation (Threshold Limit Value)*: Data not available; *Short-Term Inhalation Limits*: Data not available; *Toxicity by Ingestion*: Grade 1, LD_{50} 5 to 15 g/kg; *Late Toxicity*: Data not available; *Vapor (Gas) Irritant Characteristics*: Vapors causes a slight smarting of the eyes or respiratory system if present in high concentrations. The effect is temporary; *Liquid or Solid Irritant Characteristics*: Minimum hazard. If spilled on clothing and allowed to remain, may cause smarting and reddening of skin; *Odor Threshold*: Data not available.

Fire Hazards - *Flash Point (deg. F)*: 300 - 450 CC; *Flammable Limits in Air (%):* Data not available; *Fire Extinguishing Agents*: Dry chemical, foam, or carbon dioxide; *Fire Extinguishing Agents Not to be Used:* Water or foam may cause frothing; *Special Hazards of Combustion Products*: Not pertinent; *Behavior in Fire*: Not pertinent; *Ignition Temperature (deg. F)*: 500 - 700; *Electrical Hazard*: Not pertinent; *Burning Rate*: 4 mm/min.

Chemical Reactivity - *Reactivity with Water*: No reaction; *Reactivity with Common Materials*: No reaction; *Stability During Transport*: Stable; *Neutralizing Agents for Acids and Caustics:* Not pertinent; *Polymerization:* Not pertinent; *Inhibitor of Polymerization*: Not pertinent.

OILS, MISCELLANEOUS: MINERAL

Chemical Designations - *Synonyms*: Liquid petrolatum; White oil; *Chemical Formula*: Not applicable.

Observable Characteristics - *Physical State (as shipped)*: Liquid; *Color*: Colorless; *Odor*: Very faint.

Physical and Chemical Properties - *Physical State at 15 °C and 1 atm.*: Liquid; *Molecular Weight*: Not pertinent; *Boiling Point at 1 atm.*: Very high; *Freezing Point*: Not pertinent; *Critical Temperature*: Not pertinent; *Critical Pressure*: Not pertinent; *Specific Gravity*: 0.822 at 20 °C (liquid); *Vapor (Gas) Specific Gravity*: Not pertinent; *Ratio of Specific Heats of Vapor (Gas)*: Not pertinent; *Latent Heat of*

Vaporization: Not pertinent; *Heat of Combustion*: Data not available; *Heat of Decomposition*: Not pertinent.

Health Hazards Information - *Recommended Personal Protective Equipment*: Goggles or face shield; *Symptoms Following Exposure*: Ingestion of liquid can cause very loose bowel movements; *General Treatment for Exposure*: EYES: wash with water; *Toxicity by Inhalation (Threshold Limit Value)*: Not pertinent; *Short-Term Inhalation Limits*: Not pertinent; *Toxicity by Ingestion*: Grade 1, LD_{50} 5 to 15 g/kg; *Late Toxicity*: None; *Vapor (Gas) Irritant Characteristics*: None; *Liquid or Solid Irritant Characteristics*: None; *Odor Threshold*: Not pertinent.

Fire Hazards - *Flash Point (deg. F)*: 380 OC; *Flammable Limits in Air (%)*: Data not available; *Fire Extinguishing Agents*: Dry chemical, foam, or carbon dioxide; *Fire Extinguishing Agents Not to be Used*: Water or foam may cause frothing; *Special Hazards of Combustion Products*: Not pertinent; *Behavior in Fire*: Not pertinent; *Ignition Temperature (deg. F)*: 500 - 700; *Electrical Hazard*: Not pertinent; *Burning Rate*: 4 mm/min.

Chemical Reactivity - *Reactivity with Water*: No reaction; *Reactivity with Common Materials*: No reaction; *Stability During Transport*: Stable; *Neutralizing Agents for Acids and Caustics*: Not pertinent; *Polymerization*: Not pertinent; *Inhibitor of Polymerization*: Not pertinent.

OILS, MISCELLANEOUS: MINERAL SEAL

Chemical Designations - *Synonyms*: Long-time burning oil; Mineral colza oil; 300° oil; signal oil; *Chemical Formula*: Not applicable.

Observable Characteristics - *Physical State (as shipped)*: Liquid; *Color*: Colorless to light brown; *Odor*: Like kerosene.

Physical and Chemical Properties - *Physical State at 15 °C and 1 atm.*: Liquid; *Molecular Weight*: Not pertinent; *Boiling Point at 1 atm.*: >500, >260, >533; *Freezing Point*: 10.0, -12.2, 261; *Critical Temperature*: Not pertinent; *Critical Pressure*: Not pertinent; *Specific Gravity*: 0.811-0.825 at 15 °C (liquid); *Vapor (Gas) Specific Gravity*: Not pertinent; *Ratio of Specific Heats of Vapor (Gas)*: Not pertinent; *Latent Heat of Vaporization*: Not pertinent; *Heat of Combustion*: (est.) -18,000, -10,000, -420; *Heat of Decomposition*: Not pertinent.

Health Hazards Information - *Recommended Personal Protective Equipment*: Protective gloves; goggles or face shield; *Symptoms Following Exposure*: Vapors cause slight irritation of eyes and nose. Liquid irritates stomach; if taken into lungs causes coughing, distress, and rapidly developing pulmonary edema; *General Treatment for Exposure*: ASPIRATION: enforced bed rest; administer oxygen; call a doctor. INGESTION: do NOT induce vomiting; have victim drink water or milk. EYES: wash with copious amounts of water. SKIN: wipe off, wash with soap and water; *Toxicity by Inhalation (Threshold Limit Value)*: 200 ppm; *Short-Term Inhalation Limits*: Data not available; *Toxicity by Ingestion*: Grade 2, LD_{50} 0.5 to 5 g/kg; *Late Toxicity*: Data not available; *Vapor (Gas) Irritant Characteristics*: Vapors causes a slight smarting of the eyes or respiratory system if present in high concentrations. The effect is temporary; *Liquid or Solid Irritant Characteristics*: Minimum hazard. If spilled on clothing and allowed to remain, may cause smarting and reddening of skin; *Odor Threshold*: Data not available.

Fire Hazards - *Flash Point (deg. F)*: 170 - 275 OC; *Flammable Limits in Air (%)*: Data not available; *Fire Extinguishing Agents*: Dry chemical, foam, or carbon dioxide; *Fire Extinguishing Agents Not to be Used*: Water may be ineffective; *Special Hazards of Combustion Products*: Not pertinent; *Behavior in Fire*: Not pertinent; *Ignition Temperature*: Data not available; *Electrical Hazard*: Not pertinent; *Burning Rate*: 4 mm/min.

Chemical Reactivity - *Reactivity with Water*: No reaction; *Reactivity with Common Materials*: No reaction; *Stability During Transport*: Stable; *Neutralizing Agents for Acids and Caustics*: Not pertinent; *Polymerization*: Not pertinent; *Inhibitor of Polymerization*: Not pertinent.

OILS, MISCELLANEOUS: MOTOR

Chemical Designations - *Synonyms*: Crankcase oil; Lubricating oil; Transmission oil; *Chemical Formula*: Not applicable.

Observable Characteristics - *Physical State (as shipped)*: Liquid; *Color*: Yellow fluorescent; *Odor*: Characteristic.

Physical and Chemical Properties - *Physical State at 15 °C and 1 atm.*: Liquid; *Molecular Weight*: Not pertinent; *Boiling Point at 1 atm.*: Very high; *Freezing Point*: -29.9, 034.4, 238.8; *Critical Temperature*: Not pertinent; *Critical Pressure*: Not pertinent; *Specific Gravity*: (est.) 0.84 - 0.96 at 20 °C (liquid); *Vapor (Gas) Specific Gravity*: Not pertinent; *Ratio of Specific Heats of Vapor (Gas)*: Not pertinent; *Latent Heat of Vaporization*: Not pertinent; *Heat of Combustion*: -18,486, -19,270, -429.98; *Heat of Decomposition*: Not pertinent.

Health Hazards Information - *Recommended Personal Protective Equipment*: Protective gloves; goggles or face shield; *Symptoms Following Exposure*: INGESTION: minimal gastrointestinal tract irritation; increased frequency of bowel passage may occur. ASPIRATION: pulmonary irritation is normally minimal but may become more severe several hours after exposure; *General Treatment for Exposure*: INGESTION: do NOT lavage or induce vomiting. ASPIRATION: treatment probably not required; delayed development of pulmonary irritation can be detected by serial chest x-rays. EYES: wash with copious quantity of water. SKIN: wipe off and wash with soap and water; *Toxicity by Inhalation (Threshold Limit Value)*: Data not available; *Short-Term Inhalation Limits*: Data not available; *Toxicity by Ingestion*: Grade 1, LD_{50} 5 to 15 g/kg; *Late Toxicity*: Data not available; *Vapor (Gas) Irritant Characteristics*: Vapors causes a slight smarting of the eyes or respiratory system if present in high concentrations. The effect is temporary; *Liquid or Solid Irritant Characteristics*: Minimum hazard. If spilled on clothing and allowed to remain, may cause smarting and reddening of skin; *Odor Threshold*: Data not available.

Fire Hazards - *Flash Point (deg. F)*: 275 - 600 CC; *Flammable Limits in Air (%)*: Data not available; *Fire Extinguishing Agents*: Dry chemical, foam, or carbon dioxide; *Fire Extinguishing Agents Not to be Used*: Water may be ineffective; *Special Hazards of Combustion Products*: Not pertinent; *Behavior in Fire*: Not pertinent; *Ignition Temperature (deg. F)*: 325 - 625; *Electrical Hazard*: Not pertinent; *Burning Rate*: 4 mm/min.

Chemical Reactivity - *Reactivity with Water*: No reaction; *Reactivity with Common Materials*: No reaction; *Stability During Transport*: Stable; *Neutralizing Agents for Acids and Caustics*: Not pertinent; *Polymerization*: Not pertinent; *Inhibitor of Polymerization*: Not pertinent.

OILS, MISCELLANEOUS: NEATSFOOT

Chemical Designations - *Synonyms*: No common synonyms; *Chemical Formula*: Not applicable.

Observable Characteristics - *Physical State (as shipped)*: Liquid; *Color*: Pale yellow; *Odor*: Peculiar.

Physical and Chemical Properties - *Physical State at 15 °C and 1 atm.*: Liquid; *Molecular Weight*: Not pertinent; *Boiling Point at 1 atm.*: Very high; *Freezing Point*: 32 to 14, 0 to -10, 273 to 263; *Critical Temperature*: Not pertinent; *Critical Pressure*: Not pertinent; *Specific Gravity*: 0.915 at 16 °C (liquid); *Vapor (Gas) Specific Gravity*: Not pertinent; *Ratio of Specific Heats of Vapor (Gas)*: Not pertinent; *Latent Heat of Vaporization*: Not pertinent; *Heat of Combustion*: Not pertinent; *Heat of Decomposition*: Not pertinent.

Health Hazards Information - *Recommended Personal Protective Equipment*: Data not available; *Symptoms Following Exposure*: May cause dermatitis in sensitive individuals (humans); *General Treatment for Exposure*: Data not available; *Toxicity by Inhalation (Threshold Limit Value)*: Data not available; *Short-Term Inhalation Limits*: Data not available; *Toxicity by Ingestion*: Grade 0, LD_{50} above 15 g/kg; *Late Toxicity*: Data not available; *Vapor (Gas) Irritant Characteristics*: Data not available; *Liquid or Solid Irritant Characteristics*: Data not available; *Odor Threshold*: Data not available.

Fire Hazards - *Flash Point (deg. F)*: 430 OC; 470 CC; *Flammable Limits in Air (%)*: Not pertinent; *Fire Extinguishing Agents*: Dry chemical, foam, or carbon dioxide; *Fire Extinguishing Agents Not to be Used*: Water or foam may cause frothing; *Special Hazards of Combustion Products*: Not pertinent; *Behavior in Fire*: Not pertinent; *Ignition Temperature (deg. F)*: 828; *Electrical Hazard*: Not pertinent; *Burning Rate*: Data not available.

Chemical Reactivity - *Reactivity with Water*: No reaction; *Reactivity with Common Materials*: No reaction; *Stability During Transport*: Stable; *Neutralizing Agents for Acids and Caustics*: Not pertinent;

Polymerization: Not pertinent; *Inhibitor of Polymerization*: Not pertinent.

OILS, MISCELLANEOUS: PENETRATING

Chemical Designations - *Synonyms*: Protective oil; Water displacing oil; *Chemical Formula*: Not applicable.

Observable Characteristics - *Physical State (as shipped)*: Liquid; *Color*: Yellowish; *Odor*: Like motor oil.

Physical and Chemical Properties — *Physical State at 15 °C and 1 atm.*: Liquid; *Molecular Weight*: Not pertinent; *Boiling Point at 1 atm.*: Very high; *Freezing Point*: Not pertinent; *Critical Temperature*: Not pertinent; *Critical Pressure*: Not pertinent; *Specific Gravity*: 0.8961 at 20 °C (liquid); *Vapor (Gas) Specific Gravity*: Not pertinent; *Ratio of Specific Heats of Vapor (Gas)*: Not pertinent; *Latent Heat of Vaporization*: Not pertinent; *Heat of Combustion*: (est.) -18,000, -10,000, -420; *Heat of Decomposition*: Not pertinent.

Health Hazards Information - *Recommended Personal Protective Equipment*: Protective gloves; goggles or face shield; *Symptoms Following Exposure*: Liquid may irritate stomach and increase frequency of bowel movements; *General Treatment for Exposure*: INGESTION: have victim drink water or milk; do NOT induce vomiting. ASPIRATION: check for delayed development of pulmonary irritation by serial x-rays. EYES: wash with copious amounts of water. SKIN: wipe off, wash with soap and water; *Toxicity by Inhalation (Threshold Limit Value)*: Data not available; *Short-Term Inhalation Limits*: Data not available; *Toxicity by Ingestion*: Grade 1, LD_{50} 5 to 15 g/kg; *Late Toxicity*: Data not available; *Vapor (Gas) Irritant Characteristics*: Vapors causes a slight smarting of the eyes or respiratory system if present in high concentrations. The effect is temporary; *Liquid or Solid Irritant Characteristics*: Minimum hazard. If spilled on clothing and allowed to remain, may cause smarting and reddening of skin; *Odor Threshold*: Data not available.

Fire Hazards - *Flash Point (deg. F):* 295; *Flammable Limits in Air (%):* Data not available; *Fire Extinguishing Agents*: Dry chemical, foam, or carbon dioxide; *Fire Extinguishing Agents Not to be Used:* Water or foam may cause frothing; *Special Hazards of Combustion Products*: Not pertinent; *Behavior in Fire*: Not pertinent; *Ignition Temperature* : Data not available; *Electrical Hazard*: Not pertinent; *Burning Rate*: Data not available.

Chemical Reactivity - *Reactivity with Water*: No reaction; *Reactivity with Common Materials*: No reaction; *Stability During Transport*: Stable; *Neutralizing Agents for Acids and Caustics:* Not pertinent; *Polymerization:* Not pertinent; *Inhibitor of Polymerization*: Not pertinent.

OILS, MISCELLANEOUS: RANGE

Chemical Designations - *Synonyms*: No.1; Fuel oil; JP-1; Kerosene; Kerosine; *Chemical Formula*: Not applicable.

Observable Characteristics - *Physical State (as shipped)*: Liquid; *Color*: Colorless; *Odor*: Like kerosene.

Physical and Chemical Properties - *Physical State at 15 °C and 1 atm.*: Liquid; *Molecular Weight*: Not pertinent; *Boiling Point at 1 atm.*: 392-500, 200-260, 473-533; *Freezing Point*: -45 to -55, -43 to -48, 230 to 225; *Critical Temperature*: Not pertinent; *Critical Pressure*: Not pertinent; *Specific Gravity*: 0.80-0.85 at 15 °C (liquid); *Vapor (Gas) Specific Gravity*: Not pertinent; *Ratio of Specific Heats of Vapor (Gas)*: Not pertinent; *Latent Heat of Vaporization*: 108, 60, 2.51; *Heat of Combustion*: -18,540, -10,300, -421.24; *Heat of Decomposition*: Not pertinent.

Health Hazards Information - *Recommended Personal Protective Equipment*: Protective gloves; goggles or face shield; *Symptoms Following Exposure*: Vapors cause slight irritation of eyes and nose. Liquid irritates stomach; if taken into lungs causes coughing, distress, and rapidly developing pulmonary edema; *General Treatment for Exposure*: INGESTION: do NOT induce vomiting; call a physician. ASPIRATION: enforce bed rest; administer oxygen; call a physician. EYES: wash with plenty of water. SKIN: wipe off and wash with soap and water; *Toxicity by Inhalation (Threshold Limit Value)*: 200 ppm (suggested); *Short-Term Inhalation Limits*: Data not available; *Toxicity by Ingestion*: Grade 1; LD_{50} 5 to 15 g/kg; *Late Toxicity*: Data not available; *Vapor (Gas) Irritant Characteristics*:

Vapors causes a slight smarting of the eyes or respiratory system if present in high concentrations. The effect is temporary; *Liquid or Solid Irritant Characteristics*: Minimum hazard. If spilled on clothing and allowed to remain, may cause smarting and reddening of skin; *Odor Threshold*: 1 ppm.
Fire Hazards - *Flash Point (deg. F):* 100 CC; *Flammable Limits in Air (%):* 0.7 - 5; *Fire Extinguishing Agents*: Dry chemical, foam, or carbon dioxide; *Fire Extinguishing Agents Not to be Used:* Water may be ineffective; *Special Hazards of Combustion Products*: Not pertinent; *Behavior in Fire*: Not pertinent; *Ignition Temperature (deg. F)*: 400; *Electrical Hazard*: Not pertinent; *Burning Rate*: 4 mm/min.
Chemical Reactivity - *Reactivity with Water*: No reaction; *Reactivity with Common Materials*: No reaction; *Stability During Transport*: Stable; *Neutralizing Agents for Acids and Caustics:* Not pertinent; *Polymerization:* Not pertinent; *Inhibitor of Polymerization*: Not pertinent.

OILS, MISCELLANEOUS: RESIN
Chemical Designations - *Synonyms*: Codoil; Retinol; Rosin oil; Rosinol; *Chemical Formula*: Not applicable.
Observable Characteristics - *Physical State (as shipped)*: Liquid; *Color*: Light amber to red to black, depending on grade; *Odor*: Characteristic; like pinetree pitch.
Physical and Chemical Properties - *Physical State at 15 °C and 1 atm.*: Liquid; *Molecular Weight*: Not pertinent; *Boiling Point at 1 atm.*: 572-750, 300-400, 573-673; *Freezing Point*: Not pertinent; *Critical Temperature*: Not pertinent; *Critical Pressure*: Not pertinent; *Specific Gravity*: 0.96 at 20 °C (liquid); *Vapor (Gas) Specific Gravity*: Not pertinent; *Ratio of Specific Heats of Vapor (Gas)*: Not pertinent; *Latent Heat of Vaporization*: Not pertinent; *Heat of Combustion*: (est.) -18,000, -10,000, -420; *Heat of Decomposition*: Not pertinent.
Health Hazards Information - *Recommended Personal Protective Equipment*: Data not available; *Symptoms Following Exposure*: Data not available; *General Treatment for Exposure*: Data not available; *Toxicity by Inhalation (Threshold Limit Value)*: Data not available; *Short-Term Inhalation Limits*: Data not available, but toxicity is probably low; *Toxicity by Ingestion*: Data not available; *Late Toxicity*: Data not available; *Vapor (Gas) Irritant Characteristics*: Data not available; *Liquid or Solid Irritant Characteristics*: Data not available; *Odor Threshold*: Data not available
Fire Hazards - *Flash Point (deg. F):* 255 - 390 OC; *Flammable Limits in Air (%):* Data not available; *Fire Extinguishing Agents*: Dry chemical, foam, or carbon dioxide; *Fire Extinguishing Agents Not to be Used:* Water may be ineffective; *Special Hazards of Combustion Products*: Not pertinent; *Behavior in Fire*: Not pertinent; *Ignition Temperature (deg. F)*: 648; *Electrical Hazard*: Not pertinent; *Burning Rate*: Data not available.
Chemical Reactivity - *Reactivity with Water*: No reaction; *Reactivity with Common Materials*: No reaction; *Stability During Transport*: Stable; *Neutralizing Agents for Acids and Caustics:* Not pertinent; *Polymerization:* Not pertinent; *Inhibitor of Polymerization*: Not pertinent.

OILS, MISCELLANEOUS: ROAD
Chemical Designations - *Synonyms*: Liquid asphalt; Petroleum asphalt; Slow-curing asphalt; *Chemical Formula*: Not applicable.
Observable Characteristics - *Physical State (as shipped)*: Liquid; *Color*: Dark brown to black; *Odor*: Tarry.
Physical and Chemical Properties - *Physical State at 15 °C and 1 atm.*: Liquid; *Molecular Weight*: Not pertinent; *Boiling Point at 1 atm.*: Very high; *Freezing Point*: Not pertinent; *Critical Temperature*: Not pertinent; *Critical Pressure*: Not pertinent; *Specific Gravity*: 1.0-1.2 at 25 °C (liquid); *Vapor (Gas) Specific Gravity*: Not pertinent; *Ratio of Specific Heats of Vapor (Gas)*: Not pertinent; *Latent Heat of Vaporization*: Not pertinent; *Heat of Combustion*: (est.) -18,000, -10,000, -420; *Heat of Decomposition*: Not pertinent.
Health Hazards Information - *Recommended Personal Protective Equipment*: Protective clothing for hot asphalt; face and eye protection when hot; *Symptoms Following Exposure*: Contact with skin may cause dermatitis. Inhalation of vapors may cause moderate irritation of nose and throat. Hot liquid burns skin; *General Treatment for Exposure*: Severe burns may result from hot liquid. Cool the skin

at once with water. Cover burn with sterile dressing and seek medical attention; *Toxicity by Inhalation (Threshold Limit Value)*: 5 mg/m^3; *Short-Term Inhalation Limits*: Data not available; *Toxicity by Ingestion*: Grade 2, LD$_{50}$ 0.5 to 5 g/kg; *Late Toxicity*: None observed; *Vapor (Gas) Irritant Characteristics*: Vapors causes a slight smarting of the eyes or respiratory system if present in high concentrations. The effect is temporary; *Liquid or Solid Irritant Characteristics*: Causes smarting of the skin and first-degree burns on short exposure; may cause secondary burns on long exposure; *Odor Threshold*: Data not available.

Fire Hazards - *Flash Point (deg. F)*: 300 -550; *Flammable Limits in Air (%)*: Not pertinent; *Fire Extinguishing Agents*: Dry chemical, foam, or carbon dioxide; *Fire Extinguishing Agents Not to be Used*: Water may be ineffective; *Special Hazards of Combustion Products*: Not pertinent; *Behavior in Fire*: Not pertinent; *Ignition Temperature (deg. F)*: 400 - 700; *Electrical Hazard*: Not pertinent; *Burning Rate*: Data not available. **(vi) Chemical Reactivity -** *Reactivity with Water*: No reaction; *Reactivity with Common Materials*: No reaction; *Stability During Transport*: Stable; *Neutralizing Agents for Acids and Caustics*: Not pertinent; *Polymerization*: Not pertinent; *Inhibitor of Polymerization*: Not pertinent.

OILS, MISCELLANEOUS: ROSIN

Chemical Designations - *Synonyms*: Codoil; Retinol; Resin oil; Rosinol; *Chemical Formula*: Not applicable.

Observable Characteristics - *Physical State (as shipped)*: Liquid; *Color*: Light amber to red to black, depending on grade; *Odor*: Characteristic; like pinetree pitch.

Physical and Chemical Properties - *Physical State at 15 °C and 1 atm.*: Liquid; *Molecular Weight*: Not pertinent; *Boiling Point at 1 atm.*: 572-750, 300-400, 573-673; *Freezing Point*: Not pertinent; *Critical Temperature*: Not pertinent; *Critical Pressure*: Not pertinent; *Specific Gravity*: 0.96 at 20 °C (liquid); *Vapor (Gas) Specific Gravity*: Not pertinent; *Ratio of Specific Heats of Vapor (Gas)*: Not pertinent; *Latent Heat of Vaporization*: Not pertinent; *Heat of Combustion*: (est.) -18,000, -10,000, -420; *Heat of Decomposition*: Not pertinent.

Health Hazards Information - *Recommended Personal Protective Equipment*: Data not available; *Symptoms Following Exposure*: Data not available; *General Treatment for Exposure*: Data not available; *Toxicity by Inhalation (Threshold Limit Value)*: Data not available; *Short-Term Inhalation Limits*: Data not available, but toxicity is probably low; *Toxicity by Ingestion*: Data not available; *Late Toxicity*: Data not available; *Vapor (Gas) Irritant Characteristics*: Data not available; *Liquid or Solid Irritant Characteristics*: Data not available; *Odor Threshold*: Data not available.

Fire Hazards - *Flash Point (deg. F)*: 255 - 390 CC; *Flammable Limits in Air (%)*: Data not available; *Fire Extinguishing Agents*: Dry chemical, foam, or carbon dioxide; *Fire Extinguishing Agents Not to be Used*: Water may be ineffective; *Special Hazards of Combustion Products*: Not pertinent; *Behavior in Fire*: Not pertinent; *Ignition Temperature (deg. F)*: 648; *Electrical Hazard*: Data not available; *Burning Rate*: Data not available.

Chemical Reactivity: *Reactivity with Water*: No reaction; *Reactivity with Common Materials*: No reaction; *Stability During Transport*: Stable; *Neutralizing Agents for Acids and Caustics*: Not pertinent; *Polymerization*: Not pertinent; *Inhibitor of Polymerization*: Not pertinent.

OILS, MISCELLANEOUS: SPERM

Chemical Designations - *Synonyms*: No common synonyms; *Chemical Formula*: Not applicable.

Observable Characteristics - *Physical State (as shipped)*: Liquid; *Color*: Colorless to pale yellow; *Odor*: Characteristic.

Physical and Chemical Properties - *Physical State at 15 °C and 1 atm.*: Liquid; *Molecular Weight*: Not pertinent; *Boiling Point at 1 atm.*: Very high; *Freezing Point*: Not pertinent; *Critical Temperature*: Not pertinent; *Critical Pressure*: Not pertinent; *Specific Gravity*: 0.882 at 20 °C (liquid); *Vapor (Gas) Specific Gravity*: Not pertinent; *Ratio of Specific Heats of Vapor (Gas)*: Not pertinent; *Latent Heat of Vaporization*: Not pertinent; *Heat of Combustion*: -17,900, -9943, -416.3; *Heat of Decomposition*: Not pertinent.

Health Hazards Information - *Recommended Personal Protective Equipment*: Data not available; *Symptoms Following Exposure*: Data not available; *General Treatment for Exposure*: Data not available; *Toxicity by Inhalation (Threshold Limit Value)*: Data not available, but toxicity is probably low; *Short-Term Inhalation Limits*: Data not available; *Toxicity by Ingestion*: Data not available; *Late Toxicity*: Data not available; *Vapor (Gas) Irritant Characteristics*: Data not available; *Liquid or Solid Irritant Characteristics*: Data not available; *Odor Threshold*: Data not available.

Fire Hazards - *Flash Point (deg. F)*: 428 CC(No.1); 460 CC (No.2); 500 - 510 CC; *Flammable Limits in Air (%)*: Data not available; *Fire Extinguishing Agents*: Dry chemical, foam, or carbon dioxide; *Fire Extinguishing Agents Not to be Used:* Water or foam may cause frothing; *Special Hazards of Combustion Products*: Not pertinent; *Behavior in Fire*: Not pertinent; *Ignition Temperature (deg. F)*: 586 (No.1); *Electrical Hazard*: Not pertinent; *Burning Rate*: Data not available.

Chemical Reactivity - *Reactivity with Water*: No reaction; *Reactivity with Common Materials*: No reaction; *Stability During Transport*: Stable; *Neutralizing Agents for Acids and Caustics:* Not pertinent; *Polymerization:* Not pertinent; *Inhibitor of Polymerization*: Not pertinent.

OILS, MISCELLANEOUS: SPINDLE

Chemical Designations - *Synonyms*: Bearing oil; High-speed bearing oil; *Chemical Formula*: Not applicable.

Observable Characteristics - *Physical State (as shipped)*: Liquid; *Color*: Light brown; *Odor*: Weak, like kerosene.

Physical and Chemical Properties - *Physical State at 15 °C and 1 atm.*: Liquid; *Molecular Weight*: Not pertinent; *Boiling Point at 1 atm.*: Very high; *Freezing Point*: Not pertinent; *Critical Temperature*: Not pertinent; *Critical Pressure*: Not pertinent; *Specific Gravity*: 0.881 at 15 °C (liquid); *Vapor (Gas) Specific Gravity*: Not pertinent; *Ratio of Specific Heats of Vapor (Gas)*: Not pertinent; *Latent Heat of Vaporization*: Not pertinent; *Heat of Combustion*: Data not available; *Heat of Decomposition*: Not pertinent.

Health Hazards Information - *Recommended Personal Protective Equipment*: Protective gloves; goggles or face shield; *Symptoms Following Exposure*: Vapors cause slight irritation of eyes and nose. Liquid irritates stomach; if taken into lungs causes coughing, distress, and rapidly developing pulmonary edema; *General Treatment for Exposure*: ASPIRATION: enforced bed rest; administer oxygen; call a doctor. INGESTION: do NOT induce vomiting; have victim drink water or milk. EYES: wash with copious amounts of water. SKIN: wipe off, wash with soap and water; *Toxicity by Inhalation (Threshold Limit Value)*: Data not available; *Short-Term Inhalation Limits*: Data not available; *Toxicity by Ingestion*: Grade 1, LD_{50} 5 to 15 g/kg; *Late Toxicity*: Data not available; *Vapor (Gas) Irritant Characteristics*: Vapors causes a slight smarting of the eyes or respiratory system if present in high concentrations. The effect is temporary; *Liquid or Solid Irritant Characteristics*: Minimum hazard. If spilled on clothing and allowed to remain, may cause smarting and reddening of skin; *Odor Threshold*: Data not available.

Fire Hazards - *Flash Point (deg. F)*: 169 CC; *Flammable Limits in Air (%)*: Data not available; *Fire Extinguishing Agents*: Dry chemical, foam, or carbon dioxide; *Fire Extinguishing Agents Not to be Used:* Water may be ineffective; *Special Hazards of Combustion Products*: Not pertinent; *Behavior in Fire*: Not pertinent; *Ignition Temperature (deg. F)*: 478; *Electrical Hazard*: Not pertinent; *Burning Rate*: Data not available.

Chemical Reactivity - *Reactivity with Water*: No reaction; *Reactivity with Common Materials*: No reaction; *Stability During Transport*: Stable; *Neutralizing Agents for Acids and Caustics:* Not pertinent; *Polymerization:* Not pertinent; *Inhibitor of Polymerization*: Not pertinent.

OILS, MISCELLANEOUS: SPRAY

Chemical Designations - *Synonyms*: Dormant oil; Foliage oil; Kerosene, heavy; Plant spray oil; *Chemical Formula*: Not applicable.

Observable Characteristics - *Physical State (as shipped)*: Liquid; *Color*: Colorless to light brown; *Odor*: Like kerosene; like fuel oil.

Physical and Chemical Properties - *Physical State at 15 °C and 1 atm.*: Liquid; *Molecular Weight*: Not pertinent; *Boiling Point at 1 atm.*: 590 - 700, 310 - 371, 583 - 644; *Freezing Point*: Not pertinent; *Critical Temperature*: Not pertinent; *Critical Pressure*: Not pertinent; *Specific Gravity*: 0.82 at 20 °C (liquid); *Vapor (Gas) Specific Gravity*: Not pertinent; *Ratio of Specific Heats of Vapor (Gas)*: Not pertinent; *Latent Heat of Vaporization*: Not pertinent; *Heat of Combustion*: -18,540, -10,300, -431.24; *Heat of Decomposition*: Not pertinent.

Health Hazards Information - *Recommended Personal Protective Equipment*: Protective gloves; goggles or face shield; *Symptoms Following Exposure*: Vapors cause slight irritation of eyes and nose. Liquid irritates stomach; if taken into lungs causes coughing, distress, and rapidly developing pulmonary edema; *General Treatment for Exposure*: ASPIRATION: enforced bed rest; administer oxygen; call a doctor. INGESTION: do NOT induce vomiting; have victim drink water or milk. EYES: wash with copious amounts of water. SKIN: wipe off, wash with soap and water; *Toxicity by Inhalation (Threshold Limit Value)*: 200 ppm; *Short-Term Inhalation Limits*: Data not available; *Toxicity by Ingestion*: Grade 2, LD_{50} 0.5-5 g/kg; *Late Toxicity*: Data not available; *Vapor (Gas) Irritant Characteristics*: Vapors causes a slight smarting of the eyes or respiratory system if present in high concentrations. The effect is temporary; *Liquid or Solid Irritant Characteristics*: Minimum hazard. If spilled on clothing and allowed to remain, may cause smarting and reddening of skin; *Odor Threshold*: Data not available.

Fire Hazards - *Flash Point (deg. F)*: 140 (min.)CC; *Flammable Limits in Air (%)*: 0.6 - 4.6; *Fire Extinguishing Agents*: Dry chemical, foam, or carbon dioxide; *Fire Extinguishing Agents Not to be Used*: Water may be ineffective; *Special Hazards of Combustion Products*: Not pertinent; *Behavior in Fire*: Not pertinent; *Ignition Temperature (deg. F)*: 475; *Electrical Hazard*: Not pertinent; *Burning Rate*: 4 mm/min.

Chemical Reactivity - *Reactivity with Water*: No reaction; *Reactivity with Common Materials*: No reaction; *Stability During Transport*: Stable; *Neutralizing Agents for Acids and Caustics*: Not pertinent; *Polymerization*: Not pertinent; *Inhibitor of Polymerization*: Not pertinent.

OILS, MISCELLANEOUS: TALL

Chemical Designations - *Synonyms*: No common synonyms; *Chemical Formula*: Not applicable.

Observable Characteristics - *Physical State (as shipped)*: Liquid; *Color*: Yellow; *Odor*: Characteristic.

Physical and Chemical Properties - *Physical State at 15 °C and 1 atm.*: Liquid; *Molecular Weight*: Not pertinent; *Boiling Point at 1 atm.*: Very high; *Freezing Point*: Not pertinent; *Critical Temperature*: Not pertinent; *Critical Pressure*: Not pertinent; *Specific Gravity*: 0.951 at 16 °C (liquid); *Vapor (Gas) Specific Gravity*: Not pertinent; *Ratio of Specific Heats of Vapor (Gas)*: Not pertinent; *Latent Heat of Vaporization*: Not pertinent; *Heat of Combustion*: (est.) -18,000, -10,000, -420; *Heat of Decomposition*: Not pertinent.

Health Hazards Information - *Recommended Personal Protective Equipment*: Data not available; *Symptoms Following Exposure*: Data not available; *General Treatment for Exposure*: Data not available; *Toxicity by Inhalation (Threshold Limit Value)*: Data not available; *Short-Term Inhalation Limits*: Data not available; *Toxicity by Ingestion*: Data not available; *Late Toxicity*: Data not available; *Vapor (Gas) Irritant Characteristics*: Data not available; *Liquid or Solid Irritant Characteristics*: Data not available; *Odor Threshold*: Data not available.

Fire Hazards - *Flash Point (deg. F)*: 255; *Flammable Limits in Air (%)*: Data not available; *Fire Extinguishing Agents*: Dry chemical, foam, or carbon dioxide; *Fire Extinguishing Agents Not to be Used*: Water may be ineffective; *Special Hazards of Combustion Products*: Not pertinent; *Behavior in Fire*: Not pertinent; *Ignition Temperature* : Data not available; *Electrical Hazard*: Not pertinent; *Burning Rate*: Data not available.

Chemical Reactivity - *Reactivity with Water*: No reaction; *Reactivity with Common Materials*: No reaction; *Stability During Transport*: Stable; *Neutralizing Agents for Acids and Caustics*: Not pertinent; *Polymerization*: Not pertinent; *Inhibitor of Polymerization*: Not pertinent.

OILS, MISCELLANEOUS: TANNER'S

Chemical Designations - *Synonyms*: Sulfated neatsfoot oil, sodium salt; *Chemical Formula*: Not applicable.

Observable Characteristics - *Physical State (as shipped)*: Data not available; *Color*: Data not available; *Odor*: Data not available.

Physical and Chemical Properties - *Physical State at 15 ℃ and 1 atm.*: Liquid; *Molecular Weight*: Not pertinent; *Boiling Point at 1 atm.*: Very high; *Freezing Point*: Not pertinent; *Critical Temperature*: Not pertinent; *Critical Pressure*: Not pertinent; *Specific Gravity*: (est.) 0.85 at 20 °C (liquid); *Vapor (Gas) Specific Gravity*: Not pertinent; *Ratio of Specific Heats of Vapor (Gas)*: Not pertinent; *Latent Heat of Vaporization*: Not pertinent; *Heat of Combustion*: (est.) -18,000, -10,000, -420,; *Heat of Decomposition*: Not pertinent.

Health Hazards Information — *Recommended Personal Protective Equipment*: Data not available; *Symptoms Following Exposure*: Data not available; *General Treatment for Exposure*: Data not available; *Toxicity by Inhalation (Threshold Limit Value)*: Data not available; *Short-Term Inhalation Limits*: Data not available; *Toxicity by Ingestion*: Data not available; *Late Toxicity*: Data not available; *Vapor (Gas) Irritant Characteristics*: Data not available; *Liquid or Solid Irritant Characteristics*: Data not available; *Odor Threshold*: Data not available.

Fire Hazards - *Flash Point :* Data not available; *Flammable Limits in Air (%):* Data not available; *Fire Extinguishing Agents*: Dry chemical, foam, or carbon dioxide; *Fire Extinguishing Agents Not to be Used:* Water may be ineffective; *Special Hazards of Combustion Products*: Not pertinent; *Behavior in Fire*: Not pertinent; *Ignition Temperature :* Data not available; *Electrical Hazard*: Not pertinent; *Burning Rate*: Data not available.

Chemical Reactivity - *Reactivity with Water*: No reaction; *Reactivity with Common Materials*: No reaction; *Stability During Transport*: Stable; *Neutralizing Agents for Acids and Caustics:* Not pertinent; *Polymerization:* Not pertinent; *Inhibitor of Polymerization*: Not pertinent.

OILS, MISCELLANEOUS: TRANSFORMER

Chemical Designations - *Synonyms*: Electrical insulating oil; Insulating oil; Petroleum insulating oil; *Chemical Formula*: Not applicable.

Observable Characteristics - *Physical State (as shipped)*: Liquid; *Color*: Colorless to light brown; *Odor*: Like motor oil.

Physical and Chemical Properties - *Physical State at 15 ℃ and 1 atm.*: Liquid; *Molecular Weight*: Not pertinent; *Boiling Point at 1 atm.*: Very high; *Freezing Point*: -75, -59, 214; *Critical Temperature*: Not pertinent; *Critical Pressure*: Not pertinent; *Specific Gravity*: 0.891 at 15°C (liquid); *Vapor (Gas) Specific Gravity*: Not pertinent; *Ratio of Specific Heats of Vapor (Gas)*: Not pertinent; *Latent Heat of Vaporization*: Not pertinent; *Heat of Combustion*: Data not available; *Heat of Decomposition*: Not pertinent.

Health Hazards Information - *Recommended Personal Protective Equipment*: Protective gloves; goggles or face shield; *Symptoms Following Exposure*: Ingestion of liquid bay irritate stomach and cause increased frequency of bowel movements. If taken into lungs, delayed pulmonary irritation may occur; *General Treatment for Exposure*: INGESTION: do NOT induce vomiting. ASPIRATION: check for delayed irritation by serial X-rays. EYES: wash with copious amounts of water. SKIN: wipe off and wash with soap and water; *Toxicity by Inhalation (Threshold Limit Value)*: Data not available; *Short-Term Inhalation Limits*: Data not available; *Toxicity by Ingestion*: Grade 1, LD_{50} 5 to 15 g/kg; *Late Toxicity*: Data not available; *Vapor (Gas) Irritant Characteristics*: Vapors causes a slight smarting of the eyes or respiratory system if present in high concentrations. The effect is temporary; *Liquid or Solid Irritant Characteristics*: Minimum hazard. If spilled on clothing and allowed to remain, may cause smarting and reddening of skin; *Odor Threshold*: Data not available.

Fire Hazards - *Flash Point (deg. F):* 295 OC; *Flammable Limits in Air (%):* Data not available; *Fire Extinguishing Agents*: Dry chemical, foam, or carbon dioxide; *Fire Extinguishing Agents Not to be Used:* Water may be ineffective; *Special Hazards of Combustion Products*: Not pertinent; *Behavior in Fire*: Not pertinent; *Ignition Temperature :* Data not available; *Electrical Hazard*: Not pertinent;

Burning Rate: Data not available.
Chemical Reactivity - *Reactivity with Water*: No reaction; *Reactivity with Common Materials*: No reaction; *Stability During Transport*: Stable; *Neutralizing Agents for Acids and Caustics:* Not pertinent; *Polymerization:* Not pertinent; *Inhibitor of Polymerization*: Not pertinent.

OILS, MISCELLANEOUS: TURBINE
Chemical Designations - *Synonyms*: Steam turbine oil; Steam turbine lube oil; *Chemical Formula*: Not applicable.
Observable Characteristics - *Physical State (as shipped)*: Liquid; *Color*: Colorless to pale brown; *Odor*: Weak, like lube oil.
Physical and Chemical Properties - *Physical State at 15 °C and 1 atm.*: Liquid; *Molecular Weight*: Not pertinent; *Boiling Point at 1 atm.*: Not pertinent; *Freezing Point*: Not pertinent; *Critical Temperature*: Not pertinent; *Critical Pressure*: Not pertinent; *Specific Gravity*: 0.87 at 20 °C (liquid); *Vapor (Gas) Specific Gravity*: Not pertinent; *Ratio of Specific Heats of Vapor (Gas)*: Not pertinent; *Latent Heat of Vaporization*: Not pertinent; *Heat of Combustion*: (est.) -17,600, -9,800, -410; *Heat of Decomposition*: Not pertinent.
Health Hazards Information - *Recommended Personal Protective Equipment*: Goggles or face shield; rubber gloves; *Symptoms Following Exposure*: Contact with liquid causes slight irritation of eyes and (on prolonged contact) skin. Ingestion causes slight irritation of stomach and bowel, increased frequency of bowel movement; *General Treatment for Exposure*: EYES: wash with copious quantity of water for least 15 min. SKIN: wipe off; wash with soap and water. INGESTION: do NOT induce vomiting; do NOT lavage; 2-4 oz. Olive and 1-2 oz. Activated charcoal may be given; *Toxicity by Inhalation (Threshold Limit Value)*: 5 mg/m^3; *Short-Term Inhalation Limits*: Data not available; *Toxicity by Ingestion*: Grade 0, LD_{50} > 15 g/kg (rat); *Late Toxicity*: Data not available; *Vapor (Gas) Irritant Characteristics*: Data not available; *Liquid or Solid Irritant Characteristics*: No data; *Odor Threshold*: No data.
Fire Hazards - *Flash Point (deg. F):* 390 - 485 OC; *Flammable Limits in Air (%):* Not pertinent; *Fire Extinguishing Agents*: Dry chemical, foam, or carbon dioxide, water fog; *Fire Extinguishing Agents Not to be Used:* Water or foam may cause frothing; water may be ineffective; *Special Hazards of Combustion Products*: Not pertinent; *Behavior in Fire*: Not pertinent; *Ignition Temperature (deg. F)*: 700; *Electrical Hazard*: Not pertinent; *Burning Rate*: (approx.) 4 mm/min.
Chemical Reactivity - *Reactivity with Water*: No reaction; *Reactivity with Common Materials*: No reaction; *Stability During Transport*: Stable; *Neutralizing Agents for Acids and Caustics:* Not pertinent; *Polymerization:* Not pertinent; *Inhibitor of Polymerization*: Not pertinent.

OLEIC ACID
Chemical Designations - *Synonyms*: *cis*-8-Heptadecylene-carboxylic acid; *cis*-9-Octadecenoic acid; *cis*-9-Octadecylenic acid; Red oil; *Chemical Formula*: $CH_3(CH_2)_7CH=CH(CH_2)_7COOH$.
Observable Characteristics - *Physical State (as shipped)*: Liquid; *Color*: Pale; *Odor*: Faint.
Physical and Chemical Properties - *Physical State at 15 °C and 1 atm.*: Liquid; *Molecular Weight*: 277 (avg.); *Boiling Point at 1 atm.*: 432, 222, 495; *Freezing Point*: 57, 14, 287; *Critical Temperature*: Not pertinent; *Critical Pressure*: Not pertinent; *Specific Gravity*: 0.89 at 25 °C (liquid); *Vapor (Gas) Specific Gravity*: Not pertinent; *Ratio of Specific Heats of Vapor (Gas)*: Not pertinent; *Latent Heat of Vaporization*: 103, 57, 2.4,; *Heat of Combustion*: Data not available; *Heat of Decomposition*: Not pertinent.
Health Hazards Information - *Recommended Personal Protective Equipment*: Impervious gloves; goggles or face shield; impervious apron; *Symptoms Following Exposure*: Industrial use of compound involves no known hazards. Ingestion causes mild irritation of mouth and stomach. Contact with eyes or skin causes mild irritation; *General Treatment for Exposure*: INGESTION: give large amount of water. EYES: if eye irritation occurs, flush with water and get medical attention. SKIN: wash thoroughly with soap and water; *Toxicity by Inhalation (Threshold Limit Value)*: Data not available; *Short-Term Inhalation Limits*: Data not available; *Toxicity by Ingestion*: Grade 1, LD_{50} > 15 g/kg; *Late*

Toxicity: Data not available; *Vapor (Gas) Irritant Characteristics*: Data not available; *Liquid or Solid Irritant Characteristics*: Data not available; *Odor Threshold*: Data not available.

Fire Hazards - *Flash Point (deg. F):* 390 - 425 OC; *Flammable Limits in Air (%):* Data not available; *Fire Extinguishing Agents*: Dry chemical, foam, or carbon dioxide; *Fire Extinguishing Agents Not to be Used:* Water or foam may cause frothing; *Special Hazards of Combustion Products*: Data not available; *Behavior in Fire*: Data not available; *Ignition Temperature (deg. F)*: 685; *Electrical Hazard*: Data not available; *Burning Rate*: Data not available.

Chemical Reactivity: *Reactivity with Water*: No reaction; *Reactivity with Common Materials*: No reaction; *Stability During Transport*: Stable; *Neutralizing Agents for Acids and Caustics:* Not pertinent; *Polymerization:* Not pertinent; *Inhibitor of Polymerization*: Not pertinent.

OLEIC ACID, POTASSIUM SALT

Chemical Designations - *Synonyms*: Potassium oleate; *Chemical Formula*: $C_{17}H_{33}COOK$.

Observable Characteristics - *Physical State (as shipped)*: Solid or liquid; *Color*: Brown; *Odor*: Faint soapy.

Physical and Chemical Properties - *Physical State at 15 °C and 1 atm.*: Solid or liquid; *Molecular Weight*: 320 (solid only); *Boiling Point at 1 atm.*: Not pertinent (decomposes); *Freezing Point*: 455-464, 235-240, 508-513; *Critical Temperature*: Not pertinent; *Critical Pressure*: Not pertinent; *Specific Gravity*: > 1.1 at 20 °C (solid or liquid); *Vapor (Gas) Specific Gravity*: Not pertinent; *Ratio of Specific Heats of Vapor (Gas)*: Not pertinent; *Latent Heat of Vaporization*: Not pertinent; *Heat of Combustion*: Not pertinent; *Heat of Decomposition*: Not pertinent.

Health Hazards Information - *Recommended Personal Protective Equipment*: Chemical goggles and rubber gloves; *Symptoms Following Exposure*: Inhalation of dust causes irritation of nose and throat, coughing, and sneezing. Ingestion causes mild irritation of mouth. Contact with eyes causes irritation; *General Treatment for Exposure*: INHALATION: move to fresh air. INGESTION: give large amount of water. EYES: flush with copious quantities of tap water. SKIN: flush with water; *Toxicity by Inhalation (Threshold Limit Value)*: No data; *Short-Term Inhalation Limits*: Data not available; *Toxicity by Ingestion*: Data not available; *Late Toxicity*: Data not available; *Vapor (Gas) Irritant Characteristics*: No data; *Liquid or Solid Irritant Characteristics*: Data not available; *Odor Threshold*: Data not available.

Fire Hazards - *Flash Point (deg. F):* 140 CC; *Flammable Limits in Air (%):* No data; *Fire Extinguishing Agents*: Dry chemical, foam, or carbon dioxide; *Fire Extinguishing Agents Not to be Used:* Water may be ineffective; *Special Hazards of Combustion Products*: No data; *Behavior in Fire*: No data; *Ignition Temperature* : Data not available; *Electrical Hazard*: Not pertinent; *Burning Rate*: Not pertinent.

Chemical Reactivity - *Reactivity with Water*: No reaction; *Reactivity with Common Materials*: No reaction; *Stability During Transport*: Stable; *Neutralizing Agents for Acids and Caustics:* Not pertinent; *Polymerization:* Not pertinent; *Inhibitor of Polymerization*: Not pertinent.

OLEIC ACID, SODIUM SALT

Chemical Designations - *Synonyms*: Eunatrol; Sodium oleate; *Chemical Formula*: $C_{17}H_{33}COONa$.

Observable Characteristics - *Physical State (as shipped)*: Solid; *Color*: Light tan; *Odor*: Slight tallow-like.

Physical and Chemical Properties - *Physical State at 15 °C and 1 atm.*: Solid; *Molecular Weight*: 304 (approx.); *Boiling Point at 1 atm.*: Not pertinent (decomposes); *Freezing Point*: 450-455, 232-235, 505-508; *Critical Temperature*: Not pertinent; *Critical Pressure*: Not pertinent; *Specific Gravity*: > 1.1 at 20 °C (solid); *Vapor (Gas) Specific Gravity*: Not pertinent; *Ratio of Specific Heats of Vapor (Gas)*: Not pertinent; *Latent Heat of Vaporization*: Not pertinent; *Heat of Combustion*: Not pertinent; *Heat of Decomposition*: Not pertinent.

Health Hazards Information - *Recommended Personal Protective Equipment*: Dust mask and gloves; *Symptoms Following Exposure*: Inhalation of dust causes irritation of nose and throat, coughing, and sneezing. Ingestion causes mild irritation of mouth. Contact with eyes causes irritation; *General*

Treatment for Exposure: INHALATION: move to fresh air. INGESTION: give large amount of water. EYES: flush with copious quantities of tap water. SKIN: flush with water; *Toxicity by Inhalation (Threshold Limit Value)*: Data not available; *Short-Term Inhalation Limits*: Data not available; *Toxicity by Ingestion*: Data not available; *Late Toxicity*: Data not available; *Vapor (Gas) Irritant Characteristics*: Data not available; *Liquid or Solid Irritant Characteristics*: Data not available; *Odor Threshold*: Data not available.

Fire Hazards - *Flash Point :* Not pertinent (combustible solid;) *Flammable Limits in Air (%):* Not pertinent; *Fire Extinguishing Agents*: Dry chemical, foam, water, or carbon dioxide; *Fire Extinguishing Agents Not to be Used:* Data not available; *Special Hazards of Combustion Products*: Data not available; *Behavior in Fire*: Data not available; *Ignition Temperature :* Data not available; *Electrical Hazard*: Not pertinent; *Burning Rate*: Not pertinent.

Chemical Reactivity - *Reactivity with Water*: No reaction; *Reactivity with Common Materials*: No reaction; *Stability During Transport*: Stable; *Neutralizing Agents for Acids and Caustics:* Not pertinent; *Polymerization:* Not pertinent; *Inhibitor of Polymerization*: Not pertinent.

OLEUM

Chemical Designations - *Synonyms*: Fuming sulfuric acid; *Chemical Formula*: SO_3-H_2SO_4.

Observable Characteristics - *Physical State (as shipped)*: Liquid; *Color*: Colorless to cloudy; *Odor*: Sharp penetrating; choking.

Physical and Chemical Properties - *Physical State at 15 °C and 1 atm.*: Liquid; *Molecular Weight*: Not pertinent; *Boiling Point at 1 atm.*: Decomposes; *Freezing Point*: Not pertinent; *Critical Temperature*: Not pertinent; *Critical Pressure*: Not pertinent; *Specific Gravity*: 1.91-1.97 at 15 °C (liquid); *Vapor (Gas) Specific Gravity*: Not pertinent; *Ratio of Specific Heats of Vapor (Gas)*: Not pertinent; *Latent Heat of Vaporization*: Not pertinent; *Heat of Combustion*: Not pertinent; *Heat of Decomposition*: Not pertinent.

Health Hazards Information - *Recommended Personal Protective Equipment*: Respirator approved by U.S. Bureau of Mines for acid mists; rubber gloves; splashproof goggles; eyewash fountain and safety shower; rubber footwear; face shield; *Symptoms Following Exposure*: Acid mist is irritating to eyes, nose and throat. Liquid causes severe burns of skin and eyes; *General Treatment for Exposure*: INGESTION: have victim drink water or milk; do NOT induce vomiting. EYES: flush with plenty of water for at least 15 min.; call a doctor. SKIN: flush with plenty of water; *Toxicity by Inhalation (Threshold Limit Value)*: 1 mg/m^3; *Short-Term Inhalation Limits*: 5 mg/m^3 for 5 min.; 3 mg/m^3 for 10 min.; 2 mg/m^3 for 30 min.; 1 mg/m^3 for 60 min.; *Toxicity by Ingestion*: Severe burns of mouth and stomach; *Late Toxicity*: None; *Vapor (Gas) Irritant Characteristics*: Vapors causes a severe irritation of the eye and throat and can cause eye and lung injury. They cannot be tolerated even at low concentrations; *Liquid or Solid Irritant Characteristics*: Severe skin irritant. Causes second- and third-degree burns on short contact; very injurious to the eyes; *Odor Threshold*: 1 mg/m^3.

Fire Hazards - *Flash Point:* Not flammable; *Flammable Limits in Air (%):* Not flammable; *Fire Extinguishing Agents*: Not pertinent; *Fire Extinguishing Agents Not to be Used:* Avoid use of water on adjacent air; *Special Hazards of Combustion Products*: Toxic and irritating vapors are generated; *Behavior in Fire*: Not pertinent; *Ignition Temperature :* Not flammable; *Electrical Hazard*: Not pertinent; *Burning Rate*: Not flammable.

Chemical Reactivity - *Reactivity with Water*: Vigorous reaction with water; spatters; *Reactivity with Common Materials*: May react with cast iron with explosive violence. Attack many metals, releasing flammable hydrogen gas. Capable of igniting finely divided combustible material on contact. Extremely hazardous in contact with many materials; *Stability During Transport*: Normally stable; *Neutralizing Agents for Acids and Caustics:* Cautious dilution with water, with protection against violent spattering. Diluted acid may be neutralized with lime or soda ash; *Polymerization:* Not pertinent; *Inhibitor of Polymerization*: Not pertinent.

OXALIC ACID

Chemical Designations - *Synonyms*: Ethanedioic acid; *Chemical Formula*: $C_2H_2O_4$.

Observable Characteristics - *Physical State (as shipped)*: Solid; *Color*: White; *Odor*: Odorless.

Physical and Chemical Properties - *Physical State at 15 °C and 1 atm.*: Solid *Molecular Weight*: 126.7; *Boiling Point at 1 atm.*: Decomposes; *Freezing Point*: 214.7, 101.5, 374.7; *Critical Temperature*: Not pertinent; *Critical Pressure*: Not pertinent; *Specific Gravity*: 1.90 at 15 °C (solid); *Vapor (Gas) Specific Gravity*: Not pertinent; *Ratio of Specific Heats of Vapor (Gas)*: Not pertinent; *Latent Heat of Vaporization*: Not pertinent; *Heat of Combustion*: Not pertinent; *Heat of Decomposition*: Not pertinent.

Health Hazards Information - *Recommended Personal Protective Equipment*: Respirator for dust or mist protection; rubber, neoprene, or vinyl gloves; chemical safety glasses; rubbers, over leather or rubber safety shoes; apron or impervious clothing for splash protection; *Symptoms Following Exposure*: As dust or as a solution, can cause severe burns of eyes, skin, or mucous membranes. Ingestion of 5 grams has caused death with symptoms of nausea, shock, collapse, and convulsions coming on rapidly. Repeated or prolonged skin exposure can cause dermatitis and slow-healing ulcers; *General Treatment for Exposure*: Get medical attention for all eye exposure and any serious overexposures; treatment is symptomatic. INHALATION: rinse mouth and/or gargle repeatedly with cold water. INGESTION: dilute by drinking large amounts of water; repeat at least once and then administer milk or milk of magnesia as an emollient; do NOT induce vomiting. EYES and SKIN: flush thoroughly with water; *Toxicity by Inhalation (Threshold Limit Value)*: Not pertinent; *Short-Term Inhalation Limits*: Not pertinent; *Toxicity by Ingestion*: Grade 3, LD_{50} 50 to 500 mg/kg; *Late Toxicity*: Data not available; *Vapor (Gas) Irritant Characteristics*: Data not available; *Liquid or Solid Irritant Characteristics*: Severe skin irritant. Causes second- and third-degree burns on short contact and is very injurious to the eyes; *Odor Threshold*: Not pertinent.

Fire Hazards - *Flash Point :* Not flammable; *Flammable Limits in Air (%):* Not flammable; *Fire Extinguishing Agents*: Not pertinent; *Fire Extinguishing Agents Not to be Used:* Not pertinent; *Special Hazards of Combustion Products*: Generates poisonous gases; *Behavior in Fire*: Not pertinent; *Ignition Temperature :* Not flammable; *Electrical Hazard*: Not pertinent; *Burning Rate*: Not flammable.

Chemical Reactivity - *Reactivity with Water*: No reaction; *Reactivity with Common Materials*: No reaction; *Stability During Transport*: Stable; *Neutralizing Agents for Acids and Caustics:* Lime or soda ash; *Polymerization:* Not pertinent; *Inhibitor of Polymerization*: Not pertinent.

OXYGEN, LIQUEFIED

Chemical Designations - *Synonyms*: Liquid oxygen; LOX; *Chemical Formula*: O_2.

Observable Characteristics - *Physical State (as shipped)*: Liquefies gas; *Color*: Light blue; *Odor*: None.

Physical and Chemical Properties - *Physical State at 15 °C and 1 atm.*: Gas; *Molecular Weight*: 32.0; *Boiling Point at 1 atm.*: -297.3, -182.9, 90.3; *Freezing Point*: -361, -218, 55; *Critical Temperature*: -180, -118, 155; *Critical Pressure*: 738, 50.1, 5.09; *Specific Gravity*: 1.14 at -183 °C (liquid); *Vapor (Gas) Specific Gravity*: 1.1; *Ratio of Specific Heats of Vapor (Gas)*: 1.3962; *Latent Heat of Vaporization*: 91.6, 50.9, 2.13; *Heat of Combustion*: Not pertinent; *Heat of Decomposition*: Not pertinent.

Health Hazards Information - *Recommended Personal Protective Equipment*: Safety goggles or face shield; insulated gloves; long sleeves; trousers worn outside boots or over high-top shoes to shed spilled liquid; *Symptoms Following Exposure*: Inhalation of 100% oxygen can cause nausea, dizziness, irritation of lungs, pulmonary edema, pneumonia, and collapse. Liquid may cause frostbite of eyes and skin; *General Treatment for Exposure*: INHALATION: in all but the most severe cases (pneumonia), recovery is rapid after reduction of oxygen pressure; supportive treatment should include immediate sedation, anticonvulsive therapy if needed, and rest. EYES: treat frostbite burns. SKIN: treat frostbite; soak in lukewarm water; *Toxicity by Inhalation (Threshold Limit Value)*: Not pertinent; *Short-Term Inhalation Limits*: Not pertinent; *Toxicity by Ingestion*: Not pertinent; *Late Toxicity*: Not pertinent; *Vapor (Gas) Irritant Characteristics*: Data not available; *Liquid or Solid Irritant Characteristics*: Data not available; *Odor Threshold*: Not pertinent.

Fire Hazards - *Flash Point :* Not flammable, but supports combustion; *Flammable Limits in Air (%):* Not flammable; *Fire Extinguishing Agents*: Not pertinent; *Fire Extinguishing Agents Not to be Used:* Not pertinent; *Special Hazards of Combustion Products*: Not pertinent; *Behavior in Fire*: Increases intensity of any fire. Mixtures of liquid oxygen and any fuel are highly explosive; *Ignition Temperature* : Not pertinent; *Electrical Hazard*: Not pertinent; *Burning Rate*: Not pertinent.

Chemical Reactivity - *Reactivity with Water*: Heat of water will vigorously vaporize liquid oxygen; *Reactivity with Common Materials*: Avoid organic and combustible materials, such as oil, grease, coal dust, etc. If ignited, such mixtures can explode. Low temperature may cause brittleness in some materials; *Stability During Transport*: Stable; *Neutralizing Agents for Acids and Caustics:* Not pertinent; *Polymerization:* Not pertinent; *Inhibitor of Polymerization*: Not pertinent.

P

PARAFORMALDEHYDE

Chemical Designations - *Synonyms*: Formaldehyde polymer; Polyformaldehyde; Polyfooxymethylene; Polyoxymethylene glycol; *Chemical Formula*: $HO(CH_2O)_nH$.

Observable Characteristics - *Physical State (as shipped)*: Solid; *Color*: White; *Odor*: Pungent and irritating; like formaldehyde.

Physical and Chemical Properties - *Physical State at 15 °C and 1 atm.*: Solid; *Molecular Weight*: 600 (approx.); *Boiling Point at 1 atm.*: Decomposes; *Freezing Point*: 311 - 342, 155 - 172, 428 - 455; *Critical Temperature*: Not pertinent; *Critical Pressure*: Not pertinent; *Specific Gravity*: 1.46 at 15 °C (solid); *Vapor (Gas) Specific Gravity*: Not pertinent; *Ratio of Specific Heats of Vapor (Gas)*: Not pertinent; *Latent Heat of Vaporization*: Not pertinent; *Heat of Combustion*: -6682, -3712, -155.4; *Heat of Decomposition*: Not pertinent.

Health Hazards Information - *Recommended Personal Protective Equipment*: Goggles or face shield; protective clothing; *Symptoms Following Exposure*: Vapor or dust irritates eyes, mucous membranes, and skin; may cause dermatitis. Ingestion of solid or of a solution in water irritates mouth, throat, and stomach and may cause death; *General Treatment for Exposure*: INGESTION: give milk or white of egg beaten with water; call a doctor. SKIN OR EYES: rinse with copious amounts of water; *Toxicity by Inhalation (Threshold Limit Value)*: 5 ppm; *Short-Term Inhalation Limits*: Data not available; *Toxicity by Ingestion*: Grade 3, LD_{50} 50 to 500 mg/kg; *Late Toxicity*: Data not available; *Vapor (Gas) Irritant Characteristics*: Vapor is moderately irritating such that personnel will not usually tolerate or high vapor concentrations; *Liquid or Solid Irritant Characteristics*: Minimum hazard. If spilled on clothing and allowed to remain, may cause smarting and reddening of skin; *Odor Threshold*: Data not available.

Fire Hazards - *Flash Point (deg. F):* 199 OC; 160 CC; *Flammable Limits in Air (%):* (formaldehyde gas) 7.0 - 73.0; *Fire Extinguishing Agents*: Water, foam, dry chemical, carbon dioxide; *Fire Extinguishing Agents Not to be Used:* Data not available; *Special Hazards of Combustion Products*: Not pertinent; *Behavior in Fire*: Changes to formaldehyde gas which is highly flammable; *Ignition Temperature (deg. F)*: 572 (approx.); *Electrical Hazard*: Not pertinent; *Burning Rate*: Not pertinent.

Chemical Reactivity - *Reactivity with Water*: Forms water solution of formaldehyde gas; *Reactivity with Common Materials*: No reaction; *Stability During Transport*: Slowly decomposes to formaldehyde gas; *Neutralizing Agents for Acids and Caustics:* Not pertinent; *Polymerization:* Not pertinent; *Inhibitor of Polymerization*: Not pertinent.

PARATHION, LIQUID

Chemical Designations - *Synonyms*: O,O-Diethyl O-(p-nitrophenyl) phoshorothioate; O,O-Diethyl O-(p-nitrophenyl) thiophosphate; Ethyl Parathion; Phosphorothioic acid; O,O-diethyl O-p-nitrophenyl ester; *Chemical Formula*: $(C_2H_5O)_2PSOC_6H_4NO_2$.

Observable Characteristics - *Physical State (as shipped)*: Liquid; *Color*: Deep brown to yellow; *Odor*: Characteristic.

Physical and Chemical Properties - *Physical State at 15 °C and 1 atm.*: Liquid; *Molecular Weight*: 291.3; *Boiling Point at 1 atm.*: Very high; decomposes; *Freezing Point*: 43, 6, 279; *Critical Temperature*: Not pertinent; *Critical Pressure*: Not pertinent; *Specific Gravity*: 1.269 at 25 °C (liquid); *Vapor (Gas) Specific Gravity*: Not pertinent; *Ratio of Specific Heats of Vapor (Gas)*: Not pertinent; *Latent Heat of Vaporization*: Not pertinent; *Heat of Combustion*: -9,240, -5,140, -215; *Heat of Decomposition*: Not pertinent.

Health Hazards Information - *Recommended Personal Protective Equipment*: Neoprene-coated gloves; rubber work shoes or overshoes; latex rubber apron; goggles; respirator or mask approved for toxic dusts and organic vapors; *Symptoms Following Exposure*: Inhalation of mist, dust, or vapor (or ingestion, or absorption through the skin) cause dizziness, usually accompanied by constriction of the pupils, headache, and tightness of the chest. Nausea, vomiting, abdominal cramps, diarrhea, muscular twitching, convulsions and possibly death may follow. An increase in salivary and bronchial secretions may result which simulate severe pulmonary edema. Contact with eyes causes irritation; *General Treatment for Exposure*: Call a doctor for all exposure to this compound. INHALATION: remove victim from exposure immediately; have physician treat with atropine injections until full atropinization; 2-PAM may also be administered by physician. EYES: flush with water immediately after contact for at least 15 min. SKIN: remove all clothing and shoes immediately; quickly wipe off the affected area with a clean cloth; follow immediately with a shower, using plenty of soap. If a complete shower is impossible, wash the affected skin repeatedly with soap and water. INGESTION: if victim is conscious, induce vomiting and repeat until vomit fluid is clear; make victim drink plenty of milk or water; have him lie down and keep warm; *Toxicity by Inhalation (Threshold Limit Value)*: 0.01 mg/m³; *Short-Term Inhalation Limits*: 0.5 mg/m³ for 30 min.; *Toxicity by Ingestion*: Grade 4, oral LD_{50} = 2 mg/kg (rat); *Late Toxicity*: Birth defects in chick embryos; *Vapor (Gas) Irritant Characteristics*: Data not available; *Liquid or Solid Irritant Characteristics*: Data not available; *Odor Threshold*: 4.04 ppm.

Fire Hazards - *Flash Point :* Not flammable; *Flammable Limits in Air (%):* Not flammable; *Fire Extinguishing Agents*: Water o adjacent fires; *Fire Extinguishing Agents Not to be Used:* High pressure water hoses may scatter parathion from broken containers, increasing contamination hazard; *Special Hazards of Combustion Products*: Fumes from decomposing material may contain oxides of sulfur and nitrogen; *Behavior in Fire*: Containers may explode when heated; *Ignition Temperature :* Not pertinent; *Electrical Hazard*: Not pertinent; *Burning Rate*: Not pertinent.

Chemical Reactivity - *Reactivity with Water*: Slow reaction, not considered hazardous; *Reactivity with Common Materials*: No reaction; *Stability During Transport*: Stable; *Neutralizing Agents for Acids and Caustics:* Not pertinent; *Polymerization:* Not pertinent; *Inhibitor of Polymerization*: Not pertinent.

PENTABORANE

Chemical Designations - *Synonyms*: (9)-Pentaboron nooahydride; *Chemical Formula*: B_5H_9.

Observable Characteristics - *Physical State (as shipped)*: Liquid; *Color*: Colorless; *Odor*: Characteristic; strong, pungent; foul; sour milk.

Physical and Chemical Properties - *Physical State at 15 °C and 1 atm.*: Liquid; *Molecular Weight*: 63.2; *Boiling Point at 1 atm.*: 137.1, 58.4, 33.15; *Freezing Point*: -52.2, -46.8, 224.6; *Critical Temperature*: 441, 227, 500; *Critical Pressure*: 570, 38, 3.9; *Specific Gravity*: 0.623 at 20 °C (liquid); *Vapor (Gas) Specific Gravity*: 2.2; *Ratio of Specific Heats of Vapor (Gas)*: 1.0399; *Latent Heat of Vaporization*: 219, 122, 5.10; *Heat of Combustion*: -29,100, -16,200, -677; *Heat of Decomposition*: Not pertinent.

Health Hazards Information - *Recommended Personal Protective Equipment*: Self-contained breathing apparatus or air-line mask; goggles or face shield; rubber gloves and protective clothing; *Symptoms Following Exposure*: Inhalation of low concentrations causes dizziness, blurred vision, nausea, fatigue, light headedness or nervousness; higher concentrations also cause abnormal muscular contractions or twitching of any part of body, difficult breathing, poor muscular coordination, imperfect articulation of speech, convulsions, and (rarely) coma. Contact with liquid causes severe irritation of eyes and irritation of skin (acute local inflammation with the formation of small blisters, redness and swelling). Can be absorbed through the skin; *General Treatment for Exposure*: get medical attention following all exposures to this compound. INHALATION: remove victim to fresh air; watch for delayed symptoms for 1-2 days. EYES: wash with copious amounts of water for at least 30 min., holding eyelids apart to insure thorough flushing. SKIN: wash immediately with soap and water; rinse affected area with a 3% ammonia solution followed by additional flushing with water; *Toxicity by Inhalation (Threshold Limit Value)*: 0.005 ppm; *Short-Term Inhalation Limits*: 25 ppm for 5 min., 8 ppm for 15 min., 4 ppm for 30 min., 2 ppm for 60 min.; *Toxicity by Ingestion*: Grade 4, LD_{50} <50 mg/kg; *Late Toxicity*: Data not available; *Vapor (Gas) Irritant Characteristics*: Vapors causes a slight smarting of the eyes or respiratory system if present in high concentrations. The effect is temporary; *Liquid or Solid Irritant Characteristics*: Data not available; *Odor Threshold*: 0.8.

Fire Hazards - *Flash Point :* Not pertinent (ignites spontaneously in air); *Flammable Limits in Air (%):* 0.42 - 98; *Fire Extinguishing Agents*: Preferable let fire burn and shut off leak; extinguish with dry chemical or carbon dioxide; *Fire Extinguishing Agents Not to be Used:* Halogenated hydrocarbons, water; *Special Hazards of Combustion Products*: Toxic fumes may be formed; *Behavior in Fire*: Tends to re-ignite. Contact with water applied to adjacent fires produces flammable hydrogen gas; *Ignition Temperature :* Spontaneously flammable if impure. Approx. 35° C when pure; *Electrical Hazard*: Not pertinent; *Burning Rate*: Not pertinent.

Chemical Reactivity - *Reactivity with Water*: Reacts slowly to form flammable hydrogen gas. The reaction is not hazardous unless water is hot or unless confined; *Reactivity with Common Materials*: Corrosive to natural rubber, some synthetic rubbers, some greases and some lubricants; *Stability During Transport*: Stable below 302° F; *Neutralizing Agents for Acids and Caustics:* Not pertinent; *Polymerization:* Not pertinent; *Inhibitor of Polymerization*: Not pertinent.

PENTADECANOL

Chemical Designations - *Synonyms*: 1-Pentadecanol; Pentadecyl alcohol; *Chemical Formula*: $CH_3(CH_2)_{13}CH_2OH$.

Observable Characteristics - *Physical State (as shipped)*: Liquid; *Color*: Colorless; *Odor*: Weak alcoholic.

Physical and Chemical Properties - *Physical State at 15 °C and 1 atm.*: Liquid; *Molecular Weight*: 228.42; *Boiling Point at 1 atm.*: 572, 44, 317; *Freezing Point*: 111, 44, 317; *Critical Temperature*: 824, 440, 713; *Critical Pressure*: Not pertinent; *Specific Gravity*: 0.829 at 50 °C (liquid); *Vapor (Gas) Specific Gravity*: Not pertinent; *Ratio of Specific Heats of Vapor (Gas)*: 1.024; *Latent Heat of Vaporization*: Not pertinent; *Heat of Combustion*: Data not available; *Heat of Decomposition*: Not pertinent.

Health Hazards Information - *Recommended Personal Protective Equipment*: Goggles or face shield; *Symptoms Following Exposure*: Low toxicity. Excessive exposure produces some central nervous system depression. Prolonged contact produces skin irritation; *General Treatment for Exposure*: INHALATION: if necessary, support respiration. INGESTION: induce vomiting and call a doctor. SKIN AND EYES: wash with copious amounts of water; *Toxicity by Inhalation (Threshold Limit Value)*: Data not available; *Short-Term Inhalation Limits*: Data not available; *Toxicity by Ingestion*: Data not available; *Late Toxicity*: Data not available; *Vapor (Gas) Irritant Characteristics*: Data not available; *Liquid or Solid Irritant Characteristics*: Data not available; *Odor Threshold*: Data not available.

Fire Hazards - *Flash Point :* Data not available; *Flammable Limits in Air (%):* Data not available; *Fire Extinguishing Agents*: Foam, dry chemical, or carbon dioxide; *Fire Extinguishing Agents Not to be Used:* Water or foam may cause frothing; *Special Hazards of Combustion Products*: Not pertinent; *Behavior in Fire*: Not pertinent; *Ignition Temperature :* Data not available; *Electrical Hazard*: Not pertinent; *Burning Rate*: Data not available.

Chemical Reactivity - *Reactivity with Water*: No reaction; *Reactivity with Common Materials*: No

reaction; *Stability During Transport*: Stable; *Neutralizing Agents for Acids and Caustics:* Not pertinent; *Polymerization:* Not pertinent; *Inhibitor of Polymerization*: Not pertinent.

PENTAERYTHRITOL
Chemical Designations - *Synonyms*: Mono PE; PE; Pentaerythrite; Pentek; Tetrahydroxymethylmethane; Tetrakis (Hydroxymethyl) methane; Tetramethylolmethane; *Chemical Formula*: $C(CH_2OH)_4$.
Observable Characteristics - *Physical State (as shipped)*: Solid; *Color*: White; *Odor*: None.
Physical and Chemical Properties - *Physical State at 15 °C and 1 atm.*: Solid; *Molecular Weight*: 136.2; *Boiling Point at 1 atm.*: Not pertinent (sublimes); *Freezing Point*: 502, 261, 534; *Critical Temperature*: Not pertinent; *Critical Pressure*: Not pertinent; *Specific Gravity*: 1.39 at 25 °C (solid); *Vapor (Gas) Specific Gravity*: Not pertinent; *Ratio of Specific Heats of Vapor (Gas)*: Not pertinent; *Latent Heat of Vaporization*: Not pertinent; *Heat of Combustion*: -8,730, -4,850, -203; *Heat of Decomposition*: Not pertinent.
Health Hazards Information - *Recommended Personal Protective Equipment*: Dust mask; goggles; *Symptoms Following Exposure*: Non-toxic; no symptoms likely; *General Treatment for Exposure*: None needed; *Toxicity by Inhalation (Threshold Limit Value)*: Not pertinent ("Inert" particulate); *Short-Term Inhalation Limits*: Not pertinent; *Toxicity by Ingestion*: Grade 0, LD_{50} > 15 g/kg; *Late Toxicity*: None; *Vapor (Gas) Irritant Characteristics*: Data not available; *Liquid or Solid Irritant Characteristics*: Data not available; *Odor Threshold*: Data not available.
Fire Hazards - *Flash Point :* Not pertinent (combustible solid); *Flammable Limits in Air (%):* Not pertinent; *Fire Extinguishing Agents*: Water, dry chemical, carbon dioxide; *Fire Extinguishing Agents Not to be Used:* Not pertinent; *Special Hazards of Combustion Products*: Not pertinent; *Behavior in Fire*: Not pertinent; *Ignition Temperature (deg. F)*: 842 (dust cloud); *Electrical Hazard*: Not pertinent; *Burning Rate*: Not pertinent.
Chemical Reactivity - *Reactivity with Water*: No reaction; *Reactivity with Common Materials*: No reaction; *Stability During Transport*: Stable; *Neutralizing Agents for Acids and Caustics:* Not pertinent; *Polymerization:* Not pertinent; *Inhibitor of Polymerization*: Not pertinent.

PENTANE
Chemical Designations - *Synonyms*: No common synonyms; *Chemical Formula*: $n-C_5H_{12.}$
; **(ii) Observable Characteristics -*Physical State (as shipped)***: Liquid; *Color*: Colorless; *Odor*: Like a gasoline.
Physical and Chemical Properties - *Physical State at 15 °C and 1 atm.*: Liquid; *Molecular Weight*: 72.15; *Boiling Point at 1 atm.*: 97.0, 36.1, 309.3; *Freezing Point*: -201.0, -129.4, 143.8; *Critical Temperature*: 385.7, 196.5, 469.7; *Critical Pressure*: 490, 33.3, 3.37; *Specific Gravity*: 0.626 at 20 °C (liquid); *Vapor (Gas) Specific Gravity*: 2.5; *Ratio of Specific Heats of Vapor (Gas)*: 1.075; *Latent Heat of Vaporization*: 153.7, 85.38, 3.575; *Heat of Combustion*: -19,352, -10,751, -450; *Heat of Decomposition*: Not pertinent.
Health Hazards Information - *Recommended Personal Protective Equipment*: Goggles or face shield (as for gasoline); *Symptoms Following Exposure*: Low toxicity. Very high vapor concentrations produce narcosis. Aspiration into lungs can produce chemical pneumonitis and/or pulmonary edema; *General Treatment for Exposure*: INHALATION: remove from exposure; support respiration if needed. INGESTION: do NOT induce vomiting; call a physician; *Toxicity by Inhalation (Threshold Limit Value)*: 500 ppm; *Short-Term Inhalation Limits*: Data not available; *Toxicity by Ingestion*: Data not available; *Late Toxicity*: None; *Vapor (Gas) Irritant Characteristics*: Vapors are nonirritating to the eyes and throat; *Liquid or Solid Irritant Characteristics*: No appreciable hazard. Practically harmless to the skin; *Odor Threshold*: 10 ppm.
Fire Hazards - *Flash Point (deg. F):* -57 CC; *Flammable Limits in Air (%):* 1.4 - 8.3 (by vol.); *Fire Extinguishing Agents*: Foam, dry chemical, carbon dioxide; *Fire Extinguishing Agents Not to be Used:* Water may be ineffective; *Special Hazards of Combustion Products*: Not pertinent; *Behavior in Fire*: Containers may explode; *Ignition Temperature (deg. F)*: 544; *Electrical Hazard*: Class I, Group D;

Burning Rate: 8.6 mm/min.

Chemical Reactivity - *Reactivity with Water*: No reaction; *Reactivity with Common Materials*: No reaction; *Stability During Transport*: Stable; *Neutralizing Agents for Acids and Caustics:* Not pertinent; *Polymerization:* Not pertinent; *Inhibitor of Polymerization*: Not pertinent.

1-PENTENE

Chemical Designations - *Synonyms*: alfa-n-Amylene; Propylethylene; *Chemical Formula*: $CH_3(CH_2)_2CH=CH_2$.

Observable Characteristics - *Physical State (as shipped)*: Liquid; *Color*: Colorless; *Odor*: Like gasoline.

Physical and Chemical Properties - *Physical State at 15 °C and 1 atm.*: Liquid; *Molecular Weight*: 70.13; *Boiling Point at 1 atm.*: 85.8, 29.9, 303.1; *Freezing Point*: -265, -165, 108; *Critical Temperature*: 376.9, 191.6, 464.8; *Critical Pressure*: 588, 40, 4.5; *Specific Gravity*: 0.641 at 20 °C (liquid); *Vapor (Gas) Specific Gravity*: 2.4; *Ratio of Specific Heats of Vapor (Gas)*: 1.083; *Latent Heat of Vaporization*: 154.6, 85.87, 3.595; *Heat of Combustion*: -19.359, -10.755, -450.29; *Heat of Decomposition*: Not pertinent.

Health Hazards Information - *Recommended Personal Protective Equipment*: Goggles or face shield (as for gasoline); *Symptoms Following Exposure*: Acts as a simple asphyxiant or weak anesthetic in high vapor concentrations. Similar to effects caused by gasoline vapors; *General Treatment for Exposure*: INHALATION: remove from exposure. SKIN: wash with soap and water. EYES: flush with water; *Toxicity by Inhalation (Threshold Limit Value)*: Data not available; *Short-Term Inhalation Limits*: Data not available; *Toxicity by Ingestion*: Data not available; *Late Toxicity*: Data not available; *Vapor (Gas) Irritant Characteristics*: Not irritating; *Liquid or Solid Irritant Characteristics*: Not irritating; *Odor Threshold*: Data not available.

Fire Hazards - *Flash Point (deg. F):* - 60 CC; 0 OC; *Flammable Limits in Air (%):* 1.4 - 8.7; *Fire Extinguishing Agents*: Foam, dry chemical, or carbon dioxide. Stop flow of vapor; *Fire Extinguishing Agents Not to be Used:* Water may be ineffective; *Special Hazards of Combustion Products*: Not pertinent; *Behavior in Fire*: Containers may explode; *Ignition Temperature (deg. F)*: 527; *Electrical Hazard*: Data not available; *Burning Rate*: 9.1 mm/min.

Chemical Reactivity - *Reactivity with Water*: No reaction; *Reactivity with Common Materials*: No reaction; *Stability During Transport*: Stable; *Neutralizing Agents for Acids and Caustics:* Not pertinent; *Polymerization:* Not pertinent; *Inhibitor of Polymerization*: Not pertinent.

PERACETIC ACID

Chemical Designations - *Synonyms*: Acetyl hydroperoxide; Peroxyacetic acid; *Chemical Formula*: $CH_3COOOH-CH_3COOH$.

Observable Characteristics - *Physical State (as shipped)*: Liquid; *Color*: Colorless; *Odor*: Pungent; strong.

Physical and Chemical Properties - *Physical State at 15 °C and 1 atm.*: Liquid; *Molecular Weight*: Not pertinent; *Boiling Point at 1 atm.*: Not pertinent (mixture); *Freezing Point*: (approx.) -22, -30, 243; *Critical Temperature*: Not pertinent; *Critical Pressure*: Not pertinent; *Specific Gravity*: (est.) 1.153 at 25 °C (liquid); *Vapor (Gas) Specific Gravity*: Not pertinent; *Ratio of Specific Heats of Vapor (Gas)*: Not pertinent; *Latent Heat of Vaporization*: Data not available; *Heat of Combustion*: Data not available; *Heat of Decomposition*: Data not available.

Health Hazards Information - *Recommended Personal Protective Equipment*: Self-contained breathing apparatus; impervious gloves; full protective clothing (goggles, rubber gloves, etc.); *Symptoms Following Exposure*: inhalation causes severe irritation of mucous membrane. Contact with liquid causes severe irritation of eyes and skin. Ingestion causes severe distress, including burns of mouth and stomach; *General Treatment for Exposure*: INHALATION: remove victim to fresh air; if he is not breathing, apply artificial respiratory and oxygen; Call a doctor. EYES; *Toxicity by Inhalation (Threshold Limit Value)*: Data not available; *Short-Term Inhalation Limits*: Data not available; *Toxicity by Ingestion*: Grade 4, LD_{50} 10 mg/kg; *Late Toxicity*: Data not available; *Vapor (Gas) Irritant*

Characteristics: Data not available; *Liquid or Solid Irritant Characteristics*: Data not available; *Odor Threshold*: Data not available.

Fire Hazards - *Flash Point (deg. F)*: 104 OC; *Flammable Limits in Air (%)*: Data not available; *Fire Extinguishing Agents*: Water; *Fire Extinguishing Agents Not to be Used*: Not pertinent; *Special Hazards of Combustion Products*: Not pertinent; *Behavior in Fire*: Vapors are very flammable and explosive. Liquid will detonate if concentration rises above 56% because of evaporation of acetic acid; *Ignition Temperature (deg. F)*: 392; *Electrical Hazard*: Not pertinent; *Burning Rate*: Data not available.

Chemical Reactivity - *Reactivity with Water*: No reaction; *Reactivity with Common Materials*: May cause fire in contact with organic materials such as wood, cotton or straw. Corrosive to most metals including aluminum; *Stability During Transport*: Stable if kept cool and out of contact with most metals. At 30°C concentration decreases about 0.4% each month; *Neutralizing Agents for Acids and Caustics*: Flush with water; *Polymerization*: Not pertinent; *Inhibitor of Polymerization*: Not pertinent.

PERCHLORIC ACID

Chemical Designations - *Synonyms*: Dioxonium perchlorate solution; Perchloric acid solution; *Chemical Formula*: $HClO_4$-H_2O.

Observable Characteristics - *Physical State (as shipped)*: Liquid; *Color*: Colorless; *Odor*: None.

Physical and Chemical Properties - *Physical State at 15 °C and 1 atm.*: Liquid; *Molecular Weight*: 100.46 (solute only); *Boiling Point at 1 atm.*: Not pertinent (decomposes); *Freezing Point*: -170, -112, 161; *Critical Temperature*: Not pertinent; *Critical Pressure*: Not pertinent; *Specific Gravity*: 1.6 - 1.7 at 25 °C (liquid); *Vapor (Gas) Specific Gravity*: Not pertinent; *Ratio of Specific Heats of Vapor (Gas)*: Not pertinent; *Latent Heat of Vaporization*: Not pertinent; *Heat of Combustion*: Not pertinent; *Heat of Decomposition*: Not pertinent.

Health Hazards Information - *Recommended Personal Protective Equipment*: Rubber gloves, face shield or vapor-tight chemical-type safety goggles; rubber apron; rubber boots or shoes; *Symptoms Following Exposure*: Inhalation of vapors or mist causes burning sensation of nose and throat, and lung irritation with coughing; prolonged or excessive exposure could cause vomiting and severe coughing. Ingestion causes blistering and burns of mouth and stomach. Contact with eyes or skin causes blistering and burns; *General Treatment for Exposure*: Get medical attention following all exposures to this compound. INHALATION: move to fresh air; give oxygen if necessary. INGESTION: give large amounts of water. EYES: flush with water for at least 15 min. SKIN: flush with water; *Toxicity by Inhalation (Threshold Limit Value)*: Data not available; *Short-Term Inhalation Limits*: Data not available; *Toxicity by Ingestion*: Data not available; *Late Toxicity*: Data not available; *Vapor (Gas) Irritant Characteristics*: Data not available; *Liquid or Solid Irritant Characteristics*: Data not available; *Odor Threshold*: Data not available.

Fire Hazards - *Flash Point* : Not flammable , but may explode in fire; *Flammable Limits in Air (%)*: Not flammable; *Fire Extinguishing Agents*: Water from protected area; *Fire Extinguishing Agents Not to be Used*: Data not available; *Special Hazards of Combustion Products*: Data not available; *Behavior in Fire*: Above 160°C (320° F) will react with combustible material and increase intensity of fire. Containers may explode; *Ignition Temperature*: Not pertinent; *Electrical Hazard*: Not pertinent; *Burning Rate*: Not pertinent.

Chemical Reactivity - *Reactivity with Water*: No reaction; *Reactivity with Common Materials*: Contact with most combustible materials may cause fires and explosions. Corrosive to most metals with formation of flammable hydrogen gas, which may collect in enclosed spaces; *Stability During Transport*: Unstable if heated; *Neutralizing Agents for Acids and Caustics*: Flush with water, rinse with dilute sodium bicarbonate or soda ash solution; *Polymerization*: Not pertinent; *Inhibitor of Polymerization*: Not pertinent.

PETROLATUM

Chemical Designations - *Synonyms*: Petrolatum jelly; Petroleum jelly; Vaseline; Yellow petrolatum; *Chemical Formula*: Not applicable.

Observable Characteristics - *Physical State (as shipped)*: Liquid; *Color*: Colorless, amber, green, dark brown; *Odor*: None.

Physical and Chemical Properties - *Physical State at 15 °C and 1 atm.*: Grease; *Molecular Weight*: Not pertinent; *Boiling Point at 1 atm.*: Very high; *Freezing Point*: 100-135, 38-57, 311-330; *Critical Temperature*: Not pertinent; *Critical Pressure*: Not pertinent; *Specific Gravity*: (est.) 0.865 at 60 °C (liquid); *Vapor (Gas) Specific Gravity*: Not pertinent; *Ratio of Specific Heats of Vapor (Gas)*: Not pertinent; *Latent Heat of Vaporization*: 97-100, 54-63, 2.3-2.6; *Heat of Combustion*: Data not available; *Heat of Decomposition*: Not pertinent.

Health Hazards Information - *Recommended Personal Protective Equipment*: Goggles or face shield; *Symptoms Following Exposure*: None; *General Treatment for Exposure*: EYES: wash with water; *Toxicity by Inhalation (Threshold Limit Value)*: Not pertinent; *Short-Term Inhalation Limits*: Not pertinent; *Toxicity by Ingestion*: Grade 1, LD_{50} 5 to 5 g/kg; *Late Toxicity*: None; *Vapor (Gas) Irritant Characteristics*: None; *Liquid or Solid Irritant Characteristics*: None; *Odor Threshold*: Not pertinent.

Fire Hazards - *Flash Point (deg. F)*: 360 - 430 CC; *Flammable Limits in Air (%)*: Data not available; *Fire Extinguishing Agents*: Water, foam, dry chemical, carbon dioxide; *Fire Extinguishing Agents Not to be Used*: Not pertinent; *Special Hazards of Combustion Products*: Not pertinent; *Behavior in Fire*: Not pertinent; *Ignition Temperature* : Data not available; *Electrical Hazard*: Not pertinent; *Burning Rate*: Data not available.

Chemical Reactivity - *Reactivity with Water*: No reaction; *Reactivity with Common Materials*: No reaction; *Stability During Transport*: Stable; *Neutralizing Agents for Acids and Caustics*: Not pertinent; *Polymerization*: Not pertinent; *Inhibitor of Polymerization*: Not pertinent.

PETROLEUM NAPHTHA

Chemical Designations - *Synonyms*: Petroleum solvent; *Chemical Formula*: Not applicable.

Observable Characteristics - *Physical State (as shipped)*: Liquid; *Color*: Colorless; *Odor*: Like gasoline and kerosene.

Physical and Chemical Properties - *Physical State at 15 °C and 1 atm.*: Liquid; *Molecular Weight*: Not pertinent; *Boiling Point at 1 atm.*: 207.0, 97.2, 370.4; *Freezing Point*: Not pertinent; *Critical Temperature*: Not pertinent; *Critical Pressure*: Not pertinent; *Specific Gravity*: 0.74 at 20 °C (liquid); *Vapor (Gas) Specific Gravity*: Not pertinent; *Ratio of Specific Heats of Vapor (Gas)*: (est.) 1.030; *Latent Heat of Vaporization*: 130 - 150, 71 - 81, 3.0 - 3.4; *Heat of Combustion*: Data not available; *Heat of Decomposition*: Not pertinent.

Health Hazards Information - *Recommended Personal Protective Equipment*: Goggles or face shield (as for gasoline); *Symptoms Following Exposure*: Inhalation of concentrated vapor may cause intoxication. Liquid is not ver irritating to skin or eyes but may get into lungs by aspiration; *General Treatment for Exposure*: INHALATION: remove victim to fresh air and treat symptoms. INGESTION: have victim drink water or milk; do NOT induce vomiting. EYES: flush with water for 15 min. SKIN: wipe off and wash with soap and water; *Toxicity by Inhalation (Threshold Limit Value)*: No single TLV applicable; *Short-Term Inhalation Limits*: 500 ppm. For 30 min.; *Toxicity by Ingestion*: Grade 2, LD_{50} 0.5 - 5 g/kg; *Late Toxicity*: None; *Vapor (Gas) Irritant Characteristics*: Vapors are non-irrigating to the eyes and throat; *Liquid or Solid Irritant Characteristics*: No appreciable hazard. Practically harmless to the skin; *Odor Threshold*: Data not available.

Fire Hazards - *Flash Point (deg. F)*: 20 (approx.); *Flammable Limits in Air (%)*: 0.9 - 6.0; *Fire Extinguishing Agents*: Foam carbon dioxide, or dry chemical; *Fire Extinguishing Agents Not to be Used*: Water may be ineffective; *Special Hazards of Combustion Products*: Not pertinent; *Behavior in Fire*: Not pertinent; *Ignition Temperature (deg. F)*: 450 (approx.); *Electrical Hazard*: Not pertinent; *Burning Rate*: 4 mm/min.

Chemical Reactivity - *Reactivity with Water*: No reaction; *Reactivity with Common Materials*: No reaction; *Stability During Transport*: Stable; *Neutralizing Agents for Acids and Caustics*: Not pertinent; *Polymerization*: Not pertinent; *Inhibitor of Polymerization*: Not pertinent.

PHENOL

Chemical Designations - *Synonyms*: Carbolic acid; Hydroxybenzene; Phenic acid Phenyl hydroxide; *Chemical Formula*: C_6H_5OH.

Observable Characteristics - *Physical State (as shipped)*: Solid or liquid; *Color*: Colorless to light pink; *Odor*: Characteristically sweet; sweet, tarry; pungent, distinctive; distinct, aromatic, somewhat sickening sweet and acrid.

Physical and Chemical Properties - *Physical State at 15 °C and 1 atm.*: Solid or liquid; *Molecular Weight*: 94.11; *Boiling Point at 1 atm.*: 359.2, 181.8, 455.0; *Freezing Point*: 105.6, 40.9, 314.1; *Critical Temperature*: 790.0, 421.1, 694.3; *Critical Pressure*: 889, 60.5, 6.13; *Specific Gravity*: 1.058 at 41 °C (liquid); *Vapor (Gas) Specific Gravity*: Not pertinent; *Ratio of Specific Heats of Vapor (Gas)*: 1.089; *Latent Heat of Vaporization*: 130, 72, 3.0; *Heat of Combustion*: -13.400, -7.445, -311.7; *Heat of Decomposition*: Not pertinent.

Health Hazards Information - *Recommended Personal Protective Equipment*: Fresh-air mask for confined areas; rubber gloves; *Symptoms Following Exposure*: Will burn eyes and skin. The analgesic action may cause loss of pain sensation. Readily absorbed through skin, causing increase in heart rate, convulsions, and death; *General Treatment for Exposure*: INHALATION: if victim shows any ill effects, move him to fresh air, keep him quiet and warm, and call a doctor immediately; if breathing stops, give artificial respiration. INGESTION: do NOT induce vomiting; give milk, eggs whites, or large amounts of water and call a doctor immediately; no known antidote; treat the symptoms. EYES: immediately flush with plenty of water for at least 15 min.; continue for another 15 min. if doctor has not taken over. SKIN: immediately remove all clothing while in a shower and wash affected area with abundant flowing water or soap and water for at least 15 min.; clean clothing; *Toxicity by Inhalation (Threshold Limit Value)*: 5 ppm; *Short-Term Inhalation Limits*: Data not available; *Toxicity by Ingestion*: Grade 2, LD_{50} 0.5 to 5 g/kg; *Late Toxicity*: Carcinogenic in laboratory animals; *Vapor (Gas) Irritant Characteristics*: Vapor is moderately irritating such that personnel will not usually tolerate moderate or high vapor concentrations; *Liquid or Solid Irritant Characteristics*: Fairly severe skin irritant; may cause pain and second-degree burns after a few minutes contact; *Odor Threshold*: 0.05 ppm.

Fire Hazards - *Flash Point (deg. F)*: 185 OC; 175 CC; *Flammable Limits in Air (%)*: 1.7 - 8.6; *Fire Extinguishing Agents*: Water fog, dry chemical, carbon dioxide, foam; *Fire Extinguishing Agents Not to be Used*: Not pertinent; *Special Hazards of Combustion Products*: Toxic and irritating vapors are generated when heated; *Behavior in Fire*: Yields flammable vapors when heated which form explosive mixtures with air; *Ignition Temperature (deg. F)*: 1319; *Electrical Hazard*: Not pertinent; *Burning Rate*: 3.5 mm/min.

Chemical Reactivity - *Reactivity with Water*: No reaction; *Reactivity with Common Materials*: No reaction; *Stability During Transport*: Stable; *Neutralizing Agents for Acids and Caustics*: Not pertinent; *Polymerization*: Not pertinent; *Inhibitor of Polymerization*: Not pertinent.

PHENYLDICHLOROARSINE, LIQUID

Chemical Designations - *Synonyms*: Phenlarsenic dichloride; *Chemical Formula*: $C_6H_5AsCl_2$.

Observable Characteristics - *Physical State (as shipped)*: Liquid; *Color*: Colorless to yellow; *Odor*: Weak, but very unpleasant.

Physical and Chemical Properties - *Physical State at 15 °C and 1 atm.*: Liquid; *Molecular Weight*: 222.9; *Boiling Point at 1 atm.*: 495, 257, 530; *Freezing Point*: 3.9, -15.6, 257.6; *Critical Temperature*: Not pertinent; *Critical Pressure*: Not pertinent; *Specific Gravity*: 1.657 at 20 °C (liquid); *Vapor (Gas) Specific Gravity*: Not pertinent; *Ratio of Specific Heats of Vapor (Gas)*: Not pertinent; *Latent Heat of Vaporization*: 99, 55, 2.3; *Heat of Combustion*: (est.) -6.450, -3.600, -150; *Heat of Decomposition*: Not pertinent.

Health Hazards Information - *Recommended Personal Protective Equipment*: Full protective clothing; gas mask or self-contained breathing apparatus; *Symptoms Following Exposure*: Inhalation causes irritation of respiratory system, pulmonary edema, and systemic effects. Vapor irritates eyes. Liquid causes severe burns of eyes and severe irritation or burns of mouth and stomach; *General Treatment* *for Exposure*: *Get medical attention following all exposures to this compound*. INHALATION: remove victim from exposure; give artificial respiration if breathing has ceased. EYES: immediately wash with copious amounts of water for at least 15 min. SKIN: flush with water and wash well with soap and water; compound can be absorbed through skin and cause toxic systemic effects. INGESTION: give large amounts of water; *Toxicity by Inhalation (Threshold Limit Value)*: Data not available; *Short-Term Inhalation Limits*: Data not available; *Toxicity by Ingestion*: Data not available; *Late Toxicity*: Data not available; *Vapor (Gas) Irritant Characteristics*: Data not available; *Liquid or Solid Irritant Characteristics*: Data not available; *Odor Threshold*: Data not available.

Fire Hazards - *Flash Point :* Data not available; *Flammable Limits in Air (%):* Data not available; *Fire Extinguishing Agents*: Water; *Fire Extinguishing Agents Not to be Used:* Not pertinent; *Special Hazards of Combustion Products*: Highly toxic arsenic fumes are formed when hot; *Behavior in Fire*: Data not available; *Ignition Temperature* : Data not available; *Electrical Hazard*: Data not available; *Burning Rate*: 1.8 mm/min.

Chemical Reactivity - *Reactivity with Water*: Very slow reaction, considered non-hazardous. Hydrochloric acid is formed; *Reactivity with Common Materials*: Corrodes metals because of acid formed; *Stability During Transport*: Stable; *Neutralizing Agents for Acids and Caustics:* Not pertinent; *Polymerization:* Not pertinent; *Inhibitor of Polymerization*: Not pertinent.

PHENYLHYDRAZINE HYDROCHLORIDE

Chemical Designations - *Synonyms*: Phenyldraziniun chloride; *Chemical Formula*: $C_6H_5NHNH_2 \cdot HCl$.

Observable Characteristics - *Physical State (as shipped)*: Solid; *Color*: White to tan; *Odor*: Weak aromatic.

Physical and Chemical Properties - *Physical State at 15 °C and 1 atm.*: Solid; *Molecular Weight*: 144.6; *Boiling Point at 1 atm.*: Not pertinent (decomposes); *Freezing Point*: 469, 243, 516; *Critical Temperature*: Not pertinent; *Critical Pressure*: Not pertinent; *Specific Gravity*: >1 at 20 °C (solid); *Vapor (Gas) Specific Gravity*: Not pertinent; *Ratio of Specific Heats of Vapor (Gas)*: Not pertinent; *Latent Heat of Vaporization*: Not pertinent; *Heat of Combustion*: Not pertinent; *Heat of Decomposition*: Not pertinent.

Health Hazards Information - *Recommended Personal Protective Equipment*: Dust respirator, rubber gloves; goggles; *Symptoms Following Exposure*: Inhalation of dust irritates nose and throat; fumes from hot material may cause same symptoms as ingestion. Phenylhydrazine is a chronic poison; ingestion can cause jaundice, anorexia, nausea, and vascular thrombosis; may also cause anemia and liver injury. Contact with eyes causes irritation. Contact with skin causes irritation and dermatitis; *General Treatment for Exposure*: INHALATION: move to fresh air; get medical attention. INGESTION: give large amounts of water; induce vomiting; get medical attention. EYES: flush with water for at least 15 min.; if exposure is prolonged or repeated, get medical attention. SKIN: flush with water; *Toxicity by Inhalation (Threshold Limit Value)*: Data not available; *Short-Term Inhalation Limits*: Data not available; *Toxicity by Ingestion*: Data not available; *Late Toxicity*: Causes tumors in mice; *Vapor (Gas) Irritant Characteristics*: Data not available; *Liquid or Solid Irritant Characteristics*: Data not available; *Odor Threshold*: Data not available.

Fire Hazards - *Flash Point :* Not pertinent (combustible solid); *Flammable Limits in Air (%):* Not pertinent; *Fire Extinguishing Agents*: Water, foam, dry chemical, carbon dioxide; *Fire Extinguishing Agents Not to be Used:* Data not available; *Special Hazards of Combustion Products*: Toxic and irritating oxides of nitrogen and hydrogen chloride may form in fire; *Behavior in Fire*: The solid may sublime without melting and deposit on cool surfaces; *Ignition Temperature* : Data not available; *Electrical Hazard*: Not pertinent; *Burning Rate*: Not pertinent.

Chemical Reactivity - *Reactivity with Water*: No reaction; *Reactivity with Common Materials*: May be corrosive to metals; *Stability During Transport*: Stable; *Neutralizing Agents for Acids and Caustics:* Not pertinent; *Polymerization:* Not pertinent; *Inhibitor of Polymerization*: Not pertinent.

PHOSGENE

Chemical Designations - *Synonyms*: Carbonyl chloride; Chloroformyl chloride; *Chemical Formula*: $COCl_2$.

Observable Characteristics - *Physical State (as shipped)*: Compressed gas; *Color*: Colorless; *Odor*: Sharp, pungent odor in higher concentrations; like new-mown grass in low concentrations.

Physical and Chemical Properties - *Physical State at 15 °C and 1 atm.*: Gas; *Molecular Weight*: 98.92; *Boiling Point at 1 atm.*: 46.8, 8.2, 281.4; *Freezing Point*: -195, -126, 147; *Critical Temperature*: 360, 182, 455; *Critical Pressure*: 823, 56.0, 5.67; *Specific Gravity*: 1.38 at 20 °C (liquid); *Vapor (Gas) Specific Gravity*: 3.4; *Ratio of Specific Heats of Vapor (Gas)*: 1.170; *Latent Heat of Vaporization*: 110, 59, 2.5; *Heat of Combustion*: Not pertinent; *Heat of Decomposition*: Not pertinent.

Health Hazards Information - *Recommended Personal Protective Equipment*: Approved by U.S. Bureau of Mines respirator; protective clothing; *Symptoms Following Exposure*: Irrigates lungs, causing delayed pulmonary edema. Slight gassing produces dryness or burning sensation in the throat, numbers, pain in the chest, bronchitis, and shortness of breath; *General Treatment for Exposure*: INHALATION: remove victim from contaminated area; enforce absolute rest; call a doctor; *Toxicity by Inhalation (Threshold Limit Value)*: 0.1 ppm; *Short-Term Inhalation Limits*: 1 ppm for 5 min.; *Toxicity by Ingestion*: Data not available; *Late Toxicity*: Severe delayed pulmonary edema; *Vapor (Gas) Irritant Characteristics*: Vapors cause severe irritation of eyes and throat and can cause eye and lung injury. They cannot be tolerated even at low concentrations; *Liquid or Solid Irritant Characteristics*: Severe irritant to all tissues; *Odor Threshold*: 0.5 ppm.

Fire Hazards - *Flash Point :* Not flammable; *Flammable Limits in Air (%):* Not flammable; *Fire Extinguishing Agents*: Water to cool containers; *Fire Extinguishing Agents Not to be Used:* Not pertinent; *Special Hazards of Combustion Products*: Toxic gas is generated when heated; *Behavior in Fire*: Not pertinent; *Ignition Temperature :* Not flammable; *Electrical Hazard*: Not pertinent; *Burning Rate*: Not flammable.

Chemical Reactivity - *Reactivity with Water*: Decomposes, but not vigorously; *Reactivity with Common Materials*: No reaction; *Stability During Transport*: Stable; *Neutralizing Agents for Acids and Caustics:* Can be absorbed in caustic soda solution. One ton of phosgene requires 2,480 lbs. of caustic soda dissolved in 1000 gal. of water; *Polymerization:* Not pertinent; *Inhibitor of Polymerization*: Not pertinent.

PHOSPHORIC ACID

Chemical Designations - *Synonyms*: Orthophosphoric acid; *Chemical Formula*: H_3PO_4.

Observable Characteristics - *Physical State (as shipped)*: Liquid; *Color*: Colorless; *Odor*: Odorless.

Physical and Chemical Properties - *Physical State at 15 °C and 1 atm.*: Liquid; *Molecular Weight*: 98.00; *Boiling Point at 1 atm.*: >266, >130, >403; *Freezing Point*: Not pertinent; *Critical Temperature*: Not pertinent; *Critical Pressure*: Not pertinent; *Specific Gravity*: 1.892 at 25 °C (liquid); *Vapor (Gas) Specific Gravity*: Not pertinent; *Ratio of Specific Heats of Vapor (Gas)*: Not pertinent; *Latent Heat of Vaporization*: Not pertinent; *Heat of Combustion*: Not pertinent; *Heat of Decomposition*: Not pertinent.

Health Hazards Information - *Recommended Personal Protective Equipment*: Goggles or face shield; rubber gloves and protective clothing; *Symptoms Following Exposure*: Burns on mouth and lips, sour acrid taste, severe gastrointestinal irritation, nausea, vomiting, bloody diarrhea, difficult swallowing, severe abdominal pains, thirst, acidemia, difficult breathing, convulsion, collapse, shock, death; *General Treatment for Exposure*: INGESTION: do NOT induce vomiting; give water, milk, or vegetable oil. SKIN OR CONTACT: flush with water for at least 15 min.; *Toxicity by Inhalation (Threshold Limit Value)*: 1.0 mg/m³; *Short-Term Inhalation Limits*: Not pertinent; *Toxicity by Ingestion*: Grade 3, LD_{50} 50 to 500 mg/kg; *Late Toxicity*: None; *Vapor (Gas) Irritant Characteristics*: Not volatile; *Liquid or Solid Irritant Characteristics*: Fairly severe skin irritant; may cause pain and second-degree burns after a few minutes contact; *Odor Threshold*: Not pertinent.

Fire Hazards - *Flash Point :* Not flammable; *Flammable Limits in Air (%):* Not flammable; *Fire

Extinguishing Agents: Not pertinent; *Fire Extinguishing Agents Not to be Used:* Not pertinent; *Special Hazards of Combustion Products*: Not pertinent; *Behavior in Fire*: Not pertinent; *Ignition Temperature* : Not flammable; *Electrical Hazard*: Not pertinent; *Burning Rate*: Not flammable.

Chemical Reactivity - *Reactivity with Water*: Reacts with water to generate heat and form phosphoric acid. The reaction is not violent; *Reactivity with Common Materials*: Reacts with metals to liberate flammable hydrogen gas; *Stability During Transport*: Stable; *Neutralizing Agents for Acids and Caustics:* Flush with water, neutralize acid with lime or soda ash; *Polymerization:* Not pertinent; *Inhibitor of Polymerization*: Not pertinent.

PHOSPHORUS OXYCHLORIDE

Chemical Designations - *Synonyms*: Phosphoryl chloride; *Chemical Formula*: $POCl_3$.

Observable Characteristics - *Physical State (as shipped)*: Liquid; *Color*: Colorless to pale yellow; *Odor*: Pungent and musty; disagreeable and lingering.

Physical and Chemical Properties - *Physical State at 15 °C and 1 atm.*: Liquid; *Molecular Weight*: 153.33; *Boiling Point at 1 atm.*: 255, 107, 380; *Freezing Point*: 34, 1, 274; *Critical Temperature*: 630, 332, 605; *Critical Pressure*: Not pertinent; *Specific Gravity*: 1.675 at 20 °C (liquid); *Vapor (Gas) Specific Gravity*: Not pertinent; *Ratio of Specific Heats of Vapor (Gas)*: (est.) 1.290; *Latent Heat of Vaporization*: 97, 54, 2.3; *Heat of Combustion*: Not pertinent; *Heat of Decomposition*: Not pertinent.

Health Hazards Information - *Recommended Personal Protective Equipment*: Chemical safety goggles; face shield; self-contained or air-line respirator; hard hat; foot protection; rubber gloves and clothing; *Symptoms Following Exposure*: Vapor burns eyes and respiratory tract. Liquid is very corrosive to body tissues because of reaction with water to form hydrochloric and phosphoric acids; *General Treatment for Exposure*: *CAUTION: persons doing treatment should protect themselves against exposure.* INHALATION: remove victim from contaminated area at once; if breathing has stopped, start artificial respiration; call a doctor. INGESTION: give water or milk; do NOT induce vomiting. SKIN: remove contaminated clothing and flood exposed skin surfaces with water. EYES: retract eyelids and wash with water for at least 15 min.; call a doctor; *Toxicity by Inhalation (Threshold Limit Value)*: 0.5 ppm; *Short-Term Inhalation Limits*: Data not available; *Toxicity by Ingestion*: Grade 3, oral rat LD_{50} = 380 mg/kg; *Late Toxicity*: Data not available; *Vapor (Gas) Irritant Characteristics*: Vapor cause severe irritation of eyes and throat and can cause eye and lung injury. They cannot be tolerated even at low concentrations; *Liquid or Solid Irritant Characteristics*: Severe skin irritant. Causes second- and third-degree burns on short contact and is very injurious to the eyes; *Odor Threshold*: Data not available.

Fire Hazards - *Flash Point :* Not flammable; *Flammable Limits in Air (%):* Not flammable; *Fire Extinguishing Agents*: Sand and carbon dioxide on adjacent fires; *Fire Extinguishing Agents Not to be Used:* Water; *Special Hazards of Combustion Products*: Not pertinent; *Behavior in Fire*: Poisonous, corrosive, irritating gases are generated when heated or when in contact with water; *Ignition Temperature* : Not flammable; *Electrical Hazard*: Not pertinent; *Burning Rate*: Not flammable.

Chemical Reactivity - *Reactivity with Water*: Vigorous reaction with evolution of hydrogen chloride fumes; *Reactivity with Common Materials*: Corrosive to most metals except nickel and lead. Products of its reaction with water rapidly corrode steel and most metals with formation of flammable hydrogen gas; *Stability During Transport*: Stable; *Neutralizing Agents for Acids and Caustics:* Flush with water, neutralize acids formed with lime or soda ash; *Polymerization:* Not pertinent; *Inhibitor of Polymerization*: Not pertinent.

PHOSPHORUS PENTASULFIDE

Chemical Designations - *Synonyms*: Phosphoric sulfide; Phosphorus persulfide; Thiophosphoric anhydride; *Chemical Formula*: P_2S_5-P_4S_{10}.

Observable Characteristics - *Physical State (as shipped)*: Solid; *Color*: Light greenish yellow; greenish gray; *Odor*: Like rotten eggs. High (lethal) concentrations can paralyze the sense of smell.

Physical and Chemical Properties - *Physical State at 15 °C and 1 atm.*: Solid; *Molecular Weight*: 222.27; *Boiling Point at 1 atm.*: 957, 514, 787; *Freezing Point*: 527, 275, 548; *Critical Temperature*:

Not pertinent; *Critical Pressure*: Not pertinent; *Specific Gravity*: 2.03 at 20 °C (solid); *Vapor (Gas) Specific Gravity*: Not pertinent; *Ratio of Specific Heats of Vapor (Gas)*: Not pertinent; *Latent Heat of Vaporization*: 184, 102, 4.27; *Heat of Combustion*: -10.890, -6.050, -253.3; *Heat of Decomposition*: Not pertinent.

Health Hazards Information - *Recommended Personal Protective Equipment*: Chemical safety goggles; plastic face shielding; self-contained or air-line respirator; *Symptoms Following Exposure*: Hydrogen sulfide gas formed by reaction with moisture can cause death be respiratory failure. The gas also irritates eyes and respiratory system. The solid irritates skin and eyes; the symptoms may be delayed several hours; *General Treatment for Exposure*: INHALATION: remove victim from contaminated area; if breathing has stopped, begin artificial respiration. INGESTION: induce vomiting; call a physician. SKIN: remove contaminated clothing and wash areas with copious large amounts of water. EYES: flush with large amounts of water; *Toxicity by Inhalation (Threshold Limit Value)*: 10 ppm; *Short-Term Inhalation Limits*: 20 ppm. for 5 min. (hydrogen sulfide); *Toxicity by Ingestion*: Data not available; *Late Toxicity*: Data not available; *Vapor (Gas) Irritant Characteristics*: Hydrogen sulfide gas formed by reaction with moisture, cause severe irritation of eyes and throat and can cause eye and lung injury. It cannot be tolerated even at low concentrations; *Liquid or Solid Irritant Characteristics*: Minimum hazard. If spilled on clothing and allowed to remain, may cause smarting and reddening of skin; *Odor Threshold*: 0.0047 ppm. (hydrogen sulfide).

Fire Hazards - *Flash Point :* Flammable solid; *Flammable Limits in Air (%):* Not pertinent; *Fire Extinguishing Agents*: Sand and carbon dioxide; *Fire Extinguishing Agents Not to be Used:* Water; *Special Hazards of Combustion Products*: Products of combustion include sulfur dioxide and phosphorus pentoxide, which are irritating, toxic and corrosive; *Behavior in Fire*: Not pertinent; *Ignition Temperature (deg. F)*: 527 (liquid); *Electrical Hazard*: Not pertinent; *Burning Rate*: Not pertinent.

Chemical Reactivity - *Reactivity with Water*: Reacts with liquid water or atmospheric moisture to liberate toxic hydrogen sulfide gas; *Reactivity with Common Materials*: No reaction; *Stability During Transport*: Can be ignited by friction; *Neutralizing Agents for Acids and Caustics:* Not pertinent; *Polymerization:* Not pertinent; *Inhibitor of Polymerization*: Not pertinent.

PHOSPHORUS TRIBROMIDE

Chemical Designations - *Synonyms*: Phosphorus bromide; *Chemical Formula*: PBr_3.

Observable Characteristics - *Physical State (as shipped)*: Liquid; *Color*: Colorless or slightly yellow; *Odor*: Pungent; sharp, penetrating.

Physical and Chemical Properties - *Physical State at 15 °C and 1 atm.*: Liquid; *Molecular Weight*: 270.73; *Boiling Point at 1 atm.*: 343, 173, 446; *Freezing Point*: -42.9, -40.5, 232.2; *Critical Temperature*: Not pertinent; *Critical Pressure*: Not pertinent; *Specific Gravity*: 2.862 at 30 °C (liquid); *Vapor (Gas) Specific Gravity*: Not pertinent; *Ratio of Specific Heats of Vapor (Gas)*: Not pertinent; *Latent Heat of Vaporization*: 64.4, 35.8, 1.50; *Heat of Combustion*: Not pertinent; *Heat of Decomposition*: Not pertinent.

Health Hazards Information - *Recommended Personal Protective Equipment*: Acid-gas canister-type mask (full face type emergencies); chemical safety goggles; apron, gloves, clothing, and safety shoes all made from rubber; *Symptoms Following Exposure*: Inhalation causes severe irritation of nose, throat, and lungs. Ingestion causes burns of mouth and stomach. Contact with eyes or skin causes severe burns; *General Treatment for Exposure*: INHALATION: remove victim to clear air; if necessary, apply artificial respiration and/or administer oxygen. INGESTION: dilute by drinking water. Then neutralize with milk of magnesia, egg white, etc; do not use sodium bicarbonate. EYES: immediately flush with large amounts of water for at least 15 min. SKIN: immediately flush with large amounts of water; remove contaminated clothing; *Toxicity by Inhalation (Threshold Limit Value)*: Data not available; *Short-Term Inhalation Limits*: Data not available; *Toxicity by Ingestion*: Data not available; *Late Toxicity*: Data not available; *Vapor (Gas) Irritant Characteristics*: Data not available; *Liquid or Solid Irritant Characteristics*: Data not available; *Odor Threshold*: Data not available.

Fire Hazards - *Flash Point :* Not flammable; *Flammable Limits in Air (%):* Not flammable; *Fire

Extinguishing Agents: Not pertinent; *Fire Extinguishing Agents Not to be Used:* Do not use water on adjacent fire; *Special Hazards of Combustion Products*: Irritating hydrogen bromide and phosphoric acid vapors may form in fire; *Behavior in Fire*: Acids formed by reaction with water will attack metals and generate flammable hydrogen gas, which may form explosive mixtures in enclosed spaces; *Ignition Temperature* : Not pertinent; *Electrical Hazard*: Not pertinent; *Burning Rate*: Not pertinent.

Chemical Reactivity - *Reactivity with Water*: Reacts violently with water, evolving hydrogen bromide, an irritating and corrosive gas apparent as white fumes; *Reactivity with Common Materials*: In the presence of moisture, highly corrosive to most metals except lead and nickel; *Stability During Transport*: Unstable if heated; *Neutralizing Agents for Acids and Caustics:* Flush with water, rinse with dilute aqueous sodium bicarbonate or soda ash; *Polymerization:* Not pertinent; *Inhibitor of Polymerization*: Not pertinent.

PHOSPHORUS TRICHLORIDE

Chemical Designations - *Synonyms*: No common synonyms; *Chemical Formula*: PCl_3.

Observable Characteristics - *Physical State (as shipped)*: Liquid; *Color*: Colorless; *Odor*: Pungent, irritating, like hydrochloric acid.

Physical and Chemical Properties - *Physical State at 15 °C and 1 atm.*: Liquid; *Molecular Weight*: 137.33; *Boiling Point at 1 atm.*: 169, 76, 349; *Freezing Point*: -170, -112, 161; *Critical Temperature*: 547, 286, 559; *Critical Pressure*: Not pertinent; *Specific Gravity*: 1.575 at 20 °C (liquid); *Vapor (Gas) Specific Gravity*: 4.7; *Ratio of Specific Heats of Vapor (Gas)*: (est.) 1.290; *Latent Heat of Vaporization*: 95, 53, 2.2; *Heat of Combustion*: Not pertinent; *Heat of Decomposition*: Not pertinent.

Health Hazards Information - *Recommended Personal Protective Equipment*: Chemical safety goggles; plastic face shield; self-contained or air-line respirator; safety hat; rubber gloves and clothing; *Symptoms Following Exposure*: Vapors cause severe irritation of eyes and respiratory tract. Liquid burns eyes and skin; *General Treatment for Exposure*: CAUTION: *persons doing treatment should protect themselves against exposure.* INHALATION: remove victim from contaminated area at once; if breathing has stopped, start artificial respiration; call a doctor. INGESTION: if victim is conscious, give large quantities of water; do NOT induce vomiting. SKIN: remove contaminated clothing and flood exposed skin surfaces with water. EYES: retract eyelids and wash with water for at least 15 min.; call a doctor; *Toxicity by Inhalation (Threshold Limit Value)*: 0.5 ppm; *Short-Term Inhalation Limits*: Data not available; *Toxicity by Ingestion*: Grade 2, oral rat LD_{50} 550 mg/kg; *Late Toxicity*: None; *Vapor (Gas) Irritant Characteristics*: Vapor cause severe irritation of eyes and throat and can cause eye and lung injury. They cannot be tolerated even at low concentrations; *Liquid or Solid Irritant Characteristics*: Severe skin irritant. Causes second- and third-degree burns on short contact and is very injurious to the eyes; *Odor Threshold*: Data not available.

Fire Hazards - *Flash Point:* Not flammable; *Flammable Limits in Air (%):* Not flammable; *Fire Extinguishing Agents*: Sand, carbon dioxide and dry chemicals on adjacent fires; *Fire Extinguishing Agents Not to be Used*: Water; *Special Hazards of Combustion Products*: Not pertinent; *Behavior in Fire*: Generates toxic, irritating gases; *Ignition Temperature* : Not flammable; *Electrical Hazard*: Not pertinent; *Burning Rate*: Not flammable.

Chemical Reactivity - *Reactivity with Water*: Reacts violently and may cause flashes of fire. Hydrochloric acid fumes are formed in the reaction; *Reactivity with Common Materials*: Corrodes most common construction materials. Reacts with water to form hydrochloric acid, which reacts with most metals to form flammable hydrogen gas; *Stability During Transport*: Stable; *Neutralizing Agents for Acids and Caustics:* Flush with water; neutralize acids formed with lime or soda ash; *Polymerization:* Not pertinent; *Inhibitor of Polymerization*: Not pertinent.

PHOSPHORUS, WHITE

Chemical Designations - *Synonyms*: Yellow phosphorus; *Chemical Formula*: P.

Observable Characteristics - *Physical State (as shipped)*: Liquid; *Color*: Pale yellow to deep straw; *Odor*: Distinctive, disagreeable; pungent, sharp; like garlic.

Physical and Chemical Properties - *Physical State at 15 °C and 1 atm.*: Solid; *Molecular Weight*:

123.89; *Boiling Point at 1 atm.*: 535.5, 279.7, 552.9; *Freezing Point*: 111.4, 44.1, 317.3; *Critical Temperature*: Not pertinent; *Critical Pressure*: Not pertinent; *Specific Gravity*: 1.82 at 20 °C (solid); *Vapor (Gas) Specific Gravity*: Not pertinent; *Ratio of Specific Heats of Vapor (Gas)*: Not pertinent; *Latent Heat of Vaporization*: Not pertinent; *Heat of Combustion*: Not pertinent; *Heat of Decomposition*: Not pertinent.

Health Hazards Information - *Recommended Personal Protective Equipment*: Heavy rubber gloves and goggles or face shield; *Symptoms Following Exposure*: Solid or liquid causes severe burns of skin. If ingested, causes nausea, vomiting, jaundice, low blood pressure, depression, delirium, coma, death. Symptoms after ingestion may be delayed for from a few hours to 3 days; *General Treatment for Exposure*: INGESTION: if ingested, do NOT induce vomiting; call a doctor at once. SKIN OR EYE CONTACT: immediately flush with plenty of water for at least 15 min.; keep skin area wet until medical attention is obtained; *Toxicity by Inhalation (Threshold Limit Value)*: 0.1 mg/m^3; *Short-Term Inhalation Limits*: Data not available; *Toxicity by Ingestion*: Grade 4, LD$_{50}$ below 55 mg/kg; *Late Toxicity*: Severe attack of liver and bones; *Vapor (Gas) Irritant Characteristics*: Nonvolatile; *Liquid or Solid Irritant Characteristics*: Severe skin irritant. Causes second- and third-degree burns on short contact and is very injurious to the eyes; *Odor Threshold*: Data not available.

Fire Hazards - *Flash Point :* Ignites spontaneously in air; *Flammable Limits in Air (%):* Not pertinent; *Fire Extinguishing Agents*: Water; *Fire Extinguishing Agents Not to be Used:* Not pertinent; *Special Hazards of Combustion Products*: Fumes from burning phosphorus are highly irritating; *Behavior in Fire*: Intense white smoke is formed; *Ignition Temperature (deg. F)*: 86; *Electrical Hazard*: Not pertinent; *Burning Rate*: Not pertinent.

Chemical Reactivity - *Reactivity with Water*: No reaction; *Reactivity with Common Materials*: Ignites when exposed to air; *Stability During Transport*: Stable; *Neutralizing Agents for Acids and Caustics:* Not pertinent; *Polymerization:* Not pertinent; *Inhibitor of Polymerization*: Not pertinent.

PHTHALIC ANHYDRIDE

Chemical Designations - *Synonyms*: 1,2-Benzenedicarboxylic acid anhydride; 1,3-Dioxophthalan; PAN; Phthalandione; Phthalic acid anhydride; *Chemical Formula*: $C_6H_4(CO)_2O$.

Observable Characteristics - *Physical State (as shipped)*: Solid or liquid; *Color*: Colorless or pale yellow; *Odor*: Characteristic choking odor; choking, acrid.

Physical and Chemical Properties - *Physical State at 15 °C and 1 atm.*: Solid; *Molecular Weight*: 148.12; *Boiling Point at 1 atm.*: 544.3, 284.6, 557.8; *Freezing Point*: 268, 131, 404; *Critical Temperature*: Not pertinent; *Critical Pressure*: Not pertinent; *Specific Gravity*: 1.20 at 135 °C (liquid) 1.53 at 20 °C (solid); *Vapor (Gas) Specific Gravity*: Not pertinent; *Ratio of Specific Heats of Vapor (Gas)*: 1.080; *Latent Heat of Vaporization*: 189, 105, 4.40; *Heat of Combustion*: -9473, -526.3, -220.4; *Heat of Decomposition*: Not pertinent.

Health Hazards Information - *Recommended Personal Protective Equipment*: Coveralls and/or rubber apron; rubber shoes or boots; chemical goggles and/or face shield; Bureau of Mine organic vapor respiratory (Type AB); gauntlet-type leather or rubber gloves; *Symptoms Following Exposure*: Solid irritates skin and eyes, causing coughing and sneezing. Liquid causes severe thermal burns; *General Treatment for Exposure*: INHALATION: gargle with water and use a sedative cough mixture. INGESTION: induce vomiting and give water, milk, or vegetable oil. SKIN OR EYE CONTACT: Flush with water for at least 15 min.; if burned by molten material, remove as much solid as possible, soak off the remainder in cold water, and then treat the burn; *Toxicity by Inhalation (Threshold Limit Value)*: 2 ppm; *Short-Term Inhalation Limits*: 4 ppm for 5 min.; *Toxicity by Ingestion*: Grade 2, LD$_{50}$ 0.5-5 g/kg (rat); *Late Toxicity*: Data not available; *Vapor (Gas) Irritant Characteristics*: Vapor is moderately irritating such that personnel will not usually tolerate moderate or high vapor concentrations; *Liquid or Solid Irritant Characteristics*: Causes smarting of the skin and first-degree burns on short exposure; may cause secondary burns on long exposure; *Odor Threshold*: Data not available.

Fire Hazards - *Flash Point (deg. F):* 329 OC; 305 CC; *Flammable Limits in Air (%):* 1.7 - 10.5; *Fire Extinguishing Agents*: Water fog, dry chemical, carbon dioxide, or foam; *Fire Extinguishing Agents*

Not to be Used: Water may cause frothing; *Special Hazards of Combustion Products*: Not pertinent; *Behavior in Fire*: Not pertinent; *Ignition Temperature (deg. F)*: 1058; *Electrical Hazard*: Not pertinent; *Burning Rate*: Data not available.

Chemical Reactivity - *Reactivity with Water*: Solid has very slow reaction; no hazard. Liquid spatters when in contact with water; *Reactivity with Common Materials*: No reaction; *Stability During Transport*: Stable; *Neutralizing Agents for Acids and Caustics:* Water and sodium bicarbonate; *Polymerization:* Not pertinent; *Inhibitor of Polymerization*: Not pertinent.

PIPERAZINE

**Chemical Designations - ** *Synonyms*: Diethylenediamine; Hexahydro-1,4-diazine; Hexahydropyrazine; Lumbrical; Piperazidine; Pyrazine hexahydride; *Chemical Formula*: $NHCH_2CH_2NHCH_2CH_2$.

**Observable Characteristics - ** *Physical State (as shipped)*: Solid; *Color*: White; *Odor*: Mild, amine-like.

**Physical and Chemical Properties - ** *Physical State at 15 °C and 1 atm.*: Solid; *Molecular Weight*: 86; *Boiling Point at 1 atm.*: 299, 148, 421; *Freezing Point*: 223, 106, 379; *Critical Temperature*: Not pertinent; *Critical Pressure*: Not pertinent; *Specific Gravity*: 1.1 at 20 °C (solid); *Vapor (Gas) Specific Gravity*: Not pertinent; *Ratio of Specific Heats of Vapor (Gas)*: Not pertinent; *Latent Heat of Vaporization*: Not pertinent; *Heat of Combustion*: -14.800, -8.200, -343; *Heat of Decomposition*: Not pertinent.

**Health Hazards Information - ** *Recommended Personal Protective Equipment*: Monogoggles or face shield; rubber gloves; dust mask; *Symptoms Following Exposure*: Inhalation of dust irritates nose and throat. Ingestion causes irritation of mouth and stomach; has been known to cause severe allergic reaction. Contact with eyes causes burns. Repeated contact with skin bay cause irritation and sensitization; *General Treatment for Exposure*: INHALATION: move to fresh air. INGESTION: give large amount of water; induce vomiting; get medical attention. EYES: flush with plenty of water for at least 15 min.; get medical attention. SKIN: wash with soap and water; *Toxicity by Inhalation (Threshold Limit Value)*: Data not available; *Short-Term Inhalation Limits*: Data not available; *Toxicity by Ingestion*: Grade 2, LD_{50} 0.5-5 g/kg; *Late Toxicity*: Data not available; *Vapor (Gas) Irritant Characteristics*: Data not available; *Liquid or Solid Irritant Characteristics*: Data not available; *Odor Threshold*: Data not available.

**Fire Hazards - ** *Flash Point (deg. F):* 225 OC (molten solid); *Flammable Limits in Air (%):* Not pertinent; *Fire Extinguishing Agents*: Water, dry chemical, "alcohol" foam, carbon dioxide; *Fire Extinguishing Agents Not to be Used:* Water may cause frothing; *Special Hazards of Combustion Products*: Toxic oxides of nitrogen may form in fires; *Behavior in Fire*: Data not available; *Ignition Temperature (deg. F)*: 851; *Electrical Hazard*: Not pertinent; *Burning Rate*: Not pertinent.

**Chemical Reactivity - ** *Reactivity with Water*: No reaction; *Reactivity with Common Materials*: May be corrosive to aluminum, magnesium and zinc; *Stability During Transport*: Stable; *Neutralizing Agents for Acids and Caustics:* Flush with water; *Polymerization:* Not pertinent; *Inhibitor of Polymerization*: Not pertinent.

POLYBUTENE

**Chemical Designations - ** *Synonyms*: Butene resins; Polyisobutylene plastics; Polyisobutylene resins; Polyisobutylene waxes; *Chemical Formula*: $C(CH_3)_2CH_2$.

**Observable Characteristics - ** *Physical State (as shipped)*: Liquid; *Color*: Data not available; *Odor*: Data not available.

**Physical and Chemical Properties - ** *Physical State at 15 °C and 1 atm.*: Liquid; *Molecular Weight*: 225-230; *Boiling Point at 1 atm.*: Very high; *Freezing Point*: Not pertinent; *Critical Temperature*: Not pertinent; *Critical Pressure*: Not pertinent; *Specific Gravity*: 0.81-0.91 at 15 °C (liquid); *Vapor (Gas) Specific Gravity*: Not pertinent; *Ratio of Specific Heats of Vapor (Gas)*: Not pertinent; *Latent Heat of Vaporization*: Not pertinent; *Heat of Combustion*: (est.) -20.000, -11.000, -470; *Heat of Decomposition*: Not pertinent.

**Health Hazards Information - ** *Recommended Personal Protective Equipment*: Goggles or face shield;

Symptoms Following Exposure: Low toxicity. Vapor may act as a simple asphyxiant in high concentrations; *General Treatment for Exposure*: INHALATION: remove victim from exposure; *Toxicity by Inhalation (Threshold Limit Value)*: Data not available; *Short-Term Inhalation Limits*: Data not available; *Toxicity by Ingestion*: Grade 0, LD_{50} above 15 g/kg; *Late Toxicity*: None; *Vapor (Gas) Irritant Characteristics*: Vapors are nonirritating to the eyes and throat; *Liquid or Solid Irritant Characteristics*: No appreciable hazard. Practically harmless to the skin; *Odor Threshold*: Data not available.

Fire Hazards - *Flash Point (deg. F)*: 215 - 470 OC; *Flammable Limits in Air (%)*: Data not available; *Fire Extinguishing Agents*: Carbon dioxide, dry chemical, or foam; *Fire Extinguishing Agents Not to be Used*: Water may be ineffective; *Special Hazards of Combustion Products*: Not pertinent; *Behavior in Fire*: Not pertinent; *Ignition Temperature* : Data not available; *Electrical Hazard*: Not pertinent; *Burning Rate*: Data not available.

Chemical Reactivity - *Reactivity with Water*: No reaction; *Reactivity with Common Materials*: No reaction; *Stability During Transport*: Stable; *Neutralizing Agents for Acids and Caustics*: Not pertinent; *Polymerization*: Not pertinent; *Inhibitor of Polymerization*: Not pertinent.

POLYCHLORINATED BIPHENYL

Chemical Designations - *Synonyms*: Aroclor; Chlorinated biphenyl; Halogenated waxes; PCB; Polychloropolyphenyls; *Chemical Formula*: $(C_{12}H_{10-x})Cl_x$.

Observable Characteristics - *Physical State (as shipped)*: Liquid or solid; *Color*: Pale yellow (liquid); colorless (solid); *Odor*: Practically odorless.

Physical and Chemical Properties - *Physical State at 15 °C and 1 atm.*: Solid or liquid; *Molecular Weight*: Not pertinent; *Boiling Point at 1 atm.*: Very high; *Freezing Point*: Not pertinent; *Critical Temperature*: Not pertinent; *Critical Pressure*: Not pertinent; *Specific Gravity*: 1.3 - 1.8 at 20 °C (liquid); *Vapor (Gas) Specific Gravity*: Not pertinent; *Ratio of Specific Heats of Vapor (Gas)*: Not pertinent; *Latent Heat of Vaporization*: Not pertinent; *Heat of Combustion*: Not pertinent; *Heat of Decomposition*: Not pertinent.

Health Hazards Information - *Recommended Personal Protective Equipment*: Gloves and protective garments; *Symptoms Following Exposure*: Acne from skin contact; *General Treatment for Exposure*: SKIN: wash with soap and water; *Toxicity by Inhalation (Threshold Limit Value)*: 0.5 to 1.0 mg/m^3; *Short-Term Inhalation Limits*: Data not available; *Toxicity by Ingestion*: Grade 2, oral rat LD_{50} = 3980 mg/kg; *Late Toxicity*: Causes chromosomal abnormalities in rats, birth defects in birds; *Vapor (Gas) Irritant Characteristics*: Vapors causes severe irritation of the eyes and throat and cause eye and lung injury. They cannot be tolerated even at low concentrations; *Liquid or Solid Irritant Characteristics*: Contact with skin may cause irritation; *Odor Threshold*: Data not available.

Fire Hazards - *Flash Point (deg. F)*: >286; *Flammable Limits in Air (%)*: Data not available; *Fire Extinguishing Agents*: Water, foam, dry chemical, or carbon dioxide; *Fire Extinguishing Agents Not to be Used*: Not pertinent; *Special Hazards of Combustion Products*: Irritating gases are generating in fires; *Behavior in Fire*: Not pertinent; *Ignition Temperature* : Data not available; *Electrical Hazard*: Not pertinent; *Burning Rate*: Data not available.

Chemical Reactivity - *Reactivity with Water*: No reaction; *Reactivity with Common Materials*: No reaction; *Stability During Transport*: Stable; *Neutralizing Agents for Acids and Caustics*: Not pertinent; *Polymerization*: Not pertinent; *Inhibitor of Polymerization*: Not pertinent.

POLYMETHYLENE POLYPHENYL ISOCYANATE

Chemical Designations - *Synonyms*: PAPI; *Chemical Formula*: $C_6H_4(NCO)CH_2C_6H_4(NCO)$-and polymer.

Observable Characteristics - *Physical State (as shipped)*: Liquid; *Color*: Dark brown; *Odor*: Very weak.

Physical and Chemical Properties - *Physical State at 15 °C and 1 atm.*: Liquid; *Molecular Weight*: 400 (approx.); *Boiling Point at 1 atm.*: 392, 200, 473; *Freezing Point*: Not pertinent; *Critical*

Temperature: Not pertinent; *Critical Pressure*: Not pertinent; *Specific Gravity*: 1.20 at 20 °C (liquid); *Vapor (Gas) Specific Gravity*: Not pertinent; *Ratio of Specific Heats of Vapor (Gas)*: Not pertinent; *Latent Heat of Vaporization*: Not pertinent; *Heat of Combustion*: (est.) -13.000, -7.200, -300; *Heat of Decomposition*: Not pertinent.

Health Hazards Information - *Recommended Personal Protective Equipment*: Air-line or organic canister mask; goggles or face shield; rubber gloves and other protective clothing to prevent contact with skin; *Symptoms Following Exposure*: Inhalation causes breathless, chest discomfort, and reduces pulmonary function; wheezing, cough, and sputum may also occur. Contact with liquid irritates eyes and skin. Ingestion causes irritation of mouth and stomach; *General Treatment for Exposure*: Get *medical attention at once following all exposures to this compound*. INHALATION: remove victim to fresh air; give artificial respiration if breathing has stopped; oxygen can be given by qualified personnel. EYES: immediately wash with large amounts of water for at least 15 min. SKIN: flush immediately with water, wipe off, treat with 30% isopropyl alcohol rubbing alcohol), and wash with soap and water. INGESTION: induce vomiting at least 3 times by giving warm salt water (one tablespoon of salt per cup); follow with a quart of milk and a mild cathartic such as milk of magnesia; *Toxicity by Inhalation (Threshold Limit Value)*: 0.02 ppm; *Short-Term Inhalation Limits*: Data not available; *Toxicity by Ingestion*: Grade 1, LD_{50} 5 to 15 g/kg; *Late Toxicity*: Data not available; *Vapor (Gas) Irritant Characteristics*: Vapors are moderately irritating such that personnel will not usually tolerate moderate or high concentrations; *Liquid or Solid Irritant Characteristics*: Causes smarting of the skin and first-degree burns on short exposure; may cause second-degree burns on long exposure; *Odor Threshold*: Data not available.

Fire Hazards - *Flash Point (deg. F):* 425 OC; *Flammable Limits in Air (%):* Not pertinent; *Fire Extinguishing Agents*: Dry chemical or carbon dioxide; *Fire Extinguishing Agents Not to be Used:* Not pertinent; *Special Hazards of Combustion Products*: Not pertinent; *Behavior in Fire*: Containers may explode; *Ignition Temperature* : Data not available; *Electrical Hazard*: Not pertinent; *Burning Rate*: Data not available.

Chemical Reactivity - *Reactivity with Water*: Reacts slowly, forming heavy scum and liberating carbon dioxide gas. Dangerous pressure can build up if container is sealed; *Reactivity with Common Materials*: No hazardous reaction unless confined and wet; *Stability During Transport*: Stable if kept sealed and dry; *Neutralizing Agents for Acids and Caustics:* Not pertinent; *Polymerization:* Not pertinent; *Inhibitor of Polymerization*: Not pertinent.

POLYPROPYLENE

Chemical Designations - *Synonyms*: Propene polymer; *Chemical Formula*: $CH(CH_3)-CH_{2n}$.

Observable Characteristics - *Physical State (as shipped)*: Solid; *Color*: Tan or white; *Odor*: None.

Physical and Chemical Properties - *Physical State at 15 °C and 1 atm.*: Solid; *Molecular Weight*: Mixture; *Boiling Point at 1 atm.*: Not pertinent (decomposes); *Freezing Point*: Not pertinent; *Critical Temperature*: Not pertinent; *Critical Pressure*: Not pertinent; *Specific Gravity*: 0.90 at 20 °C (solid); *Vapor (Gas) Specific Gravity*: Not pertinent; *Ratio of Specific Heats of Vapor (Gas)*: Not pertinent; *Latent Heat of Vaporization*: Not pertinent; *Heat of Combustion*: (est.) -19.600, -10.900, -456; *Heat of Decomposition*: Not pertinent.

Health Hazards Information - *Recommended Personal Protective Equipment*: Filter respirator; *Symptoms Following Exposure*: No apparent toxicity; *General Treatment for Exposure*: None required; *Toxicity by Inhalation (Threshold Limit Value)*: Data not available; *Short-Term Inhalation Limits*: Data not available; *Toxicity by Ingestion*: Data not available; *Late Toxicity*: Causes central; *Vapor (Gas) Irritant Characteristics*: Data not available; *Liquid or Solid Irritant Characteristics*: Data not available; *Odor Threshold*: Data not available.

Fire Hazards - *Flash Point (deg. F)*: -162 CC; *Flammable Limits in Air (%):* 2.0 - 11; *Fire Extinguishing Agents:* Stop the flow of gas; *Fire Extinguishing Agents Not To Be Used:* Not pertinent; *Special Hazards of Combustion Products:* Not pertinent; *Behavior in Fire:* Containers may explode. Vapors are heavier than air and can travel to a source of ignition and flash back; *Ignition Temperature (deg. F):* 927; *Electrical Hazard:* Class I, Group D; *Burning Rate:* 8 mm/min as liquid.

Chemical Reactivity - *Reactivity with Water:* No reaction; *Reactivity with Common Materials:* No reactions; *Stability During Transport:* Stable; *Neutralizing Agents for Acids and Caustics:* Not pertinent; *Polymerization:* Not pertinent; *Inhibitor of Polymerization:* Not pertinent.

POLYPROPYLENE GLYCOL

Chemical Designations - *Synonyms*: Polyoxipropylene glycol; Polyoxpropylene ether, PPG; Pluracol polyol; Polyproplene glycols P400 to P4000; Thanol PPG; *Chemical Formula*: $HOCH(CH3)CH2O[CH2CH(CH3)O]_n$-H n averages 2-34.

Observable Characteristics - *Physical State (as shipped)*: Liquid; *Color*: Colorless; clear, lightly colored; *Odor*: None; slight sweet; faint ether-like.

Physical and Chemical Properties - *Physical State at 15 °C and 1 atm.*: Liquid; *Molecular Weight*: Variable — 200 to 2000; *Boiling Point at 1 atm.*: Not pertinent (decomposes); *Freezing Point*: -22 to -58, -30 to -50, -243 to 223; *Critical Temperature*: Not pertinent; *Critical Pressure*: Not pertinent; *Specific Gravity*: 1.012 at 20 °C (liquid); *Vapor (Gas) Specific Gravity*: Not pertinent; *Ratio of Specific Heats of Vapor (Gas)*: Not pertinent; *Latent Heat of Vaporization*: Not pertinent; *Heat of Combustion*: -14.200, -7.900, -330; *Heat of Decomposition*: Not pertinent.

Health Hazards Information - *Recommended Personal Protective Equipment*: Safety glasses or face shield; rubber gloves; *Symptoms Following Exposure*: The compound has a very low toxicity; few, if any, symptoms will be observed. Contact of liquid with eyes causes slight transient pain and irritation similar to that caused by a mild soap; *General Treatment for Exposure*: EYES: flush with water until mild irritation is gone; *Toxicity by Inhalation (Threshold Limit Value)*: Data not available; *Short-Term Inhalation Limits*: Not pertinent; *Toxicity by Ingestion*: (depends on molecular wt.) Grade 2, oral LD_{50} = 2.150-5 mg/kg (rat); Grade 1, LD_{50} 5 to 15 g/kg; Grade 1, LD_{50} >15 g/kg; *Late Toxicity*: Data not available; *Vapor (Gas) Irritant Characteristics*: Data not available; *Liquid or Solid Irritant Characteristics*: Data not available; *Odor Threshold*: Not pertinent.

Fire Hazards - *Flash Point (deg. F)*: 210 CC, 225 OC; *Flammable Limits in Air (%):* 2.6 - 12.5; *Fire Extinguishing Agents:* Water fog, alcohol foam, carbon dioxide, or dry chemical; *Fire Extinguishing Agents Not To Be Used:* Not pertinent; *Special Hazards of Combustion Products:* Not pertinent; *Behavior in Fire:* Not pertinent; *Ignition Temperature (deg. F):* 790; *Electrical Hazard:* Not pertinent; *Burning Rate:* 1.5 mm/min.

Chemical Reactivity - *Reactivity with Water:* No reaction; *Reactivity with Common Materials:* No reactions; *Stability During Transport:* Stable; *Neutralizing Agents for Acids and Caustics:* Not pertinent; *Polymerization:* Not pertinent; *Inhibitor of Polymerization:* Not pertinent.

POTASSIUM, METALLIC

Chemical Designations - *Synonyms*: No common synonyms; *Chemical Formula*: K.

Observable Characteristics - *Physical State (as shipped)*: Solid; *Color*: Silvery white; *Odor*: None.

Physical and Chemical Properties - *Physical State at 15 °C and 1 atm.*: Solid; *Molecular Weight*: 39; *Boiling Point at 1 atm.*: 1,425, 774, 1,0047; *Freezing Point*: 145, 63, 336; *Critical Temperature*: Not pertinent; *Critical Pressure*: Not pertinent; *Specific Gravity*: 0.86 at 20 °C (solid); *Vapor (Gas) Specific Gravity*: Not pertinent; *Ratio of Specific Heats of Vapor (Gas)*: Not pertinent; *Latent Heat of Vaporization*: Not pertinent; *Heat of Combustion*: -2,003, -1.113, -46.57; *Heat of Decomposition*: Not pertinent.

Health Hazards Information - *Recommended Personal Protective Equipment*: Goggles or face shield; rubber gloves; *Symptoms Following Exposure*: Contact with eyes or skin causes severe burns; *General Treatment for Exposure*: EYES OR SKIN: flush with water; treat caustic burns; *Toxicity by Inhalation (Threshold Limit Value)*: Data not available; *Short-Term Inhalation Limits*: Not pertinent; *Toxicity by Ingestion*: Not pertinent; *Late Toxicity*: Data not available; *Vapor (Gas) Irritant Characteristics*: Data not available; *Liquid or Solid Irritant Characteristics*: Data not available; *Odor Threshold*: Not pertinent.

Fire Hazards - *Flash Point :* Not pertinent. This is a combustible solid; *Flammable Limits in Air (%):* Not pertinent; *Fire Extinguishing Agents*: Graphite, sand, sodium chloride; *Fire Extinguishing Agents*

Not to be Used: Water, foam, carbon dioxide, or halogenated hydrocarbons; *Special Hazards of Combustion Products*: No data; *Behavior in Fire*: Reacts violently with water, forming flammable and explosive hydrogen gas. This product may spontaneously ignite in air; *Ignition Temperature* : No data; *Electrical Hazard*: Not pertinent; *Burning Rate*: Not pertinent.

Chemical Reactivity - *Reactivity with Water*: Reacts violently forming flammable hydrogen gas and a strong caustic solution; *Reactivity with Common Materials*: May ignite combustible materials if they are damp or moist; *Stability During Transport*: Stable if protected from air and moisture; *Neutralizing Agents for Acids and Caustics:* Caustic that is formed by the reaction with water should be flushed with water and then can be rinsed with dilute acetic acid solution; *Polymerization:* Not pertinent; *Inhibitor of Polymerization*: Not pertinent.

POTASSIUM ARSENATE

Chemical Designations - *Synonyms*: Macquer's salt; *Chemical Formula*: KH_2AsO_4

Observable Characteristics - *Physical State (as shipped)*: Solid; *Color*: White; *Odor*: None.

Physical and Chemical Properties - *Physical State at 15 °C and 1 atm.*: Solid; *Molecular Weight*: 180.0; *Boiling Point at 1 atm.*: Not pertinent (decomposes); *Freezing Point*: 550, 288, 561; *Critical Temperature*: Not pertinent; *Critical Pressure*: Not pertinent; *Specific Gravity*: 2.8 at 20 °C (solid); *Vapor (Gas) Specific Gravity*: Not pertinent; *Ratio of Specific Heats of Vapor (Gas)*: Not pertinent; *Latent Heat of Vaporization*: Not pertinent; *Heat of Combustion*: Not pertinent; *Heat of Decomposition*: Not pertinent.

Health Hazards Information - *Recommended Personal Protective Equipment*: Dust respirator; rubber gloves; *Symptoms Following Exposure*: Dust may irritates eyes. Ingestion or severe exposure by inhalation can cause burning of throat and mouth, abdominal pain, vomiting, diarrhea with hemorrhage, dehydration, jaundice, and collapse; *General Treatment for Exposure*: EYES: flush with water to remove dust. INGESTION: immediately induce evacuation of intestinal tract by inducing vomiting, giving gastric lavage and saline cathartic; see physician at once; consider possible development of arsenic poisoning; *Toxicity by Inhalation (Threshold Limit Value)*: 0.5 mg/m^3; *Short-Term Inhalation Limits*: Data not available; *Toxicity by Ingestion*: Data not available; *Late Toxicity*: May be carcinogenic; arsenic poisoning may develop; *Vapor (Gas) Irritant Characteristics*: Data not available; *Liquid or Solid Irritant Characteristics*: Data not available; *Odor Threshold*: Not pertinent.

Fire Hazards - *Flash Point :* Not flammable; *Flammable Limits in Air (%):* Not flammable; *Fire Extinguishing Agents*: Not pertinent; *Fire Extinguishing Agents Not to be Used:* Not pertinent; *Special Hazards of Combustion Products*: Not pertinent; *Behavior in Fire*: Not pertinent; *Ignition Temperature* : Not pertinent; *Electrical Hazard*: Not pertinent; *Burning Rate*: Not pertinent.

Chemical Reactivity - *Reactivity with Water*: No reaction; *Reactivity with Common Materials*: No reactions; *Stability During Transport*: Stable; *Neutralizing Agents for Acids and Caustics:* Not pertinent; *Polymerization:* Not pertinent; *Inhibitor of Polymerization*: Not pertinent.

POTASSIUM BINOXALATE

Chemical Designations - *Synonyms*: Potassium acid oxalate; Salt acetosella; Salt of sorrel; *Chemical Formula*: KHC_2O_4.

Observable Characteristics - *Physical State (as shipped)*: Solid; *Color*: White; *Odor*: None.

Physical and Chemical Properties - *Physical State at 15 °C and 1 atm.*: Solid; *Molecular Weight*: 128.11; *Boiling Point at 1 atm.*: Not pertinent (decomposes); *Freezing Point*: Not pertinent; *Critical Temperature*: Not pertinent; *Critical Pressure*: Not pertinent; *Specific Gravity*: 2.0 at 20 °C (solid); *Vapor (Gas) Specific Gravity*: Not pertinent; *Ratio of Specific Heats of Vapor (Gas)*: Not pertinent; *Latent Heat of Vaporization*: Not pertinent; *Heat of Combustion*: Not pertinent; *Heat of Decomposition*: Not pertinent.

Health Hazards Information - *Recommended Personal Protective Equipment*: Dust mask; goggles or face shield; protective gloves; *Symptoms Following Exposure*: Inhalation of dust causes irritation of nose and throat. Ingestion causes burning pain in throat, esophagus, and stomach; exposed areas of mucous membrane turn white; vomiting, severe purging, weak pulse, and cardiovascular collapse; if death is delayed, neuromuscular symptoms develop. Contact with dust irritates eyes and may cause mild irritation of skin; *General Treatment for Exposure*: INHALATION: move to fresh air; if exposure to dust is severe, get medical attention. INGESTION: give immediately by mouth a dilute solution of any soluble calcium salt (calcium lactate, lime water, chalk solution; or even milk); large amounts of Ca are required; administer gastric lavage with dilute lime water consult physician; watch for edema of the glottis and delayed constriction of esophagus. EYES: flush with water for at least 15 min.; *Toxicity by Inhalation (Threshold Limit Value)*: Data not available; *Short-Term Inhalation Limits*: Data not available; *Toxicity by Ingestion*: Grade 3, LD_{50} 50-500 mg/kg; *Late Toxicity*: Data not available; *Vapor (Gas) Irritant Characteristics*: Data not available; *Liquid or Solid Irritant Characteristics*: Data not available; *Odor Threshold*: Odorless.

Fire Hazards - *Flash Point :* Not flammable; *Flammable Limits in Air (%):* Not flammable; *Fire Extinguishing Agents*: Not pertinent; *Fire Extinguishing Agents Not to be Used:* Not pertinent; *Special Hazards of Combustion Products*: Not pertinent; *Behavior in Fire*: Not pertinent; *Ignition Temperature* : Not pertinent; *Electrical Hazard*: Not pertinent; *Burning Rate*: Not pertinent.

Chemical Reactivity - *Reactivity with Water*: Below 50 °C product dissolves in water and reacts to form the precipitate potassium tetraoxalate; *Reactivity with Common Materials*: No reactions; *Stability During Transport*: Stable; *Neutralizing Agents for Acids and Caustics:* Not pertinent; *Polymerization:* Not pertinent; *Inhibitor of Polymerization*: Not pertinent.

POTASSIUM CYANIDE

Chemical Designations - *Synonyms*: Cyanide; *Chemical Formula*: KCN.

Observable Characteristics - *Physical State (as shipped)*: Solid; *Color*: White; *Odor*: Like hydrogen cyanide; almond-like.

Physical and Chemical Properties - *Physical State at 15 °C and 1 atm.*: Solid; *Molecular Weight*: 65.12; *Boiling Point at 1 atm.*: Very high; *Freezing Point*: 1174, 634.5, 907.7; *Critical Temperature*: Not pertinent; *Critical Pressure*: Not pertinent; *Specific Gravity*: 1.52 at 16 °C (solid); *Vapor (Gas) Specific Gravity*: Not pertinent; *Ratio of Specific Heats of Vapor (Gas)*: Not pertinent; *Latent Heat of Vaporization*: Not pertinent; *Heat of Combustion*: Not pertinent; *Heat of Decomposition*: Not pertinent.

Health Hazards Information - *Recommended Personal Protective Equipment*: Wear dry cotton gloves and U.S. Bureau of Mines approved dust respirator when handling solid potassium cyanide. Wear rubber gloves and approved Chemical safety goggles when handling solutions; *Symptoms Following Exposure*: Is rapidly fatal poison when taken into the digestive system. Dust may cause toxic symptoms when inhaled, and prolonged contact with the skin may cause irritation and possibly poisoning if skin is broken. Strong solutions are corrosive to skin and may cause deep ulcers that heal slowly; *General Treatment for Exposure*: INGESTION: call physician immediately; have victim lie down and keep him quiet and warm. If he is conscious, induce vomiting by having him drink warm salt water (1 tablespoon per cup of water); repeat until vomit fluid is clear; then give orally 1 pint of 1% solution of sodium thiosulfate, to be repeated in 15 min. Victim is not breathing, give artificial respiration until breathing starts. If victim is unconscious but breathing, give oxygen from an inhalator if he does not respond to treatment. In all cases, break an amyl nitrite pearl in a cloth and hold lightly under victim's nose for 15 sec., repeating 5 times at about 15-sec. intervals; if necessary, repeat procedure every 3 min. with fresh pearls until 3 or 4 have been used. *Amyl nitrite pearls must not be over 2 years old. Avoid breathing the vapor while administering it to the victim*; *Toxicity by Inhalation (Threshold Limit Value)*: Not pertinent; *Short-Term Inhalation Limits*: Data not available; *Toxicity by Ingestion*: Grade 4, LD_{50} below 50 mg/kg (mice); *Late Toxicity*: None; *Vapor (Gas) Irritant Characteristics*: Non-volatile, but moisture in air can liberate come lethal hydrogen cyanide gas; *Liquid or Solid Irritant Characteristics*: Moist solid can cause caustic-type irritation of skin and formation of ulcers; *Odor Threshold*: Not pertinent.

Fire Hazards - *Flash Point :* Not flammable; *Flammable Limits in Air (%):* Not flammable; *Fire Extinguishing Agents*: Not pertinent; *Fire Extinguishing Agents Not to be Used:* Not pertinent; *Special Hazards of Combustion Products*: Not pertinent; *Behavior in Fire*: Not flammable; *Ignition Temperature* : Not flammable; *Electrical Hazard*: Not pertinent; *Burning Rate*: Not flammable.

Chemical Reactivity - *Reactivity with Water*: When potassium cyanide dissolves in water, a mild reaction occurs and poisonous hydrogen cyanide gas is released. The gas readily dissipates, however if it collects in a confined space, then workers may be exposed to toxic levels. If the water is acidic, toxic amounts of the gas will form instantly; *Reactivity with Common Materials*: Contact with even weak acids will result in the formation of deadly hydrogen cyanide gas; *Stability During Transport*: Stable; *Neutralizing Agents for Acids and Caustics:* Not pertinent; *Polymerization:* Not pertinent; *Inhibitor of Polymerization*: Not pertinent.

POTASSIUM DICHLORO-S-TRIAZINETRIONE
Chemical Designations - *Synonyms*: Potassium; *Chemical Formula*: $Kcl_2(NCO)_3$.
Observable Characteristics - *Physical State (as normally shipped)*: Solid; *Color*: White; *Odor*: Like chlorine.
Physical and Chemical Properties - *Physical State at 15 °C and 1 atm.*: Solid; *Molecular Weight*: 236,1; *Boiling Point at 1 atm.*: Not pertinent (decomposes); *Freezing Point*: Not pertinent; *Critical Temperature*: Not pertinent; *Critical Pressure*: Not pertinent; *Specific Gravity*: 0.96 at 20 °C (solid); *Vapor (Gas) Density*: Not pertinent; *Ratio of Specific Heats of Vapor (Gas)*: Not pertinent; *Latent Heat of Vaporization*: Not pertinent; *Heat of Combustion*: Not pertinent; *Heat of Decomposition*: Not pertinent.
Health Hazards Information - *Recommended Personal Protective Equipment*: Dust mask or chorine canister mask; goggles; rubber gloves and other protective clothing to prevent contact with skin; *Symptoms Following Exposure*: Dust causes sneezing; is moderately irritating to the eyes and causes itching and redness of skin. Ingestion causes burn of mouth and stomach; *General Treatment for Exposure:* INHALATION: remove victim to fresh air. EYES: irrigate with running water fir 15 min.; call a physician. SKIN: flush with water. INGESTION: induce vomiting and call physician *Toxicity by Inhalation (Threshold Limit Value):* Not pertinent; *Short-Term Inhalation Limits:* Not pertinent; *Toxicity by Ingestion:* Grade 3;50 to 500 mg/kg (human); *Late Toxicity:* Some suggestion of lung cancer; *Vapor (Gas) Irritant Characteristics:* Dust or mists may severe irritation of eyes and throat and can cause eye and lung injury. They cannot be tolerated even at low concentrations.; *Liquid or Solid Irritant Characteristics:* Severe skin irritant; causes second- and third-degree burns on short contact and is very injurious to the eyes; *Odor Threshold:* Not pertinent.
Fire Hazards - *Flash Point :* Not flammable but may cause fires upon contact with ordinary combustibles; *Flammable Limits in Air (%):* Not pertinent; *Fire Extinguishing Agents*: Water; *Fire Extinguishing Agents Not to be Used:* Not pertinent; *Special Hazards of Combustion Products*: May form toxic chlorine and other gases in fires; *Behavior in Fire*: Decomposition can be initiated with a heat source and can propagate throughout the mass with the evolution of dense fumes. Containers may also explode when exposed to the heat from adjacent fires; *Ignition Temperature :* Not pertinent; *Electrical Hazard*: Not pertinent; *Burning Rate*: Not pertinent.
Chemical Reactivity - *Reactivity with Water*: A non violent reaction occurs resulting in the formation of a bleach solution; *Reactivity with Common Materials*: Contact with most foreign materials, organic matter, or easily chlorinated or oxidized materials may result in fire. Avoid contact with oils, greases, sawdust, floor sweepings, and other easily oxidized organic compounds; *Stability During Transport*: Stable if kept dry; *Neutralizing Agents for Acids and Caustics:* Not pertinent; *Polymerization:* Not pertinent; *Inhibitor of Polymerization*: Not pertinent.

POTASSIUM DICHROMATE
Chemical Designations - *Synonyms*: Potassium bichromate, Bichromate; *Chemical Formula*: $K_2Cr_2O_7$.
Observable Characteristics - *Physical State (as normally shipped)*: Solid; *Color*: Orange-red; *Odor*: Odorless.
Physical and Chemical Properties - *Physical State at 15 °C and 1 atm.*: Solid; *Molecular Weight*: 294,19; *Boiling Point at 1 atm.*: Decomposes; *Freezing Point*: 748, 398, 671; *Critical Temperature*: Not pertinent; *Critical Pressure*: Not pertinent; *Specific Gravity*: 2.676 at 25 °C (solid); *Vapor (Gas) Density*: Not pertinent; *Ratio of Specific Heats of Vapor (Gas)*: Not pertinent; *Latent Heat of*

Vaporization: Not pertinent; *Heat of Combustion*: Not pertinent; *Heat of Decomposition*: Not pertinent.
Health Hazards Information - *Recommended Personal Protective Equipment*: Approved dust mask; protective gloves; goggles or face shield; *Symptoms Following Exposure*: Highly corrosive to skin and mucous membranes. If ingested, causes violent gastroenteritis, peripheral vascular collapse, vertigo, muscle cramps, coma, and (later) toxic nephritis with glycosuria. Allergic reaction may also occur; *General Treatment for Exposure*: INGESTION: have victim drink water or milk; do NOT induce vomiting. SKIN: treat like acid burns; external lesions may be scrubbed with 2% solution of sodium thiosulfate; *Toxicity by Inhalation (Threshold Limit Value)*: Not pertinent; *Short-Term Exposure Limits*: Not pertinent; *Toxicity by Ingestion*: Grade 3;50 to 500 mg/kg (human); *Late Toxicity*: Some suggestion of lung cancer; *Vapor (Gas) Irritant Characteristics*: Dust or mists may severe irritation of eyes and throat and can cause eye and lung injury. They cannot be tolerated even at low concentrations; *Liquid or Solid Irritant Characteristics:* Severe skin irritant; causes second- and third-degree burns on short contact and is very injurious to the eyes; *Odor Threshold*: Not pertinent.
 Fire Hazards - *Flash Point :* Not flammable; *Flammable Limits in Air (%):* Not flammable; *Fire Extinguishing Agents*: Flood the spill area with water; *Fire Extinguishing Agents Not to be Used:* Not pertinent; *Special Hazards of Combustion Products*: Not pertinent; *Behavior in Fire*: May decompose, generating oxygen and hence supports the combustion of other materials; *Ignition Temperature :* Not flammable; *Electrical Hazard*: Not pertinent; *Burning Rate*: Not pertinent.
Chemical Reactivity - *Reactivity with Water*: No reaction; *Reactivity with Common Materials*: Ignition may occur when the product is in contact with finely divided combustibles, such as sawdust; *Stability During Transport*: Stable; *Neutralizing Agents for Acids and Caustics:* Not pertinent; *Polymerization:* Not pertinent; *Inhibitor of Polymerization*: Not pertinent.

POTASSIUM HYDROXIDE
Chemical Designations - *Synonyms:* Caustic potash; *Chemical Formula:* KOH.
Observable Characteristics - *Physical State (as normally shipped):* Solid; *Color:* White; *Odor:* None.
Physical and Chemical Properties - *Physical State at 15 °C and 1 atm:* Solid; *Molecular Weight:* 56,11; *Boiling Point at 1 atm:* Very high; *Freezing Point:* 716, 380, 653; *Critical Temperature:* Not pertinent; *Critical Pressure:* Not pertinent; *Specific Gravity:* 2,04 at 15°C (solid); *Vapor (Gas) Density:* Not pertinent; *Ratio of Specific Heats of Vapor (Gas):* Not pertinent; *Latent Heat of Vaporization:* Not pertinent; *Heat of Combustion:* Not pertinent; *Heat of Decomposition:* Not pertinent.
Health Hazards Information - *Recommended Personal Protective Equipment:* Wide brimmed hat and close-fitting safety goggles with rubber side; respirator for dust; long-sleeved cotton shirt or jacket with buttoned collar and buttoned sleeves; rubber or rubber-coated canvas gloves (shift sleeves should be buttoned over the gloves); rubber shoes or boots; cotton coveralls (with trouser cuffs worn over boots); rubber apron.; *Symptoms Following Exposure:* Causes severe burn of eyes, skin, and mucous membranes.; *General Treatment for Exposure:* (Act quickly!) Call a physician at once, even when injury seems to be slight. INGESTION: give water and milk; do NOT induce vomiting. EYES: flush with water at once for at least 15 min. SKIN: flush with water, then rinse with dilute vinegar; *Toxicity by Inhalation (Threshold Limit Value):* Not pertinent; *Short-Term Exposure Limits:* Not pertinent; *Toxicity by Ingestion:* Grade 3; oral rat LD_{50} = 364 mg/kg; *Late Toxicity:* None; *Vapor (Gas) Irritant Characteristics:* Not pertinent; *Liquid or Solid Irritant Characteristics:* Severe skin irritant; causes second- and third-degree burns on short contact and is very injurious to the eyes; *Odor Threshold:* Not pertinent.
Fire Hazards - *Flash Point :* Not flammable; *Flammable Limits in Air (%):* Not flammable; *Fire Extinguishing Agents*: Not pertinent; *Fire Extinguishing Agents Not to be Used:* Not pertinent; *Special Hazards of Combustion Products*: Not pertinent; *Behavior in Fire*: Not pertinent; *Ignition Temperature :* Not flammable; *Electrical Hazard*: Not pertinent; *Burning Rate*: Not flammable.
Chemical Reactivity - *Reactivity with Water*: Dissolves with the liberation of much heat. Steam and violent agitation can be observed in the reaction; *Reactivity with Common Materials*: When wet, this material attacks metals such as aluminum, tin, lead, and zinc, producing flammable hydrogen gas; *Stability During Transport*: Stable; *Neutralizing Agents for Acids and Caustics:* Flush with water and

rinse with a dilute solution of acetic acid; *Polymerization:* Not pertinent; *Inhibitor of Polymerization*: Not pertinent.

POTASSIUM IODIDE
Chemical Designations - *Synonyms:* No common synonyms; *Chemical Formula:* KI.
Observable Characteristics - *Physical State (as normally shipped):* Solid; *Color:* White; *Odor:* Odorless.
Physical and Chemical Properties - *Physical State at 15 °C and 1 atm:* Solid; *Molecular Weight:* 166,01; *Boiling Point at 1 atm:* Very high; *Freezing Point:* 1258, 681, 954; *Critical Temperature:* Not pertinent; *Critical Pressure:* Not pertinent; *Specific Gravity:* 3,13 at 15°C (solid); *Vapor (Gas) Density:* Not pertinent; *Ratio of Specific Heats of Vapor (Gas):* Not pertinent; *Latent Heat of Vaporization:* Not pertinent; *Heat of Combustion:* Not pertinent; *Heat of Decomposition:* Not pertinent.
Health Hazards Information - *Recommended Personal Protective Equipment:* Goggles or face shield; *Symptoms Following Exposure:* May irritate eyes or open cuts; *General Treatment for Exposure:* Flush all affected areas with water; *Toxicity by Inhalation (Threshold Limit Value):* Not pertinent; *Short-Term Exposure Limits:* Not pertinent; *Toxicity by Ingestion:* Grade 3; oral rat LD_{50} = 364 mg/kg; *Late Toxicity:* Data not available; *Vapor (Gas) Irritant Characteristics:* Non-volatile; *Liquid or Solid Irritant Characteristics:* None; *Odor Threshold:* Not pertinent.
Fire Hazards - *Flash Point :* Not flammable; *Flammable Limits in Air (%):* Not flammable; *Fire Extinguishing Agents*: Not pertinent; *Fire Extinguishing Agents Not to be Used:* Not pertinent; *Special Hazards of Combustion Products*: Not pertinent; *Behavior in Fire*: Not pertinent; *Ignition Temperature* : Not flammable; *Electrical Hazard*: Not pertinent; *Burning Rate*: Not flammable.
Chemical Reactivity - *Reactivity with Water*: No reaction; *Reactivity with Common Materials*: Corrosive in all concentrations to most metals, except stainless steels, titanium, and tantalum; *Stability During Transport*: Stable; *Neutralizing Agents for Acids and Caustics:* Not pertinent; *Polymerization:* Not pertinent; *Inhibitor of Polymerization*: Not pertinent.

POTASSIUM OXALATE
Chemical Designations - *Synonyms*: Potassium oxalate monohydrate; *Chemical Formula*: $K_2C_2O_4 \cdot H_2O$.
Observable Characteristics - *Physical State (as normally shipped)*: Solid; *Color*: Colorless; White; *Odor*: None.
Physical and Chemical Properties - *Physical State at 15 °C and 1 atm.*: Solid; *Molecular Weight*: 184.24; *Boiling Point at 1 atm.*: Not pertinent (decomposes); *Freezing Point*: Not pertinent; *Critical Temperature*: Not pertinent; *Critical Pressure*: Not pertinent; *Specific Gravity*: 2.13 at 18.5 °C (solid); *Vapor (Gas) Density*: Not pertinent; *Ratio of Specific Heats of Vapor (Gas)*: Not pertinent; *Latent Heat of Vaporization*: Not pertinent; *Heat of Combustion*: Not pertinent; *Heat of Decomposition*: Not pertinent.
Health Hazards Information - *Recommended Personal Protective Equipment*: Approved dust respirator, chemical goggles: rubber or plastic-coated gloves; *Symptoms Following Exposure*: Inhalation of dust can cause systemic poisoning. Ingest ion causes burning pain in throat, esophagus, and stomach; exposed areas of mucous membrane turn white; vomiting, severe purging, weak pulse, and cardiovascular collapse may result; if death is delayed, neuromuscular symptoms develop. Contact with eyes or skin causes irritation; *General Treatment for Exposure*: Act promptly! INHALATION: remove victim to fresh air; if exposure to dust is severe, get medical attention. INGESTION: call physician immediately: have victim drink dilute calcium lactate, lime water, chalk soln, or even milk; large amounts of calcium arc required; administer gastric lavage with dilute lime water; watch for edema of the glottis and delayed constriction of esophagus. EYES: flush with water and seek medical attention. SKIN: flush with; *Toxicity by Inhalation (Threshold Limit Value)*: Data not available; *Short-Term Exposure Limits*: Data not available; *Toxicity by Ingestion*: Grade 3, LD_{50} 50-500 mg/kg; *Late Toxicity*: Data not available; *Vapor (Gas) Irritant Characteristics*: Data not available; *Liquid or Solid Irritant Characteristics*: Data not available; *Odor Threshold*: Odorless.
Fire Hazards - *Flash Point :* Not flammable; *Flammable Limits in Air (%):* Not flammable; *Fire

Extinguishing Agents: Not pertinent; *Fire Extinguishing Agents Not to be Used:* Not pertinent; *Special Hazards of Combustion Products*: Not pertinent; *Behavior in Fire*: Loses water at about 160 °C and decomposes to carbonate with no charring. The reaction is considered non hazardous; *Ignition Temperature* : Not flammable; *Electrical Hazard*: Not pertinent; *Burning Rate*: Not flammable.

Chemical Reactivity - *Reactivity with Water*: No reaction; *Reactivity with Common Materials*: No reactions; *Stability During Transport*: Stable; *Neutralizing Agents for Acids and Caustics:* Not pertinent; *Polymerization:* Not pertinent; *Inhibitor of Polymerization*: Not pertinent.

POTASSIUM PERMANGANATE

Chemical Designations - *Synonyms:* No common synonyms; *Chemical Formula:* $KmnO_4$.

Observable Characteristics - *Physical State (as normally shipped):* Solid; *Color:* Dark purple or bronze-like; *Odor:* Odorless.

Physical and Chemical Properties - *Physical State at 15 °C and 1 atm:* Solid; *Molecular Weight:* 158,04; *Boiling Point at 1 atm:* Decomposes; *Freezing Point:* >464, >240, >513; *Critical Temperature:* Not pertinent; *Critical Pressure:* Not pertinent; *Specific Gravity:* 2,70 at 15°C (solid); *Vapor (Gas) Density:* Not pertinent; *Ratio of Specific Heats of Vapor (Gas):* Not pertinent; *Latent Heat of Vaporization:* Not pertinent; *Heat of Combustion:* Not pertinent; *Heat of Decomposition:* Not pertinent.

Health Hazards Information - *Recommended Personal Protective Equipment:* Goggles or face shield, rubber gloves; *Symptoms Following Exposure:* Burns and stains the dark brown. If ingested will cause severe distress of Castro-intestinal system. May be fatal if over 4 oz. Are consumed; *General Treatment for Exposure:* INGESTION: induce vomiting and with thorough gastric lavage, demulcent, glucose I.V., fluid therapy, and antibiotics. Tracheotomy may be lifesaving; *Toxicity by Inhalation (Threshold Limit Value):* Not pertinent; *Short-Term Exposure Limits:* Not pertinent; *Toxicity by Ingestion:* Grade 3; LD_{50} 50 to 500 mg/kg; *Late Toxicity:* None; *Vapor (Gas) Irritant Characteristics:* Non-volatile; *Liquid or Solid Irritant Characteristics:* Can burn skin if not flushed with water; *Odor Threshold:* Not pertinent.

Fire Hazards - *Flash Point :* Not flammable; *Flammable Limits in Air (%):* Not flammable; *Fire Extinguishing Agents*: Flood spill area with water; *Fire Extinguishing Agents Not to be Used:* Not pertinent; *Special Hazards of Combustion Products*: Not pertinent; *Behavior in Fire*: May cause fire on contact with combustibles. Also containers may e; *Ignition Temperature* : Not flammable; *Electrical Hazard*: Not pertinent; *Burning Rate*: Not flammable.

Chemical Reactivity - *Reactivity with Water*: No reaction; *Reactivity with Common Materials*: Attacks rubber and most fibrous materials. May cause ignition of organic materials such as wood. Some acids, such as sulfuric acid, may result in explosion; *Stability During Transport*: Stable; *Neutralizing Agents for Acids and Caustics:* Not pertinent; *Polymerization:* Not pertinent; *Inhibitor of Polymerization*: Not pertinent.

POTASSIUM PEROXIDE

Chemical Designations - *Synonyms:* Potassium superoxide; *Chemical Formula:* K_2O_2.

Observable Characteristics - *Physical State (as normally shipped):* Powder; *Color:* Yellow; *Odor:* None.

Physical and Chemical Properties - *Physical State at 15 °C and 1 atm:* Solid; *Molecular Weight:* 110; *Boiling Point at 1 atm:* Not pertinent (decomposes); *Freezing Point:* 914, 490, 763; *Critical Temperature:* Not pertinent; *Critical Pressure:* Not pertinent; *Specific Gravity:* >1 at 20°C (solid); *Vapor (Gas) Density:* Not pertinent; *Ratio of Specific Heats of Vapor (Gas):* Not pertinent; *Latent Heat of Vaporization:* Not pertinent; *Heat of Combustion:* Not pertinent; *Heat of Decomposition:* Not pertinent.

Health Hazards Information - *Recommended Personal Protective Equipment:* Dust mask; goggles or face shield; protective gloves; *Symptoms Following Exposure:* Inhalation causes respiratory irritation. Ingest ion causes severe burn of mouth and stomach. Contact with eyes or skin causes irritation and caustic burns; *General Treatment for Exposure:* INHALATION: remove from exposure; support respiration. INGESTION: give large amount of water; do NOT induce vomiting; get medical attention EYES: irrigate with large quantities of water for at least 15 min.; get medical attention for caustic burn. SKIN: flush with water; treat caustic burns; *Toxicity by Inhalation (Threshold Limit Value):* Data not available; *Short-Term Exposure Limits:* Data not available; *Toxicity by Ingestion:* Data not available; *Late Toxicity:* Data not available; *Vapor (Gas) Irritant Characteristics:* Data not available; *Liquid or Solid Irritant Characteristics:* Data not available; *Odor Threshold:* Data not available.

Fire Hazards - *Flash Point* : Not pertinent; *Flammable Limits in Air (%):* Not pertinent; *Fire Extinguishing Agents:* Flood with water from a protected area; *Fire Extinguishing Agents Not To Be Used:* A small amount of water may cause an explosion; *Special Hazards of Combustion Products:* No data; *Behavior in Fire:* Increases intensity of fire and can start fires when in contact with organic materials; *Ignition Temperature :* Not pertinent; *Electrical Hazard:* Not pertinent; *Burning Rate:* Not pertinent.

Chemical Reactivity - *Reactivity with Water:* Reacts violently with liberation of heat and oxygen and the formation of caustic solution; *Reactivity with Common Materials:* Forms explosive and self-igniting mixtures with wood and other combustible materials; *Stability During Transport:* Stable if kept dry; *Neutralizing Agents for Acids and Caustics:* Following the reaction with water, the caustic solution formed can be flushed away with water and area rinsed with dilute acetic acid; *Polymerization:* Not pertinent; *Inhibitor of Polymerization:* Not pertinent.

PROPANE
Chemical Designations - *Synonyms:* Dimethylmethane; *Chemical Formula:* $CH_3CH_2CH_3$.

Observable Characteristics - *Physical State (as normally shipped):* Liquefied compressed gas; *Color:* Colorless; *Odor:* Faint gassy.

Physical and Chemical Properties - *Physical State at 15 °C and 1 atm:* Gas; *Molecular Weight:* 44.09; *Boiling Point at 1 atm:* -49, -42.1, 231.1; *Freezing Point:* -305.9, -108.7, 85.5; *Critical Temperature:* -142.01, -96.67, 176.53; *Critical Pressure:* 616.5, 41.94, 4.249; *Specific Gravity:* 0.590 at -50°C; *Vapor (Gas) Density:* 1.5; *Ratio of Specific Heats of Vapor (Gas):* 1.130; *Latent Heat of Vaporization:* 183.2, 101.8, 4.262; *Heat of Combustion:* -19.782, -10.990, -460.13; *Heat of Decomposition:* Not pertinent.

Health Hazards Information - *Recommended Personal Protective Equipment:* Self-contained breathing apparatus for high concentrations of gas; *Symptoms Following Exposure:* Vaporizing liquid may cause frostbite. Concentrations in air greater than 10% cause dizziness in a few minutes. 1% concentrations give the effect in 10 min. High concentration cause asphyxiation; *General Treatment for Exposure:* Remove to open air. If victim is by gas apply artificial respiration. Guard against self-injury if confused; *Toxicity by Inhalation (Threshold Limit Value):* 1000 ppm; *Short-Term Exposure Limits:* Data not available; *Toxicity by Ingestion:* Not pertinent; *Late Toxicity:* None; *Vapor (Gas) Irritant Characteristics:* Vapors are nonirritating to the eyes and throat; *Liquid or Solid Irritant Characteristics:* No appreciable hazard. Practically harmless to the skin because it evaporates quickly; *Odor Threshold:* 5.000-20.000 ppm.

Fire Hazards - *Flash Point (deg. F):* -156 CC; *Flammable Limits in Air (%):* 2.1 - 9.5; *Fire Extinguishing Agents:* Stop the flow of gas. For small fires, use dry chemicals. Cool adjacent areas with water spray; *Fire Extinguishing Agents Not To Be Used:* Water; *Special Hazards of Combustion Products:* Not pertinent; *Behavior in Fire:* Containers may explode. Vapor is heavier than air and can travel considerable distances to a source of ignition and flash back; *Ignition Temperature (deg. F):* 842; *Electrical Hazard:* Class I, Group D; *Burning Rate:* 8.2 mm/min.

Chemical Reactivity - *Reactivity with Water:* No reaction; *Reactivity with Common* Materials: No reactions; *Stability During Transport:* Stable; *Neutralizing Agents for Acids and Caustics:* Not pertinent; *Polymerization:* Not pertinent; *Inhibitor of Polymerization:* Not pertinent.

BETA-PROPIOLACTONE
Chemical Designations - *Synonyms:* Betaprone; Hydracrylic acid, beta-lactone; 2-Oxetanone; Propanolide; beta-Propionolactone; *Chemical Formula:* OCH_2CH_2CO.

Observable Characteristics - *Physical State (as normally shipped):* Liquid; *Color:* Colorless; *Odor:* Pungent, acrylic; irrigating.

Physical and Chemical Properties - *Molecular Weight:* 72.1; *Boiling Point at 1 atm:* Not pertinent; *Freezing Point:* -28.1, -33.4, 239.8; *Critical Temperature:* Not pertinent; *Critical Pressure:* Not pertinent; *Specific Gravity:* 1.148 at 20°C; *Vapor (Gas) Density:* 2.5; *Ratio of Specific Heats of Vapor (Gas):* 1.1089; *Latent Heat of Vaporization:* Not pertinent; *Heat of Combustion:* -8.510, -4.730, -198; *Heat of Decomposition:* Not pertinent.

Health Hazards Information - *Recommended Personal Protective Equipment:* Air mask or organic canister mask; goggles or face shield; rubber gloves; protective clothing to prevent all contact with skin; *Symptoms Following Exposure:* Inhalation causes irritation of nose, throat, and respiratory tract. Contact of liquid with eyes causes irritation and tears. Contact with skin causes irritation and blistering. Ingestion causes burn of mouth and stomach; *General Treatment for Exposure:* Get medical attention following all exposures to this compound. INHALATION: move victim to fresh air; if breathing has stopped, give artificial respiration. EYES: flush continuously with water for at least 15 min. SKIN: flush with water; if blistering occurs, alert physician to fact than fluid from blister will cause additional blistering of adjacent skin. INGESTION: give large amount of water and induce vomiting; *Toxicity by Inhalation (Threshold Limit Value):* 0.07 ppm. Because of the high incidence of cancer, either in man or animal, no exposure or contact by any route — respiratory, oral, or skin — should be permitted; *Short-Term Exposure Limits:* Data not available; *Toxicity by Ingestion:* Grade 3; oral LDL_0 = 50 mg/kg (rat); *Late Toxicity:* Because of the high incidence of cancer, either in man or animal, no exposure or contact by any route — respiratory, oral, or skin — should be permitted; *Vapor (Gas) Irritant Characteristics:* Vapors are moderately irritating such that personnel will not usually tolerate moderate or high concentrations; *Liquid or Solid Irritant Characteristics:* Fairly severe skin irritant. May cause pain and second-degree burn after a few minutes' contact; *Odor Threshold:* Data not available.

Fire Hazards - *Flash Point (deg. F):* 165 CC; *Flammable Limits in Air (%):* 2.9 (LEL); *Fire Extinguishing Agents:* Water, dry chemical, foam, or carbon dioxide; *Fire Extinguishing Agents Not To Be Used:* Not pertinent; *Special Hazards of Combustion Products:* Vapors of unburned material are highly toxic; *Behavior in Fire:* Containers may explode; *Ignition Temperature :* No data; *Electrical Hazard:* No data; *Burning Rate:* No data.

Chemical Reactivity - *Reactivity with Water:* A slow, non-hazardous reaction occurs forming beta-hydroxypropionic acid; *Reactivity with Common Materials:* No reactions; *Stability During Transport:* Stable; *Neutralizing Agents for Acids and Caustics:* Not pertinent; *Polymerization:* Can polymerize and rupture containers especially at elevated temperatures. At 22 °C, approximately 0.04 % polymerizes per day; *Inhibitor of Polymerization:* None reported in the literature.

PROPIONALDEHYDE

Chemical Designations - *Synonyms:* Melhylacelaldehyde; Propaldehyde; Propanal; Propionic aldehyde; Propylaldehyde; *Chemical Formula:* CH_3CH_2CHO.

Observable Characteristics - *Physical State (as normally shipped):* Liquid; *Color:* Colorless; *Odor:* Pungent, unpleasant; suffocating.

Physical and Chemical Properties - *Physical State at 15 °C and 1 atm:* Liquid; *Molecular Weight:* 58.08; *Boiling Point at 1 atm:* 118.4, 48.08, 321.2; *Freezing Point:* -112, -80, 193; *Critical Temperature:* 433, 223, 496; *Critical Pressure:* 690, 47, 4.8; *Specific Gravity:* 0.805 at 20°C; *Vapor (Gas) Density:* 2.0; *Ratio of Specific Heats of Vapor (Gas):* 1.120; *Latent Heat of Vaporization:* 211, 117, 4.90; *Heat of Combustion:* -12.470, -6.930, -290.1,; *Heat of Decomposition:* Not pertinent.

Health Hazards Information - *Recommended Personal Protective Equipment:* Air-supplied mask for high vapor concentrations; plastic glove; goggles; *Symptoms Following Exposure:* Vapors will irritate nose and throat, and may cause nausea and vomiting. Liquid causes eye irritation; *General Treatment for Exposure:* INHALATION: remove victim to fresh air; give oxygen if breathing is difficult; call a physician. EYES: flush with plenty of water for at least 15 min., and call a physician. SKIN: flush with water; *Toxicity by Inhalation (Threshold Limit Value):* Data not available; *Short-Term Exposure Limits:* Data not available; *Toxicity by Ingestion:* Grade 3; oral LDL_0 = 50 mg/kg (rat); *Late Toxicity:* Data not available; *Vapor (Gas) Irritant Characteristics:* Vapors are moderate irritation, such that personnel will find high concentrations unpleasant. The effect is temporary; *Liquid or Solid Irritant Characteristics:* Minimum hazard. If spilled on clothing and allowed to remain, may cause smarting and reddening of the skin; *Odor Threshold:* 1 ppm.

Fire Hazards - *Flash Point (deg. F):* -22 OC; *Flammable Limits in Air (%):* 2.6 - 16.1; *Fire Extinguishing Agents:* On small fires use carbon dioxide or dry chemical. For large fires use alcohol type foam; *Fire Extinguishing Agents Not To Be Used:* Water may be ineffective; *Special Hazards of Combustion Products:* Not pertinent; *Behavior in Fire:* Vapor is heavier than air and may travel considerable distance to source of ignition and flash back; *Ignition Temperature (deg. F):* 405; *Electrical Hazard:* Not pertinent; *Burning Rate:* 4.4 mm/min.

Chemical Reactivity - *Reactivity with Water:* No reaction; *Reactivity with Common Materials:* No reactions; *Stability During Transport:* Stable; *Neutralizing Agents for Acids and Caustics:* Not pertinent; *Polymerization:* Polymerizes in the presence of acids and caustics; *Inhibitor of Polymerization:* Not pertinent.

PROPIONIC ANHYDRIDE
Chemical Designations - *Synonyms:* Methylacetic anhydride; Propanoic anhydride; Propionyl oxide; *Chemical Formula:* $(CH_3CH_2CO)_2O$.

Observable Characteristics - *Physical State (as normally shipped):* Liquid; *Color:* Colorless; *Odor:* Pungent.

Physical and Chemical Properties - *Physical State at 15 °C and 1 atm:* Liquid; *Molecular Weight:* 130.1; *Boiling Point at 1 atm:* 336, 169, 442; *Freezing Point:* -45, -43, 230; *Critical Temperature:* 660, 349, 622; *Critical Pressure:* 490, 33, 3.3; *Specific Gravity:* 1.01 at 20°C (liquid); *Vapor (Gas) Density:* 4.5; *Ratio of Specific Heats of Vapor (Gas):* 1.0543; *Latent Heat of Vaporization:* 149, 83, 3.5; *Heat of Combustion:* (at 15 °C) -10.320, -5.740, -240; *Heat of Decomposition:* Not pertinent.

Health Hazards Information - *Recommended Personal Protective Equipment:* Organic canister mask; goggles or face shield; rubber gloves; *Symptoms Following Exposure:* Inhalation causes irritation of eyes and respiratory tract. Contact with liquid causes burns of eyes and skin. Ingestion causes burns of mouth and stomach; *General Treatment for Exposure:* INHALATION: move victim to fresh air; if breathing has stopped, give artificial respiration. EYES: immediately flush with plenty of water for at least 15 min.; get medical attention. SKIN: immediately flush with plenty of water for at least 15 min. INGESTION: give large amount of water; do NOT induce vomiting; *Toxicity by Inhalation (Threshold Limit Value):* Data not available; *Short-Term Exposure Limits:* Data not available; *Toxicity by Ingestion:* Grade 1; LD_{50} 5 to 15 g/kg; Grade 2; LD_{50} 0.5 to 5 g/kg; *Late Toxicity:* Data not available; *Vapor (Gas) Irritant Characteristics:* Vapors are moderatory irritating such that personnel will not usually tolerate moderate or high concentrations; *Liquid or Solid Irritant Characteristics:* Causes smarting of the skin and first-degree burns on short exposure; my cause second-degree burns on long exposure; *Odor Threshold:* Data not available.

Fire Hazards - *Flash Point (deg. F):* 136 OC; 145 CC; *Flammable Limits in Air (%):* 1.48 - 11.9; *Fire Extinguishing Agents:* Water, dry chemical, alcohol foam, or carbon dioxide; *Fire Extinguishing Agents Not To Be Used:* Not pertinent; *Special Hazards of Combustion Products:* Not pertinent; *Behavior in Fire:* Not pertinent; *Ignition Temperature (deg. F):* 545; *Electrical Hazard:* No data; *Burning Rate:* 3.0 mm/min.

Chemical Reactivity - *Reactivity with Water:* Reacts slowly forming weak propionic acid. The reaction is non-violent and non-hazardous; *Reactivity with Common Materials:* Slowly forms a corrosive material if wet; *Stability During Transport:* Stable; *Neutralizing Agents for Acids and Caustics:* Flush with water and rinse with sodium bicarbonate or lime solution; *Polymerization:* Not pertinent; *Inhibitor of Polymerization:* Not pertinent.

N-PROPYL ACETATE
Chemical Designations - *Synonyms:* Acetic acid; Propyl ester; Methylacetic anhydride; Propanoic anhydride; Propionyl oxide; *Chemical Formula:* $CH_3COOCH_2CH_2CH_3$.

Observable Characteristics - *Physical State (as normally shipped)*: Liquid; *Color:* Colorless; *Odor:* Mild fruity.

Physical and Chemical Properties - *Physical State at 15 °C and 1 atm:* Liquid; *Molecular Weight:* 102.13; *Boiling Point at 1 atm:* 214.9, 101.6, 374.8; *Freezing Point:* -139, -95.0, 178.2; *Critical Temperature:* 529, 276, 549; *Critical Pressure:* 485, 33, 3.3; *Specific Gravity:* 0.886 at 20°C (liquid); *Vapor (Gas) Density:* Not pertinent; *Ratio of Specific Heats of Vapor (Gas):* 1.071; *Latent Heat of Vaporization:* 145, 80.3, 3.36; *Heat of Combustion:* (at 15 °C) -10.320, -5.740, -240; *Heat of Decomposition:* Not pertinent.

Health Hazards Information - *Recommended Personal Protective Equipment:* Air-supplied mask or chemical canister; goggles or face shield; protective gloves; *Symptoms Following Exposure:* Contact with skin and eyes causes no serious injury. High vapor concentrations will be irritating and will cause nausea, vomiting, and dizziness, with final loss of consciousness; *General Treatment for Exposure:* INHALATION: remove victim to fresh air; give artificial respiration if breathing has stopped; give oxygen if will breathing is difficult. SKIN AND EYES: flush with water; *Toxicity by Inhalation (Threshold Limit Value):* 200 ppm; *Short-Term Exposure Limits:* 200 ppm for 60 min.; *Toxicity by Ingestion:* Grade 2; LD_{50} 0.5 to 5 g/kg; *Late Toxicity:* None; *Vapor (Gas) Irritant Characteristics:* Vapors causes a slight smarting of the eyes or respiratory system if present in high concentrations. The effect is temporary; *Liquid or Solid Irritant Characteristics:* Minimum hazard. If spilled on clothing and allowed to remain, may cause smarting and reddening of the skin; *Odor Threshold:* 70 mg/m³.

Fire Hazards - *Flash Point (deg. F):* 58 CC, 65 OC; *Flammable Limits in Air (%):* 2.0 - 8.0; *Fire Extinguishing Agents:* For small fires use carbon dioxide or dry chemical. For large fires, use alcohol foam; *Fire Extinguishing Agents Not To Be Used:* Water may be ineffective; *Special Hazards of Combustion Products:* Not pertinent; *Behavior in Fire:* Not pertinent; *Ignition Temperature (deg. F):* 842; *Electrical Hazard:* Not pertinent; *Burning Rate:* No data.

Chemical Reactivity - *Reactivity with Water:* No reaction; *Reactivity with Common Materials:* No reactions; *Stability During Transport:* Stable; *Neutralizing Agents for Acids and Caustics:* Not pertinent; *Polymerization:* Not pertinent; *Inhibitor of Polymerization:* Not pertinent.

N-PROPYL ALCOHOL

Chemical Designations - *Synonyms:* Ethylcarbinol; l-Propanol; Propylalcohol; *Chemical Formula:* $CH_3CH_2CH_2OH$.

Observable Characteristics - *Physical State (as normally shipped):* Liquid; *Color:* Colorless; *Odor:* Resembles that of ethyl alcohol.

Physical and Chemical Properties - *Physical State at 15 °C and 1 atm:* Liquid; *Molecular Weight:* 60.10; *Boiling Point at 1 atm:* 207.0, 97.2, 370.4; *Freezing Point:* -195.2, -126.2, 147.0; *Critical Temperature:* 506.5, 263.6, 536.8; *Critical Pressure:* 750, 51, 5.2; *Specific Gravity:* 0.803 at 20°C (liquid); *Vapor (Gas) Density:* Not pertinent; *Ratio of Specific Heats of Vapor (Gas):* 2.1; *Latent Heat of Vaporization:* 292.7, 162.6, 6.808; *Heat of Combustion:* -13.130, -7.296, -305.5; *Heat of Decomposition:* Not pertinent.

Health Hazards Information - *Recommended Personal Protective Equipment:* Air-supplied respirator for high concentrations; goggles or face shield; plastic gloves; *Symptoms Following Exposure:* Contact with eyes extremely irritating and may cause burns. Vapors irritate nose and throat. In high concentrations, may cause nausea, dizziness, headache, and stupor; *General Treatment for Exposure:* INHALATION: remove victim to fresh air; call a physician. SKIN: OR EYE CONTACT: flush at once with plenty of water; get medical care for eyes; *Toxicity by Inhalation (Threshold Limit Value):* 200 ppm; *Short-Term Exposure Limits:* 400 ppm for 30 min.; *Toxicity by Ingestion:* Grade 2; LD_{50} 0.5 to 5 g/kg (rat); *Late Toxicity:* None; *Vapor (Gas) Irritant Characteristics:* Vapors causes a slight smarting of the eyes or respiratory system if present in high concentrations. The effect is temporary; *Liquid or Solid Irritant Characteristics:* No appreciable hazard. Practically harmless to the skin; *Odor Threshold:* 30 ppm.

Fire Hazards - *Flash Point (deg. F):* 81 OC, 77 CC; *Flammable Limits in Air (%):* 2.1 - 13.5; *Fire Extinguishing Agents:* Carbon dioxide for small fires, and alcohol foam for large fires; *Fire*

Extinguishing Agents Not To Be Used: Water may be ineffective; *Special Hazards of Combustion Products:* Not pertinent; *Behavior in Fire:* Not pertinent; *Ignition Temperature (deg. F):* 700; *Electrical Hazard:* Class I, Group D; *Burning Rate:* 2.9 mm/min.

Chemical Reactivity - *Reactivity with Water:* No reaction; *Reactivity with Common Materials:* No reactions; *Stability During Transport:* Stable; *Neutralizing Agents for Acids and Caustics:* Not pertinent; *Polymerization:* Not pertinent; *Inhibitor of Polymerization:* Not pertinent.

PROPYLENE OXIDE

Chemical Designations - *Synonyms:* 1,2-Epoxypropane; Methyloxirane propene oxide; *Chemical Formula:* CH_3CHCH_2O.

Observable Characteristics - Physical State (as normally shipped): Liquid; *Color:* Colorless; *Odor:* Ethereal; characteristic; sweet, alcoholyc; like natural gas.

Physical and Chemical Properties - *Physical State at 15 °C and 1 atm:* Liquid; *Molecular Weight:* 58.08; *Boiling Point at 1 atm:* 93.7, 34.3, 307.5; *Freezing Point:* -169.4, -11.9, 161.3; *Critical Temperature:* 408.4, 209.1 482.3; *Critical Pressure:* 714, 48.6, 4.92; *Specific Gravity:* 0.830 at 20°C; *Vapor (Gas) Density:* 2.0; *Ratio of Specific Heats of Vapor (Gas):* 1.133; *Latent Heat of Vaporization:* 205, 114, 4.77; *Heat of Combustion:* -13.000, -7.221, -302.3; *Heat of Decomposition:* Not pertinent.

Health Hazards Information - *Recommended Personal Protective Equipment:* Air-supplied mask; rubber or plastic gloves; vapor-proof goggles; *Symptoms Following Exposure:* Inhalation may introduce headache, nausea, vomiting, and unconsciousness; mild depression of central nervous system; lung irritation. Slightly irritating to skin, but covered contact may cause burn. Very irritating to eyes; *General Treatment for Exposure:* INHALATION: remove person to fresh air immediately, keep quit and warm; call a physician; if breathing stop, start artificial respiration. SKIN OR EYE CONTACT: immediately flush with plenty of water for at 15 min.; immediately remove contaminated clothing, watch bands, rings, etc. to prevent confiding product to skin; for eyes get medical attention; *Toxicity by Inhalation (Threshold Limit Value):* 100 ppm; *Short-Term Exposure Limits:* Data not available; *Toxicity by Ingestion:* Grade 2; LD_{50} 0.5 to 5 g/kg (rat); *Late Toxicity:* Data not available; *Vapor (Gas) Irritant Characteristics:* Vapors is moderatory irritating such that personnel will not short exposure; may cause secondary burn on long exposure; *Liquid or Solid Irritant Characteristics:* Causes smarting on the skin and first-degree burns on short exposure; may causes secondary burns on long exposure; *Odor Threshold:* 200 ppm.

Fire Hazards - *Flash Point (deg. F):* -35 CC, -20 OC; *Flammable Limits in Air (%):* 2.1 - 38.5; *Fire Extinguishing Agents:* Carbon dioxide or dry chemical for small fires. Alcohol or polymer foam for large fires.; *Fire Extinguishing Agents Not To Be Used:* Water may be ineffective; *Special Hazards of Combustion Products:* Not pertinent; *Behavior in Fire:* Containers may explode. Vapors are heavier than air and can travel to a source of ignition and flash back; *Ignition Temperature (deg. F):* 869; *Electrical Hazard:* Class I, Group B; *Burning Rate:* 3.3 mm/min.

Chemical Reactivity - *Reactivity with Water:* No reaction; *Reactivity with Common Materials:* No reactions; *Stability During Transport:* Stable; *Neutralizing Agents for Acids and Caustics:* Not pertinent; *Polymerization:* Polymerization can occur when this product is exposed to high temperatures or is contaminated with alkalies, aqueous acids, amines, and acidic alcohols; *Inhibitor of Polymerization:* Not pertinent.

PROPYLENE TETRAMER

Chemical Designations - *Synonyms:* Dodecene (nonelinear); Tetrapropylene; *Chemical Formula:* $CH_{12}H_{14.}$

Observable Characteristics - Physical State (as normally shipped): Liquid; *Color:* Colorless; *Odor:* Data not available.

Physical and Chemical Properties - *Physical State at 15 °C and 1 atm:* Liquid; *Molecular Weight:* 168.3; *Boiling Point at 1 atm:* 365-368, 185-196, 458-469; *Freezing Point:* Not pertinent; *Critical Temperature:* Data not available; *Critical Pressure:* Not pertinent; *Specific Gravity:* 0.2937 at 20°C (liquid); *Vapor (Gas) Density:* Not pertinent; *Ratio of Specific Heats of Vapor (Gas):* Not pertinent;

Latent Heat of Vaporization: 205, 58.6, 2.45; *Heat of Combustion:* -19.100, -10.600, -444; *Heat of Decomposition:* Not pertinent.

Health Hazards Information - *Recommended Personal Protective Equipment:* Goggles or face shield; *Symptoms Following Exposure:* No inhalation hazard expected. Aspiration hazard if ingested; *General Treatment for Exposure:* INHALATION: remove victim to fresh air. INGESTION: do NOT lavage or induce vomiting; give vegetable oil demulcent; call physician. EYES: flush with water for 15 min. SKIN: wash with soap and water; *Toxicity by Inhalation (Threshold Limit Value):* 200 ppm; *Short-Term Exposure Limits:* Data not available; *Toxicity by Ingestion:* Grade 0; LD_{50} above 15 g/kg; *Late Toxicity:* Data not available; *Vapor (Gas) Irritant Characteristics:* Vapors causes a slight smarting of the eyes or respiratory system if present in high concentrations. The effect is temporary; *Liquid or Solid Irritant Characteristics:* Minimum hazard. If spilled on clothing and allowed to remain, may cause smarting and reddening of skin; *Odor Threshold:* Data not available.

Fire Hazards - *Flash Point (deg. F):* 120 CC, 134 OC; *Flammable Limits in Air (%):* No data; *Fire Extinguishing Agents:* Water fog, foam, carbon dioxide or dry chemical; *Fire Extinguishing Agents Not To Be Used:* Not pertinent; *Special Hazards of Combustion Products:* Not pertinent; *Behavior in Fire:* Not pertinent; *Ignition Temperature (deg. F):* 400; *Electrical Hazard:* Not pertinent; *Burning Rate:* No data.

Chemical Reactivity - *Reactivity with Water:* No reaction; *Reactivity with Common Materials:* No reactions; *Stability During Transport:* Stable; *Neutralizing Agents for Acids and Caustics:* Not pertinent; *Polymerization:* Not pertinent; *Inhibitor of Polymerization:* Not pertinent.

N-PROPYL MERCAPTAN

Chemical Designations - *Synonyms*: 1-Propanethiol; Propane-1-thiol; *Chemical Formula*: $CH_3CH_2CH_2SH$.

Observable Characteristics - *Physical State (as normally shipped)*: Liquid; *Color*: Colorless; *Odor*: Skunky.

Physical and Chemical Properties - *Physical State at 15 °C and 1 atm.*: Liquid; *Molecular Weight*: 76.2; *Boiling Point at 1 atm.*: 153, 67, 340; *Freezing Point*:-171, -113, 160; *Critical Temperature*: 495, 257, 530; *Critical Pressure*: 667, 45.3, 4.60; *Specific Gravity*: at 20 °C (solid): 0.841 at 20°C; *Vapor (Gas) Density*: 2.6; *Ratio of Specific Heats of Vapor (Gas)*: 1.0984; *Latent Heat of Vaporization*: 179, 99, 4.16; *Heat of Combustion*: -15.990, -8.890, -372; *Heat of Decomposition*: Not pertinent.

Health Hazards Information - *Recommended Personal Protective Equipment*: Goggles or face shield; rubber gloves; self-contained breathing apparatus or organic canister mask; *Symptoms Following Exposure*: Inhalation causes muscular weakness, convulsion, and respiratory paralysis; high concentrations may cause pulmonary irritation. Contact with liquid causes irritation of eyes and skin. Ingestion causes irritation of mouth and stomach; *General Treatment for Exposure*: INHALATION: remove victim from contaminated atmosphere; give artificial respiratory paralysis; *Toxicity by Inhalation (Threshold Limit Value)*: Data not available; *Short-Term Exposure Limits*: Data not available; *Toxicity by Ingestion*: Grade 2, LD_{50} = 1.790 mg/kg; *Late Toxicity*: Data not available; *Vapor (Gas) Irritant Characteristics*: Data not available; *Liquid or Solid Irritant Characteristics*: Data not available; *Odor Threshold*: 0.00075 ppm.

Fire Hazards - *Flash Point (deg. F)*: 5 OC; *Flammable Limits in Air (%):* No data; *Fire Extinguishing Agents:* Dry chemical, foam, or carbon dioxide; *Fire Extinguishing Agents Not To Be Used:* Water may be ineffective; *Special Hazards of Combustion Products:* Toxic vapors of sulfur dioxide are generated; *Behavior in Fire:* Not pertinent; *Ignition Temperature :* No data; *Electrical Hazard:* No data; *Burning Rate:* 5.1 mm/min.

Chemical Reactivity - *Reactivity with Water:* No reaction; *Reactivity with Common Materials:* No reactions; *Stability During Transport:* Stable; *Neutralizing Agents for Acids and Caustics:* Not pertinent; *Polymerization:* Not pertinent; *Inhibitor of Polymerization:* Not pertinent.

PROPYLENEIMINE, INHIBITED

Chemical Designations - *Synonyms*: 2-Methylaziridine; 2-Methylethynimine; Propylenimine; *Chemical*

Formula: CH₃CHCH₂NH.

Observable Characteristics - *Physical State (as normally shipped)*: Liquid; *Color*: Colorless; *Odor*: Strong, ammonia-like.

Physical and Chemical Properties - *Physical State at 15 °C and 1 atm.*: Liquid; *Molecular Weight*: 57.1; *Boiling Point at 1 atm.*: 151, 66, 339; *Freezing Point*: Not pertinent; *Critical Temperature*: Not pertinent; *Critical Pressure*: Not pertinent; *Specific Gravity*: 0.802 at 25 °C (Liquid); *Vapor (Gas) Density*: 2; *Ratio of Specific Heats of Vapor (Gas)*: Data not available; *Latent Heat of Vaporization*: 250, 139, 5.82; *Heat of Combustion*: (est.) -15.500, -8.600, -360; *Heat of Decomposition*: Not pertinent.

Health Hazards Information - *Recommended Personal Protective Equipment*: Self-contained breathing apparatus; goggles or face shield; *Symptoms Following Exposure*: Inhalation causes vomiting, breathing difficulty, and irritation of eyes, nose, and throat; on prolonged exposure, vapors tend to redden the whites of the eyes. Contact with liquid causes eye irritation, like that caused by strong ammonia. Liquid causes skin burns, which are slow lo heal. Ingestion causes burns of mouth and stomach; *General Treatment for Exposure*: INHALATION: move victim to fresh air; if he is not breathing, apply artificial respiration, oxygen; if breathing is difficult, administer oxygen; call physician. EYES: flush with plenty of water for at least 30 min. and obtain prompt medical attention. SKIN: remove all contaminated clothing and flush with water; rinse with vinegar and water. INGESTION: drink large amounts of milk or water; get prompt medical attention; *Toxicity by Inhalation (Threshold Limit Value)*:2 ppm; *Short-Term Exposure Limits*: Data not available; *Toxicity by Ingestion*: Grade 4, oral LD₅₀ = 19 mg/kg (rat); *Late Toxicity*: Data not available; *Vapor (Gas) Irritant Characteristics*: Data not available; *Liquid or Solid Irritant Characteristics*: Data not available; *Odor Threshold*: Data not available.

Fire Hazards - *Flash Point (deg. F)*: 25 OC; *Flammable Limits in Air (%)*: No data; *Fire Extinguishing Agents*: Dry chemical or carbon dioxide; *Fire Extinguishing Agents Not To Be Used*: Water or foam may be ineffective; *Special Hazards of Combustion Products*: Irritating nitrogen oxides are generated in fires; *Behavior in Fire*: Containers may explode when exposed to heat; *Ignition Temperature*: No data; *Electrical Hazard*: No data; *Burning Rate*: 4.1 mm/min.

Chemical Reactivity - *Reactivity with Water*: A slow, non-hazardous reaction occurs, forming propanolamine; *Reactivity with Common Materials*: No reactions; *Stability During Transport*: The product is stable if it is kept in contact with solid caustic soda (sodium hydroxide); *Neutralizing Agents for Acids and Caustics*: Dilute with water and rinse with vinegar solution; *Polymerization*: This material will polymerize explosively when in contact with any acid; *Inhibitor of Polymerization*: Solid sodium hydroxide (caustic soda).

PYRIDINE

Chemical Designations - *Synonyms*: No common synonyms; *Chemical Formula*: C₅H₅N.

Observable Characteristics - *Physical State (as normally shipped)*: Liquid; *Color*: Yellow or colorless; *Odor*: Disagreeable: strong unpleasant, characteristic unpleasant: sharp penetrating, unpleasant.

Physical and Chemical Properties - *Physical State at 15 °C and 1 atm.*: Liquid; *Molecular Weight*: 79.1; *Boiling Point at 1 atm.*: 239.5, 115.3, 388.5; *Freezing Point*: -44, -42, 231; *Critical Temperature*: 656.2, 346.8, 620; *Critical Pressure*: 817.3, 55.6, 5.63; *Specific Gravity*: 0.983 at 20 °C (solid); *Vapor (Gas) Density*: 2.73; *Ratio of Specific Heats of Vapor (Gas)*: 1.123; *Latent Heat of Vaporization*: 193, 107, 4.48; *Heat of Combustion*: -14.390, -7992, -334.6; *Heat of Decomposition*: Not pertinent.

Health Hazards Information - *Recommended Personal Protective Equipment*: Air-supplied mask or organic canister; vapor-proof goggles; rubber gloves and protective clothing; *Symptoms Following Exposure*: Vapor irritates eyes and nose. Liquid irritates skin and is absorbed through the skin. Overexposure causes nausea, headache, nervous symptoms, increased urinary frequency; *General Treatment for Exposure*: INHALATION: remove victim from contaminated area; give artificial respiration and oxygen if necessary; treat symptomatically. INGESTION: induce vomiting and follow with gastric lavage. SKIN: wash thoroughly with water. EYES: irrigate with water for at least 15 min.

Toxicity by Inhalation (Threshold Limit Value): 5 ppm; *Short-Term Exposure Limits*: No data; *Toxicity by Ingestion*: LD_{50} 0.5 to 5 g/kg (rat); *Late Toxicity*: Liver and kidney damage after ingestion; *Vapor (Gas) Irritant Characteristics*: Vapors cause moderate irritation; high concentrations are unpleasant. Effect are temporary; *Liquid or Solid Irritant Characteristics*: Causes smarting of the skin and first-degree burns on short exposure; causes secondary burns on long exposure; *Odor Threshold*: 0.021 ppm.
Fire Hazards - *Flash Point (deg. F)*: 68 CC; *Flammable Limits in Air (%)*: 1.8 - 12.4; *Fire Extinguishing Agents:* Alcohol foam, dry chemical, or carbon dioxide; *Fire Extinguishing Agents Not To Be Used:* Water may be ineffective; *Special Hazards of Combustion Products:* Not pertinent; *Behavior in Fire:* Vapor is heavier than air and may travel to a source of ignition and flash back; *Ignition Temperature (deg. F):* 900; *Electrical Hazard:* Class I, Group D; *Burning Rate:* 4.3 mm/min.
Chemical Reactivity - *Reactivity with Water:* No reaction; *Reactivity with Common Materials:* No reactions; *Stability During Transport:* Stable; *Neutralizing Agents for Acids and Caustics:* Flush with water; *Polymerization:* Not pertinent; *Inhibitor of Polymerization:* Not pertinent.

PYROGALLIC ACID

Chemical Designations - *Synonyms*: 1,2,3-Benzenetriol; Pyrogallol; 1,2,3-Trihydroxybenzene; *Chemical Formula*: $1,2,3-C_6H_3(OH)_3$.
Observable Characteristics - *Physical State (as normally shipped)*:Solid; *Color*: White to gray; *Odor*: None.
Physical and Chemical Properties - *Physical State at 15 °C and 1 atm.*: Solid; *Molecular Weight*: 126; *Boiling Point at 1 atm.*: 588, 309, 582; *Freezing Point*: 268, 309, 582; *Critical Temperature*: Not pertinent; *Critical Pressure*: Not pertinent; *Specific Gravity*: 1.45 at 20 °C (solid); *Vapor (Gas) Density*: Not pertinent; *Ratio of Specific Heats of Vapor (Gas)*: Not pertinent; *Latent Heat of Vaporization*: Not pertinent; *Heat of Combustion*: -9.130, -5.070, -212; *Heat of Decomposition*: Not pertinent.
Health Hazards Information - *Recommended Personal Protective Equipment*: Rubber gloves: safety goggles, dust mask; *Symptoms Following Exposure*: Inhalation of dust causes irritation of nose and throat. Ingestion may cause severe gastrointestinal irritation, convulsions, circulatory collapse, and death. Contact with eyes causes irritation. Skin contact can cause local discoloration, irritation. eczema, and death: repeated contact can cause sensitization; *General Treatment for Exposure*: INHALATION: remove victim lo fresh air. INGESTION: give large amount of water: induce vomiting immediately: consult a physician. EYES: flush with water for at least 15 min.: consult a physician. SKIN: wash immediately with soap and water: consult a physician if exposure has been severe; *Toxicity by Inhalation (Threshold Limit Value)*: Data not available; *Short-Term Exposure Limits*: Data not available; *Toxicity by Ingestion*: Grade 2, oral LD_{50} =719 mg/kg (rat); *Late Toxicity*: Depresses growth in chicks; *Vapor (Gas) Irritant Characteristics*: Data not available; *Liquid or Solid Irritant Characteristics*: Data not available; *Odor Threshold*: Odorless.
Fire Hazards - *Flash Point* : Not pertinent; this is a combustible solid; *Flammable Limits in Air (%):* Not pertinent; *Fire Extinguishing Agents:* Water, foam, dry chemical, or carbon dioxide; *Fire Extinguishing Agents Not To Be Used:* Not pertinent; *Special Hazards of Combustion Products:* Not pertinent; *Behavior in Fire:* Not pertinent; *Ignition Temperature :* No data; *Electrical Hazard:* Not pertinent; *Burning Rate:* Not pertinent.
Chemical Reactivity - *Reactivity with Water:* No reactions; *Reactivity with Common Materials:* No reactions; *Stability During Transport:* Stable; *Neutralizing Agents for Acids and Caustics:* Not pertinent; *Polymerization:* Not pertinent; *Inhibitor of Polymerization:* Not pertinent.

Q

QUINOLINE

Chemical Designations - *Synonyms*: 1-Azanapthalene; 1-Benzazine; Benzo(b)pyridine; Chinoline; Leucol; *Chemical Formula*: C_9H_7N.

Observable Characteristics - *Physical State (as normally shipped)*: Liquid; *Color*: Colorless to brown; *Odor*: Strong, unpleasant.

Physical and Chemical Properties - *Physical State at 15 °C and 1 atm.*: Liquid; *Molecular Weight*: 129; *Boiling Point at 1 atm.*: 459, 237, 510; *Freezing Point*: 5, -15, 258; *Critical Temperature*: 948, 509, 782; *Critical Pressure*: Data not available; *Specific Gravity*: 1.095 at 20 °C (liquid); *Vapor (Gas) Density*: 4.5; *Ratio of Specific Heats of Vapor (Gas)*: Not pertinent; *Latent Heat of Vaporization*: (est.) 155, 86, 3.6; *Heat of Combustion*: -15.700, -8.710, -365; *Heat of Decomposition*: Not pertinent.

Health Hazards Information - *Recommended Personal Protective Equipment*: U. S. Bu. Mines approved respirator: rubber gloves; safety glasses with side shields or chemical goggles: coveralls or rubber apron; *Symptoms Following Exposure*: Inhalation of vapors or dust causes irritation of respiratory tract, ingestion causes burns of mucous membranes, severe diarrhea, pallor, sweating, weakness, headache, dizziness, tinnitus, shock, and severe convulsions: may also cause siderosis of the spleen and tubular injury to the kidney. Contact with eyes causes irritation. Can be absorbed from wounds or through unbroken skin, producing severe dermatitis, methemoglobinemia, cyanosis, convulsions, tachycardia, dyspnea, and death; *General Treatment for Exposure*: INHALATION: remove victim to fresh air; if he is not breathing, give artificial respiration, preferably mouth-to-mouth; if breathing is difficult, give oxygen, call a physician. INGESTION: give activated charcoal; administer gastric lavage with water; consult physician. EYES: flush with water for IS min. SKIN: flush with water; *Toxicity by Inhalation (Threshold Limit Value)*: Data not available; *Short-Term Exposure Limits*: Data not available; *Toxicity by Ingestion*: Grade 3, oral LD_{50} =460 mg/kg (rat); *Late Toxicity*: Data not available; *Vapor (Gas) Irritant Characteristics*: Data not available; *Liquid or Solid Irritant Characteristics*: Data not available; *Odor Threshold*: 71 ppm.

Fire Hazards - *Flash Point (deg. F)*: 225 CC; *Flammable Limits in Air (%):* No data; *Fire Extinguishing Agents:* Water, dry chemical, foam, or carbon dioxide; *Fire Extinguishing Agents Not To Be Used:* Not pertinent; *Special Hazards of Combustion Products:* Toxic oxides of nitrogen form in fires; *Behavior in Fire:* Exposure to heat can result in pressure build-up in closed containers, resulting in bulging or even explosion; *Ignition Temperature (deg. F):* 896; *Electrical Hazard:* No data; *Burning Rate:* 4.1 mm/min.

Chemical Reactivity - *Reactivity with Water:* No reaction; *Reactivity with Common Materials:* Attacks some forms of plastics; *Stability During Transport:* Stable; *Neutralizing Agents for Acids and Caustics:* Not pertinent; *Polymerization:* Not pertinent; *Inhibitor of Polymerization:* Not pertinent..

S

SALICYLIC ACID

Chemical Designations - *Synonyms*: o-Hydroxybenzoic acid; Retarder W; *Chemical Formula*: 1,2-HOC_6H_4COOH.

Observable Characteristics - *Physical State (as normally shipped)*: Solid; *Color*: White to very light; *Odor*: None.

Physical and Chemical Properties - *Physical State at 15 °C and 1 atm.*: Solid; *Molecular Weight*: 138.13; *Boiling Point at 1 atm.*: Not pertinent (decomposes); *Freezing Point*: 315, 157, 430; *Critical Temperature*: Not pertinent; *Critical Pressure*: Not pertinent; *Specific Gravity*: 1.44 at 20 °C (solid);

Vapor (Gas) Density: Not pertinent; *Ratio of Specific Heats of Vapor (Gas)*: Not pertinent; *Latent Heat of Vaporization*: Not pertinent; *Heat of Combustion*: -9.420, -5.230, -219; *Heat of Decomposition*: Not pertinent.

Health Hazards Information - *Recommended Personal Protective Equipment*: Gloves: goggles: respirator For dust: clean body-covering clothing; *Symptoms Following Exposure*: Inhalation of dust irritates nose and throat. Vomiting may occur spontaneously if large amounts are swallowed. Contact with eyes causes irritation, marked pain, and corneal injury which should heal. Prolonged or repeated skin contact may cause marked irritation or even a mild burn; *General Treatment for Exposure*: INHALATION: move to fresh air. INGESTION: induce vomiting and get medical attention promptly. EYES: promptly flush with water for 15 min. and get medical attention. SKIN: wash with soap and water; *Toxicity by Inhalation (Threshold Limit Value)*: Data not available; *Short-Term Exposure Limits*: Data not available; *Toxicity by Ingestion*: Grade 2, LD_{50} 0.5-5 g/kg; *Late Toxicity*: Data not available; *Vapor (Gas) Irritant Characteristics*: Data not available; *Liquid or Solid Irritant Characteristics*: Data not available; *Odor Threshold*: Data not available.

Fire Hazards - *Flash Point* : Not pertinent; this is a combustible solid; *Flammable Limits in Air (%):* Not pertinent; *Fire Extinguishing Agents:* Water, foam, dry chemical, or carbon dioxide; *Fire Extinguishing Agents Not To Be Used:* Application of water or foam may cause frothing; *Special Hazards of Combustion Products:* Irritating vapors of unburned product and phenol form during fires; *Behavior in Fire:* This product sublimes and forms vapor or dust that can explode; *Ignition Temperature* : No data; *Electrical Hazard:* Not pertinent; *Burning Rate:* Not pertinent.

Chemical Reactivity - *Reactivity with Water:* No reaction; *Reactivity with Common Materials:* No reactions; *Stability During Transport:* Stable; *Neutralizing Agents for Acids and Caustics:* Not pertinent; *Polymerization:* Not pertinent; *Inhibitor of Polymerization:* Not pertinent.

SELENIUM DIOXIDE

Chemical Designations - *Synonyms*: Selenious anhydride; Selenium oxide; *Chemical Formula*: SeO_2.

Observable Characteristics - *Physical State (as normally shipped)*: Solid; *Color*: White; *Odor*: Pungent; sour.

Physical and Chemical Properties - *Physical State at 15 °C and 1 atm.*: Solid; *Molecular Weight*: 111; *Boiling Point at 1 atm.*: Not pertinent (decomposes); *Freezing Point*: Not pertinent; *Critical Temperature*: Not pertinent; *Critical Pressure*: Not pertinent; *Specific Gravity*: 3.95 at 20 °C (solid); *Vapor (Gas) Density*: Not pertinent; *Ratio of Specific Heats of Vapor (Gas)*: Not pertinent; *Latent Heat of Vaporization*: Not pertinent; *Heat of Combustion*: Not pertinent; *Heat of Decomposition*: Not pertinent.

Health Hazards Information - *Recommended Personal Protective Equipment*: Dust mask; rubber gloves; protective clothing; *Symptoms Following Exposure*: Absorption of selenium may be demonstrated by presence of the element in the urine and by a garlic-like odor of the breath, inhalation of dust can cause bronchial spasms, symptoms of asphyxiation, and pneumonitis. Acute symptoms of ingestion include sternal pain, cough, nausea, pallor, coated tongue, gastrointestinal disorders, nervousness and conjunctivitis. Contact with eyes causes irritation; *General Treatment for Exposure*: Consult physician after all exposures or his compound. INHALATION remove victim to fresh air: give oxygen if needed. INGESTION: induce vomiting; follow with gastric lavage and saline cathartics. EYES: flush immediately and thoroughly with water. SKIN: flush with water; *Toxicity by Inhalation (Threshold Limit Value)*: 0.2 mg/m^3 (as selenium); *Short-Term Exposure Limits*: 0.3 mg/m^3, 30 min. (as selenium); *Toxicity by Ingestion*: Data not available; *Late Toxicity*: Data not available; *Vapor (Gas) Irritant Characteristics*: Data not available; *Liquid or Solid Irritant Characteristics*: Data not available; *Odor Threshold*: 0.0002 mg/m^3.

Fire Hazards - *Flash Point* : Not flammable; *Flammable Limits in Air (%):* Not flammable; *Fire Extinguishing Agents:* Not pertinent; *Fire Extinguishing Agents Not To Be Used:* Not pertinent; *Special Hazards of Combustion Products:* This product sublimes and forms toxic vapors when heated in fires; *Behavior in Fire:* Not pertinent; *Ignition Temperature* : Not pertinent; *Electrical Hazard:* Not pertinent; *Burning Rate:* Not pertinent.

Chemical Reactivity - *Reactivity with Water:* No reaction; *Reactivity with Common Materials:* In presence of water will corrode most metals; *Stability During Transport:* Stable; *Neutralizing Agents for Acids and Caustics:* Not pertinent; *Polymerization:* Not pertinent; *Inhibitor of Polymerization:* Not pertinent.

SELENIUM TRIOXIDE

Chemical Designations - *Synonyms*: Selenic anhydride; *Chemical Formula*: SeO_3.

Observable Characteristics - *Physical State (as normally shipped)*: Solid; *Color*: White; *Odor*: Data not available.

Physical and Chemical Properties - *Physical State at 15 °C and 1 atm.*: Solid; *Molecular Weight*: 126.9; *Boiling Point at 1 atm.*: Not pertinent (decomposes); *Freezing Point*: 244, 118, 391; *Critical Temperature*: Not pertinent; *Critical Pressure*: Not pertinent; *Specific Gravity*: 3.6 at 20 °C (solid); *Vapor (Gas) Density*: Not pertinent; *Ratio of Specific Heats of Vapor (Gas)*: Not pertinent; *Latent Heat of Vaporization*: Not pertinent; *Heat of Combustion*: Not pertinent; *Heat of Decomposition*: Not pertinent.

Health Hazards Information - *Recommended Personal Protective Equipment*: Dust mask: goggles or Face shield: rubber gloves; *Symptoms Following Exposure*: Absorption of selenium may be demonstrated by presence of the element in the urine and by a garlic-like odor of breath, inhalation can cause bronchial spasms symptoms of asphyxiation, and pneumonitis. Acute symptoms of ingestion include sternal pain, cough, nausea, pallor, coated tongue, gastrointestinal disorders, nervousness and conjunctivitis. Contact with eyes or skin causes irritation; *General Treatment for Exposure*: INHALATION: remove victim to fresh air; give oxygen if necessary. INGESTION: induce vomiting: follow with gastric lavage and saline cathartics. EYES: flush with water; *Toxicity by Inhalation (Threshold Limit Value)*: 0.2 mg/m³, (as selenium); *Short-Term Exposure Limits*: 0.3 mg/m³, 30 min. (as selenium); *Toxicity by Ingestion*: Data not available; *Late Toxicity*: Data not available; *Vapor (Gas) Irritant Characteristics*: Data not available; *Liquid or Solid Irritant Characteristics*: Data not available; *Odor Threshold*: Data not available.

Fire Hazards - *Flash Point* : Not flammable; *Flammable Limits in Air (%):* Not flammable; *Fire Extinguishing Agents:* Not pertinent; *Fire Extinguishing Agents Not To Be Used:* Not pertinent; *Special Hazards of Combustion Products:* Not pertinent; *Behavior in Fire:* Not pertinent; *Ignition Temperature* : Not pertinent; *Electrical Hazard:* Not pertinent; *Burning Rate:* Not pertinent.

Chemical Reactivity - *Reactivity with Water:* Reacts vigorously with water forming selenic acid solution; *Reactivity with Common Materials:* Corrodes all metals in the presence of water; *Stability During Transport:* Stable; *Neutralizing Agents for Acids and Caustics:* Flush with water and rinse with dilute solution of sodium bicarbonate or soda ash; *Polymerization:* Not pertinent; *Inhibitor of Polymerization:* Not pertinent.

SILICON TETRACHLORIDE

Chemical Designations - *Synonyms*: Silicon chloride; *Chemical Formula*: $SiCl_4$.

Observable Characteristics - *Physical State (as normally shipped)*: Liquid; *Color*: Colorless to pale yellow; *Odor*: Suffocating.

Physical and Chemical Properties - *Physical State at 15 °C and 1 atm.*:Liquid; *Molecular Weight*: 169.9; *Boiling Point at 1 atm.*: 135.7, 57.6, 330.8; *Freezing Point*: -94, -70, 203; *Critical Temperature*: 472.5, 233.6, 506.8; *Critical Pressure*: 542, 36.8, 3.74; *Specific Gravity*: 1.48 at 20 °C (liquid); *Vapor (Gas) Density*: 5.86; *Ratio of Specific Heats of Vapor (Gas)*: Data not available; *Latent Heat of Vaporization*: 74.2, 41.2, 1.73; *Heat of Combustion*: Not pertinent; *Heat of Decomposition*: Not pertinent.

Health Hazards Information - *Recommended Personal Protective Equipment*: Add-canister-type gas mask or self-contained breathing apparatus; goggles or face shield, rubber gloves, other protective clothing to prevent contact with skin; *Symptoms Following Exposure*: Inhalation causes severe irritation of upper respiratory tract resulting in coughing, choking, and a feeling of suffocation; continued inhalation may produce ulceration of the nose, throat, and larynx; if inhaled deeply, edema of the lungs may occur. Contact of liquid with eyes causes severe irritation and painful burns; may cause permanent visual impairment. Liquid may cause severe burns of skin. Repeated skin contact with dilute solutions or exposure to concentrated vapors may cause dermatitis, ingestion causes severe internal injury with pain in the throat and stomach, intense thirst, difficulty in swallowing. nausea, vomiting, and diarrhea; in severe cases, collapse and unconsciousness may result; *General Treatment for Exposure*: *Get medical attention at once following any exposure to this compound*, INHALATION: remove victim from contaminated atmosphere; if breathing has ceased, start mouth-to-mouth resuscitation; oxygen should only be administered by an experienced person when authorized by a physician; keep patient warm and comfortable. EYES: immediately flush with large quantities of running water for a minimum of 15 min.; continue irrigation for an additional 15 min. if physician is not available. SKIN: immediately flush affected area with water; severe or extensive burns may be caused by silicon tetrachloride, producing shock symptoms (rapid pulse, sweating and collapse); keep patient comfortably warm. INGESTION: if patient is conscious give large amounts of lime water or milk of magnesia; plain water should be given if neither of these is available; do NOT give sodium bicarbonate or make any attempt to induce vomiting, if patient is unconscious, do not give anything but ensure there is no obstruction to breathing (tongue should be kept forward and false teeth removed); victim will be less likely to aspirate vomitus if placed in a face-downward position; *Toxicity by Inhalation*: Data not available; *Short-Term Exposure Limits*: Data not available; *Toxicity by Ingestion*: Grade 4, $LD_{50} < 50$ mg/kg; *Late Toxicity*: Data not available; *Vapor (Gas) Irritant Characteristics*: Vapors cause severe irritation of eyes and throat and can cause eye and lung injury. Vapors cannot be tolerated even at low concentrations; *Liquid or Solid Irritant Characteristics*: Data not available; *Odor Threshold*: Data not available.

Fire Hazards - *Flash Point* : Not flammable; *Flammable Limits in Air (%):* Not flammable; *Fire Extinguishing Agents:* Not pertinent; *Fire Extinguishing Agents Not To Be Used:* Do not apply water or foam on adjacent fires; *Special Hazards of Combustion Products:* Not pertinent; *Behavior in Fire:* Contact with water or foam applied to adjacent fires results in the formation of toxic and irritating fumes of hydrogen chloride; *Ignition Temperature:* Not pertinent; *Electrical Hazard:* Not pertinent; *Burning Rate:* Not pertinent.

Chemical Reactivity - *Reactivity with Water:* Reacts vigorously with water forming hydrogen chloride (hydrochloric acid); *Reactivity with Common Materials:* In the presence of moisture, will corrode metals. The reaction is generally non-hazardous; *Stability During Transport:* Stable; *Neutralizing Agents for Acids and Caustics:* Flush with water and rinse with sodium bicarbonate or lime solution; *Polymerization:* Not pertinent; *Inhibitor of Polymerization:* Not pertinent.

SILVER ACETATE

Chemical Designations - *Synonyms*: No common synonyms; *Chemical Formula*: CH_3COOAg.

Observable Characteristics - *Physical State (as normally shipped)*: Solid; *Color*: White to gray; *Odor*: None.

Physical and Chemical Properties - *Physical State at 15 °C and 1 atm.*:Solid; *Molecular Weight*: 166.9; *Boiling Point at 1 atm.*: Not pertinent (decomposes); *Freezing Point*: Not pertinent; *Critical Temperature*: Not pertinent; *Critical Pressure*: Not pertinent; *Specific Gravity*: 3.26 at 20 °C (solid); *Vapor (Gas) Density*: Not pertinent; *Ratio of Specific Heats of Vapor (Gas)*: Not pertinent; *Latent Heat of Vaporization*: Not pertinent; *Heat of Combustion*: Not pertinent; *Heat of Decomposition*: Not pertinent.

Health Hazards Information - *Recommended Personal Protective Equipment*: Dust mask, goggles or face shield, protective gloves; *Symptoms Following Exposure*: Inhalation of dust irritates nose and throat. Contact with eyes or skin causes irritation. If continued for a long period, ingestion or inhalation of silver compounds can cause permanent discoloration of skin (argyria); *General Treatment for Exposure*: INHALATION: move to fresh air. INGESTION: give large amount of water; induce vomiting. EYES: flush with water for at least 15 min. SKIN: flood with water; *Toxicity by Inhalation (Threshold Limit Value)*:0.01 mg/m^3; *Short-Term Exposure Limits*: Data not available; *Toxicity by Ingestion*: Data not available; *Late Toxicity*: Data not available; *Vapor (Gas) Irritant Characteristics*: Data not available; *Liquid or Solid Irritant Characteristics*: Data not available; *Odor Threshold*: Data not available.

Fire Hazards - *Flash Point* : Not flammable; *Flammable Limits in Air (%):* Not flammable; *Fire Extinguishing Agents:* Not pertinent; *Fire Extinguishing Agents Not To Be Used:* Not pertinent; *Special Hazards of Combustion Products:* Not pertinent; *Behavior in Fire:* Not pertinent; *Ignition Temperature :* Not pertinent; *Electrical Hazard:* Not pertinent; *Burning Rate:* Not pertinent.

Chemical Reactivity - *Reactivity with Water:* No reaction; *Reactivity with Common Materials:* No reactions; *Stability During Transport:* Stable; *Neutralizing Agents for Acids and Caustics:* Not pertinent; *Polymerization:* Not pertinent; *Inhibitor of Polymerization:* Not pertinent.

SILVER CARBONATE

Chemical Designations - *Synonyms*: No common synonyms; *Chemical Formula*: Ag_2CO_3.

Observable Characteristics - *Physical State (as normally shipped)*: Solid; *Color*: Yellow to brown; *Odor*: None.

Physical and Chemical Properties - *Physical State at 15 °C and 1 atm.*: Solid; *Molecular Weight*: 275.75; *Boiling Point at 1 atm.*: Not pertinent (decomposes); *Freezing Point*: Not pertinent; *Critical Temperature*: Not pertinent; *Critical Pressure*: Not pertinent; *Specific Gravity*: 6.1 at 20 °C (solid); *Vapor (Gas) Density*: Not pertinent; *Ratio of Specific Heats of Vapor (Gas)*: Not pertinent; *Latent Heat of Vaporization*: Not pertinent; *Heat of Combustion*: Not pertinent; *Heat of Decomposition*: Not pertinent.

Health Hazards Information - *Recommended Personal Protective Equipment*: Dust mask, goggles or face shield, rubber gloves; *Symptoms Following Exposure*: Contact with eyes causes irritation. If continued for a long period, ingestion or inhalation of silver compounds can cause permanent discoloration of the skin (argyria); *General Treatment for Exposure*: INHALATION: move to fresh air. INGESTION: give large amount of water, induce vomiting. EYES: flush with water for at least 15 min. SKIN: flush with water, wash with soap and water; *Toxicity by Inhalation (Threshold Limit Value)*: 0.01 mg/m^3; *Short-Term Exposure Limits*: Data not available; *Toxicity by Ingestion*: Data not available; *Late Toxicity*: Data not available; *Vapor (Gas) Irritant Characteristics*: Data not available; *Liquid or Solid Irritant Characteristics*: Data not available; *Odor Threshold*: Data not available.

Fire Hazards - *Flash Point* : Not flammable; *Flammable Limits in Air (%):* Not flammable; *Fire Extinguishing Agents:* Not pertinent; *Fire Extinguishing Agents Not To Be Used:* Not pertinent; *Special Hazards of Combustion Products:* Not pertinent; *Behavior in Fire:* Decomposes to silver oxide, silver, and carbon dioxide. The reaction is non violent; *Ignition Temperature :* Not pertinent; *Electrical Hazard:* Not pertinent; *Burning Rate:* Not pertinent.

Chemical Reactivity - *Reactivity with Water:* No reaction; *Reactivity with Common Materials:* No reactions; *Stability During Transport:* Stable; *Neutralizing Agents for Acids and Caustics:* Not pertinent; *Polymerization:* Not pertinent; *Inhibitor of Polymerization:* Not pertinent.

SILVER FLUORIDE

Chemical Designations - *Synonyms*: Argentous fluoride; Silver monofluoride; *Chemical Formula*: AgF.

Observable Characteristics - *Physical State (as normally shipped)*: Solid; *Color*: Yellow to gray; *Odor*: None.

Physical and Chemical Properties - *Physical State at 15 °C and 1 atm.*: Solid; *Molecular Weight*: 126.9; *Boiling Point at 1 atm.*: 2.118, 1.159, 1.432; *Freezing Point*: Not pertinent; *Critical Temperature*: Not pertinent; *Critical Pressure*: Not pertinent; *Specific Gravity*: 5.82 at 20 °C (solid); *Vapor (Gas) Density*: Not pertinent; *Ratio of Specific Heats of Vapor (Gas)*: Not pertinent; *Latent Heat of Vaporization*: Not pertinent; *Heat of Combustion*: Not pertinent; *Heat of Decomposition*: Not pertinent.

Health Hazards Information - *Recommended Personal Protective Equipment*: Dust mask; goggles or face shield; protective gloves; *Symptoms Following Exposure*: inhalation of dust causes irritation of nose and throat. Ingestion may cause vomiting, salty taste, abdominal pain, diarrhea, convulsions, collapse, thirst, disturbed color vision, and acute toxic nephritis. Contact with eyes causes irritation.

Skin may be blackened on prolonged exposure; *General Treatment for Exposure*: INHALATION: move to fresh air. INGESTION: get medical attention at once: give large amount of water and induce vomiting. EYES: flush with water for at least 15 min. SKIN: flush with water; *Toxicity by Inhalation (Threshold Limit Value)*: 0.01 mg/m^3; *Short-Term Exposure Limits*: Data not available; *Toxicity by Ingestion*: Data not available; *Late Toxicity*: Data not available; *Vapor (Gas) Irritant Characteristics*: Data not available; *Liquid or Solid Irritant Characteristics*: Data not available; *Odor Threshold*: Data not available.

Fire Hazards - *Flash Point* : Not flammable; *Flammable Limits in Air (%):* Not flammable; *Fire Extinguishing Agents:* Not pertinent; *Fire Extinguishing Agents Not To Be Used:* Not pertinent; *Special Hazards of Combustion Products:* Not pertinent; *Behavior in Fire:* Not pertinent; *Ignition Temperature :* Not pertinent; *Electrical Hazard:* Not pertinent; *Burning Rate:* Not pertinent.

Chemical Reactivity - *Reactivity with Water:* No reaction; *Reactivity with Common Materials:* No reactions; *Stability During Transport:* Stable; *Neutralizing Agents for Acids and Caustics:* Not pertinent; *Polymerization:* Not pertinent; *Inhibitor of Polymerization:* Not pertinent.

SILVER IODATE

Chemical Designations - *Synonyms*: No common synonyms; *Chemical Formula*: AgIO$_3$.

Observable Characteristics - *Physical State (as normally shipped)*: Solid; *Color*: White; *Odor*: None.

Physical and Chemical Properties - *Physical State at 15 °C and 1 atm.*: Solid; *Molecular Weight*: 282.1; *Boiling Point at 1 atm.*: Not pertinent (decomposes); *Freezing Point*: Not pertinent; *Critical Temperature*: Not pertinent; *Critical Pressure*: Not pertinent; *Specific Gravity*: 5.53 at 20 °C (solid); *Vapor (Gas) Density*: Not pertinent; *Ratio of Specific Heats of Vapor (Gas)*: Not pertinent; *Latent Heat of Vaporization*: Not pertinent; *Heat of Combustion*: Not pertinent; *Heat of Decomposition*: Not pertinent.

Health Hazards Information - *Recommended Personal Protective Equipment*: Dust mask: goggles or face shield, protective gloves; *Symptoms Following Exposure*: Contact with eyes causes irritation, if continued for a long period, ingestion or inhalation of silver compounds can cause permanent discoloration of the skin (argyria); *General Treatment for Exposure*: INHALATION: move to fresh air. INGESTION: give large amount of water: induce vomiting. EYES: flush with water for at least 15 min. SKIN: flush with water, wash with soap and water; *Toxicity by Inhalation (Threshold Limit Value)*: 0.01 mg/m^3; *Short-Term Exposure Limits*: Data not available; *Toxicity by Ingestion*: Data not available; *Late Toxicity*: Data not available; *Vapor (Gas) Irritant Characteristics*: Data not available; *Liquid or Solid Irritant Characteristics*: Data not available; *Odor Threshold*: Data not available.

Fire Hazards - *Flash Point* : Not flammable; *Flammable Limits in Air (%):* Not flammable; *Fire Extinguishing Agents:* Not pertinent; *Fire Extinguishing Agents Not To Be Used:* Not pertinent; *Special Hazards of Combustion Products:* Not pertinent; *Behavior in Fire:* Not pertinent; *Ignition Temperature :* Not pertinent; *Electrical Hazard:* Not pertinent; *Burning Rate:* Not pertinent.

Chemical Reactivity - *Reactivity with Water:* No reaction; *Reactivity with Common Materials:* No reactions; *Stability During Transport:* Stable; *Neutralizing Agents for Acids and Caustics:* Not pertinent; *Polymerization:* Not pertinent; *Inhibitor of Polymerization:* Not pertinent.

SILVER NITRATE

Chemical Designations - *Synonyms*: Lunar caustic; *Chemical Formula*: AgNO$_3$.

Observable Characteristics - *Physical State (as normally shipped)*: Solid; *Color*: Colorless; *Odor*: Odorless.

Physical and Chemical Properties - *Physical State at 15 °C and 1 atm.*: Solid; *Molecular Weight*: 169.87; *Boiling Point at 1 atm.*: Decomposes; *Freezing Point*: 414, 212, 485; *Critical Temperature*: Not pertinent; *Critical Pressure*: Not pertinent; *Specific Gravity*: 4.35 at 19 °C (solid); *Vapor (Gas) Density*: Not pertinent; *Ratio of Specific Heats of Vapor (Gas)*: Not pertinent; *Latent Heat of Vaporization*: Not pertinent; *Heat of Combustion*: Not pertinent; *Heat of Decomposition*: Not pertinent.

Health Hazards Information - *Recommended Personal Protective Equipment*: Goggles or face shield; rubber gloves; *Symptoms Following Exposure*: Concentrated solution will produce irritation, ulceration, and discoloration of the skin; also causes severe irritation of the eyes. Ingestion will produce violent abdominal pain and other gastroenteric symptoms; *General Treatment for Exposure*: INGESTION: gastric lavage with dilute solution of chloride, followed by cathartics and demulcents. Other treatment is symptomatic. SKIN: wash promptly; *Toxicity by Inhalation (Threshold Limit Value)*: Not pertinent; *Short-Term Exposure Limits*: Not pertinent; *Toxicity by Ingestion*: Grade 3, LD_{50} 50-500 mg/kg; *Late Toxicity*: Data not available; *Vapor (Gas) Irritant Characteristics*: Non-volatile; *Liquid or Solid Irritant Characteristics*: Burns skin on prolonged contact; *Odor Threshold*: Not pertinent.

Fire Hazards - *Flash Point* : Not flammable; *Flammable Limits in Air (%):* Not flammable; *Fire Extinguishing Agents:* Not pertinent; *Fire Extinguishing Agents Not To Be Used:* Not pertinent; *Special Hazards of Combustion Products:* Not pertinent; *Behavior in Fire:* Increases the flammability of combustible materials; *Ignition Temperature :* Not pertinent; *Electrical Hazard:* Not pertinent; *Burning Rate:* Not pertinent.

Chemical Reactivity - *Reactivity with Water:* No reaction; *Reactivity with Common Materials:* No reactions; *Stability During Transport:* Stable; *Neutralizing Agents for Acids and Caustics:* Not pertinent; *Polymerization:* Not pertinent; *Inhibitor of Polymerization:* Not pertinent.

SILVER OXIDE

Chemical Designations - *Synonyms*: Argentous oxide; *Chemical Formula*: Ag_2O.

Observable Characteristics - *Physical State (as normally shipped)*: Solid; *Color*: Brown-black; *Odor*: None.

Physical and Chemical Properties - *Physical State at 15 °C and 1 atm.*: Solid; *Molecular Weight*: 231.8; *Boiling Point at 1 atm.*: Not pertinent (decomposes); *Freezing Point*: Not pertinent; *Critical Temperature*: Not pertinent; *Critical Pressure*: Not pertinent; *Specific Gravity*: 7.14 at 20 °C (solid); *Vapor (Gas) Density*: Not pertinent; *Ratio of Specific Heats of Vapor (Gas)*: Not pertinent; *Latent Heat of Vaporization*: Not pertinent; *Heat of Combustion*: Not pertinent; *Heat of Decomposition*: Not pertinent.

Health Hazards Information - *Recommended Personal Protective Equipment*: Dust mask; goggles or face shield; protective gloves; *Symptoms Following Exposure*: Contact with eyes causes mild irritation. If continued for a long period, ingestion or inhalation of silver compounds can cause permanent discoloration of the skin (argyria); *General Treatment for Exposure*: EYES: flush with water. SKIN: flush with water; wash soap and water; *Toxicity by Inhalation (Threshold Limit Value)*: 0.01 mg/m³; *Short-Term Exposure Limits*: Data not available; *Toxicity by Ingestion*: Grade 2, LD_{50} 0.5-5 g/kg; *Late Toxicity*: Data not available; *Vapor (Gas) Irritant Characteristics*: Data not available; *Liquid or Solid Irritant Characteristics*: Data not available; *Odor Threshold*: Odorless.

Fire Hazards - *Flash Point* : Not flammable; *Flammable Limits in Air (%):* Not flammable; *Fire Extinguishing Agents:* Not pertinent; *Fire Extinguishing Agents Not To Be Used:* Not pertinent; *Special Hazards of Combustion Products:* Not pertinent; *Behavior in Fire:* Decomposes into metallic silver and oxygen. If large amounts of the product are involved in a fire, the oxygen liberated may increase the intensity of the fire; *Ignition Temperature :* Not pertinent; *Electrical Hazard:* Not pertinent; *Burning Rate:* Not pertinent.

Chemical Reactivity - *Reactivity with Water:* No reaction; *Reactivity with Common Materials:* No reactions; *Stability During Transport:* Stable; *Neutralizing Agents for Acids and Caustics:* Not pertinent; *Polymerization:* Not pertinent; *Inhibitor of Polymerization:* Not pertinent.

SILVER SULFATE

Chemical Designations - *Synonyms*: No common synonyms; *Chemical Formula*: Ag_2SO_4.

Observable Characteristics - *Physical State (as normally shipped)*: Solid; *Color*: White to gray; *Odor*: None.

Physical and Chemical Properties - *Physical State at 15 °C and 1 atm.*: Solid; *Molecular Weight*: 311.80; *Boiling Point at 1 atm.*: Not pertinent; *Freezing Point*: Not pertinent; *Critical Temperature*: Not pertinent; *Critical Pressure*: Not pertinent; *Specific Gravity*: 5.45 at 20 °C (solid); *Vapor (Gas) Density*: Not pertinent; *Ratio of Specific Heats of Vapor (Gas)*: Not pertinent; *Latent Heat of*

Vaporization: Not pertinent; *Heat of Combustion*: Not pertinent; *Heat of Decomposition*: Not pertinent.
Health Hazards Information - *Recommended Personal Protective Equipment*: Dust mask; goggles or face shield; protective gloves; *Symptoms Following Exposure*: Contact with eyes causes irritation. If continued for long period, ingestion or inhalation of silver compounds can cause permanent discoloration of the skin (argyria); *General Treatment for Exposure*: INHALATION: move to fresh air. INGESTION: give large amount of water; induce vomiting; EYES flush with water for at least 15 min. SKIN: flush with water, wash with soap and water; *Toxicity by Inhalation (Threshold Limit Value)*: 0.01 mg/m^3; *Short-Term Exposure Limits*: Data not available; *Toxicity by Ingestion*: Data not available; *Late Toxicity*: Data not available; *Vapor (Gas) Irritant Characteristics*: Data not available; *Liquid or Solid Irritant Characteristics*: Data not available; *Odor Threshold*: Data not available.
Fire Hazards - *Flash Point* : Not flammable; *Flammable Limits in Air (%)*: Not flammable; *Fire Extinguishing Agents:* Not pertinent; *Fire Extinguishing Agents Not To Be Used:* Not pertinent; *Special Hazards of Combustion Products:* Not pertinent; *Behavior in Fire:* Not pertinent; *Ignition Temperature : Not pertinent; *Electrical Hazard:* Not pertinent; *Burning Rate:* Not pertinent.
Chemical Reactivity - *Reactivity with Water:* No reaction; *Reactivity with Common Materials:* No reactions; *Stability During Transport:* Stable; *Neutralizing Agents for Acids and Caustics:* Not pertinent; *Polymerization:* Not pertinent; *Inhibitor of Polymerization:* Not pertinent.

SODIUM
Chemical Designations - *Synonyms*: No common synonyms; *Chemical Formula*: Na.
Observable Characteristics - *Physical State (as normally shipped)*: Soft solid or liquid; *Color*: Silvery white, changing to gray on exposure to air; *Odor*: Odorless.
Physical and Chemical Properties - *Physical State at 15 °C and 1 atm.*: Solid; *Molecular Weight*: 22.49; *Boiling Point at 1 atm.*: 1621, 883, 1156; *Freezing Point*: 207.5, 97.5, 370.7; *Critical Temperature*: 3632, 2000, 2273; *Critical Pressure*: 5040, 343, 34.8; *Specific Gravity*: 0.971 at 20 °C (solid); *Vapor (Gas) Density*: Not pertinent; *Ratio of Specific Heats of Vapor (Gas)*: Not pertinent; *Latent Heat of Vaporization*: Not pertinent; *Heat of Combustion*: Not pertinent; *Heat of Decomposition*: Not pertinent.
Health Hazards Information - *Recommended Personal Protective Equipment*: Maximum protective clothing; goggles and face shield; *Symptoms Following Exposure*: Severe burns caused by burning metal or by caustic soda formed by reaction with moisture on skin; *General Treatment for Exposure*: SKIN: brush off any metal, then flood with water for at least 15 min.; treat as heat or caustic burn; call a doctor; *Toxicity by Inhalation (Threshold Limit Value)*: Not pertinent; *Short-Term Exposure Limits*: Not pertinent; *Toxicity by Ingestion*: Not pertinent; *Late Toxicity*: None; *Vapor (Gas) Irritant Characteristics*: Non-volatile; *Liquid or Solid Irritant Characteristics*: Severe skin irritant. Cause second- and third-degree burns on short contact and is very injurious to the eyes; *Odor Threshold*: Not pertinent.
Fire Hazards - *Flash Point* : Not pertinent; *Flammable Limits in Air (%)*: Not pertinent; *Fire Extinguishing Agents:* Dry soda ash, graphite, salt, or other approved dry powder such as dry limestone; *Fire Extinguishing Agents Not To Be Used:* Water, carbon dioxide, or halogenated extinguishing agents; *Special Hazards of Combustion Products:* The fumes of burning sodium are highly irritating to the eyes, skin, and mucous membranes.; *Behavior in Fire:* Not pertinent; *Ignition Temperature (deg. F):* 250; *Electrical Hazard:* Not pertinent; *Burning Rate:* Not pertinent.
Chemical Reactivity - *Reactivity with Water:* Sodium reacts violently with water, forming flammable hydrogen gas, and caustic soda solution. Fire often accompanies the reaction; *Reactivity with Common Materials:* No reactions; *Stability During Transport:* Stable; *Neutralizing Agents for Acids and Caustics:* After the reaction with water, the caustic soda formed as a by-product can be diluted with water and then neutralized with acetic acid; *Polymerization:* Not pertinent; *Inhibitor of Polymerization:* Not pertinent.

SODIUM ALKYLBENZENESULFONATES
Chemical Designations - *Synonyms*: Alkylbenzenesulfonic acid, sodium salt; Sulfonated alkylbenzene;

Chemical Formula: $C_nH_{2n}+_1C_6H_4SO_3Na$.

Observable Characteristics - *Physical State (as normally shipped)*: Powder or thick liquid; *Color*: Pale yellow; *Odor*: Faint detergent.

Physical and Chemical Properties - *Physical State at 15 °C and 1 atm.*: Liquid or solid; *Molecular Weight*: Not pertinent; *Boiling Point at 1 atm.*: Decomposes; *Freezing Point*: Not pertinent; *Critical Temperature*: Not pertinent; *Critical Pressure*: Not pertinent; *Specific Gravity*: 1.0 at 20 °C (solid); *Vapor (Gas) Density*: Not pertinent; *Ratio of Specific Heats of Vapor (Gas)*: Not pertinent; *Latent Heat of Vaporization*: Not pertinent; *Heat of Combustion*: Not pertinent; *Heat of Decomposition*: Not pertinent.

Health Hazards Information - *Recommended Personal Protective Equipment*: Goggles or face shield; rubber gloves; *Symptoms Following Exposure*: In general, these chemical have a moderate order of toxicity. Repeated skin contact with concentrated solutions may cause dermatitis. Ingestion may cause gastrointestinal, vomiting, and diarrhea; *General Treatment for Exposure*: INGESTION: induce vomiting and call a doctor. EYES OR SKIN: flush with copious amounts of water; *Toxicity by Inhalation (Threshold Limit Value)*: Not pertinent; *Short-Term Exposure Limits*: Not pertinent; *Toxicity by Ingestion*: Grade 2, LD_{50} 0.5-5 g/kg; *Late Toxicity*: None; *Vapor (Gas) Irritant Characteristics*: Non-volatile; *Liquid or Solid Irritant Characteristics*: Minimum hazard. If spilled on clothing and allowed to remain, may cause smarting and reddening of skin; *Odor Threshold*: Not pertinent.

Fire Hazards - *Flash Point* : Not flammable; *Flammable Limits in Air (%)*: Not flammable; *Fire Extinguishing Agents*: Not pertinent; *Fire Extinguishing Agents Not To Be Used*: Not pertinent; *Special Hazards of Combustion Products*: Not pertinent; *Behavior in Fire*: Irritating vapors form in fires; *Ignition Temperature* : Not pertinent; *Electrical Hazard*: Not pertinent; *Burning Rate*: Not pertinent.

Chemical Reactivity - *Reactivity with Water*: No reaction; *Reactivity with Common Materials*: No reactions; *Stability During Transport*: Stable; *Neutralizing Agents for Acids and Caustics*: Not pertinent; *Polymerization*: Not pertinent; *Inhibitor of Polymerization*: Not pertinent.

SODIUM ALKYL SULFATES

Chemical Designations - *Synonyms*: Sodium hydrogen alkyl sulfate; *Chemical Formula*: $C_nH_{2n+1}OSO_2ONa$.

Observable Characteristics - *Physical State (as normally shipped)*: Liquid; *Color*: Colorless to pale yellow; *Odor*: Weak Detergent.

Physical and Chemical Properties - *Physical State at 15 °C and 1 atm.*: Liquid or solid; *Molecular Weight*: Not pertinent; *Boiling Point at 1 atm.*: Decomposes; *Freezing Point*: Not pertinent; *Critical Temperature*: Not pertinent; *Critical Pressure*: Not pertinent; *Specific Gravity*: Data not available; *Vapor (Gas) Density*: Not pertinent; *Ratio of Specific Heats of Vapor (Gas)*: Not pertinent; *Latent Heat of Vaporization*: Not pertinent; *Heat of Combustion*: Not pertinent; *Heat of Decomposition*: Not pertinent.

Health Hazards Information - *Recommended Personal Protective Equipment*: Goggles or face shield; rubber gloves; *Symptoms Following Exposure*: In general, these chemical have a moderate order of toxicity. Repeated skin contact with concentrated solutions may cause dermatitis. Ingestion may cause gastrointestinal, vomiting, and diarrhea; *General Treatment for Exposure*: INGESTION: induce vomiting and follow with gastric lavage. SKIN: wash off with water; *Toxicity by Inhalation (Threshold Limit Value)*: Not pertinent; *Short-Term Exposure Limits*: Not pertinent; *Toxicity by Ingestion*: Grade 2, LD_{50} 0.5-5 g/kg; *Late Toxicity*: None; *Vapor (Gas) Irritant Characteristics*: Non-volatile; *Liquid or Solid Irritant Characteristics*: Minimum hazard. If spilled on clothing and allowed to remain, may cause smarting and reddening of the skin; *Odor Threshold*: Not pertinent.

Fire Hazards - *Flash Point* : Not flammable; *Flammable Limits in Air (%)*: Not flammable; *Fire Extinguishing Agents*: Not pertinent; *Fire Extinguishing Agents Not To Be Used*: Not pertinent; *Special Hazards of Combustion Products*: Irritating vapors are generated in fires; *Behavior in Fire*: Not pertinent; *Ignition Temperature* : Not pertinent; *Electrical Hazard*: Not pertinent; *Burning Rate*: Not pertinent.

Chemical Reactivity - *Reactivity with Water*: No reaction; *Reactivity with Common Materials*: No

reactions; *Stability During Transport:* Stable; *Neutralizing Agents for Acids and Caustics:* Not pertinent; *Polymerization:* Not pertinent; *Inhibitor of Polymerization:* Not pertinent.

SODIUM AMIDE

Chemical Designations - *Synonyms*: Sodamite; *Chemical Formula*: $NaNH_2$

Observable Characteristics - *Physical State (as normally shipped)*: Solid; *Color*: Gray; *Odor*: Like a ammonia.

Physical and Chemical Properties - *Physical State at 15 °C and 1 atm.*: Solid; *Molecular Weight*: 39.01; *Boiling Point at 1 atm.*: 752, 400, 673; *Freezing Point*: 410, 210, 483; *Critical Temperature*: Not pertinent; *Critical Pressure*: Not pertinent; *Specific Gravity*: 1.39 at 20 °C (solid); *Vapor (Gas) Density*: Not pertinent; *Ratio of Specific Heats of Vapor (Gas)*: Not pertinent; *Latent Heat of Vaporization*: Not pertinent; *Heat of Combustion*: Not pertinent; *Heat of Decomposition*: Not pertinent.

Health Hazards Information - *Recommended Personal Protective Equipment*: Goggles or face shield, dust respirator; rubber gloves and shoes; *Symptoms Following Exposure*: Ammonia gas formed by reaction of solid with moisture irritates eyes and skin. Solid causes caustic burns of eyes and skin. Ingestion burns mouth and stomach in same way as caustic soda and may cause perforation of tissue; *General Treatment for Exposure*: INGESTION: give water or milk followed by dilute vinegar or fruit juice; do NOT induce vomiting; call a doctor. SKIN OR EYES: flood all affect areas with copious amounts of water; *Toxicity by Inhalation (Threshold Limit Value)*: Not pertinent; *Short-Term Exposure Limits*: Not pertinent; *Toxicity by Ingestion*: Data not available; *Late Toxicity*: None; *Vapor (Gas) Irritant Characteristics*: Only that of ammonia formed by reaction of solid with moisture in air; *Liquid or Solid Irritant Characteristics*: Burns skin and eyes just like caustic soda; *Odor Threshold*: Not pertinent.

Fire Hazards - *Flash Point* : Flammable solid; *Flammable Limits in Air (%)*: Not pertinent; *Fire Extinguishing Agents*: Dry soda ash, graphite, salt, or other recommended dry powder such as dry limestone; *Fire Extinguishing Agents Not To Be Used*: Water; *Special Hazards of Combustion Products*: Toxic and irritating ammonia gas may be formed in fires; *Behavior in Fire*: No data; *Ignition Temperature* : Not pertinent; *Electrical Hazard*: No data; *Burning Rate*: Not pertinent.

Chemical Reactivity - *Reactivity with Water:* Reacts violently and often bursts into flames. Also forms caustic soda solution; *Reactivity with Common Materials:* No data; *Stability During Transport:* Stable; *Neutralizing Agents for Acids and Caustics:* The caustic solution formed by the reaction with water can be diluted with water and then neutralized by acetic acid; *Polymerization:* Not pertinent; *Inhibitor of Polymerization:* Not pertinent.

SODIUM ARSENATE

Chemical Designations - *Synonyms*: Disodium arsenate heptahydrate; Sodium arsenate, dibasic; *Chemical Formula*: $Na_2HasO_4 \cdot 7H_2O$.

Observable Characteristics - *Physical State (as normally shipped)*: Solid; *Color*: White; *Odor*: None.

Physical and Chemical Properties - *Physical State at 15 °C and 1 atm.*: Solid; *Molecular Weight*: 312; *Boiling Point at 1 atm.*: (decomposes) 356, 180, 453; *Freezing Point*: 135, 57, 330; *Critical Temperature*: Not pertinent; *Critical Pressure*: Not pertinent; *Specific Gravity*: 1.87 at 20 °C (solid); *Vapor (Gas) Density*: Not pertinent; *Ratio of Specific Heats of Vapor (Gas)*: Not pertinent; *Latent Heat of Vaporization*: Not pertinent; *Heat of Combustion*: Not pertinent; *Heat of Decomposition*: Not pertinent.

Health Hazards Information - *Recommended Personal Protective Equipment*: Dust mask, goggles or face shield, protective gloves; *Symptoms Following Exposure*: Inhalation of massive doses can cause laryngitis, bronchitis. Ingestion cause concentration in throat and difficulty in swallowing; also causes burning and pain, vomiting, profuse diarrhea. Cyanosis, coma, convulsion, and death. Contact with eyes causes irritation. Contact with skin causes various skin eruptions, more after as a late manifestation, or chronic poisoning; *General Treatment for Exposure*: INHALATION: remove victim from exposure; support respiration; INGESTION: gastric lavage with water, followed by 1 glass of milk; consult physician. EYES: flush with water for at least 15 min. SKIN: flush with water; *Toxicity* by Inhalation (Threshold Limit Value): 0.5 mg/m^3 (as arsenic); Short-Term Exposure Limits: Data not available; Toxicity by Ingestion: Grade 4, LD$_{50}$ < 50 mg/kg; Late Toxicity: Possible carcinogenic effects on skin and lungs; Vapor (Gas) Irritant Characteristics: Data not available; Liquid or Solid Irritant Characteristics: Data not available; Odor Threshold: Data not available.

Fire Hazards - Flash Point : Not flammable; Flammable Limits in Air (%): Not flammable; Fire Extinguishing Agents: Not pertinent; Fire Extinguishing Agents Not To Be Used: Not pertinent; Special Hazards of Combustion Products: Not pertinent; Behavior in Fire: No information; Ignition Temperature : Not pertinent; Electrical Hazard: Not pertinent; Burning Rate: Not pertinent.

Chemical Reactivity - Reactivity with Water: No reaction; Reactivity with Common Materials: No reactions; Stability During Transport: Stable; Neutralizing Agents for Acids and Caustics: Not pertinent; Polymerization: Not pertinent; Inhibitor of Polymerization: Not pertinent.

SODIUM ARSENITE

Chemical Designations - Synonyms: Sodium metaarsenite; Chemical Formula: Na_3AsO_3-$NaAsO_2$.

Observable Characteristics - Physical State (as normally shipped): Solid; Color: White to gray; Odor: None.

Physical and Chemical Properties - Physical State at 15 °C and 1 atm.: Solid; Molecular Weight: Not pertinent; Boiling Point at 1 atm.: Not pertinent (decomposes); Freezing Point: 1.139, 615, 888; Critical Temperature: Not pertinent; Critical Pressure: Not pertinent; Specific Gravity: 1.87 at 20 °C (solid); Vapor (Gas) Density: Not pertinent; Ratio of Specific Heats of Vapor (Gas): Not pertinent; Latent Heat of Vaporization: Not pertinent; Heat of Combustion: Not pertinent; Heat of Decomposition: Not pertinent.

Health Hazards Information - Recommended Personal Protective Equipment: Dust mask; rubber gloves, goggles or face shield; Symptoms Following Exposure: Dust may irritate eyes. Ingestion or excessive inhalation of dust causes irritation of stomach and intestines with nausea, vomiting, and diarrhea: bloody stools, shock, rapid pulse, coma; General Treatment for Exposure: EYES: flush with water for at least 15 min. SKIN: wash with large amounts of water. INGESTION: immediately induce evacuation of intestinal tract by gastric lavage and saline cathartic, sec physician immediately, consider possible development of arsenic poisoning; Toxicity by Inhalation (Threshold Limit Value): 0.5 mg/m^3 (as arsenic); Short-Term Exposure Limits: Data not available; Toxicity by Ingestion: Grade 4, LD$_{50}$ = 42 mg/kg; Late Toxicity: May be carcinogenic. Arsenic poisoning may develop; Vapor (Gas) Irritant Characteristics: Data not available; Liquid or Solid Irritant Characteristics: Data not available; Odor Threshold: Not pertinent.

Fire Hazards - Flash Point : Not flammable; Flammable Limits in Air (%): Not flammable; Fire Extinguishing Agents: Not pertinent; Fire Extinguishing Agents Not To Be Used: Not pertinent; Special Hazards of Combustion Products: Toxic arsenic fumes may form. The use of an SCBA is recommended; Behavior in Fire: Not pertinent; Ignition Temperature : Not pertinent; Electrical Hazard: Not pertinent; Burning Rate: Not pertinent.

Chemical Reactivity - Reactivity with Water: No reaction; Reactivity with Common Materials: No reactions; Stability During Transport: Stable; Neutralizing Agents for Acids and Caustics: Not pertinent; Polymerization: Not pertinent; Inhibitor of Polymerization: Not pertinent.

SODIUM AZIDE

Chemical Designations - Synonyms: Hydrazoic acid, sodium salt; Chemical Formula: NaN_3.

Observable Characteristics — Physical State (as normally shipped): Solid; Color: White; Odor: None.

Physical and Chemical Properties — Physical State at 15 °C and 1 atm.: Solid; Molecular Weight: 65; Boiling Point at 1 atm.: Not pertinent (decomposes); Freezing Point: Not pertinent; Critical Temperature: Not pertinent; Critical Pressure: Not pertinent; Specific Gravity: 1.85 at 20 °C (solid); Vapor (Gas) Density: Not pertinent; Ratio of Specific Heats of Vapor (Gas): Not pertinent; Latent Heat of Vaporization: Not pertinent; Heat of Combustion: Not pertinent; Heat of Decomposition: Not pertinent.

Health Hazards Information - Recommended Personal Protective Equipment: Dust mask, protective clothing, goggles; *Symptoms Following Exposure*: Inhalation or ingestion causes dizziness, weakness, blurred vision, slight shortness of breath, and reeling of going to faint; moderate reduction of blood pressure and bardycardia. Contact with eyes or skin causes irritation; *General Treatment for Exposure*: Give oxygen if weakness, pallor, or low blood pressure is observed. INHALATION: remove victim lo fresh air, enforce rest; call a doctor. EYES: flush with water for at least 15 min. SKIN: flush with water, wash with soap and water. INGESTION: give large amount of water and induce vomiting at once, get medical attention; *Toxicity by Inhalation (Threshold Limit Value)*: Data not available; *Short-Term Exposure Limits*: Data not available; *Toxicity by Ingestion*: Grade 4, oral rat LD_{50} = 27 mg/kg (technical); *Late Toxicity*: Potent mutagen of salmon-sperm DNA; *Vapor (Gas) Irritant Characteristics*: Data not available; *Liquid or Solid Irritant Characteristics*: Data not available; *Odor Threshold*: Not pertinent.

Fire Hazards - *Flash Point* : Not flammable; *Flammable Limits in Air (%):* Not flammable; *Fire Extinguishing Agents:* Not pertinent; *Fire Extinguishing Agents Not To Be Used:* Not pertinent; *Special Hazards of Combustion Products:* May form toxic hydrazoic acid fumes in fires; *Behavior in Fire:* Containers may explode; *Ignition Temperature :* Not pertinent; *Electrical Hazard:* Not pertinent; *Burning Rate:* Not pertinent.

Chemical Reactivity - *Reactivity with Water:* Dissolves to form an alkaline solution. The reaction is non-violent; *Reactivity with Common Materials:* Forms explosion-sensitive materials with some metals such as lead, silver, mercury, and copper; *Stability During Transport:* Stable but must not be in contact with acids; *Neutralizing Agents for Acids and Caustics:* Not pertinent; *Polymerization:* Not pertinent; *Inhibitor of Polymerization:* Not pertinent.

SODIUM BISULFITE

Chemical Designations - *Synonyms*: Sodium acid sulfite; Sodium metebisulfite; Sodium pyrosulfite; *Chemical Formula*: $NaHSO_4$-$Na_2S_2O_5$.

Observable Characteristics - *Physical State (as normally shipped)*: Solid; *Color*: White; *Odor*: Pungent odor of sulfur dioxide when moist.

Physical and Chemical Properties - *Physical State at 15 °C and 1 atm.*: Solid; *Molecular Weight*: 104.06; *Boiling Point at 1 atm.*: Decomposes; *Freezing Point*: Not pertinent; *Critical Temperature*: Not pertinent; *Critical Pressure*: Not pertinent; *Specific Gravity*: 1.48 at 20 °C (solid); *Vapor (Gas) Density*: Not pertinent; *Ratio of Specific Heats of Vapor (Gas)*: Not pertinent; *Latent Heat of Vaporization*: Not pertinent; *Heat of Combustion*: Not pertinent; *Heat of Decomposition*: Not pertinent.

Health Hazards Information - *Recommended Personal Protective Equipment*: Dust mask; goggles or face shield; *Symptoms Following Exposure*: Powder is irritating to eyes, nose, and throat and can irritate skin. Ingestion may cause irritation of stomach. Very large doses cause violent colic, diarrhea, depression, and death; *General Treatment for Exposure*: INHALATION OR INGESTION: get medical attention at once. SKIN: wash with plenty of water. EYES: flush with plenty of water for at least 15 min., and get medical attention at once; *Toxicity by Inhalation (Threshold Limit Value)*: Not pertinent; *Short-Term Exposure Limits*: Not pertinent; *Toxicity by Ingestion*: Grade 2, LD_{50} 0.5 to 5 g/kg; *Late Toxicity*: None; *Vapor (Gas) Irritant Characteristics*: Non-volatile; *Liquid or Solid Irritant Characteristics*: Irritates skin and mucous membranes; *Odor Threshold*: Not pertinent.

Fire Hazards - *Flash Point* : Not flammable; *Flammable Limits in Air (%):* Not flammable; *Fire Extinguishing Agents:* Not pertinent; *Fire Extinguishing Agents Not To Be Used:* Not pertinent; *Special Hazards of Combustion Products:* Not pertinent; *Behavior in Fire:* Not pertinent; *Ignition Temperature :* Not pertinent; *Electrical Hazard:* Not pertinent; *Burning Rate:* Not pertinent.

 (vi) Chemical Reactivity - *Reactivity with Water:* No reaction; *Reactivity with Common Materials:* No reactions; *Stability During Transport:* Stable; *Neutralizing Agents for Acids and Caustics:* Not pertinent; *Polymerization:* Not pertinent; *Inhibitor of Polymerization:* Not pertinent.

SODIUM BORATE

Chemical Designations - *Synonyms*: Borax, anhydrous; Sodium biborate; Sodium pyroborate; Sodium tetraborate,anhydrous; *Chemical Formula*: $Na_2B_4O_7$.

Observable Characteristics - *Physical State (as normally shipped)*: Solid; *Color*: White; *Odor*: None.
Physical and Chemical Properties - *Physical State at 15 °C and 1 atm.*: Solid; *Molecular Weight*: 201.26; *Boiling Point at 1 atm.*: Not pertinent (decomposes); *Freezing Point*: Not pertinent; *Critical Temperature*: Not pertinent; *Critical Pressure*: Not pertinent; *Specific Gravity*: 2.367 at 20 °C (solid); *Vapor (Gas) Density*: Not pertinent; *Ratio of Specific Heats of Vapor (Gas)*: Not pertinent; *Latent Heat of Vaporization*: Not pertinent; *Heat of Combustion*: Not pertinent; *Heat of Decomposition*: Not pertinent.
Health Hazards Information - *Recommended Personal Protective Equipment*: Dust mask and goggles or face shield; *Symptoms Following Exposure*: No adverse effects from inhaling borax have been reported. Ingestion may cause acute or chronic effects: initial symptoms are nausea, vomiting, and diarrhea: these may be followed by weakness, depression, headaches, skin rashes, drying skin, cracked lips, and loss of hair; shock may follow ingestion of large doses and may interfere with breathing. Eye contact with powder or solutions may cause irritation; no chronic effects have been recognized, but continued contact should be avoided. Local skin irritation may result from contact with powder or strong solutions: the latter may cause chronic dermatitis on prolonged contact, and if skin is broken. enough boron may be absorbed lo cause boron poisoning (symptoms are similar to those for ingestion); *General Treatment for Exposure*: INHALATION: move to fresh air: call physician immediately; give mouth-to-mouth resuscitation if breathing has ceased: give oxygen if authorized by physician: keep victim warm. INGESTION: get medical attention quickly: if victim is conscious, give warm salty or soapy water to induce vomiting: repeat until vomitus is clear: additional water may be given lo wash out stomach. EYES: get medical attention quickly: Hush with copious amounts of water for at least 15 min. (30 min. if physician is not available), holding eyelids apart. SKIN: Hush with water: remove contaminated clothing under shower: do not use chemical neulralizers: get medical attention unless burn is minor; *Toxicity by Inhalation (Threshold Limit Value)*: Data not available; *Short-Term Exposure Limits*: Data not available; *Toxicity by Ingestion*: Grade 2, LD_{50} 0.5-5 g/kg; *Late Toxicity*: Data not available; *Vapor (Gas) Irritant Characteristics*: Data not available; *Liquid or Solid Irritant Characteristics*: Data not available; *Odor Threshold*: Data not available.
Fire Hazards - *Flash Point* : Not flammable; *Flammable Limits in Air (%):* Not flammable; *Fire Extinguishing Agents:* Not pertinent; *Fire Extinguishing Agents Not To Be Used:* Not pertinent; *Special Hazards of Combustion Products:* Not pertinent; *Behavior in Fire:* The compound melts into a glassy material that may flow in large quantities and ignite combustible materials it comes in contact with; *Ignition Temperature :* Not pertinent; *Electrical Hazard:* Not pertinent; *Burning Rate:* Not pertinent.
Chemical Reactivity - *Reactivity with Water:* No reaction; *Reactivity with Common Materials:* No reactions; *Stability During Transport:* Stable; *Neutralizing Agents for Acids and Caustics:* Not pertinent; *Polymerization:* Not pertinent; *Inhibitor of Polymerization:* Not pertinent.

SODIUM BOROHYDRIDE

Chemical Designations - *Synonyms*: No common synonyms; *Chemical Formula*: $NaBH_4$.
Observable Characteristics - *Physical State (as normally shipped)*: Solid or solution in caustic; *Color*: White; *Odor*: Odorless.
Physical and Chemical Properties - *Physical State at 15 °C and 1 atm.*:Solid; *Molecular Weight*: 37.83; *Boiling Point at 1 atm.*: Decomposes; *Freezing Point*: Not pertinent; *Critical Temperature*: Not pertinent; *Critical Pressure*: Not pertinent; *Specific Gravity*: 1.074 at 20 °C (solid); *Vapor (Gas) Density*: Not pertinent; *Ratio of Specific Heats of Vapor (Gas)*: Not pertinent; *Latent Heat of Vaporization*: Not pertinent; *Heat of Combustion*: Not pertinent; *Heat of Decomposition*: Not pertinent.
Health Hazards Information - *Recommended Personal Protective Equipment*: Goggles, rubber gloves, and protective clothing; *Symptoms Following Exposure*: Solid irritates skin lf ingested can form large volume of gas and lead to a gas embolism; *General Treatment for Exposure*: INGESTION: do NOT induce vomiting: give dilute vinegar, lemon juice, milk. or olive oil: call a doctor. SKIN AND EYES: flood with large amount of water; *Toxicity by Inhalation (Threshold Limit Value)*: Not pertinent; *Short-Term Exposure Limits*: Not pertinent; *Toxicity by Ingestion*: Violent reaction with acid in stomach. Considered toxic because of boron content; *Late Toxicity*: Data not available; *Vapor (Gas) Irritant*

Characteristics: Non-volatile; *Liquid or Solid Irritant Characteristics*: Irritates skin; *Odor Threshold*: Not pertinent.

Fire Hazards - *Flash Point* : Not flammable; *Flammable Limits in Air (%):* Not pertinent; *Fire Extinguishing Agents:* Graphite, limestone, soda ash, sodium chloride powders; *Fire Extinguishing Agents Not To Be Used:* Water, carbon dioxide, or halogenated extinguishing agents; *Special Hazards of Combustion Products:* Not pertinent; *Behavior in Fire:* Decomposes and produces highly flammable hydrogen gas; *Ignition Temperature :* Not pertinent; *Electrical Hazard:* Not pertinent; *Burning Rate:* Not pertinent.

Chemical Reactivity - *Reactivity with Water:* Reacts to form flammable hydrogen gas; *Reactivity with Common Materials:* Reacts with acids to form toxic, flammable diborane gas. Slowly attacks and destroys glass; *Stability During Transport:* Stable unless contaminated with acids or is overheated, thereby forming flammable hydrogen gas; *Neutralizing Agents for Acids and Caustics:* Caustic formed by the reaction with water can be diluted with water and then neutralized with acetic acid; *Polymerization:* Not pertinent; *Inhibitor of Polymerization:* Not pertinent.

SODIUM CACODYLATE

Chemical Designations - *Synonyms*: Arsecodile; Arsicodile; Arsycodile; Phytar 160; Phytar 560; Sodium dimethylarsonate; *Chemical Formula*: $(CH_3)_2AsOONa$.

Observable Characteristics - *Physical State (as normally shipped)*: Solid or water solution; *Color*: Colorless to light yellow; *Odor*: None.

Physical and Chemical Properties - *Physical State at 15 °C and 1 atm.*: Solid; *Molecular Weight*: 160.0; *Boiling Point at 1 atm.*: Not pertinent (decomposes); *Freezing Point*: Not pertinent; *Critical Temperature*: Not pertinent; *Critical Pressure*: Not pertinent; *Specific Gravity:* (est.) > 1 at 20 °C (solid); *Vapor (Gas) Density*: Not pertinent; *Ratio of Specific Heats of Vapor (Gas)*: Not pertinent; *Latent Heat of Vaporization*: Not pertinent; *Heat of Combustion*: Not pertinent; *Heat of Decomposition*: Not pertinent.

Health Hazards Information - *Recommended Personal Protective Equipment*: Goggles or face shield: dust mask: rubber gloves; *Symptoms Following Exposure*: Dust may irritate eyes. Ingestion or excessive inhalation causes irritation of stomach and intestines with nausea, vomiting, diarrhea, shock, rapid pulse, coma; *General Treatment for Exposure*: INHALATION: remove victim from exposure: call physician. EYES: flush with water. SKIN: flush with water and wash well with soap and water. INGESTION: call physician: induce evacuation of intestinal tract by inducing vomiting, giving gastric lavage and a saline cathartic. Do NOT use BAL as an antidote; *Toxicity by Inhalation*: No data; *Short-Term Exposure Limits*: Data not available; *Toxicity by Ingestion*: Grade 2, oral LD_{50} = 2,600 mg/kg (rat); *Late Toxicity*: Data not available; *Vapor (Gas) Irritant Characteristics*: Data not available; *Liquid or Solid Irritant Characteristics*: Data not available; *Odor Threshold*: Not pertinent.

Fire Hazards - *Flash Point* : Not flammable; *Flammable Limits in Air (%):* Not flammable; *Fire Extinguishing Agents:* Not pertinent; *Fire Extinguishing Agents Not To Be Used:* Not pertinent; *Special Hazards of Combustion Products:* Arsenic containing fumes are formed in fires; *Behavior in Fire:* Not pertinent; *Ignition Temperature :* Not pertinent; *Electrical Hazard:* Not pertinent; *Burning Rate:* Not pertinent.

Chemical Reactivity - *Reactivity with Water:* No reaction; *Reactivity with Common Materials:* Corrodes many common metals but the reaction is non-hazardous; *Stability During Transport:* Stable; *Neutralizing Agents for Acids and Caustics:* Not pertinent; *Polymerization:* Not pertinent; *Inhibitor of Polymerization:* Not pertinent.

SODIUM CHLORATE

Chemical Designations - *Synonyms*: Chlorate of soda; *Chemical Formula*: $NaClO_3$.

Observable Characteristics - *Physical State (as normally shipped)*: Solid; *Color*: Pale yellow to white; *Odor*: Odorless.

Physical and Chemical Properties - *Physical State at 15 °C and 1 atm.*: Solid; *Molecular Weight*: 106.45; *Boiling Point at 1 atm.*: Decomposes; *Freezing Point*: 478, 248, 521; *Critical Temperature*:

Not pertinent; *Critical Pressure*: Not pertinent; *Specific Gravity*: 2.49 at 15 °C (solid); *Vapor (Gas) Density*: Not pertinent; *Ratio of Specific Heats of Vapor (Gas)*: Not pertinent; *Latent Heat of Vaporization*: Not pertinent; *Heat of Combustion*: Not pertinent; *Heat of Decomposition*: Not pertinent.
Health Hazards Information - *Recommended Personal Protective Equipment*: Clean work clothing (must be washed well with water after each exposure): rubber gloves and shoes: where dusty, goggles and an approved dust respirator. Do NOT use oils, greases, or protective creams on skin; *Symptoms Following Exposure*: ingestion of a toxic dose (at least ½ oz.) leads to severe gastroenteric pain, vomiting, and diarrhea. Possible respiratory difficulties, including failure of respiration. Kidney and liver injury may also be produced. The lethal oral dose For an adult is approximately 15 gm. Contact with eyes causes irritation; *General Treatment for Exposure*: INGESTION: induce vomiting and follow with gastric lavage, saline cathartics. Fluid therapy, and oxygen. EYES: wash thoroughly with water; *Toxicity by Inhalation (Threshold Limit Value)*: Not pertinent; *Short-Term Exposure Limits*: Not pertinent; *Toxicity by Ingestion*: Grade 3, LD_{50} 50 to 500 mg/kg; *Late Toxicity*: Data not available; *Vapor (Gas) Irritant Characteristics*: Non-volatile; *Liquid or Solid Irritant Characteristics*: Prolonged exposure to solid or dust may irritate skin; *Odor Threshold*: Not pertinent.
Fire Hazards - *Flash Point* : Not flammable but the product will support combustion; *Flammable Limits in Air (%)*: Not pertinent; *Fire Extinguishing Agents*: Water; *Fire Extinguishing Agents Not To Be Used:* Fire blankets; *Special Hazards of Combustion Products:* In fire situations, oxygen is liberated which can increase the intensity of fires; *Behavior in Fire:* The product melts and then decomposes giving off oxygen gas that increases the intensity of fires. This product reacts explosively, either as a solid or liquid with all organic matter and some metals; *Ignition Temperature :* Not pertinent; *Electrical Hazard:* Not pertinent; *Burning Rate:* Not pertinent.
Chemical Reactivity - *Reactivity with Water:* No reaction; *Reactivity with Common Materials:* Chlorates are powerful oxidizing agents and can cause explosions when heated or rubbed with wood, organic matter, sulfur, and many metals. Even water solutions react in this manner if the solution is more than 30% concentrated, especially when warm; *Stability During Transport:* This product begins decomposing at 572 of with the evolution of oxygen gas. The decomposition may become self-sustaining. Oxygen liberation will increase the intensity of fires; *Neutralizing Agents for Acids and Caustics:* Not pertinent; *Polymerization:* Not pertinent; *Inhibitor of Polymerization:* Not pertinent.

SODIUM CHROMATE
Chemical Designations - *Synonyms*: Neutral sodium chromate, anhydrous; Sodium chromate (VI); *Chemical Formula*: Na_2CrO_4.
Observable Characteristics - *Physical State (as normally shipped)*: Solid; *Color*: Yellow; *Odor*: None.
Physical and Chemical Properties - *Physical State at 15 °C and 1 atm.*: Solid; *Molecular Weight*: 162; *Boiling Point at 1 atm.*: Not pertinent (decomposes); *Freezing Point*: Not pertinent; *Critical Temperature*: Not pertinent; *Critical Pressure*: Not pertinent; *Specific Gravity*: 2.723 at 25 °C (solid); *Vapor (Gas) Density*: Not pertinent; *Ratio of Specific Heats of Vapor (Gas)*: Not pertinent; *Latent Heat of Vaporization*: Not pertinent; *Heat of Combustion*: Not pertinent; *Heat of Decomposition*: Not pertinent.
Health Hazards Information - *Recommended Personal Protective Equipment*: U. S. Bu. Mines approved respirator, rubber gloves; chemical safely goggles; rubber apron and sleeves, face shield, rubber shoes, protective clothing; *Symptoms Following Exposure*: Inhalation causes irritation and may ulcerate mucous membranes: continued irritation of the nose may lead to perforation of the septum, ingestion causes severe circulatory collapse and toxic nephritis, may be fatal. Contact with eyes causes severe irritation and possible conjunctivitis. Irritates skin and can cause ulcers: if skin is broken, prolonged contact may cause "chrome sores" (slow-healing, hard-rimmed ulcers), which leave the area vulnerable lo infection as a secondary effect; *General Treatment for Exposure*: INHALATION: remove victim to fresh air: get medical attention. INGESTION: get immediate medical help: if vomiting is not spontaneous, give an emetic such as soapy water followed by copious water in lake. EYES: immediately flush with plenty of water for at least 15 min.; consult physician promptly. SKIN: immediately flush with plenty of water for at least 15 min.; persistent dermatitis should be referred to

a physician: wash contaminated skin or clothing until chromate color disappears; *Toxicity by Inhalation (Threshold Limit Value)*: Data not available; *Short-Term Exposure Limits*: Data not available; *Toxicity by Ingestion*: Grade 3, LD$_{50}$ 50-500 mg/kg; *Late Toxicity*: Possible lung cancer; *Vapor (Gas) Irritant Characteristics*: Data not available; *Liquid or Solid Irritant Characteristics*: Data not available; *Odor Threshold*: Data not available.

Fire Hazards - *Flash Point* : Not flammable; *Flammable Limits in Air (%):* Not flammable; *Fire Extinguishing Agents:* Not pertinent; *Fire Extinguishing Agents Not To Be Used:* Not pertinent; *Special Hazards of Combustion Products:* Toxic chromium oxide fumes may form in fires; *Behavior in Fire:* Can increase the intensity of fires when in contact with combustible materials; *Ignition Temperature :* Not pertinent; *Electrical Hazard:* Not pertinent; *Burning Rate:* Not pertinent.

Chemical Reactivity - *Reactivity with Water:* No reaction; *Reactivity with Common Materials:* Causes fire when in contact with combustible materials; *Stability During Transport:* Stable; *Neutralizing Agents for Acids and Caustics:* Not pertinent; *Polymerization:* Not pertinent; *Inhibitor of Polymerization:* Not pertinent.

SODIUM CYANIDE

Chemical Designations - *Synonyms*: Hydrocyanic acid; *Chemical Formula*: NaCN.

Observable Characteristics - *Physical State (as normally shipped)*: Solid; *Color*: White; *Odor*: Odorless when dry. When moist it has a slight odor of hydrocyanic acid.

Physical and Chemical Properties - *Physical State at 15 °C and 1 atm.*: Solid; *Molecular Weight*: 49.01; *Boiling Point at 1 atm.*: Very high; *Freezing Point*: 1047, 564, 837; *Critical Temperature*: Not pertinent; *Critical Pressure*: Not pertinent; *Specific Gravity*: 1.60 at 25 °C (solid); *Vapor (Gas) Density*: Not pertinent; *Ratio of Specific Heats of Vapor (Gas)*: Not pertinent; *Latent Heat of Vaporization*: Not pertinent; *Heat of Combustion*: Not pertinent; *Heat of Decomposition*: Not pertinent.

Health Hazards Information - *Recommended Personal Protective Equipment*: Protective gloves when handling solid sodium cyanide; rubber gloves when handling cyanide solutions (wash hands and rubber gloves thoroughly with running water after handling cyanides); U. S. Bureau of Mines approved dust respirator; approved chemical safety goggles; *Symptoms Following Exposure*: As little as 180 milligrams is a rapidly fatal poison if ingested non-lethal doses may cause toxic symptoms. Strong water solutions, or the solid itself, can be absorbed by the skin and cause deep ulcers which heal slowly; *General Treatment for Exposure*: INGESTION: start treatment immediately; call a physician; carry victim to fresh air; have him lie down; keep him quiet and warm until physician arrives. If victim is conscious and breathing: induce vomiting by giving emetic of warm salt water (1 tablespoon salt/cup water) repeat until vomit fluid is clear; then have victim drink one pint of 1 % solution of sodium thiosulfate, to be repealed in 15 min. If victim has stopped breathing: give artificial respiration until breathing starts. If victim is unconscious but breathing: give oxygen from an inhalator. For all of above conditions, have victim breathe amyl nitrite. Break nitrite pearl in a cloth and hold lightly under victim's nose for 15 sec., repeating 5 times at about 15-sec. intervals, If necessary repeat this procedure every 3 min. with fresh pearls until 3 or 4 have been given. (Pearls must not be over 2 years old. Avoid breathing amyl nitrite while administering it to victim.); *Toxicity by Inhalation (Threshold Limit Value)*: Not pertinent; *Short-Term Exposure Limits*: Not pertinent; *Toxicity by Ingestion*: Grade 4, bellow 50 mg/kg; *Late Toxicity*: None; *Vapor (Gas) Irritant Characteristics*: Non-volatile, but moisture in air liberate some lethal hydrogen cyanide gas; *Liquid or Solid Irritant Characteristics*: Fairly severe skin irritant; may cause pain and second-degree burns after a few minutes contact; *Odor Threshold*: Not pertinent.

Fire Hazards - *Flash Point* : Not flammable; *Flammable Limits in Air (%):* Not flammable; *Fire Extinguishing Agents:* Not pertinent; *Fire Extinguishing Agents Not To Be Used:* Not pertinent; *Special Hazards of Combustion Products:* Not pertinent; *Behavior in Fire:* Not pertinent; *Ignition Temperature :* Not flammable; *Electrical Hazard:* Not pertinent; *Burning Rate:* Not pertinent.

Chemical Reactivity - *Reactivity with Water:* When sodium cyanide dissolves in water, a mild reaction occurs and some poisonous hydrogen cyanide gas is liberated. The gas is not generally a concern unless it is generated in an enclosed space. If the water is acidic, then large amounts of the toxic gas forms rapidly; *Reactivity with Common Materials:* No reactions; *Stability During Transport:* Stable; *Neutralizing Agents for Acids and Caustics:* Not pertinent; *Polymerization:* Not pertinent; *Inhibitor of Polymerization:* Not pertinent.

SODIUM DICHROMATE

Chemical Designations - *Synonyms*: No common synonyms; *Chemical Formula*: $Na_2Cr_2O_7$.
Observable Characteristics - *Physical State (as normally shipped)*: Solid; *Color*: Bright orange red; *Odor*: Odorless.
Physical and Chemical Properties - *Physical State at 15 °C and 1 atm.*: Solid; *Molecular Weight*: 262.01; *Boiling Point at 1 atm.*: Decomposes; *Freezing Point*: 675, 357, 630; *Critical Temperature*: Not pertinent; *Critical Pressure*: Not pertinent; *Specific Gravity*: 2.35 at 25 °C (solid); *Vapor (Gas) Density*: Not pertinent; *Ratio of Specific Heats of Vapor (Gas)*: Not pertinent; *Latent Heat of Vaporization*: Not pertinent; *Heat of Combustion*: Not pertinent; *Heat of Decomposition*: Not pertinent.
Health Hazards Information - *Recommended Personal Protective Equipment*: Approved dust mask, protective gloves; goggles or face shield; *Symptoms Following Exposure*: Inhalation of dust or mist causes respiratory irritation sometimes resembling asthma, nasal septal perforation may occur. Ingestion causes vomiting, diarrhea, and (rarely) stomach and kidney complications. Contact with eyes or skin produces local irritation: repeated skin exposure causes dermatitis; *General Treatment for Exposure*: INGESTION: have victim drink water or milk; do NOT induce vomiting; call a doctor. SKIN OR EYE CONTACT: treat like acid burns; flush eyes with water for at least 15 min.; external lesions can be scrubbed with a 2% solution of sodium thiosulfate; *Toxicity by Inhalation (Threshold Limit Value)*: Not pertinent; *Short-Term Exposure Limits*: Data not available; *Toxicity by Ingestion*: Grade 3, LD_{50} 50-500 mg/kg; *Late Toxicity*: Some suggestion of lung cancer; *Vapor (Gas) Irritant Characteristics*: Dusts or mists may cause severe irritation of eye and throat and can cause eye and lung injury. They cannot be tolerated even at low concentrations; *Liquid or Solid Irritant Characteristics*: Severe skin irritant. Causes second- and third-degree burns on short contact and is very injurious to the eyes; *Odor Threshold*: Odorless.
Fire Hazards - *Flash Point* : Not flammable; *Flammable Limits in Air (%)*: Not flammable; *Fire Extinguishing Agents:* Flood with large amounts of water; *Fire Extinguishing Agents Not To Be Used:* Not pertinent; *Special Hazards of Combustion Products:* Not pertinent; *Behavior in Fire:* Decomposes to produce oxygen upon heating. May ignite other combustibles upon contact; *Ignition Temperature :* Not flammable; *Electrical Hazard:* Not pertinent; *Burning Rate:* Not flammable.
Chemical Reactivity - *Reactivity with Water:* No reaction; *Reactivity with Common Materials:* When in contact with finely divided combustibles, such as sawdust, ignition may occur; *Stability During Transport:* Stable; *Neutralizing Agents for Acids and Caustics:* Not pertinent; *Polymerization:* Not pertinent; *Inhibitor of Polymerization:* Not pertinent.

SODIUM HYDRIDE

Chemical Designations - *Synonyms*: No common synonyms; *Chemical Formula*: NaH.
Observable Characteristics - *Physical State (as normally shipped)*: Solid; *Color*: Gray; *Odor*: Odorless.
Physical and Chemical Properties - *Physical State at 15 °C and 1 atm.*:Solid; *Molecular Weight*: Not applicable; *Boiling Point at 1 atm.*: Very high; *Freezing Point*: Not pertinent; *Critical Temperature*: Not pertinent; *Critical Pressure*: Not pertinent; *Specific Gravity*: Data not available; *Vapor (Gas) Density*: Not pertinent; *Ratio of Specific Heats of Vapor (Gas)*: Not pertinent; *Latent Heat of Vaporization*: Not pertinent; *Heat of Combustion*: Not pertinent; *Heat of Decomposition*: Not pertinent.
Health Hazards Information - *Recommended Personal Protective Equipment*: Face shield; rubber gloves; *Symptoms Following Exposure*: Moisture of body converts compound to caustic soda, which irritates all tissues; *General Treatment for Exposure*: INGESTION: do NOT induce vomiting; neutralize alkali in stomach by drinking dilute vinegar, lemon juice, or orange juice; call a physician. SKIN CONTACT: brush off all particles at once and flood the affected area with water; *Toxicity by Inhalation (Threshold Limit Value)*: Not pertinent; *Short-Term Exposure Limits*: Not pertinent; *Toxicity by Inhalation (Threshold Limit Value)*: Not pertinent; *Short-Term Exposure Limits*: Not pertinent; *Toxicity*

by Ingestion: Data not available; *Late Toxicity*: None; *Vapor (Gas) Irritant Characteristics*: Non-volatile; *Liquid or Solid Irritant Characteristics*: Severe skin irritant. Causes second- and third-degree burns on short contact and is very injurious ti the EYES; *Odor Threshold*: Not pertinent.

Fire Hazards - *Flash Point* : Oil is flammable; *Flammable Limits in Air (%):* Not pertinent; *Fire Extinguishing Agents:* Powdered limestone and nitrogen-propelled dry powder; *Fire Extinguishing Agents Not To Be Used:* Water, soda ash, chemical foam, or carbon dioxide; *Special Hazards of Combustion Products:* Not pertinent; *Behavior in Fire:* Accidental contact with water used to extinguish surrounding fires will result in the release of hydrogen gas and possible explosion; *Ignition Temperature :* No data; *Electrical Hazard:* Not pertinent; *Burning Rate:* Not pertinent.

Chemical Reactivity - *Reactivity with Water:* Reacts vigorously with water with the release of flammable hydrogen gas; *Reactivity with Common Materials:* No reactions; *Stability During Transport:* Stable at temperatures below 225 °C; *Neutralizing Agents for Acids and Caustics:* Neutralize only when accidental reaction with water is complete. Do not neutralize the flammable solid with aqueous solutions. Spent reaction solution may be neutralized with dilute solutions of acetic acid.; *Polymerization:* Not pertinent; *Inhibitor of Polymerization:* Not pertinent.

SODIUM HYDROSULFIDE SOLUTION

Chemical Designations - *Synonyms*: Sodium bisulfide; Sodium hydrogen sulfide; Sodium sulfhydrate; *Chemical Formula*: To be developed.

Observable Characteristics - *Physical State (as normally shipped)*: Liquid; *Color*: Light lemon; pale yellow; amber to dark red; *Odor*: Rotten eggs.

Physical and Chemical Properties - *Physical State at 15 °C and 1 atm.*:Liquid; *Molecular Weight*: Not pertinent; *Boiling Point at 1 atm.*: (approx.) 212, 100, 373; *Freezing Point:* (approx.) 63, 17, 290; *Critical Temperature*: Not pertinent; *Critical Pressure*: Not pertinent; *Specific Gravity*: 1.3 at 15 °C (liquid); *Vapor (Gas) Density*: Not pertinent; *Ratio of Specific Heats of Vapor (Gas)*: Not pertinent; *Latent Heat of Vaporization*: Not pertinent; *Heat of Combustion*: Not pertinent; *Heat of Decomposition*: Not pertinent.

Health Hazards Information - *Recommended Personal Protective Equipment*: Rubber protective equipment, such as apron, boots, splash-proof goggles, gloves; canister-type respirator or self-contained breathing apparatus; *Symptoms Following Exposure*: Inhalation of mist causes irritation of respiratory tract and possible systemic poisoning, hydrogen sulfide gas, which may be given off when acid is present, causes headache, dizziness, nausea, vomiting: continued exposure can lead to loss of consciousness, respiratory failure, and death. Liquid causes marked eye irritation, itching, lachrymation, swelling, and corneal injury causing blurring of vision are the most common effects exposure to light may increase the painful effects. Contact of liquid with skin causes irritation and corrosion of tissue, continued exposure may cause dermatitis. Ingestion causes severe burning and corrosion of all portions of the gastro-intestinal tract, pain in the throat and abdomen, nausea, and vomiting, followed by diarrhea, in severe cases, collapse, unconsciousness, and paralysis of respiration may be expected; *General Treatment for Exposure*: INHALATION: move victim from contaminated atmosphere, call physician, if breathing has ceased, start mouth-to-mouth resuscitation. EYES: immediately flush with large quantities of running water for a minimum of 15 min., obtain medical attention as soon as possible; while awaiting instructions from physician, patient may be kept in a dark room and ice compresses applied to the eyes and forehead. SKIN: immediately flush affected areas with water: obtain medical attention if irritation persists. INGESTION: obtain medical attention as soon as possible; if patient is conscious, induce vomiting by giving large amounts of water or warm salty water (2 tablespoons of table sail to a pint of water), if this measure is unsuccessful, vomiting may be induced by tickling back of patient's throat with a finger. Vomiting should be encouraged until the vomitus is clear. if patient is unconscious, do not give anything but ensure there is no obstruction to breathing (his tongue should be kept forward and false teeth removed). He will be less likely to aspirate vomitus if he is placed in a face-down position; *Toxicity by Inhalation (Threshold Limit Value)*: Data not available; *Short-Term Exposure Limits*: Data not available; *Toxicity by Ingestion*: Grade 2, LD_{50} 0.5 to 5 g/kg; *Late Toxicity*: Data not available; *Vapor (Gas) Irritant Characteristics*: Vapors cause moderate irritation such that personnel will find high concentrations unpleasant. The effect is temporary; *Liquid or Solid Irritant Characteristics*: Fairly severe skin irritant. May cause pain and second-degree burns after a few minutes contact; *Odor Threshold*: 0.0047 ppm.

Fire Hazards - *Flash Point* : Not flammable; *Flammable Limits in Air (%):* Not flammable; *Fire Extinguishing Agents:* Not pertinent; *Fire Extinguishing Agents Not To Be Used:* Not pertinent; *Special Hazards of Combustion Products:* Not pertinent; *Behavior in Fire:* Not pertinent; *Ignition Temperature :* Not pertinent; *Electrical Hazard:* Not pertinent; *Burning Rate:* Not pertinent.

Chemical Reactivity - *Reactivity with Water:* No reaction; *Reactivity with Common Materials:* Corrodes most metals, but the reactions are generally non-hazardous; *Stability During Transport:* No reaction; *Neutralizing Agents for Acids and Caustics:* Flood with water; *Polymerization:* Not pertinent; *Inhibitor of Polymerization:* Not pertinent.

SODIUM HYDROXIDE

Chemical Designations - *Synonyms*: Caustic soda; *Chemical Formula*: NaON.

Observable Characteristics - *Physical State (as normally shipped)*: Solid; *Color*: White; *Odor*: Odorless.

Physical and Chemical Properties - *Physical State at 15 °C and 1 atm.*:Solid; *Molecular Weight*: 40.00; *Boiling Point at 1 atm.*: Very high; *Freezing Point*: 604, 318, 591; *Critical Temperature*: Not pertinent; *Critical Pressure*: Not pertinent; *Specific Gravity*: 2.13 at 20 °C (solid); *Vapor (Gas) Density*: Not pertinent; *Ratio of Specific Heats of Vapor (Gas)*: Not pertinent; *Latent Heat of Vaporization*: Not pertinent; *Heat of Combustion*: Not pertinent; *Heat of Decomposition*: Not pertinent.

Health Hazards Information - *Recommended Personal Protective Equipment*: Chemical safely goggles, lace shield: tiller or dust-type respirator: rubber boots: rubber gloves; *Symptoms Following Exposure*: Strong corrosive, destroys tissues. INHALATION: dust may cause damage to upper respiratory tract and lung itself, producing from mild nose irritation to pneumonitis. INGESTION: severe damage to mucous membranes; severe scar formation or perforation may occur. EYE CONTACT: produces severe damage; *General Treatment for Exposure*: INHALATION: remove from exposure; support respiration; call physician. INGESTION: give water or milk followed by dilute vinegar or fruit juice: do NOT induce vomiting. SKIN: wash immediately with large quantities of water under emergency safety shower while removing clothing: continue washing until medical help arrives, call physician. EYES: irrigate immediately with copious amounts of water for at least 15 min.; call physician; *Toxicity by Inhalation (Threshold Limit Value)*: Not pertinent; *Short-Term Exposure Limits*: Not pertinent; *Toxicity by Ingestion*: (10% solution) oral Grade 2, LDL_0 = 500 mg/kg; *Late Toxicity*: None; *Vapor (Gas) Irritant Characteristics*: Non-volatile; *Liquid or Solid Irritant Characteristics*: Severe skin irritant. Causes second-and third-degree burns on short contact and is very injurious to the eyes; *Odor Threshold*: Not pertinent.

Fire Hazards - *Flash Point* : Not flammable; *Flammable Limits in Air (%):* Not flammable; *Fire Extinguishing Agents:* Not pertinent; *Fire Extinguishing Agents Not To Be Used:* Not pertinent; *Special Hazards of Combustion Products:* Not pertinent; *Behavior in Fire:* Not pertinent; *Ignition Temperature :* Not flammable; *Electrical Hazard:* Not pertinent; *Burning Rate:* Not flammable.

Chemical Reactivity - *Reactivity with Water:* Dissolves with the liberation of considerable heat. The reaction violently produces steam and agitation; *Reactivity with Common Materials:* When wet, attacks metals such as aluminum , tin, lead, and zinc to produce flammable hydrogen gas; *Stability During Transport:* Stable; *Neutralizing Agents for Acids and Caustics:* Flush with water, rinse with dilute acetic acid; *Polymerization:* Not pertinent; *Inhibitor of Polymerization:* Not pertinent.

SODIUM HYPOCHLORITE

Chemical Designations - *Synonyms*: Clorox; *Chemical Formula*: NaOCI-H₂O.

Observable Characteristics - *Physical State (as normally shipped)*: Liquid; *Color*: Green-yellow; *Odor*: Like bleach solution.

Physical and Chemical Properties - *Physical State at 15 °C and 1 atm.*: Liquid; *Molecular Weight*: Not applicable; *Boiling Point at 1 atm.*: Decomposes; *Freezing Point*: Not pertinent; *Critical*

Temperature: Not pertinent; *Critical Pressure*: Not pertinent; *Specific Gravity*: 1.06 at 20 °C (liquid); *Vapor (Gas) Density*: Not pertinent; *Ratio of Specific Heats of Vapor (Gas)*: Not pertinent; *Latent Heat of Vaporization*: Not pertinent; *Heat of Combustion*: Not pertinent; *Heat of Decomposition*: Not pertinent.

Health Hazards Information - *Recommended Personal Protective Equipment*: Rubber gloves: goggles; *Symptoms Following Exposure*: Liquid can be irritating to skin and eyes; *General Treatment for Exposure*: INGESTION: induce vomiting, give water, and repeat. SKIN: wash off contacted skin area. EYES: flush with plenty of water for 15 min. and consult a physician; *Toxicity by Inhalation (Threshold Limit Value)*: Data not available; *Short-Term Exposure Limits*: Data not available; *Toxicity by Ingestion*: Data not available; *Late Toxicity*: Data not available; *Vapor (Gas) Irritant Characteristics*: Not pertinent; *Liquid or Solid Irritant Characteristics*: Data not available; *Odor Threshold*: Not pertinent.

Fire Hazards - *Flash Point* : Not flammable; *Flammable Limits in Air (%):* Not flammable; *Fire Extinguishing Agents:* Not pertinent; *Fire Extinguishing Agents Not To Be Used:* Not pertinent; *Special Hazards of Combustion Products:* Not pertinent; *Behavior in Fire:* May decompose, generating irritating chlorine gas; *Ignition Temperature :* Not flammable; *Electrical Hazard:* Not pertinent; *Burning Rate:* Not flammable.

Chemical Reactivity - *Reactivity with Water:* No reaction; *Reactivity with Common Materials:* No reactions; *Stability During Transport:* Stable; *Neutralizing Agents for Acids and Caustics:* Destroy with sodium bisulfite or hypo and water, then neutralize with soda ash; *Polymerization:* Not pertinent; *Inhibitor of Polymerization:* Not pertinent.

SODIUM METHYLATE

Chemical Designations - *Synonyms*: Sodium methoxide; *Chemical Formula*: CH_3ONa.

Observable Characteristics - *Physical State (as normally shipped)*: Solid; *Color*: White; *Odor*: None.

Physical and Chemical Properties - *Physical State at 15 °C and 1 atm.* :Solid; *Molecular Weight*: 54.0; *Boiling Point at 1 atm.*: Not pertinent (decomposes); *Freezing Point:* Not pertinent; *Critical Temperature:* Not pertinent; *Critical Pressure:* Not pertinent; *Specific Gravity:* > 1 at 20°C (solid); *Vapor (Gas) Density:* Not pertinent; *Ratio of Specific Heats of Vapor (Gas):* Not pertinent; *Latent Heat of Vaporization:* Not pertinent; *Heat of Combustion*: Data not available; *Heat of Decomposition:* Not pertinent.

Health Hazards Information - *Recommended Personal Protective Equipment:* Self-contained breathing apparatus, rubber gloves and apron; goggles or face shield; *Symptoms Following Exposure:* Inhalation of dust causes severe irritation of nose and throat. Contact with eyes or skin causes severe irritation and burns. Ingestion causes irritation of mouth and stomach; *General Treatment for Exposure:* Get medical attention at once following all exposures to this compound. INHALATION: remove victim from contamination and keep him quiet and warm. Rest is essential. Hot tea or coffee may be given as a stimulant if patient is conscious, if breathing has apparently ceased, give artificial respiration, if available, oxygen should be administered by experienced personnel. EYES: wash well with water, then with 3% boric acid solution and additional water washes. SKIN: wash well with water, then with dilute vinegar. INGESTION: if victim is conscious, induce vomiting by administering a glassful of warm water containing a teaspoon full of salt; repeat until vomit is clear, then give two teaspoons of baking soda every 15 min.; keep victim's eyes covered until all visual and retinal changes have disappeared, alert physician to possibility of methyl alcohol poisoning; *Toxicity by Inhalation (Threshold Limit Value):* Data not available; *Short-Term Exposure Limits:* Data not available; *Toxicity by Ingestion:* Data not available; *Late Toxicity:* Data not available; *Vapor (Gas) Irritant Characteristics:* Not pertinent; *Liquid or Solid Irritant Characteristics:* Data not available; *Odor Threshold:* Not pertinent.

Fire Hazards - *Flash Point* : Not pertinent; this is a flammable solid; *Flammable Limits in Air (%):* Not pertinent; *Fire Extinguishing Agents:* Dry chemical, inert powders such as sand or limestone, or carbon dioxide; *Fire Extinguishing Agents Not To Be Used:* Water, foam; *Special Hazards of Combustion Products:* Not pertinent; *Behavior in Fire:* Contact with water or foam adjacent to fires will produce flammable mathanol; *Ignition Temperature :* Not pertinent; *Electrical Hazard:* Not pertinent; *Burning Rate:* Not pertinent.

Chemical Reactivity - *Reactivity with Water:* Produces a caustic soda solution and a solution of methyl alcohol. The reaction is not violent; *Reactivity with Common Materials:* Attacks some polymers such as nylon and polyesters; *Stability During Transport:* Stable if kept dry; *Neutralizing Agents for Acids and Caustics:* Water followed by dilute acetic acid or vinegar; *Polymerization:* Not pertinent; *Inhibitor of Polymerization:* Not pertinent.

SODIUM NITRITE

Chemical Designations - *Synonyms*: Erinitrit; Filmerine; *Chemical Formula*: $NaNO_2$.
Observable Characteristics - *Physical State (as shipped)*: Solid; *Color*: White; *Odor*: None.
Physical and Chemical Properties - *Physical State at 15 °C and 1 atm.*: Solid; *Molecular Weight*: 69; *Boiling Point at 1 atm.*: (decomposes) >608, >320, >593; *Freezing Point*: 520, 271, 544; *Critical Temperature*: Not pertinent; *Critical Pressure*: Not pertinent; *Specific Gravity*: 2.17 at 20°C (solid); *Vapor (Gas) Specific Gravity*: Not pertinent; *Ratio of Specific Heats of Vapor (Gas)*: Not pertinent; *Latent Heat of Vaporization*: Not pertinent; *Heat of Combustion*: Not pertinent; *Heat of Decomposition*: Not pertinent.
Health Hazards Information - *Recommended Personal Protective Equipment*: Dust mask; goggles or face shield; protective gloves; *Symptoms Following Exposure*: Ingestion (or inhalation of excessive amounts of dust) causes rapid drop in blood pressure, persistent and throbbing headache, vertigo, palpitations, and visual disturbances; skin becomes flushed and sweaty, later cold and cyanotic; other symptoms include nausea, vomiting, diarrhea (sometimes), fainting, methemoglobinemia. Contact with eyes causes irritation; *General Treatment for Exposure*: INHALATION: move to fresh air; if exposure is severe, get medical attention. INGESTION: keep patient recumbent in a shock position and comfortably warm; administer gastric lavage; consult a physician. EYES or SKIN: flush with water; *Toxicity by Inhalation (Threshold Limit Value)*: Data not available; *Short-Term Inhalation Limits*: Data not available; *Toxicity by Ingestion*: Grade 3, LD_{50} 50-500 mg/kg; *Late Toxicity*: Data not available; *Vapor (Gas) Irritant Characteristics*: Data not available; *Liquid or Solid Irritant Characteristics*: Data not available; *Odor Threshold*: Data not available.
Fire Hazards - *Flash Point* : Not flammable, but may intensify fires; *Flammable Limits in Air (%):* Not flammable; *Fire Extinguishing Agents:* Apply large amounts of water to adjacent fires. Cool exposed containers with water; *Fire Extinguishing Agents Not To Be Used:* Not pertinent; *Special Hazards of Combustion Products:* Toxic oxides of nitrogen may form in fires; *Behavior in Fire:* May increase the intensity of fires if the chemical is in contact with combustible materials. This product may melt and flow at elevated temperatures; *Ignition Temperature :* Not pertinent; *Electrical Hazard:* Not pertinent; *Burning Rate:* Not pertinent.
Chemical Reactivity - *Reactivity with Water:* No reaction; *Reactivity with Common Materials:* No reactions; *Stability During Transport:* Stable; *Neutralizing Agents for Acids and Caustics:* Not pertinent; *Polymerization:* Not pertinent; *Inhibitor of Polymerization:* Not pertinent.

SODIUM OXALATE

Chemical Designations - *Synonyms*: Ethanedioic acid; Disodium salt; *Chemical Formula*: $Na_2C_2O_4$.
Observable Characteristics - *Physical State (as shipped)*: Solid; *Color*: White; *Odor*: None.
Physical and Chemical Properties - *Physical State at 15 °C and 1 atm.*: Solid; *Molecular Weight*: 134.0; *Boiling Point at 1 atm.*: Not pertinent (decomposes); *Freezing Point*: Not pertinent; *Critical Temperature*: Not pertinent; *Critical Pressure*: Not pertinent; *Specific Gravity*: 2.27 at 20 °C (solid); *Vapor (Gas) Specific Gravity*: Not pertinent; *Ratio of Specific Heats of Vapor (Gas)*: Not pertinent; *Latent Heat of Vaporization*: Not pertinent; *Heat of Combustion*: Not pertinent; *Heat of Decomposition*: Not pertinent.
Health Hazards Information - *Recommended Personal Protective Equipment*: Dust mask; goggles or face shield; rubber gloves; *Symptoms Following Exposure*: Inhalation or ingestion causes pain in throat, esophagus, and stomach; mucous membranes turn white; other symptoms include vomiting, severe purging, weak pulse, cardiovascular collapse, neuromuscular symptoms, and kidney damage. Contact with eyes or skin causes irritation; *General Treatment for Exposure*: *Act promptly!* INHALATION:

move to fresh air; if exposure to dust is severe, get medical attention. INGESTION: give dilute calcium lactate, lime water, or milk; administer gastric lavage; consult physician; watch for edema of the glottis and delayed constriction of esophagus. EYES or SKIN: flush with water; *Toxicity by Inhalation (Threshold Limit Value)*: Data not available; *Short-Term Inhalation Limits*: Data not available; *Toxicity by Ingestion*: Grade 3, LD_{50} 50-500 mg/kg; *Late Toxicity*: Data not available; *Vapor (Gas) Irritant Characteristics*: Data not available; *Liquid or Solid Irritant Characteristics*: Data not available; *Odor Threshold*: Data not available.

Fire Hazards - *Flash Point* : Not flammable; *Flammable Limits in Air (%):* Not flammable; *Fire Extinguishing Agents:* Not pertinent; *Fire Extinguishing Agents Not To Be Used:* Not pertinent; *Special Hazards of Combustion Products:* Not pertinent; *Behavior in Fire:* Not pertinent; *Ignition Temperature :* Not flammable; *Electrical Hazard:* Not pertinent; *Burning Rate:* Not flammable.

Chemical Reactivity - *Reactivity with Water:* No reaction; *Reactivity with Common Materials:* No reactions; *Stability During Transport:* Stable; *Neutralizing Agents for Acids and Caustics:* Not pertinent; *Polymerization:* Not pertinent; *Inhibitor of Polymerization:* Not pertinent.

SODIUM PHOSPHATE

Chemical Designations - *Synonyms*: "Sodium phosphate" is generic term and includes the following: (1) monosodium phosphate (MSP; sodium phosphate, monobasic), (2) disodium phosphate (DSP; sodium phosphate dibasic), (3) trisodium phosphate (TSP; sodium phosphate, tribasic), (4) sodium acid pyrophosphate (ASPP; SAPP; disodium pyrophosphate (TSPP), (6) sodium metaphosphate (insoluble sodium metaphosphate), (7) sodium trimetaphosphate, and (9) sodium tripolyphosphate (STPP; TPP); *Chemical Formula*: (1) NaH_2PO_4; (2) Na_2HPO_4; (3) Na_3PO_4; (4) $Na_2H_2P_2O_7$; (5) $Na_4P_2O_7$; (6) $(NaPO_3)_n$; (7) $(NaPO_3)_3$; (8) $(NaPO_3)_n \cdot NaO$; (9) $Na_5P_3O_{10}$.

Observable Characteristics - *Physical State (as shipped)*: Granular or powdered solid; some may appear glassy; *Color*: White; *Odor*: None.

Physical and Chemical Properties - *Physical State at 15 °C and 1 atm.*: Solid; *Molecular Weight*: Values for anhydrous salt run from 120 to high polymer values; *Boiling Point at 1 atm.*: Not pertinent (decomposes); *Freezing Point*: Not pertinent; *Critical Temperature*: Not pertinent; *Critical Pressure*: Not pertinent; *Specific Gravity*: 1.8-2.5 at 25 °C (solid); *Vapor (Gas) Specific Gravity*: Not pertinent; *Ratio of Specific Heats of Vapor (Gas)*: Not pertinent; *Latent Heat of Vaporization*: Not pertinent; *Heat of Combustion*: Not pertinent; *Heat of Decomposition*: Not pertinent.

Health Hazards Information - *Recommended Personal Protective Equipment*: U.S. Bu. Mines toxic dust mask; protective gloves; chemical-type goggles; full-cover clothing; *Symptoms Following Exposure*: Inhalation of heavy dust may irritate nose and throat. Ingestion may injure mouth, throat, and gastrointestinal tract, resulting in nausea, vomiting, cramps and diarrhea; pain and burning in moth may occur. Contact with eyes produces local irritation; can lead to chronic damage. Contact with skin produces local irritation; repeated or prolonged contact can lead to dermatitis; *General Treatment for Exposure*: If the following measures do not eliminate the symptoms, see a physician. INHALATION: give large amounts of water or warm salty water to induce vomiting; repeat until vomitus is clear; milk, eggs, or olive oil may then be given to soothe stomach. EYES: immediately flush with large amounts of water for at least 15 min., holding eyelids to ensure flushing or entire surface; avoid chemical neutralizers. SKIN: flush with water; avoid chemical neutralizers; *Toxicity by Inhalation (Threshold Limit Value)*: Data not available; *Short-Term Inhalation Limits*: Data not available; *Toxicity by Ingestion*: Data not available; *Late Toxicity*: Data not available; *Vapor (Gas) Irritant Characteristics*: Data not available; *Liquid or Solid Irritant Characteristics*: Data not available; *Odor Threshold*: Data not available.

Fire Hazards - *Flash Point* : Not flammable; *Flammable Limits in Air (%):* Not flammable; *Fire Extinguishing Agents:* Not pertinent; *Fire Extinguishing Agents Not To Be Used:* Not pertinent; *Special Hazards of Combustion Products:* Not pertinent; *Behavior in Fire:* May melt with the loss of steam; *Ignition Temperature :* Not pertinent; *Electrical Hazard:* Not pertinent; *Burning Rate:* Not pertinent.

Chemical Reactivity - *Reactivity with Water:* All variations or grades of this chemical readily dissolve in water. ASPP and MSP form weakly acidic solutions. TSP forms a strong caustic solution, similar to soda lye: TSPP forms weekly alkali solution; *Reactivity with Common Materials:* When wet, MSP, ASPP, and TSP corrodes mild steel or brass. Others are not considered corrosive; *Stability During Transport:* All forms of sodium phosphate are stable. TSP tends to be hygroscopic and will form a hard cake; *Neutralizing Agents for Acids and Caustics:* For those grades of sodium hydroxide that form acidic or alkali solutions, dilution by water is recommended; *Polymerization:* Not pertinent; *Inhibitor of Polymerization:* Not pertinent.

SODIUM SILICATE

Chemical Designations - *Synonyms*: Water glass; Soluble glass; *Chemical Formula*: Na_2SiO_3-$NaSiO_4$-H_2O.

Observable Characteristics - *Physical State (as shipped)*: High-viscosity liquid; *Color*: Colorless; *Odor*: Odorless.

Physical and Chemical Properties - *Physical State at 15 °C and 1 atm.*: Liquid; *Molecular Weight*: Not applicable; *Boiling Point at 1 atm.*: Decomposes; *Freezing Point*: Not pertinent; *Critical Temperature*: Not pertinent; *Critical Pressure*: Not pertinent; *Specific Gravity*: 1.1 - 1.7 at 20 °C (liquid); *Vapor (Gas) Specific Gravity*: Not pertinent; *Ratio of Specific Heats of Vapor (Gas)*: Not pertinent; *Latent Heat of Vaporization*: Not pertinent; *Heat of Combustion*: Not pertinent; *Heat of Decomposition*: Not pertinent.

Health Hazards Information - *Recommended Personal Protective Equipment*: Goggles or face shield; *Symptoms Following Exposure*: If large doses are ingested, some irritation of mucous membranes may occur, similar to that caused by caustic solution; *General Treatment for Exposure*: INGESTION (large doses): give water or milk; do NOT induce vomiting; *Toxicity by Inhalation (Threshold Limit Value)*: Not pertinent; *Short-Term Inhalation Limits*: Not pertinent; *Toxicity by Ingestion*: Grade 2, LD_{50} 0.5 to 5 g/kg (human); *Late Toxicity*: None; *Vapor (Gas) Irritant Characteristics*: Non-volatile; *Liquid or Solid Irritant Characteristics*: None; *Odor Threshold*: Not pertinent.

Fire Hazards - *Flash Point* : Not flammable; *Flammable Limits in Air (%):* Not flammable; *Fire Extinguishing Agents:* Not pertinent; *Fire Extinguishing Agents Not To Be Used:* Not pertinent; *Special Hazards of Combustion Products:* Not pertinent; *Behavior in Fire:* Not pertinent; *Ignition Temperature* : Not flammable; *Electrical Hazard:* Not pertinent; *Burning Rate:* Not flammable.

Chemical Reactivity - *Reactivity with Water:* No reaction; *Reactivity with Common Materials:* No reactions; *Stability During Transport:* Stable; *Neutralizing Agents for Acids and Caustics:* Not pertinent; *Polymerization:* Not pertinent; *Inhibitor of Polymerization:* Not pertinent.

SODIUM SILICOFLUORIDE

Chemical Designations - *Synonyms*: Salufer; Sodium fluosilicate; Sodium hexafluorosilicate; *Chemical Formula*: Na_2SiF_6.

Observable Characteristics - *Physical State (as shipped)*: Solid; *Color*: White; *Odor*: None.

Physical and Chemical Properties - *Physical State at 15 °C and 1 atm.*: Solid; *Molecular Weight*: 188; *Boiling Point at 1 atm.*: Not pertinent (decomposes); *Freezing Point*: Not pertinent; *Critical Temperature*: Not pertinent; *Critical Pressure*: Not pertinent; *Specific Gravity*: 2.68 at 20 °C (solid); *Vapor (Gas) Specific Gravity*: Not pertinent; *Ratio of Specific Heats of Vapor (Gas)*: Not pertinent; *Latent Heat of Vaporization*: Not pertinent; *Heat of Combustion*: Not pertinent; *Heat of Decomposition*: Not pertinent.

Health Hazards Information - *Recommended Personal Protective Equipment*: Dust respirator; goggles or face shield; protective gloves; *Symptoms Following Exposure*: Inhalation of dust may irritate nose and throat. Ingestion causes symptoms similar to fluoride poisoning; compound is highly toxic; initial symptoms include nausea, cramps, vomiting, diarrhea, and dehydration; in severe cases, convulsions, shock, and cyanosis are followed by death in 2-4 hr. Contact with eyes causes irritation. Contact with skin causes rash, redness, and burning, sometimes followed by ulcer formation; *General Treatment for Exposure*: INHALATION: move to fresh air. INGESTION: seek medical attention; administer gastric lavage with lime water, then give lime water or milk at frequent intervals. EYES: flush with water for at least 15 min. SKIN: flush with water; wash with soap and water; *Toxicity by Inhalation (Threshold*

Limit Value): 2.5 mg/m³ (as fluoride); *Short-Term Inhalation Limits*: Data not available; *Toxicity by Ingestion*: Grade 3; LD₅₀ 50-500 mg/kg; *Late Toxicity*: Data not available; *Vapor (Gas) Irritant Characteristics*: Data not available; *Liquid or Solid Irritant Characteristics*: Data not available; *Odor Threshold*: Data not available.

Fire Hazards - *Flash Point* : Not flammable; *Flammable Limits in Air (%):* Not flammable; *Fire Extinguishing Agents:* Not pertinent; *Fire Extinguishing Agents Not To Be Used:* Not pertinent; *Special Hazards of Combustion Products:* Not pertinent; *Behavior in Fire:* Decomposes at red heat; *Ignition Temperature :* Not pertinent; *Electrical Hazard:* Not pertinent; *Burning Rate:* Not pertinent.

Chemical Reactivity - *Reactivity with Water:* No reaction; *Reactivity with Common Materials:* No reaction; *Stability During Transport:* Stable; *Neutralizing Agents for Acids and Caustics:* Not pertinent; *Polymerization:* Not pertinent; *Inhibitor of Polymerization:* Not pertinent.

SODIUM SULFIDE

Chemical Designations - *Synonyms*: No common synonyms; *Chemical Formula*: Na₂S.

Observable Characteristics - *Physical State (as shipped)*: Solid; *Color*: Yellow or light buff; *Odor*: Like rotten eggs.

Physical and Chemical Properties - *Physical State at 15 °C and 1 atm.*: Solid; *Molecular Weight*: 78.4; *Boiling Point at 1 atm.*: Very high; *Freezing Point*: Not pertinent; *Critical Temperature*: Not pertinent; *Critical Pressure*: Not pertinent; *Specific Gravity*: 1.856 at 20 °C (solid); *Vapor (Gas) Specific Gravity*: Not pertinent; *Ratio of Specific Heats of Vapor (Gas)*: Not pertinent; *Latent Heat of Vaporization*: Not pertinent; *Heat of Combustion*: Not pertinent; *Heat of Decomposition*: Not pertinent.

Health Hazards Information - *Recommended Personal Protective Equipment*: Goggles or face shield; *Symptoms Following Exposure*: Caustic action on skin and eyes. If ingested may liberate hydrogen sulfide in stomach; *General Treatment for Exposure*: INGESTION: give water; induce vomiting; call a doctor. SKIN OR EYE CONTACT: wash with water for at least 15 min.; *Toxicity by Inhalation (Threshold Limit Value)*: Not pertinent; *Short-Term Inhalation Limits*: Not pertinent; *Toxicity by Ingestion*: Grade 3; LD₅₀ 50 - 500 mg/kg (human); *Late Toxicity*: None; *Vapor (Gas) Irritant Characteristics*: Non-volatile; *Liquid or Solid Irritant Characteristics*: Irritates skin and mucous membranes; *Odor Threshold*: Not pertinent.

Fire Hazards - *Flash Point* : Moderately flammable solid; *Flammable Limits in Air (%):* Not pertinent; *Fire Extinguishing Agents:* Water; *Fire Extinguishing Agents Not To Be Used:* Not pertinent; *Special Hazards of Combustion Products:* Irritating fumes of sulfur dioxide are generated in fires; *Behavior in Fire:* Not pertinent; *Ignition Temperature :* Not pertinent; *Electrical Hazard:* Not pertinent; *Burning Rate:* Not pertinent.

Chemical Reactivity - *Reactivity with Water:* No reaction; *Reactivity with Common Materials:* No reactions; *Stability During Transport:* Stable; *Neutralizing Agents for Acids and Caustics:* Not pertinent; *Polymerization:* Not pertinent; *Inhibitor of Polymerization:* Not pertinent.

SODIUM SULFITE

Chemical Designations - *Synonyms*: No common synonyms; *Chemical Formula*: Na₂SO₃.

Observable Characteristics - *Physical State (as shipped)*: Solid; *Color*: White; *Odor*: Odorless.

Physical and Chemical Properties - *Physical State at 15 °C and 1 atm.*: Solid; *Molecular Weight*: 126.04; *Boiling Point at 1 atm.*: Decomposes; *Freezing Point*: Not pertinent; *Critical Temperature*: Not pertinent; *Critical Pressure*: Not pertinent; *Specific Gravity*: 2.633 at 15 °C (solid); *Vapor (Gas) Specific Gravity*: Not pertinent; *Ratio of Specific Heats of Vapor (Gas)*: Not pertinent; *Latent Heat of Vaporization*: Not pertinent; *Heat of Combustion*: Not pertinent; *Heat of Decomposition*: Not pertinent.

Health Hazards Information - *Recommended Personal Protective Equipment*: Dust mask; goggles or face shield; *Symptoms Following Exposure*: When ingested, solutions cause gastric irritation by the liberation of sulfurous acid. Because of rapid oxidation to sulfate, sulfites are well tolerated until large doses are reached; than violent colic and diarrhea, circulatory disturbances, central nervous depression, and death can occur; *General Treatment for Exposure*: INGESTION: treatment is symptomatic and supportive; call a doctor; *Toxicity by Inhalation (Threshold Limit Value)*: Not pertinent; *Short-Term*

Inhalation Limits: Not pertinent; *Toxicity by Ingestion*: Grade 2, LD_{50} 0.5 to 5 g/kg; *Late Toxicity*: None; *Vapor (Gas) Irritant Characteristics*: Non-volatile; *Liquid or Solid Irritant Characteristics*: Not very irritating; *Odor Threshold*: Not pertinent.

Fire Hazards - *Flash Point* : Not flammable; *Flammable Limits in Air (%):* Not flammable; *Fire Extinguishing Agents:* Not pertinent; *Fire Extinguishing Agents Not To Be Used:* Not pertinent; *Special Hazards of Combustion Products:* Not pertinent; *Behavior in Fire:* Not pertinent; *Ignition Temperature :* Not flammable; *Electrical Hazard:* Not pertinent; *Burning Rate:* Not flammable.

Chemical Reactivity - *Reactivity with Water:* No reaction; *Reactivity with Common Materials:* No reactions; *Stability During Transport:* Stable; *Neutralizing Agents for Acids and Caustics:* Not pertinent; *Polymerization:* Not pertinent; *Inhibitor of Polymerization:* Not pertinent.

SODIUM THIOCYANATE

Chemical Designations - *Synonyms*: Rhodanate; Sodium rhodanide; Sodium sulfocyanate; *Chemical Formula*: NaSCN.

Observable Characteristics - *Physical State (as shipped)*: Solid; *Color*: White; *Odor*: None.

Physical and Chemical Properties - *Physical State at 15 °C and 1 atm.*: Solid; *Molecular Weight*: 81.08; Boiling Point at 1 atm.: Not pertinent (decomposes); *Freezing Point*: 572, 300, 573; *Critical Temperature*: Not pertinent; *Critical Pressure*: Not pertinent; *Specific Gravity*: >1 at 20 °C (solid); *Vapor (Gas) Specific Gravity*: Not pertinent; *Ratio of Specific Heats of Vapor (Gas)*: Not pertinent; *Latent Heat of Vaporization*: Not pertinent; *Heat of Combustion*: Not pertinent; *Heat of Decomposition*: Not pertinent.

Health Hazards Information - *Recommended Personal Protective Equipment*: Rubber or plastic gloves; standard goggles; rubber or plastic apron; *Symptoms Following Exposure*: Inhalation of dust causes irritation of nose and throat. Ingestion of large doses causes vomiting, extreme cerebral excitement, convulsions, and death in 10-48 hrs.; chronic poisoning can cause flu-like symptoms, skin rashes, weakness, fatigue, vertigo, nausea, vomiting, diarrhea, confusion. Contact with eyes causes irritation. Prolonged contact with skin may produce various skin eruptions, dizziness, cramps, nausea, and mild to severe disturbance of the nervous system; *General Treatment for Exposure*: INHALATION: move to fresh air; if exposure has been great, get medical attention. INGESTION: consult physician; hemodialysis is recommended as the treatment of choice. EYES or SKIN: flush with water for 15 min.; *Toxicity by Inhalation (Threshold Limit Value)*: Data not available; *Short-Term Inhalation Limits*: Data not available; *Toxicity by Ingestion*: Grade 2, LD_{50} 0.5 - 5 g/kg; *Late Toxicity*: Causes birth defects in chick embryos; *Vapor (Gas) Irritant Characteristics*: Data not available; *Liquid or Solid Irritant Characteristics*: Data not available; *Odor Threshold*: Data not available.

Fire Hazards - *Flash Point* : Not flammable; *Flammable Limits in Air (%):* Not flammable; *Fire Extinguishing Agents:* Not pertinent; *Fire Extinguishing Agents Not To Be Used:* Not pertinent; *Special Hazards of Combustion Products:* Irritating oxides of sulfur and nitrogen form in fires; *Behavior in Fire:* Not pertinent; *Ignition Temperature :* Not flammable; *Electrical Hazard:* Not pertinent; *Burning Rate:* Not flammable.

Chemical Reactivity - *Reactivity with Water:* No reaction; *Reactivity with Common Materials:* No reactions; *Stability During Transport:* Stable; *Neutralizing Agents for Acids and Caustics:* Not pertinent; *Polymerization:* Not pertinent; *Inhibitor of Polymerization:* Not pertinent.

SORBITOL

Chemical Designations - *Synonyms*: D-Glucitol; Hexahydric alcohol; 1,2,3,4,5,6-Hexanehexol; Sorbit; Sorbo; Sorbol; *Chemical Formula*: $CH_2OH(CHOH)_4CH_2OH$.

Observable Characteristics - *Physical State (as shipped)*: Liquid; *Color*: Colorless; *Odor*: Odorless.

Physical and Chemical Properties - *Physical State at 15 °C and 1 atm.*: Solid; *Molecular Weight*: 182.17; *Boiling Point at 1 atm.*: Very high; *Freezing Point*: 230, 110, 383; *Critical Temperature*: Not pertinent; *Critical Pressure*: Not pertinent; *Specific Gravity*: 1.49 at 15 °C (liquid); *Vapor (Gas) Specific Gravity*: Not pertinent; *Ratio of Specific Heats of Vapor (Gas)*: Not pertinent; *Latent Heat of Vaporization*: Not pertinent; *Heat of Combustion*: (est.) -6,750, -3,750, -157; *Heat of Decomposition*:

Not pertinent.

Health Hazards Information - *Recommended Personal Protective Equipment*: Goggles or face shield; protective clothing for hot liquid; *Symptoms Following Exposure*: Hot liquid will burn skin; *General Treatment for Exposure*: Only for burns caused by hot liquid; *Toxicity by Inhalation (Threshold Limit Value)*: Not pertinent; *Short-Term Inhalation Limits*: Not pertinent; *Toxicity by Ingestion*: None; *Late Toxicity*: None; *Vapor (Gas) Irritant Characteristics*: Non-volatile; *Liquid or Solid Irritant Characteristics*: Data not available; *Odor Threshold*: Not pertinent.

Fire Hazards - *Flash Point (deg. F)*: 542; *Flammable Limits in Air (%):* Not flammable; *Fire Extinguishing Agents:* Water; *Fire Extinguishing Agents Not To Be Used:* Not data; *Special Hazards of Combustion Products:* Not pertinent; *Behavior in Fire:* Not pertinent; *Ignition Temperature :* No data; *Electrical Hazard:* Not pertinent; *Burning Rate:* No data.

Chemical Reactivity - *Reactivity with Water:* No reaction; *Reactivity with Common Materials:* No reactions; *Stability During Transport:* Stable; *Neutralizing Agents for Acids and Caustics:* Not pertinent; *Polymerization:* Not pertinent; *Inhibitor of Polymerization:* Not pertinent.

STEARIC ACID

Chemical Designations - *Synonyms*: 1-Heptadecancecarboxylic acid; Octadecanoic acid; n-Octadecylic acid; Stearophanic acid; *Chemical Formula*: $CH_3(CH_2)_{16}CO_2H$.

Observable Characteristics - *Physical State (as shipped)*: Solid; *Color*: White; *Odor*: Fatty.

Physical and Chemical Properties - *Physical State at 15 °C and 1 atm.*: Solid; *Molecular Weight*: (avg.) 282; *Boiling Point at 1 atm.*: Not pertinent (decomposes); *Freezing Point*: 157, 70, 343; *Critical Temperature*: Not pertinent; *Critical Pressure*: Not pertinent; *Specific Gravity*: 0.86 at 20°C (solid); *Vapor (Gas) Specific Gravity*: Not pertinent; *Ratio of Specific Heats of Vapor (Gas)*: Not pertinent; *Latent Heat of Vaporization*: Not pertinent; *Heat of Combustion*: -17,310, -9,616, -402.3; *Heat of Decomposition*: Not pertinent.

Health Hazards Information - *Recommended Personal Protective Equipment*: For prolonged exposure to vapors, use air-supplied mask of chemical cartridge respirator; impervious gloves; goggles; impervious apron; *Symptoms Following Exposure*: Compound is generally considered nontoxic. Inhalation of dust irritates nose and throat. Dust causes mild irritation of eyes; *General Treatment for Exposure*: INGESTION: drink large volume of water; induce vomiting; call a physician. EYES: flush with water; if irritation persists, get medical attention. SKIN: wash thoroughly with soap and water; *Toxicity by Inhalation (Threshold Limit Value)*: Data not available; *Short-Term Inhalation Limits*: Data not available; *Toxicity by Ingestion*: Grade 0; $LD_{50} > 15$ g/kg; *Late Toxicity*: Data not available; *Vapor (Gas) Irritant Characteristics*: Data not available; *Liquid or Solid Irritant Characteristics*: Data not available; *Odor Threshold*: 20 ppm.

Fire Hazards - *Flash Point (deg. F)*: 410 ~ 435 OC, 365 CC (as molten solid); *Flammable Limits in Air (%):* Not pertinent; *Fire Extinguishing Agents:* Foam, dry chemical, or carbon dioxide; *Fire Extinguishing Agents Not To Be Used:* Water or foam may cause frothing; *Special Hazards of Combustion Products:* No data; *Behavior in Fire:* No data; *Ignition Temperature (deg. F):* 743; *Electrical Hazard:* Not pertinent; *Burning Rate:* Not pertinent.

Chemical Reactivity - *Reactivity with Water:* No reaction; *Reactivity with Common Materials:* No reactions; *Stability During Transport:* Stable; *Neutralizing Agents for Acids and Caustics:* Not pertinent; *Polymerization:* Not pertinent; *Inhibitor of Polymerization:* Not pertinent.

STYRENE

Chemical Designations - *Synonyms*: Phenethylene; Phenylethylene; Styrol; Styrolene; Vinylbenzene; *Chemical Formula*: $C_6H_5CH=CH_2$.

Observable Characteristics - *Physical State (as shipped)*: Liquid; *Color*: Colorless; *Odor*: Sweet at low concentrations; characteristic pungent; sharp; disagreeable.

Physical and Chemical Properties - *Physical State at 15 °C and 1 atm.*: Liquid; *Molecular Weight*: 104.15; *Boiling Point at 1 atm.*: 293.4, 145.2, 418.4; *Freezing Point*: -23.1, -30.6, 242.6; *Critical Temperature*: 703, 373, 646; *Critical Pressure*: 580, 39.46, 4.00; *Specific Gravity*: 0.906 at 20 °C (liquid); *Vapor (Gas) Specific Gravity*: Not pertinent; *Ratio of Specific Heats of Vapor (Gas)*: 1.074; *Latent Heat of Vaporization*: 156, 86.8, 3.63; *Heat of Combustion*: Not pertinent; *Heat of Decomposition*: Not pertinent.

Health Hazards Information - *Recommended Personal Protective Equipment*: Air-supplied mask or approved canister; rubber or plastic gloves; boots; goggles or face shield; *Symptoms Following Exposure*: Moderate irritation of eyes and skin. High vapor concentrations cause dizziness, drunkenness, and anesthesia; *General Treatment for Exposure*: INHALATION: remove to fresh air; keep warm and quiet; use artificial respiration if needed. INGESTION: do NOT induce vomiting; call physician; no known antidote. SKIN OR EYE CONTACT: flush with plenty of water; for eyes get medical attention; *Toxicity by Inhalation (Threshold Limit Value)*: 100 ppm; *Short-Term Inhalation Limits*: 100 ppm for 30 min.; *Toxicity by Ingestion*: Grade 2, LD_{50} 0.5 to 5 g/kg; *Late Toxicity*: Data not available; *Vapor (Gas) Irritant Characteristics*: Vapors cause moderate irritation such that personnel will find high concentrations unpleasant. The effect is temporary; *Liquid or Solid Irritant Characteristics*: Causes smarting of the skin and first-degree burns on short exposure; may cause second-degree burns on long exposure; *Odor Threshold*: 0.148 ppm.

Fire Hazards - *Flash Point (deg. F)*: 93 OC, 88 CC; *Flammable Limits in Air (%)*: 1.1 - 6.1; *Fire Extinguishing Agents:* Water fog, foam, carbon dioxide, or dry chemical; *Fire Extinguishing Agents Not To Be Used:* Water may be ineffective; *Special Hazards of Combustion Products:* Not pertinent; *Behavior in Fire:* Vapor is heavier than air and may travel considerable distance to a source of ignition and flash back. At elevated temperatures as under fire conditions, polymerization may occur, resulting in containers exploding; *Ignition Temperature (deg. F):* 914; *Electrical Hazard:* Class I, Group D; *Burning Rate:* 5.2 mm/min.

Chemical Reactivity - *Reactivity with Water:* No reaction; *Reactivity with Common Materials:* No reactions; *Stability During Transport:* Stable; *Neutralizing Agents for Acids and Caustics:* Not pertinent; *Polymerization:* Polymerization can occur if the product's temperature is raised above 150 of. This can cause the rupture of containers. Avoid contact with metal salts, peroxides, and strong acids, which can cause polymerization to occur; *Inhibitor of Polymerization:* Tertiarybutylcatechol (10 ~ 15 ppm).

SUCROSE

Chemical Designations - *Synonyms*: Beet sugar; Cane sugar; Saccharose; Saccharum; Sugar; *Chemical Formula*: $C_{12}H_{22}O_{11}$.

Observable Characteristics - *Physical State (as shipped)*: Solid; *Color*: White; *Odor*: None.

Physical and Chemical Properties - *Physical State at 15 °C and 1 atm.*: Solid; *Molecular Weight*: 342.3; *Boiling Point at 1 atm.*: Not pertinent (decomposes); *Freezing Point*: (decomposes) 320-367, 160-186,433-459; *Critical Temperature*: Not pertinent; *Critical Pressure*: Not pertinent; *Specific Gravity*: 1.59 at 20 °C (solid); *Vapor (Gas) Specific Gravity*: Not pertinent; *Ratio of Specific Heats of Vapor (Gas)*: Not pertinent; *Latent Heat of Vaporization*: Not pertinent; *Heat of Combustion*: -6,400, -3,600, -150; *Heat of Decomposition*: Not pertinent.

Health Hazards Information - *Recommended Personal Protective Equipment*: Dust mask and goggles or face shield; *Symptoms Following Exposure*: None; *General Treatment for Exposure*: EYES: flush with water; *Toxicity by Inhalation (Threshold Limit Value)*: Data not available; *Short-Term Inhalation Limits*: Data not available; *Toxicity by Ingestion*: Grade 0; oral LD_{50} (100 days) = 28,500 mg/kg (rat); *Late Toxicity*: None; *Vapor (Gas) Irritant Characteristics*: Data not available; *Liquid or Solid Irritant Characteristics*: Data not available; *Odor Threshold*: Not pertinent.

Fire Hazards - *Flash Point* : Not pertinent; this is a combustible solid; *Flammable Limits in Air (%):* Not pertinent; *Fire Extinguishing Agents:* Water; *Fire Extinguishing Agents Not To Be Used:* Not pertinent; *Special Hazards of Combustion Products:* Irritating fumes may form in fire situations; *Behavior in Fire:* The product melts and chars; *Ignition Temperature :* Not pertinent; *Electrical Hazard:* Not pertinent; *Burning Rate:* Not pertinent.

Chemical Reactivity - *Reactivity with Water:* No reaction; *Reactivity with Common Materials:* No reactions; *Stability During Transport:* Stable; *Neutralizing Agents for Acids and Caustics:* Not pertinent; *Polymerization:* Not pertinent; *Inhibitor of Polymerization:* Not pertinent.

SULFOLANE
Chemical Designations - *Synonyms*: Sulfolane-W; Tetrahydrothiophene-1,1-dioxide; Tetramethylene sulfone; *Chemical Formula*: $CH_2CH_2CH_2CH_2SO_2$.
Observable Characteristics - *Physical State (as shipped)*: Liquid; *Color*: Colorless; *Odor*: Weak oily.
Physical and Chemical Properties - *Physical State at 15 °C and 1 atm.*: Solid; *Molecular Weight*: 120.17; *Boiling Point at 1 atm.*: 545, 285, 558; *Freezing Point*: 79, 26, 299; *Critical Temperature*: Not pertinent; *Critical Pressure*: Not pertinent; *Specific Gravity*: 1.26 at 20 °C (liquid); *Vapor (Gas) Specific Gravity*: Not pertinent; *Ratio of Specific Heats of Vapor (Gas)*: Not pertinent; *Latent Heat of Vaporization*: Not pertinent; *Heat of Combustion*: (est.) -9,500, -5,300, -220; *Heat of Decomposition*: Not pertinent.
Health Hazards Information - *Recommended Personal Protective Equipment*: Goggles or face shield; rubber gloves; *Symptoms Following Exposure*: Very mildly irritating to the eyes; *General Treatment for Exposure*: INGESTION: induce vomiting. SKIN OR EYE CONTACT: flush with water; *Toxicity by Inhalation (Threshold Limit Value)*: Not pertinent; *Short-Term Inhalation Limits*: Not pertinent; *Toxicity by Ingestion*: Grade 2, LD_{50} 0.5 to 5 g/kg (rat, mouse); *Late Toxicity*: Data not available; *Vapor (Gas) Irritant Characteristics*: Vapors are nonirritating to eyes and throat; *Liquid or Solid Irritant Characteristics*: No appreciable hazard. Practically harmless to the skin; *Odor Threshold*: Not pertinent.
Fire Hazards - *Flash Point (deg. F)*: 330 CC; *Flammable Limits in Air (%):* No data; *Fire Extinguishing Agents:* Water, foam, dry chemicals, or carbon dioxide; *Fire Extinguishing Agents Not To Be Used:* Not pertinent; *Special Hazards of Combustion Products:* Toxic and irritating gases may form in fire situations; *Behavior in Fire:* Not pertinent; *Ignition Temperature :* No data; *Electrical Hazard:* Not pertinent; *Burning Rate:* No data.
Chemical Reactivity - *Reactivity with Water:* No reaction; *Reactivity with Common Materials:* No reactions; *Stability During Transport:* Stable; *Neutralizing Agents for Acids and Caustics:* Not pertinent; *Polymerization:* Not pertinent; *Inhibitor of Polymerization:* Not pertinent.

SULFUR DIOXIDE
Chemical Designations - *Synonyms*: No common synonyms; *Chemical Formula*: SO_2.
Observable Characteristics - *Physical State (as shipped)*: Liquefied gas; *Color*: Colorless; *Odor*: Sharp, pungent; characteristic; like burning sulfur.
Physical and Chemical Properties - *Physical State at 15 °C and 1 atm.*: Gas; *Molecular Weight*: 64.06; *Boiling Point at 1 atm.*: 14, -10, 263.2; *Freezing Point*: -103.9, -75.5, 197.7; *Critical Temperature*: 315, 157, 430; *Critical Pressure*: 1142, 77.69, 7.870; *Specific Gravity*: 1.45 at -10 °C (liquid); *Vapor (Gas) Specific Gravity*: 2.2; *Ratio of Specific Heats of Vapor (Gas)*: 1.265; *Latent Heat of Vaporization*: 171, 94.8, 3.97; *Heat of Combustion*: Not pertinent; *Heat of Decomposition*: Not pertinent.
Health Hazards Information - *Recommended Personal Protective Equipment*: Air-supplied mask or approved canister; goggles or face shield; rubber gloves; rubber clothing where contact with liquid is possible; *Symptoms Following Exposure*: Vapors cause irritation of eyes and lungs, with severe choking; *General Treatment for Exposure*: INHALATION: remove from exposure; support respiration; administer oxygen; call a doctor. SKIN: flush with water after exposure to liquid. EYES: wash promptly for at least 15 min.; call physician; *Toxicity by Inhalation (Threshold Limit Value)*: 5 ppm; *Short-Term Inhalation Limits*: 20 ppm for 5 min.; *Toxicity by Ingestion*: Not pertinent; *Late Toxicity*: Data not available; *Vapor (Gas) Irritant Characteristics*: Vapors cause severe irritation of eye and throat and can cause eye and lung injury. They cannot be tolerated even at low concentrations; *Liquid or Solid Irritant Characteristics*: Liquid can cause frostbite; *Odor Threshold*: 3 ppm.
Fire Hazards - *Flash Point :* Not flammable; *Flammable Limits in Air (%):* Not flammable; *Fire Extinguishing Agents:* Not pertinent; *Fire Extinguishing Agents Not To Be Used:* Not pertinent; *Special Hazards of Combustion Products:* Not pertinent; *Behavior in Fire:* Containers may rupture, releasing toxic and irritating sulfur dioxide; *Ignition Temperature :* Not flammable; *Electrical Hazard:* Not

pertinent; *Burning Rate:* Not pertinent.

Chemical Reactivity - *Reactivity with Water:* Reacts non-violently with water to form corrosive acid; *Reactivity with Common Materials:* Corrodes aluminum; *Stability During Transport:* Stable; *Neutralizing Agents for Acids and Caustics:* The mild acidity of water solution may be neutralized by dilute caustic soda; *Polymerization:* Not pertinent; *Inhibitor of Polymerization:* Not pertinent.

SULFURIC ACID

Chemical Designations - *Synonyms*: Battery acid; Chamber acid; Fertilizer acid; Oil of vitriol; *Chemical Formula*: H_2SO_4.

Observable Characteristics - *Physical State (as shipped)*: Liquid; *Color*: Colorless (pure) to dark brown; *Odor*: Odorless unless hot, then choking.

Physical and Chemical Properties - *Physical State at 15 °C and 1 atm.*: Liquid; *Molecular Weight*: 98.08; *Boiling Point at 1 atm.*: 644, 340, 613; *Freezing Point*: Not pertinent; *Critical Temperature*: Not pertinent; *Critical Pressure*: Not pertinent; *Specific Gravity*: 1.84 at 20 °C (liquid); *Vapor (Gas) Specific Gravity*: Not pertinent; *Ratio of Specific Heats of Vapor (Gas)*: Not pertinent; *Latent Heat of Vaporization*: Not pertinent; *Heat of Combustion*: Not pertinent; *Heat of Decomposition*: Not pertinent.

Health Hazards Information - *Recommended Personal Protective Equipment*: Safety shower; eyewash fountain; safety goggles; face shield; approved respirator (self-contained or air-line); rubber safety shoes; rubber apron; *Symptoms Following Exposure*: Inhalation of vapor from hot, concentrated acid may injure lungs. Swallowing may cause severe injury or death. Contact with skin or eyes causes severe burns; *General Treatment for Exposure*: Call a doctor. INHALATION: observe victim for delayed pulmonary reaction. INGESTION: have victim drink water if possible; do NOT induce vomiting. EYES AND SKIN: wash with large amounts of water for at least 15 min.; do not use oils or ointments in eyes; treat skin burns; *Toxicity by Inhalation (Threshold Limit Value)*: 1 mg/m^3; *Short-Term Inhalation Limits*: 10 mg/m^3 for 5 min.; 5 mg/m^3 for 10 min.; 2 mg/m^3 for 30 min.; 1 mg/m^3 for 60 min.; *Toxicity by Ingestion*: No effects except those secondary to tissue damage; *Late Toxicity*: None; *Vapor (Gas) Irritant Characteristics*: Vapors from hot acid (77-98%) cause moderate irritation of eyes and respiratory system. The effect is temporary; *Liquid or Solid Irritant Characteristics*: 77-98% acid causes severe second- and third-degree burns on short contact and is very injurious to the eyes; *Odor Threshold*: Greater than 1 mg/m^3.

Fire Hazards - *Flash Point* : Not flammable; *Flammable Limits in Air (%):* Not flammable; *Fire Extinguishing Agents:* Not pertinent; *Fire Extinguishing Agents Not To Be Used:* Water used on adjacent fires should be carefully handled; *Special Hazards of Combustion Products:* Not pertinent; *Behavior in Fire:* Not flammable; *Ignition Temperature :* Not flammable; *Electrical Hazard:* None; *Burning Rate:* Not flammable.

Chemical Reactivity - *Reactivity with Water:* Reacts violently with the evolution of heat (exothermic reaction). Significant agitation and spattering occurs when water is added to the chemical; *Reactivity with Common Materials:* Sulfuric acid is extremely hazardous in contact with many materials, particularly metals and combustibles. Dilute acid reacts with most metals, releasing hydrogen which can form explosive mixtures with air in confined spaces; *Stability During Transport:* Stable; *Neutralizing Agents for Acids and Caustics:* Dilute with large amounts of water, then neutralize with lime, limestone, or soda ash; *Polymerization:* Not flammable; *Inhibitor of Polymerization:* Not flammable.

SULFURIC ACID, SPENT

Chemical Designations - *Synonyms*: No common synonyms; *Chemical Formula*: $H_2SO_4 \cdot H_2O$.

Observable Characteristics - *Physical State (as shipped)*: Liquid; *Color*: Colorless to dark brown; *Odor*: Odorless.

Physical and Chemical Properties - *Physical State at 15 °C and 1 atm.*: Liquid; *Molecular Weight*: Not pertinent; *Boiling Point at 1 atm.*: 212, 100, 373; *Freezing Point*: Not pertinent; *Critical Temperature*: Not pertinent; *Critical Pressure*: Not pertinent; *Specific Gravity*: 1.39 at 20 °C (liquid); *Vapor (Gas) Specific Gravity*: Not pertinent; *Ratio of Specific Heats of Vapor (Gas)*: Not pertinent;

Latent Heat of Vaporization: Not pertinent; *Heat of Combustion*: Not pertinent; *Heat of Decomposition*: Not pertinent.

Health Hazards Information - *Recommended Personal Protective Equipment*: Chemical safety goggles and face shield; rubber gloves, boots, and apron; *Symptoms Following Exposure*: Contact with eyes or skin causes severe burns, the severity depending on the strength of the acid. Ingestion can cause severe irritation of mouth and stomach; *General Treatment for Exposure*: Call a doctor. INGESTION: do NOT induce vomiting. SKIN OR EYES: flush affected parts with large amounts of water for at least 15 min.; do NOT use oils or ointments in eyes; treat burns; *Toxicity by Inhalation (Threshold Limit Value)*: Not pertinent; *Short-Term Inhalation Limits*: Not pertinent; *Toxicity by Ingestion*: No effects except those stemming from tissue damage; *Late Toxicity*: None; *Vapor (Gas) Irritant Characteristics*: Non-volatile; *Liquid or Solid Irritant Characteristics*: Severe skin irritant. Causes second- and third-degree burns on short contact and is very injurious to the eyes; *Odor Threshold*: Not pertinent.

Fire Hazards - *Flash Point* : Not flammable; *Flammable Limits in Air (%):* Not flammable; *Fire Extinguishing Agents:* Not pertinent; *Fire Extinguishing Agents Not To Be Used:* Not pertinent; *Special Hazards of Combustion Products:* Not pertinent; *Behavior in Fire:* Not pertinent; *Ignition Temperature :* Not flammable; *Electrical Hazard:* Not pertinent; *Burning Rate:* Not flammable

Chemical Reactivity - *Reactivity with Water:* No reaction, unless strength is above 80 ~ 90 %, in which case an exothermic reaction will occur. See sulfuric acid; *Reactivity with Common Materials:* Attacks many metals, releasing flammable hydrogen gas; *Stability During Transport:* Stable; *Neutralizing Agents for Acids and Caustics:* Neutralize with limestone, lime, or soda ash after further dilution with water; *Polymerization:* Not pertinent; *Inhibitor of Polymerization:* Not pertinent.

T

TITANIUM TETRACHLORIDE

Chemical Designations - *Synonyms*: No common synonyms; *Chemical Formula*: $TiCl_4$.

Observable Characteristics - *Physical State (as normally shipped)*: Liquid; *Color*: Colorless; *Odor*: Acrid; choking.

Physical and Chemical Properties - *Physical State at 15 °C and 1 atm.*: Liquid; *Molecular Weight*: 189.71; *Boiling Point at 1 atm.*: 277, 136, 249; *Freezing Point*: -11, -24, 249; *Critical Temperature*: Not pertinent; *Critical Pressure*: Not pertinent; *Specific Gravity*: 1.726 at 20 °C (liquid); *Vapor (Gas) Density*: Not pertinent; *Ratio of Specific Heats of Vapor (Gas)*: (est.) 1.221; *Latent Heat of Vaporization*: 79.7, 44.3, 1.86; *Heat of Combustion*: Not pertinent; *Heat of Decomposition*: Not pertinent.

Health Hazards Information - *Recommended Personal Protective Equipment*: Goggles and face shield; air-supplied mask or approved canister; rubber gloves; protective clothing; *Symptoms Following Exposure*: Vapors can cause severe irritation and damage to eyes, coughing, headache, dizziness, lung damage, bronchial pneumonia. Liquid causes thermal and acid burns of eyes, skin, throat, and stomach. If ingested, causes nausea, vomiting, cramps, diarrhea, and possible tissue ulceration; *General Treatment for Exposure*: INHALATION: remove victim to fresh air; if symptoms persist, call a doctor. INGESTION: give large amounts of water, then induce vomiting; give milk, eggs ir olive oil; call a doctor. EYES: immediately flush with copious amounts of water for at least 15 min.; call a doctor. SKIN: flush with water; obtain medical attention if irritation persists; *Toxicity by Inhalation (Threshold Limit Value)*: 5 ppm in moist air; *Short-Term Exposure Limits*: Data not available; *Toxicity by Ingestion*: Data not available; *Late Toxicity*: Disturbances of upper respiratory and nervous system in man; *Vapor (Gas) Irritant Characteristics*: Vapor is moderately irritating such that personnel will not tolerate moderate or high vapor concentrations; *Liquid or Solid Irritant Characteristics*: Fairly Severe skin irritant; may cause pain and second- degree burns after a few minutes' contact; *Odor Threshold*: Data not available.

Fire Hazards - *Flash Point* : Not flammable; *Flammable Limits in Air (%):* Not flammable; *Fire Extinguishing Agents:* Dry powder or carbon dioxide on adjacent fires; *Fire Extinguishing Agents Not To Be Used:* Do not use water if it can directly contact this chemical; *Special Hazards of Combustion Products:* Not pertinent; *Behavior in Fire:* If containers leak, a very dense white fume can form and obscure operations; *Ignition Temperature :* Not flammable; *Electrical Hazard:* Not pertinent; *Burning Rate:* Not flammable.

Chemical Reactivity - *Reactivity with Water:* Reacts with moisture in air forming a dense white fume. Reaction with liquid water gives off heat and forms hydrochloric acid; *Reactivity with Common Materials:* The acid formed by reaction with moisture attacks metals, forming flammable hydrogen gas; *Stability During Transport:* Stable; *Neutralizing Agents for Acids and Caustics:* Acid formed by the reaction with water can be neutralized by limestone, lime, or soda ash; *Polymerization:* Not pertinent; *Inhibitor of Polymerization:* Not pertinent.

TOLUENE

Chemical Designations - *Synonyms*: Methylbenzen, Methylbenzol, Toluol; *Chemical Formula*: $C_6H_6CH_3$.

Observable Characteristics - *Physical State (as normally shipped)*: Liquid; *Color*: Colorless; *Odor*: Pungent; aromatic, benzene-like; distinct, pleasant.

Physical and Chemical Properties - *Physical State at 15 °C and 1 atm.*: Liquid; *Molecular Weight*: 92.14; *Boiling Point at 1 atm.*: 231.1, 110.6, 383.8; *Freezing Point*: -139, -95.0, 178.2; *Critical Temperature*: 605.4, 318.6, 591.8; *Critical Pressure*: 596.1, 40.55, 4.108; *Specific Gravity*: 0.867 at 20 °C (liquid); *Vapor (Gas) Density*: Not pertinent; *Ratio of Specific Heats of Vapor (Gas)*: 1.089; *Latent Heat of Vaporization*: 155, 86.1, 3.61; *Heat of Combustion*: -17,430, -9686, -405.5; *Heat of Decomposition*: Not pertinent.

Health Hazards Information - *Recommended Personal Protective Equipment*: Air-supplied mask; goggles and face shield; plastic gloves; *Symptoms Following Exposure*: Vapors irritate eyes and upper respiratory tract; cause dizziness, headache, anesthesia, respiratory arrest. Liquid irritates eyes and causes coughing, gagging distress, and rapidly developing pulmonary edema. If ingested causes vomiting, griping, diarrhea, depressed respiration; *General Treatment for Exposure*: INHALATION: remove victim to fresh air, give artificial respiration and oxygen if needed; call a doctor. INGESTION: do NOT induce vomiting; call a doctor. EYES: flush with water for at least 15 min. SKIN: wipe off, wash with soap and water; *Toxicity by Inhalation (Threshold Limit Value)*: 100 ppm; *Short-Term Exposure Limits*: 600 ppm for 30 min.; *Toxicity by Ingestion*: Grade 2, LD_{50} 0.5 to 5 g/kg; *Late Toxicity*: Kidney and liver damage may follow ingestion; *Vapor (Gas) Irritant Characteristics*: Vapors cause a slight smarting of the eyes or respiratory system if present in high concentration. The effect is temporary; *Liquid or Solid Irritant Characteristics*: Minimum hazard. If spilled on clothing and allowed to remain, may cause smarting and reddening of the skin; *Odor Threshold*: 0.17 ppm.

Fire Hazards - *Flash Point (deg. F)*: 40 CC, 55 OC; *Flammable Limits in Air (%):* 1.27 - 7.0; *Fire Extinguishing Agents:* Carbon dioxide or dry chemical for small fire; ordinary foam for large fires; *Fire Extinguishing Agents Not To Be Used:* Water may be ineffective; *Special Hazards of Combustion Products:* Not pertinent; *Behavior in Fire:* Vapors are heavier than air and may travel considerable distances to a source of ignition and flash back; *Ignition Temperature (deg. F):* 997; *Electrical Hazard:* Class I, Group D; *Burning Rate:* 5.7 mm/min.

Chemical Reactivity - *Reactivity with Water:* No reaction; *Reactivity with Common Materials:* No reactions; *Stability During Transport:* Stable; *Neutralizing Agents for Acids and Caustics:* Not pertinent; *Polymerization:* Not pertinent; *Inhibitor of Polymerization:* Not pertinent.

TOLUENE 2,4-DIISOCYANATE

Chemical Designations - *Synonyms*: Hylene T; Mondur TDS; Nacconate 100; 2,4-Tolylene diisocyanate; TDE; *Chemical Formula*: $1-CH_3C_6H_3(NCO)_2-2,4$.

Observable Characteristics - *Physical State (as normally shipped)*: Liquid; *Color*: Colorless to light yellow; *Odor*: Sweet, fruity, pungent.

Physical and Chemical Properties - *Physical State at 15 °C and 1 atm.*:Solid; *Molecular Weight*: 174.16; *Boiling Point at 1 atm.*: 482, 250, 523; *Freezing Point*: 68 - 72, 20 - 22, 293 - 295; *Critical Temperature*: Not pertinent; *Critical Pressure*: Not pertinent; *Specific Gravity*: 1.22 at 25 °C (liquid); *Vapor (Gas) Density*: Not pertinent; *Ratio of Specific Heats of Vapor (Gas)*: Not pertinent; *Latent Heat of Vaporization*: Not pertinent; *Heat of Combustion*: (est.) -10,300, -5720, -239; *Heat of Decomposition*: Not pertinent.

Health Hazards Information - *Recommended Personal Protective Equipment*: Organic vapor canister; goggles and face shield; rubber gloves; boots and apron; *Symptoms Following Exposure*: Irritates eyes and skin. Potent sensitizer and lung irritant if inhaled. May produce bronhospasm (asthma), pneumonitis, bronchitis, and pulmonary edema. Nocturnal cough and shortness of breath are common. Repeated low-level exposure may produce chronic lung disease. Oral toxicity is low; *General Treatment for Exposure*: INHALATION: remove victim to fresh air; administer artificial respiration and oxygen if needed; call a doctor. EYES: flush with water; wipe off; with rubbing alcohol; wash with soap and water; *Toxicity by Inhalation (Threshold Limit Value)*: 0.02 ppm; *Short-Term Exposure Limits*: 0.02 ppm for 5 min.; *Toxicity by Ingestion*: Grade 2, LD_{50} 0.5 to 5 g/kg; *Late Toxicity*: Data not available; *Vapor (Gas) Irritant Characteristics*: Vapors is moderately irritating such that personnel will not usually tolerate moderate or high vapor concentration; *Liquid or Solid Irritant Characteristics*: Fairly Severe skin irritant; may cause pain and second-degree burns after a few minutes' contact; *Odor Threshold*: 0.4-2.14 ppm.

Fire Hazards - *Flash Point (deg. F)*: 270 OC; *Flammable Limits in Air (%)*: 0.9 - 9.5; *Fire Extinguishing Agents:* Water, foam, dry chemical, or carbon dioxide; *Fire Extinguishing Agents Not To Be Used:* Water or foam may cause frothing; *Special Hazards of Combustion Products:* Irritating vapors are generated upon heating; *Behavior in Fire:* Not pertinent; *Ignition Temperature (deg. F):* > 300; *Electrical Hazard:* Not pertinent; *Burning Rate:* No data.

Chemical Reactivity - *Reactivity with Water:* A non violent reaction occurs forming carbon dioxide gas and an organic base; *Reactivity with Common Materials:* No reactions; *Stability During Transport:* Stable; *Neutralizing Agents for Acids and Caustics:* Not pertinent; *Polymerization:* Slow polymerization occurs at temperatures above 113 of. The reaction is not hazardous; *Inhibitor of Polymerization:* Not pertinent.

P-TOLUENESULFONIC ACID

Chemical Designations - *Synonyms*: Methylbenzenesulfonic acid; Tosic acid; p-TSA; *Chemical Formula*: $CH_3C_6H_4SO_3H$.

Observable Characteristics - *Physical State (as normally shipped)*: Solid; *Color*: White to brown to black; yellow to amber; *Odor*: None when pure; technical grade has slight aromatic odor.

Physical and Chemical Properties - *Physical State at 15 °C and 1 atm.*: Solid; *Molecular Weight*: 172.2; *Boiling Point at 1 atm.*: Not pertinent (decomposes); *Freezing Point*: 219 - 221, 104 - 105, 377 - 378; *Critical Temperature*: Not pertinent; *Critical Pressure*: Not pertinent; *Specific Gravity*: 1.45 at 25 °C (solid); *Vapor (Gas) Density*: Not pertinent; *Ratio of Specific Heats of Vapor (Gas)*: Not pertinent; *Latent Heat of Vaporization*: Not pertinent; *Heat of Combustion*: Not pertinent; *Heat of Decomposition*: Not pertinent.

Health Hazards Information - *Recommended Personal Protective Equipment*: Chemical goggles and face shield; rubber gloves; *Symptoms Following Exposure*: Contact with eyes or skin causes severe irritation. Ingestion causes irritation of mouth and stomach; *General Treatment for Exposure*: EYES: wash thoroughly with copious amounts of water for at least 15 min.; call physician if irritation persists. SKIN: wash thoroughly with large amounts of water for at least 15 min. INGESTION: give large amount of water; *Toxicity by Inhalation (Threshold Limit Value)*: Data not available; *Short-Term Exposure Limits*: Not pertinent; *Toxicity by Ingestion*: Grade 3, oral LD_{50} =400 mg/kg (rat); *Late Toxicity*: Data not available; *Vapor (Gas) Irritant Characteristics*: Data not available; *Liquid or Solid Irritant Characteristics*: Data not available; *Odor Threshold*: Not pertinent.

Fire Hazards - *Flash Point* : Not pertinent; this is a solid with low flammability; *Flammable Limits in Air (%):* Not pertinent; *Fire Extinguishing Agents:* Water; *Fire Extinguishing Agents Not To Be*

Used: Not pertinent; *Special Hazards of Combustion Products:* Irritating oxides of sulfur may be formed; *Behavior in Fire:* Not pertinent; *Ignition Temperature :* No data; *Electrical Hazard:* Not pertinent; *Burning Rate:* Not pertinent.

Chemical Reactivity - *Reactivity with Water:* No reaction; *Reactivity with Common Materials:* This is a strong acid that can react with common materials; *Stability During Transport:* Stable; *Neutralizing Agents for Acids and Caustics:* Flush with water and rinse with sodium bicarbonate or lime solution; *Polymerization:* Not pertinent; *Inhibitor of Polymerization:* Not pertinent.

O-TOLUIDINE

Chemical Designations - *Synonyms*: 2-Amino-1-methyl-benzene; 2-Aminotoluene; 2-Methylaniline; o- Methylaniline; *Chemical Formula*: $1,2\text{-}CH_3C_6H_4NH$.

Observable Characteristics - *Physical State (as normally shipped)*: Liquid; *Color*: Clear to light yellow; turns yellow, brown or deep red on exposure to air and light; *Odor*: Aromatic, aniline-like.

Physical and Chemical Properties - *Physical State at 15 °C and 1 atm.*:Liquid; *Molecular Weight*: 107.2; *Boiling Point at 1 atm.*: 392, 200, 473; *Freezing Point*: -11, -24, 249; *Critical Temperature*: 790, 421, 694; *Critical Pressure*: 544, 37.0, 3.75; *Specific Gravity*: 0.998 at 20 °C (liquid); *Vapor (Gas) Density*: Not pertinent; *Ratio of Specific Heats of Vapor (Gas)*: Not pertinent; *Latent Heat of Vaporization*: 179.1, 99.5, 4.16; *Heat of Combustion*: -16,180, -8,990, -376; *Heat of Decomposition*: Not pertinent.

Health Hazards Information - *Recommended Personal Protective Equipment*: Chemical safety goggles; face shield; Bu. Mines approved respirator; leather or rubber safety shoes; butyl rubber gloves; *Symptoms Following Exposure*: Absorption of toxic quantities by any route causes cyanosis (blue discoloration of lips, nails, skin); nausea, vomiting, and coma may follow. Repeated inhalation of low concentration may cause pallor, low-grade secondary anemia, and loss of appetite. Contact with eyes causes irritation; *General Treatment for Exposure*: Get medical attention following all exposure to this compound. INHALATION: move to fresh air. INGESTION: if victim is conscious, promptly induce vomiting by giving lukewarm soapy water or mustard and water. EYES: flush with copious amounts of water for at least 15 min., holding lids apart. SKIN: remove all contaminated clothing; wash affected areas immediately and thoroughly with plenty of warm water and soap; *Toxicity by Inhalation (Threshold Limit Value)*: 5 ppm; *Short-Term Exposure Limits*: Data not available; *Toxicity by Ingestion*: Grade 2, oral LD_{50}=900 mg/kg (rat); *Late Toxicity*: Causes tumors in urinary bladder of rats; *Vapor (Gas) Irritant Characteristics*: Data not available; *Liquid or Solid Irritant Characteristics*: Data not available; *Odor Threshold*: Data not available.

Fire Hazards - *Flash Point (deg. F)*: 167 OC, 85 CC; *Flammable Limits in Air (%):* No data; *Fire Extinguishing Agents:* Foam, dry chemical, or carbon dioxide; *Fire Extinguishing Agents Not To Be Used:* Water may be ineffective; *Special Hazards of Combustion Products:* Toxic oxides of nitrogen and flammable vapors may form; *Behavior in Fire:* No data; *Ignition Temperature (deg. F):* 900; *Electrical Hazard:* No data; *Burning Rate:* 3.6 mm/min.

Chemical Reactivity - *Reactivity with Water:* No reaction; *Reactivity with Common Materials:* No reactions; *Stability During Transport:* Stable; *Neutralizing Agents for Acids and Caustics:* Not pertinent; *Polymerization:* Not pertinent; *Inhibitor of Polymerization:* Not pertinent.

TOXAPHENE

Chemical Designations - *Synonyms*: Octachlorocamphene; *Chemical Formula*: $C_{10}H_8Cl_8$.

Observable Characteristics - *Physical State (as normally shipped)*: Waxy solid; *Color*: Amber; *Odor*: Mild turpentine odor.

Physical and Chemical Properties - *Physical State at 15 °C and 1 atm.*: Waxy solid; *Molecular Weight*: 414 (avg.); *Boiling Point at 1 atm.*: Decomposes; *Freezing Point*: 149-194, 65-90, 338-363; *Critical Temperature*: Not pertinent; *Critical Pressure*: Not pertinent; *Specific Gravity*: 1.6 at 15 °C (solid); *Vapor (Gas) Density*: Not pertinent; *Ratio of Specific Heats of Vapor (Gas)*: Not pertinent; *Latent Heat of Vaporization*: Not pertinent; *Heat of Combustion*: Not pertinent; *Heat of Decomposition*: Not pertinent.

Health Hazards Information - *Recommended Personal Protective Equipment*: Chemical-type respirator; rubber gloves; chemical Goggles and face shield; *Symptoms Following Exposure*: May be absorbed through skin, lungs, or intestinal tract. Symptoms include salivation, leg and back muscle spasms, nausea, vomiting, hyper excitability, tremors, shivering, clonic convulsions, then titanic contraction of all skeletal muscles. Lethal doses cause respiratory failure. Respiration, affected as a result of the exertion from vomiting or convulsions, is first arrested because of titanic muscular contraction, then increased in both amplitude and rate as the muscles relax; *General Treatment for Exposure*: If symptoms of poisoning appear, promptly remove the unabsorbed pesticide from the stomach by inducing vomiting with warm salty or soapy water (if the patient is conscious) or from the skin with soap and water. Keep patient warm and quiet. Call a physician; *Toxicity by Inhalation (Threshold Limit Value)*: Not pertinent; *Short-Term Exposure Limits*: Not pertinent; *Toxicity by Ingestion*: Grade 4, LD_{50} below 50 mg/kg (dot); *Late Toxicity*: Data not available; *Vapor (Gas) Irritant Characteristics*: The solid is non-volatile. For solutions, see meta-xylene; *Liquid or Solid Irritant Characteristics*: Minimum hazard. If spilled on clothing and allowed to remain, may cause smarting and reddening of the skin; *Odor Threshold*: Not pertinent.

Fire Hazards - *Flash Point (deg. F)*: 84 CC (solution); *Flammable Limits in Air (%)*: 1.1 - 6.4; *Fire Extinguishing Agents:* Foam, dry chemical, or carbon dioxide; *Fire Extinguishing Agents Not To Be Used:* Water may be ineffective; *Special Hazards of Combustion Products:* Toxic vapors are generated when heated; *Behavior in Fire:* Solution in xylene may produce corrosive products when heated; *Ignition Temperature (deg. F):* 986 (solution); *Electrical Hazard:* Not pertinent; *Burning Rate:* 5.8 mm/min.

Chemical Reactivity - *Reactivity with Water:* No reaction; *Reactivity with Common Materials:* No reactions; *Stability During Transport:* Stable; *Neutralizing Agents for Acids and Caustics:* Not pertinent; *Polymerization:* Not pertinent; *Inhibitor of Polymerization:* Not pertinent.

TRICHLOROETHYLENE

Chemical Designations - *Synonyms*: Algylen; Clorilen; Gemalgene; Threthylene; Trethylene; Tri; Trihloran; Trihloroethelene; TriClene; Trielene; Triline; Trimar; *Chemical Formula*: $CHCl=CCl_2$.

Observable Characteristics - *Physical State (as shipped)*: Liquid; *Color*: Colorless; *Odor*: Chloroform-like; ethereal.

Physical and Chemical Properties - *Physical State at 15 °C and 1 atm.*: Liquid; *Molecular Weight*: 131.39; *Boiling Point at 1 atm.*: 189, 87, 360; *Freezing Point*: -123.5, -86.4, 186.8; *Critical Temperature*: Not pertinent; *Critical Pressure*: Not pertinent; *Specific Gravity*: Not pertinent; *Vapor (Gas) Specific Gravity*: 4.5; *Ratio of Specific Heats of Vapor (Gas)*: 1.116; *Latent Heat of Vaporization*: 103, 57.2, 2.40; *Heat of Combustion*: Not pertinent; *Heat of Decomposition*: Not pertinent.

Health Hazards Information - *Recommended Personal Protective Equipment*: Organic vapor-acid gas canister; self-contained breathing apparatus for emergencies; neoprene or vinyl gloves; chemical safety goggles; face-shield; neoprene safety shoes; neoprene suit or apron for splash protection; *Symptoms Following Exposure*: INHALATION: symptoms range from irritation of the nose and throat to nausea, an attitude of irresponsibility, blurred vision, and finally disturbance of central nervous system resulting in cardiac failure. Chronic exposure may cause organic injury. INGESTION: symptoms similar to inhalation. SKIN: defatting action can cause dermatitis. EYES: slightly irritating sensation and lachrymation; *General Treatment for Exposure*: Do NOT administer adrenaline or epinephrine; get medical attention for all cases of overexposure. INHALATION: remove victim to fresh air; if necessary, apply artificial respiration and/or administer oxygen. INGESTION: have victim drink water and induce vomiting; repeat three times; then give 1 tablespoon Epsom salts in water. EYES: flush thoroughly with water. SKIN: wash thoroughly with soap and warm water; *Toxicity by Inhalation (Threshold Limit Value)*: 100 ppm; *Short-Term Inhalation Limits*: 200 ppm for 30 min.; *Toxicity by Ingestion*: Grade 3, LD_{50} 50 to 500 mg/kg; *Late Toxicity*: Data not available; *Vapor (Gas) Irritant Characteristics*: Vapor cause slight smarting of the eyes or respiratory system if present in high concentration. The effect is temporary; *Liquid or Solid Irritant Characteristics*: Minimum hazard. If spilled on clothing and allowed to remain, may cause smarting and reddening of the skin; *Odor Threshold*: 50 ppm.

Fire Hazards - *Flash Point (deg. F)*: 90 CC; practically nonflammable; *Flammable Limits in Air (%):* 8.0 - 10.5; *Fire Extinguishing Agents:* Water fog; *Fire Extinguishing Agents Not To Be Used:* Not pertinent; *Special Hazards of Combustion Products:* Toxic and irritating vapors are produced in fire situations; *Behavior in Fire:* Not pertinent; *Ignition Temperature (deg. F):* 770; *Electrical Hazard:* Not pertinent; *Burning Rate:* Not pertinent.

Chemical Reactivity - *Reactivity with Water:* No reaction; *Reactivity with Common Materials:* No reaction; *Stability During Transport:* Stable; *Neutralizing Agents for Acids and Caustics:* Not pertinent; *Polymerization:* Not pertinent; *Inhibitor of Polymerization:* Not pertinent.

TRICHLOROFLUOROMETHANE

Chemical Designations - *Synonyms*: Arcton 9; Freon 11; Isceon 11; Eskimon 11; Frigen 11; Isotron 11;F-11; Genetron 11; Ucon 11; *Chemical Formula*: $CFCl_3$.

Observable Characteristics - *Physical State (as shipped)*: Liquid; *Color*: Colorless; *Odor*: Odorless; weak chlorinated solvent.

Physical and Chemical Properties - *Physical State at 15 °C and 1 atm.*: Liquid; *Molecular Weight*: 137.4; *Boiling Point at 1 atm.*: 4.8, 23.8, 297.0; *Freezing Point*: -168, -111, 162; *Critical Temperature*: 388, 198, 471; *Critical Pressure*: 639.4, 43.5, 4.41; *Specific Gravity*: 1.49 at 20°C (liquid); *Vapor (Gas) Specific Gravity*: 4.7; *Ratio of Specific Heats of Vapor (Gas):*(est.) 1.128; *Latent Heat of Vaporization*: 78.3, 43.5, 1.82; *Heat of Combustion*: Not pertinent; *Heat of Decomposition*: Not pertinent.

Health Hazards Information - *Recommended Personal Protective Equipment*: Air line respirator; rubber gloves; monogoggles; *Symptoms Following Exposure*: Breathing concentration approaching 10% in air will cause dizziness and drowsiness. Contact with tissues may cause frostbite; *General Treatment for Exposure*: INHALATION: remove victim to not-contaminated area and apply artificial respiration if breathing has stopped; call a physician immediately; oxygen inhalation may be utilized. SKIN: if frostbite has occurred, flush areas with warm water; *Toxicity by Inhalation (Threshold Limit Value)*: 1000 ppm; *Short-Term Inhalation Limits*: Data not available; *Toxicity by Ingestion*: Data not available; *Late Toxicity*: Data not available; *Vapor (Gas) Irritant Characteristics*: Non-irritating; *Liquid or Solid Irritant Characteristics*: May cause frostbite; *Odor Threshold*: Data not available.

Fire Hazards - *Flash Point* : Not flammable; *Flammable Limits in Air (%):* Not flammable; *Fire Extinguishing Agents:* Not pertinent; *Fire Extinguishing Agents Not To Be Used:* Not pertinent; *Special Hazards of Combustion Products:* Produces toxic and irritating vapors when heated to its decomposition temperature; *Behavior in Fire:* Not pertinent; *Ignition Temperature :* Not flammable; *Electrical Hazard:* Not pertinent; *Burning Rate:* Not pertinent.

Chemical Reactivity - *Reactivity with Water:* No reaction; *Reactivity with Common Materials:* No reactions; *Stability During Transport:* Stable; *Neutralizing Agents for Acids and Caustics:* Not pertinent; *Polymerization:* Not pertinent; *Inhibitor of Polymerization:* Not pertinent.

TRICHLOROPHENOL

Chemical Designations - *Synonyms*: Dowicide-2; Omal; Phenachlor; 2,4,5-Trihlorophenol; *Chemical Formula*: $1-HOC_6H_2Cl_3-2,4,5$.

Observable Characteristics - *Physical State (as shipped)*: Solid; *Color*: Colorless to gray; *Odor*: Strong disinfectant.

Physical and Chemical Properties - *Physical State at 15 °C and 1 atm.*: Solid; *Molecular Weight*: 197.5; *Boiling Point at 1 atm.*: 485, 252, 525; *Freezing Point*: 135, 57, 330; *Critical Temperature*: Not pertinent; *Critical Pressure*: Not pertinent; *Specific Gravity*: 1.7 at 25 °C (solid); *Vapor (Gas) Specific Gravity*: Not pertinent; *Ratio of Specific Heats of Vapor (Gas)*: Not pertinent; *Latent Heat of Vaporization*: Not pertinent; *Heat of Combustion*: Not pertinent; *Heat of Decomposition*: Not pertinent.

Health Hazards Information - *Recommended Personal Protective Equipment*: Approved dust respirator for toxic dust; goggles; protective clothing to prevent contact with skin. *Symptoms Following Exposure*: INHALATION: Dust may cause swelling of eyes and eye injury, irritation of nose and throat. Solid irritates skin on prolonged contact; *General Treatment for Exposure*: INHALATION: remove to fresh air; get medical attention; EYES: flush with water for at least 15 min.; get medical attention. SKIN: wash with soap and water; *Toxicity by Inhalation (Threshold Limit Value)*: Not pertinent; *Short-Term Inhalation Limits*: Data not available; *Toxicity by Ingestion*: 20% solution in fuel oil: Grade 2, LD_{50} 0.5 to 5 g/kg (rat); *Late Toxicity*: Data not available; *Vapor (Gas) Irritant Characteristics*: Essentially non-volatile at ordinary temperatures; *Liquid or Solid Irritant Characteristics*: May cause injury to eye. Prolonged contact with skin causes a slight burn. Dust irritates nose and throat; *Odor Threshold*: Data not available.

 Fire Hazards - *Flash Point* : Not flammable; *Flammable Limits in Air (%):* Not flammable; *Fire Extinguishing Agents:* Not pertinent; *Fire Extinguishing Agents Not To Be Used:*; *Special Hazards of Combustion Products:*; *Behavior in Fire:*; *Ignition Temperature :* Not pertinent; *Electrical Hazard:* Not pertinent; *Burning Rate:* Not pertinent.

Chemical Reactivity - *Reactivity with Water:* No reaction; *Reactivity with Common Materials:* No reactions; *Stability During Transport:* Stable; *Neutralizing Agents for Acids and Caustics:* Not pertinent; *Polymerization:* Not pertinent; *Inhibitor of Polymerization:* Not pertinent.

2,4,5-TRICHLOROPHENOXY ACETIC ACID

Chemical Designations - *Synonyms*: 2,4,5-T; *Chemical Formula*: 2,4,5-$Cl_3C_6H_2OCH_2COOH$.

Observable Characteristics - *Physical State (as shipped)*:Solid; *Color*: White; *Odor*: None.

Physical and Chemical Properties - *Physical State at 15 °C and 1 atm.*: Solid; *Molecular Weight*: 255.5; *Boiling Point at 1 atm.*: Not pertinent; *Freezing Point*: 316, 158, 431; *Critical Temperature*: Not pertinent; *Critical Pressure*: Not pertinent; *Specific Gravity*: 1.803 at 20 °C (solid); *Vapor (Gas) Specific Gravity*: Not pertinent; *Ratio of Specific Heats of Vapor (Gas)*: Not pertinent; *Latent Heat of Vaporization*: Not pertinent; *Heat of Combustion*: -6,500, -3,600, -150; *Heat of Decomposition*: Not pertinent.

Health Hazards Information - *Recommended Personal Protective Equipment*: Dust mask and rubber gloves. *Symptoms Following Exposure*: INHALATION: Overexposure to dust by inhalation or ingestion may cause fatigue, nausea, vomiting, lowered blood pressure, convulsions, coma. Dust may irritate eyes and skin; *General Treatment for Exposure*: INHALATION: remove victim to fresh air; if required, give artificial respiration. EYES: flush with water until irritating dust is removed. SKIN: wash with soap and water. INGESTION: call physician at once; induce vomiting and administer gastric lavage; *Toxicity by Inhalation (Threshold Limit Value)*: 10 mg/m^5; *Short-Term Inhalation Limits*: Data not available; *Toxicity by Ingestion*: Grade 3, oral LD_{50} 500 mg/kg (rat); *Late Toxicity*: Birth defects in rats and mice. Causes an acne-like skin eruption among human workers; *Vapor (Gas) Irritant Characteristics*: Not pertinent; *Liquid or Solid Irritant Characteristics*: Data not available; *Odor Threshold*: Not pertinent.

Fire Hazards - *Flash Point* : Not pertinent (solid); *Flammable Limits in Air (%):* Not pertinent; *Fire Extinguishing Agents:* Water, foam, dry chemical, or carbon dioxide; *Fire Extinguishing Agents Not To Be Used:* Not pertinent; *Special Hazards of Combustion Products:* Toxic hydrogen chloride and phosgene gases; *Behavior in Fire:* Not pertinent; *Ignition Temperature :* No data; *Electrical Hazard:* Not pertinent; *Burning Rate:* Not pertinent.

Chemical Reactivity - *Reactivity with Water:* No reaction; *Reactivity with Common Materials:* Can be corrosive to common metals; *Stability During Transport:* Stable; *Neutralizing Agents for Acids and Caustics:* Not pertinent; *Polymerization:* Not pertinent; *Inhibitor of Polymerization:* Not pertinent.

TRICHLOROSILANE

Chemical Designations - *Synonyms*: Silicochloroform, Trichloromonosilane; *Chemical Formula*: $SiHCl_3$.

Observable Characteristics - *Physical State (as shipped)*:Liquid; *Color*: Colorless; *Odor*: Sharp, choking, like hydrochloric acid.

Physical and Chemical Properties - *Physical State at 15 °C and 1 atm.*: liquid; *Molecular Weight*: 135.5; *Boiling Point at 1 atm.*: 90, 32, 305; *Freezing Point*: -197, -127, 146; *Critical Temperature*: Not pertinent; *Critical Pressure*: Not pertinent; *Specific Gravity*: 1.344 at 20 °C (liquid); *Vapor (Gas) Specific Gravity*: 4.9; *Ratio of Specific Heats of Vapor (Gas)*: Data not available; *Latent Heat of Vaporization*: 85, 47, 2.0; *Heat of Combustion*: -6,500, -3,600, -150; *Heat of Decomposition*: Data not available.

Health Hazards Information - *Recommended Personal Protective Equipment*: Acid-vapor-type respiratory protection; rubber gloves; chemical worker's goggles; other protective equipment as necessary to protect skin and eyes; *Symptoms Following Exposure*: Inhalation causes severe irritation of respiratory system. Liquid causes severe burns of eyes and skin. Ingestion causes severe burns of mouth and stomach; *General Treatment for Exposure*: INHALATION: remove victim from exposure; if breathing is difficult or stopped, give artificial respiration; call physician. EYES or SKIN: flush with plenty of water immediately for at least 15 min., and get medical attention. INGESTION: do NOT induce vomiting; give large amount of water; get medical attention; *Toxicity by Inhalation (Threshold Limit Value)*: Data not available; *Short-Term Inhalation Limits*: Data not available; *Toxicity by Ingestion*: Grade 2, oral LD_{50} = 1,000 mg/kg (rat); *Late Toxicity*: Data not available; *Vapor (Gas) Irritant Characteristics*: Vapors cause severe irritation of eyes and throat and can cause eye and lung injure. They cannot be tolerated even at low concentrations; *Liquid or Solid Irritant Characteristics*: Severe skin irritation. Causes second-and third-degree burns on short contact and is very injurious to the eyes; *Odor Threshold*: Data not available.

Fire Hazards - *Flash Point (deg. F)*: -18 OC, > -58 CC; *Flammable Limits in Air (%)*: 1.2 - 90.5; *Fire Extinguishing Agents:* Dry chemical, carbon dioxide; *Fire Extinguishing Agents Not To Be Used:* Water, foam; *Special Hazards of Combustion Products:* Toxic hydrogen chloride and phosgene gases may form; *Behavior in Fire:* Difficult to extinguish; reignition may occur. Also, vapor is heavier than air and can travel to a source of ignition and flash back; *Ignition Temperature (deg. F):* 220; *Electrical Hazard:* No data; *Burning Rate:* No data.

Chemical Reactivity - *Reactivity with Water:* Reacts violently to form hydrogen chloride fumes (hydrochloric acid); *Reactivity with Common Materials:* Reacts with surface moisture to form hydrochloric acid which corrodes metals and generates flammable hydrogen gas; *Stability During Transport:* Stable; *Neutralizing Agents for Acids and Caustics:* Flush with water and rinse with sodium bicarbonate or lime solution; *Polymerization:* Not pertinent; *Inhibitor of Polymerization:* Not pertinent.

TRICHLORO-S-TRIAZINETRIONE

Chemical Designations - *Synonyms*: Trichloroiminoisocyanuris acid; Trichloroisocyanuric acid; Trichloro-s-Triazine-2,4,6-(1H,3H,5H)-trion; Trichlorotriazinetrion; 1,3,5- Trichloro- 2,4,6-trioxo-1,3,5-triazine; *Chemical Formula*: $Cl_3(NCO)_3$.

Observable Characteristics - *Physical State (as shipped)*: Solid; *Color*: White; *Odor*: Like chlorine.

Physical and Chemical Properties - *Physical State at 15 °C and 1 atm.*: solid; *Molecular Weight*: 232.5; *Boiling Point at 1 atm.*: Not pertinent; *Freezing Point*: Not pertinent; *Critical Temperature*: Not pertinent; *Critical Pressure*: Not pertinent; *Specific Gravity*: (est.) > 1 at 20 °C (solid); *Vapor (Gas) Specific Gravity*: Not pertinent; *Ratio of Specific Heats of Vapor (Gas)*: Not pertinent; *Latent Heat of Vaporization:* Not pertinent; *Heat of Combustion*: Not pertinent; *Heat of Decomposition*: Not pertinent.

Health Hazards Information - *Recommended Personal Protective Equipment*: Dust mask or chlorine canister mask, goggles; rubber gloves; *Symptoms Following Exposure*: Inhalation causes sneezing and coughing. Contact with dust causes moderate irritation of eyes and itching and redness of skin. Ingestion causes burns of mouth and stomach; *General Treatment for Exposure*: INHALATION: remove victim to fresh air. EYES: irrigate with running water for 15 min.; call physician. SKIN: flush with water INGESTION: induce vomiting and call physician. *Toxicity by Inhalation (Threshold Limit Value)*: Data not available; *Short-Term Inhalation Limits*: Data not available; *Toxicity by Ingestion*: Grade 2, oral LD_{50} = 750 mg/kg (rat); *Late Toxicity*: Data not available; *Vapor (Gas) Irritant Characteristics*: Data not available; *Liquid or Solid Irritant Characteristics*: Data not available; *Odor Threshold*: Data not available.

Fire Hazards - *Flash Point* : Not flammable but may cause fire on contact with ordinary combustible materials; *Flammable Limits in Air (%):* Not pertinent; *Fire Extinguishing Agents:* Water in large amounts; *Fire Extinguishing Agents Not To Be Used:* Not pertinent; *Special Hazards of Combustion Products:* Toxic chlorine or nitrogen trichloride may be formed in fires; *Behavior in Fire:* Containers may explode when heated; *Ignition Temperature* : Not pertinent; *Electrical Hazard:* Not pertinent; *Burning Rate:* Not pertinent.

Chemical Reactivity - *Reactivity with Water:* A non-hazardous reaction occurs forming a bleach solution; *Reactivity with Common Materials:* Contact with most foreign material, organic matter, or easily chlorinated or oxidized materials may result in fires. Avoid contacting this product with oil, sawdust, floor sweepings, other easily oxidized organic compounds; *Stability During Transport:* Stable; *Neutralizing Agents for Acids and Caustics:* Not pertinent; *Polymerization:* Not pertinent; *Inhibitor of Polymerization:* Not pertinent.

TRICRESYL PHOSPHATE

Chemical Designations - *Synonyms*: TCP; Tri-p-tolil phosphate; Tri-p-cresyl phosphate; *Chemical Formula*: $(p\text{-}CH_3C_6H_4O)_3PO$.

Observable Characteristics - *Physical State (as shipped)*:Liquid; *Color*: Colorless; *Odor*: Odorless.

Physical and Chemical Properties - *Physical State at 15 °C and 1 atm.*: liquid; *Molecular Weight*: 368; *Boiling Point at 1 atm.*: 770, 410, 683; *Freezing Point*: -27, -33, 240; *Critical Temperature*: Not pertinent; *Critical Pressure*: Not pertinent; *Specific Gravity*: 1.16 at 20 °C (liquid); *Vapor (Gas) Specific Gravity*: Not pertinent; *Ratio of Specific Heats of Vapor (Gas)*: Not pertinent; *Latent Heat of Vaporization*: (est.) 80.0, 44.5, 1.86; *Heat of Combustion*: Not pertinent; *Heat of Decomposition*: Not pertinent.

Health Hazards Information - *Recommended Personal Protective Equipment*: Goggles or face shield; *Symptoms Following Exposure*: Vapors may irritate eyes, but only at high temperatures. Ingestion of liquid may cause severe damage to central nervous system and death if significant amounts of the toxic ortho-isomer are present. *General Treatment for Exposure*: INGESTION: induce vomiting and call a doctor. EYES: flush with water for at least 15 min. SKIN: wipe off, wash with soap and water; *Toxicity by Inhalation (Threshold Limit Value)*: Not pertinent; *Short-Term Inhalation Limits*: Not pertinent; *Toxicity by Ingestion*: Grade 2, LD_{50} 0.5 to 5 g/kg (chicken LD_{50} > 2 g/kg); *Late Toxicity*: Data not available; *Vapor (Gas) Irritant Characteristics*: Vapors cause a slight smarting of the eyes or respiratory system if present in high concentration. The effect is temporary. The compound is not-volatile for all practical purposes; *Liquid or Solid Irritant Characteristics*: No appreciable hazard. Practically harmless to the skin; *Odor Threshold*: Not pertinent.

Fire Hazards - *Flash Point (deg. F)*: 410 CC; *Flammable Limits in Air (%):* No data; *Fire Extinguishing Agents:* Foam, dry chemical, or carbon dioxide; *Fire Extinguishing Agents Not To Be Used:* Water or foam may cause frothing; *Special Hazards of Combustion Products:* Not pertinent; *Behavior in Fire:* Not pertinent; *Ignition Temperature* : No data; *Electrical Hazard:* Not pertinent; *Burning Rate:* No data.

Chemical Reactivity - *Reactivity with Water:* No reaction; *Reactivity with Common Materials:* No reactions; *Stability During Transport:* Stable; *Neutralizing Agents for Acids and Caustics:* Not pertinent; *Polymerization:* Not pertinent; *Inhibitor of Polymerization:* Not pertinent.

TRIDECANOL

Chemical Designations - *Synonyms*: Isotridecanol; Isotridecyl alcohol; 1-tridecanol; *Chemical Formula*: $C_{12}H_{25}CH_2OH$.

Observable Characteristics - *Physical State (as shipped)*:Liquid; *Color*: Colorless; *Odor*: Mild alcoholic.

Physical and Chemical Properties - *Physical State at 15 °C and 1 atm.*: liquid; *Molecular Weight*: 200.37; *Boiling Point at 1 atm.*: 525, 274, 547; *Freezing Point*: Not pertinent; *Critical Temperature*: Not pertinent; *Critical Pressure*: Not pertinent; *Specific Gravity*: 0.846 at 20 °C (liquid); *Vapor (Gas) Specific Gravity*: Not pertinent; *Ratio of Specific Heats of Vapor (Gas)*: 1.027; *Latent Heat of*

Vaporization: 120, 64, 2.7; *Heat of Combustion*:: -12,200, -6,790, -284; *Heat of Decomposition*: Not pertinent.

Health Hazards Information - *Recommended Personal Protective Equipment*: Synthetic rubber gloves; chemical goggles. *Symptoms Following Exposure*: Inhalator hazard slight. Skin contact results in moderate irritation. Liquid contact with eyes causes severe irritation and possible eye damage. *General Treatment for Exposure*: EYES: promptly flush with clean water for at least 15 min. and see a physician. SKIN: wash exposed area with soap and water; *Toxicity by Inhalation (Threshold Limit Value)*: Data not available; *Short-Term Inhalation Limits*: Data not available; *Toxicity by Ingestion*: Data not available; *Late Toxicity*: Data not available; *Vapor (Gas) Irritant Characteristics*: Vapors are nonirritating to the eye and throat; *Liquid or Solid Irritant Characteristics*: No appreciable hazard. Practically harmless to the skin; *Odor Threshold*: Data not available.

Fire Hazards - *Flash Point (deg. F)*:250 OC; *Flammable Limits in Air (%):* No data; *Fire Extinguishing Agents:* Alcohol, dry chemical, water fog; *Fire Extinguishing Agents Not To Be Used:* Water or foam may cause frothing; *Special Hazards of Combustion Products:* Not pertinent; *Behavior in Fire:* Not pertinent; *Ignition Temperature :* No data; *Electrical Hazard:* Not pertinent; *Burning Rate:* No data.

Chemical Reactivity - *Reactivity with Water:* No reaction; *Reactivity with Common Materials:* No reactions; *Stability During Transport:* Stable; *Neutralizing Agents for Acids and Caustics:* Not pertinent; *Polymerization:* Not pertinent; *Inhibitor of Polymerization:* Not pertinent.

1-TRIDECENE

Chemical Designations - *Synonyms*: Undecylethylene; *Chemical Formula*: $CH_3(CH_2)_{10}CH=CH_2$.

Observable Characteristics - *Physical State (as shipped)*:Liquid; *Color*: Colorless; *Odor*: Mild, pleasant.

Physical and Chemical Properties - *Physical State at 15 °C and 1 atm.*: liquid; *Molecular Weight*: 182.35; *Boiling Point at 1 atm.*: 451, 233, 506; *Freezing Point*: -11, -24, 249; *Critical Temperature*: Not pertinent; *Critical Pressure*: Not pertinent; *Specific Gravity*: 0.765 at 20 °C (liquid); *Vapor (Gas) Specific Gravity*: Not pertinent; *Ratio of Specific Heats of Vapor (Gas)*: 1.029; *Latent Heat of Vaporization*: 110, 59, 2.5; *Heat of Combustion*:: -19,048, -10,582, -443.05; *Heat of Decomposition*: Not pertinent.

Health Hazards Information - *Recommended Personal Protective Equipment*: Goggles of face shield. *Symptoms Following Exposure*: Liquid may irritate eyes. *General Treatment for Exposure*: EYES: flush with water for 15 min.; *Toxicity by Inhalation (Threshold Limit Value)*: Data not available; *Short-Term Inhalation Limits*: Not pertinent; *Toxicity by Ingestion*: Data not available; *Late Toxicity*: Data not available; *Vapor (Gas) Irritant Characteristics: Non-volatile; *Liquid or Solid Irritant Characteristics*: Data not available; *Odor Threshold*: Not pertinent.

Fire Hazards - *Flash Point (deg. F)*: 175 (est.); *Flammable Limits in Air (%):* No data; *Fire Extinguishing Agents:* Dry chemical, foam, or carbon dioxide; *Fire Extinguishing Agents Not To Be Used:* Water may be ineffective; *Special Hazards of Combustion Products:* Not pertinent; *Behavior in Fire:* Not pertinent; *Ignition Temperature :* No data; *Electrical Hazard:* Not pertinent; *Burning Rate:* No data.

Chemical Reactivity - *Reactivity with Water:* No reaction; *Reactivity with Common Materials:* No reactions; *Stability During Transport:* Stable; *Neutralizing Agents for Acids and Caustics:* Not pertinent; *Polymerization:* Not pertinent; *Inhibitor of Polymerization:* Not pertinent.

TRIETHANOLAMINE

Chemical Designations - *Synonyms*: 2,2'2''-Nitrilotriethanol; Triethilolamine; Trihydroxy-triethylamine; Tris(hydroxyethy)amine; *Chemical Formula*: $(HOCH_2CH_2)_3N$.

Observable Characteristics - *Physical State (as shipped)*:Liquid; *Color*: Colorless; *Odor*: Mild ammoniacal.

Physical and Chemical Properties - *Physical State at 15 °C and 1 atm.*: liquid; *Molecular Weight*: 149.19; *Boiling Point at 1 atm.*: decomposes; *Freezing Point*: 70.9, 21.6, 294.8; *Critical Temperature*:

Not pertinent; *Critical Pressure*: Not pertinent; *Specific Gravity*: 1.13 at 20 °C (liquid); *Vapor (Gas) Specific Gravity*: Not pertinent; *Ratio of Specific Heats of Vapor (Gas)*: Not pertinent; *Latent Heat of Vaporization*: 176, 97.8, 4.10; *Heat of Combustion*:: -11,050, -6140, -257; *Heat of Decomposition*: Not pertinent.

Health Hazards Information - *Recommended Personal Protective Equipment*: Goggles of face shield; rubber gloves and boots; *Symptoms Following Exposure*: Liquid may irritate eyes and skin; *General Treatment for Exposure*: EYES: flush with water for 15 min.; call a doctor. SKIN: wipe off, wash with soap and water; *Toxicity by Inhalation (Threshold Limit Value)*: Not pertinent; *Short-Term Inhalation Limits*: Not pertinent; *Toxicity by Ingestion*: Data not available; *Late Toxicity*: Data not available; *Vapor (Gas) Irritant Characteristics: Non-volatile*; *Liquid or Solid Irritant Characteristics*: Data not available; *Odor Threshold*: Not pertinent.

Fire Hazards - *Flash Point (deg. F)*: 355 CC, 375 OC; *Flammable Limits in Air (%):* No data; *Fire Extinguishing Agents:* Alcohol foam, dry chemical, or carbon dioxide; *Fire Extinguishing Agents Not To Be Used:* Water or foam may cause frothing; *Special Hazards of Combustion Products:* Not pertinent; *Behavior in Fire:* Not pertinent; *Ignition Temperature :* No data; *Electrical Hazard:* Not pertinent; *Burning Rate:* No data.

Chemical Reactivity - *Reactivity with Water:* No reaction; *Reactivity with Common Materials:* No reactions; *Stability During Transport:* Stable; *Neutralizing Agents for Acids and Caustics:* Dilute with water; *Polymerization:* Not pertinent; *Inhibitor of Polymerization:* Not pertinent.

TRIETHYLALUMINUM

Chemical Designations - *Synonyms*: ATE; Aluminum triethyl; TEA; *Chemical Formula*: $(C_2H_5)_3Al$.

Observable Characteristics - *Physical State (as shipped)*:Liquid; *Color*: Colorless; *Odor*: Not pertinent.

Physical and Chemical Properties - *Physical State at 15 °C and 1 atm.*: liquid; *Molecular Weight*: 114.2; *Boiling Point at 1 atm.*: 367.9, 186.6, 459.8; *Freezing Point*: -51, -46, 227; *Critical Temperature*: 761, 405, 678; *Critical Pressure*: 1,970, 134, 13.6; *Specific Gravity*: 0.836 at 20 °C (liquid); *Vapor (Gas) Specific Gravity*: Not pertinent; *Ratio of Specific Heats of Vapor (Gas)*: Data not available; *Latent Heat of Vaporization*: 216, 120, 5.02; *Heat of Combustion*:: -18,364, -10,202, -426.85; *Heat of Decomposition*: Not pertinent.

Health Hazards Information - *Recommended Personal Protective Equipment*: Full protection clothing, preferably of aluminized glass cloth; goggles; face shield; gloves. In case of fire, all purpose canister or self-contained breathing apparatus; *Symptoms Following Exposure*: Exposure to smoke from fire causes metal-fume (flu-like symptoms). Since liquid ignites spontaneously, contact with eyes or skin causes several burns; *General Treatment for Exposure*: INHALATION: only fumes from fire need be considered; metal-fume fever is not critical, lasting less then 36 hrs. EYES: flush gently with copious quantities of water for 15 min. with lids held open; treat burns if fire occurred; get medical attention. SKIN: wash with water; treat burns if fire occurred; get medical attention. *Toxicity by Inhalation (Threshold Limit Value)*: Not pertinent; *Short-Term Inhalation Limits*: Not pertinent; *Toxicity by Ingestion*: Not pertinent; *Late Toxicity*: Data not available; *Vapor (Gas) Irritant Characteristics:* Not pertinent; *Liquid or Solid Irritant Characteristics*: Severe skin irritant. Causes second- and third-degree burns on short contact and very injurious to the see; *Odor Threshold*: Not pertinent.

Fire Hazards - *Flash Point :* Spontaneously ignites in air at all temperatures; *Flammable Limits in Air (%):* Not pertinent; *Fire Extinguishing Agents:* Inert powders such as limestone or sand, or dry chemical; *Fire Extinguishing Agents Not To Be Used:* Water, foam, halogenated extinguishing agents; *Special Hazards of Combustion Products:* Intense smoke may cause metal-fume fever; *Behavior in Fire:* Dense smoke of aluminum oxide is formed. Contact with water on adjacent fires causes violent reaction producing toxic and flammable gases; *Ignition Temperature :* Not pertinent - product spontaneously ignites at ambient temperature; *Electrical Hazard:* Not pertinent; *Burning Rate:* Not pertinent.

Chemical Reactivity - *Reactivity with Water:* Reacts violently to form flammable ethane gas; *Reactivity with Common Materials:* No significant reactions reported; *Stability During Transport:* Stable; *Neutralizing Agents for Acids and Caustics:* Not pertinent; *Polymerization:* Not pertinent; *Inhibitor of*

Polymerization: Not pertinent.

TRIETHYLAMINE

Chemical Designations - *Synonyms*: TEN; *Chemical Formula*: $(C_2H_5)_3N$.

Observable Characteristics - *Physical State (as shipped)*:Liquid; *Color*: Colorless; *Odor*: Fishy.

Physical and Chemical Properties - *Physical State at 15 °C and 1 atm.*: liquid; *Molecular Weight*: 101.19; *Boiling Point at 1 atm.*: 193.1, 89.5, 362.7; *Freezing Point*: -174.5, -114.7, 158.5; *Critical Temperature*: 504, 262, 535; *Critical Pressure*: 440, 30, 3.0; *Specific Gravity*: 0.729 at 20 °C (liquid); *Vapor (Gas) Specific Gravity*: 3.5; *Ratio of Specific Heats of Vapor (Gas)*: 1.055; *Latent Heat of Vaporization*: 140, 80, 3.3; *Heat of Combustion*:: -17,040, -9,466, -369.3; *Heat of Decomposition*: Not pertinent.

Health Hazards Information - *Recommended Personal Protective Equipment*: Air-supplied mask; goggles or face shield; rubber gloves; *Symptoms Following Exposure*: Vapors irritate nose, throat, and lungs, causing coughing, choking and difficult breathing. Contact with eyes causes several burns. Clothing wet with chemical causes burns; *General Treatment for Exposure*: INHALATION: remove victim to fresh air; give artificial respiration in needed; call a doctor. INGESTION: induce vomiting if patient is conscious. EYES: flush with water for at least 30 min.; call a doctor. SKIN: flush with water for at least 30 min.; *Toxicity by Inhalation (Threshold Limit Value)*: 25 ppm; *Short-Term Inhalation Limits*: 100 ppm for 30 min.; *Toxicity by Ingestion*: Grade 3, LD_{50} 50 to 500 mg/kg (rat - LD_{50} 460 mg/kg); *Late Toxicity*: Data not available; *Vapor (Gas) Irritant Characteristics:* Vapors cause moderate irritation, such that personnel will find high concentration unpleasant. The effect is temporary; *Liquid or Solid Irritant Characteristics*: Causes smarting of the skin and first-degree burns on short exposure; may cause secondary burns on long exposure; *Odor Threshold*: Data not available.

Fire Hazards - *Flash Point (deg. F)*: 20 OC; *Flammable Limits in Air (%):* 1.2 - 8.0; *Fire Extinguishing Agents:* Carbon dioxide or dry chemicals for small fires; alcohol foam for large fires; *Fire Extinguishing Agents Not To Be Used:* Water may be ineffective; *Special Hazards of Combustion Products:* Not pertinent; *Behavior in Fire:* Not pertinent; *Ignition Temperature (deg. F):* 842; *Electrical Hazard:* Not pertinent; *Burning Rate:* 6.2 mm/min.

Chemical Reactivity - *Reactivity with Water:* No reaction; *Reactivity with Common Materials:* No reactions; *Stability During Transport:* Stable; *Neutralizing Agents for Acids and Caustics:* Dilute with water; *Polymerization:* Not pertinent; *Inhibitor of Polymerization:* Not pertinent.

TRIETHYLBENZENE

Chemical Designations - *Synonyms*: 1,3,5-Triethylbenzene; sym Triethylbenzene; *Chemical Formula*: $C_6H_3(C_2H_5)_3$-1,3,5.

Observable Characteristics - *Physical State (as shipped)*:Liquid; *Color*: Colorless; *Odor*: Weak aromatic.

Physical and Chemical Properties - *Physical State at 15 °C and 1 atm.*: liquid; *Molecular Weight*: 162.27; *Boiling Point at 1 atm.*: 421, 216, 489; *Freezing Point*: Not pertinent; *Critical Temperature*: Not pertinent; *Critical Pressure*: Not pertinent; *Specific Gravity*: 0.861 at 20 °C (liquid); *Vapor (Gas) Specific Gravity*: Not pertinent; *Ratio of Specific Heats of Vapor (Gas)*: 1.039; *Latent Heat of Vaporization*: 120, 65, 2.7; *Heat of Combustion*: Data not available; *Heat of Decomposition*: Not pertinent.

Health Hazards Information - *Recommended Personal Protective Equipment*: Goggles or face shield; rubber gloves; *Symptoms Following Exposure*: Eye irritation by vapors or liquid. Central nervous system depression. Prolonged skin contact with liquid can cause dermatitis; *General Treatment for Exposure*: EYES: flush with water for at least 15 min.; call a doctor. SKIN: wipe off, wash with soap and water; *Toxicity by Inhalation (Threshold Limit Value)*: Data not available; *Short-Term Inhalation Limits*: Data not available; *Toxicity by Ingestion*: Data not available; *Vapor (Gas) Irritant Characteristics:* Vapors cause a slight smarting of the eyes or respiratory system if present in high concentration. The effect is temporary; *Liquid or Solid Irritant Characteristics*: Minimum hazard. If spilled on clothing and allowed to remain, may be cause smarting and reddening of the skin; *Odor*

Threshold: Data not available.

Fire Hazards - *Flash Point (deg. F)*: 181 OC; *Flammable Limits in Air (%):* No data; *Fire Extinguishing Agents:* Dry chemical, foam, or carbon dioxide; *Fire Extinguishing Agents Not To Be Used:* Water may be ineffective; *Special Hazards of Combustion Products:* Not pertinent; *Behavior in Fire:* Not pertinent; *Ignition Temperature :* No data; *Electrical Hazard:* Not pertinent; *Burning Rate:* Not pertinent.

Chemical Reactivity - *Reactivity with Water:* No reaction; *Reactivity with Common Materials:* No reactions; *Stability During Transport:* Stable; *Neutralizing Agents for Acids and Caustics:* Not pertinent; *Polymerization:* Not pertinent; *Inhibitor of Polymerization:* Not pertinent.

TRIETHYLENE GLYCOL

Chemical Designations - *Synonyms*: Di-beta-hydroxyethoxy-ethan; 2,2'-Ethylenedioxydiethanol; Ethylene glycol dihydroxydiethyl ether; TEG; Triglycol; *Chemical Formula*: $HO(CH_2CH_2O)_3CH$.

Observable Characteristics - *Physical State (as shipped)*:Liquid; *Color*: Colorless; *Odor*: Very mild, sweet.

Physical and Chemical Properties - *Physical State at 15 ℃ and 1 atm.*: liquid; *Molecular Weight*: 150.17; *Boiling Point at 1 atm.*: 550, 288, 561; *Freezing Point*: 24.3, -4.3, 268.9; *Critical Temperature*: Not pertinent; *Critical Pressure*: Not pertinent; *Specific Gravity*: 1.125 at 20 ℃ (liquid); *Vapor (Gas) Specific Gravity*: Not pertinent; *Ratio of Specific Heats of Vapor (Gas)*: 1.039; *Latent Heat of Vaporization*: 180, 99, 4.1; *Heat of Combustion*:: -10,190, -5,660, -237.0; *Heat of Decomposition*: Not pertinent.

Health Hazards Information - *Recommended Personal Protective Equipment*: goggles; plastic gloves; *Symptoms Following Exposure*: Vapors and liquid are unlikely to cause harm; *General Treatment for Exposure*: flush eyes and skin with water; *Toxicity by Inhalation (Threshold Limit Value)*: Not pertinent; *Short-Term Inhalation Limits*: Not pertinent; *Toxicity by Ingestion*: Grade 1, LD_{50} 5 to 15 g/kg (guinea pig); *Late Toxicity*: Data not available; *Vapor (Gas) Irritant Characteristics:* Vapors are non-irritation to the eyes and skin; *Liquid or Solid Irritant Characteristics*: No appreciable hazard. Practically harmless to the skin; *Odor Threshold*: Not pertinent.

Fire Hazards - *Flash Point (deg. F)*: 350 CC, 330 OC; *Flammable Limits in Air (%):* 0.9 - 9.2; *Fire Extinguishing Agents:* Alcohol foam, dry chemical, or carbon dioxide; *Fire Extinguishing Agents Not To Be Used:* Water or foam may cause frothing; *Special Hazards of Combustion Products:* Not pertinent; *Behavior in Fire:* Not pertinent; *Ignition Temperature (deg. F):* 700; *Electrical Hazard:* Not pertinent; *Burning Rate:* 1.7 mm/min.

Chemical Reactivity - *Reactivity with Water:* No reaction; *Reactivity with Common Materials:* No reactions; *Stability During Transport:* Stable; *Neutralizing Agents for Acids and Caustics:* Not pertinent; *Polymerization:* Not pertinent; *Inhibitor of Polymerization:* Not pertinent.

TRIETHYLENETETRAMINE

Chemical Designations - *Synonyms*: N,N'-Bis(2-aminoethyl)-ethylenediamine; TETA; Trien; *Chemical Formula*: $NH_2(CH_2)_2NH(CH_2)_2NH(CH_2)_2NH_2$.

Observable Characteristics - *Physical State (as normally shipped)*: Liquid; *Color*: Light straw; amber; *Odor*: Ammoniacal.

Physical and Chemical Properties - *Physical State at 15 ℃ and 1 atm.*: Liquid; *Molecular Weight*: 146.24; *Boiling Point at 1 atm.*: 531.3, 277.4, 550.6; *Freezing Point*: -31, -35, 238; *Critical Temperature*: 860, 460, 733; *Critical Pressure*: 470, 32, 3.2; *Specific Gravity*: 0.982 at 20 °C (liquid); *Vapor (Gas) Density*: Not pertinent; *Ratio of Specific Heats of Vapor (Gas)*: 1.037; *Latent Heat of Vaporization*: Not pertinent; *Heat of Combustion*: -13,500, -7,530, -315.0; *Heat of Decomposition*: Not pertinent.

Health Hazards Information - *Recommended Personal Protective Equipment*: Amine-type canister; goggles or face shield; rubber gloves; *Symptoms Following Exposure*: Vapors from hot liquid can irritate eyes and upper respiratory system. Liquid burns eyes and skin. May cause sensitization of skin; *General Treatment for Exposure*: INHALATION: remove victim to fresh air; INGESTION: do NOT induce vomiting; give large quantities of water; give at least one ounce of vinegar in equal amount of water; get medical attention. SKIN: flush with plenty of water. EYES: flush with plenty of water for at least 15 min. and get medical attention; *Toxicity by Inhalation (Threshold Limit Value)*: Data not available; *Short-Term Exposure Limits*: Data not available; *Toxicity by Ingestion*: Grade 2, LD_{50} 0.5 to 5 g/kg (rat); *Late Toxicity*: May cause dermatitis, asthma and other allergic reaction in man; *Vapor (Gas) Irritant Characteristics*: Vapors cause moderate irritation such that personnel will find high concentration unpleasant. The effect is temporary; *Liquid or Solid Irritant Characteristics*: Causes smarting of the skin and first-degree burns on short exposure; may cause secondary burns on long exposure; *Odor Threshold*: Data not available.

Fire Hazards - *Flash Point (deg. F)*: 275 CC, 290 OC; *Flammable Limits in Air (%):* No data; *Fire Extinguishing Agents:* Dry chemical, alcohol foam, or carbon dioxide; *Fire Extinguishing Agents Not To Be Used:* Application of water or foam may cause frothing; *Special Hazards of Combustion Products:* Not pertinent; *Behavior in Fire:* Not pertinent; *Ignition Temperature (deg. F):* 640; *Electrical Hazard:* Not pertinent; *Burning Rate:* No data.

Chemical Reactivity - *Reactivity with Water:* No reaction; *Reactivity with Common Materials:* No reactions; *Stability During Transport:* Stable; *Neutralizing Agents for Acids and Caustics:* After dilution with water, can be stabilized with acetic acid; *Polymerization:* Not pertinent; *Inhibitor of Polymerization:* Not pertinent.

TRIFLUOROCHLOROETHYLENE

Chemical Designations - *Synonyms*: Chlorotrifluoroethylene; Kel F monomer; Plascon monomer; Trifluoromonochloroethylene; Trifluoroevinil chloride; *Chemical Formula*: $F_2C=CFCl$.

Observable Characteristics - *Physical State (as normally shipped)*:Compressed liquefied gas; *Color*: Colorless; *Odor*: None faint ethereal.

Physical and Chemical Properties - *Physical State at 15 °C and 1 atm.*: Gas; *Molecular Weight*: 116.5; *Boiling Point at 1 atm.*: -18, -28, 245; *Freezing Point*: Not pertinent; *Critical Temperature*: (est.) 223.2, 106.2, 379.4; *Critical Pressure*: 592, 40.2, 4.08; *Specific Gravity*: 1.307 at 20 °C (liquid); *Vapor (Gas) Density*: 4.02; *Ratio of Specific Heats of Vapor (Gas)*: Data not available; *Latent Heat of Vaporization*: 83, 46, 1.92; *Heat of Combustion*: Data not available; *Heat of Decomposition*: Not pertinent.

Health Hazards Information - *Recommended Personal Protective Equipment*: Self-contained breathing apparatus; goggles; rubber gloves; *Symptoms Following Exposure*: Inhalation causing dizziness, nausea, vomiting; liver and kidney injury may develop after several hours and cause jaundice and necrosis of the kidney; *General Treatment for Exposure*: Call a physician after all exposures to this compound; it is more toxic then most oh the closely related propellant gases. INHALATION: remove victim to fresh air; enforce bed rest; administer oxygen for 30 min. of very hour oh 6 hours, even if no symptoms appear. SKIN: if frostbite has occurred, apply warm and tread burn; *Toxicity by Inhalation (Threshold Limit Value)*: 20 ppm (suggested); *Short-Term Exposure Limits*: Data not available; *Toxicity by Ingestion*: Not pertinent (TFC is a gas at normal temperatures); *Late Toxicity*: Data not available; *Vapor (Gas) Irritant Characteristics*: Data not available; *Liquid or Solid Irritant Characteristics*: Data not available; *Odor Threshold*: Data not available.

Fire Hazards - *Flash Point* : Not pertinent; this is a gas; *Flammable Limits in Air (%):* 16 - 34; *Fire Extinguishing Agents:* Let fire burn; stop the flow of gas; cool containers with water; *Fire Extinguishing Agents Not To Be Used:* Not pertinent; *Special Hazards of Combustion Products:* Toxic hydrogen chloride and hydrogen fluoride gases are formed; *Behavior in Fire:* Vapor is heavier than air and can travel considerable distance to a source of ignition and flash back; *Ignition Temperature* : No data; *Electrical Hazard:* Not pertinent; *Burning Rate:* Not pertinent.

Chemical Reactivity - *Reactivity with Water:* No reaction; *Reactivity with Common Materials:* No reactions; *Stability During Transport:* Stable; *Neutralizing Agents for Acids and Caustics:* Not pertinent; *Polymerization:* Polymerization can occur; *Inhibitor of Polymerization:* Terpenes or Tributylamine (1%).

TRIFLURALIN
Chemical Designations - *Synonyms*: 2,6-Dinitro-N,N-dipropyl-4-trifluoromethylaniline; 2,6-Dinitro-N,N-dipropyl-alpha, alpha-trifluoro-p-toluidine; N,N-Dipropyl-2,6-dinitro-4-trifluoro-methylaniline; Treflan; alpha, alpha-Trifluoro-2,6-dinitro-N,N-dipropyl-p-toluidine; *Chemical Formula*: $C_{13}H_{16}F_3N_3O_4$.
Observable Characteristics - *Physical State (as normally shipped)*: Solid; *Color*: Yellow-orange; *Odor*: Data not available.
Physical and Chemical Properties - *Physical State at 15 °C and 1 atm.*: Solid; *Molecular Weight*: 335.3; *Boiling Point at 1 atm.*: Not pertinent (decomposes); *Freezing Point*: 108, 42, 315; *Critical Temperature*: Not pertinent; *Critical Pressure*: Not pertinent; *Specific Gravity*: 1.294 at 20 °C; *Vapor (Gas) Density*: Not pertinent; *Ratio of Specific Heats of Vapor (Gas)*: Not pertinent; *Latent Heat of Vaporization*: Not pertinent; *Heat of Combustion*: (est.) -9,040, -5,020, -210; *Heat of Decomposition*: Not pertinent.
Health Hazards Information - *Recommended Personal Protective Equipment*: Protective gloves; goggles; dust mask; *Symptoms Following Exposure*: Dust may irritate eyes. No toxic symptoms have been observed during the manufacture and use of this compound; *General Treatment for Exposure*: INHALATION: move to fresh air. EYES: wash with running water; call physician if irritation persist. SKIN: wash with soap and running water. INGESTION: induce vomiting; call physician; *Toxicity by Inhalation (Threshold Limit Value)*: Data not available; *Short-Term Exposure Limits*: Data not available; *Toxicity by Ingestion*: Grade 3, oral LD_{50} = 500 mg/kg (rat); *Late Toxicity*: Data not available; *Vapor (Gas) Irritant Characteristics*: Data not available; *Liquid or Solid Irritant Characteristics*: Data not available; *Odor Threshold*: Data not available.
Fire Hazards - *Flash Point (deg. F)*: >185 OC; *Flammable Limits in Air (%):* Not pertinent; *Fire Extinguishing Agents:* Water, foam, dry chemical, or carbon dioxide; *Fire Extinguishing Agents Not To Be Used:* Not pertinent; *Special Hazards of Combustion Products:* Toxic and hazardous hydrogen fluoride gas may be formed in fires; *Behavior in Fire:* Not pertinent; *Ignition Temperature :* Not pertinent; *Electrical Hazard:* Not pertinent; *Burning Rate:* Not pertinent.
Chemical Reactivity - *Reactivity with Water:* No reaction; *Reactivity with Common Materials:* No reactions; *Stability During Transport:* Stable; *Neutralizing Agents for Acids and Caustics:* Not pertinent; *Polymerization:* Not pertinent; *Inhibitor of Polymerization:* Not pertinent.

TRIISOBUTYLALUMINUM
Chemical Designations - *Synonyms*: Aluminium triisobutyl; TIBA; TIBAL; *Chemical Formula*: (iso-$C_4H_9)_3Al$.
Observable Characteristics - *Physical State (as normally shipped)*: Liquid; *Color*: Colorless; *Odor*: Not pertinent.
Physical and Chemical Properties - *Physical State at 15 °C and 1 atm.*: Liquid; *Molecular Weight*: 198.3; *Boiling Point at 1 atm.*: 414, 212, 485; *Freezing Point*: 33.8, 1.0, 274.2; *Critical Temperature*: Not pertinent; *Critical Pressure*: Not pertinent; *Specific Gravity*: 0.788 at 20 °C (liquid); *Vapor (Gas) Density*: Not pertinent; *Ratio of Specific Heats of Vapor (Gas)*: Not pertinent; *Latent Heat of Vaporization*: 101, 56, 2.3; *Heat of Combustion*: -18,423, -10,235, -428.23; *Heat of Decomposition*: Not pertinent.
Health Hazards Information - *Recommended Personal Protective Equipment*: Full protective clothing, preferably of aluminized glass cloth; goggles; face shield; gloves. In case of fire, all-purpose canister or self-contained breathing apparatus; *Symptoms Following Exposure*: Inhalation of smoke from fire causes metal-fume fever (flu-like symptoms). Contact with liquid can cause severe burns of eyes and skin because of spontaneous ignition; *General Treatment for Exposure*: INHALATION: only fumes from fire need be considered; methal-fume fever lasts less than 36 hrs. and is not critical. EYES: flush gently with copious quantities of water for 15 min. with lids open; treat burns, if fire occurred; get medical attention. SKIN: wash with water; treat burns caused by fire; get medical attention. INGESTION: Not pertinent; *Toxicity by Inhalation (Threshold Limit Value)*: Not pertinent; *Short-Term Exposure Limits*: Not pertinent; *Toxicity by Ingestion*: Not pertinent; *Late Toxicity*: Metal fume fever may develop following exposure to smoke from fire; *Vapor (Gas) Irritant Characteristics*: Not

pertinent; *Liquid or Solid Irritant Characteristics*: Severe skin irritant. Causes second- and third-degree burns on short contact and is very injurious to the eyes; *Odor Threshold*: Not pertinent.

Fire Hazards - *Flash Point* : Not pertinent; this product ignites spontaneously; *Flammable Limits in Air (%):* Not pertinent; *Fire Extinguishing Agents:* Inert powder such as sand or limestone, or dry chemical; *Fire Extinguishing Agents Not To Be Used:* Water, foam, halogenated extinguishing agents; *Special Hazards of Combustion Products:* Dense smoke may cause metal-fume fever; *Behavior in Fire:* Dense smoke of aluminum oxide forms in fires; *Ignition Temperature :* Ignites spontaneously under ambient conditions; *Electrical Hazard:* Not pertinent; *Burning Rate:* Not pertinent.

Chemical Reactivity - *Reactivity with Water:* Reacts violently to form flammable hydrocarbon gases; *Reactivity with Common Materials:* Not compatible with silicone rubber or urethane rubbers; *Stability During Transport:* Stable; *Neutralizing Agents for Acids and Caustics:* Not pertinent; *Polymerization:* Not pertinent; *Inhibitor of Polymerization:* Not pertinent.

TRIMETHYLAMINE

Chemical Designations - *Synonyms*: No common synonyms; *Chemical Formula*: $(CH_3)_3N$.

Observable Characteristics - *Physical State (as normally shipped)*: Liquefied compressed gas; *Color*: Colorless; *Odor*: Ammonical.

Physical and Chemical Properties - *Physical State at 15 °C and 1 atm.*: Gas; *Molecular Weight*: 59.11; *Boiling Point at 1 atm.*: 37.2, 2.9, 276.1; *Freezing Point*: -178.8, -117.1, -156.1; *Critical Temperature*: 320.2, 160.1, 433.3; *Critical Pressure*: 591, 40.2, 4.07; *Specific Gravity*: 0.633 at 20 °C (liquid); *Vapor (Gas) Density*: 2.0; *Ratio of Specific Heats of Vapor (Gas)*: 1.139; *Latent Heat of Vaporization*: 174, 96.5, 4.04; *Heat of Combustion*: -17,660, -9,810, -410.7; *Heat of Decomposition*: Not pertinent.

Health Hazards Information - *Recommended Personal Protective Equipment*: Vapor-proof goggles and face shield; rubber gloves; air-supplied mask; *Symptoms Following Exposure*: Vapor irritates eyes, nose, and throat; high concentrations can cause pulmonary edema. Liquid burns eyes and skin; *General Treatment for Exposure*: INHALATION: remove victim to fresh air and call a doctor; give artificial respiration and oxygen if needed. EYES: flush with water for at least 15 min.; consult an eye doctor. SKIN: flush with water, wash with soap and water; *Toxicity by Inhalation (Threshold Limit Value)*: Data not available; *Short-Term Exposure Limits*: Data not available; *Toxicity by Ingestion*: Data not available; *Late Toxicity*: Data not available; *Vapor (Gas) Irritant Characteristics*: Vapor is moderately irritating such that personnel will not usually tolerate moderate or high concentrations; *Liquid or Solid Irritant Characteristics*: Causes smarting of the skin and first-degree burns on short exposure; may cause secondary burns on long exposure; *Odor Threshold*: less than 100 ppm.

Fire Hazards - *Flash Point* : Not pertinent; this is a gas; *Flammable Limits in Air (%):* 2.0 - 11.6; *Fire Extinguishing Agents:* Stop flow of gas. Use water, alcohol foam, dry chemical, or carbon dioxide on water solution fires; *Fire Extinguishing Agents Not To Be Used:* Not pertinent; *Special Hazards of Combustion Products:* Not pertinent; *Behavior in Fire:* Vapor is heavier than air and may travel considerable distance to a source of ignition and flash back; *Ignition Temperature (deg. F):* 374; *Electrical Hazard:* Not pertinent; *Burning Rate:* 8.0 mm/min.

Chemical Reactivity - *Reactivity with Water:* No reaction; *Reactivity with Common Materials:* No reactions; *Stability During Transport:* Stable; *Neutralizing Agents for Acids and Caustics:* Although water solutions may be neutralized with acetic acid, simple evaporation will remove all the compound; *Polymerization:* Not pertinent; *Inhibitor of Polymerization:* Not pertinent.

TRIMETHYLCHLOROSILANE

Chemical Designations - *Synonyms*: Chlorotrimethylsilane; Trimethylsilyl chloride; *Chemical Formula*: $(CH_3)_3SiCl$.

Observable Characteristics - *Physical State (as normally shipped)*: Liquid; *Color*: Colorless; *Odor*: Sharp, hydrochloric acid-like; acrid.

Physical and Chemical Properties - *Physical State at 15 °C and 1 atm.*: Liquid; *Molecular Weight*: 108.7; *Boiling Point at 1 atm.*: 135, 57, 330; *Freezing Point*: Not pertinent; *Critical Temperature*: Not pertinent; *Critical Pressure*: Not pertinent; *Specific Gravity*: 0.846 at 20 °C (liquid); *Vapor (Gas) Density*: 3.7; *Ratio of Specific Heats of Vapor (Gas)*: (est.) 1.0683; *Latent Heat of Vaporization*: 126, 70, 2.9; *Heat of Combustion*: (est.) -10,300, -5,700, -240; *Heat of Decomposition*: Not pertinent.

Health Hazards Information - *Recommended Personal Protective Equipment*: Acid-vapor-type respiratory protection; rubber gloves; chemical worker's goggles; other protective equipment as necessary to protect skin and eyes; *Symptoms Following Exposure*: Inhalation of vapor irritates mucous membranes. Contact of liquid with eyes or skin causes severe burns of mouth and stomach; *General Treatment for Exposure*: Get medical attention all exposures to this compound. INHALATION: remove victim from exposure; if breathing is difficult or stopped, give artificial respiration. EYES: flush with water for 15 min. SKIN: flush with water. INGESTION: do NOT induce vomiting; give large amount of water; *Toxicity by Inhalation (Threshold Limit Value)*: Data not available; *Short-Term Exposure Limits*: Data not available; *Toxicity by Ingestion*: Grade 3, LD_{50} 0.5 to 5 g/kg; *Late Toxicity*: Data not available; *Vapor (Gas) Irritant Characteristics*: Vapors cause severe irritation of eyes and throat and can cause eye and lung injury. They cannot be tolerated even at low concentrations; *Liquid or Solid Irritant Characteristics*: Severe skin irritant. Causes second- and third-degree burns on short contact and is very injurious to the eyes; *Odor Threshold*: Data not available.

Fire Hazards - *Flash Point (deg. F)*: 0 OC; *Flammable Limits in Air (%):* 1.8 (LEL); *Fire Extinguishing Agents:* Dry chemical; *Fire Extinguishing Agents Not To Be Used:* Water, foam; *Special Hazards of Combustion Products:* Toxic and irritating hydrogen chloride and phosgene may form in fires; *Behavior in Fire:* Difficult to extinguish; material easily re-ignites. Contact with water on adjacent fires should be avoided as irritating and toxic hydrogen chloride gas will form; *Ignition Temperature (deg. F):* 743; *Electrical Hazard:* No data; *Burning Rate:* 5.3 mm/min.

Chemical Reactivity - *Reactivity with Water:* Reacts vigorously forming hydrogen chloride (hydrochloric acid); *Reactivity with Common Materials:* Reacts with surface moisture evolving hydrogen chloride, which will corrode common metals and form flammable hydrogen gas; *Stability During Transport:* Stable; *Neutralizing Agents for Acids and Caustics:* Flush with water and rinse with sodium bicarbonate or lime solution; *Polymerization:* Not pertinent; *Inhibitor of Polymerization:* Not pertinent.

TRIPROPYLENE GLYCOL

Chemical Designations - *Synonyms*: No common synonyms; *Chemical Formula*: $HO(C_3H_6)_2C_3H_6OH$.

Observable Characteristics — *Physical State (as normally shipped)*: Liquid; *Color*: Colorless; *Odor*: Characteristic.

Physical and Chemical Properties - *Physical State at 15 °C and 1 atm.*: Liquid; *Molecular Weight*: 192.26; *Boiling Point at 1 atm.*: Not pertinent (decomposes) 523, 273, 546; *Freezing Point*: (sets to glass) -49, -45, 228; *Critical Temperature*: Not pertinent; *Critical Pressure*: Not pertinent; *Specific Gravity*: 1.022 at 20 °C (liquid); *Vapor (Gas) Density*: Not pertinent; *Ratio of Specific Heats of Vapor (Gas)*: Not pertinent; *Latent Heat of Vaporization*: Data not available; *Heat of Combustion*: (est.) -13,700, -7,6610, -318; *Heat of Decomposition*: Not pertinent.

Health Hazards Information - *Recommended Personal Protective Equipment*: Plastic gloves; safety glasses of face shield; *Symptoms Following Exposure*: Non-irritation; no symptoms observed by any exposure route; *General Treatment for Exposure*: INGESTION: if large amounts are swallowed, induce vomiting; treat symptomatically. EYES: or SKIN: flush with water, get medical attention if ill effects develop; *Toxicity by Inhalation (Threshold Limit Value)*: Data not available; *Short-Term Exposure Limits*: Data not available; *Toxicity by Ingestion*: Grade 2, oral LD_{50} =3,000 mg/kg (rat); *Late Toxicity*: Vapors are nonirritating to eyes and throat; *Vapor (Gas) Irritant Characteristics*: No appreciable hazard. Practically harmless to the skin; *Liquid or Solid Irritant Characteristics*: Data not available; *Odor Threshold*: Odorless.

Fire Hazards - *Flash Point (deg. F)*: 285 OC; *Flammable Limits in Air (%)*: 0.8 - 5.0; *Fire Extinguishing Agents:* Alcohol foam, dry chemical, or carbon dioxide; *Fire Extinguishing Agents Not To Be Used:* Water may be ineffective; *Special Hazards of Combustion Products:* Acrid fumes of acids and aldehydes may form in fires; *Behavior in Fire:* No data; *Ignition Temperature :* No data; *Electrical*

Hazard: No data; *Burning Rate:* No data.

Chemical Reactivity - *Reactivity with Water:* No reaction; *Reactivity with Common Materials:* May attack some forms of plastics and elastomers; *Stability During Transport:* Stable; *Neutralizing Agents for Acids and Caustics:* Not pertinent; *Polymerization:* Not pertinent; *Inhibitor of Polymerization:* Not pertinent.

TRIS(AZIRIDINYL)PHOSPHINE OXIDE

Chemical Designations - *Synonyms*: APO; Phosphoric acid triethileneimide; Triethylenephosphor-amide; Tris (l-aziridinyl) phosphin oxide; *Chemical Formula*: $(CH_2CH_2N)_3PO$ or $C_6H_{12}N_3PO$.

Observable Characteristics - *Physical State (as normally shipped)*: Solid; *Color*: White; *Odor*: Data not available.

Physical and Chemical Properties - *Physical State at 15 °C and 1 atm.*: Solid; *Molecular Weight*: 173.16; *Boiling Point at 1 atm.*: Not pertinent (decomposes); *Freezing Point*: 106, 41, 314; *Critical Temperature*: Not pertinent; *Critical Pressure*: Not pertinent; *Specific Gravity*: (est.) >1 at 20 °C (solid); *Vapor (Gas) Density*: Not pertinent; *Ratio of Specific Heats of Vapor (Gas)*: Not pertinent; *Latent Heat of Vaporization*: Not pertinent; *Heat of Combustion*: Not pertinent; *Heat of Decomposition*: Not pertinent.

Health Hazards Information - *Recommended Personal Protective Equipment*: Protective clothing and gloves to prevent contact with skin; goggles; *Symptoms Following Exposure*: Inhalation (unlikely unless a heave mist is formed) causes symptoms similar to the those observed after ingestion. Contact with liquid or powder causes irritation of eyes and (on prolonged contact) irritation and burns of skin. Burns are slow to develop and slow to heal. May sensitize on repeated contact. Ingestion causes depression, anorexia and diarrhea, appearing 2-3 days before death, followed by terminal dyspnea, incoordination, epistaxis, salivation and cyanosis; *General Treatment for Exposure*: INHALATION: remove victim to fresh air. EYES: flush with water at once for at least 15 min.; get medical attention. SKIN: flush with water at once, followed by vinegar and dilute hydrogen peroxide. INGESTION: only symptomatic and supportive measured are available; *Toxicity by Inhalation (Threshold Limit Value)*: Data not available; *Short-Term Exposure Limits*: Data not available; *Toxicity by Ingestion*: Grade 4, oral rat LD_{50}=37 mg/kg; *Late Toxicity*: None observed; *Vapor (Gas) Irritant Characteristics*: Data not available; *Liquid or Solid Irritant Characteristics*: Data not available; *Odor Threshold*: Data not available.

Fire Hazards - *Flash Point* : Not flammable; *Flammable Limits in Air (%):* Not flammable; *Fire Extinguishing Agents:* Not pertinent; *Fire Extinguishing Agents Not To Be Used:* Not pertinent; *Special Hazards of Combustion Products:* Phosphoric acid mist may form in fires. Toxic oxide of nitrogen may form; *Behavior in Fire:* No data; *Ignition Temperature :* Not pertinent; *Electrical Hazard:* Not pertinent; *Burning Rate:* Not pertinent.

Chemical Reactivity - *Reactivity with Water:* No reaction unless in the presence of acids and caustics; *Reactivity with Common Materials:* Slow decomposition occurs, but generally the reactions are not hazardous; *Stability During Transport:* Stable if cool; *Neutralizing Agents for Acids and Caustics:* Not pertinent; *Polymerization:* Violent, exothermic polymerization occurs at about 225 of. Acid fumes will also cause polymerization at ordinary temperatures; *Inhibitor of Polymerization:* None reported.

TURPENTINE

Chemical Designations - *Synonyms*: D.D. turpentine, Gum turpentine, Spirits of turpentine, Sulfate turpentine, Turps, Wood turpentine; *Chemical Formula*: $C_{10}H_{16.}$

Observable Characteristics - *Physical State (as normally shipped)*: Liquid; *Color*: Colorless; *Odor*: Aromatic, rather unpleasant, penetrating.

Physical and Chemical Properties - *Physical State at 15 °C and 1 atm.*: Liquid; *Molecular Weight*: Not pertinent; *Boiling Point at 1 atm.*: 302-320, 150-160, 423-433; *Freezing Point*: Not pertinent; *Critical Temperature*: Not pertinent; *Critical Pressure*: Not pertinent; *Specific Gravity*: 0.86 at 15 °C; *Vapor (Gas) Density*: Not pertinent; *Ratio of Specific Heats of Vapor (Gas)*: Not pertinent; *Latent Heat of Vaporization*: Not pertinent; *Heat of Combustion*: Data not available; *Heat of Decomposition*: Data not available.

Health Hazards Information - *Recommended Personal Protective Equipment*: Organic canister or air-supplied mask; goggles or face shield; rubber gloves; *Symptoms Following Exposure*: Vapors cause headache, confusion, respiratory distress. Liquid irritates skin. If ingested, can irritate the entire digestive system and may injure kidneys. If liquid is taken into lungs, causes several pneumonitis; *General Treatment for Exposure*: INHALATION: remove victim to fresh air; call a doctor; administer artificial respiration and oxygen if required. INGESTION: give water and induce vomiting; call a doctor. EYES: flush with water for at least 15 min. SKIN: wipe off, wash with soap and water; *Toxicity by Inhalation (Threshold Limit Value)*: 100 ppm; *Short-Term Exposure Limits*: 200 ppm for 30 min.; *Toxicity by Ingestion*: Grade 2, LD_{50} 0.5 to 5 g/kg; *Late Toxicity*: None; *Vapor (Gas) Irritant Characteristics*: Vapor causes a slight smarting of the eyes or respiratory system if present in high concentration. The effect is temporary; *Liquid or Solid Irritant Characteristics*: Minimum hazard. If spilled on clothing and allowed to remain, may be cause smarting and reddening of the skin; *Odor Threshold*: Data not available.

Fire Hazards - *Flash Point (deg. F):* 95 CC; *Flammable Limits in Air (%):* 0.8 (LEL); *Fire Extinguishing Agents:* Foam, dry chemical, or carbon dioxide; *Fire Extinguishing Agents Not to be Used:* Water may be ineffective; *Special Hazards of Combustion Products:* Not pertinent; *Behavior in Fire:* Forms heavy black smoke and soot; *Ignition Temperature (deg. F):* 488; *Electrical Hazard:* Not pertinent; *Burning Rate:* 2.4 mm/min.

Chemical Reactivity - *Reactivity with Water* No reaction; *Reactivity with Common Materials:* No reactions; *Stability During Transport*: Stable; *Neutralizing Agents for Acids and Caustics:* Not pertinent; *Polymerization:* Not pertinent; *Inhibitor of Polymerization:* Not pertinent.

U

UNDECANOL
Chemical Designations - *Synonyms*: Hendecanoic alcohol; 1-Hendecanol; n-Hendecylenic alcohol; 1-Undecanol; Undecyl alcohol; Undecylic alcohol; *Chemical Formula*: $CH_3(CH_2)_9CH_2OH$.

Observable Characteristics - *Physical State (as normally shipped)*: Liquid; *Color*: Colorless; *Odor*: Faint alcohol.

Physical and Chemical Properties - *Physical State at 15 °C and 1 atm.*: Liquid; *Molecular Weight*: 172.30; *Boiling Point at 1 atm.*: 473, 245, 518; *Freezing Point*: 60.6, 15.9, 289.1; *Critical Temperature*: 739, 393, 666; *Critical Pressure*: 308, 21, 2.1; *Specific Gravity*: 0.835 at 20 °C (liquid); *Vapor (Gas) Density*: Not pertinent; *Ratio of Specific Heats of Vapor (Gas)*: 1.032; *Latent Heat of Vaporization*: Not pertinent; *Heat of Combustion*: (est.) -18,000, -10,000, -419; *Heat of Decomposition*: Not pertinent.

Health Hazards Information - *Recommended Personal Protective Equipment*: Goggles and face shield; *Symptoms Following Exposure*: Liquid can irritate eyes; *General Treatment for Exposure*: Wash eyes with water for at least 15 min.; *Toxicity by Inhalation (Threshold Limit Value)*: Not pertinent; *Short-Term Exposure Limits*: Not pertinent; *Toxicity by Ingestion*: Grade 2, LD_{50} 0.5 to 5 g/kg; *Late Toxicity*: Data not available; *Vapor (Gas) Irritant Characteristics*: None; *Liquid or Solid Irritant Characteristics*: No appreciable hazard. Practically harmless to the skin; *Odor Threshold*: Not pertinent.

Fire Hazards - *Flash Point (deg. F):* 200 OC; *Flammable Limits in Air (%):* No data; *Fire Extinguishing Agents*: Foam, carbon dioxide, or dry chemical; *Fire Extinguishing Agents Not to be Used:* Water or foam may cause frothing; *Special Hazards of Combustion Products*: Not pertinent; *Behavior in Fire*: Not pertinent; *Ignition Temperature*: No data; *Electrical Hazard*: Not pertinent; *Burning Rate*: No data.

Chemical Reactivity - *Reactivity with Water*: No reaction; *Reactivity with Common Materials*: No reactions; *Stability During Transport*: Stable; *Neutralizing Agents for Acids and Caustics:* Not pertinent; *Polymerization:* Not pertinent; *Inhibitor of Polymerization*: Not pertinent.

1-UNDECENE

Chemical Designations - *Synonyms*: n-Nonylethylene; *Chemical Formula*: $CH_3(CH_2)_8CH=CH_2$.
Observable Characteristics - *Physical State (as normally shipped)*: Liquid; *Color*: Colorless; *Odor*: Mild, pleasant.
Physical and Chemical Properties - *Physical State at 15 °C and 1 atm.*: Liquid; *Molecular Weight*: 154.2; *Boiling Point at 1 atm.*: 378.9, 192.7, 465.9; *Freezing Point*: -56, 49, 224; *Critical Temperature*: Not pertinent; *Critical Pressure*: Not pertinent; *Specific Gravity*: 0.750 at 20 °C (solid); *Vapor (Gas) Density*: Not pertinent; *Ratio of Specific Heats of Vapor (Gas)*: 1.035; *Latent Heat of Vaporization*: 154, 85.8, 3.59; *Heat of Combustion*: -19.084, -10.602, -443.89; *Heat of Decomposition*: Not pertinent.
Health Hazards Information - *Recommended Personal Protective Equipment*: Goggles or face shield; rubber gloves; *Symptoms Following Exposure*: Aspiration hazard if ingested. Slight skin and eye irritation. No inhalation hazard expected; *General Treatment for Exposure*: INHALATION: remove victim to fresh air. INGESTION: do NOT lavage or induce vomiting; give vegetable oil and demulcents; call a doctor. EYES: flush with water for 15 min. SKIN: wipe off, wash with soap and water; *Toxicity by Inhalation (Threshold Limit Value)*: Data not available; *Short-Term Exposure Limits*: Data not available; *Toxicity by Ingestion*: Data not available; *Late Toxicity*: Data not available; *Vapor (Gas) Irritant Characteristics*: Slight smarting of eyes and respiratory system at high concentrations. The effect is temporary; *Liquid or Solid Irritant Characteristics*: Minimum hazard. If spilled on clothing and allowed to remain, may cause smarting and reddening of the skin; *Odor Threshold*: Data not available.
Fire Hazards - *Flash Point (deg. F)*: 160 OC; *Flammable Limits in Air (%)*: No data; *Fire Extinguishing Agents*: Foam, dry chemical, or carbon dioxide; *Fire Extinguishing Agents Not to be Used*: Water may be ineffective; *Special Hazards of Combustion Products*: Not pertinent; *Behavior in Fire*:;Not pertinent *Ignition Temperature* : No data; *Electrical Hazard*: Not pertinent; *Burning Rate*: 4.8 mm/min.
Chemical Reactivity - *Reactivity with Water*: No reaction; *Reactivity with Common Materials*: No reactions; *Stability During Transport*: Stable; *Neutralizing Agents for Acids and Caustics:* Not pertinent; *Polymerization:* Not pertinent; *Inhibitor of Polymerization*: Not pertinent.

N-UNDECYLBENZENE

Chemical Designations - *Synonyms*: 1-Phenylundecane; *Chemical Formula*: $C_6H_5(CH_2)_{10}CH_3$.
Observable Characteristics — *Physical State (as normally shipped)*: Liquid; *Color*: Colorless; *Odor*: Mild.
Physical and Chemical Properties - *Physical State at 15 °C and 1 atm.*: Liquid; *Molecular Weight*: 232.4; *Boiling Point at 1 atm.*: 601, 316, 589; *Freezing Point*: 23, -5, 268; *Critical Temperature*: 918.1, 492.3, 765.5; *Critical Pressure*: 234, 15.9, 1.61; *Specific Gravity*: 0.855 at 20 °C (liquid); *Vapor (Gas) Density*: Not pertinent; *Ratio of Specific Heats of Vapor (Gas)*: Not pertinent; *Latent Heat of Vaporization*: 101.27, 56.26, 2.354; *Heat of Combustion*: -19.490, -10.830, -453.1; *Heat of Decomposition*: Not pertinent.
Health Hazards Information - *Recommended Personal Protective Equipment*: Goggles or face shield; rubber gloves; *Symptoms Following Exposure*: Ingestion may cause intestinal disturbances. Contact with eyes causes mild irritation; *General Treatment for Exposure*: INGESTION: induce vomiting if large amount has been swallowed. EYES: flush with water. SKIN: remove spills on skin or clothing by washing with soap and water; *Toxicity by Inhalation (Threshold Limit Value)*: Data not available; *Short-Term Exposure Limits*: Data not available; *Toxicity by Ingestion*: Data not available; *Late Toxicity*: Data not available; *Vapor (Gas) Irritant Characteristics*: Data not available; *Liquid or Solid Irritant Characteristics*: Data not available; *Odor Threshold*: Data not available.
Fire Hazards - *Flash Point (deg. F)*: 285 CC; *Flammable Limits in Air (%)*: No data; *Fire Extinguishing Agents*: Foam, dry chemical, or carbon dioxide; *Fire Extinguishing Agents Not to be Used:* Water may be ineffective; *Special Hazards of Combustion Products*: No data; *Behavior in Fire*:

No data; *Ignition Temperature* : No data; *Electrical Hazard*: No data; *Burning Rate*: No data.
Chemical Reactivity - *Reactivity with Water*: No reaction; *Reactivity with Common Materials*: May attack some forms of plastics; *Stability During Transport*: Stable; *Neutralizing Agents for Acids and Caustics:* Stable; *Polymerization:* Not pertinent; *Inhibitor of Polymerization*: Not pertinent.

URANYL ACETATE
Chemical Designations - *Synonyms*: Bis(acetato)dixouranium; Uranium acetate; Uranium acetate dihydrate; Uranium oxyacetate dihydrate; Uranyl acetate dihydrate; *Chemical Formula*: $UO_2(C_2H_3O_2)_2 \cdot 2H_2O$.
Observable Characteristics - *Physical State (as normally shipped)*: Solid; *Color*: Yellow; *Odor*: Slight vinegar.
Physical and Chemical Properties - *Physical State at 15 °C and 1 atm.*: Solid; *Molecular Weight*: 424.2; *Boiling Point at 1 atm.*: Not pertinent (decomposes); *Freezing Point*: Not pertinent; *Critical Temperature*: Not pertinent; *Critical Pressure*: Not pertinent; *Specific Gravity*: 2.89 at 20 °C (solid); *Vapor (Gas) Density*: Not pertinent; *Ratio of Specific Heats of Vapor (Gas)*: Not pertinent; *Latent Heat of Vaporization*: Not pertinent; *Heat of Combustion*: Not pertinent; *Heat of Decomposition*: Not pertinent.
Health Hazards Information - *Recommended Personal Protective Equipment*: Approved dust respirator; goggles or face shield; protective clothing; *Symptoms Following Exposure*: Inhalation of dust may irritate nose and throat. Contact with eyes causes irritation; *General Treatment for Exposure*: *Get medical attention after all exposures to this compound.* INHALATION: move to fresh air. INGESTION: give large amount of water; induce vomiting. EYES: flush with water for at least 15 min. SKIN: flush with water; *Toxicity by Inhalation (Threshold Limit Value)*: 0.2 mg/m^3 (as uranium); *Short-Term Exposure Limits*: Data not available; *Toxicity by Ingestion*: Data not available; *Late Toxicity*: Data not available; *Vapor (Gas) Irritant Characteristics*: Data not available; *Liquid or Solid Irritant Characteristics*: Data not available; *Odor Threshold*: Data not available.
Fire Hazards - *Flash Point :* Not flammable; *Flammable Limits in Air (%)*: Not flammable; *Fire Extinguishing Agents*: Not pertinent; *Fire Extinguishing Agents Not to be Used:* Not pertinent; *Special Hazards of Combustion Products*: Not pertinent; *Behavior in Fire*: Not pertinent; *Ignition Temperature* : Not pertinent; *Electrical Hazard*: Not pertinent; *Burning Rate*: Not pertinent.
Chemical Reactivity - *Reactivity with Water*: Dissolves and reacts producing a milky like solution. The reaction is non hazardous; *Reactivity with Common Materials*: No reactions; *Stability During Transport*: Stable; *Neutralizing Agents for Acids and Caustics:* Not pertinent; *Polymerization:* Not pertinent; *Inhibitor of Polymerization*: Not pertinent.

URANYL NITRATE
Chemical Designations - *Synonyms*: Uranium nitrate; *Chemical Formula*: $UO_2(NO_3)_2 \cdot 6H_2O$.
Observable Characteristics - *Physical State (as normally shipped)*: Solid; *Color*: Pale yellow; *Odor*: None.
Physical and Chemical Properties - *Physical State at 15 °C and 1 atm.*: Solid; *Molecular Weight*: 502.13; *Boiling Point at 1 atm.*: Not pertinent (decomposes); *Freezing Point*: 140.4, 60.2, 333.4; *Critical Temperature*: Not pertinent; *Critical Pressure*: Not pertinent; *Specific Gravity*: 2.81 at 13 °C (solid); *Vapor (Gas) Density*: Not pertinent; *Ratio of Specific Heats of Vapor (Gas)*: Not pertinent; *Latent Heat of Vaporization*: Not pertinent; *Heat of Combustion*: Not pertinent; *Heat of Decomposition*: Not pertinent.
Health Hazards Information - *Recommended Personal Protective Equipment*: Dust mask, gloves, goggles; *Symptoms Following Exposure*: Excessive inhalation of dust may cause irritation of lungs and delayed symptoms similar to those observed after ingestion. Dust irritates eyes and skin and may be absorbed through skin on prolonged exposure, ingestion causes irritation of mouth and stomach; inflammation of kidney and liver develops 1 to 4 days after exposure; *General Treatment for Exposure*: INHALATION: remove victim to fresh air. EYES: Hush with water for at least 15 min.: see physician if irritation persists. SKIN: wash thoroughly with soap and water. INGESTION: administer large doses

of sodium bicarbonate.(This will convert the uranium salt lo the bicarbonate, which is much less toxic.) Additional treatment is symptomatic get medical attention; *Toxicity by Inhalation (Threshold Limit Value)*: 0.05 mg/m³; *Short-Term Exposure Limits*: Data not available; *Toxicity by Ingestion*: Grade 3, LD_{50} 50 to 500 mg/kg; *Late Toxicity*: Delayed inflammation of kidneys. Airborne radioactive particles have apparent been responsible for a significantly increased death rate from lung cancer among long-term uranium miners; *Vapor (Gas) Irritant Characteristics*: Data not available; *Liquid or Solid Irritant Characteristics*: Data not available; *Odor Threshold*: Not pertinent.

Fire Hazards - *Flash Point:* Not flammable but may cause fire on contact with combustible materials; *Flammable Limits in Air (%):* Not flammable; *Fire Extinguishing Agents*: Apply flooding amounts of water; *Fire Extinguishing Agents Not to be Used:* Not pertinent; *Special Hazards of Combustion Products*: Toxic oxides of nitrogen are formed in fires; *Behavior in Fire*: Intensifies fires. When large quantities are involved, nitrate may fuse or melt. The application of water may then cause extensive scattering of the molten material; *Ignition Temperature* : Not pertinent; *Electrical Hazard*: Not pertinent; *Burning Rate*: Not pertinent.

Chemical Reactivity - *Reactivity with Water*: Dissolves in water forming a weak solution of nitric acid. The reaction is nonhazardous; *Reactivity with Common Materials*: When in contact with easily oxidizable materials, this chemical may react rapidly enough to cause ignition, violent combustion, or explosion. Water solutions are acidic and can corrode metals; *Stability During Transport*: Stable; *Neutralizing Agents for Acids and Caustics:* Flush with water; *Polymerization*: Not pertinent; *Inhibitor of Polymerization*: Not pertinent.

URANYL SULFATE

Chemical Designations - *Synonyms*: Uranium sulfate; Uranium sulfate trihydrate; Uranyl sulfate trihydrate; *Chemical Formula*: $UO_2SO_4 \cdot 3H_2O$.

Observable Characteristics - *Physical State (as normally shipped)*: Solid; *Color*: Yellow; *Odor*: None.

Physical and Chemical Properties - *Physical State at 15 °C and 1 atm.*: Solid; *Molecular Weight*: 420.2; *Boiling Point at 1 atm.*: Not pertinent (decomposes); *Freezing Point*: Not pertinent; *Critical Temperature*: Not pertinent; *Critical Pressure*: Not pertinent; *Specific Gravity*: 3.28 at 20 °C (solid); *Vapor (Gas) Density*: Not pertinent; *Ratio of Specific Heats of Vapor (Gas)*: Not pertinent; *Latent Heat of Vaporization*: Not pertinent; *Heat of Combustion*: Not pertinent; *Heat of Decomposition*: Not pertinent.

Health Hazards Information - *Recommended Personal Protective Equipment*: Approved dust respirator; goggles or face shield; protective clothing; *Symptoms Following Exposure*: irritates eyes and skin, stomach; *General Treatment for Exposure*: *Get medical attention after all exposure to this compound.* INGESTION: give large amounts of water; induce vomiting. EYES: flush with water for at least 15 min. SKIN: flush with water; *Toxicity by Inhalation (Threshold Limit Value)*: 0.2 mg/m³ (as uranium); *Short-Term Exposure Limits*: Data not available; *Toxicity by Ingestion*: Grade 1, LD_{50} 5-15 g/kg; *Late Toxicity*: Data not available; *Vapor (Gas) Irritant Characteristics*: Data not available; *Liquid or Solid Irritant Characteristics*: Data not available; *Odor Threshold*: Data not available.

Fire Hazards - *Flash Point :* Not flammable; *Flammable Limits in Air (%)*: Not flammable; *Fire Extinguishing Agents*: Not pertinent; *Fire Extinguishing Agents Not to be Used:* Not pertinent; *Special Hazards of Combustion Products*: No data; *Behavior in Fire*: No data; *Ignition Temperature* : Not pertinent; *Electrical Hazard*: Not pertinent; *Burning Rate*: Not pertinent.

Chemical Reactivity - *Reactivity with Water*: No reaction; *Reactivity with Common Materials*: No data; *Stability During Transport*: Stable; *Neutralizing Agents for Acids and Caustics:* Not pertinent; *Polymerization:* Not pertinent; *Inhibitor of Polymerization*: Not pertinent.

UREA

Chemical Designations - *Synonyms*: Carbamide, Carbonyldiamide; *Chemical Formula*: NH_2CONH_2.

Observable Characteristics - *Physical State (as normally shipped)*: Solid; *Color*: White; *Odor*: Odorless, or slight ammonia odor.

Physical and Chemical Properties - *Physical State at 15 °C and 1 atm.*: Solid; *Molecular Weight*:

60.06; *Boiling Point at 1 atm.*: Decomposes; *Freezing Point*: 271, 133, 406; *Critical Temperature*: Not pertinent; *Critical Pressure*: Not pertinent; *Specific Gravity*: 1.34 at 20°C (solid); *Vapor (Gas) Density*: Not pertinent; *Ratio of Specific Heats of Vapor (Gas)*: Not pertinent; *Latent Heat of Vaporization*: Not pertinent; *Heat of Combustion*: -3913, -2174, -91.02; *Heat of Decomposition*: Not pertinent.

Health Hazards Information - *Recommended Personal Protective Equipment*: Goggles or face shield; dust mask; *Symptoms Following Exposure*: May irritate eyes; *General Treatment for Exposure*: Wash eyes with water; *Toxicity by Inhalation (Threshold Limit Value)*: Not pertinent; *Short-Term Exposure Limits*: Not pertinent; *Toxicity by Ingestion*: Data not available; *Late Toxicity*: None; *Vapor (Gas) Irritant Characteristics*: Non-volatile; *Liquid or Solid Irritant Characteristics*: None; *Odor Threshold*: Not pertinent.

Fire Hazards - *Flash Point :* Not flammable; *Flammable Limits in Air (%):* Not flammable; *Fire Extinguishing Agents*: Water; *Fire Extinguishing Agents Not to be Used:* Not pertinent; *Special Hazards of Combustion Products*: Not pertinent; *Behavior in Fire*: Melts and decomposes, generating ammonia; *Ignition Temperature :* Not flammable; *Electrical Hazard*: Not pertinent; *Burning Rate*: Not flammable.

Chemical Reactivity - *Reactivity with Water*: No reaction; *Reactivity with Common Materials*: No reactions; *Stability During Transport*: Occurs only above melting point (132°C), yielding ammonia and other products. The decomposition is not explosive; *Neutralizing Agents for Acids and Caustics:* Not pertinent; *Polymerization:* Not pertinent; *Inhibitor of Polymerization*: Not pertinent.

UREA PEROXIDE

Chemical Designations - *Synonyms*: Carbamide peroxide, Carbonyldiamine, Hydrogen peroxide carbamide, Percarbamide, Perhydrol-Urea, Urea hydrogen peroxide; *Chemical Formula*: $CO(NH_2)_2 \cdot H_2O_2$.

Observable Characteristics - *Physical State (as normally shipped)*: Solid; *Color*: White; *Odor*: None.

Physical and Chemical Properties - *Physical State at 15 °C and 1 atm.*: Solid; *Molecular Weight*: 94.1; *Boiling Point at 1 atm.*: Not pertinent (decomposes); *Freezing Point*: Not pertinent; *Critical Temperature*: Not pertinent; *Critical Pressure*: Not pertinent; *Specific Gravity*: 0.8 at 20 °C (solid); *Vapor (Gas) Density*: Not pertinent; *Ratio of Specific Heats of Vapor (Gas)*: Not pertinent; *Latent Heat of Vaporization*: Not pertinent; *Heat of Combustion*: Not pertinent; *Heat of Decomposition*: -540, -300, -12.5.

Health Hazards Information - *Recommended Personal Protective Equipment*: Rubber gloves and protective goggles; *Symptoms Following Exposure*: Inhalation of dust causes irritation of nose From hydrogen peroxide formed when heated. Contact with eyes causes severe damage. Contact with moist skin causes temporary itching or burning sensation. Ingestion causes irritation of mouth and stomach; *General Treatment for Exposure*: INHALATION: remove victim from exposure; call physician. EYES: wash thoroughly with large quantities of water for at least 15 min., call physician; *Toxicity by Inhalation (Threshold Limit Value)*: Data not available; *Short-Term Exposure Limits*: Data not available; *Toxicity by Ingestion*: Data not available; *Late Toxicity*: Data not available; *Vapor (Gas) Irritant Characteristics*: Data not available; *Liquid or Solid Irritant Characteristics*: Data not available; *Odor Threshold*: Not pertinent.

Fire Hazards - *Flash Point :* Not pertinent. This is a combustible solid that may cause fire upon contact with ordinary combustible materials; *Flammable Limits in Air (%):* Not pertinent; *Fire Extinguishing Agents*: Inert powders such as sand and limestone, or water; *Fire Extinguishing Agents Not to be Used:* Not pertinent; *Special Hazards of Combustion Products*: Irritating ammonia gas may be formed in fires; *Behavior in Fire*: Melts and decomposes, giving off oxygen and ammonia. Increases the severity of fires. Containers may explode; *Ignition Temperature (deg. F)*: > 680; *Electrical Hazard*: No data; *Burning Rate*: Not pertinent.

Chemical Reactivity - *Reactivity with Water*: Forms solution of hydrogen peroxide. The reaction is nonhazardous; *Reactivity with Common Materials*: There are no significant reactions under ordinary conditions and temperatures. At 50 °C (122 of) the chemical reacts with dust and rubbish; *Stability During Transport*: Stable below 60 °C (140 of); *Neutralizing Agents for Acids and Caustics:* Not pertinent; *Polymerization:* Not pertinent; *Inhibitor of Polymerization*: Not pertinent.

V

VANADIUM OXYTRICHLORIDE

Chemical Designations - *Synonyms*: Trichloroxo vanadium; Vanadyl chloride; Vanadyl trichloride; *Chemical Formula*: VOCl$_3$.

Observable Characteristics - *Physical State (as normally shipped)*: Liquid; *Color*: Lemon yellow; *Odor*: Acrid.

Physical and Chemical Properties - *Physical State at 15 °C and 1 atm.*: Liquid; *Molecular Weight*: 173.3; *Boiling Point at 1 atm.*: 259, 126, 399; *Freezing Point*: -107, -77, 196; *Critical Temperature*: Data not available; *Critical Pressure*: Data not available; *Specific Gravity*: 1.83 at 20°C (liquid); *Vapor (Gas) Density*: 5.98; *Ratio of Specific Heats of Vapor (Gas)*: Data not available; *Latent Heat of Vaporization*: Data not available; *Heat of Combustion*: Not pertinent; *Heat of Decomposition*: Not pertinent.

Health Hazards Information - *Recommended Personal Protective Equipment*: Acid vapor mask. rubber gloves: face shield: acid-resistant clothing; *Symptoms Following Exposure*: Inhalation of vapor causes irritation of nose and throat. Ingestion causes irritation of mouth and stomach. Contact with eyes or skin causes severe irritation; *General Treatment for Exposure*: *Consult a physician after all exposures to this compound.* INHALATION: move to fresh air: give artificial respiration if necessary. INGESTION: give large amount of water. EYES: flush with water for 15 min. SKIN: wipe exposed areas free of the chemical with a dry cloth, then flush thoroughly with water; *Toxicity by Inhalation (Threshold Limit Value)*: Data not available; *Short-Term Exposure Limits*: 5 ppm (HCl) ceiling level, based on fact that compound decomposes in moist air into vanadic acid and HCl; *Toxicity by Ingestion*: Grade 3, LD$_{50}$ = 140 mg/kg; *Late Toxicity*: Repeated exposures may cause discoloration of tongue, loss of appetite, anemia, kidney disorders, and blindness; *Vapor (Gas) Irritant Characteristics*: Data not available; *Liquid or Solid Irritant Characteristics*: Data not available; *Odor Threshold*: 10 ppm HCl, based on decomposition of compound in moist air.

Fire Hazards - *Flash Point :* Not flammable; *Flammable Limits in Air (%):* Not flammable; *Fire Extinguishing Agents*: Not pertinent; *Fire Extinguishing Agents Not to be Used:* Water, unless in flooding amounts should not be used on adjacent fires; *Special Hazards of Combustion Products*: Irritating fumes of hydrogen chloride form during fires;*Behavior in Fire*: No data; *Ignition Temperature* : Not pertinent; *Electrical Hazard*: Not pertinent; *Burning Rate*: Not pertinent.

Chemical Reactivity - *Reactivity with Water*: Reacts forming a solution of hydrochloric acid; *Reactivity with Common Materials*: In presence of moisture will corrode most metals; *Stability During Transport*: Stable; *Neutralizing Agents for Acids and Caustics:* Flush with water and sprinkle with powdered limestone or rinse with dilute solution of sodium bicarbonate or soda ash; *Polymerization:* Not pertinent; *Inhibitor of Polymerization*: Not pertinent.

VANADIUM PENTOXIDE

Chemical Designations - *Synonyms*: Vanadic anhydride, Vanadium pentaoxide; *Chemical Formula*: V$_2$O$_5$.

Observable Characteristics - *Physical State (as normally shipped)*: Solid; *Color*: Yellow-orange (powder), dark gray (flakes), yellow brown; *Odor*: None.

Physical and Chemical Properties - *Physical State at 15 °C and 1 atm.*: Solid; *Molecular Weight*: 181.88; *Boiling Point at 1 atm.*: Not pertinent (decomposes); *Freezing Point*: Not pertinent; *Critical Temperature*: Not pertinent; *Critical Pressure*: Not pertinent; *Specific Gravity*: 3.36 at 20°C (solid); *Vapor (Gas) Density*: Not pertinent; *Ratio of Specific Heats of Vapor (Gas)*: Not pertinent; *Latent Heat of Vaporization*: Not pertinent; *Heat of Combustion*: Not pertinent; *Heat of Decomposition*: Not pertinent.

Health Hazards Information - *Recommended Personal Protective Equipment*: Bu. Mines approved respirator: rubber gloves: goggles for prolonged exposure; *Symptoms Following Exposure*: Inhalation of dust irritates nose and throat. Ingestion causes irritation of mouth and stomach. Contact with eyes or skin causes irritation: eczema may develop; *General Treatment for Exposure*: INHALATION: move to fresh air: if exposure to dust has been severe, get medical attention. INGESTION: induce vomiting: get medical attention. EYES: flush w water for at least 15 min. SKIN: flush with water; wash with soap and water; *Toxicity by Inhalation (Threshold Limit Value)*: 0.5 mg/m^3; *Short-Term Exposure Limits*: Data not available; *Toxicity by Ingestion*: Grade 4, LD$_{50}$ = 23 mg/kg (mouse); *Late Toxicity*: Repeated exposures may cause discoloration of tongue, loss of appetite, kidney disorders, and blindness; *Vapor (Gas) Irritant Characteristics*: Data not available; *Liquid or Solid Irritant Characteristics*: Data not available; *Odor Threshold*: Data not available.

Fire Hazards - *Flash Point* : Not flammable; *Flammable Limits in Air (%)*: Not flammable; *Fire Extinguishing Agents*: Not pertinent; *Fire Extinguishing Agents Not to be Used:* Not pertinent; *Special Hazards of Combustion Products*: Not pertinent; *Behavior in Fire*: May increase the intensity of fires; *Ignition Temperature* : Not pertinent; *Electrical Hazard*: Not pertinent; *Burning Rate*: Not pertinent.

Chemical Reactivity - *Reactivity with Water*: No reaction; *Reactivity with Common Materials*: No reactions; *Stability During Transport*: Stable; *Neutralizing Agents for Acids and Caustics:* Not pertinent; *Polymerization:* Not pertinent; *Inhibitor of Polymerization*: Not pertinent.

VANADYL SULFATE

Chemical Designations - *Synonyms*: Vanadium oxysulfate, Vanadyl sulfate dihydrate; *Chemical Formula*: VOSO$_4$•2H$_2$O.

Observable Characteristics - *Physical State (as normally shipped)*: Solid; *Color*: Pale blue; *Odor*: None.

Physical and Chemical Properties - *Physical State at 15 °C and 1 atm.*: Solid; *Molecular Weight*: 199.1; *Boiling Point at 1 atm.*: Not pertinent (decomposes); *Freezing Point*: Not pertinent; *Critical Temperature*: Not pertinent; *Critical Pressure*: Not pertinent; *Specific Gravity*: 2.5 at 20 °C (solid); *Vapor (Gas) Density*: Not pertinent; *Ratio of Specific Heats of Vapor (Gas)*: Not pertinent; *Latent Heat of Vaporization*: Not pertinent; *Heat of Combustion*: Not pertinent; *Heat of Decomposition*: Not pertinent.

Health Hazards Information - *Recommended Personal Protective Equipment*: Dust mask, goggles or face shield: protective gloves; *Symptoms Following Exposure*: Inhalation of dust causes irritation of nose and throat, ingestion is irritation of mouth and stomach. Contact with eyes or skin causes irritation; *General Treatment for Exposure*: INHALATION: move to fresh air if exposure to dust has been severe. INGESTION: give large amount of water; induce vomiting; get medical attention. EYES: flush with water for at least 15 min. SKIN: flush with water; *Toxicity by Inhalation (Threshold Limit Value)*: Data not available; *Short-Term Exposure Limits*: Data not available; *Toxicity by Ingestion*: Grade 3, LD$_{50}$ 500-500 mg/kg; *Late Toxicity*: Repeated exposures may cause discoloration of tongue, loss of appetite, anemia, and blindness; *Vapor (Gas) Irritant Characteristics*: Data not available; *Liquid or Solid Irritant Characteristics*: No data; *Odor Threshold*: Data not available.

Fire Hazards - *Flash Point:* Not flammable; *Flammable Limits in Air:* Not flammable; *Fire Extinguishing Agents*: Not pertinent; *Fire Extinguishing Agents Not to be Used:* Not pertinent; *Special Hazards of Combustion Products*: Not pertinent; *Behavior in Fire*: No data; *Ignition Temperature* : Not pertinent; *Electrical Hazard*: Not pertinent; *Burning Rate*: Not pertinent.

Chemical Reactivity - *Reactivity with Water*: No reaction; *Reactivity with Common Materials*: No reactions; *Stability During Transport*: Stable; *Neutralizing Agents for Acids and Caustics:* Not pertinent; *Polymerization:* Not pertinent; *Inhibitor of Polymerization*: Not pertinent.

VINYL ACETATE

Chemical Designations - *Synonyms*: VAM, Vinyl A monomer, Vy Ac; *Chemical Formula*: CH$_3$COOCH=CH$_2$.

Observable Characteristics - *Physical State (as normally shipped)*: Liquid; *Color*: Colorless; *Odor*: Not unpleasant, sweet smell in small quantities; pleasant fruity; characteristic.

Physical and Chemical Properties - *Physical State at 15 °C and 1 atm.*: Liquid; *Molecular Weight*:

86.09; *Boiling Point at 1 atm.*: 163.2, 72.9, 346.1; *Freezing Point*: -135.0, -92.8, 180.4; *Critical Temperature*: 486, 252, 525; *Critical Pressure*: 617, 42, 4.25; *Specific Gravity*: 0.934 at 20 °C (liquid); *Vapor (Gas) Density*: Not pertinent; *Ratio of Specific Heats of Vapor (Gas)*: 1.103; *Latent Heat of Vaporization*: 163, 90.6, 3.79; *Heat of Combustion*: -9754, -5419, -226.9; *Heat of Decomposition*: Not pertinent.

Health Hazards Information - *Recommended Personal Protective Equipment*: Approved canister or air-supplied mask; goggles or face shield; rubber or plastic gloves; *Symptoms Following Exposure*: High vapor concentrations cause narcosis. Liquid irritates eyes and may irritate skin; *General Treatment for Exposure*: INHALATION: remove victim to fresh air; give artificial respiration if required. EYES: flush with water for at least 15 min.; *Toxicity by Inhalation (Threshold Limit Value)*: 10 ppm; *Short-Term Exposure Limits*: Data not available; *Toxicity by Ingestion*: Grade 2, LD_{50} 0.5 to 5 g/kg (rat); *Late Toxicity*: Data not available; *Vapor (Gas) Irritant Characteristics*: Vapors cause a slight smarting of the eyes or respiratory system if present in high concentrations. The effect is temporary; *Liquid or Solid Irritant Characteristics*: Minimum hazard. If spilled on clothing and allowed to remain, may cause smarting and reddening of the skin; *Odor Threshold*: 0.12 ppm.

Fire Hazards - *Flash Point (deg. F):* 18 CC, 23 OC; *Flammable Limits in Air (%):* 2.6 - 13.4; *Fire Extinguishing Agents*: Carbon dioxide or dry chemical for small fires, and ordinary foam for large fires; *Fire Extinguishing Agents Not to be Used:* Water may be ineffective; *Special Hazards of Combustion Products*: Not pertinent; *Behavior in Fire*: Vapor is heavier than air and may travel to a source of ignition and flash back, causing product to polymerize and burst or explode containers; *Ignition Temperature (deg. F)*: 800; *Electrical Hazard*: Class I, Group D; *Burning Rate*: 3.8 mm/min.

Chemical Reactivity - *Reactivity with Water*: No reaction; *Reactivity with Common Materials*: No reactions; *Stability During Transport*: Stable; *Neutralizing Agents for Acids and Caustics:* Not pertinent; *Polymerization:* Polymerization can occur when the product is in contact with peroxides and strong acids, but only under extreme conditions; *Inhibitor of Polymerization*: Hydroquinone and or Diphenylamine.

VINYL CHLORIDE

Chemical Designations - *Synonyms*: Chloroethene, Chloroethylene, Vinyl C Monomer, VCL,VCM; *Chemical Formula*: $CH_2=CHCl$.

Observable Characteristics - *Physical State (as normally shipped)*: Liquefied gas; *Color*: Colorless; *Odor*: Pleasant, sweet.

Physical and Chemical Properties - *Physical State at 15 °C and 1 atm.*: Gas; *Molecular Weight*: 62.50; *Boiling Point at 1 atm.*: 7.2, -13.8, 259.4; *Freezing Point*: -244.8, -153.8, 259.4; *Critical Temperature*: 317.1, 158.4, 431.6; *Critical Pressure*: 775, 52.7, 5.34; *Specific Gravity*: 0.969 at-13 °C (liquid); *Vapor (Gas) Density*: 2.2; *Ratio of Specific Heats of Vapor (Gas)*: 1.186; *Latent Heat of Vaporization*: 160, 88, 3.7; *Heat of Combustion*: -8136, -4520,-189.1; *Heat of Decomposition*: Not pertinent.

Health Hazards Information - *Recommended Personal Protective Equipment*: Rubber gloves and shoes; gas-tight goggles, organic vapor canister or self-contained breathing apparatus; *Symptoms Following Exposure*: INHALATION: high concentrations cause dizziness, anesthetic lung irritation. SKIN: may cause frostbite; phenol inhibitor may be absorbed through skin if large amounts of liquid evaporate; *General Treatment for Exposure*: INHALATION: remove patient lo fresh air arid keep him quiet and warm, call a doctor; give artificial respiration if breathing stops. EYES AND SKIN: flush with plenty of water for at least 15 min.; for eyes, get medical attention; remove contaminated clothing; *Toxicity by Inhalation (Threshold Limit Value)*: 200 ppm; *Short-Term Exposure Limits*: 500 ppm for 5 min.; *Toxicity by Ingestion*: Data not available; *Late Toxicity*: Chronic exposure may cause liver damage; *Vapor (Gas) Irritant Characteristics*: Vapors cause moderate irritation such that personnel will find high concentrations unpleasant. The effect is temporary; *Liquid or Solid Irritant Characteristics*: Minimum hazard. If spilled on clothing and allowed to remain, may cause smarting and reddening of skin. May cause frostbite; *Odor Threshold*: 260 ppm.

Fire Hazards - *Flash Point (deg. F):* -110 OC; *Flammable Limits in Air (%):* 4 - 26; *Fire*

Extinguishing Agents: For small fires use dry chemical or carbon dioxide. For large fires stop the flow of gas if feasible. Cool exposed containers with water; *Fire Extinguishing Agents Not to be Used:* Not pertinent; *Special Hazards of Combustion Products*: Forms highly toxic combustion products such as hydrogen chloride, phosgene, and carbon monoxide; *Behavior in Fire*: Container may explode in fire. Gas is heavier than air and may travel to a source of ignition and flash back; *Ignition Temperature (deg. F)*: 882; *Electrical Hazard*: Class I, Group D; *Burning Rate*: 4.3 mm/min.

Chemical Reactivity - *Reactivity with Water*: No reaction; *Reactivity with Common Materials*: No reactions; *Stability During Transport*: Stable; *Neutralizing Agents for Acids and Caustics:* Not pertinent; *Polymerization:* Polymerizes when exposed to sunlight, air, or heat unless stabilized by inhibitors; *Inhibitor of Polymerization*: Not normally used except when high temperatures are expected. Normally phenol can be used (typically 40 to 100 ppm).

VINYL FLUORIDE, INHIBITED

Chemical Designations - *Synonyms*: Fluoroethylene, Monofluoro ethylene; *Chemical Formula*: $CH_2=CHF$.

Observable Characteristics - *Physical State (as normally shipped)*: Liquefied compressed gas; *Color*: Colorless; *Odor*: Faint ethereal.

Physical and Chemical Properties - *Physical State at 15 °C and 1 atm.*:Gas; *Molecular Weight*: 46.1; *Boiling Point at 1 atm.*: -98, -72, 201; *Freezing Point*: -258, -161, 112; *Critical Temperature*: 130.5, 54.7, 327.9; *Critical Pressure*: 760, 51.6, 5.24; *Specific Gravity*: 0.707 at 0°C (liquid); *Vapor (Gas) Density*: 1.6; *Ratio of Specific Heats of Vapor (Gas)*: 1.2097; *Latent Heat of Vaporization*: 156, 86.5, 3.62; *Heat of Combustion*: (est.) -6.500, -3.600, -150; *Heat of Decomposition*: Not pertinent.

Health Hazards Information - *Recommended Personal Protective Equipment*: Protective gloves; safety glasses; self-contained breathing apparatus; *Symptoms Following Exposure*: Inhalation of vapor causes slight intoxication, some shortness of breath. Liquid may cause frostbite of eyes or skin; *General Treatment for Exposure*: INHALATION: remove victim to fresh air. SKIN: if frostbite has occurred, immerse in warm water, treat burn; *Toxicity by Inhalation (Threshold Limit Value)*: Data not available; *Short-Term Exposure Limits*: Data not available; *Toxicity by Ingestion*: Not pertinent (gas at normal temperatures); *Late Toxicity*: Data not available; *Vapor (Gas) Irritant Characteristics*: Data not available; *Liquid or Solid Irritant Characteristics*: Data not available; *Odor Threshold*: Data not available.

Fire Hazards - *Flash Point :* Not pertinent. This is a flammable, compressed liquified gas; *Flammable Limits in Air (%)*: 2.6 - 21.7; *Fire Extinguishing Agents*: Allow fire to burn out; stop the flow of gas if feasible. Cool adjacent containers with water; *Fire Extinguishing Agents Not to be Used:* Not pertinent; *Special Hazards of Combustion Products*: Toxic hydrogen fluoride gas is generated in fires; *Behavior in Fire*: Vapor is heavier than air and can travel to a source of ignition and flash back. Containers may explode; *Ignition Temperature (deg. F)*: 725; *Electrical Hazard*: No data; *Burning Rate*: Not pertinent.

Chemical Reactivity - *Reactivity with Water*: No reaction; *Reactivity with Common Materials*: No reactions; *Stability During Transport*: Stable; *Neutralizing Agents for Acids and Caustics:* Not pertinent; *Polymerization:* Polymerization can occur in the absence of inhibitor; *Inhibitor of Polymerization*: Terpene B (0.2%).

VINYLIDENE CHLORIDE, INHIBITED

Chemical Designations - *Synonyms*: 1,1-Dichloroethylene, unsym-Dichloroethylene; *Chemical Formula*: $CH_2=CCl_2$.

Observable Characteristics - *Physical State (as normally shipped)*: Liquid; *Color*: Colorless; *Odor*: Sweet, like carbon tetrachloride or chloroform.

Physical and Chemical Properties - *Physical State at 15 °C and 1 atm.*: Liquid; *Molecular Weight*: 96.95; *Boiling Point at 1 atm.*: 88.9, 31.6, 304.8; *Freezing Point*: -187.6, -122.0, 151.2; *Critical Temperature*: Not pertinent; *Critical Pressure*: Not pertinent; *Specific Gravity*: 1.21 at 20 °C (liquid); *Vapor (Gas) Density*: 3.3; *Ratio of Specific Heats of Vapor (Gas)*: Data not available; *Latent Heat of*

Vaporization: 130, 72, 3.0; *Heat of Combustion*: -4860, -2700, -113.0; *Heat of Decomposition*: Not pertinent.

Health Hazards Information - *Recommended Personal Protective Equipment*: Approved canister or air-supplied mask: goggles or face shield, rubber gloves and boots; *Symptoms Following Exposure*: Vapor can cause dizziness and drunkenness: high levels cause anesthesia. Liquid irritates eyes and skin; *General Treatment for Exposure*: INHALATION: if any illness develops, remove person to fresh air pr keep warm and quiet, and get medical attention; if breathing stops, start artificial respiration. INGESTION: not likely a problem, no known antidote, treat symptomatically. EYES OR SK flush with plenty of water for at least 15 min.; get medical attention for eyes; remove contaminated clothing and wash before reuse; *Toxicity by Inhalation (Threshold Limit Value)*: 25 ppm (suggested); *Short-Term Exposure Limits*: Data not available; *Toxicity by Ingestion*: Grade 3, Oral LD_{50} 24 hr = 84 mg/kg; *Late Toxicity*: Data not available; *Vapor (Gas) Irritant Characteristics*: Vapors cause moderate irritation such that personnel will find high concentrations unpleasant. The effect is temporary; *Liquid or Solid Irritant Characteristics*: Causes smarting of the skin and first-degree burns oi short exposure, may cause secondary burns on long exposure; *Odor Threshold*: Data not available.

Fire Hazards - *Flash Point (deg. F)*: 0 OC; *Flammable Limits in Air (%):* 7.3 - 16.0; *Fire Extinguishing Agents*: Foam, carbon dioxide, or dry chemical; *Fire Extinguishing Agents Not to be Used:* Water may be ineffective; *Special Hazards of Combustion Products*: Toxic hydrogen chloride and phosgene form in fires;*Behavior in Fire*: May explode in fires due to polymerization. Vapor is heavier than air and can travel to a source of ignition and flash back; *Ignition Temperature (deg. F)*: 955 - 1,031; *Electrical Hazard*: Not pertinent; *Burning Rate*: 2.7 mm/min.

Chemical Reactivity - *Reactivity with Water*: No reaction; *Reactivity with Common Materials*: Contact with copper or aluminum can cause polymerization; *Stability During Transport*: Stable; *Neutralizing Agents for Acids and Caustics:* Not pertinent; *Polymerization*: Can occur if the product is exposed to sunlight, air, copper, aluminum, or heat; *Inhibitor of Polymerization*: Methyl Ether of Hydroquinone (200 ppm) and or phenol (0.6 to 0.8 %).

VINYL METHYL ETHER, INHIBITED

Chemical Designations - *Synonyms*: Methoxyethylene, Methyl vinyl ether; *Chemical Formula*: $CH_2=CH-O-CH_3$.

Observable Characteristics - *Physical State (as normally shipped)*: Liquefied compressed gas; *Color*: Colorless; *Odor*: Sweet, pleasant.

Physical and Chemical Properties - *Physical State at 15 °C and 1 atm.*: Gas; *Molecular Weight*: 58.1; *Boiling Point at 1 atm.*: Not pertinent (decomposes) 41.9, 5.5, 278.7; *Freezing Point*: -188, -122, 151; *Critical Temperature*: Not pertinent; *Critical Pressure*: Not pertinent; *Specific Gravity*: 0.777 at 0 °C (liquid); *Vapor (Gas) Density*: 2.0; *Ratio of Specific Heats of Vapor (Gas)*: (est.) 1.1473; *Latent Heat of Vaporization*: (est.) 180, 100, 4.2; *Heat of Combustion*: (est.) -14.200, -7.900, -330; *Heat of Decomposition*: Not pertinent.

Health Hazards Information - *Recommended Personal Protective Equipment*: Organic-vapor musk: plastic or rubber gloves, safety glasses; *Symptoms Following Exposure*: Inhalation causes intoxication, blurring of vision, headache, dizziness, excitation, loss of consciousness. Liquid or concentrated vapor irritates eyes and causes frostbite of skin. Aspiration of the liquid will cause chemical pneumonitis; *General Treatment for Exposure*: INHALATION: remove victim to fresh air if breathing is difficult, administer oxygen; call physician. EYES: wash with copious quantities of water; consult an eye specialist. SKIN: wash with copious quantities of water; treat frostbite by use of warm water or blankets. INGESTION: do NOT induce vomiting: get medical attention; *Toxicity by Inhalation (Threshold Limit Value)*: Data not available; *Short-Term Exposure Limits*: Data not available; *Toxicity by Ingestion*: Grade 2, LD_{50} 0.5 to 5 g/kg; *Late Toxicity*: Data not available; *Vapor (Gas) Irritant Characteristics*: Data not available; *Liquid or Solid Irritant Characteristics*: Data not available; *Odor Threshold*: Data not available.

Fire Hazards - *Flash Point (deg. F):* -69 OC; *Flammable Limits in Air (%)*: 2.6 - 39; *Fire Extinguishing Agents*: Allow fire to burn and shut off the flow of gas if feasible. Extinguish small fires with dry chemical or carbon dioxide; *Fire Extinguishing Agents Not to be Used:* Water may be ineffective; *Special Hazards of Combustion Products*: Not pertinent; *Behavior in Fire*: Containers may explode. Vapor is heavier than air and can travel to a source of ignition and flash back; *Ignition Temperature* : No data; *Electrical Hazard*: No data; *Burning Rate*: Not pertinent.

Chemical Reactivity - *Reactivity with Water*: Reacts slowly to form acetaldehyde. The reaction is generally not hazardous unless occurring in hot water or acids are present; *Reactivity with Common Materials*: Acids cause polymerization; *Stability During Transport*: Stable but must be segregated from acids; *Neutralizing Agents for Acids and Caustics:* Not pertinent; *Polymerization:* Can polymerize in the presence of acids; *Inhibitor of Polymerization*: Dioctylamine; Triethanolamine; Solid Potassium Hydroxide.

VINYLTOLUENE

Chemical Designations - *Synonyms*: Methylystyrene; *Chemical Formula*: $CH_3C_6H_4CH=CH_2$.

Observable Characteristics — *Physical State (as normally shipped)*: Liquid; *Color*: Colorless; *Odor*: Disagreeable.

Physical and Chemical Properties - *Physical State at 15 °C and 1 atm.*:Liquid; *Molecular Weight*: 118.18; *Boiling Point at 1 atm.*: 333.9, 167.7, 440.9; *Freezing Point*: -106.6, -77.0, 196; *Critical Temperature*: Not pertinent; *Critical Pressure*: Not pertinent; *Specific Gravity*: 0.897 at 20 °C (liquid); *Vapor (Gas) Density*: Not pertinent; *Ratio of Specific Heats of Vapor (Gas)*: (est.) 1.060; *Latent Heat of Vaporization*: 150, 83.5, 3.50; *Heat of Combustion*: -17.710, -9840, -412.0; *Heat of Decomposition*: Not pertinent.

Health Hazards Information - *Recommended Personal Protective Equipment*: Air-supplied mask, goggles or lace shield, plastic gloves; *Symptoms Following Exposure*: Vapors irritate eyes and nose, high levels cause dizziness drunkenness, and anesthesia. Liquid irritation eyes and may irritate skin; *General Treatment for Exposure*: INHALATION: remove person to fresh air, give artificial respiration and oxygen if needed; call a doctor. INGESTION: do NOT induce vomiting; no known antidote. Call a doctor. EYES: flush with water for at least 15 min. SKIN: wipe off, wash with soap and water; *Toxicity by Inhalation (Threshold Limit Value)*: 100 ppm; *Short-Term Exposure Limits*: 400 ppm for 5 min.; *Toxicity by Ingestion*: Grade 2, LD_{50} 0.5 to 5 g/kg (rat); *Late Toxicity*: Data not available; *Vapor (Gas) Irritant Characteristics*: Vapors cause moderate irritation such that personnel will find high concentrations unpleasant. The effect is temporary; *Liquid or Solid Irritant Characteristics*: Minimum hazard. If spilled on clothing and allowed to remain, may cause smarting and reddening of the skin; *Odor Threshold*: 50 ppm.

Fire Hazards - *Flash Point (deg. F)*: 137 OC, 125 CC; *Flammable Limits in Air (%):* 0.8 - 11; *Fire Extinguishing Agents*: Water fog, foam, carbon dioxide, or dry chemical; *Fire Extinguishing Agents Not to be Used:* Not pertinent; *Special Hazards of Combustion Products*: Not pertinent; *Behavior in Fire*: Containers may explode or rupture in fires due to polymerization; *Ignition Temperature (deg. F)*: 914; *Electrical Hazard*: Not pertinent; *Burning Rate*: 6.0 mm/min.

Chemical Reactivity - *Reactivity with Water*: No reaction; *Reactivity with Common Materials*: No reactions; *Stability During Transport*: Stable; *Neutralizing Agents for Acids and Caustics:* Not pertinent; *Polymerization:* Slow at ordinary temperatures but when hot may rupture container. Also polymerized by metal salts such as those of iron or aluminum; *Inhibitor of Polymerization*: Tertiary Butylcatechol (typically 10 to 50 ppm).

VINYLTRICHLOROSILANE

Chemical Designations - *Synonyms*: Trichlorovinylsilane, Trichlorovinylsilicane, Vinylsilicon trichloride; *Chemical Formula*: $CH_2=CHSiCl_3$.

Observable Characteristics - *Physical State (as normally shipped)*: Liquid; *Color*: Colorless or pale yellow; *Odor*: Sharp, choking, like hydrochloric acid.

Physical and Chemical Properties - *Physical State at 15 °C and 1 atm.*: Liquid; *Molecular Weight*: 161.5; *Boiling Point at 1 atm.*: 195.1, 90.6, 363.8; *Freezing Point*: -139, -95, 178; *Critical Temperature*: Not pertinent; *Critical Pressure*: Not pertinent; *Specific Gravity*: 1.26 at 20°C (liquid);

Vapor (Gas) Density: 5.61; *Ratio of Specific Heats of Vapor (Gas)*: Data not available; *Latent Heat of Vaporization*: 88, 49, 2.0; *Heat of Combustion*: (est.) -4.300, -2.400, -100; *Heat of Decomposition*: Not pertinent.

Health Hazards Information - *Recommended Personal Protective Equipment*: Acid-vapor-type respiratory protection, rubber gloves, chemical worker's goggles; other protective equipment as necessary lo protect skin and eyes; *Symptoms Following Exposure*: Inhalation causes irritation of mucous membranes. Vapor irritates eyes. Contact with liquid causes severe burns of eyes and skin. Ingestion causes burns of mouth and stomach; *General Treatment for Exposure*: *Get medical attention following all exposures to this compound.* INHALATION: remove victim from exposure; give artificial respiration if required. EYES: flush with water for 15 min. SKIN: Hush with water. INGESTION: do NOT induce vomiting: give large amount of water; *Toxicity by Inhalation (Threshold Limit Value)*: Data not available; *Short-Term Exposure Limits*: Data not available; *Toxicity by Ingestion*: Grade 2, oral LD_{50} = 1.280 mg/kg (rat); *Late Toxicity*: Data not available; *Vapor (Gas) Irritant Characteristics*: Vapors cause severe irritation of eyes and throat and can cause eye and lung injury. They cannot be tolerated even at low concentrations; *Liquid or Solid Irritant Characteristics*: Severe skin irritant. Causes second- and third-degree burns on short contact and is very injurious to the eyes; *Odor Threshold*: Data not available.

Fire Hazards - *Flash Point (deg. F)*: 60 OC, 52 CC; *Flammable Limits in Air (%)*: 3 (LEL); *Fire Extinguishing Agents*: Dry chemical or carbon dioxide; *Fire Extinguishing Agents Not to be Used*: Water, foam; *Special Hazards of Combustion Products*: Toxic chlorine and phosgene gases are formed; *Behavior in Fire*: Fire is difficult to extinguish because of ease in re-ignition. Contact with water applied to fight adjacent fires will result in the formation of irritating hydrogen chloride gas; *Ignition Temperature (deg. F)*: 505; *Electrical Hazard*: No data; *Burning Rate*: 2.9 mm/min.

Chemical Reactivity - *Reactivity with Water*: Reacts vigorously, producing hydrogen chloride (hydrochloric acid); *Reactivity with Common Materials*: Reacts with surface moisture to evolve hydrogen chloride, which will corrode common metals and form flammable hydrogen gas; *Stability During Transport*: Stable if protected from moisture; *Neutralizing Agents for Acids and Caustics:* Flush with water and rinse with sodium bicarbonate or lime solution; *Polymerization:* May occur in absence of inhibitor; *Inhibitor of Polymerization*: Diphenylamine, Hydroquinone.

WAXES: CARNAUBA

Chemical Designations - *Synonyms*: No common synonyms; *Chemical Formula*: Not pertinent.

Observable Characteristics - *Physical State (as normally shipped)*: Liquid; *Color*: Yellow to dark brownish green; *Odor*: None.

Physical and Chemical Properties - *Physical State at 15 °C and 1 atm.*: Liquid; *Molecular Weight*: Not pertinent; *Boiling Point at 1 atm.*: Very high; *Freezing Point*: 176-187, 80-86, 353-359; *Critical Temperature*: Not pertinent; *Critical Pressure*: Not pertinent; *Specific Gravity*: 0.998 at 20°C (solid); *Vapor (Gas) Density*: Not pertinent; *Ratio of Specific Heats of Vapor (Gas)*: Not pertinent; *Latent Heat of Vaporization*: Not pertinent; *Heat of Combustion*: Data not available; *Heat of Decomposition*: Not pertinent.

Health Hazards Information - *Recommended Personal Protective Equipment*: Goggles or face shield; protective gloves and clothing for hot liquid wax; *Symptoms Following Exposure*: Hot wax can burn skin and eyes; *General Treatment for Exposure*: SKIN OR EYE CONTACT: remove solidified wax from skin, wash with soap and water; if in eyes, or if skin is burned, call a doctor; *Toxicity by Inhalation (Threshold Limit Value)*: Not pertinent; *Short-Term Exposure Limits*: Not pertinent; *Toxicity by Ingestion*: Data not available; *Late Toxicity*: No data; *Vapor (Gas) Irritant Characteristics*: Non-volatile; *Liquid or Solid Irritant Characteristics*: Hot wax can burn skin and eyes; *Odor Threshold*: Not pertinent.

Fire Hazards - *Flash Point (deg. F):* 540 CC; *Flammable Limits in Air (%):* Not pertinent; *Fire Extinguishing Agents*: Water, foam, dry chemical, or carbon dioxide; *Fire Extinguishing Agents Not to be Used:* Water or foam may cause frothing; *Special Hazards of Combustion Products*: Not pertinent; *Behavior in Fire*: Not pertinent; *Ignition Temperature* : Data not available; *Electrical Hazard*: Not pertinent; *Burning Rate*: Not pertinent.

Chemical Reactivity - *Reactivity with Water*: No reaction; *Reactivity with Common Materials*: No reaction; *Stability During Transport*: Stable; *Neutralizing Agents for Acids and Caustics:* Not pertinent; *Polymerization:* Not pertinent; *Inhibitor of Polymerization*: Not pertinent.

WAXES: PARAFFIN

Chemical Designations - *Synonyms*: Petroleum wax; *Chemical Formula*: Not pertinent.

Observable Characteristics - *Physical State (as normally shipped)*: Liquid to hard solid; *Color*: Yellow to white; *Odor*: Very weak.

Physical and Chemical Properties - *Physical State at 15 °C and 1 atm.*: Solid; *Molecular Weight*: Not pertinent; *Boiling Point at 1 atm.*: Very high; *Freezing Point*: 118 - 149, 48 - 65, 321 - 338; *Critical Temperature*: Not pertinent; *Critical Pressure*: Not pertinent; *Specific Gravity*: 0.78-0.79 at 20°C (liquid); *Vapor (Gas) Density*: Not pertinent; *Ratio of Specific Heats of Vapor (Gas)*: Not pertinent; *Latent Heat of Vaporization*: Not pertinent; *Heat of Combustion*: -18.000, -10.000, -430; *Heat of Decomposition*: Not pertinent.

Health Hazards Information - *Recommended Personal Protective Equipment*: Goggles or face shield; protective gloves and clothing for hot liquid wax; *Symptoms Following Exposure*: Hot wax can burn skin and eyes; *General Treatment for Exposure*: SKIN OR EYE CONTACT: remove solidified wax, wash with soap and water; if in eyes, call a doctor; *Toxicity by Inhalation (Threshold Limit Value)*: Not pertinent; *Short-Term Exposure Limits*: Not pertinent; *Toxicity by Ingestion*: Grade 1, LD_{50} 5 to 15 g/kg; *Late Toxicity*: None; *Vapor (Gas) Irritant Characteristics*: Non-volatile; *Liquid or Solid Irritant Characteristics*: None; *Odor Threshold*: Not pertinent.

Fire Hazards - *Flash Point (deg. F):* 390 CC; 380 - 465 OC; *Flammable Limits in Air (%):* Not pertinent; *Fire Extinguishing Agents*: Water, foam, dry chemical, or carbon dioxide; *Fire Extinguishing Agents Not to be Used:* Water or foam may cause frothing; *Special Hazards of Combustion Products*: Not pertinent; *Behavior in Fire*: Not pertinent; *Ignition Temperature (deg. F)*: 473; *Electrical Hazard*: Not pertinent; *Burning Rate*: Not pertinent.

Chemical Reactivity - *Reactivity with Water*: No reaction; *Reactivity with Common Materials*: No reaction; *Stability During Transport*: Stable; *Neutralizing Agents for Acids and Caustics:* Not pertinent; *Polymerization:* Not pertinent; *Inhibitor of Polymerization*: Not pertinent.

X

M-XYLENE

Chemical Designations - *Synonyms*: 1,3-Dimethilbenzene, Xylol; *Chemical Formula*: m-$C_6H_4(CH_3)_2$.

Observable Characteristics - *Physical State (as normally shipped)*: Liquid; *Color*: Colorless; *Odor*: Like benzene; characteristic aromatic.

Physical and Chemical Properties - *Physical State at 15 °C and 1 atm.*: Liquid; *Molecular Weight*: 106,16; *Boiling Point at 1 atm.*: 269.4, 131.9, 405.1; *Freezing Point*: -54.2, -47.9, 225.3; *Critical Temperature*: 680.5, 343.8, 617.0; *Critical Pressure*: 34.95, 513.8, 3.540; *Specific Gravity*: 0.864 at 20 °C; *Vapor (Gas) Density*: 36.4 dynes/cm =0.0364 n/m at 30 °C; *Ratio of Specific Heats of Vapor (Gas)*: 1.071; *Latent Heat of Vaporization*: 147, 81.9, 3.43; *Heat of Combustion*: -17,554, -9752, -408.31; *Heat of Decomposition*: Not pertinent.

Health Hazards Information - *Recommended Personal Protective Equipment*: Approved canister or

Air-supplied mask; goggles and face shield; plastic gloves and boots; *Symptoms Following Exposure*: Vapors cause headache and dizziness. Liquid irritates eyes and skin. If taken into lungs, causes severe coughing, distress, and coma; can be fatal. Kidney and liver damage can occur; *General Treatment for Exposure*: INHALATION: remove victim to fresh air; administer artificial respiration and oxygen if required; call a doctor. INGESTION: do NOT induce vomiting; call a doctor. EYES: flush with water for at least 15 min. SKIN: wipe off, wash with soap and water; *Toxicity by Inhalation (Threshold Limit Value)*: 100 ppm; *Short-Term Exposure Limits*: 300 ppm for 30 min.; *Toxicity by Ingestion*: Grade 3, LD_{50} 50 to 500 g/kg; *Late Toxicity*: Kidney and liver damage; *Vapor (Gas) Irritant Characteristics*: Vapor causes a slight smarting of the eyes or respiratory system if present in high concentration; *Liquid or Solid Irritant Characteristics*: Minimum hazard. If spilled on clothing and allowed to remain, may be cause smarting and reddening of the skin; *Odor Threshold*: 0.05 ppm.
Fire Hazards - *Flash Point (deg. F)*: 84 CC; *Flammable Limits in Air (%)*: 1.1 - 6.4; *Fire Extinguishing Agents*: Foam, dry chemical, or carbon dioxide; *Fire Extinguishing Agents Not to be Used*: Water may be ineffective; *Special Hazards of Combustion Products*: Not pertinent; *Behavior in Fire*: Vapor is heavier than air and may travel a considerable distance to a source of ignition and flash back; *Ignition Temperature (deg. F)*: 986; *Electrical Hazard*: Class I, Group D; *Burning Rate*: 5.8 mm/min.
Chemical Reactivity - *Reactivity with Water*: No reaction; *Reactivity with Common Materials*: No reaction; *Stability During Transport*: Stable; *Neutralizing Agents for Acids and Caustics:* Not pertinent; *Polymerization:* Not pertinent; *Inhibitor of Polymerization*: Not pertinent.

O-XYLENE
Chemical Designations - *Synonyms*: 1,2-Dimethilbenzene; *Chemical Formula*: o-$C_6H_4(CH_3)_2$.
Observable Characteristics - *Physical State (as normally shipped)*: Liquid; *Color*: Colorless; *Odor*: Benzene-like; characteristic aromatic.
Physical and Chemical Properties - *Physical State at 15 °C and 1 atm.*: Liquid; *Molecular Weight*: 106,16; *Boiling Point at 1 atm.*: 291.9, 144.4, 417.6; *Freezing Point*: -13.3, -25.2, 248.0; *Critical Temperature*: 674.8, 357.1, 630.3; *Critical Pressure*: 36.84, 541.5, 3.732; *Specific Gravity*: 0.880 at 20°C; *Vapor (Gas) Density*: 30.53 dynes/cm = 0.03053 N/m at 15.5°C; *Ratio of Specific Heats of Vapor (Gas)*: 1.068; *Latent Heat of Vaporization*: 149, 82.9, 3.47; *Heat of Combustion*: -17,558, -9754, -408.41; *Heat of Decomposition*: Not pertinent.
Health Hazards Information - *Recommended Personal Protective Equipment*: Approved canister or air-supplied mask; goggles and face shield; plastic gloves and boots; *Symptoms Following Exposure*: Vapors cause headache and dizziness. Liquid irritates eyes and skin. If taken into lungs, causes severe coughing, distress, and coma; can be fatal. Kidney and liver damage can occur; *General Treatment for Exposure*: INHALATION: remove victim to fresh air; administer artificial respiration and oxygen if required; call a doctor. INGESTION: do NOT induce vomiting; call a doctor. EYES: flush with water for at least 15 min. SKIN: wipe off, wash with soap and water; *Toxicity by Inhalation (Threshold Limit Value)*: 100 ppm; *Short-Term Exposure Limits*: 300 ppm for 30 min.; *Toxicity by Ingestion*: Grade 3, LD_{50} 50 to 500 mg/kg; *Late Toxicity*: Kidney and liver damage; *Vapor (Gas) Irritant Characteristics*: Vapor causes a slight smarting of the eyes or respiratory system if present in high concentration. The effect is temporary; *Liquid or Solid Irritant Characteristics*: Minimum hazard. If spilled on clothing and allowed to remain, may be cause smarting and reddening of the skin; *Odor Threshold*: 0.05 ppm.
Fire Hazards - *Flash Point (deg. F)*: 63 CC; 75 OC; *Flammable Limits in Air (%)*: 1.1 - 7.0; *Fire Extinguishing Agents*: Foam, dry chemical, or carbon dioxide; *Fire Extinguishing Agents Not to be Used*: Water may be ineffective; *Special Hazards of Combustion Products*: Not pertinent; *Behavior in Fire*: Vapor is heavier than air and may travel considerable distance to a source of ignition and flash back; *Ignition Temperature (deg. F)*: 869; *Electrical Hazard*: Class I, Group D; *Burning Rate*: 5.8 mm/min.
Chemical Reactivity - *Reactivity with Water*: No reaction; *Reactivity with Common Materials*: No reaction; *Stability During Transport*: Stable; *Neutralizing Agents for Acids and Caustics:* Not pertinent; *Polymerization:* Not pertinent; *Inhibitor of Polymerization*: Not pertinent.

P-XYLENE
Chemical Designations- *Synonyms*: 1,4-Dimethilbenzene; *Chemical Formula*: p-$C_6H_4(CH_3)_2$.
Observable Characteristics - *Physical State (as normally shipped)*: Liquid; *Color*: Colorless; *Odor*: Like benzene; characteristic aromatic.
Physical and Chemical Properties - *Physical State at 15 °C and 1 atm.*: Liquid; *Molecular Weight*: 106,16; *Boiling Point at 1 atm.*: 280.9, 138.3, 411.5; *Freezing Point*: 55.9, 13.3, 286.5; *Critical Temperature*: 649.4, 343.0, 616.2; *Critical Pressure*: 34.65, 509.4, 3.510; *Specific Gravity*: 0.861 at 20°C; *Vapor (Gas) Density*: 28.3 dynes/cm = 0.0283 N/m at 20°C; *Ratio of Specific Heats of Vapor (Gas)*: 1.071; *Latent Heat of Vaporization*: 150, 81, 3.4; *Heat of Combustion*: -17,559, -9754.7, -408.41; *Heat of Decomposition*: Not pertinent.
Health Hazards Information - *Recommended Personal Protective Equipment*: Approved canister or air-supplied mask; goggles and face shield; plastic gloves and boots; *Symptoms Following Exposure*: Vapors cause headache and dizziness. Liquid irritates eyes and skin. If taken into lungs, causes severe coughing, distress, and coma; can be fatal. Kidney and liver damage can occur; *General Treatment for Exposure*: INHALATION: remove victim to fresh air; administer artificial respiration and oxygen if required; call a doctor. INGESTION: do NOT induce vomiting; call a doctor. EYES: flush with water for at least 15 min. SKIN: wipe off, wash with soap and water; *Toxicity by Inhalation (Threshold Limit Value)*: 100 ppm; *Short-Term Exposure Limits*: 300 ppm for 30 min.; *Toxicity by Ingestion*: Grade 3, LD_{50} 50 to 500 mg/kg; *Late Toxicity*: Kidney and liver damage; *Vapor (Gas) Irritant Characteristics*: Vapor causes a slight smarting of the eyes or respiratory system if present in high concentration. The effect is temporary; *Liquid or Solid Irritant Characteristics*: Minimum hazard. If spilled on clothing and allowed to remain, may be cause smarting and reddening of the skin; *Odor Threshold*: 0.05 ppm.
Fire Hazards - *Flash Point (deg. F):* 81 CC; *Flammable Limits in Air (%)*: 1.1 - 6.6; *Fire Extinguishing Agents*: Foam, dry chemical, or carbon dioxide; *Fire Extinguishing Agents Not to be Used:* Water may be ineffective; *Special Hazards of Combustion Products*: Not pertinent; *Behavior in Fire*: Vapor is heavier than air and may travel considerable distance to a source of ignition and flash back; *Ignition Temperature (deg. F)*: 870; *Electrical Hazard*: Class I, Group D; *Burning Rate*: 5.8 mm/min.
Chemical Reactivity - *Reactivity with Water*: No reaction; *Reactivity with Common Materials*: No reaction; *Stability During Transport*: Stable; *Neutralizing Agents for Acids and Caustics:* Not pertinent; *Polymerization:* Not pertinent; *Inhibitor of Polymerization*: Not pertinent.

XYLENOL
Chemical Designations - *Synonyms*: Cresylic acid; 2,6-Dimetuylphenol; 2-Hydroxy-m-xylene; 2,6-Xylenol; vic-m-Xylenol; *Chemical Formula*: 2,6-$(CH_3)_2C_6H_3OH$.
Observable Characteristics - *Physical State (as normally shipped)*: Solid or liquid; *Color*: Light yellow-brown; *Odor*: Sweet tarry.
Physical and Chemical Properties - *Physical State at 15 °C and 1 atm.*:Solid or liquid; *Molecular Weight*: 122.2; *Boiling Point at 1 atm.*: 413, 212, 485; *Freezing Point*: -40 to +106, -40 to +45, 233 to 318; *Critical Temperature*: Not pertinent; *Critical Pressure*: Not pertinent; *Specific Gravity*: 1.01 at 20 °C (liquid); *Vapor (Gas) Density*: Not pertinent; *Ratio of Specific Heats of Vapor (Gas)*: Not pertinent; *Latent Heat of Vaporization*: 212.74, 118.19, 4.9451 at 25 °C; *Heat of Combustion*: -15,310, -8,500, -356 at 25 °C; *Heat of Decomposition*: Not pertinent.
Health Hazards Information - *Recommended Personal Protective Equipment*: Organic canister mask; goggles and face shield; rubber gloves; other protective clothing to prevent contact with skin; *Symptoms Following Exposure*: Vapor irritates eyes, nose, and throat and readily absorbed through mucous membranes and lungs; producing general toxic symptoms (weakness, dizziness, headache, difficult breathing, twitching). Contact with skin causes temporary prickling and intense burning, then local anesthesia. Affected areas initially show white discoloration, wrinkling, and softening, then become red, then brown or black (sings of gangrene). Extensive burns may permit absorption of chemical to produce toxic symptoms described above. Ingestion causes irritation of mouth and stomach, nausea, abdominal pain, weakness, dizziness, headache, difficult breathing, and twitching; *General Treatment*

for Exposure: Get medical attention at once following exposure to this compound. INHALATION: remove patient immediately to fresh air; irritation of nose or throat may be somewhat relieved by spraying or gargling with water until all odor is gone; 100% oxygen inhalation is indicated for cyanosis or respiratory distress; keep patient warm, but not hot. EYES: flood with running water for 15 min.; if physician is not immediately available, continue irritation for another 15 min.; do not use oils or oily ointments unless ordered by physician. SKIN: wash affected areas with large quantities of water or soapy water until all odor is gone; then wash with alcohol or 20% glycerin solution and more water; keep patient warm, but not hot; cover chemical burns continuously with compresses wet with saturated solution of sodium thiosulphate; apply no salves or ointments for 24 hrs after injury. INGESTION: give large quantities of liquid (salt water, weak sodium bicarbonate solution, milk, or gruel) followed by demulcent such as raw egg white or corn starch paste; if profuse vomiting does not follow immediately, give a mild emetic (such as 1 tbsp. mustard in glass of water), or tickle back of throat. Repeat procedure until vomitus is free of the odor. Some demulcent should be left in stomach after vomiting. Keep patient comfortably warm; *Toxicity by Inhalation (Threshold Limit Value)*: 45 ppm; *Short-Term Exposure Limits*: Data not available; *Toxicity by Ingestion*: Grade 2, oral $LD_{50} = 1,070$ mg/kg (mouse); *Late Toxicity*: Damage to heart muscle, and changes in liver, kidney in rats; *Vapor (Gas) Irritant Characteristics*: Data not available; *Liquid or Solid Irritant Characteristics*: Data not available; *Odor Threshold*: Data not available.

Fire Hazards - *Flash Point (deg. F)*: 186; *Flammable Limits in Air (%)*: 1.4 (LEL); *Fire Extinguishing Agents*: Water, dry chemical, carbon dioxide, foam; *Fire Extinguishing Agents Not to be Used:* Not pertinent; *Special Hazards of Combustion Products*: Toxic vapor of unburned material may form in fire; *Behavior in Fire*: Not pertinent; *Ignition Temperature (deg. F)*: 1110; *Electrical Hazard*: Data not available; *Burning Rate*: Data not available.

Chemical Reactivity - *Reactivity with Water*: No reaction; *Reactivity with Common Materials*: No reaction; *Stability During Transport*: Stable; Neutralizing Agents for Acids and Caustics: Not pertinent; *Polymerization:* Not pertinent; *Inhibitor of Polymerization*: Not pertinent.

Z

ZINC ACETATE

Chemical Designations - *Synonyms*: Acetic acid, zinc salt; Dicarbomethoxyzine; Zinc acetate dehydrate; Zinc diacetate; *Chemical Formula*: $Zn(C^2H^3O^2)_2$ or $Zn(C^2H^3O^2)_2 2H^2O$.

Observable Characteristics - *Physical State (as normally shipped)*: Solid; *Color*: White; *Odor*: Faint acetic acid.

Physical and Chemical Properties - *Physical State at 15 °C and 1 atm.*: Solid; *Molecular Weight*: 219.49; *Boiling Point at 1 atm.*: Not pertinent; *Freezing Point*: Not pertinent; *Critical Temperature*: Not pertinent; *Critical Pressure*: Not pertinent; *Specific Gravity*: 1.74 at 20 °C (solid); *Vapor (Gas) Density*: Not pertinent; *Ratio of Specific Heats of Vapor (Gas)*: Not pertinent; *Latent Heat of Vaporization*: Not pertinent; *Heat of Combustion*: Not pertinent; *Heat of Decomposition*: Not pertinent.

Health Hazards Information - *Recommended Personal Protective Equipment*: MSA respirator; rubber gloves; chemical goggles; *Symptoms Following Exposure*: Inhalator causes mild irritation of nose and throat, coughing, and sneezing. Ingestion can cause irritation or corrosion of the alimentary tract, resulting in vomiting. Contact with dust causes irritation of eyes and mild irritation of skin; *General Treatment for Exposure*: INHALATION: move to fresh air; if exposure is severe, get medical attention. INGESTION: induce vomiting; followed by prompt and complete gastric lavage, cathartics, and demulcents. EYES: flush with water for at least 15 min.; consult physician. SKIN: wash with soap and water; *Toxicity by Inhalation*: No data; *Short-Term Exposure Limits*: No data; *Toxicity by Ingestion*: Grade 2, LD_{50} 0.5-5 g/kg; *Late Toxicity*: Data not available; *Vapor (Gas) Irritant Characteristics*: No data; *Liquid or Solid Irritant Characteristics*: Data not available; *Odor Threshold*: Data not available.

Fire Hazards - *Flash Point:* Not flammable; *Flammable Limits in Air:* Not flammable; *Fire Extinguishing Agents*: Not pertinent; *Fire Extinguishing Agents Not to be Used:* Not pertinent; *Special Hazards of Combustion Products*: Not pertinent; *Behavior in Fire*: Not pertinent; *Ignition Temperature*: Not pertinent; *Electrical Hazard*: Not pertinent; *Burning Rate*: Not pertinent.

Chemical Reactivity - *Reactivity with Water*: No reaction; *Reactivity with Common Materials*: No reaction; *Stability During Transport*: Stable; *Neutralizing Agents for Acids and Caustics:* Not pertinent; *Polymerization:* Not pertinent; *Inhibitor of Polymerization*: Not pertinent.

ZINC AMMONIUM CHLORIDE

Chemical Designation - *Synonyms*: Ammonium pentachlorozincate; Ammonium zinc chloride; *Chemical Formula*: $ZnCl_2 3NH_4Cl$.

Observable Characteristics - *Physical State (as normally shipped)*: Solid; *Color*: White; *Odor*: None.

Physical and Chemical Properties - *Physical State at 15 °C and 1 atm.*: Solid; *Molecular Weight*: 296.8; *Boiling Point at 1 atm.*: (sublimes) 644, 340, 613; *Freezing Point*: Not pertinent; *Critical Temperature*: Not pertinent; *Critical Pressure*: Not pertinent; *Specific Gravity*: 1.81 at 20 °C (solid); *Vapor (Gas) Density*: Not pertinent; *Ratio of Specific Heats of Vapor (Gas)*: Not pertinent; *Latent Heat of Vaporization*: Not pertinent; *Heat of Combustion*: Not pertinent; *Heat of Decomposition*: Not pertinent.

Health Hazards Information - *Recommended Personal Protective Equipment*: Dust mask; Goggles and face shield; protective gloves; *Symptoms Following Exposure*: Inhalation of dust irritates nose and throat. Ingestion can cause irritation or corrosion of the alimentary tract. Contact with eyes or skin causes irritation; *General Treatment for Exposure*: INHALATION: remove dust. INGESTION: immediately induce evacuation of intestinal tract by inducing vomiting and giving gastric lavage and saline cathartic; see physician at once; consider development of arsenic poisoning; *Toxicity by Inhalation (Threshold Limit Value)*: 0.5 mg/m$_3$; *Short-Term Exposure Limits*: Data not available; *Toxicity by Ingestion*: Data not available; *Late Toxicity*: May be carcinogenic. Arsenic poisoning may develop; *Vapor (Gas) Irritant Characteristics*: Data not available; *Liquid or Solid Irritant Characteristics*: Data not available; *Odor Threshold*: Not pertinent.

Fire Hazards - *Flash Point :* Not flammable; *Flammable Limits in Air (%):* Not flammable; *Fire Extinguishing Agents*: Not pertinent; *Fire Extinguishing Agents Not to be Used:* Not pertinent; *Special Hazards of Combustion Products*: Not pertinent; *Behavior in Fire*: Not pertinent; *Ignition Temperature* : Not pertinent; *Electrical Hazard*: Not pertinent; *Burning Rate*: Not pertinent.

Chemical Reactivity - *Reactivity with Water*: No reaction; *Reactivity with Common Materials*: No reaction; *Stability During Transport*: Stable; *Neutralizing Agents for Acids and Caustics:* Not pertinent; *Polymerization:* Not pertinent; *Inhibitor of Polymerization*: Not pertinent.

ZINC BROMIDE

Chemical Designations - *Synonyms*: No common synonyms; *Chemical Formula*: $ZnBr_2$.

Observable Characteristics - *Physical State (as normally shipped)*: Solid; *Color*: White; *Odor*: None.

Physical and Chemical Properties - *Physical State at 15 °C and 1 atm.*: Solid; *Molecular Weight*: 225.18; *Boiling Point at 1 atm.*: Not pertinent (decomposes); *Freezing Point*: Not pertinent; *Critical Temperature*: Not pertinent; *Critical Pressure*: Not pertinent; *Specific Gravity*: 4.22 at 20 °C; *Vapor (Gas) Density*: Not pertinent; *Ratio of Specific Heats of Vapor (Gas)*: Not pertinent; *Latent Heat of Vaporization*: Not pertinent; *Heat of Combustion*: Not pertinent; *Heat of Decomposition*: Not pertinent.

Health Hazards Information - *Recommended Personal Protective Equipment*: Chemical goggles and face shield; dust mask; *Symptoms Following Exposure*: Inhalation of dust may irritate nose and throat. Ingestion can cause irritation or corrosion of the alimentary tract; if large amount is swallowed and not thrown up, drowsiness and other symptoms of bromide poisoning may occur. Contact with eyes or skin causes irritation; *General Treatment for Exposure*: INHALATION: move to fresh air. INGESTION: give large amount off water; induce vomiting, followed by prompt and complete gastric lavage, catharsis, and demulcents. EYES or SKIN: wash immediately with large volumes of water; *Toxicity by Inhalation (Threshold Limit Value)*: Data not available; *Short-Term Exposure Limits*: Data not available; *Toxicity by Ingestion*: Grade 2, LD_{50} 0.5 - 15 g/kg; *Late Toxicity*: No data; *Vapor (Gas) Irritant Characteristics*: No data; *Liquid or Solid Irritant Characteristics*: No data; *Odor Threshold*: No data.

Fire Hazards - *Flash Point :* Not flammable; *Flammable Limits in Air (%):* Not flammable; *Fire Extinguishing Agents*: Not pertinent; *Fire Extinguishing Agents Not to be Used:* Not pertinent; *Special Hazards of Combustion Products*: Not pertinent; *Behavior in Fire*: Not pertinent; *Ignition Temperature*: Not pertinent; *Electrical Hazard*: Not pertinent; *Burning Rate*: Not pertinent.

Chemical Reactivity - *Reactivity with Water*: No reaction; *Reactivity with Common Materials*: No reaction; *Stability During Transport*: Stable; *Neutralizing Agents for Acids and Caustics:* Not pertinent; *Polymerization:* Not pertinent; *Inhibitor of Polymerization*: Not pertinent.

ZINC CHLORIDE

Chemical Designations - *Synonyms*: No common synonyms; *Chemical Formula*: $ZnCl_2$

Observable Characteristics - *Physical State (as normally shipped)*: Solid; *Color*: White; *Odor*: Odorless.

Physical and Chemical Properties - *Physical State at 15 °C and 1 atm.*:Solid; *Molecular Weight*: 136.28; *Boiling Point at 1 atm.*: Very high; *Freezing Point*: 541, 283, 556; *Critical Temperature*: Not pertinent; *Critical Pressure*: Not pertinent; *Specific Gravity*: 2.91 at 25 °C; *Vapor (Gas) Density*: Not pertinent; *Ratio of Specific Heats of Vapor (Gas)*: Not pertinent; *Latent Heat of Vaporization*: Not pertinent; *Heat of Combustion*: Not pertinent; *Heat of Decomposition*: Not pertinent.

Health Hazards Information - *Recommended Personal Protective Equipment*: Goggles and face shield; *Symptoms Following Exposure*: Solid or water solution is astringent and can irritate the eyes. When ingested, can cause intoxication, several irritation of stomach, nausea, vomiting, and diarrhea; *General Treatment for Exposure*: INGESTION: give large volumes of water and induce vomiting; repeat process; call a doctor. EYES: wash with water for at least 15 min.; *Toxicity by Inhalation (Threshold Limit Value)*:Not pertinent; *Short-Term Exposure Limits*: Not pertinent; *Toxicity by Ingestion*: Grade 3, LD_{50} 50 to 500 mg/kg; *Late Toxicity*: Data not available; *Vapor (Gas) Irritant Characteristics*: Non-volatile; *Liquid or Solid Irritant Characteristics*: Solid irritates skin on prolonged contact; *Odor Threshold*: Not pertinent.

Fire Hazards - *Flash Point :* Not flammable; *Flammable Limits in Air (%):* Not flammable; *Fire Extinguishing Agents*: Not pertinent; *Fire Extinguishing Agents Not to be Used:* Not pertinent; *Special Hazards of Combustion Products*: Not pertinent; *Behavior in Fire*: Not pertinent; *Ignition Temperature* : Not pertinent; *Electrical Hazard*: Not pertinent; *Burning Rate*: Not pertinent.

Chemical Reactivity - *Reactivity with Water*: No reaction; *Reactivity with Common Materials*: No reaction; *Stability During Transport*: Stable; *Neutralizing Agents for Acids and Caustics:* Not pertinent; *Polymerization:* Not pertinent; *Inhibitor of Polymerization*: Not pertinent.

ZINC CHROMATE

Chemical Designations - *Synonyms*: Buttercup yellow; Zinc chromate (VI) hydroxide; Zinc yellow; *Chemical Formula*: $4ZnO \cdot K_2O \cdot 4CrO \cdot 3H_2O$.

Observable Characteristics - *Physical State (as normally shipped)*: Solid; *Color*: Yellow; *Odor*: None.

Physical and Chemical Properties - *Physical State at 15 °C and 1 atm.*: Solid; *Molecular Weight*: 874 (approx.); *Boiling Point at 1 atm.*: Not pertinent (decomposes); *Freezing Point*: Not pertinent; *Critical Temperature*: Not pertinent; *Critical Pressure*: Not pertinent; *Specific Gravity*: 3.43 at 20 °C (solid); *Vapor (Gas) Density*: Not pertinent; *Ratio of Specific Heats of Vapor (Gas)*: Not pertinent; *Latent Heat of Vaporization*: Not pertinent; *Heat of Combustion*: Not pertinent; *Heat of Decomposition*: Not pertinent.

Health Hazards Information - *Recommended Personal Protective Equipment*: Suitable respirator (For dust): rubber gloves: chemical goggles or face shield; *Symptoms Following Exposure*: Inhalation of dust causes irritation of nose and throat. Ingestion can cause irritation or corrosion of the alimentary tract, circulatory collapse, and toxic nephritis. Contact with eyes or skin causes irritation; *General Treatment for Exposure*: INHALATION: move to fresh air: if exposure has been severe, gel medical attention.

INGESTION: induce vomiting, followed by prompt and complete gastric lavage. catharsis, and demulcents. EYES: flush with water. SKIN: wash thoroughly with soap and water; *Toxicity by Inhalation (Threshold Limit Value)*: 0.1 mg/m^3; *Short-Term Exposure Limits*: Data not available; *Toxicity by Ingestion*: Grade 2, LD$_{50}$ 0.5-5 g/kg; *Late Toxicity*: Possible lung cancer; *Vapor (Gas) Irritant Characteristics*: Data not available; *Liquid or Solid Irritant Characteristics*: Data not available; *Odor Threshold*: Data not available.

Fire Hazards - *Flash Point :* Not flammable; *Flammable Limits in Air (%):* Not flammable; *Fire Extinguishing Agents*: Not pertinent; *Fire Extinguishing Agents Not to be Used:* Not pertinent; *Special Hazards of Combustion Products*: Not pertinent; *Behavior in Fire*: Not pertinent; *Ignition Temperature* : Not pertinent; *Electrical Hazard*: Not pertinent; *Burning Rate*: Not pertinent.

Chemical Reactivity - *Reactivity with Water*: No reaction; *Reactivity with Common Materials*: No reaction; *Stability During Transport*: Stable; *Neutralizing Agents for Acids and Caustics:* Not pertinent; *Polymerization:* Not pertinent; *Inhibitor of Polymerization*: Not pertinent.

ZINC DIALKYLDITHIOPHOSPHATE

Chemical Designations - *Synonyms*: ZincO.O-di-n-butylphos: phorodithioale; Zinc dihexyldithio-phosphalc; Zinc dihexylphosphoro-dithioate; *Chemical Formula*: [(RO)$_2$PSS]$_2$Zn where R=C$_4$H$_9$.

Observable Characteristics - *Physical State (as normally shipped)*: Solid or liquid; *Color*: Straw yellow; yellow-green; *Odor*: Sweet, alcohol-like.

Physical and Chemical Properties - *Physical State at 15 °C and 1 atm.*: Solid; *Molecular Weight*: 548 (approx.); *Boiling Point at 1 atm.*: Not pertinent (decomposes); *Freezing Point*: Not pertinent; *Critical Temperature*: Not pertinent; *Critical Pressure*: Not pertinent; *Specific Gravity*: 1.12-1.26 at 20 °C (liquid) 1.6 at 20 °C (solid); *Vapor (Gas) Density*: Not pertinent; *Ratio of Specific Heats of Vapor (Gas)*: Not pertinent; *Latent Heat of Vaporization*: Not pertinent; *Heat of Combustion*: Not pertinent; *Heat of Decomposition*: Not pertinent.

Health Hazards Information - *Recommended Personal Protective Equipment*: Rubber gloves: safety glasses or face shield: dust respirator for solid form; *Symptoms Following Exposure*: (All commercially available members of this class have about the same health hazards.) Inhalation of dust can cause respiratory discomfort. Ingestion causes irritation of mouth and stomach. Contact with eyes causes moderately severe irritation. Contact with skin causes mild irritation; *General Treatment for Exposure*: INHALATION: move from exposure. INGESTION: if large amounts have been ingested, induce vomiting. EYES: flush with copious amounts of water, if irritation persists, consult a physician. SKIN: wash affected areas with soap and water; *Toxicity by Inhalation (Threshold Limit Value)*: Data not available; *Short-Term Exposure Limits*: Data not available; *Toxicity by Ingestion*: Grade 2, LD$_{50}$ 0.5-5 g/kg; *Late Toxicity*: Data not available; *Vapor (Gas) Irritant Characteristics*: Data not available; *Liquid or Solid Irritant Characteristics*: Data not available; *Odor Threshold*: Data not available.

Fire Hazards - *Flash Point (deg. F):* 360 CC; *Flammable Limits in Air (%):* Data not available; *Fire Extinguishing Agents*: Water, dry chemical, foam, carbon dioxide; *Fire Extinguishing Agents Not to be Used:* Data not available; *Special Hazards of Combustion Products*: Irritating oxides of sulfur and phosphorus may form in fires; *Behavior in Fire*: Data not available; *Ignition Temperature* : Data not available; *Electrical Hazard*: Not pertinent; *Burning Rate*: Not pertinent.

Chemical Reactivity - *Reactivity with Water*: No reaction at ordinary temperatures; *Reactivity with Common Materials*: No reaction; *Stability During Transport*: Stable; Neutralizing Agents for Acids and Caustics: Not pertinent; *Polymerization:* Not pertinent; *Inhibitor of Polymerization*: Not pertinent.

ZINC FLUOROBORATE

Chemical Designations - *Synonyms*: Zinc fluoborate solution; *Chemical Formula*: Zn(SF$_4$)$_2$-H$_2$O.

Observable Characteristics - *Physical State (as normally shipped)*: Liquid; *Color*: Colorless; *Odor*: None.

Physical and Chemical Properties - *Physical State at 15 °C and 1 atm.*: Liquid; *Molecular Weight*: 238.98 (solute only); *Boiling Point at 1 atm.*: (approx.) 212, 100, 373; *Freezing Point*: Data not available; *Critical Temperature*: Not pertinent; *Critical Pressure*: Not pertinent; *Specific Gravity*: 1.45

at 20 °C (liquid); *Vapor (Gas) Density*: Not pertinent; *Ratio of Specific Heats of Vapor (Gas)*: Not pertinent; *Latent Heat of Vaporization*: Not pertinent; *Heat of Combustion*: Not pertinent; *Heat of Decomposition*: Not pertinent.

Health Hazards Information - *Recommended Personal Protective Equipment*: Rubber gloves; safely glasses or face shield; *Symptoms Following Exposure*: Ingestion may cause irritation or corrosion tract. Contact with eyes or skin causes irritation; *General Treatment for Exposure*: INGESTION: give gastric lavage, cathartics, and demulcents. EYES: flush with plenty of water: get medical attention. SKIN: flush with plenty of water; *Toxicity by Inhalation (Threshold Limit Value)*: Data not available; *Short-Term Exposure Limits*: Data not available; *Toxicity by Ingestion*: Grade 2, LD$_{50}$ 0.5-5 g/kg; *Late Toxicity*: Data not available; *Vapor (Gas) Irritant Characteristics*: Data not available; *Liquid or Solid Irritant Characteristics*: Data not available; *Odor Threshold*: Data not available.

Fire Hazards - *Flash Point :* Not flammable; *Flammable Limits in Air (%):* Not flammable; *Fire Extinguishing Agents*: Not pertinent; *Fire Extinguishing Agents Not to be Used:* Not pertinent; *Special Hazards of Combustion Products*: Not pertinent; *Behavior in Fire*: Not pertinent; *Ignition Temperature* : Not pertinent; *Electrical Hazard*: Not pertinent; *Burning Rate*: Not pertinent.

Chemical Reactivity - *Reactivity with Water*: No reaction; *Reactivity with Common Materials*: No reaction; *Stability During Transport*: Stable; *Neutralizing Agents for Acids and Caustics:* Not pertinent; *Polymerization:* Not pertinent; *Inhibitor of Polymerization*: Not pertinent.

ZINC NITRATE

Chemical Designations - *Synonyms*: Zinc nitrate hexahydrate; *Chemical Formula*: $Zn(NO_3)_2 \cdot 6H_2O$.
Observable Characteristics - *Physical State (as normally shipped)*: Solid; *Color*: White; colorless; *Odor*: None.
Physical and Chemical Properties - *Physical State at 15 °C and 1 atm.*:Solid; *Molecular Weight*: 297.47; *Boiling Point at 1 atm.*: Not pertinent (decomposes); *Freezing Point*: 97, 36, 309; *Critical Temperature*: Not pertinent; *Critical Pressure*: Not pertinent; *Specific Gravity*: 2.07 at 20 °C (solid); *Vapor (Gas) Density*: Not pertinent; *Ratio of Specific Heats of Vapor (Gas)*: Not pertinent; *Latent Heat of Vaporization*: Not pertinent; *Heat of Combustion*: Not pertinent; *Heat of Decomposition*: None.
Health Hazards Information - *Recommended Personal Protective Equipment*: Dust mask; goggles or face shield: protective gloves; *Symptoms Following Exposure*: Inhalation of dust may irritate nose and throat. Ingestion can cause irritation or corrosion of the alimentary tract. Contact with eyes causes irritation, which may be delayed. Contact with skin causes irritation; *General Treatment for Exposure*: INHALATION: move to fresh air. INGESTION: induce vomiting, followed by prompt and complete gastric lavage, cathartics, and demulcents. EYES: flush with water; consult a physician. SKIN: wash with soap and water.; *Toxicity by Inhalation (Threshold Limit Value)*: Data not available; *Short-Term Exposure Limits*: Data not available; *Toxicity by Ingestion*: Grade 2, oral LD$_{50}$ = 2.500 mg/kg; *Late Toxicity*: Causes enlarged liver, spleen, and bone marrow in rabbits; *Vapor (Gas) Irritant Characteristics*: Data not available; *Liquid or Solid Irritant Characteristics*: No data; *Odor Threshold*: Odorless.
Fire Hazards - *Flash Point :* Not flammable; *Flammable Limits in Air (%):* Not flammable; *Fire Extinguishing Agents*: Not pertinent; *Fire Extinguishing Agents Not to be Used:* Not pertinent; *Special Hazards of Combustion Products*: Toxic oxides of nitrogen may form in fire; *Behavior in Fire*: May increase intensity of fire when in contact with combustible material; *Ignition Temperature* : Not pertinent; *Electrical Hazard*: Not pertinent; *Burning Rate*: Not pertinent.
Chemical Reactivity - *Reactivity with Water*: No reaction; *Reactivity with Common Materials*: No reaction; *Stability During Transport*: Stable; *Neutralizing Agents for Acids and Caustics:* Not pertinent; *Polymerization:* Not pertinent; *Inhibitor of Polymerization*: Not pertinent.

ZINC PHENOLSULFONATE

Chemical Designations - *Synonyms*: p-Hydroxybcnzenesulfonic acid, zinc salt; Zinc p-phenolsulfonate: Zinc sulfocarbolate; Zinc sulfophenalc; *Chemical Formula*: (1,4-HOC$_6$H$_4$SO$_3$)$_2$Zn•8H20.

Observable Characteristics - *Physical State (as normally shipped)*: Solid; *Color*: White; *Odor*: None.
Physical and Chemical Properties - *Physical State at 15 °C and 1 atm.*: Solid; *Molecular Weight*:
555.8; *Boiling Point at 1 atm.*: (decomposes) 248, 120, 393; *Freezing Point*: Not pertinent; *Critical
Temperature*: Not pertinent; *Critical Pressure*: Not pertinent; *Specific Gravity*: >1 at 20 °C (solid);
Vapor (Gas) Density: Not pertinent; *Ratio of Specific Heats of Vapor (Gas)*: Not pertinent; *Latent Heat
of Vaporization*: Not pertinent; *Heat of Combustion*: Not pertinent; *Heat of Decomposition*: Not
pertinent.
Health Hazards Information - *Recommended Personal Protective Equipment*: Dust mask: goggles or
Face shield, protective gloves; *Symptoms Following Exposure*: Inhalation of dust may irritate nose and
throat. Ingestion of large doses has emetic and astringent effects, can cause irritation or corrosion of
the alimentary tract. Contact with eyes causes irritation. Contact with skin causes mild irritation;
General Treatment for Exposure: INHALATION: move to fresh air. INGESTION: if large amount has
been swallowed, induce vomiting, followed by prompt and complete gastric lavage, cathartics, and
demulcents. EYES or SKIN: flush with water; *Toxicity by Inhalation (Threshold Limit Value)*: Data
not available; *Short-Term Exposure Limits*: Data not available; *Toxicity by Ingestion*: Grade 2, LD_{50}
0.5-5 g/kg; *Late Toxicity*: Data not available; *Vapor (Gas) Irritant Characteristics*: Data not available;
Liquid or Solid Irritant Characteristics: Data not available; *Odor Threshold*: Odorless.
Fire Hazards - *Flash Point :* Not flammable; *Flammable Limits in Air (%):* Not flammable; *Fire
Extinguishing Agents*: Not pertinent; *Fire Extinguishing Agents Not to be Used:* Not pertinent; *Special
Hazards of Combustion Products*: Irritating oxides of sulfur may form in fires; *Behavior in Fire*: Not
pertinent; *Ignition Temperature :* Not pertinent; *Electrical Hazard*: Not pertinent; *Burning Rate*: Not
pertinent.
Chemical Reactivity - *Reactivity with Water*: No reaction; *Reactivity with Common Materials*: No
reaction; *Stability During Transport*: Stable; *Neutralizing Agents for Acids and Caustics:* Not pertinent;
Polymerization: Not pertinent; *Inhibitor of Polymerization*: Not pertinent.

ZINC PHOSPHIDE
Chemical Designations - *Synonyms*: No common synonyms; *Chemical Formula*: Zn_3P_2.
Observable Characteristics - *Physical State (as normally shipped)*: Solid; *Color*: Gray or gray-black;
Odor: Faint phosphorus.
Physical and Chemical Properties - *Physical State at 15 °C and 1 atm.*: Solid; *Molecular Weight*:
258.10; *Boiling Point at 1 atm.*: 2.012, 1.110, 1.373; *Freezing Point*: (sublimes) 788, 420, 693;
Critical Temperature: Not pertinent; *Critical Pressure*: Not pertinent; *Specific Gravity*: 4.55 at 13 °C
(solid); *Vapor (Gas) Density*: Not pertinent; *Ratio of Specific Heats of Vapor (Gas)*: Not pertinent;
Latent Heat of Vaporization: Not pertinent; *Heat of Combustion*: -4.100, -2.270, -95; *Heat of
Decomposition*: Not pertinent.
Health Hazards Information - *Recommended Personal Protective Equipment*: Dust mask or self-
contained breathing apparatus; goggles or face shield: protective gloves; *Symptoms Following Exposure*:
When inhaled or ingested, compound releases phosphine, which causes faintness, weakness, nausea,
vomiting, dyspnea, fall in blood pressure, change in pulse rate, diarrhea, intense thirst, convulsions,
paralysis, and coma. Contact with eyes or skin causes irritation; *General Treatment for Exposure*:
INHALATION: move lo fresh air; give artificial respiration if required; get medical attention for
phosphine poisoning. INGESTION: give one tablespoonful of mustard in a glass of warm water; repeat
until vomit fluid is clear; avoid use of all oils; call a physician immediately; have patient lie down and
keep warm. EYES: flush with water for at least 15 min. SKIN: flush with water, wash with soap and
water; *Toxicity by Inhalation (Threshold Limit Value)*: Data not available; *Short-Term Exposure Limits*:
Data not available; *Toxicity by Ingestion*: Grade 4, oral LD_{50} = 40 mg/kg (rat); *Late Toxicity*: Data not
available; *Vapor (Gas) Irritant Characteristics*: Data not available; *Liquid or Solid Irritant
Characteristics*: Data not available; *Odor Threshold*: Data not available.
Fire Hazards - *Flash Point :* Not flammable; *Flammable Limits in Air (%):* Not flammable; *Fire
Extinguishing Agents*: Use water, foam, or dry chemical on adjacent fires; *Fire Extinguishing Agents
Not to be Used:* Any agent with an acid reaction (e.g. carbon dioxide or halogenated agents) will

liberate phosphine, a toxic and spontaneously flammable gas; *Special Hazards of Combustion Products*: Irritating oxides of phosphorus may be formed in fires; *Behavior in Fire*: Not pertinent; *Ignition Temperature* : Not pertinent; *Electrical Hazard*: Not pertinent; *Burning Rate*: No data.

Chemical Reactivity - *Reactivity with Water*: Reacts slowly with water, more rapidly with dilute acid, to form phosphine gas, which is toxic and spontaneously flammable; *Reactivity with Common Materials*: No reaction; *Stability During Transport*: Stable unless exposed to moisture; toxic phosphine gas may then be released and collect in closed spaces; *Neutralizing Agents for Acids and Caustics:* Not pertinent; *Polymerization:* Not pertinent; *Inhibitor of Polymerization*: Not pertinent.

ZINC SILICOFLUORIDE
Chemical Designations - *Synonyms*: Zinc fluosilicate; Zinc hexafluorosilicate; Zinc silicofluoride hexahydrate; *Chemical Formula*: $ZnSiF_6 \cdot 6H_2O$.

Observable Characteristics - *Physical State (as normally shipped)*: Solid; *Color*: White; transparent; *Odor*: None.

Physical and Chemical Properties - *Physical State at 15 °C and 1 atm.*: Solid; *Molecular Weight*: 315.5; *Boiling Point at 1 atm.*: 122 - 158, 50 - 70, 232 - 343; *Freezing Point*: Not pertinent; *Critical Temperature*: Not pertinent; *Critical Pressure*: Not pertinent; *Specific Gravity*: 2.10 at 20 °C (solid); *Vapor (Gas) Density*: Not pertinent; *Ratio of Specific Heats of Vapor (Gas)*: Not pertinent; *Latent Heat of Vaporization*: Not pertinent; *Heat of Combustion*: Not pertinent; *Heat of Decomposition*: Not pertinent.

Health Hazards Information - *Recommended Personal Protective Equipment*: Dust respirator: chemical goggles or face shield: protective gloves; *Symptoms Following Exposure*: Inhalation of dust irritates nose and throat, excessive inhalation may cause severe pulmonary inflammation. Ingestion causes nausea, cramps, vomiting, shock, convulsions, cyanosis, and other symptoms of fluoride poisoning. Contact with eyes or skin causes irritation: skin ulcers may develop; *General Treatment for Exposure*: INHALATION: move to fresh air. INGESTION: cause vomiting by giving soapy water or mustard water: have patient drink large quantities of lime water: if necessary, give stimulant such as strong coffee. EYES: flush with water: call physician as necessary. SKIN: wash with soap and water; *Toxicity by Inhalation (Threshold Limit Value)*: 2.5 mg/m^3 (as fluoride); *Short-Term Exposure Limits*: Data not available; *Toxicity by Ingestion*: LD_{LO} = 100 mg/kg (rat); *Late Toxicity*: Data not available; *Vapor (Gas) Irritant Characteristics*: Data not available; *Liquid or Solid Irritant Characteristics*: Data not available; *Odor Threshold*: Data not available.

Fire Hazards - *Flash Point :* Not flammable; *Flammable Limits in Air (%):* Not flammable; *Fire Extinguishing Agents*: Not pertinent; *Fire Extinguishing Agents Not to be Used:* Not pertinent; *Special Hazards of Combustion Products*: Toxic and irritating hydrogen fluoride and silicon tetrafluoride are formed in fires; *Behavior in Fire*: Not pertinent; *Ignition Temperature* : Not pertinent; *Electrical Hazard*: Not pertinent; *Burning Rate*: Not pertinent.

Chemical Reactivity - *Reactivity with Water*: No reaction; *Reactivity with Common Materials*: No reaction; *Stability During Transport*: Stable; *Neutralizing Agents for Acids and Caustics:* Not pertinent; *Polymerization:* Not pertinent; *Inhibitor of Polymerization*: Not pertinent.

ZINC SULFATE
Chemical Designations - *Synonyms*: White vitroil; Zinc sulfate heptahydrate; Zinc vitriol; *Chemical Formula*: $ZnSO_4 \cdot 7H_2O$.

Observable Characteristics - *Physical State (as normally shipped)*: Solid; *Color*: Colorless; *Odor*: None.

Physical and Chemical Properties - *Physical State at 15 °C and 1 atm.*: Solid; *Molecular Weight*: 287.54; *Boiling Point at 1 atm.*: Not pertinent (decomposes); *Freezing Point*: (decomposes) 122 -212, 50 -100, 323 -373; *Critical Temperature*: Not pertinent; *Critical Pressure*: Not pertinent; *Specific Gravity*: 1.96 at 20 °C (solid); *Vapor (Gas) Density*: Not pertinent; *Ratio of Specific Heats of Vapor (Gas)*: Not pertinent; *Latent Heat of Vaporization*: Not pertinent; *Heat of Combustion*: Not pertinent; *Heat of Decomposition*: Not pertinent.

Health Hazards Information - *Recommended Personal Protective Equipment*: Dust mask, goggles or face shield, protective gloves; *Symptoms Following Exposure*: Inhalation of dust causes irritation of nose and throat. Ingestion can cause irritation or corrosion of the alimentary tract. Contact with eyes or skin causes irritation; *General Treatment for Exposure*: INHALATION: move to fresh air. INGESTION: induce vomiting, followed by prompt and complete gastric lavage, cathartics, and demulcents. EYES or SKIN: flush with water; *Toxicity by Inhalation (Threshold Limit Value)*: Data not available; *Short-Term Exposure Limits*: Data not available; *Toxicity by Ingestion*: Grade 2, LD_{50} 0.5-5 g/kg; *Late Toxicity*: Data not available; *Vapor (Gas) Irritant Characteristics*: Data not available; *Liquid or Solid Irritant Characteristics*: Data not available; *Odor Threshold*: Data not available.

Fire Hazards - *Flash Point :* Not flammable; *Flammable Limits in Air (%):* Not flammable; *Fire Extinguishing Agents*: Not pertinent; *Fire Extinguishing Agents Not to be Used:* Not pertinent; *Special Hazards of Combustion Products*: Not pertinent; *Behavior in Fire*: Not pertinent; *Ignition Temperature* : Not pertinent; *Electrical Hazard*: Not pertinent; *Burning Rate*: Not pertinent.

Chemical Reactivity - *Reactivity with Water*: No reaction; *Reactivity with Common Materials*: No reaction; *Stability During Transport*: Stable; *Neutralizing Agents for Acids and Caustics:* Not pertinent; *Polymerization:* Not pertinent; *Inhibitor of Polymerization*: Not pertinent.

ZIRCONIUM ACETATE

Chemical Designations - *Synonyms*: Zirconium acetate solution; *Chemical Formula*: $Zr(C_2H_3O_2)_4 \cdot H_2O$.

Observable Characteristics - *Physical State (as normally shipped)*: Liquid; *Color*: Colorless; *Odor*: Weak vinegar.

Physical and Chemical Properties - *Physical State at 15 °C and 1 atm.*:Liquid; *Molecular Weight*: 327 (solute only); *Boiling Point at 1 atm.*: Not pertinent; *Freezing Point*: Not pertinent; *Critical Temperature*: Not pertinent; *Critical Pressure*: Not pertinent; *Specific Gravity*: 1.37 at 20°C (liquid); *Vapor (Gas) Density*: Not pertinent; *Ratio of Specific Heats of Vapor (Gas)*: Not pertinent; *Latent Heat of Vaporization*: Not pertinent; *Heat of Combustion*: Not pertinent; *Heat of Decomposition*: Not pertinent.

Health Hazards Information - *Recommended Personal Protective Equipment*: Rubber gloves; chemical goggles or face shield; *Symptoms Following Exposure*: Has only a mild pharmacological action. Contact with eyes or skin may cause irritation; *General Treatment for Exposure*: INGESTION: give large amount of water. EYES: flush with water for at least 15 min.; consult a physician if irritation persist. SKIN: flush with water; *Toxicity by Inhalation (Threshold Limit Value)*: 5 mg/m^3 (as zirconium); *Short-Term Exposure Limits*: Data not available; *Toxicity by Ingestion*: Grade 2, LD_{50} 0.5 - 5 g/kg (rat); *Late Toxicity*: Data not available; *Vapor (Gas) Irritant Characteristics*: Data not available; *Liquid or Solid Irritant Characteristics*: Data not available; *Odor Threshold*: Data not available.

Fire Hazards - *Flash Point :* Not flammable; *Flammable Limits in Air (%):* Not flammable; *Fire Extinguishing Agents*: Not pertinent; *Fire Extinguishing Agents Not to be Used:* Not pertinent; *Special Hazards of Combustion Products*: Not pertinent; *Behavior in Fire*: Not pertinent; *Ignition Temperature* : Not pertinent; *Electrical Hazard*: Not pertinent; *Burning Rate*: Not pertinent.

Chemical Reactivity - *Reactivity with Water*: No reaction; *Reactivity with Common Materials*: No reaction; *Stability During Transport*: Stable; *Neutralizing Agents for Acids and Caustics:* Not pertinent; *Polymerization:* Not pertinent; *Inhibitor of Polymerization*: Not pertinent.

ZIRCONIUM NITRATE

Chemical Designations - *Synonyms*: Zirconium nitrate; *Chemical Formula*: $Zr(NO_3) \cdot 5H_2O$.

Observable Characteristics - *Physical State (as normally shipped)*: Solid; *Color*: White; *Odor*: Nine.

Physical and Chemical Properties - *Physical State at 15 °C and 1 atm.*: Solid; *Molecular Weight*: 429.3; *Boiling Point at 1 atm.*: Not pertinent (decomposes); *Freezing Point*: Not pertinent; *Critical Temperature*: Not pertinent; *Critical Pressure*: Not pertinent; *Specific Gravity*: >1 at 20 °C (solid); *Vapor (Gas) Density*: Not pertinent; *Ratio of Specific Heats of Vapor (Gas)*: Not pertinent; *Latent Heat of Vaporization*: Not pertinent; *Heat of Combustion*: Not pertinent; *Heat of Decomposition*: Not pertinent.

Health Hazards Information - *Recommended Personal Protective Equipment*: Dust mask; goggles or face shield; protective gloves; *Symptoms Following Exposure*: Has only a mild pharmacological action. Inhalation of dust may irritate nose and throat. Contact with eyes or skin causes irritation; *General Treatment for Exposure*: INHALATION: move to fresh air. INGESTION: give large amount of water. EYES or SKIN: flush with water; *Toxicity by Inhalation (Threshold Limit Value)*: 5 mg/m$_3$ (as zirconium); *Short-Term Exposure Limits*: Data not available; *Toxicity by Ingestion*: Grade 2, LD$_{50}$ = 25 g/kg (rat); *Late Toxicity*: Data not available; *Vapor (Gas) Irritant Characteristics*: Data not available; *Liquid or Solid Irritant Characteristics*: Data not available; *Odor Threshold*: Data not available.

Fire Hazards - *Flash Point :* Not flammable but may intensify fire; *Flammable Limits in Air (%):* Not flammable; *Fire Extinguishing Agents*: Not pertinent; *Fire Extinguishing Agents Not to be Used:* Not pertinent; *Special Hazards of Combustion Products*: Toxic oxides of nitrogen may form in fire; *Behavior in Fire*: May increase intensity of fire when in contact with combustible materials; *Ignition Temperature :* Not pertinent; *Electrical Hazard*: Not pertinent; *Burning Rate*: Not pertinent.

Chemical Reactivity - *Reactivity with Water*: Dissolves to give an acid solution; *Reactivity with Common Materials*: Will corrode most metals; *Stability During Transport*: Stable; *Neutralizing Agents for Acids and Caustics:* Flush with water; *Polymerization:* Not pertinent; *Inhibitor of Polymerization*: Not pertinent.

Synonyms Index

A

B

C

D

E

F

G

H

I

K

L

M

N

O

P

Q

R

S

T

U

V

W

X

Y

Z